Selenium

Selenium

edited by **Ralph A. Zingaro**

Department of Chemistry
Texas A & M University
College Station, Texas

and

W. Charles Cooper

United Nations Development Programme
Santiago, Chile

 Van Nostrand Reinhold Company
New York/Cincinnati/Toronto/London/Melbourne

Van Nostrand Reinhold Company Regional Offices:
New York Cincinnati Chicago Millbrae Dallas

Van Nostrand Reinhold Company International Offices:
London Toronto Melbourne

Copyright © 1974 by Litton Educational Publishing, Inc.
Library of Congress Catalog Card Number: 74-1246
ISBN: 0-442-29575-8

Manufactured in the United States of America
Published by Van Nostrand Reinhold Company
450 West 33rd Street, New York, N.Y. 10001

Published simultaneously in Canada by Van Nostrand Reinhold Ltd.
15   14   13   12   11   10   9   8   7   6   5   4   3   2   1

Library of Congress Cataloging in Publication Data
Zingaro, Ralph A
Selenium.
Includes bibliographies.
1. Selenium. I. Cooper, W. Charles, 1921-
joint author. II. Title.
QD181.S5Z56   546'.724'1   74-1246
ISBN 0-442-29575-8

Preface

Following the discovery of selenium by Berzelius in 1817, it was some time before a number of the physical properties of the element were noted, in particular, its photoconductivity. However, the erratic and unpredictable performance of the early photoelectric devices designed around this element led investigators to conclude that selenium had indeed been most appropriately named, being as fickle as the moon. One of the obvious characteristics of this element is, in fact, its schizophrenic chemical personality, behaving as a metallic non-metal and a non-metallic metal. This behavior is what makes this element interesting, but difficult to work with in the chemical laboratory.

Actually, it is only in more recent years that the electrical and optical properties of selenium have come to be reasonably well-understood. Thus, following the introduction of the silicon rectifier in the mid-1950's, considerable improvements in the manufacture and quality of selenium rectifiers were made. Although possibly not of commercial interest, the expitaxial growth of selenium single crystals has led to the development of a single crystal selenium rectifier and to a much better knowledge of the asymmetric conduction of selenium. The use of selenium in the xerographic process, introduced during the mid-1950s, stimulated a great deal of research on the structural, optical, and electrophotographic properties of selenium. This research has been extended to the various allotropic modifications, with major advances in selenium physics being presented at conferences in London in 1964 (*Recent Advances in Selenium Physics*, Pergamon Press, 1965) and Montreal in 1967 (*The Physics of Selenium and Tellurium*, Pergamon Press, 1969).

The discovery of the role of selenium in animal nutrition in 1957 together with the well-known toxic effects of the ingestion of abnormal amounts of selenium such as are found in certain seleniferous plants has led to a greatly expanded interest in the biochemistry of selenium. The vital function of selenium as a nutrient is now so well-established that the Food and Drug Administration has been formally requested to establish minimum acceptable levels for this element in commercial livestock feeds. Extraordinary progress is currently being made in understanding the biochemical role of this element and several selenoproteins have been shown to be present as components of various enzymes.

Also noteworthy is the continued and expanding use of selenium in such areas as glass and ceramics, rubber and plastics, ferrous and non-ferrous metallurgy, as well as the extensive research on semiconducting selenium compounds and alloys which hold promise of new device applications. Of unusual interest, especially in view of increasing concern with alternate energy sources, is the potential of the cadmium chalcogenides in the field of solar energy conversion. Thus, CdS, CdSe, and CdTe have been referred to as "the optimum semiconductors" and "the only present day hope for truly low cost (solar) cells."

The scientific and technical literature on selenium and its sister element, tellurium, as compiled by *Chemical Abstracts*, has grown markedly since 1960 and has averaged three thousand publications per annum during the period 1968–1973, with the majority of these works pertaining to selenium.

The rapid growth in the accumulation of knowledge about selenium and the lack of a monograph in which there is a comprehensive treatment of the science and technology of the element, prompted the publication of the present volume (recently, the organic chemistry of selenium has been very thoroughly treated, see *Organic Selenium Compounds: Their Chemistry and Biology*, D. L. Klayman and W. H. H. Günther, eds., Wiley-Interscience, New York, N.Y., 1973). Although it has not been possible to cover all topics related to selenium, those subjects which are presented cover areas of major interest as well as areas in which new and interesting research is in progress. In all cases the authors have been selected on the basis of their direct knowledge and working experience in the field.   The spatial limitations in a publication of this kind should be partly obviated by the extensive literature references.

It has been the objective of the editors to provide a treatise which would meet the needs of a broad spectrum of readers, including teachers and students, research workers, and industrialists.   In this undertaking the editors wish to express their sincere appreciation to the authors for their contributions and to the Selenium–Tellurium Development Association for its support of this publication.

Ralph A. Zingaro
Department of Chemistry
Texas A&M University
College Station, Texas

W. Charles Cooper
United Nations Development Programme
Santiago, Chile

Contents

Preface, v

1. **The History, Occurrence, and Properties of Selenium**—*W. Charles Cooper/Kathryn G. Bennett/ Frank C. Croxton*

   History, 1

   Occurrence, 2
   Minerals, 2   Meteorites and Volcanic Matter, 5   Soils, 8   Plants, 9   Fossil Fuels, 10   Water and Air, 11

   Properties, 12

   Applications, 25

   Production, 26

2. **Recovery and Refining of Selenium**—*P. H. Jennings/J. C. Yannopoulos*

   Part I: Recovery of Selenium, 31
   Introduction, 31
   Recovery of Selenium from Ores, 37

Recovery of Selenium from Copper Refinery Slimes, 38
Smelting Processes, 39   Roasting of Copper Refinery Slimes, 40
Sulfatizing of Copper Refinery Slimes, 46   Leaching of Copper
Refinery Slimes, 48   Other Methods of Treating Copper Refinery
Slimes, 52

Treatment of Sulfuric Acid Plant Sludge, 52

Recovery of Selenium from Miscellaneous Metallurgical Products, 54

Secondary Selenium, 56

Recovery of Selenium from Aqueous Solutions, 56

Part II: Refining of Selenium, 60

Introduction, 60

Chemical Methods of Refining, 61
Selective Precipitation, 61   Selective Leaching and Recrystallization,
62   Oxide Purification, 61   Hydride Purification, 63   Chloride Purification, 66   Purification of Selenium Solutions by Ion-Exchange
and Solvent Extraction, 67   Electrolytic Methods, 69   Fusion with
a Nitrate, 69

Physical Methods of Refining, 69
Refining of Selenium solutions by Absorption Methods, 69   Zone
Refining of Selenium, 70   Distillation of Selenium, 70

3. **The Structure of Selenium—*W. Charles Cooper/R. A. Westbury***

Introduction, 87

The Vapor Phase, 88
Dissociation Constants for Species in the Vapor Phase, 95

Liquid Selenium, 96
Magnetic Susceptibility and Electron Paramagnetic Resonance, 100
Viscosity, 102   Theoretical Analysis, 106   Summary, 107

The Amorphous Allotropes of Solid Selenium, 109
Red Amorphous Selenium and Black Amorphous Selenium, 109
Vitreous Selenium, 113   Thin Amorphous Films of Selenium, 117

The Crystalline Allotropes of Selenium, 119
α-Monoclinic and β-Monoclinic Selenium, 120   Trigonal Selenium,
123   Miscellaneous Crystalline Isotopes, 124

Crystallization of Selenium, 125
Factors Affecting the Crystallization Behavior of Selenium, 127

The Kinetics of Selenium Crystallization, 133

The Growth of Selenium Single Crystals, 137
Trigonal Selenium, 137   Monoclinic Selenium, 141

4. **The Interaction with Light of Phonons in Selenium**—*Richard Zallen/ Gerald Lucovsky*

Introduction, 148

Symmetry and Selection Rules for Trigonal Se, 149

The Photon–Phonon Interaction in Trigonal Se, 152
Reststrahlen in Elemental Crystals, 152   Dynamic Charge in Trigonal Se, 153   The Reststrahlen and Polariton Spectra, 156

Higher-Order Optical Interactions in Trigonal Se, 159
Higher-Order Phonon-Involving Optical Processes, 159   Raman Scattering, 160   Two-Phonon Absorption, 162   Phonon-Assisted Electronic Optical Processes, 162

The Spectra of Monoclinic Se, 163
Structure, 163   Infrared Absorption, 164   Raman Scattering, 165   Vibrational Assignments, 166

The Spectra and Structure of Amorphous Se, 167
Introduction, 167   Infrared Absorption, 167   Raman Scattering, 168   The Molecular Structure of Amorphous Se, 169

5. **Optical and Electrical Properties of Selenium**—*Josef Stuke*

Introduction, 174

Trigonal Selenium, 177
Band Structure, 177   Optical Properties, 190   Electrical Properties, 217   Photoconductivity, 248   Selenium Rectifiers and Photovoltaic Cells, 270

Monoclinic Selenium, 287
Crystal Structure, 287   Optical Properties, 288   Electrical Properties, 291   Photoconductivity, 293

6. **The Structural Aspects of Selenium Chemistry**—*A. W. Cordes*
Introduction, 298

Elemental Selenium, 299

Structures Containing Noncyclic X—Se—X Linkages, 301

The Diselenide Ion and Structures Containing Noncyclic X—Se—Se—X Linkages, 303

Compounds Containing Noncyclic X—Se—Se—Se—X Linkages, 305

Selenium Oxide Structures, 307

Oxyacids and Oxyanions of Selenium, 309

Selenium Tetrahalides and Alkyl and Aryl Selenium Halides, 312

Seleninyl Dihalides and Related Structures, 314

Selenium Hexafluoride and the Hexahaloselenates, 320

Cyclic Systems Involving Selenium, 321
Structures Related to 1,4-Diselenane, 321   Other Cyclic Systems:
Five-Membered Rings, 325   Other Cyclic Systems, 329

Structures with Singly Bonded Selenium Atoms, 332
Summary of Bond Distances Involving Selenium, 333

7. **Coordination Compounds in which Selenium Functions as the Donor Atom—
V. Krishnan/Ralph A. Zingaro**

Introduction, 337

Inorganic Ligands, 338
Selenium Analogues of the Tetraoxometallates, 338   Heteropoly Acids
Containing Selenium, 340   Hydrogen Selenide, 341   Selenocyanates,
342

Ligands which Contain Phosphorus-Selenium Bonds, 354
Dialkyldiselenophosphates, 354   Diselenophosphoric Diamides, 358
Diselenophosphinates and Related Mixed Ligands, 358   Phosphine
Selenides, 363

Organic Ligands, 364
Neutral Ligands, 364   Uninegative Ligands, 378   Binegatively Charged
Ligands, 387

Organometallic Compounds, 391
Carbonyls, 392   Cyclopentadienyls, 394   Organometallic Nontransition Metal Compounds Bonded to Selenium, 395

Miscellaneous Complexes, 398

8. **The Organic Chemistry of Selenium—Kurt J. Irgolic/Mohan V. Kudchadker**

Introduction, 408

Selenols, RSeH, 409
Selenols from Elemental Selenium, 410   Selenols from Sodium Hydro-
gen Selenide, Hydrogen Selenide, Carbon Diselenide and Carbon
Oxide Selenide, 411   Selenols by Reduction of Diselenides, Seleno-

cyanates and Seleninic Acids, 413   Physical Properties and Reactions, 414

Selenium Compounds of the General Formula R—Se—X, 414
Selenenic Acids, R—Se—OH, 415   Organyl Selenium Halides, 415
Organyl Selenium Isocyanates, Thiocyanates, Selenocyanates and
and Organyl Selenosulfates, 417   Esters, Anhydrides, and Amides of
Selenic Acids, 417   Reactions, 418

Organyl Selenocyanates and Isoselenocyanates, 418
Organyl Selenocyanates, 418   Organyl Isoselenocyanates, 421

Seleninic Acids, Selenonic Acids and Their Derivatives, 423
Seleninic Acids, 423   Selenonic Acids, $RSeO_3H$, 426

Aliphatic and Aromatic Selenium Trihalides, 427

Diorganyl Diselenides, R_2Se_2, 427
Symmetric Diselenides, 428   Unsymmetric Diselenides R—Se—Se—R′, 435   Reactions of Diselenides, 435

Diorganyl Selenide Sulfides, R—Se—S—R, 436

Diorganyl Triselenides, R_2Se_3, 437

Diorganyl Selenides, 437
Symmetric Dialkyl Selenides, 438   Unsymmetric Aliphatic Diorganyl
Selenides, 441   Symmetric Diaryl Selenides, 450   Unsymmetric Diaryl
Selenides, 452   Aromatic Aliphatic Selenides, 456   Aryl Trifluoro-
methyl Selenides, 475   Diorganyl Selenides, R—Se—$(CH_2)_n$—Se—R, 476

Se-Esters of Selenocarboxylic Acids, 479

Selenium Containing Carbohydrates, 480

Diorganyl Selenium Dihalides and Realted Compounds, 497
The Reaction of Diorganyl Selenides with Halogens, 497   The Condensation of Selenium Tetrachloride and Oxychloride with Hydrocarbons,
498   The Addition of Selenium Tetrachloride and Oxychloride to
Carbon-Carbon Multiple Bonds, 499   Diorganyl Selenium Dihalides
by Halogen Exchange and from Diorganyl Selenoxides, 500   Physical
Properties and Reactions of Diorganyl Selenium Dihalides, 500   Diorganyl Selenium Dialkoxides, Dicarboxylates and Imines, 501

Diorganyl Selenoxides, 502

Diorganyl Selenones, 504

Triorganyl Selenonium Compounds, 505

Tetraorganyl Selenium Compounds, 509

Compounds with a Formal Carbon-Selenium Double Bond: Selenoaldehydes, Selenoketones, Selenocarboxylic Acid Amides and Related Compounds,   509
  Selenoaldehydes, 509   Selenoketones, 511   Selenocarboxylic Acid Amides, 512   Other Selenocarboxylic Acid Derivatives, 515
Selenium Compounds of Biological Importance, 515
  Organylselenoaminocarboxylic Acids, 516   Se-Aryl Aminomonoselenocarboxylates, 519   Bis(aminocarboxyalkyl) Diselenides and Selenides, 520   Selenium-Containing Peptides, 521   Other Organic Selenium Compounds of Biological Importance, 525

9. **Biochemistry of Selenium**—*Howard E. Ganther*

Introduction, 546

Total Content and Distribution of Selenium, 547
  Plants and Microorganisms, 548   Animal Tissues, 548

Forms of Selenium, 550
  Low Molecular Weight Compounds, 550   Proteins, 555

Metabolism of Selenium, 567
  Intermediary Metabolism of Selenite, 568   Studies with Selenomethionine, 572

Properties of Selenium and Effects on Biological Systems, 574
  Oxidation-Reduction Reactions of Selenium Compounds, 574   Substitution of Selenium for Sulfur, 580   Selenium and Cellular Processes, 584

Interrelationships of Selenium with Other Substances, 588
  Nutritional Interrelationships with Vitamin E, Sulfur Amino Acids, and Heavy Metals, 588   Sulfate, 592   Arsenic, 593   Heavy Metals, 594   Other Substances, 597

On the Biological Role of Selenium, 597

Addendum, 609
  Selenoproteins, 609   Protection by Selenium Against Heavy Metals, 611

10. **Analytical Chemistry of Selenium**—*W. Charles Cooper*

Introduction, 615

Separation and Isolation of Selenium, 616
  Decomposition, Dissolution and Other Preliminary Treatment of

Inorganic Materials, 616   Decomposition of Organic Materials Containing Selenium, 618   Selective Separation Methods, 619

Detection and Identification, 621

Determination of Selenium, 623
Precipitation and Gravimetric Methods of Determination, 623   Volumetric Methods, 624   Polarography and Other Electrochemical Methods, 625   Photometric Methods, 627   Neutron Activation Analysis, 631

Analysis of Selenium (and Tellurium), 632
Analysis of Selenium, 632   Determination of Impurities in Selenium, 632 Analysis of Selenium Compounds, 637

Determination of Selenium in Specific Materials, 640
Iron and Steel, 640   Other Metals and Alloys, 641   Semiconductor Materials, 641   Rocks, Minerals, Ores and Anode Slimes, 642   Biological Materials, 644   Air and Water, 647

11. **The Toxicology of Selenium and its Compounds—*W. Charles Cooper/ J. R. Glover***

Introduction, 654

Selenium Intoxication in Man, 655

Selenium in the Animal Body, 658
Retention and Distribution of Selenium, 658   Pathology of Selenium Intoxication, 662   Selenium Detoxification and Elimination, 662

Toxicity of Selenium, Selenium Compounds, and Selenium-Bearing Natural Products, 664

Elemental Selenium, 664
Selenium Dioxide, Selenites and Selenates, 666   Hydrogen Selenide, 667   Selenium Oxychloride, 668   Organoselenium Compounds and Selenium-Bearing Natural Products, 669

Conclusion, 671

12. **Selenium in Agriculture—*Alvin L. Moxon/Oscar E. Olson***

Introduction, 675

Selenium Poisoning in Livestock, 676
Distribution of Seleniferous Soils and Vegetation Over the World, 677 Forms of Selenium in Soils, 678   Selenium in Plants, 679   Selenium in Waters, 681   Symptoms of Selenium Poisoning, 681   Control of Selenium Poisoning in Livestock, 685

Selenium as a Nutrient, 687
  Selenium-Responsive Diseases, 687   Distribution of Selenium-Responsive Diseases, 694   Control Measures, 695
Biological Functions of Selenium, 697
  Selenium as an Essential Element for Plants, 698   Selenium as an Insecticide, 699

13. **Selenium in Glass**—*George B. Hares*

Introduction, 708

Selenium Pink Glasses, 709

Selenium as a Decolorizer for Glass, 711

Selenium Ruby Glasses, 716

Selenium in Chalcogenide Glasses, 721

14. **Selenium and Selenium Compounds in Rubber and Plastics**—*William J. Mueller*

Introduction, 728

Curing-Accelerator Systems, 729
  Natural Rubber, 729   SBR (Styrene-Butadiene Rubber), 733   Nitrile Rubber, 737   Butyl Rubber, 737   Neoprene, 742   Halogenated Butyl, 742   Hypalon, 742   Viton, 746   Acrylate Polymer, 746   Hard Rubber, 746   Surface Vulcanization, 746

Antioxidants, 750
  SBR, 750   Butyl Rubber, 750   Natural Rubber, 750   Staining and Discoloration, 754   Miscellaneous Effects, 754

UV Stabilizers, 754

Bonding Agents, 757

Polymerization Additives, 758

Modification of Butyl Rubber, 758

Activating Carbon Black, 760

Latex Preservative, 760

Flameproofing, 760

15. **Metallurgical Aspects and Uses of Selenium**—*S. C. Carapella, Jr./R. H. Aborn/L. R. Cornwell*

Introduction, 762

Crystal Growth, 763

Mechanical Behavior, 765

Alloying Behavior, 766
  Selenium Binary Systems, 766   Selenium Ternary Systems, 769

Oxidation and Corrosion, 769

Diffusion, 770

Metallurgical Uses, 771
  Ferrous Alloys, 771   Nonferrous Metals and Alloys, 779   Miscellaneous Applications, 782

16. **Selenium in Electrophotography**—*Gerald Lucovsky/Mark D. Tabak*

Introduction: The Xerographic Process and the Ideal Photoreceptor, 788

Amorphous Selenium as a Xerographic Photoreceptor, 791

Electrical and Optical Properties of Amorphous Se, 794
  Optical Properties Due to Electronic Transitions, 795   Electrical and Photoelectrical Properties of Amorphous Selenium, 797

Models for the Photoelectronic Properties of Amorphous Se, 799
  Model for Nonphotoconductive Absorption, 800   Molecular Origin of Nonphotoconductive States: Theory and Discussion, 803   Trapping Kinetics, 804

Recent Developments in Xerographic Photoreceptors Employing Se, 805
  Alloy Photoreceptors, 805   Sensitized Organic Photoconductors, 806

Index, 809

1

The History, Occurrence, and Properties of Selenium

W. CHARLES COOPER*

Noranda Research Centre, Pointe Claire, Quebec, Canada

KATHRYN G. BENNETT

Department of Chemistry, Sir George Williams University, Montreal, Canada

FRANK C. CROXTON

Battelle Memorial Institute, Columbus, Ohio

HISTORY

It is possible that a man by the name of Arnold de Villanova was the first to observe and describe the element which came to be known as selenium. In his writings, *Rosarius Philosophorum*, around the beginning of the fourteenth century,[35] he described the vaporization of sulfur and the appearance on the walls of the container of a deposit which he called *sulphur rubeum*. It was not until 1817, however, that a reliable account of the isolation and identification of the new element, named selenium, was published,[9] by J. J. Berzelius. In a letter to Berthollet, dated February 9, 1818, he spoke of the discovery of lithium by his student, Arfredson. He continued, then, to tell of his and J. G. Gahn's discovery of selenium during the burning of sulfur from Falun pyrites at Gripsholm, Sweden. In this process, a reddish deposit was recovered which, when

* Present address: United Nations Development Programme, Casilla 197-D, Santiago, Chile.

1

burning, produced a very strong odor of horseradish.   It was believed at first that it was tellurium, but further examination showed it to be readily volatile and reducible.   Berzelius named it selenium, enclosed a small sample for Berthollet and asked him to inform Gay-Lussac of his letter.   The new element was given the name of selenium, for the moon, because of its similarity to tellurium which, 35 years before, had been named for the earth. It is interesting to note[62] that both Berzelius and Gahn had a financial interest in the Gripsholm sulfuric acid plant.   Gahn was also the discoverer of manganese, the dioxide of which was used in the eighteenth century for the decolorizing of glass.   Now it is selenium itself which is employed almost exclusively for this purpose.

OCCURRENCE

Selenium is an element which is widely distributed in small concentrations in the earth's crust, having an abundance around 7×10^{-5} weight per cent, which approximates that of cadmium and antimony.[17]   The selenium content of a wide range of minerals, meteorites and volcanic matter, soils, plants, fossil fuels, and various waters has been determined.   The accuracy of these determinations has been enhanced by the development, in recent years, of methods of analysis which are sensitive to 0.05 ppm or less of selenium.

Because of the preponderance of igneous rocks in the earth's crust and of the association between selenium and sulfur in their behavior during the crystallization of the magma, an indication of the abundance of selenium can be obtained from selenium analyses of sulfide minerals.   Unfortunately, no extensive analytical data are available from a broad sampling of such minerals. In the Soviet Union, where the most comprehensive recent studies on selenium mineralogy, particularly in sulfide deposits, have been undertaken, Sindeeva[47] has reported a Clarke index for selenium in igneous rocks of 1.4×10^{-5}.   In a comparison of the relative abundance of sulfur, selenium, and tellurium, Vinogradov[60] has given the Clarke indices for these elements as follows: S, 0.05; Se, 6×10^{-5}; and Te, 10^{-6}.   These data give a S : Se : Te ratio of 50,000 : 60 : 1 and a S : Se ratio of approximately 800 : 1.   There is an obvious need that research similar to that done in the Soviet Union be undertaken in other countries to provide a more complete picture of the occurrence, mineralogy, and geochemistry of selenium.   In this connection the detailed studies on the incidence of selenium in soils and plants in the seleniferous regions of the United States are specially noteworthy.

Minerals

Since Berzelius' discovery of selenium and his identification of the selenium minerals Berzelianite, Cu_2Se — and, Eucairite, AgCuSe, some 40 different selenium minerals have been analyzed (see Table 1-1). The advances in ore microscopy have resulted in the recent discovery of a number of selenium minerals. Although x-ray diffraction studies have yet to be undertaken on numerous selenides, there are indications to date that a number of the complex selenium minerals are indeed mixtures.

The occurrence of independent selenium minerals requires a considerably increased concentration of the element. It appears that this process takes place during the later stages of mineralization with endogenic selenium minerals being formed during the hydrothermal stage of mineralization in association with oxide minerals such as barite, quartz–barite, and barite–carbonate. Consequently, selenium minerals are usually found in the upper horizons of mineral deposits. The resulting segregations of selenium are characterized by their very small size and the absence of any economically meaningful concentrations.

The dissemination of selenium in sulfide minerals, such as are found in hydrothermal sulfide deposits,[59] is related to the association of selenium with sulfur in the form of isomorphous replacement of sulfur by selenium in the sulfide crystal lattice. This similar crystallochemistry of the two elements explains why selenium, and not selenium minerals are found in pyrite and pyrrhotite–chalcopyrite deposits.[22,24] The deposits of chalcopyrite–pentlandite–pyrrhotite and of quartz–pyrite–chalcopyrite–polymetallic formations[43] are of particular importance as to their selenium content, with the ores of chalcopyrite and bornite being claimed as richest in selenium and tellurium.[42] In a study of copper molybdenum deposits, Pidzhyan[40] found the selenium concentration to be maximal in quartz–chalcopyrite–molybdenite, quartz–molybdenite, quartz–chalcopyrite, and especially in the quartz–tennantite–enargite stage of mineralization.

The similar geochemistry of selenium and uranium, particularly as regards their favorable oxidation behavior and their inertness under reducing conditions has resulted in an association in nature between the two elements.[1,4,10,54]

The abundance of selenium cannot be related to the concentration of copper, nickel, sulfur, or tellurium. Also, selenium minerals do not occur in sufficient quantity to provide a commercial source for the element. The dissemination of selenium in copper sulfide minerals generates considerable commercial excitement, since it is from these minerals that selenium is eventually recovered through the treatment of anode slimes produced during the electrolytic refining of copper. It must be noted, however, that there is a wide variation in the selenium content of sulfide minerals with such factors as the mineralogical composition and type of deposit and the geological setting of the ore body playing an important role in deciding the selenium level. The wide variation of selenium in sulfide ores results in the principal sources

TABLE 1-1   MINERALS OF SELENIUM[47]

Cation	Name	Formula	Synonyms	Reference

I. Native elements and Intermetallic Compounds

	Native selenium	Se		
	Tellurium selenide	TeSe		

II. Selenides

	Hydrogen selenide	H_2Se		
Ag	Naumannite	Ag_2Se	Cacheutaite	
	Aguilarite	Ag_4SeS		
Ag—Bi	Bohdanowiczite	$AgBiSe_2$		5
Ag—Cu	Eucairite	$AgCuSe$		
Ag—Cu—Tl	Crookesite	$(Cu,Tl,Ag)_2Se$		
Pb	Clausthalite	$PbSe$	Lead Selenide	
Zn	Stilleite	$ZnSe(?)$		
Cd	Cadmoselite	$CdSe$		
Hg	Tiemannite	$HgSe$		
	Onofrite	$Hg(S,Se)$		
Fe	Achavalite	$FeSe$		
	Ferroselite	$FeSe_2$		
Cu	Klockmannite	$CuSe$		
	Berzelianite	Cu_2Se	Selenocuprite, copper selenide, berzeline	
	Umangite	$Cu_{4-x}Se_2$		
Cu—Fe	Eskebornite	$CuFeSe_2$		31
Ni	Blockite	$NiSe_2$	Penroseite	
	Wilkmanite	Ni_3Se_4		61
	Trüstedtite	Ni_3Se_4		61
	Kullerudite	$NiSe_2$		61
	Sederholmite	β-NiSe		61
	Mäkinenite	γ-NiSe		61
Ni—Cu—Co	Tyrrellite	$(Cu,Co,Ni)_2Se_4$		
Co	Trögtalite	$CoSe_2$		
	Hastite	$CoSe_2$		
	Bornhardtite	Co_3Se_4		
	Freboldite	$CoSe(?)$		
Bi	Guanajuatite	$Bi_2(Se, S)_3$	Frenzelite, castillite, bismuth selenide, selenobismuthite	
	Paraguanajuatite	Bi_2Se_2S		
	Laitakariite	$Bi_8(S,Se)_7(?)$		
As	Jermoite	$As(S,Se)_2$		
Pd	Palladium selenide			

III. Sulfosalts

Pb—Bi	Weibullite	$PbBi_2(S,Se)_4$		
	Platynite	$PbBi_2(Se,S)_3$		
	Wittite	$[Pb_5Bi_6(Se,S)_{14}(?)]$ or $5PbS \cdot 3Bi_2(S,S)_3$		

ABLE 1-1 (*Continued*)

Cation	Name	Formula	Synonyms	Reference
		IV. *Oxides*		
	Selenolite	SeO_2		
		V. *Oxyacid Salts*		
Pb	Kerstenite	$PbSeO_4 \cdot 2H_2O$	Glasbachite, selenocerussite	
	Molybdomenite	$PbSeO_3 \cdot nH_2O$		
Cu	Chalcomenite	$CuSeO_3 \cdot 2H_2O$		
Pe	Iron selenites	?		
Hg	Mercury selenite	?		
Ni	Ahlfeldite	$NiSeO_3 \cdot 2H_2O$		
Co	Cobaltomenite			

of selenium being the sulfidic copper ores in Canada, (mainly in the Canadian Shield), the United States, and the Soviet Union.

The various types of deposits in which selenium is found are summarized in Table 1-2. Of special interest are the extensive studies on selenium mineralization particularly in sulfide deposits which have been undertaken in the Soviet Union.[15,16,20,25,30,34,52,53,58,63] Unfortunately, there are no comparable data from other countries.

Meteorites and Volcanic Matter

Various types of meteorites have shown a wide variation in selenium content within each kind. A value as high as 0.23% was reported for an iron meteorite from Bohumilitz, Bohemia.[47] In the estimation of the cosmic abundance of the elements, considerable reliance has been placed on the analyses of chondritic materials.[48] However, it should be noted that the various groups of chondrites are not chemically homogeneous.[2] According to Greenland[21] who analyzed representative samples of all chondrite groups for various elements, including selenium, by a neutron activation technique, the carbonaceous Type one and possibly, the enstatite Type A chondrites give the closest approximation to the composition of the primordial nonvolatile material. The selenium contents of these two types were found to be 27 and 11 to 34 ppm, respectively, with atomic abundances relative to $Si = 10^6$, of 89 and 48. These values should be compared with the cosmic abundance for selenium of 68 given by Suess and Urey.[48] Apparently the selenium content decreases throughout the carbonaceous chondrites and is the least in ordinary chondrites for which average selenium values of 6.5,[21] 7, 9,[14] and

TABLE 1-2 TYPES OF SELENIUM DEPOSITS[47]

Genetic Type of Deposit	Type of Mineral Association	Form of Deposition of Se	Morphology of Ore Bodies	Relation to Intrusive Rocks	Examples of Deposits	Age of Deposit
Magmatic	Pyrrhotite-chalcopyrite pentlandite	Selenium in disseminated form in sulfides	Zone of mineral deposits and breccia zones situated at the bottom of intrusives, intersecting veins	Basic and ultrabasic rocks	Noril'sk, Pechenga Monchegorsk, Sudbury	Prepaleozoic, seldom Lower and upper Paleozoic
Volcanic	Native sulfur	Mixed crystals with sulfur	Flows and veins	—	Kamchatka, Kurile Islands, Hawaiian Islands	—
	—	Selenite, selenate, and natural selenium	Layers of volcanic tuff and bentonite	—	Tuff deposits in Wyoming, U.S.A. (Shoshone, Riverton)	Oligocene
Hydro-thermal	Chalcopyrite-molybdenite	Selenium in disseminated form in sulfides chiefly in molybdenite	Stockworks, zones of small veins and scattering, more rarely quartz ore veins	Moderately silicic granitoids	Kadzharan, Dastakert Agarak, Paragachay-skoye	Tertiary Variscan
	Pyritic	Selenium in disseminated form in sulfide	Lenses, deposits, bodies of irregular shape	Deposited in effusives; para-genetically, probably, connected with subvolcanic intrusives	Pyrite deposits in the Urals (Blyava, Sibay, Degtyarka), Añtai	Paleozoic, Mesozoic
	Cobaltite-selenite	Forms its own minerals (clausthalite, guanajuatite)	Stockwork zones, veins	Moderately silicic granite	Akdzhil'ga	Mesozoic
	Selenide	Selenium forms its own minerals (blockite, naumannite, clausthalite and others)	Siderite-hematite-barite veins	?	Pakajaca in Bolivia; Tilkerode Lerbach, and Zorge in Harz, and San-Andreasberg, Germany	Tertiary Paleozoic

Pitchblende-selenide	Selenium forms its own minerals (selenides of copper, lead, mercury, and silver)	Veins	High to moderately silicic rocks	Katanga in Belgian Congo; Athabasca in Canada	Proterozoic, Paleozoic
Gold-silver quartz-adularia	Selenium forms its own minerals, silver and lead selenides	Stockwork zones	Deposited in extrusives: associated probably with subvolcanic intrusives ?	Redjang-Lebong, Lebong-Donok in Sumatra, Nagyag in Romania	Tertiary (possibly significant)
Gold-selenium	Selenium in the form of clausthalite	Spall-type faults; spallation zones filled with calcite		Eisenberg	Paleozoic
Galena-sphalerite-quartz-carbonate (polymetallic)	Selenium only isomorphically in sulfides	Veins, lenses, metasomatic bodies	Paragenetically, deposits are related to small intrusives of granite and divided into three types according to condition of deposition: 1) In granite-shale, etc. 2) In volcanic sedimentary rocks 3) In carbonate rocks, acid granites and their alkaline derivatives	Sadon, Zavodinskoye, Smirnovskoye, Zyryanovskoye	Variscan, Kimmeridgian, Alpine
Cinnabar-stibnite	Selenium chiefly forms its own minerals. Selenium partially enters isomorphically	Bedded deposits, veins and stockworks	—	Kwei-Chou (China) Khaydarkan (Central Asia), Monte-Amiata	Paleozoic, Mesozoic
Carnotite-uranium black	In the form of native selenium and selenides of iron, copper, lead, and silver	Beds, metasomatic bodies		—	Paleozoic, Mesozoic
Exogenic (type of iron cap)	In dispersed condition chiefly in elemental form	Iron cap on sulfide deposits		Colorado plateau, Maykain, Kul'yurt Tau, Uchaly, Buribay Blyava, Chudak	—

7

9.8 ppm[44] have been reported. It is of some significance to note that a selenium content of 0.2 ppm has been reported for lunar soil.[36]

The separation of selenium and sulfur from the magma during crystallization and the release of these elements in gaseous form in volcanic exhalations have prompted studies of the geochemistry of selenium in such emissions. Although the particular chemical composition of the selenium in volcanic gases has not as yet been determined, from a study of the equilibrium constants for four possible reactions involving Se, SeO_2, SO_3, H_2 and H_2O in the temperature range 400 to 1,000°K, Suzuoki[49] concluded that the selenium in volcanic gases near the outlet of a fumarole is present as gaseous elemental selenium possibly in such forms as Se_2 and Se_6. On the other hand, most of the sulfur is present as sulfur dioxide and hydrogen sulfide. According to Suzuoki the S : Se ratio in fumarolic gases varies inversely as the temperature. The actual selenium content (approximately 0.1 ppm) increased slightly with temperatures in the range of 400 to 700°K. Suzuoki[50] found a selenium content ranging between 0.01 and 0.31 ppm for the fumarolic gases from the Nasudaki volcano in Japan. The content of selenium as well as that of other elements such as chlorine, fluorine, and boron showed a considerable variation from year to year.

Lakin[32] in an examination of 36 samples of volcanic sulfur found that 12 samples contained less than 2 ppm selenium, 16 samples less than 15 ppm selenium with the highest selenium content being 1,600 ppm. As much as 5.18 % selenium has been reported for seleniferous sulfur (Kilauea, Hawaii).[39]

Soils

The occurrence of selenium in soils is a topic of considerable interest particularly in view of its toxic effects on cattle through plants which can accumulate the element from seleniferous soils. There are, besides, deficiency diseases prevalent in animals on a selenium-deficient diet, especially animals grazing in areas in which the soils are low in selenium.

As indicated by Lakin and Davidson,[33] the selenium level in soils is related to the whole geochemical cycle of selenium and the solution chemistry of the element. Factors of importance include the selenium content of host rocks, redox potentials, pH, and the nature of the drainage waters.

Rosenfeld and Beath[42] have pointed out that selenium in the soil may be derived from: (1) rock formations and outcrops, (2) formations lying below the soil mantle, (3) decomposition through the weathering of the host rock and subsequent transport of selenium by wind or surface waters, (4) indicator plants, and (5) enrichment of the soil with selenium resulting from mining operations, and as stated by Lakin and Davidson,[33] the combustion of fossil fuels.

According to Sindeeva,[47] soils have a selenium content of around 0.01 ppm. While concentrations are found in soils resulting from the disintegration of rocks, selenium enrichment has occurred as a result of intensive volcanic activity. The results of the analysis of several thousand soil samples in the United States have indicated a maximum concentration of selenium of less than 100 ppm, with the majority of seleniferous soils containing on the average less than 2 ppm.[42] The selenium content of most soils, as estimated by Swaine[51] is between 0.1 and 0.2 ppm.

The selenium in soils may occur as selenites, selenates, elemental selenium, and as selenium in association with pyrite and other minerals. It should be remembered that in seleniferous soils a significant amount of selenium in the form of soluble organic or inorganic compounds is present as a result of the decay of seleniferous plant materials.

Although the most extensive examination of soils for their selenium content has been undertaken in the United States, especially in relation to the production of toxic selenium vegetation, the selenium in the soils of other countries such as Australia, Canada, Colombia, Israel, Ireland, Mexico, and South Africa has been studied also. The high selenium content of soils in Ireland is particularly noteworthy, and the study of soil profiles has shown a selenium content as high as 1,200 ppm. Although the source of selenium is not really known, the rock formations in all of the toxic areas of Ireland are of the Upper and Middle Carboniferous limestones associated with shale and sandstones. The soils with the greater selenium content occur in low-lying areas which are poorly drained and high in organic residues.

Plants

The accumulation of selenium by certain plants has been known for some time and has been the subject of considerable research in regions of the western and southwestern United States. The ingestion of seleniferous plants by cattle grazing in these areas has given rise to the well-known toxic conditions of blind staggers and alkali disease.

According to Rosenfeld and Beath,[42] there are 24 species and varieties of the plant *Astragalus* which grow in seleniferous soils and for which selenium is an essential nutrient. Such plants have been designated as selenium indicator plants and strangely enough their occurrence outside of the United States, Canada, and Mexico is relatively unknown. However, seleniferous vegetation arising from seleniferous soils has been noted in other countries, notably Colombia. Thus, although there are no known selenium indicator plants in Colombia, grain and forage crops from seleniferous regions have shown a high selenium content; notably, wheat (115 ppm), barley (137 ppm), corn (40 ppm), and grass (280 ppm).[8]   In contrast to the location of seleniferous vegetation in the United States in arid or semi-arid areas, such flora are encountered in Colombia in humid regions of high rainfall.

There is a considerable variation in the capacity of different forms of vegetation to accumulate selenium as shown by the several species of plants grown on soils from the Niobrara formation in Wyoming:[42]

Astragalus bisulcatus	5,530 ppm Se
Stanleya pinnata	1,190
Atriplex nuttallii	300
Grasses	23

Beath and co-workers[7] studied the correlation between the distribution of these so-called selenium indicator plants and the nature of the soil and host rocks.   It was noted that most of the seleniferous formations are found to be of the Cretaceous period.   Thus, soil resulting from the Niobrara formation in Wyoming gave rise to a notably high maximum selenium content in indicator plants of 3,939 ppm with minimum and average values of 199 and 689 ppm, respectively.   The plants in question were principally of the genera *Astragalus* and *Stanleya*.

The particular form in which the selenium is available has a bearing on the selenium uptake by plants.   Thus, selenates are more readily absorbed than selenites and calcium selenate more so than sodium selenate.[37]   It should be understood that an important source of the selenium available to plants is the selenium accumulated from previous seleniferous vegetation.

The somewhat offensive odor of seleniferous vegetation can be related to the release of volatile selenium compounds of which four have been shown to be produced consistently, including dimethyl diselenide.[27]

It is also interesting to note that although the geological mapping of seleniferous vegetation has been of major importance in agriculture, the presence of selenium indicator plants has been used as a geobotanical prospecting tool for uranium.[11]

The geobotany of selenium and particularly the occurrence and growth of selenium indicator and accumulator plants have been discussed in considerable detail by Rosenfeld and Beath.[42]   The distribution of selenium in plants has been reviewed more recently by Johnson, Asher, and Broyer.[27]

Fossil Fuels

It is of the utmost significance to examine the selenium content of fossil fuels and thereby obtain an indication not only of the distribution of selenium in these materials but also of the amount of selenium released to the atmosphere from this source.   The selenium present in 120 fossil fuels samples, both coal and oil, in the United States, has been determined by neutron activation analysis.[41]   In the majority of the coal samples, the selenium ranged between 1 and 5 ppm with an average value of 3.2 ppm.   In the case of crude oil[40] (samples mainly from Texas), the selenium content was much lower, 0.06 to 0.35 ppm with an average value of 0.17 ppm.

In this particular study the S : Se ratios were calculated.   For the majority of the coal samples the ratio was in the range 20,000 to 2000 with an average of 3500.   In the crude oil samples the ratio extended from 160,000 to 8000 with an average value of 20,000.   Although lower S : Se ratios were evidenced in a number of the Texas crude oil samples and with selected coal samples from Montana and New Mexico, the results are not sufficiently conclusive to be related to the accumulation of selenium in soils and plants which is so well-known in the western and southern United States.

The selenium in fossil fuels is no doubt a source of the selenium which is present in airborne dust.[37]   In accordance with the above S : Se ratios and an estimated annual SO_2 emission in the United States of approximately 30,000,000 tons (90% from coal and 10% from oil combustion), the annual release of selenium would be in the order of 4,000 tons.[41]

Water and Air

Selenium has been denoted as a minor constituent in potable water in a concentration range of 0.1 to 100 μg/liter.[12]   According to the United States Department of Health,[57] the safe upper limit for selenium in drinking water is considered to be 10 μg/liter.   In the case of surface waters much higher concentrations of selenium have been found, notably in seleniferous regions. Thus, a maximum value of 400 μg/liter was noted by Scott and Voegeli[46] in an examination of 42 surface waters in Colorado.   In this study, however, the median value was 1 μg/liter.   The selenium content of ground waters in seleniferous areas is subject to considerable variation with an exceptional sample of well water from Colorado analyzing as high as 9000 μg/liter.[6]   In most instances the selenium content of ground waters is well below 100 μg/liter.

Hashimoto and Winchester[23] determined selenium in 22 samples of snow and rain during the winter of 1964–65 in Cambridge, Massachussetts as well as in 7 air samples during the following spring.   The average selenium content of the precipitation samples was 0.21 μg/liter with the S : Se ratio being 10,000.   Not unexpectedly, a significantly lower selenium value was found for well and tap water, viz., 0.1 μg/liter.   In the case of air samples, the average selenium content was 0.09 μg/100 m^3 air.   Approximately the same S : Se ratio was found as for the precipitation samples.

Although the previous S : Se ratio is typical for terrestrial materials, in the case of sea water the abundance of sulfur is much greater relative to selenium. Thus, Goldschmidt[18] has estimated a S : Se ratio in sea water of 232,000. An average value of 0.09 μg/liter was found for selenium in the principal oceans with a marked, but not consistent, variation in depth at two locations in the eastern Pacific Ocean.[45]   Values on the order of 4 μg/liter have been reported for samples from the North Sea[19] and the coast of Japan.[26]

Considering the amount of selenium carried into the oceans, there must be a substantial removal of selenium from aqueous solution by co-precipitation with basic oxides such as hydrous ferric oxide and the incorporation of these precipitates into marine sediments.

Properties

As a member of Group VIA of the periodic table, selenium displays a number of similarities to sulfur and tellurium in many of its properties.   The increasing metallic nature of the elements with increasing atomic weight is seen by the fact that oxygen and sulfur are insulators, selenium and tellurium, semiconductors, and polonium, a metal.   The resemblance between selenium and sulfur is more pronounced in a number of respects than that between selenium and tellurium.   The periodicity in properties, similar to that found in neighboring groups of the periodic system, is shown in Tables 1-3 and 1-4.

TABLE 1-3   BOND ENERGIES OF GROUP VIA ELEMENTS[a]

Bond	Energies (kcal/mole)
O—O	33.2
S—S	50.9
Se—Se	44.0
Te—Te	33.0

[a] Pauling, Linus, *The Nature of the Chemical Bond*, 3rd ed., Cornell University Press, Ithaca, N.Y., 1960.

TABLE 1-4   PERIODICITY AMONG GROUP VIA ELEMENTS[a]

	Oxygen	Sulfur	Selenium	Tellurium
Electron configuration	2-6	2-8-6	2-8-18-6	2-8-18-18-6
Covalent radius, Å	0.74	1.04	1.17	1.37
Electronegativity (Pauling scale)	3.5	2.5	2.4	2.1

[a] Pauling, Linus, *The Nature of the Chemical Bond*, 3rd ed., Cornell University Press, Ithaca, N.Y., 1960.

TABLE 1-5   ATOMIC AND NUCLEAR PROPERTIES OF SELENIUM

Property	Allotropic Modification or Physical State	Value	References and Remarks
Atomic number		34	
Atomic weight		78.96	
Periodic position		Group VIA Period 4	
Valence states		2, 4, 6	
Electronic configuration		$1s^2, 2s^2, 2p^6, 3s^2, 3p^6,$ $3d^{10}, 4s^2, 4p^4$	
Allotropic forms	Trigonal (grey) α-monoclinic (red) β-monoclinic (red) Vitreous (black) Red amorphous Black amorphous		
Stable isotopes and natural abundance		74 (0.87%) 76 (9.02%) 77 (7.58%) 78 (23.52%) 80 (49.82%) 82 (9.19%)	R. L. Heath, *Handbook of Chemistry and Physics*, 50th ed., Chemical Rubber Publishing Company, Cleveland, Ohio, 1969–70.
Radioactive isotopes and half-lives		70 (44 min) 71 (5 min) 72 (8.4 days) 73 (7.1 hr) 73 (42 min) 75 (120.4 days) 77 (17.5 sec) 79 (6.5×10^4 yr) 81 (57 min)	R. L. Heath, *Handbook of Chemistry and Physics*, 50th ed., Chemical Rubber Publishing Company, Cleveland, Ohio, 1969–70.

(Continued)

TABLE 1-5 ATOMIC AND NUCLEAR PROPERTIES OF SELENIUM (*Continued*)

Property	Allotropic Modification or Physical State	Value	References and Remarks
		81 (18.6 min)	
		83 (70 sec)	
		83 (23 min)	
		84 (3.3 min)	
		85 (39 sec)	
Ionization potentials		I 9.75 eV	R. L. Heath, *Handbook of Chemistry and Physics*,
		II 21.5 eV	50th ed., Chemical Rubber Publishing
		III 32 eV	Company, Cleveland, Ohio, 1969–70.
		IV 43 eV	
		V 68 eV	
		VI 82 eV	
		VII 155 eV	
Crystal ionic radii		Se^{-2} 1.91 Å	R. L. Heath, *Handbook of Chemistry and Physics*,
		Se^{-1} 2.32 Å	50th ed., Chemical Rubber Publishing
		Se^{+1} 0.66 Å	Company, Cleveland, Ohio, 1969–70.
		Se^{+4} 0.50 Å	
		Se^{+6} 0.42 Å	
Bond lengths: Nearest neighbor— same ring or chain	Trigonal	2.373 ± 0.0055 Å	P. Unger and P. Cherin, *The Physics of Selenium and Tellurium*, ed. W. Charles Cooper, Pergamon Press, Inc., New York, 1969, p. 224.
	α-monoclinic	2.32 Å	*Ibid.*, p. 226.
	β-monoclinic	2.34 Å	*Ibid.*, p. 227.
	Vitreous	2.34 Å (value varies with quench temperature)	R. Kaplow, T. A. Rowe, and B. L. Averbach, *Phys. Rev.* **168**(1), 1068 (1968).

Property		Value	Reference
Nearest neighbor—adjacent ring or chain	Liquid	2.36 Å	R. C. Buschert, Ph.D. Thesis, Purdue University, Lafayette, Indiana, 1957.
	Trigonal	3.436 Å	P. Unger and P. Cherin, *The Physics of Selenium and Tellurium*, ed. W. Charles Cooper, Pergamon Press, Inc., New York, 1969, p. 224.
	α-monoclinic	3.48 Å	*Ibid.*, p. 226.
	β-monoclinic	3.44 Å	*Ibid.*, p. 227.
Bond angles	Trigonal	103.1 ± 0.2°	P. Unger and P. Cherin, *The Physics of Selenium and Tellurium*, ed. W. Charles Cooper, Pergamon Press, Inc., New York, 1969, p. 224.
	α-monoclinic	105.9 ± 1.7°	*Ibid.*, p. 226.
	β-monoclinic	105.5°	*Ibid.*, p. 227.
Lattice constants	Trigonal (space group $P3_121$ or $P3_221$)	a: 4.3662 Å c: 4.9536 Å c/a: 1.134_5 β: 120°	H. E. Swanson, N. T. Gilfrich, and G. M. Ugrinic, N. B. S. Circular 539, Vol. V, 1955.
	α-monoclinic (space group $P2_1/n$)	a: 9.054 Å b: 9.083 Å c: 11.601 Å β: 90.81°	P. Unger and P. Cherin, *The Physics of Selenium and Tellurium*, ed. W. Charles Cooper, Pergamon Press, Inc., New York, 1969, p. 223.
	β-monoclinic (space group $P2_1/a$)	a: 12.85 Å b: 8.07 Å c: 9.31 Å β: 93.13°	*Ibid.*, p. 223.
Thermal neutron cross sections: Absorption	Se (normal isotopic mixture)	12.2 ± 0.6 barns	R. L. Heath, *Handbook of Chemistry and Physics*, 50th ed., Chemical Rubber Publishing Company, Cleveland, Ohio, 1969–70.
Scattering	Se (normal isotopic mixture)	11 ± 2 barns	R. L. Heath, *Handbook of Chemistry and Physics*, 50th ed., Chemical Rubber Publishing Company, Cleveland, Ohio, 1969–70.

TABLE 1-6 THERMAL PROPERTIES OF SELENIUM

Property	Allotropic Modification or Physical State	Value	References and Remarks
Melting point	Trigonal	217°C	W. Klement, Jr., L. H. Cohen, G. C. Kennedy, *J. Phys. Chem. Solids* **27**, 171–7 (1966). Purity of Se: 99.999+ % Reference also includes variation of the melting point with pressure. (10 kbar–40 kbar)
Normal boiling point	Liquid-Vapor	684.9 ± 0.1°C	M. de Selincourt, *Proc. Phys. Soc.* **52**, 348–52 (1940). Purity of Se: 99.9+ %
Vapor pressure	Liquid → Vapor	1 mm Hg (344.6°C) 10 mm Hg (431.5°C) 40 mm Hg (496.7°C) 100 mm Hg (547.4°C) 400 mm Hg (636.5°C) 760 mm Hg (684.7°C) 2 atm (742.8°C) 5 atm (831.3°C) 10 atm (909.3°C) 20 atm (999.1°C) 40 atm (1103.6°C) 100 atm (1271.4°C) 150 atm (1359.5°C)	A. A. Kudryatsev and G. P. Ustyugov, *Tr. Mosk. Khim. Tekhnol. Inst.* No **38**, 42–6 (1962). As cited in *Chem. Abs.* **60**, 1134g (1964). The equation quoted in *Chem. Abs.* is incorrect. It should read: $$\log P \text{ (mm)} \simeq -\frac{5010.7}{T} + 8.1105$$ E. H. Baker, *J. Chem. Soc.* **A3**, 1089–92 (1968). Purity of Se: 99.999+ %

Property	Transition	Value	Reference
Heat of dissociation	$Se_2(g) \rightarrow 2\ Se(g)$	72.94 ± 0.05 kcal/mole	G. G. Chandler, R. F. Barrow, and B. Meyer, *Phil. Trans. Roy. Soc. London* **260A**, 395 (1966).
Heat of fusion	Trigonal → Liquid	1.49 ± 0.04 kcal/mole	C. T. Moynihan and U. E. Schnaus, *Nat. Sci. Eng.* **6** (4), 277–8 (1970).
Heat of evaporation	Liquid → Vapor	22.92 kcal per mole	A. A. Kudryatsev and G. P. Ustyugov (cf. vapor pressure reference).
Heat of combustion	Trigonal	−56.6 kcal/mole (298°K)	L. L. Andreevna and A. A. Kudryatsev, *Tr. Mosk. Khim. Tekhnol. Inst.* **No 49**, 25–7 (1965).
Heat capacity	Trigonal (polycrystalline)	5.86 cal/deg g-atom (calculated for 298.16°K)	C. T. Moynihan and U. E. Schnaus, *Mat. Sci Eng.* **6** (4), 277–8 (1970). These authors coupled their data with those of W. Desorbo, *J. Chem. Phys.* **21** (7), 1144–8 (1953), and those of C. T. Anderson, *J. Am. Chem. Soc.* **59**, 1036–7 (1937) to obtain the equation $\bar{C}_p = 5.40 + 0.00154T$ (T in °K) Valid over the temperature range 200–480°K. The heat capacity value, at 298.16°K, was calculated from this equation.
	Vitreous	6.125 cal/deg-g-atom (298.16°K)	C. T. Anderson, *J. Am. Chem. Soc.* **59**, 1036–7 (1937). Data in range 49.9–299.1°K cf. also J. C. Lasjaunias, *Compt. Rend.* **269B** 763–6 (1969) for range 1.5–22°K, and P. Chaudhari, P. Beardmore and M. B. Bever, *Phys. Chem. Glasses* **7** (5), 157 (1966) for range 298–358°K.

(Continued)

TABLE 1-6 THERMAL PROPERTIES OF SELENIUM (*Continued*)

Property	Allotropic Modification or Physical State	Value	References and Remarks
	Liquid	7.0 cal/deg g-atom	O. Kubaschewski, Z. *Metalkunde* **41**, 445 (1950). Valid from the melting point to 790°C.
Debye temperature	Trigonal single crystal	$152.5 \pm 0.7°$K	J. C. Lasjaunias, *Compt. Rend.* **269B**, 763 (1969).
Entropy	Trigonal	10.15 ± 0.05 cal/deg g-atom at 298.16°K	W. Desorbo, *J. Chem. Phys.* **21** (7), 114–8 (1953).
Coefficient of thermal expansion	Trigonal	$(\perp C)$ 74.09×10^{-6}/deg $(\parallel C)$ -17.89×10^{-6}/deg	M. E. Straumanis, *J. Appl. Phys.* **20**, 726–34 (1949).
	Vitreous	$(<Tg)$ 1.397×10^{-4}/deg $(>Tg)$ 4.115×10^{-4}/deg $(Tg) = 30.2°$C	G. Gattow and B. Buss, *Naturwissenschaften* **56** (1), 35–6 (1969).
	α-monoclinic	a_0: -1.5×10^{-6}/°C b_0: $+84.7 \times 10^{-6}$/°C c_0: $+63.3 \times 10^{-6}$/°C	C. J. Newton and M. Y. Colby, *Acta Cryst.* **4**, 477 (1951).

Property	Form	Value	Reference
Thermal conductivity	Trigonal	5.93×10^{-3} cal/cm-sec-°K (room temp)	G. B. Abdullaev, S. I. Mekhtieva, D. S. Abdinov, G. M. Aliev, and S. G. Alieva, *Phys. Status Solidi* **13**, 315 (1966). (Purity of Se given as 99.99999%)
	Vitreous	1.21×10^{-3} cal/cm-sec-°K (room temp)	*Ibid.*
		1.72×10^{-3} cal/cm-sec-°K (304°K)	*Ibid.*
Grüneisen parameter		1.0 ± 0.1	R. K. Kirby and B. D. Rothrock, *J. Amer. Ceram. Soc.* **51** (9), 535 (1968). Valid over the temperature range 83–293°K
Critical constants		T_c: 1757°K P_c: 249 atm V_c: 168.6 cc/g-atom	As calculated by D. S. Gates and G. Thodos, *A.I.Ch.E. J.* **6** (1), 50–4 (1960).

TABLE 1-7 VISCOELASTIC PROPERTIES OF SELENIUM

Property	Allotropic Modification or Physical State	Value	References and Remarks
Viscosity	Liquid	22.10 poise 220°C 14.67 ,, 230°C 9.87 ,, 240°C 6.03 ,, 260°C 3.71 ,, 280°C 2.38 ,, 300°C 1.54 ,, 320°C 1.06 ,, 340°C 0.70 ,, 360°C	Kh. M. Khalilov, *Izv. Akad. Nauk. Azerb. SSR, Ser. Fiz. Mat. i Tekhn. Nauk.* (6), 67 (1959).
Glass transition temperature	Vitreous	$303.4 \pm 0.1°K$ (varies with purity and heating rate)	G. Gattow and G. Buss, *Naturwiss.* **56** (1), 37 (1969).
Compressibility	Trigonal single-crystal	1.06×10^{-11} cm²/kg	Calculated by J. Mort, *J. Appl. Phys.* **38**, 341 (1961).
	Trigonal polycrystalline	0.19×10^{-5} cm²/kg	Calculated by D. B. Sirdeshmukh, *Acta Cryst.* **A24**, 318 (1968).
	α-monoclinic	0.064×10^{-5} cm²/kg	*Ibid.*
	Vitreous	1.04×10^{-5} cm²/kg	*Ibid.*
	Liquid	1.47×10^{-5} cm²/kg	Calculated by D. E. Harrison, *J. Chem. Phys.* **41** (3), 844 (1964).
Microhardness	Vitreous	29 kg/mm² (room temp)	V. N. Lange, T. I. Lange, and V. A. Titov, *Sov. Phys.–Cryst.* **12** (3), 456 (1967).
		45.8 kg/mm² (0°C)	N. A. Korpatskii, *Referat. Zh. Met.* 1960, Abst. No. 7600, cited in *Chem. Abs.* **55**, 32376 (1961).
		11 kg/mm² (58°C)	*Ibid.*

Property	Form	Value	Reference
Plastic constants	Trigonal single-crystal	C_{11} 1.87 × 10¹¹ dyne/cm² — C_{11} 1.87 × 10^{11} dyne/cm², C_{12} 0.71 × ,, C_{13} 2.62 × ,, C_{14} 0.62 × ,, C_{33} 7.41 × ,, C_{44} 1.49 × ,, (All at room temp)	J. Mort, *J. Appl. Phys.* **38**, 3414 (1967).
	Vitreous	C_{11} 1.436 × 10^{11} dyne/cm², C_{12} 0.698 × 10^{11} dyne/cm² (both at 25°C)	K. Vedam, D. L. Miller, and R. Roy, *J. Appl. Phys.* **37** (9), 3432 (1966).
Young's modulus	Trigonal single-crystal	2.34 × 10^{11} dyne/cm²	Calculated by D. B. Sirdeshmukh and K. G. Subhadra, *J. Appl. Phys.* **40** (13), 5404 (1969).
	Vitreous	1.021 × 10^{11} dyne/cm²	Calculated by K. Vedam, D. L. Miller, and R. Roy, *J. Appl. Phys.* **37** (9), 3432 (1966).
Shear modulus	Trigonal single-crystal	0.92 × 10^{11} dyne/cm²	O. L. Anderson, *Physical Acoustics*, v. IIIB, p. 43, Academic Press, Inc., N.Y., 1965.
	Vitreous	0.369 × 10^{11} dyne/cm²	Calculated by K. Vedam, D. L. Miller, and R. Roy, *J. Appl. Phys.* **37** (9), 3432 (1966).
Bulk modulus	Trigonal single-crystal	1.74 × 10^{11} dyne/cm²	O. L. Anderson, *Physical Acoustics*, v. IIIB, p. 43, Academic Press, Inc., N.Y., 1965.
Poisson's ratio	Trigonal single-crystal	0.27	Calculated by D. B. Sirdeshmukh and K. G. Subhadra, *J. Appl. Phys.* **40** (13), 5404 (1969).
	Vitreous	0.327	Calculated by K. Vedam, D. L. Miller, and R. Roy, *J. Appl. Phys.* **37** (9), 3432 (1966).

TABLE 1-8   VARIOUS PHYSICAL AND OTHER PROPERTIES OF SELENIUM

Property	Allotropic Modification or Physical State	Value	References and Remarks
Density	Trigonal	4.819 g/cc (298°K)	L. A. Nisel'son and V. M. Glazov, *Izv. Akad. Nauk. SSSR, Neorg. Mater.* **4** (11), 1849 (1968). Equation is: Trigonal density $= 4.833 - 0.542 \times 10^{-3}t$ (t in °C)
	α-monoclinic	4.389 ± 0.015 g/cc (room temp)	Valid in range 20°C to melting point J. D. Taynai and M. A. Nicolet, *J. Phys. Chem. Solids* **31** (7), 1651 (1970).
	β-monoclinic	4.4 g/cc (calculated)	H. P. Klug, *Z. Krist.* **88**, 128 (1934).
	Liquid	3.975 g/cc (490°K)	L. A. Nisel'son and V. M. Glazov, *Izv. Akad. Nauk SSSR, Neorg. Mater.* **4** (11), 1849 (1968). Equation is: Liquid density $= 4.203 - 1.050 \times 10^{-3}t$ (t in °C)
	Vitreous	4.285 ± 0.0025 g/cc	Valid in range melting point to 820° G. Gattow and G. Buss, *Naturwiss.* **56** (1), 35 (1969). Equations are: Below glass transition temperature Vitreous density $= 4.46296 - 5.97 \times 10^{-4}T$ T in deg abs. Above glass transition temperature Vitreous density $= 4.81460 - 1.756 \times 10^{-3}T$ T in deg abs
Dielectric constant	Liquid	5.40 (250°C)	Nat. Bur. Standards, Circ. 514, U.S. Govt. Printing Office, Washington, D.C., 1951.
	Solid (allotrope unknown)	6.6 (room temp)	E. C. Gregg, Jr., *Handbook of Chemistry and Physics*, 50th ed., Chemical Rubber Publishing Company, Cleveland, Ohio, 1969–70.

Property	Condition	Value	Reference
Electrical resistivity	Trigonal	10^{10} ohm cm	F. Eckhart, *Ann. Physik.* **14**, 233 (1954). Depends upon purity of sample.
	Vitreous	10^{13} ohm cm	A. D. Andreev, *Sov. Phys.—Semicond.* **4** (1) 26 (1970). Depends upon purity of sample: this author used a thin film sample.
	Liquid	1.3×10^5 ohm cm (400°C)	E. H. Baker, *J. Chem. Soc. A.*, 1089 (1968). This value calculated from Baker's equation $$\log(\text{elec. rest.}) = \frac{6000}{T} - 3.80$$ (T in deg abs)
Magnetic susceptibility	Trigonal	$-12.6 \times 10^{-7}/\text{cm}^3$	M. Risi and S. Yuan, *Helv. Phys. Acta* **33** (9) 1002 (1960).
	Monoclinic	$-14.1 \times 10^{-7}/\text{cm}^3$	
	Vitreous	$-13.4 \times 10^{-7}/\text{cm}^3$ (all at 300°C)	
Molar volume	Trigonal	16.39 cc/g-atom	Calculated by author (KGB). (H. E. Straumanis *J. Appl. Phys.* **20**, 726 (1494), reports 16.420 ± 0.007 cc/g-atom; but value inconsistent with density of trig Se).
Photoelectric work function		4.62 eV	R. Hamer, *J. Opt. Soc. Am.*, **9**, 251 (1924). (R. Schulze, *Z. Physik*, **92**, 212 (1934) reports 5.11 eV)
Refractive index Ordinary	Trigonal single crystal 1.06 microns	2.790 ± 0.008	L. Gampel and F. M. Johnson, *J. Opt. Soc. Am.* **59** (1), 72 (1969).
	1.15 ,,	2.737 ± 0.008	
	3.39 ,,	2.65 ± 0.01	All values at 23 ± 2°C
	10.6 ,,	2.64 ± 0.01	
Extraordinary	Trigonal single crystal 1.06 microns	3.608 ± 0.008	
	1.15 ,,	3.573 ± 0.008	
	3.39 ,,	3.46 ± 0.01	
	10.6 ,,	3.41 ± 0.01	
	Vitreous 1.152 microns	2.4969 (21°C)	W. C. Schneider and K. Vedam, *J. Opt. Soc. Am.* **60** (6), 800 (1970).

(*Continued*)

TABLE 1-8 VARIOUS PHYSICAL AND OTHER PROPERTIES OF SELENIUM *(Continued)*

Property	Allotropic Modification or Physical State	Value	References and Remarks
Self-diffusion coefficient	Trigonal (polycrystalline)	3.8×10^{-12} cm²/sec (35°C) $4.1 \times$ " (40°C) $6.3 \times$ " (50°C) $8.9 \times$ " (81°C) $21.1 \times$ " (100°C) $88.5 \times$ " (140°C) 7.7×10^{-12} cm²/sec (35°C) $24.0 \times$ " (40°C) $320 \times$ " (50°C) $2200 \times$ " (81°C)	B. J. Boltaks and B. T. Plachenov, *Sov. Phys.—Tech. Phys.* **2**, 2071 (1957). *Chem. Abs.* **52**, 17883h (1958). Values can be expressed by the equation $$D_T = 1.4 \times 10^{-4}\, \frac{e^{-11,700}}{RT}$$
	Trigonal single-cryst. (‖c-axis)	1.0×10^{-13} cm²/sec (217°C)	*Ibid.*
	(⊥c-axis)	6.5×10^{-13} cm²/sec (217°C)	P. Bratter and H. Gobrecht, *Phys. Status Solidi* **37**, 869 (1970). *Ibid.*
Standard reduction potential	$Se + 2e^- \rightarrow Se^{2-}$ $Se + 2H^+ + 2e^- \rightarrow H_2Se(aq)$	-0.78v -0.36v	J. F. Hunsberger, *Handbook of Chemistry and Physics*, 50th ed., Chemical Rubber Publishing Company, 1969–70.
Surface tension	Liquid	105.5 dynes/cm (220°C) 100.5 dynes/cm (250°C) 98.0 dynes/cm (280°C) 95.2 dynes/cm (310°C)	K. V. Astakhov, N. A. Penin and E. Dobkina, *J. Phys. Chem.* (*USSR*) **20**, 403 (1946). *Chem. Abs.* **40**, 6915[8] (1946). Purity of Se: 99.95%
Critical surface tension	Trigonal	32 dynes/cm	D. A. Olsen, R. W. Moravec and A. J. Osteraas, *J. Phys. Chem.* **71** (13), 4464 (1967).
Thermoelectric emf	Trigonal	0.9 mv/deg	I. K. Bandrovskaya and I. P. Klyus, *Izv. Vysshidk. Uchebn. Zavdenii, Fiz.* **11** (12), 121 (1968). As cited in *Chem. Abs.* **70**, 71933n (1969).

Since the important chemical and electrical properties of selenium are dealt with elsewhere in this volume, the major physical properties and certain other properties of the element are tabulated here for convenient reference (see Tables 1-5 through 1-8). The atomic and nuclear properties as well as the thermal and viscoelastic ones are included. The structural aspects of selenium in the solid, liquid, and gaseous states are treated in detail in Chapter 3.

APPLICATIONS

Selenium has a very broad range of applications because of the variety of physical, chemical, and electronic properties of the element. However, it was not until quite recently that the utilization of these properties led to a significant consumption of selenium. Although it is difficult to indicate the percentage distribution of the uses of selenium and its compounds in industry, the principal applications are found in glass, pigments, steel, xerography, and rectifiers. In light of the extensive research on the electronic propreties of selenium and of semiconducting selenium compounds, new and expanded uses of selenium in solid state devices can be anticipated.

It was not until 1873, more than fifty years after the discovery of selenium, that its photoconductivity was noted. This resulted in the development of the selenium photocell and the first practical application of the substance. This photocell, which has undergone considerable improvement over the years, has found numerous applications, particularly in photography. Of even greater significance has been the development of the photocopying process of xerography which is based on the reproduction of images induced by light falling on a film of amorphous selenium having a static electric charge. The important xerographic properties of selenium are discussed in Chapter 16.

The asymmetric conduction of selenium has been used for some time in selenium rectifiers. These rectifiers which were prominent in the 1950s and gave rise to a strong demand for selenium, have now been supplanted in many instances by silicon rectifiers. The lower cost of selenium rectifiers and their ability to withstand power surges are their principal advantages.

It was during World War I that selenium replaced manganese dioxide as a glass decolorizer to neutralize the green tint imparted by iron, particularly in bottle glass. This application is now one of the major uses of selenium.

Cadmium sulfoselenide pigments are widely used and are considered to be superior to the mercury red and the organic red pigments. As the selenium content is increased relative to sulfur, the pigment changes color progressively from orange to deep red. Selenium pigments are used in numerous materials including paints, enamels, plastics, printing inks, glass, ceramics, and rubber.

The metallurgical applications of selenium are centered on the addition of selenium, usually in the form of ferroselenium or iron selenide, to improve the machinability of stainless steels.   This and other metallurgical applications are discussed in Chapter 15.

There are numerous minor uses of selenium.   As discussed in Chapter 14, selenium compounds such as selenium diethyldithiocarbamate are used in curing-accelerator systems for a number of rubbers.   A suspension of selenium disulfide is used to combat seborrheic dermatitis of the scalp.   The introduction of sodium selenate into plant tissues has been employed successfully in protecting certain greenhouse plants, such as carnations, against attack by thrips and aphids.   The nutritional value of trace amounts of selenium in numerous animal species including sheep, cattle, hogs, and poultry has been clearly established.[38]   Animal deficiency diseases such as white muscle disease and ill thrift in sheep in New Zealand have been counteracted successfully by the injection of small amounts of sodium selenite.   These diseases have also been combatted through the use of a preparation of vitamin E and selenium.   A selenium deficiency in livestock can also be overcome through the use of grain feeds grown in seleniferous regions.   Selenium compounds have been used for some time as photographic toners and in metal finishing operations such as steel bluing.   In addition they have shown promise as additives in plating, notably in chrome plating, where an addition of sodium selenate induces microcracking and leads to superior corrosion resistance. Selenides of refractory metals such as $MoSe_2$, $NbSe_2$, and WSe_2 have been effective as high temperature vacuum-stable solid lubricants.   Numerous organic reactions including oxidation, hydrogenation, isomerization and polymerization are catalyzed by selenium and its compounds such as selenium dioxide.   A comprehensive review of the catalytic activity of selenium has been presented.[29]   Selenium catalysts have been used commercially in catalytic liquid phase oxidations and in the isomerization of unsaturated oils. A number of selenium compounds have been proposed as antioxidants for lubricating oil and extreme pressure lubricants.   Numerous patents have been issued for these applications but the actual use of selenium in this area appears to be limited.

PRODUCTION

Since selenium is produced principally as a by-product of electrolytic copper refining, its availability is closely related to copper production.

The association of selenium with sulfur in copper sulfide minerals has resulted in the world production of selenium being related largely to those countries processing such minerals in significant tonnages *viz*, Canada, Japan, the United States and the Soviet Union.   Smaller quantities of selenium are

produced in Belgium, Chile, Finland, Peru, Sweden, West Germany, Yugo-
slavia, and Zambia.

Although it is difficult to obtain accurate data on selenium production, it
is estimated that the annual free world production is of the order of 2.7 million
pounds with a demand that approximates this level.   A substantial part of
this production comes from the United States (1,229,800 lbs in 1969 and
993,000 lbs in 1970) followed by Canada and Japan.[13,28,55,56]   The 1970
production in the United States declined from the record set in 1969 due principally to the loss of selenium in exported ores and to lower selenium values
in the ores processed.[3]   Data on the world production of selenium by countries are presented in Table 1-9.   No data are available on the production in
the Soviet Union.   The demand for selenium appears to be strong, and its
increased availability will depend on an appropriate increase in copper production combined with a higher recovery of selenium from sulfide ores.

Selenium is generally marketed as commercial grade selenium (99.5% Se
minimum) in the form of powder, small lumps, or as high purity selenium in
the form of pellets, assaying 99.99$^+$% Se.   Actually, the composition of high
purity selenium is best expressed in terms of the impurity levels since there
are a number of critical impurities which bear an important relation to the
end use of the selenium.   These include mercury, tellurium, arsenic, and
chlorine in rectifier applications.   In high purity form, selenium is used in the
manufacture of photocells and rectifiers and in xerography.   Selenium is also

TABLE 1-9 WORLD PRODUCTION OF SELENIUM BY
COUNTRIES[56] (IN THOUSAND POUNDS OF
CONTAINED SELENIUM)

Country	1967	1968	1969 p
Australia e	4	4	4
Belgium-Luxembourg			
(exports)	72	54	55
Canada	725	636	711
Finland	15	16	15
Japan	422	399	435
Mexico	5	2	42
Peru	11	13	15
Sweden	r 132	e 176	176
United States	598	633	1229
Yugoslavia	10	22	20
Zambia[a] e	57	57	57
Total[b]	r 2,051	2,012	2,759

e Estimate   p Preliminary   r revised
[a] Contained in copper refiners slimes exported for treatment.
[b] Total is of listed figures only.

available as ferroselenium and iron selenide, selenium dioxide, sodium selenite and sodium selenate.

The price of selenium has fluctuated over the years reflecting the changes in demand and availability of the element. Thus, in the case of high purity selenium, the price has ranged from $5.50 per pound in 1960 to a high of $18.00 per pound in 1956 then declining to $6.00 by 1963 and rising again in more recent years to a price of $12.50 per pound in 1973. Commercial grade selenium has followed a similar pattern; $3.50 per pound in 1950, a high of $15.50 in 1956, dropping to $4.50 per pound in 1963 and rising again to a level of $9.00–10.00 per pound in 1973.

Data are available on the annual production and consumption of selenium and these publications should be consulted for up-to-date information.[13,55,56]

REFERENCES

1. Agrinier, H., Jacques, G., and Raoul, F., *Compt. Rend., Serv.* D263 (5), 465 (1966).
2. Anders, E., *Space Sci, Rev.* 3, 583 (1964).
3. Baltrusaitis, V. A., *Eng. Min. J.*, p. 101, March 1971.
4. Bana, M., and Mochnacka, K. Freiberg, *Forschungsh.* C230, 327 (1968).
5. Banas, M., and Ottemann, J., *Przegl. Geol.* 15 (5), 240 (1967).
6. Beath, O. A., *Science News Letter*, 81, 254 (1962).
7. Beath, O. A., Draize, J. H., Eppson, H. F., Gilbert, C. S. and McCreary, O. C., *J. Am. Pharm. Assoc. Sci. Ed.* 23, 94 (1934).
8. Benavides, S. T., and Mojica, R. F. S. *Inst. Geograf. Colombia Publ.* No. IT 3, 1 (1959).
9. Berzelius, J. J., *Annales de Chimie et de Physique*, Paris, Serie 2, Tome 7, 199 (1817).
10. Boitsov, V. E., *At. Energ,* (*USSR*) 20 (1), 46 (1966).
11. Cannon, H. L., *U.S. Geol. Sur. Bull.* No. 1030, 399 (1957).
12. Davis, S. N. and De Wiest, R. J. M., *Hydrogeology*, John Wiley & Sons, New York, 1966.
13. *Yearly Review, Selenium and Tellurium.* Department of Energy, Mines and Resources, Mineral Resources Branch, Ottawa, Canada.
14. Dufresne A., *Geochim Cosmochim. Acta*, 20, 141 (1960).
15. Dzhandzhgava, M. I., *Geokhimiya* 5, 579 (1966).
16. Efendiev, G. Kh., Kislyakova, L. E., and Zul'fugarly, N. D., *Geokim. Redk. Elem.*, 9, (1966).
17. Fleischer, M., *J. Chem. Ed.*, 31, 446 (1954).
18. Goldschmidt, V. M., *Fortschr. Mineral. Krist. Petrog.* 19, 183 (1935).
19. Goldschmidt, V. M., and Strock, L. W., Nachr. Geo. Wiss. Gottingen, Jahresber, *Geschaftsjahr Math.—Physik. Kl, Fachgruppen II*, 1, 123 (1935).
20. Gorbunov, G. I., *Mater. Mineral. Kol'sk. Polnostrova, Akad. Nauk. SSR, Kol'sh Filial* No. 4, 5 (1965).
21. Greenland, L., *Geochimica et Cosmochimica Acta*, 31, 849 (1967).
22. Hamada, S., Sato, R., and Shinai, T., *Bull. Chem. Soc. Jap.*, 41 (4), 850 (1968).
23. Hashimoto, Y., and Winchester, J. W., *Environ. Sci. and Tech.*, 1, 338 (1967).

24. Inin, V. D., *Tr. Inst. Geol. Nauk. Akad. Nauk. Kaz. SSR*, **17**, 110 (1966).
25. Inin, V. D. and Litvinovich, A. N., *Tr. Inst. Geol. Nauk. Akad. Nauk. Kaz. SSR*, **17**, 125 (1966).
26. Ishibashi, M., Shigemetsu, T., and Nakagawa, Y., *Oceanog. Works Japan Rec.* **1**, 44 (1953).
27. Johnson, C. M., Asher, C. J., and Broyer, T. C., *Symposium-Selenium in Biomedicine*, Muth, O. H., ed., Avi Publishing Company, Westport, Conn., 1967, p. 57.
28. Killin, A. F., *Selenium and Tellurium Report* No.43, Mineral Resources Branch, Department of Energy, Mines and Resources, Ottawa, Canada, 1968.
29. Kollonitsch, V., and Kline, C. H., *Ind. Eng. Chem.* **55** (12), 18 (1963).
30. Kozyrev, V. V., *Sb. Nauchin. Tr., Inst. Geol. i Geofiz., Akad. Nauk. Uz. SSR* (3), 14 (1967).
31. Kvacek, M., Suran, J., and Ambroz, F., *Casopis Mineral Geol.*, **10** (4), 411 (1965).
32. Lakin, H. W., in *Symposium-Selenium in Biomedicine*, Muth, O. H. ed., Avi Publishing Company, Westport, Conn., 1967, p. 30.
33. Lakin, H. W. and Davidson, D. F., *Symposium-Selenium in Biomedicine*, Muth, O. H., ed., Avi Publishing Company, Westport, Connecticut, 1967, p. 27.
34. Litvinovich, A. N., *Tr. Inst. Geol. Nauk, Akad. Nauk. Kaz. SSR*, **17**, 98 (1966).
35. Mellor, J. W., *A Comprehensive Treatise on Inorganic and Theoretical Chemistry*, John Wiley & Sons, Inc., New York.
36. Morrison, G. H., *Chem. Eng. News*, Jan. 19, 1970.
37. Moxon, A. L., Olson, D. E., and Searight, W. V., *South Dakota Agr. Expt. Stn. Revised Techn. Bull.* No. 2, 1 (1950).
38. Muth, O. H., ed., *Symposium-Selenium in Biomedicine*, Avi Publishing Company, Westport, Conn., 1967.
39. Palachi, C., Berman, H., and Frondel, C., *System of Mineralogy*, 7th ed., John Wiley & Sons, New York, 1951.
40. Pidzhyan, G. O., *Izv. Akad. Nauk. Am. SSR, Nauki Zemle*, **20** (5–6), 81 (1967).
41. Pillay, K. K. S., Thomas, C. C., Jr., and Kaminski, J. W., *Nucl. Applns and Tech.*, **7**, 478 (1969).
42. Rosenfeld, I., and Beath, O. A., *Selenium-Geobotany, Biochemistry, Toxicity, and Nutrition*, Academic Press, Inc., New York, 1964.
43. Ruzmatov, S. R., *Uzb. Geol. Zh.*, **10** (6), 55 (1966).
44. Schindewolf, U., *Geochim. Cosmochim. Acta* **19**, 134 (1960).
45. Schutz, D. F., and Turekiam, K. K., *Geochim et Cosmochim. Acta*, **29**, 259 (1965).
46. Scott, R. C., and Voegeli, P. T., *Colo.-Water Conservation Bd., Basic Data Report* No. 7, 1961.
47. Sindeeva, N. D., *Mineralogy and Types of Deposits of Selenium and Tellurium*, Interscience Publishers, John Wiley & Sons, New York, 1964.
48. Suess, H. E., and Urey, H. C., *Rev. Mod. Phys.* **28**, 53 (1956).
49. Suzuoki, T., *Chem. Soc. Japan Bull.* **38**, 1940 (1965).
50. Suzuoki, T., *Chem. Soc. Japan Bull.* **38**, 1946 (1965).
51. Swaine, D. J., *Tech. Comm.* **48**, Commonwealth Bur. Soil Sci., Rothamsted Expt. Stn., Harpenden, England, 1955.
52. Timofeeva, T. S., *Tr. Sredneaziat. Nauch.—Issled. Inst. Geol. Miner. Syr'ya*, **7**, 23 (1966).

53. Tugarinov, A. I., and D'yachkova, I. B., *Geokhimiya*, **9**, 1035 (1966).
54. Tugarinov, A. I., and D'yachkova, I. B., *Vop. Prikl. Radiogeol.* **2**, 380 (1967).
55. *Selenium*, U.S. Bureau of Mines, Quarterly Mineral Industry Survey.
56. *Minerals Yearbook*. U.S. Bureau of Mines.
57. *Drinking Water Standards*, U.S. Dept. of Health, Education and Welfare, Public Health Service Publication 956 (1962).
58. Vakhrushev, V. A., and Dorosh, V. M., *Geokhimiya*, (11), 1349 (1966).
59. Velikii, A. S., and Sindeeva, N. D., *Geokhim. Mineral Genet. Tipy Mestorozhd. Redk. Elem., Akad. Nauk. SSSR. Gos. Geol. Kom. SSSR. Inst. Mineral, Geokim, Kristallokhim, Redk. Elem.* **3**, 534, 832 (1966).
60. Vinogradov, A. P., in Sindeeva *Mineralogy and Types of Deposits of Selenium and Tellurium*, Interscience Publishers, John Wiley & Sons, New York, 1964, p. 162.
61. Vuorelainen, Y., Huhma, A., and Häkli, A., *Bull. Comm. Geol. Finlande*, **36** (215), 113 (1964).
62. Weeks, M. E., *J. Chem. Ed.*, **9**, 477 (1932).
63. Zul'fugarly, N. D., and Geidarov, A. S., *Izv. Akad. Nauk, Azub. SSR, Ser Nauk Zemle*, **3**, 52 (1966).

2

Recovery and Refining of Selenium

P. H. JENNINGS

Noranda Research Centre, Pointe Claire, Quebec

J. C. YANNOPOULOS

Newmont Exploration Company, Ltd., Danbury, Connecticut

Part I: Recovery of Selenium

INTRODUCTION

Naturally occurring concentrations of selenium are, for the most part, too lean for economic recovery to be possible, and probably all the world's selenium is a by-product of some other industry. The largest source of selenium is the anode slimes that are formed during the electrolytic refining of copper, but significant quantities are also recovered from the sludge accumulating in sulfuric acid plants and from electrostatic precipitator dust collected during the processing of copper and lead ores. The selenium in each of these three intermediate products is derived from minerals such as chalcopyrite, galena, and pyrite, where it substitutes for sulfur.[209]

During pyrometallurgical processing, some selenium follows the sulfur into the gaseous products, but being less volatile and less readily oxidized than its sister element, the greater part of it ultimately ends up in condensed phases such as matte, slag, and dust.   Studies have been made of the distribution of selenium between the products of copper smelting,[42,44,51,170,189] and the selenium distributes itself roughly as follows:

	%	%
Roaster dust and gases	25–30	20
Reverberatory dust and gases ⎫	15–30	4
Reverberatory slag ⎭		9
Converter dust and gases	5–15	4
Blister copper	25–54	63
Reference	44	189

The very limited degree of volatilization of selenium in the converter has been confirmed in small-scale experiments.[19]   In one plant,[132] the overall loss of selenium in processing ore to blister copper has been reported to be only 25%.   In electrorefining, selenium in the anodes is found almost quantitatively in the slimes along with other elements that do not dissolve in the electrolyte, such as tellurium, gold, silver, lead, and the platinum metals.

Copper Refinery Slimes.   There are wide variations amongst copper refineries both in the yield of anode slimes and in their composition.   Table 2-1 lists some typical data.   Besides the compositional analysis, the phase analysis is of interest in achieving an understanding of the formation of slimes and in the development of physical methods of separation such as flotation.   Methods for the phase analysis of slimes have been developed by several investigators[62,64,102,108,158] based mainly on the use of selective solvents.   Some results of research of this kind are given in Table 2-2.

While some constituent phases in copper anodes may enter the anode slimes without undergoing chemical change, the majority are probably the result of precipitation reactions occurring in the electrolyte.   Greiver[95] made a composite anode by clamping together an alloy of copper and selenium and an alloy of copper and silver.   The slimes formed during electrolysis contained the compound CuAgSe which could not have been present in the anode.   This compound is generally found in slimes that contain silver and selenium.   An excess of either element may precipitate as silver or as copper selenide ($Cu_{2-x}Se$), respectively.   Silver selenide, Ag_2Se, may be formed as a result of the slow decomposition of CuAgSe in the electrolyte through the action of sulfuric acid and dissolved oxygen.   A similar action on $Cu_{2-x}Se$ may result in the formation of CuSe.

The particle size of most of the constituents of anode slimes is extremely small.   The slow settling of these particles is responsible for their occlusion onto the cathode.   The use of polyacrylamide increases the particle size by coagulation.[135]

Dusts from Copper and Lead Smelters.   The operations of roasting and smelting eliminate substantial quantities of selenium along with sulfur, probably in the form of selenium dioxide.   The dioxide has a sublimation temperature of 315°C, and it is unlikely to condense completely ahead of the precipitators. However elemental selenium, formed as a result of reduction by sulfur dioxide, condenses and reports in the dust.   An important part of the selenium in the dust is derived from the reaction between volatile metals and their oxides with selenium or selenium dioxide, forming selenides and selenites.   By selective dissolution, it is possible to make separate determinations of selenium, selenium dioxide, and the selenides and selenites of lead, zinc, and mercury.[69]

The selenium content of precipitator dust varies from a few ppm up to 1% or more.   Roaster dusts contain large amounts of entrained ore particles which dilute the condensation products, but smelter and copper converter dusts have higher proportions of condensed solids and are more correctly referred to as fume.   Roaster dusts, therefore, are a less promising source of selenium than the dusts derived from smelting and converting.

Sulfuric Acid Plant Sludge.   Comparatively little selenium is derived from acid plants that use elemental sulfur, but the use of roaster gases derived from pyrites or nonferrous base metal ores generally leads to the accumulation of selenium-bearing sludges.   In the chamber process, part of the selenium forms a sludge in the Glover tower, but most of it collects in the lead chambers.   In the contact process, in which the gas is thoroughly cleaned before entering the converter, the selenium reports in the scrubber mud.   In each case, the selenium is present mainly in elemental form, arising from the reduction of selenious acid by sulfur dioxide.   A method has been prescribed for the identification of the forms in which selenium occurs in sulfuric acid plant sludge.[63]

Other Sources.   The development of wet-oxidation processes for treating sulfide ores opens up the possibility of more efficient recovery of selenium. In some of these processes the sulfur is recovered in the elemental form instead of being converted to the dioxide.   In a process developed by the International Nickel Company,[45] nickel sulfide anodes are electrolyzed with the production of a sulfur-rich anode mud which can be fractionally distilled to remove selenium.   In this way selenium can be upgraded from 0.15% in the mud to 20%.   A notable source of secondary selenium arises in those industries where the element is used in the uncombined state, as in selenium rectifiers and xerography.

TABLE 2-1 TYPICAL COMPOSITIONS OF COPPER REFINERY SLIMES

Refinery	Yield lb Slimes per Ton Anodes	Composition %																References
		Cu	Ag	Au	Se	Te	As	Sb	Bi	Pb	Ni	Fe	S	Al$_2$O$_3$	CaO	SiO$_2$	Others	
A. S. and R., Baltimore, Md.	12.5	26.68	10.5	0.77	9.65	7.34	3.78	3.10	0.25	5.15	0.49	0.08	—	—	—	—		272, 273
A. S. and R., Perth Amboy, N.J.	—	20	20.6*		7.5	1.5	5.0	4.0	0.7	12.0	1.0	1.5	4	0.6	0.7	1.0	Sn 0.7	272, 273
Anaconda, Great Falls, Mont.	5.6	19.3	30.3	0.26	3.4	8.1	5.1	4.2	0.2	4.6	0.17	0.07	2.0	—	—	—	Zn 0.1	272, 273
Anglo American, Nkana, Zambia	2.8	36	4.0	.00031	7.9	1.0	—	—	0.2	0.9	—	—	—	2.8	—	6.9		271, 272
Baia Mare, Romania	—	13.1	23.2	0.42	2.08	0.09	1.3	4.0	—	14.0	—	0.5	—	—	—	15.2	Zn 0.6	90
Bolidens, Sweden	17	40	11.3	1.54	21	1	0.8	1.5	0.8	10.0	0.5	0.04	3.5	—	0.05	—	Co 0.02, Zn < 1	272, 273
Candian Copper Refiners, Montreal E., Que.	16	31.8	15.7	1.73	18.28	4.26	0.26	0.48	—	4.73	0.10	—	2.64	—	—	4.14		30, 272, 273
Cerro de Pasco Corp., Oroya, Peru	45	2.74	26.5	0.17	0.64	0.45	2.80	16.0	0.47	13.8	—	2.10	5.10	0.90	—	16.3	Zn 0.4	272, 273
Csepel, Hungary	—	—	—	—	1.49	—	—	—	—	—	—	—	—	—	—	—		76

* Total of Ag and Au contents.

34

Electrolytic Refining and Smelting, Port Kembla, N.S.W.	20	16.91	4.5	1.65	2.70	0.98	9.75	17.0	0.52	6.50	2.20	—	4.6	—	—	—		272, 273
Furukawa Electric Company, Japan	—	10–22	6–20	0.1–0.9	2–5	—	0.3–2	0.9–2.7	0.2–0.4	18–33	0.4–1	—	—	—	—	—		221
International Nickel, Copper Cliff, Ont.,	8.3	25.4	—	—	13.2	1.80	0.45	0.15	—	2.25	6.65	0.12	7.25	—	—	—		272, 273
Kennecott Copper Corp., Garfield, Utah	15	26.4	15.2	1.39	11.59	4.84	1.40	0.49	0.14	8.64	—	—	—	—	—	—		272, 273
Mansfeld A.G., East Germany	20–24	18–20	35–40	—	6–7	—	1	3–6	—	3–6	5–10	—	2	—	—	—		152
Mount Lyell, Tasmania	35–40	75.6	0.88	0.16	2.01	0.33	0.05	0.40	Tr.*	0.30	0.09	0.23	11.15	—	—	—	Insol. 1.2	272, 273
Nippon Mining Company, Japan	—	19.12	9.46	0.49	4.71	2.07	—	4.6	0.76	17.9	—	—	—	—	—	—		222
Outokumpu, Pori, Finland	9–13	11	9.6	0.51	4.33	—	0.70	0.04	—	2.62	45.21	0.60	2.32	—	—	—	Sn 1.0, Insol. 2.25	272, 273
Roan Selection Trust, Mufulira, Zambia	2.55	34	9.6	.00043	4.6	0.7	—	—	9.30	1.35	—	—	—	—	10.1	13.6		271, 272
Roan Selection Trust, Ndola, Zambia	—	82	1.8	.00018	2.9	0.1	—	—	0.4	0.25	—	—	—	—	2.0	1.57		271
Shen'yan, China	—	11.5	—	—	1.3–1.4	0.14	3.0	10.0	—	—	5.0	—	—	—	—	—		266

* Trace.

35

TABLE 2-2   ELEMENTARY AND PHASE ANALYSIS OF COPPER REFINERY SLIMES

No.	Slimes Analysis, %						Atomic Ratio Ag/Se	Selenium-Bearing Constituents Identified				References[a]
	Cu	Ag	Au	Se	Te	Others		Ag₂Se	CuAgSe	Cu₂₋ₓSe	Others	
1	38	33	3	21	4	Pb 7	1.15		Major	Small Minor		196
2	34	27	2	17	5	Pb 14	1.16		Major	Trace(?)	PbSO₄ major	196
3	45	20	<1	21	13	Pb <1	0.70		Major	Major		196
4	32.8	5.87	0.11	8.64	5.39	Pb 0.48 S 5.1	0.50	Yes	Yes	Major	Ag, Ag₂S, Ag₂Te, CuSe, NiS₂, PbSO₄, PbSe, Au, (Au, Ag)Te₂, Cu₂Te	67
5	20.3	13.7	0.2	7.3	1.5	Pb 23.5 Bi 0.8 As 5.4 Sb 6.0 Sn 1.2 Ni 0.7 S 3.7	1.37	Yes	Yes		PbSO₄, SnO₂, PbSb₂O₆, etc	102
6	6.3	33.5	0.01	0.7	0.3	Pb 11.3 Bi 1.1 As 4.8 Sb 8.6 Sn 7.7 Ni 0.3 S 1.9	35	Yes	Yes		Ag, SnO₂, PbSb₂O₆, etc.	102
7	18–20	45		7		Pb 3–6 As 1 Sb 3–6 Ni 5–10 S 2	4.7	Yes	No	Possibly		152

RECOVERY OF SELENIUM FROM ORES

In the recovery of selenium from intermediate products, its production is, of course, governed by the demand for the major end product, which is generally copper. There exist, in certain locations, ores of selenium whose development not only appears to be feasible but would provide an independent source of the element. The methods that have been proposed for treating selenium ores include flotation and leaching, the principal leaching agents being cyanides and calcium oxychloride. Gold ores containing selenium or tellurium are generally roasted to sublime these elements, and their recovery from sublimates has been proposed.[268]

Flotation. The recovery of selenium by flotation from certain sandstone formations of New Mexico has been shown to be both practical and economical.[23] The ores in the Morrison formation contain up to about 0.4% selenium in the form of the native element and as selenides of iron, copper, silver, lead, and bismuth. Some of the selenium-bearing deposits are currently being exploited for uranium, the selenium discarded in the tailing. Test work[23] has shown that the selenium minerals are liberated at 48-mesh, and after conditioning with soda ash and sodium silicate to depress slimy gangue minerals, the selenium values can be recovered by the use of kerosene (0.1 lb/ton), potassium amyl xanthate (0.05 lb/ton), and Dowfroth 250 (0.1 lb/ton). Concentrates ranging from 3 to 16% selenium can be made with overall recoveries between 71 and 93%. By a slightly modified procedure, the lower grade Galisteo deposits can be worked up to a concentrate containing 4% selenium with a recovery of 75%. The calculated cutoff grades are 0.12 and 0.04% selenium for underground and open pit mining operations respectively, when selenium is the only value recovered. As little as 0.01% selenium, however, is enough to justify its recovery as a by-product in the treatment of a uranium ore.

Leaching. The solubility of elemental selenium in solutions of alkali cyanides is well known and it is utilized in the treatment of certain intermediate products (see the following discussion). The ability of cyanide solutions to dissolve selenium from selenides is the basis of a patented method[31] for treating Mexican ores containing Aguilerite (Ag_4SeS), Naumanite (Ag_2Se), Eucairite (CuAgSe), Clausthalite (PbSe), and other minerals. The method is claimed to be effective for the recovery of germanium as well as any associated silver or gold. The procedure is similar to that used in the cyanidation of gold ores except that more concentrated cyanide solutions are used (0.1–0.3%). The precipitation of selenium from the clarified leach solution, like that of silver and gold, is claimed to be effected by means of zinc or aluminum powder, though other methods are required for the recovery of germanium, if it is present.

Plaksin et al.[177] found that cyanidation was not suitable for ores containing both gold and selenium because of interference by selenium in the precipitation of gold on zinc dust, the selenium coating the zinc particles. This difficulty does not seem to have been encountered in the work described in the preceding paragraph in which, despite the claim that gold, silver, and selenium were all precipitated by zinc, no procedure is given for separating the selenium from the associated elements. Plaksin[177] developed a method for treating gold-selenium ores in which the selenium is first leached in calcium oxychloride solution from which it can be recovered by means of sulfur dioxide or iron, and the leach residue is then cyanided to recover the gold. Selenium extractions of 95% or higher were achieved by agitation leaching or percolation leaching with a 1–3% calcium oxychloride solution, which oxidized selenides to selenious acid.

Methods have been patented for the recovery of selenium from carbonate leach solutions arising from the treatment of uranium ores.[101,191] One involves precipitating selenium as the sulfide and coagulating the precipitate with copper ammine sulfate. Alternatively, the selenium may be precipitated with the uranium and then volatilized by roasting at about 800°C.

RECOVERY OF SELENIUM FROM COPPER REFINERY SLIMES

In the early days of electrolytic copper refining there was little interest in recovering selenium, and the quick recovery of the precious metals was accomplished by smelting processes in which the unwanted elements were slagged off. Although all refineries still employ smelting for gathering the precious metals into a bullion which is subsequently parted, today nearly all precede this operation with one or more hydrometallurgical or roasting steps. Smelting treatment is facilitated by the removal of copper, which can be accomplished by means of leaching in dilute sulfuric acid and providing oxidizing conditions either with nitre or air, but the process is slow. Selenium compounds can be broken down by roasting, the selenium being driven off as the dioxide or, if soda is added to the charge, retained as water-soluble sodium selenite and selenate. Roasting processes tend to be characterized by dusting losses, and they have been replaced in many refineries by sulfatizing treatments using concentrated sulfuric acid. This method of treatment results in the oxidation of copper, selenium, tellurium, and other elements, and if the temperature is high enough selenium can be expelled and recovered from the gases. The use of sulfuric acid as an oxidizing agent has the disadvantage that a reaction product is sulfur dioxide.

There is a resemblance between the methods employed for treating base-metal sulfide ores and those used for processing copper refinery slimes, in that the acidic elements (sulfur, selenium, tellurium, arsenic, etc.) are removed by oxidation. The methods are primarily oxidizing roasting or smelting with oxidizing fluxes. The need to recover some of the acidic elements has led to the adoption of processes in which closer control can be exercised over the chemistry. The development of pressure-leaching methods for treating base-metal sulfide ores constitutes a parallel trend, though less far advanced, which also results in the more effective recovery of the acidic element, sulfur, either in the elemental form or as ammonium sulfate, etc.

Smelting Processes

In the most widely used smelting process for slimes treatment, the slimes are fused in a small reverberatory furnace, known as a Doré furnace, and subjected to fluxing treatments which are varied to suit the nature of the charge. Selenium and tellurium are generally collected in a soda slag, being oxidized by blowing air into the bath, and the slag is subsequently leached to dissolve the selenium. The treatment of selenium-bearing slimes now almost invariably begins with other processes designed to make a more effective recovery of this element, and Doré furnace smelting serves to recover residual amounts of selenium besides providing a ready means for treating reverts.

The cupellation process can be used for treating copper refinery slimes when a lead refinery is located in the same plant. The slimes are introduced onto the surface of a bath of molten lead, the metal serving to collect the silver and gold, and the resulting litharge containing the other slimes constituents is returned to the lead smelter. This method tends to give a poor recovery of selenium owing to the fact that the element is distributed between slag, dross, bullion, and fume. The slimes are first treated with dilute sulfuric acid at a temperature near the boiling point and sparged with air to dissolve most of the copper, which would otherwise contaminate the litharge. This process is used by the Métallurgie Hoboken N.V., Olen, Belgium, and the Cerro Corporation, Oroya, Peru. It was formerly used by the Norddeutsche Affinerie, Hamburg.

At Hoboken,[230] a lead bath is obtained by fusing lead containing precious metals (e.g., crusts from the Parkes desilverization process). The slimes, which are first decopperized by leaching in sulfuric acid, are introduced onto the surface of the bath, selenium is oxidized and is recovered from the fume, while tellurium and precious metals pass into the bullion. Tellurium is subsequently removed during the cupellation and finally collected in the pig lead resulting from the reduction of tellurium-bearing lead oxide dross. Contacting the pig lead with molten caustic soda in the absence of an oxidizing agent effects the extraction of the tellurium into the slag as sodium telluride. Its recovery from the slag is accomplished by the usual method of water leaching and aeration.

The Cerro process[20,132] treats a mixture of anode slimes from the company's lead and copper refineries. This is first smelted and then converted, whereupon lead is slagged and antimony and arsenic are fumed off. The resulting metal is cupelled to yield a gold–silver alloy, a bismuth-rich slag, a tellurium-rich soda slag, and selenium-bearing fume. The soda slag is leached with water, and the resulting solution is used to leach the cupellation furnace fume. The final solution contains about 30 gpl (grams per liter) selenium and 60–80 gpl tellurium and is treated by conventional methods. The recovery of selenium from the slimes is 45–50%. Direct treatment of slimes, rather than fume, for selenium recovery has not so far been economically practical, owing to their low selenium content (1.8%).

A modification of soda-smelting has been patented in which the valence state of the selenium in the slag can be controlled.[186] Selenium is generally present in soda slags in the +4 and +6 valence states, and its recovery from slag leach solutions is incomplete unless special measures are adopted to reduce the selenium from the higher valence state. Controlled additions of carbonaceous material are made to the molten slag, so as to reduce the hexavalent selenium, the degree of oxidation being shown by the color of solidified samples.

Roasting of Copper Refinery Slimes

In the roasting of copper refinery slimes, atmospheric oxygen is used to convert selenides and tellurides into oxidized forms. Selenium may be expelled as selenium dioxide vapor and subsequently recovered from the gases, or retained as the selenite of copper and silver. Tellurium dioxide is not sufficiently volatile to be vaporized and is usually recovered from the roasted product by leaching. By roasting a mixture of slimes and soda ash, the selenium can be converted to water-soluble forms, thus facilitating its recovery. Sulfatizing of slimes differs from roasting in that sulfuric acid is used as the oxidant; this is considered in the following section.

While the fine particulate form of slimes might be considered an advantage for promoting oxidation by air, complete oxidation is by no means easy. Silver selenite, Ag_2SeO_3, is very readily formed. This compound melts at 530°C, frequently causing the formation of a liquid film which obstructs the contact of oxygen with the unoxidized material. The success of roasting processes, therefore, depends upon the maintenance of a porous texture that allows the continued penetration of the solid aggregate by oxygen. Before considering the ways in which this can be achieved, the reactions that occur during roasting will be discussed.

The Chemistry of Roasting. Several studies have been made of the roasting of slimes and of their synthesized constituents,[3,117,152,247] while the dissociation of the intermediate products $CuSeO_3$ and Ag_2SeO_3 was the subject of separate investigations.[18,34]

The oxidation of Cu_2Se takes place in air above 300°C forming $CuSeO_3$ and CuO. At about 580°C, the normal selenite changes to a basic compound, $2CuO \cdot SeO_2$, with evolution of selenium dioxide, and the basic selenite dissociates at 640°C. Pure Ag_2Se is resistant to oxidation by air up to 400°C, but in the presence of catalysts such as elemental silver, CuO, Ag_2O, and sodium carbonate, it oxidizes at lower temperatures. The dissociation of silver selenite takes place according to the reaction:

$$Ag_2SeO_3 \longrightarrow 2Ag + SeO_2 + \tfrac{1}{2}O_2$$

which occurs energetically at about 550°C and goes to completion above 700°C.

The double selenide, CuAgSe, behaves rather like a mixture of the simple selenides:[117] basic copper selenite is formed at 400–450°C, and silver selenite at 500–550°C. At still higher temperatures, first the copper selenite and then the silver selenite dissociate with the evolution of selenium dioxide.

In the presence of sodium carbonate, the expulsion of selenium dioxide vapor is largely prevented, and the selenium is converted to a mixture of sodium selenite and sodium selenate. Tishchenko[227–229] studied the effects of heating pure synthesized selenides with sodium carbonate in a current of air, and postulated the following reactions which he demonstrated to be thermodynamically possible:

$$Ag_2Se + Na_2CO_3 + O_2 \longrightarrow 2Ag + Na_2SeO_3 + CO_2$$
$$Ag_2Se + Na_2CO_3 + \tfrac{3}{2}O_2 \longrightarrow 2Ag + Na_2SeO_4 + CO_2$$
$$Cu_2Se + Na_2CO_3 + 2O_2 \longrightarrow 2CuO + Na_2SeO_3 + CO_2$$
$$Cu_2Se + Na_2CO_3 + \tfrac{5}{2}O_2 \longrightarrow 2CuO + Na_2SeO_4 + CO_2$$

He found that the best temperature for soda-roasting was in the range 650–700°C, since the selenium reported mainly in the form of selenite (rather than selenate) and losses by volatilization were low. The preference for sodium selenite over the selenate arises from the fact that the former can be reduced in acid solution by sulfur dioxide, whereas the latter compound cannot unless a halide is present. Tishchenko's experimental results apply to pure selenides, and no experiments with mixtures of selenides or the double selenide CuAgSe were done.

The behavior of tellurium during soda-roasting is important, since it is frequently desirable to minimize the dissolution of tellurium in water leaching of the roasted product. Sodium tellurite is water-soluble, and most of the tellurium contained in pure gold telluride ($AuTe_2$) and silver telluride (Ag_2Te) is converted to this form on roasting with an excess of soda at 400–500°C,[212] particularly if some nitre is present. At higher temperatures, the sodium tellurite tends to oxidize to the tellurate, which is sparingly soluble in water. The yield of water-soluble tellurium is lower if copper compounds are present, owing to the formation of water-insoluble sodium copper tellurites and tellurates. Thus, the solubility of tellurium after soda-roasting tends to be depressed by high roasting temperatures, by the presence of copper in the slimes, and by the addition of nitre to the charge.

Roasting processes fall into the following three groups according to the form in which the selenium reports:

1. Selenium is retained as alkali-soluble selenites.
2. Selenium is volatilized as the dioxide.
3. Selenium is retained as water-soluble sodium selenite and selenate.

These methods are considered separately in the following sections.

Roasting to Form Selenites. There are several descriptions of processes, or development work, in which the object was to oxidize the selenides in the slimes to selenites and thus render them soluble in alkaline solutions.[22,24,57,90,94,148–150,192] Much work has been done at Mansfeld AG, in East Germany, on the development of suitable roasting techniques.[149,150,152] In 1940, this refinery was using a smelting process for treating slimes, which gave way to a process in which partially decopperized slimes were roasted at 500°C with soda. This, in turn, was superseded by a volatilization process at 700°C, but the tendency of the slimes to fuse together resulted in an overall recovery of selenium of only 42% and the process was abandoned in 1957.

The Mansfeld slimes have an average composition as follows:

	%		%
Cu	18–20	Sb	3–6
Ag	35–40	Pb	3–6
Se	6–7	Ni	5–10
As	1	S	2

A study of the roasting of copper selenide and silver selenide[152] led to the development of a low temperature roasting process which was introduced into the refinery in 1957. The slimes are not decopperized before roasting, because copper oxide is a catalyst for the oxidation of silver selenide. The oxidation step is carried out in muffle furnaces at 350–400°C, and the roasted product is pulverized and leached in sodium hydroxide solution. After precipitating the selenium, the overall recovery is stated to be 91.4%. In a patented modification of this process[148] the slimes are mixed with 25% of their weight of magnesium oxide before roasting. This is claimed to give certain advantages in operation. The use of muffle furnaces is apparently not entirely satisfactory since it is difficult to maintain close control over the temperature. Optimal results are obtained between 350 and 380°C with the Mansfeld slimes, and with the unusually low sintering temperature of these slimes (460°C), overheating soon leads to difficulties.

In an attempt to find a roasting method in which more accurate control of temperature could be exercised, research was done on the application of fluid-bed roasting to Mansfeld slimes.[192]   This technique was found to work best at near sintering temperatures, for lower temperatures resulted in the elimination of the finer particles from the reactor before they were completely oxidized.   By operating at 420°C, it was found that sufficient agglomeration occurred to ensure adequate residence times for the smaller particles, and an average residence time of about 1 h resulted in the solubilization of more than 90% of the selenium.   In spite of its apparent success, there has been no indication of the adoption of this process.

The Romanian practice in the treatment of anode slimes for selenium recovery apparently differs little from the Mansfeld process.[24,90]   The composition of the Baia Mare slimes is as follows:

	%		%
Cu	13.1	Sb	4.0
Ag	23.2	Pb	14.0
Au	0.42	Fe	0.5
Se	2.08	Zn	0.6
Te	0.09	Si	15.2
As	1.3		

The dried slimes are roasted in an oxidizing atmosphere at 270–320°C, and the product is leached in 20% sodium hydroxide solution. After the reclaiming of selenium from the solution, the overall yield is about 75% at a purity of 98–99.5%.

Although sodium hydroxide is normally used to extract selenium from the product of roasting, alkali metal chloride solution may be employed, with the formation of silver chloride which has a lower solubility product than silver selenite.[22]

Roasting to Expel Selenium Dioxide.   At the high temperatures needed to dissociate silver selenite (700°C or higher) and to eliminate selenium from the slimes, the problem caused by the tendency of the slimes to fuse is, of course, much more acute than in the lower temperature roasting processes. The exothermic oxidation reactions tend to proceed so fast that the heat developed cannot readily be dissipated. Two patented processes have been developed which claim to have solved this problem,[100,146] and in a notable development known as the Leningrad Mining Institute process, a new approach has been made to the reclaiming of the evolved selenium dioxide.

In a patent assigned to American Metal Climax Inc.,[100] which is applicable to other selenium-bearing materials besides anode slimes, the material is formed into pellets or other suitable shapes, such as briquettes or noodles, so that roasting can be satisfactorily performed in beds of far greater thickness than is practicable with a powdery charge. In making the pellets, the moisture content is adjusted to 9–12%, and 5–7% bentonite is added as a binder and thoroughly mixed into the slimes. Roasting can be carried out on a moving belt roasting furnace, in a rotary kiln, or other devices, at a temperature in the range 650–760°C, and preheated air is circulated freely through the bed which may be as much as six inches deep. The elimination of over 98% of the selenium is said to take place, and the selenium dioxide is scrubbed from the effluent gases by conventional means. (This patent also covers a soda-roasting process which is described later.)

A different approach from pelletizing is followed in a patented procedure assigned to the American Smelting and Refining Company[146]: a thin layer of slimes is passed on a continuous belt through a gas-heated muffle having two temperature zones. In the first zone the temperature is 540°C and the selenides are oxidized by a current of air without selenium being volatilized. In the second zone, at 600–700°C, selenium dioxide is expelled and is subsequently scrubbed from the gases. The volatilization can be accelerated by adding to the charge either the oxides or the sulfides of copper or iron.

Pelletizing of slimes is also applied in the Leningrad Mining Institute (LGI) method,[86,87,92,93,97,199] which has been in commercial operation for several years.[87,199] The slimes are partially dried to about 16% moisture, made into pellets 10–30 mm in diameter, and roasted in an electrically heated vertical kiln. Preheated air is passed upwards through the bed, which is maintained at 600–650°C for 16–18 h. The selenium dioxide, equivalent to at least 98% of the selenium in the charge, is conveyed in the roaster gases to heated chambers containing layers of sodium carbonate, where it is quantitatively absorbed. The resulting sodium selenite and selenate are reduced to the selenide by heating the soda with charcoal or fuel oil.[174] The product is dissolved in water and sparged with air. Elemental selenium forms according to the reaction:

$$Na_2Se + H_2O + \tfrac{1}{2}O_2 \longrightarrow Se + 2NaOH$$

and the sodium carbonate is regenerated by passing carbon dioxide into the filtered solution.   This process is said to give higher purity products than the sulfatizing method (see later section), to consume less reagent, and to result in improved operating conditions.

Selenium dioxide can also be recovered from the gases by absorption in zinc oxide or sodium oxide.   When zinc oxide is used, the surfaces of the granules are converted to zinc selenite.   The selenium can be extracted by leaching in sodium carbonate solution, and the resulting zinc carbonate recycled.[41,96,181,182]

Zambian copper refinery slimes[271] do not fuse below 750°C, and the development of a roasting process for treating them proved to be relatively straightforward.   A rotary kiln was employed to ensure freedom from hot spots and free access of air.   By operating it with a discharge temperature of 830°C, it was possible to eliminate more than 99% of the selenium in 1.5 h.   At this temperature mild sintering occurred which reduced dust losses without affecting selenium elimination.   Copper in the roasted slimes could be dissolved in dilute sulfuric acid at 60°C, leaving a residue from which silver and gold were recoverable by conventional methods.   Zambian slimes are now shipped abroad for treatment, but the process developed for their domestic treatment is likely to become economically feasible in the future.

Roasting with Soda.   The purpose of soda-roasting is to convert selenium to a water-soluble compound—sodium selenite or selenate.   When tellurium is present, an attempt is usually made to minimize its solubility in water. It is often recovered by an acid leach after selenium has been extracted. This sequence of operations—soda-roasting, water leaching, and acid leaching—facilitates the separate recovery of selenium and tellurium from the leach solutions.   The conditions required for the conversion of selenium to a water-soluble form are: (1) intimate mixing of the slimes with soda ash, and (2) an adequate supply of air, which necessitates maintaining an open texture.   Incipient fusion results in the blocking of pores so that air cannot reach the unreacted material.   Close control of temperature is, therefore, crucial.   The maximum safe temperature depends on the fusion point of the lowest melting constituent, and hence on the composition.   The highest practicable temperature is often employed in order to minimize the yield of water-soluble tellurium.

Soda-roasting processes are used at Bolidens Gruvaktiebolag, Rönnskar, Sweden,[203] at Norddeutsche Affinerie, Hamburg, Germany[61], and at plants in the Soviet Union.[35,154]   At Bolidens, roasting is carried out in multiple hearth furnaces at the unusually low temperature of 450°C.   The slimes are very low in tellurium (1%), and the use of higher temperatures to render tellurium insoluble in water is probably unnecessary. The ratio of soda to slimes is 0.45. A rotary kiln is employed at Norddeutsche, where the soda: slimes ratio is 0.20. The slimes are first decopperized by leaching in dilute sulfuric acid. The roasting temperature is 650–680°C, the slimes being initially treated to destroy glue arising from the refinery electrolyte which would otherwise cause overheating. This is achieved by mixing them with 4% of nitre and giving them a preliminary low temperature pass through the furnace. Shaft furnaces and pelletized charges are in use in the Soviet Union[154] where the roasting temperature is controlled so as not to exceed 575°C. Electrical heating is employed in all these plants to secure the necessary control of temperature.

Other methods of preparing charges having suitable physical characteristics for soda-roasting include the extrusion of noodles[100] and the careful drying of a paste made with sodium carbonate monohydrate above 35°C, yielding a "macroporous bed of microporous masses."[238] Fluidized beds have been used experimentally for roasting,[169] but no plants operating this kind of roaster have been studied.

Sulfatizing of Copper Refinery Slimes

Roasting processes have been replaced in many refineries by concentrated sulfuric acid treatment. This reagent sulfatizes the base metals in the slimes and oxidizes selenium and tellurium. At suitable temperatures, selenium is expelled almost wholly as the dioxide with little or no elimination of tellurium. The process has the advantage over the roasting of slimes in air in that dusting losses are negligible, but the use of sulfuric acid as an oxidizing agent gives rise to a pollution problem by the emission of sulfur dioxide. The mixture of the dioxides of selenium and sulfur usually passes to a scrubber system where selenious acid is formed, but considerable quantities of amorphous selenium are precipitated by the action of sulfur dioxide. This causes difficulties in reclaiming the product.

The sulfatizing treatment is usually carried out in two stages, referred to, respectively, as digestion and roasting. The first stage is carried out below the boiling point of sulfuric acid (330°C) in heated kettles, the formation of copper sulfate and other base metal sulfates providing bulk and enabling the proper consistency to be attained for charging to the roasters. Without this preliminary step, slimes rich in selenium would not be able to hold enough sulfuric acid in a paste of the right consistency. The second stage takes place in a roasting furnace through which a thin layer of the digested slimes is passed; its temperature is raised sufficiently to expel selenium dioxide. The combustion gases are generally confined to separate compartments.

Chemistry of the Sulfatizing Process.   Studies in which slimes or their constituents have been treated with concentrated sulfuric acid have shown that at the temperature of the digestion stage selenium is oxidized to $SeSO_3$ and tellurium to $TeSO_3$[119,120,183] by reactions such as the following:

$$Se + H_2SO_4 \longrightarrow SeSO_3 + H_2O \qquad (\sim 100°C)$$
$$Cu_2Se + 5H_2SO_4 \longrightarrow 2CuSO_4 + SeSO_3 + 2SO_2 + 5H_2O \quad (170–230°C)$$
$$Ag_2Se + 3H_2SO_4 \longrightarrow Ag_2SO_4 + SeSO_3 + SO_2 + 3H_2O \quad (170–230°C)$$

The compound $SeSO_3$ forms a green solution which, on dilution with water, precipitates selenium. The decomposition of $SeSO_3$ takes place slowly at 200°C but rapidly at 300°C according to the reaction:

$$2SeSO_3 \longrightarrow SeO_2 + Se + 2SO_2$$

In practice, considerably higher temperatures are employed in the roasting stage and most of the selenium is probably expelled as the dioxide.

The reactions of tellurium and tellurides are analogous to those of selenium and selenides; the corresponding reaction of tellurium in the roasting stage is stated to be:

$$5TeSO_3 \longrightarrow 2TeO_2 \cdot SO_3 + 3Te + 4SO_2$$

the tellurium oxysulfate being non-volatile and remaining in the solid product.

Sulfatizing Processes.   The sulfatizing process we have described is the most popular method for the recovery of selenium from anode slime. It is practiced in the U.S., Canada,[30,53] the Soviet Union,[226] Finland,[203] Hungary,[76] China,[266] and Japan.[222] Digestion is normally carried out between 200 and 300°C for times ranging from 1 to 12 h, the vessels being made of cast iron or mild steel which, despite the need for frequent replacement, is less expensive than more corrosion-resistant materials.

Roasting temperatures are generally between 400 and 500°C, but may be as high as 700°C.[30] The expulsion of selenium exceeds 90% in every reported case except one[53] where half of the selenium is recovered from Doré furnace products. Roasting furnaces are generally of a type that permits a thin layer to be passed through the hot zone continuously; a moving belt made up of overlapping plates constitutes a commonly used arrangement. Some refineries use rotary kilns[222] or retorts.[76]

The proportion of sulfuric acid in the charge to the digesters varies according to the selenium and base metal contents of the slimes. One part slimes requires from 0.5 to 2.0 parts sulfuric acid (66°Bé).

The recovery of selenium from the roaster gases is usually accomplished by passing them through scrubber systems or wet Cottrell precipitators. A series of bubblers is used in one plant,[266] but such units have the disadvantage of requiring considerable back pressure to operate them. The collection of amorphous selenium from scrubber systems may be facilitated by the use of anionic surface-active agents such as lignosulfonic acid (Goulac) which maintains the reduced selenium in a dispersed state.[188]

A modified sulfatizing process was developed by the Kennecott Copper Corporation[134] in which the slimes are fused with a mixture of sodium sulfate or bisulfate and sulfuric acid; the elimination of selenium is thus achieved in a single step. The fusion kettle is in the form of a rotary kiln operated at 650°C, the residence time of the charge being 8 h. The weight of the charge can be reduced by a preliminary treatment with dilute sulfuric acid to remove part of the copper. Such treatment, however, would probably not be permissible before the normal process, on account of the part played by copper sulfate in promoting satisfactory physical properties.

A more radical departure from normal sulfatizing practices has been patented[176] in which the slimes or other selenium-bearing material are mixed with sulfuric acid and heated to 730°C in a retort. The evolved selenium dioxide passes through a bed of charcoal nearer the mouth of the retort, where it is reduced to the element, and liquid selenium is collected in a condenser held at 230°C.

Where sulfatizing is used in the treatment of slimes, it is normally the first operation after dewatering. It is usually followed by leaching in dilute sulfuric acid to extract copper together with variable proportions of the silver and tellurium. The latter two elements are cemented with copper powder or scrap and the precipitate is treated for their recovery. The leach residue is normally treated in a Doré furnace for the collection of precious metals. The alternative method of recovering gold and silver by cyanidation is claimed to permit the recovery of lead and antimony, which are normally lost in Doré furnace slags.[55,221]

Leaching of Copper Refinery Slimes

Leaching is adjunct to some of the processes that have already been discussed concerning the treatment of copper refinery slimes. In these processes, the components of the slimes are oxidized prior to leaching, for example, by roasting in air or digesting with concentrated sulfuric acid, and leaching serves merely to extract the elements that have been solubilized. Oxidative leaching, on the other hand, is a fairly new approach to slimes treatment, apart from the well-established method for decopperizing slimes in aerated sulfuric acid solution. It owes its inception to the establishment of pressure hydrometallurgy in a number of fields, particularly for the direct leaching of sulfides.

Slimes constitutents, particularly selenides, are not readily broken down by leaching below 100°C when the oxidant is atmospheric oxygen, but high rates of attack can be attained at somewhat higher temperatures and under hyperatmospheric pressures of oxygen. Despite considerable interest in pressure leaching using oxygen in either acid or alkaline media, only one such process has yet been reported to be in operation for the treatment of the slimes.[221]

Interest has also been shown in the use of oxidants that are sufficiently potent to break down the slimes constituents at lower temperatures. These include gaseous chlorine,[68,127] hydrochloric acid,[231] and ferric chloride solution. The chlorination of silver and gold, which takes place when gaseous chlorine is used, is likely to cause volatilization losses if pyrometallurgical methods are employed to obtain the precious metals. This necessitates the use of wet processing throughout the flow-sheet.

Chemistry of Leaching. Powerful oxidants such as gaseous chlorine attack the slimes constituents more or less indiscriminately, while the use of less potent leachants permits the selective oxidation of the elements. Gaseous oxygen, although thermodynamically capable of oxidizing all the usual slimes components except gold, may be used to selectively oxidize copper and tellurium. It has been shown[37] that various copper(II) species such as cupric ion and cupric hydroxide are theoretically capable of oxidizing elemental selenium and tellurium to the tetravalent states but not to the hexavalent states.

A general approach to the chemistry of leaching is afforded by the use of potential pH diagrams and solubility data. According to the compilation of Pourbaix,[187] the oxidation potentials of the usual slimes components increase in the order:

$$Cu < Te < Se < Ag < Au$$

at 25°C and at all pH values except those in strongly alkaline media, when copper and tellurium are reversed. This is a simplified picture because slimes constituents are, for the most part, compounds rather than elements. That it is generally applicable is, however, shown by the fact that the elements tend to be oxidized in the order prescribed. When oxygen is the oxidant, the attack may be confined to the earlier members of the series by limiting the time or temperature.

The leach may be conducted in acid, neutral, or alkaline media, and the selection of a suitable pH range permits different results depending on the solubility of the oxidized species. The main considerations on which to base the choice of pH (apart from practical limitations such as corrosion) for media based on sulfuric acid and sodium hydroxide are as follows:

1. Copper(II) is soluble in acid media, insoluble in neutral media, and slightly soluble in strongly alkaline media.
2. Selenium(IV) and selenium(VI) are soluble at all pH values.
3. Tellurium(IV) is sparingly soluble in acid media, insoluble in neutral media, and soluble in alkaline media.
4. Tellurium(VI) is soluble in acid media and insoluble in alkaline media.
5. Silver(I) is soluble in acid media and insoluble in neutral and alkaline media.

The use of strongly complexing acids or bases, such as hydrochloric acid and ammonia will, of course, lead to quite different solubilities. The situation may be further complicated by side-reactions between various species, such as the precipitation of silver selenite in a certain range of pH, and by the effects of the leaching medium on consituents such as lead sulfate and silica, which dissolve in strongly alkaline conditions.

Leaching Processes.   The investigations of Tamura and his associates[133,219-221] led to the introduction of the first all-hydrometallurgical process for treating copper refinery slimes.   The composition of the slimes is given in Table 2-1, and the process is outlined in Figure 2-1.   The process is based on the decomposition of the slimes by pressure leaching in dilute

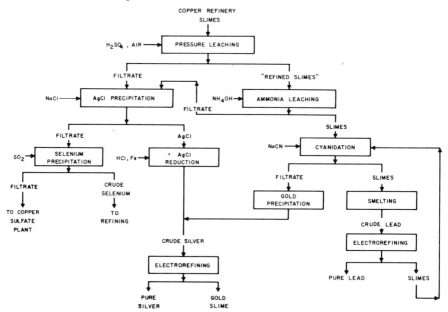

Figure 2-1 Slimes treatment process operated by the Furukawa Electric Company (Nikko Electric Copper Works), Japan (221).

sulfuric acid at 180°C using air at 700 psi.   At this stage substantially all the copper and nickel, 90% of the selenium, and about half the silver dissolve. The residue is leached in ammonia to dissolve most of the remaining silver. Silver in both acid and ammonia solutions is precipitated as the chloride which is filtered off and reduced to the metal by iron powder in a concentrated hydrochloric acid solution.   The crude silver is melted together with crude gold (from a later stage) and treated for the recovery of the pure metals by conventional electrorefining processes.   The residue from ammoniacal leaching contains all the gold, the remaining silver (about 10% of the silver in the slimes), and lead sulfate.   It is cyanided to collect the silver and gold, and lead is reclaimed from the residue.

Selenium is recovered from the pressure leach solution, in which it is present in the selenate form, by reduction with sulfur dioxide.   The reduction of copper, which would lead to the formation of copper selenide, is prevented by maintaining a sulfuric acid concentration of 1.5 N.

The process is not only simpler than the pyrometallurgical one it replaced, but it gives an improved selenium recovery (90%), permits the treatment of slimes varying widely in composition, and saves 30–40% in the cost of labor.

Leaching in solutions of ferric or cupric salts, particularly ferric chloride, permits selenides and tellurides to be broken down at temperatures below the boiling point.   Tellurides leach more readily than selenides, but these reagents appear not to be selective.[16,17,233]

Leaching in alkaline media has attracted more attention than acid or saline leaching.   Alkaline leaching, using sodium hydroxide or occasionally sodium carbonate solutions, has been the subject of a number of investigations in which either oxygen[28,29,33,39,59,60,162,242] or cupric oxide[36,38,40,66] has been used to effect oxidation.   The earliest investigations are those of Vaaler[242] employing air as the source of oxygen.   The leaching of slimes at 200–250°C in 10–40% sodium hydroxide solution with an oxygen partial pressure of 60–120 psi resulted in the solubilization of all the selenium as the selenate, and the evolved solution was practically free from tellurium.

Many later studies have attempted to obtain selenium in the form of the more easily reduced selenite.   This may be done either by the use of milder leaching conditions or by the substitution of cupric oxide for oxygen.   In the second approach, the ability to achieve clean separations between selenium and tellurium is lost, the insoluble sodium tellurate not being formed. Moreover, silver selenide is reported[38] to be scarcely attacked even at 250°C.

By conducting the leaching operation at lower temperatures or under lower pressures of oxygen, Elkin et al.[59,60] have shown that the oxidation of tellurium(IV) to tellurium(VI) proceeds under milder conditions than that of selenium(IV) to selenium(VI) and that it is possible to recover most of the selenium from the slimes in the form of a sodium selenite solution free from tellurium. The oxidation of selenium to the higher valence state, however, may be catalyzed by other constituents of the slimes, for while it has been noted[29] that pure copper selenide yields all its selenium in the form of sodium selenite on leaching at 100–300°C, a process evaluation on a pilot plant scale[162] indicated that between 10 and 15% of the selenium was oxidized to sodium selenate.

Other Methods of Treating Copper Refinery Slimes

Experimental studies have been made of distillation and chlorination in the treatment of anode slimes, but no commercial processes have been described in the literature. Attempts have been made to volatilize selenium and tellurium from slimes by mixing them with sulfur and heating *in vacuo*.[113,114] Although the sulfides of copper and silver are more stable than the selenides and tellurides, the formation of sulfides is hindered by the tendency to volatilize the sulfur before it has had time to react. Some of the selenium and tellurium, but not all, can be volatilized from decopperized slimes by heating *in vacuo* at about 700°C, especially if the slimes are mixed with soda ash or cupric oxide.[65]

Two chlorination processes have been patented. In each process, the slimes are treated with gaseous chlorine, or a mixture of chlorine and carbon monoxide, at a temperature at which the chlorides of selenium, tellurium, arsenic, tin, etc. are volatilized, while those of the precious metals remain in the residue. In one process, the slimes are pelletized with a binder such as bentonite.[105] In the other, the slimes are suspended in a molten bath consisting of the chlorides of alkali and alkaline earth metals, and chlorine is bubbled through the melt.[156,157] If chlorination is carried out at 800°C in the presence of carbon, silver is volatilized as its chloride together with selenium, tellurium etc., while gold and platinum group metals remain in the residue.[122]

TREATMENT OF SULFURIC ACID PLANT SLUDGE

While the concentration of selenium in sulfuric acid plant sludge may be as high as 30%,[7] it is more usual for the selenium content to be about 1%.[80,184] Even such low concentrations may be from 300 to 1500 times greater than those in the raw materials, such as zinc or pyrite concentrates.[6,80] The development of methods for treating sulfuric acid plant sludge for the recovery of selenium has been discouraged, due to the low percentages of selenium, but considerable attention has been accorded in recent years in the Soviet Union to the exploitation of this source.[143]

The composition of sulfuric acid plant sludge is very variable and depends on the relative percentages of volatile elements in the raw material. In zinc roasting plants the sludge may contain as much as 52% lead in the form of lead sulfate.[184] Arsenic and mercury are also common impurities. This variability probably accounts for the large number of approaches to the problem of selenium recovery and for the fact that no single method appears to have been generally adopted.

Flotation has proved effective for the concentration of selenium from sludges containing less than about 2% of this element.[184] The flotation is carried out in an acid pulp (pH below 3), and the use of reagents is not always necessary as the sludge may contain suitable collectors and frothers. Red (monoclinic) selenium floats more readily than the gray variety, and exposure of the sludge to high temperatures that would cause conversion to the gray form should therefore be avoided. Flotation results in the concentration of mercury as well as selenium.

Sublimation and vacuum distillation have been studied[7,116,142,232] and are sometimes found to give poor yields, possibly because the selenium vapors tend to form selenides by reaction with other constituents of the sludge. Mercury selenide can be effectively decomposed by iron shavings at 800°C.[116,232]

Attempts have been made to use leachants that will dissolve elemental selenium, such as alkali metal sulfides,[21,125] sulfites,[91,184,201] or cyanides.[118,244] The equilibrium:

$$Se + SO_3^{-2} \rightleftharpoons SeSO_3^{-2}$$

has been studied[129] and selenium concentrations of 0.5 M and above are attainable if the solution is kept on the alkaline side. One attempt to utilize the reaction for extracting selenium from sludge resulted in incomplete extraction and the resulting solutions were unstable.[184] On the other hand, the method is claimed[91] to be capable of yielding a sulfur-free product from a mud containing 2% selenium and up to 20% sulfur. Cyanide leaching, in which the selenium forms the seleno-cyanide complex, has been employed, but it is disfavored on account of the hazard attending the precipitation of selenium by acidification.[7]

The more successful methods of treatment involve oxidation of the selenium. As in the treatment of copper refinery slimes, roasting in air with or without soda ash[244] and sulfatizing with sulfuric acid[171] enjoy considerable popularity. Roasting with soda ash results in the formation of sodium selenite and selenate which can be extracted with water and the selenium precipitated.[44] In the sulfatizing method, selenium dioxide is evolved and is recovered in scrubbers. A modification of the sulfatizing method has been patented[176] in which the operation is conducted in a retort and the selenium dioxide vapour passes through a bed of carbonaceous material.   In this way it is reduced to the element and may be condensed to liquid selenium.

The use of nitric acid has been studied in this connection,[7] but a large excess of the reagent is required, the reaction is a vigorous one, and protection from nitrous fumes is essential.   Other oxidants that have been tried include potassium chlorate[210], and chlorine.[58,128]

Sludges that contain much arsenic require a preliminary treatment to remove this impurity.   Leaching in dilute hydrochloric acid is the usual method,[144] but leaching first in hot water and then in hot sodium carbonate solution is claimed to offer advantages in reagent cost.[110,111]   Another attempt was made to develop a method for upgrading sludges containing high concentrations of lead by leaching the lead sulfate in brine,[184] but the consumption of the reagent proved to be too high for economic application.

RECOVERY OF SELENIUM FROM MISCELLANEOUS METALLURGICAL PRODUCTS

Along with other elements that have volatile species, selenium tends to concentrate in various intermediate metallurgical products, notably Cottrell precipitator dust in plants for the roasting of pyrites, zinc sulfide, or lead sulfide,[80,248] as well as dusts collected from copper and lead smelter gases. Selenium accompanies copper into the sulfide dross formed in the drossing of lead bullion.[81]   Sometimes the selenium occurs in the elemental form, permitting its recovery by flotation and a method has been patented[88] in which the material is given a low temperature densifying roast, followed by water-leaching to remove soluble salts and flotation of selenium from the repulped residue.   Cyanidation has also been used for recovering elemental selenium.[239]

When the selenium is wholly or partially in the form of heavy metal selenides, the usual method of treatment is sulfatization with concentrated sulfuric acid, the process resembling that which is applied to copper refinery slimes.[75,82,83,164,185,215]   Arsenic remains in the residue unless coke is added to the charge,[75] while indium, thallium, tellurium, and germanium are always left in the residue.   Sulfatization is unsuitable for dusts containing recoverable amounts of germanium since the solubility of this element is lowered during the treatment.[75]

Selenium and tellurium, either as the elements or as selenides and tellurides, can be reclaimed by treatment with 1% sodium amalgam in the presence of water.[240]   Sodium selenide and telluride are formed and dissolve in the water, while the base metals unite in an amalgam.   By passing air through the solution, the selenium and tellurium are precipitated.

Electrolytic leaching in sodium hydroxide solution has been studied as a means for recovering selenium from lead sinter-plant dust.[48,138] From a dust containing 49% lead, 12% zinc, 15% arsenic, 2.2% sulfur, 1.0% cadmium, 0.7% copper, and 0.58% selenium, the selenium, arsenic, and sulfur were dissolved and selenium was recovered by acidifying with hydrochloric acid followed by reduction with sulfur dioxide.

Smelting processes have been described for the treatment of lead cupellation dusts[180] and sulfide dross[14] and a sintering method has been outlined for treating sludges containing arsenic and selenium.[2]

In the aqueous electrolytic process for the recovery of nickel from nickel sulfide anodes,[45] the anodic residue of elemental sulfur contains any selenium originally present in the anodes.   In the Port Colborne refinery of the International Nickel Company, the anode residue contains 0.15% selenium, and after filtering off the metallic impurities above the melting point of sulfur, the elements are separated by rectification, yielding a bottom product containing 20% selenium.

Efforts have been made to develop methods for removing selenium and tellurium from copper either during the conversion of white metal or subsequently by treatment during fire refining.   The applications of such processes would be limited because electrolytic refining, which effectively removes selenium and tellurium is generally necessitated by the presence of precious metals.   Research[19] has shown that selenium and tellurium cannot be removed by volatilization during the conversion of matte, but by selective conversion it is possible to retain considerable proportions of the selenium and tellurium in a relatively small quantity of white metal.   There are practical difficulties, however, in separating white metal and copper in a normal converter.   Treatment of molten copper in the fire refining furnace with a mixture of soda ash, lime, and coal was effective in removing selenium only if reducing conditions were maintained, which would complicate the normal refining operation.   Additions of elemental calcium to molten deoxidized copper were shown to extract selenium into a slag so long as a proportion of the calcium remained dissolved in the copper.   Subsequently, the selenium returned to the melt.   The removal of the small quantity of selenium-rich slag at the right stage was considered to be impracticable. In a separate study the addition of lime to a copper melt containing oxygen and selenium was shown to cause a redistribution of selenium in favor of the slag.[79]

An electrolytic procedure for the removal of oxygen, sulfur, selenium, and tellurium from molten copper has been demonstrated on a laboratory scale.[259] The electrolyte was barium chloride and the operation was carried out at 1150°C.   Sulfur and selenium entered the electrolyte as barium sulfide and selenide, while tellurium escaped from the cell as the vapor.   Although the power requirements were shown not to be excessive, the practical difficulties in operating a cell using molten barium chloride as the electrolyte were considerable.

SECONDARY SELENIUM

Where selenium is used in the elemental form, as in rectifiers and xerography, its recovery from secondary sources is feasible. In selenium rectifiers, it is associated with aluminum and a bismuth-cadmium alloy, and in xerography, with an aluminum alloy only. Mechanical means such as machining, hammer-milling, or shot-blasting with particles of a soft material, are usually employed to detach the selenium from the aluminum base[104,194,211] and a method has been patented[141] which uses a liquid that wets the molten selenium but not the bismuth–cadmium alloy.

A well-known process for the recovery of selenium involves leaching in fused caustic soda[126,211,254] in which disproportionation occurs:

$$6NaOH + (2x + 1)Se \longrightarrow 2Na_2Se_x + Na_2SeO_3 + 3H_2O$$

The solidified mass is leached with water, and selenium is precipitated by the addition of an acid (such as hydrochloric, acetic) or bubbling with carbon dioxide, causing the reversal of the reaction. The conditions must be controlled so as to prevent the evolution of hydrogen selenide, which is highly toxic.

Sulfite leaching is an alternative method which avoids the hazards associated with the caustic fusion method. The leachant is a solution of sodium sulfite, and the selenium may be reprecipitated from the resulting selenosulfite solution by lowering the pH to 5–6 with a mineral acid or chlorine gas.[104,141]

Other methods that have been described include vacuum distillation, leaching in nitric acid,[211] and leaching in alkaline phosphate solutions.[260]

RECOVERY OF SELENIUM FROM AQUEOUS SOLUTIONS

Selenium-bearing solutions are the penultimate stage of most of the processes for the recovery of the element. The preceding operations depend upon the nature of the starting material; once in solution, however, selenium recovery depends upon two main factors: (1) the valence state of the selenium, and (2) the presence of impurities that may contaminate the product.

Dissolved selenium may have a valence of -2, $+4$ or $+6$. Many industrial solutions contain the element in more than one valence state, and solutions are often purified before precipitating selenium. The selenium can generally be obtained in the elemental state by oxidation from the -2 valence state or by reduction from the $+4$ and $+6$ valence states.

Recovery from the −2 *Valence State.* In the Bolidens soda roast process for treating copper refinery slimes,[203] the resulting solution contains the carbonate, selenite, and selenate of sodium. It is evaporated to dryness, reduced with coke in an electric furnace to form sodium selenide, and the resulting product is then redissolved in water. This new solution is then sparged with air, and 90% of the selenium is precipitated by the reaction:

$$2Na_2Se + O_2 + 2H_2O \longrightarrow 2Se + 4NaOH$$

This reaction is also used in the Leningrad Mining Institute process, referred to previously (see Roasting of Copper Refinery Slimes). Following the separation of the precipitated selenium, the solution is carbonated to precipitate the remaining selenium and to regenerate sodium carbonate:

$$2Na_2Se + O_2 + 2CO_2 \longrightarrow 2Se + 2Na_2CO_3$$

Elemental selenium, when treated with concentrated sodium hydroxide solutions or with molten sodium hydroxide, disproportionates into polyselenide and selenite. These species can coexist only in strongly alkaline solutions, and upon acidification selenium is precipitated by the reaction:

$$2Na_2Se_x + Na_2SeO_3 + 3H_2SO_4 \longrightarrow (2x + 1)Se + 3Na_2SO_4 + 3H_2O$$

The electrodeposition of selenium from selenide solutions is feasible, and it has been claimed to give a product particularly suitable for the manufacture of rectifiers and photocells.[25]

Recovery from the +4 *Valence State.* Most of the selenium produced industrially is recovered from solutions where it is found predominantly in the tetravalent form. Such solutions may be derived from the dissolution of selenium dioxide in weakly acidic aqueous media. More often, these solutions contain an excess of sodium carbonate, as when soda slags from Doré furnacing or the products of soda-roasting are leached in water. Selenium is precipitated by the use of sulfur dioxide or, occasionally, sodium sulfite,[132] by a reaction such as:

$$H_2SeO_3 + 2SO_2 + H_2O \longrightarrow Se + 2H_2SO_4$$

Although sulfuric acid is a product of the reaction, the initial alkaline solution is always neutralized before introducing sulfur dioxide, so as to precipitate impurities such as tellurium, lead, and silica. Lead precipitates as the sulfate at about pH 12, silica precipitates slowly and incompletely over a wide pH range, and tellurium precipitates below pH 10, reaching a minimum solubility between pH 5 and 6. Since separation of the selenium product from tellurium is important, the pH of the solution is generally carefully adjusted to about 5 or 6, and the solids are filtered off before further acidi-fication and gassing with sulfur dioxide.[30,53,132]   This procedure generally results in a small loss of selenium as lead selenite, which could probably be prevented by carbonation at a pH above 9 where basic lead carbonate has a lower solubility than lead selenite.   The neutral solution after filtration, generally contains between 0.1 and 1 gpl tellurium.[30,53]   The selenium content may be as high as 125 gpl.

The precipitation of selenium by sulfur dioxide is often carried out below 30°C, or between 30 and 35°C[253], to ensure the formation of the red, amor-phous form which readily remains in suspension and thus does not clog solution lines.   Precipitation of the gray hexagonal form is practical only above about 85°C.   At intermediate temperatures, the selenium may form plastic or coke-like masses which foul cooling coils.   Completion of the reaction may require from 2 to 10 h, depending on the initial concentration of selenium.   In practice, a small amount of selenium is allowed to remain unreduced in order to prevent the contamination of the product with tellur-ium.   An analysis of the kinetics of the reduction of selenious acid and tellurous acid by sulfur dioxide in a sulfuric acid medium[99] showed that the rate constant for the precipitation of selenium is directly proportional to $[H^+]^2$, while that for the precipitation of tellurium is inversely proportional to $[H^+]^2$.   Thus, the increasing concentration of sulfuric acid during the precipitation of selenium tends to inhibit the precipitation of tellurium. Accordingly, careful process control should enable practically all the selen-ium to be precipitated without contamination.

The roaster gases in the sulfatizing of anode slimes contain both selenium dioxide and sulfur dioxide, resulting in the precipitation of part of the selenium in the scrubber system.   Since the temperature cannot be kept below 30°C, the precipitated selenium tends to coagulate and to deposit on the surfaces of the equipment, necessitating shutdowns for cleaning.   By maintaining a concentration of at least one gpl lignosulfonic acid, it is claimed[188] that the precipitated selenium can be maintained in the form of a dispersion.   Animal glue or other flocculents are used subsequently to recover the selenium.

For the recovery of selenium and tellurium from dilute solutions metallic copper can be used[124] in the case of those solutions which contain the elements in their hexavalent forms.   The precipitate is a mixture of copper selenide and copper telluride.   The precipitation of selenium(IV) by metallic copper in sulfuric and nitric acid media forms the compounds Cu_2Se and CuSe.   In a hydrochloric acid medium, the formation of elemental selenium is favored.[78]   Selenium can also be precipitated from acid solutions by metal-lic iron, and the reaction is greatly accelerated by the presence of Cu^{2+} ions.[145]

In the production of sulfuric acid, traces of selenium may be removed and reclaimed from the acid by means of sulfur dioxide,[98] elemental sulfur at 125–130°C,[252] or sodium silicate.[160]

The use of zinc amalgam is reputed[85,251] to effect the precipitation of both selenium and tellurium from solutions containing sulfuric acid. In a 5% sodium carbonate solution, selenium is not precipitated but tellurium, and lead are reduced and removed from the solution.

In the carbonate leach process for extracting uranium from its ores, any selenium present in the solution can be precipitated by the addition of a soluble sulfide and flocculated by means of cuprammonium sulfate.[101]

Solvent extraction is a well-known analytical tool for separating tetravalent selenium and tellurium from base metals as well as from the hexavalent forms of these elements. Solvents such as tributyl phosphate are used, but the method has not been applied to industrial separations. On the other hand, the extraction of platinum metals by amines is purported to effect a clean separation from selenium.[26]

Recovery from the +6 Valence State.   Selenate solutions cannot be reduced by treatment with sulfur dioxide at atmospheric pressure except in the presence of a catalyst such as a halide or thiourea. Solutions in which a significant fraction of the selenium is present as selenate, particularly those involved in soda-roast processes, are generally acidified with hydrochloric acid (2–20% of concentrated HCl) before gassing with sulfur dioxide.[24,90,132] However, a mixture of sodium chloride and sulfuric acid can be used equally well and is cheaper. Since lead-lined vessels cannot withstand acid chloride solutions, expensive equipment with rubber or glass linings is necessary. The selenium obtained in this way is generally less pure than that which is obtained from selenite solutions, and therefore separate precipitation of selenium from the two species has been proposed.[162]

In order to catalyze the reduction of selenate solutions, a mixture of chloride and ferrous ion is more effective than chloride ion alone.[162] Thiourea is capable of reducing selenate solutions without the aid of sulfur dioxide, but the product contains much sulfur.[106] A substantially sulfur-free product is obtained less expensively when sulfur dioxide is used in conjunction with thiourea. Under these conditions one g of thiourea will reduce from 3 to 10 g of selenium in a sulfuric acid solution (40–200 gpl) heated to 80°C or above.[106]

A patented method[249] for the collection of selenium from waste solutions containing sulfate utilizes the large difference in solubility between calcium sulfate and calcium selenate. The selenium values are first oxidized to the selenate form by means of sodium hypochlorite, and milk of lime is added to precipitate most of the sulfate as calcium sulfate. Selenium remains in solution and the filtrate is treated with barium chloride to precipitate barium selenate together with a small amount of barium sulfate. Selenate values in acid solutions may also be precipitated by boiling with copper powder, the product being copper selenide. The low solubility of magnesium tellurate (Mg_3TeO_6) can be utilized in the separation of tellurium(VI) from selenium(VI). The selenate remains in solution.[130]

Alkaline solutions containing selenium in the hexavalent form are generally acidified before being treated with reducing agents. The Bolidens process (see Recovery from the -2 Valence State) is unique in that it employs reduction of selenide in an alkaline medium. Precipitation as barium selenate has been proposed[242,243] for the recovery of selenium from caustic pressure leaching of anode slimes and other materials, followed by dissolution of the precipitated selenate in hydrochloric acid and gassing with sulfur dioxide. Reduction with hydrogen under pressure has been studied.[172] Using a pressure of 40–60 atm, hydrogen, and a temperature of 225°C, reduction was almost complete in 3 h, resulting in the formation of a selenide solution which when treated with air and carbon dioxide at atmospheric pressure, precipitated selenium. Elemental selenium will also react with selenate at 225°C with the resulting production of selenite. The use of hydrogen at high pressure can thus be avoided, but a temperature of 225°C is also inconveniently high for an industrial pressure reactor.

Recovery of Selenium from Other Solutions. When elemental selenium is leached from raw materials by means of sodium sulfite solution, reprecipitation of the selenium is achieved by acidification, owing to the instability of the selenosulfite ion in acidic solutions.[129] Recovery of selenium from ammonium sulfate mother liquors, where it may be present in small concentrations, may be effected by treatment with sulfur dioxide[216] or by adsorption on ferric or aluminum hydroxide.[103]

Part II: Refining of Selenium

INTRODUCTION

One of the important new aspects of technology is the influence of minute amounts of impurities or crystalline imperfections on the performance of many vital materials. Semiconductor materials are notorious for their sensitivity to impurities and imperfections. High purity has become increasingly necessary for progress in the science and application of solid-state materials.   The role of impurities in solid state research has been reviewed by Fuller,[72] and more recently by Sazhin.[202]

The most common and most abundant impurity in selenium is tellurium. Due to the similarities in the chemical properties of the two elements, tellurium is also difficult to remove from these mixtures.   Among the contaminations commonly found in selenium are Cu, Pb, Hg, Ag, Au, Fe, Si, S, Cl, O, and a number of other elements, depending on the source and method of selenium recovery.   The achievement of ultrapurity and the maintenance of high purity levels depend also on effective anlytical detection techniques.

Two grades of selenium are recognized commercially: refined selenium assaying 99.5–99.8% and marketed usually as powder, and high-purity grade assaying 99.95–99.9999% and sold mostly as selenium shot, but also as pellets and sticks.   Ultrahighpurity selenium (99.9999$^+$%) has no appreciable market but it is manufactured in minute quantities, mostly for research purposes.   What is known as "doped" selenium is made from high-purity material with the controlled addition of minute amounts of added components and is also sold in shot form.   Ferroselenium is used in the steel industry and contains 50–58% Se.   It is made from refined selenium.

The purification of selenium has been reviewed briefly by Scholten[204], Lumbroso,[153] Pleteneva,[178] and more extensively by Chizhikov and Schastlivyi.[44]

The procedures of selenium refining are very diversified but can be divided broadly into the chemical and physical methods of purification. The processes are usually sequences of chemical and physical purification steps.

CHEMICAL METHODS OF REFINING

In general, chemical methods of selenium purification have as a common objective, the preparation of an intermediate compound amenable to high purity and from which selenium can be easily recovered.   Thus, the formation of volatile selenium compounds—selenium dioxide, selenium hydride, selenium chlorides—is often the main step in selenium refining.

Selective Precipitation

Precipitation of selenium from solutions mainly through reduction by SO_2 has been used as a refining step.   Wagenmann et al.[255,257] have proposed the precipitation of $MgSeO_3 \cdot 6H_2O$ as a refining procedure.   The precipitate is dissolved in acid, reprecipitated and redissolved.   Selenium is finally precipitated in elemental form with sulfur dioxide from a selenite solution.

Selective Leaching and Recrystallization

A method[44] which has found industrial application, is based on the solulity of elemental selenium in sodium sulfite solution and the formation of soluble selenosulfate:

$$Na_2SO_3 + Se \rightleftharpoons Na_2SeSO_3$$

The equilibrium constant of this reaction is significantly influenced by temperature ($K = 4.35$ at 20°C, $K = 0.80$ at 97.5°C according to Chizhikov *et al.*).[44] Technical grade selenium is dissolved in boiling sodium sulfite solution and the pregnant solution is filtered and clarified while still hot. Selenium is recrystallized from the clean solution by cooling. The purity of the selenium can be increased with a number of cyclic leachings and recrystallizations. Due to the minor solubility of tellurium and other contaminations in hot sulfite solutions, a final retorting of the recrystallized selenium is required in order to obtain an impurity level of less than 80 ppm in the final product.

Oxide Purification

Selenium dioxide is a common and one of the most stable oxidized forms of selenium. This oxide is volatile (sp 315°C) and can be easily purified by sublimation. Oxidation of selenium followed by sublimation of the resulting selenium dioxide is a well-established method of purification. Divers and Shimose[52] as well as Metzner[159] at the end of the last century, used boiling sulphuric acid as the oxidizing medium and purified the resulting selenium dioxide by sublimation. Hugot[107] and Threlfall[224] used dilute nitric acid to transform selenium into selenious acid.

These early methods of selenium refining were not extended to an industrial scale. The first industrial selenium refining process involving the use of the oxide was the one developed by Canadian Copper Refiners Limited.[43] Selenium vapors are oxidized by air over a catalyst at 700–800°F. The resulting selenium dioxide is collected as a finely divided crystalline material after cooling the reaction gases. High purity selenium is produced by dissolving the selenium dioxide in de-ionized water and reducing the selenious acid with chemically pure sulphur dioxide.[30] The selenium obtained is finally purified by simple distillation in silica glass retorts. The advantage of this method lies in the simplicity of using air oxidation at moderate temperatures. It involves, nevertheless, a series of operations and requires the handling of solid selenium dioxide which is an obnoxious skin irritant.

Roseman, Neptune, and Allan[195] introduced the use of hydrogen peroxide for the oxidation of an aqueous slurry of selenium at 35°C. Only 69% of the expected selenium dioxide was collected as solid product after filtration.

The method involves a filtration step and requires the use of excess peroxide, as well as a rather lengthy reaction and sedimentation rate.

Ato and Ichioka[12] dissolved crude Se in nitric acid. The SeO_2 formed was dried by evaporation, dissolved in concentrated sulfuric acid, and distilled at about 330°C. The distillate (H_2SeO_3 in H_2SO_4) was reduced by sulfurous acid. A Russian patent by Shebunin et al.[207] also claimed nitric acid as the oxidizing agent for the production of SeO_2. The resulting selenium dioxide is sublimed and then dissolved in water to form selenious acid. Selenium is obtained through reduction by formic acid in the presence of ammonium carbonate. After a final distillation, selenium of 99.998% purity is produced.

Almassy et al.[8] dissolved crude Se in nitric and sulfuric acids. The solution was treated with sulfur dioxide until precipitation started, and then following decantation, all the selenium was precipitated. This second precipitate was redissolved in nitric acid and the solution obtained was finally purified through ion-exchange resins.

Pleteneva et al.[179] obtained high-purity Se (99.999%) by burning technical selenium ($\sim 97.3\%$) in a stream of oxygen at 500–550°C and subliming the resulting SeO_2 at 320–350°C. The pure SeO_2 was dissolved in water to form selenious acid from which elemental selenium was precipitated by sulfur dioxide. The latter was finally purified by distillation under vacuum.

Ivashentsev[121] proposed the purification of selenium by sublimation of SeO_2 at 320–340°C and reduction of the sublimate with graphite at 800–850°C.

High-purity selenium dioxide is prepared, according to a recent Russian patent,[214] by oxidizing elemental selenium with oxygen during heating and subsequently condensing it. Before condensation, the selenium dioxide is filtered through a dense filter, such as glass wool, at 400°–500°C.

Hydride Purification

The purification of a number of elements, e.g., silicon, germanium, antimony, selenium, making use of the respective hydrides, is an established procedure.

Hydrogen selenide is a gas (bp -42°C) and as such is amenable to purification. There is a wide discrepancy among the published thermochemical properties of hydrogen selenide. Rawling and Toguri,[190] in a recent study, presented the following values for the reaction:

$$H_{2(g)} + Se_{(1)} \longrightarrow H_2Se_{(g)} \quad (525–625°C) \quad (1)$$

$$\Delta H_T^\circ = 10107 - 7.33T + 1.36 \times 10^{-3}T^2 + 0.43 \times 10^5 T^{-1} \text{ cal./mole}$$

$$\Delta G_T^\circ = -59.80T + 10107 + 7.33T \ln T - 1.36 \times 10^{-3}T^2 + 0.22 \times 10^5 T^{-1}$$

$$\text{cal/mole}$$

Their results are in agreement with the extrapolation of Bodenstein's[27] old results obtained in the temperature range 254 to 493°C. The equilibrium constant for the reaction represented by reaction (1), $K = P_{H_2Se}/P_{H_2}$, increases, in a pronounced way, with temperature (Figure 2-2) and it reaches $K = 0.87$ at 600°C.[190] Therefore, even at 600°C, where the corrosion problems are acute, a large excess of hydrogen must be used.

Devyatykh and Yushin[50] calculated the equilibrium constants for the thermal decomposition reactions of the simplest volatile hydrides of Group III–VI elements. Their calculated results are in serious disagreement with the experimental results obtained by Rawling and Toguri.[190]

Purification of selenium through the formation and decomposition of hydrogen selenide was reported by Nielsen and Heritage.[165] The apparatus was built on a laboratory scale—10 g Se/h. A product of high purity was claimed, although the analytical technique employed did not possess low detection limits (Sn 10 ppm, Bi 20, Fe 10, Si 20, As 300, Sb 50, Te 200). Mostecký, Weisser, Landa and Ruprych[163] proposed refining of selenium

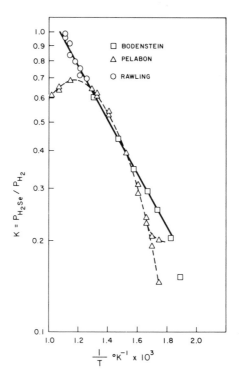

Figure 2-2 Equilibrium constant for the reaction: $H_{2(g)} + Se_{(1)} = H_2Se_{(g)}$ versus 1/T. (190) (reproduced by permission of the National Research Council of Canada from the *Canadian Journal of Chemistry*, **44**, p. 454 (1966)).

by hydrogenation and subsequent oxidation of the hydrogen selenide with 30% hydrogen peroxide, selenious acid, or air. The hydrogenation took place at 350°C in the presence of 20–30% MoS$_2$. Starting from heavily contaminated selenium (ash > 1.5%, Hg 1.2 ppm and Te 1%, a high-purity material was produced in which Te, As, Fe could not be detected and Si and Mg were reduced to trace levels (<5 ppm).

Giesekus and Schnoering,[84] in a patent granted to Farbenfabriken Bayer A. G., claim the production of high-purity selenium through its hydride. Liquid selenium is treated in a countercurrent process at 550°C (Figure 2-3) with small volumes of hydrogen. Impurities such as halides and sulphur are easily hydrogenated and carried away, while selenium is not hydrogenated because of the unfavorable equilibrium. The prepurified selenium flows on to an evaporator to be hydrogenated at 650–685°C with pure, dry hydrogen. Hydrogen selenide passes to a cooling section where dust, traces of mercury

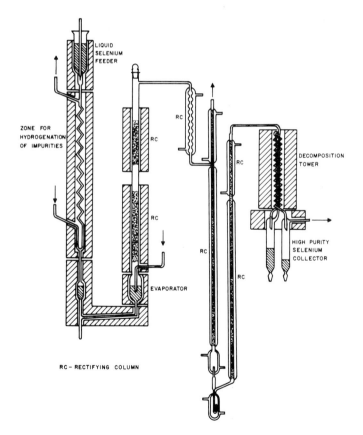

Figure 2-3   Apparatus for the production of high purity selenium (Giesekus and Schnoering (84)).

and water, and high-boiling contaminating hydrides are removed at $-40°C$. The purified gas then enters a two-stage rectifying column where it is liquefied at $-60°C$, and low-boiling contaminating hydrides are eliminated. The high-purity hydrogen selenide is then decomposed at $1000°C$ in a decomposition tower, where about 75% is reconverted into the elements. Selenium leaves the hot zone as a liquid and is collected in a storage vessel. Any fraction which did not decompose is recirculated to the tower and is finally oxidized to selenium by hydrogen peroxide. This method is adaptable to the production of high-purity selenides by reaction of the purified hydrogen selenide with pure metals:

$$Me + H_2Se \longrightarrow MeSe + H_2 \tag{2}$$

The method may be suitable for the production of ultrahigh-purity selenium in small quantities. It appears, nevertheless, that it cannot be competitive with other methods for the industrial production of high-purity selenium, since it requires the use of a material which can resist selenium corrosion and be properly sealed at $1000°C$.

Chloride Purification

Selenium with chlorine forms selenium monochloride (Se_2Cl_2) which, in the presence of an excess of chlorine, is transformed to selenium tetrachloride ($SeCl_4$). Simons[208] and Yost and Kircher[265] have proved the existence of selenium dichloride ($SeCl_2$) in the vapor of the thermally dissociated tetrachloride.

Selenium monochloride is a dark yellow oil (mp $-85°C$), selenium dichloride is a gas, and selenium tetrtachloride is a solid (mp $305°C$) which sublimes at $196°C$. The volatility of the selenium chlorides allows purification by distillation or sublimation.

The chlorination of selenium:

$$2Se + Cl_2 \longrightarrow Se_2Cl_2 \tag{3}$$

is a highly exothermic reaction. The standard heat of formation of selenium monochloride is reported to vary from $-19,990$ cal/mole to $-22,400$ cal/mole, depending on the form of selenium used.[175,223]

Selenium monochloride is dissociated during distillation:

$$2Se_2Cl_2 \longrightarrow SeCl_4 + 3Se \tag{4}$$

and can be decomposed by steam, ethyl alcohol, and ether:[153]

$$2Se_2Cl_2 + 2H_2O \longrightarrow SeO_2 + 3Se + 4HCl \tag{5}$$

Selenium tetrachloride is hydrolyzed even by traces of moisture:

$$SeCl_4 + H_2O \longrightarrow SeOCl_2 + 2HCl \tag{6}$$

$$SeCl_4 + 3H_2O \longrightarrow H_2SeO_3 + 4HCl \tag{7}$$

Artamonov[11] patented the purification of technical grade selenium by chlorination with gaseous chlorine to Se_2Cl_2. The chloride is purified by evaporation and hydrolyzed in the vapor phase by steam in the presence of air. The solid product is washed free of chlorine, dried, and fused.

Refining through chlorination has the obvious disadvantage of producing selenium which contains residual traces of chlorine. These are detrimental for certain uses of high-purity selenium.

Purification of Selenium Solutions by Ion-Exchange and Solvent Extraction

The ability of ion-exchange resins to react selectively with ionic species has stimulated much interest in the field of hydrometallurgy. Apart from water treatment, one of the major fields of application of ion-exchange has been the concentration, recovery, and purification of valuable metals in solution.

Selenium in aqueous solutions forms anionic species. Therefore, the purification of selenium solutions can be expected by either anion-exchange resins, where the selenium species are retained, or cation-exchange resins, where cationic contaminations are retained, or a combination of the two.

A number of attempts to apply ion-exchange techniques for purification of selenium solutions have been reported. These can be divided into analytical treatments,[10,13,15,70,109,147,161,198,246,264] general investigations, and industrial purifications.

The possible separation of selenium from tellurium by the use of a cation-exchanger (KU-1) was studied by Zelyanskaya et al.[269] Experiments with solutions containing both elements show that at pH 1.4, selenium passes completely into the filtrate, but tellurium is absorbed on the cation-exchanger and may be eluted with ammonium hydroxide. At pH 1.5–3.7, Cu, Fe, Pb and Zn are completely absorbed.

Gaibakyan and Darbinyan studied extensively[46,47,73,74] the ion-exchange separation of selenium from tellurium. The complete separation, under certain concentration limits, is possible with cation-exchange resins (KU-2, KU-1) at pH 2.5.[73] The separation of selenium from tellurium with anion-exchange resins is only possible with solutions which contain high concentrations of hydrochloric acid (4–12N). At low acid concentration, tellurium can be absorbed as TeO_3^{-2} simultaneously with selenium as SeO_3^{-2}. However, with increasing acidity, TeO^{+2} forms which is poorly absorbed on the anion-exchanger. In a very strong acid medium, absorption of tellurium again increases because of the formation of the complex anion $TeOCl_4^{-2}$.

Static and dynamic methods of separation of selenium and tellurium on the cation-exchanger (KU-2) in an acid form were studied[46] with different concentrations of acetic, citric, and tartaric acids. Tellurium in $0.5N$ solution is absorbed well.

The ions of Se(IV) and Te(IV) in solution were studied by the same investigators[73] with electrophoresis on chromatographic paper. Tellurium is present as a cation (probably TeO^{+2}) in $0.005-2N$ HCl. In very concentrated hydrochloric acid, tellurium forms an anionic complex. Selenium and tellurium can be separated in $0.1N$ HCl by electrophoresis at 350 V. Selenium moves to the anode and tellurium to the cathode.

Bueker and Kofron[32] have patented a selenium purification procedure involving a cation exchange resin. Impure selenium is converted to selenious acid, and the pH of this solution is adjusted between 0.8 to 2 by dilution. Freshly precipitated selenium ($\sim 1\%$ of the total selenium content) is dispersed in the solution in order to remove any heavy contamination by mercury. The acidic solution is passed through a cation-exchanger which retains trace impurities of Ni, Zn, Te, Cd, Hg, Pb, Sn, Fe, Bi, Cu and Ag.

Van Goetsenhoven[89] obtained pure selenium from a commercial solution by purifying through cation (Amberlite 1R 120) and anion (Amberlite 1RA 400) exchange resins. The first resin absorbs Pb, Cu, and partly absorbs Ca and Fe. The second resin absorbs Hg, (as $HgCl_3^-$), Ag, and $HSeO_3^-$, but only the latter is eluted with hydrochloric acid.

Purification of selenium solutions through the use of ion-exchange resins has also been claimed in a Japanese patent by Tsuji[235] and in an East German patent by Schumann[205].

Almassy and his collaborators[8] reported the purification of selenious acid solutions through a cation-exchange resin (Dowex 50). To remove the last traces of mercury the effluent was passed through an anion-exchange resin (Dowex-2 OH-form), and the absorbed Se (SeO_3^{-2}) was eluted with $2N$ NaOH.

Selezneva[206] described the production of high-purity selenium from a solution contaminated with Te, Hg, Sb, Cu, Fe, As, Sn and Ag by treating it through a cation and an anion ion exchange columns. Ruprych[197] patented the production of high-purity selenium by dissolving crude Se in nitric acid, subjecting the selenious acid obtained to ion-exchange treatment and reducing the pure, pregnant solution with sulfur dioxide.

A rapid extraction of Se(IV) from $4M$ HBr with benzene containing 1% phenol has been described by McGee et al.[155] The separation is highly selective for selenium. Attempts to refine selenium solutions by solvent extraction have also been reported by Timofeeva[225] and Tserekova and Giganov.[234]

Electrolytic Methods

Electrorefining has not been applied successfully in the purification of selenium. Alekperov and Mirzoeva[4,5] reported an electrochemical separation of small amounts of tellurium from selenium. Tellurium ($<0.2\%$) was separated from selenium by electrolysis of a solution containing 20% NaCl, $\sim 3.6\%$ NaOH, and 51 g Na_2SeO_3/liter at 18–20°C using a current density of 2–3 mA/cm^2 and platinum screen electrodes. The selenium obtained from this solution contained 0.2 ppm Te.

The electrolytic reduction of selenium(IV) and tellurium(IV) on a mercury cathode has been studied by Zarechanskaya et al.[267] The reduction is feasible from acidic but not from alkaline solutions, and the method cannot lead to high-purity selenium. Pacauskas et al.[173] have described various cells and results obtained from the electrolysis of selenium dioxide solutions in concentrated sulfuric acid.

Fusion with a Nitrate

Impurities more easily oxidized than selenium can be removed by heating the impure selenium with an alkali metal nitrate. Kozmin and Kupcheno [137,139] patented a process involving the fusion of selenium with ammonium nitrate at 200–230°C. The fusion was found to be effective for the removal of S, Fe, and Te. Technical grade selenium (Se: 96.82%, Te: 0.88%, Fe: 0.09%) was purified to a product containing Se: 99.26%, Te: 0.44%, Fe: 0.018%. A German patent[256] claims the use of an alkali metal nitrate with or without an alkali metal hydroxide. The method is able to remove low concentrations of arsenic and antimony.

Although the purification of selenium by oxidative fusion may be very useful in a refining process, it cannot lead, by itself, to high-purity selenium.

PHYSICAL METHODS OF REFINING

Physical methods of selenium refining are based on mass-transfer between two phases. These methods have the advantage of obtaining purification without the addition of any reacting material which, in general, will introduce undesirable trace components into the system. Physical methods of refining constitute, as a rule, final steps in long processes of purification.

Refining of Selenium Solutions by Adsorption Methods

The phenomenon of selective adsorption can explain the efficiency of freshly precipitated selenium for the removal of mercury from selenium solutions[32] as mentioned earlier. The purification of selenium solutions from mercury contaminations by stepwise precipitation of selenium has also been recently reported.[213] The precipitation of hydrated aluminum hydroxide for the adsorption of impurities from selenite solutions has been patented by Voight and Krebs.[140,250] Wallden and Edenwall[258] proposed the precipitation of a number of gelatinous hydroxides (Fe, Al, Ni, Co, Mg, or Mn) for the adsorption of impurities from aqueous selenium solutions.

The purification of industrial aqueous selenium solutions by continuous adsorption on activated carbon is feasible.[261] Commercial selenium solutions were passed through columns packed with granular activated carbon where up to 90% of the tellurium, 50% of the iron, and 80% of the copper contaminations were removed without significant selenium adsorption. The use of adsorbents such as Al_2O_3 in molten selenium has been claimed in a Belgian patent[49] as a refining step prior to vacuum distillation.

Zone Refining of Selenium

The zone-refining process was applied to selenium of 99.994% purity by Tsuzuki.[236] It has not been confirmed that the application of zone refining to selenium is effective. The impurities Ga, Cu, Bi and Te, separate at the starting end while Al, Fe, Mg, Mn, Sb and Si move to the finishing end.

Distillation of Selenium

Under the general term of distillation of selenium, processes employing simple retorting (a still without a rectification column at atmospheric pressure or under vacuum) as well as rectification are assembled.

Selenium contaminated with metallic impurities (any metal except mercury and cadmium) can be effectively separated by a simple distillation process, since these impurities form selenides which remain in solution in the liquid phase. During the distillation of impure selenium, mercury is removed with the first fraction of the distillate because of the pronounced volatility of mercury selenide. Therefore, the main problem in the purification of selenium by distillation involves the separation of tellurium from a binary selenium–tellurium mixture with an extremely high selenium content.

The vapor-liquid equilibria of the sulphur–selenium system, were investigated by Zhuravleva and Chufarov.[270] The more industrially important selenium–tellurium system has only recently been investigated. Sato and Kaneko[200] reported isobaric and isothermal measurements of vapor-liquid equilibria of selenium–tellurium solutions containing up to 22% Te. These authors claimed that an azeotrope existed at about 3% Te and that the purification of such solutions by simple distillation would be impossible.

Yannopoulos and Themelis[263] studied the vapor-liquid equilibria of the selenium–tellurium system in the composition range from 40 to 99.999% Se by weight, under isobaric conditions.   Their report is detailed in the extremely high Se concentration ranges, and therefore is relevant to the production of high-purity selenium.   The results presented (Figures 2-4 and 2-5) are in disagreement with those of Sato and Kaneko[200] since they establish the absence of an azeotropic solution.   These results[263] are supported by the fact that selenium purification is carried out in a number of industrial distillation units.

Abdullaev et al.[1] measured the vapor pressure of solid selenium–tellurium solutions at low temperatures (290°C).   The object of this work was to apply vacuum sublimation to the separation of Se–Te mixtures and it was, therefore, concerned only with the solid–gas equilibrium of this system.   Two Russian studies on the vapor-liquid equilibrium of the tellurium–selenium system have been published recently.[136,241]   Both are concerned with the refining of tellurium from selenium contaminations.

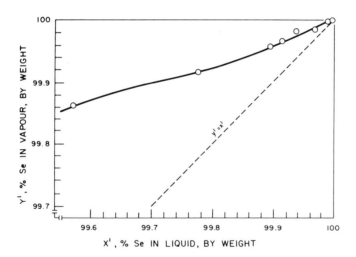

Figure 2-4   Vapor-liquid equilibria for the selenium–tellurium system at very high selenium concentrations (263) (reproduced from the *Canadian Journal of Chemical Engineering*).

Rectification of Selenium.   The enrichment of the vapor stream, as it passes through a column in contact with the liquid stream (reflux) by simultaneous heat and mass transfer between the two phases, is called rectification.   According to the vapor-liquid equilibria of the selenium–tellurium system,[263] rectification presents a possible method of selenium purification.   Theoretically, heavily contaminated selenium can be brought to a level of high-purity by rectification through a contacting column.   This rectification should be able to replace a number of chemical and physical steps for the production

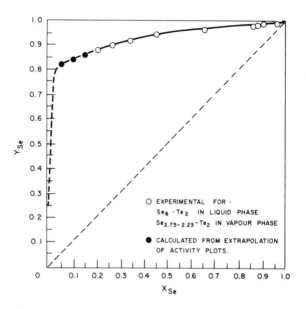

Figure 2-5 Equilibrium curve of the selenium–tellurium system (263) (reproduced from the *Canadian Journal of Chemical Engineering*).

of high-purity selenium.   Higher levels of purity could be achieved or more contaminated selenium charges could be handled by simply increasing the height of the rectification column.

The purification of selenium by rectification, nevertheless, has not proceeded beyond the laboratory stage, because the problem of selenium rectification on the industrial scale depends on finding a suitable material for the construction of the apparatus.   Such a material would be required to maintain its essential physical properties (corrosion resistance, strength, impermeability, heat conductivity) over a wide temperature range with a relatively high degree of efficiency.   It must absolutely resist selenium attack since, otherwise, the rectification product would be contaminated.

Janjua, Yannopoulos and Cooper[123,262] have experimentally demonstrated the vigorous corrosivity of selenium toward a number of refractory materials at high temperatures.   At relatively low temperature (400°C), high chromium content stainless steel has displayed an appreciable resistance to selenium corrosion.[262]   As a result, an alternative to rectification under reduced pressure was considered, but because an increased viscosity in the liquid phase is encountered at decreased temperatures both the purifying mass-transfer phenomenon and the flow through the column are hindered.   To minimize the drop in a column operating at reduced pressure, extremely low vaporization rates would be necessary.

Rectification under reduced pressure (10 mm, 450°C) was performed in the laboratory by Eckart[56] in an attempt to produce high-purity selenium for conductivity measurements. The apparatus consisted of a flask, a column, and a column head all made from Yenaer-Geratel glass 20. The product obtained after five consecutive rectifications did not contain any spectrally detectable contaminations.

Yannopoulos and Cooper[262] have used a laboratory rectification still made from Vycor glass and operated at atmospheric pressure. This still consisted of a three-neck flask, a column packed with Vycor Raschig rings, and a reflux head (Figure 2-6).

Nisel'son et al.[167] presented preliminary experimental data on the rectification of selenium in a quartz sieve-plate column, 3 cm in diameter. The possibility of refining selenium from a mixture containing 75% Se, 25% Te, and other trace additions was demonstrated. Comparison of selenium refining by rectification with selenium obtained by simple distillation demonstrates the obvious advantages of rectification.

Isakova et al.[112] used a laboratory plate-type still (Figure 2-7) under a

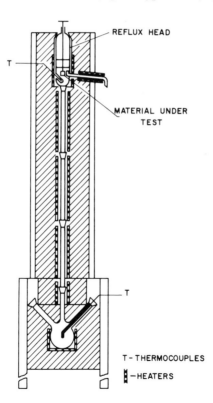

Figure 2-6   Silica glass selenium rectification column (262).

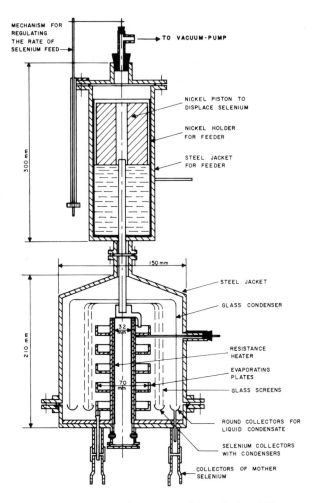

Figure 2-7   Apparatus for vacuum refining selenium (112).

vacuum (370–430°C, 0.2–1.0 mm Hg). The use of a countercurrent flow of vapor and liquid and decreasing the temperature improved the condensate purity.

Koch[131] patented an apparatus for the purification of cadmium and selenium in which the reflux or condensation area is kept just below the mp of the refined substance. Koch claimed that through this process another refining step, crystallization, takes place in addition to distillation. The practicality of this apparatus for charges larger than a few g is questionable. Attempts at selenium rectification have also been published by Reznyakov et al. [115,193]

Simple Distillation (Retorting) of Selenium. Although chemical methods of purification contribute substantially to the refining of selenium, they are not able, as a rule, to achieve purity levels with a total contamination content less than 2 ppm. Therefore, the distillation of selenium represents an almost compulsory final purifying step in most refining processes producing high-purity selenium. The simple distillation of selenium has been performed either at atmospheric pressure or preferably, under reduced pressure, in an attempt to minimize the corrosivity of selenium toward the materials of construction.

A distillation step is often used as an intermediate refining procedure as well. Crude selenium is refined by simple distillation in cast iron retorts in a number of metallurgical plants.[30,266] The retorting of crude selenium decreases the tellurium and metal contamination and recovers in the nonvolatile residue any silver and gold that may have been in the crude selenium. Since the product of retorting is not high-purity selenium, the corrosion of the cast iron does not interfere significantly.

Doucot[54] has patented a retort for the distillation of selenium closed at one end by a seal of liquid selenium to be purified and at the other end by a seal of the condensate. (Figure 2-8.) The retort is usually made from borosilicate or quartz glass.

Flemming[71] suggested the use of a quartz apparatus for the separation of selenium through distillation to top and bottom fractions. Simple distillation of selenium under reduced pressure is applied more often than atmospheric distillation. Reduced pressure is achieved either by operation under vacuum or in a gas stream (reduced partial pressure of the vapor phase).[7,8,57,166,218]

Vacuum distillation has been advocated by Isakova and Nesterov[113] for the recovery of technical grade selenium from materials high in selenium and low in mercury. The role of mercury impurities in the vacuum distillation of selenium was studied by Tsvetkov et al.[237] A comparison of mercury distributions at various condenser temperatures was made.

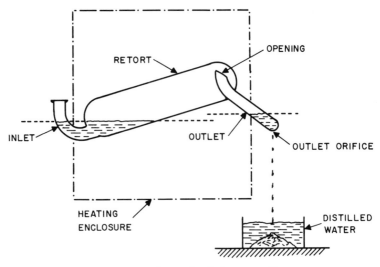

Figure 2-8   Doucot's selenium retort.[54]

Takahashi[217] reduced vapor phase pressure by passing 5–8% SO_2 through the retort.   A distillate containing 98.5% Se was obtained from a raw material containing only 16% Se.   The sulphur dioxide functioned simultaneously as diluent and reducing agent for any form of oxidized selenium.

Attempts to produce high-purity selenium by distillation in the Mansfeld refinery in East Germany were reported by Loeschau.[149–151]   Distillation of selenium in a purified stream of nitrogen gave poor results in an apparatus made from stainless steel due to metal corrosion, but better results were obtained with an electrically heated quartz apparatus.   The main fraction obtained assayed at 99.995% Se.

Distillation of Selenium with the Addition of Another Compound.   In industrial practice, multi-component mixtures are frequently encountered in which normal distillation methods are not practical, either due to the formation of azeotropes or the very low relative volatility over a wide concentration region.   Among a large number of such mixtures, it has been found possible to modify the relative volatility of the original components by the addition of another component.   This is called extractive and azeotropic distillation. The added compound can alter the activity coefficients of the various components, thereby altering their relative volatilities.   The addition of another component has been tried in the distillation of crude selenium.   Gardner[77] has patented the addition of powdered carbon to the granular starting material, heating it to 615–680°C to remove impurities more volatile than selenium, and then to 690–1000°C to vaporize the selenium.   The operation is carried out in an inert atmosphere (Ar, He, N_2, CO_2). Takei et al.[218] suggested fusion with $NaNO_3$–KNO_3 with addition of Fe_3O_4, followed by vacuum distillation for the purification of selenium. Oberbacher and Schlier[168] claimed the production of high-purity selenium by distillation, under vacuum, of crude selenium to which 0.05–2% magnesium powder was added. The addition of adsorbents, such as Al_2O_3, to the molten selenium before distillation has also been claimed.[49]

The effect of iron additions on the behavior of sulfur and tellurium during the distillation of selenium was also studied.[245] Introduction of iron into the selenium–sulfur system enhances the formation of nonvolatile ferrous sulfide.

REFERENCES

1. Abdullaev, G. B., Shakhtakhtinskii, M. G., and Kuliev, A. A., *Dokl. Akad. Nauk Azerb. SSR* **16**, 219 (1960); *Chem. Abs.* **55**: 20591,.
2. Aidarova, P. I., and Koz'min, Yu. A., *Sbornik Nauch. Trudov Vsesoyuz. Nauch.—Issledovatel. Gornomet. Inst. Tsvetnoi. Met.* **6**, 402 (1960); *Chem. Abs.* **56**: 4430.
3. Akhmetov, S. F., et al., *Tr. Khim.—Met. Inst., Akad. Nauk Kaz. SSR* **4**, 20 (1967); *Chem. Abs.* **69**: 40899.
4. Alekperov, A. I., and Mirzoeva, A. A., *Azerb. Khim. Zh.* **1**, 133 (1962); *Chem. Abs.* **58**: 4158.
5. Alekperov, A. I., and Mirzoeva, A. A., *Zh. Prikl. Khim.* **38** (3), 555 (1965); *Chem. Abs.* **62**: 15756.
6. Alekperov, R. A., Nazirova, T. M., and Alekperova,. A. A., *Geokhim. Redk Elem., Akad. Nauk Azerb. SSR, Inst. Neorg. Fiz. Khim* **1966**, 24; *Chem. Abs.*. **68**: 15023.
7. Almassy, G., *Magy. Kem. Lapja* **17**, 165 (1962).
8. Almassy, G., et al., Hungarian Patent 148,585, Nov. 30, 1961; *Chem. Abs.* **58**: 5287.
9. Almassy, G., et al., Hungarian Patent 150,464; *Chem. Abs.* **60**: 3755.
10. Aoki, F., *Bull. Chem. Soc. Japan* **26**, 480 (1953); *Chem. Abs.* **49**: 1012.
11. Artamonov, N. S., Russ. Patent 66,128, Apr. 30, 1946; *Chem. Abs.* **41**: 1818.
12. Ato, S., and Ichioka, K., *Repts. Sci. Research Inst. (Japan)* **32**, 27 (1956); *Chem. Abs.* **50**: 10352.
13. Attebery, R. W., Larson, Q. V., and Boyd, G. E., *Abstract Am. Chem. Soc.,* 118th Meeting 8G (1950).
14. Auezov, Zh., *Met. i Khim. Prom. Kazakhstana, Nauch. Tekhn. Sb.* **4**, 49 (1961); *Se–Te Abs.* **5**: 50.
15. Ayres, J. A., *Ind. Eng. Chem.* **43**, 1526 (1951); *Chem. Abs.* **45**: 9205.
16. Azerbaeva, R. G., and Tseft, A. L., *Met. i Khim. Prom. Kazakhstana, Nauchn.—Tekhn. Sb.* (6), 92 (1962); *Chem. Abs.* **60**: 233.
17. Azerbaeva, R. G., and Tseft, A. L., *Zh. Prikl. Khim.* **37** (11), 2367 (1964); *Chem. Abs.* **62**: 4688.
18. Bakeeva, S. S., Buketov, E. A., and Pashinkin, A. S., *Zh. Neorg. Khim.* **13** (1), 32 (1968).

19. Baker, W. A., and Hallowes, A. P. C., *Bull. Inst. Mining Met.* No. **513**, 1 (1949).
20. Barker, I. L., *J. Metals* **8**, 1058 (1956).
21. Bartosevich, N. K., *et al.*, *Zh. Vses. Khim. Obshchestva im. D. I. Mendeleeva* **8** (5), 584 (1963); *Chem. Abs.* **60**: 5197.
22. Bastius, H., Ger. (East) Patent No. 11,394, March 17, 1956.
23. Bhappu, R. B., *J. Metals* **14**, 429 (1962).
24. Blankenburg, A., *Rev. chim. (Bucharest)* **9**, 96 (1958); *Chem. Abs.* **53**: 6553.
25. Bloom, M. C., U.S. Patent 2,414,438, Jan. 21, 1947; *Chem. Abs.* **41**: 3383
26. Bobikov, P. I., *et al.*, *Tsvetn. Metal.* **36** (12), 54 (1963); *Chem. Abs.* **61**: 342.
27. Bodenstein, M., *Z. Physik. Chem. Leipzig* **29**, 429 (1899).
28. Bondin, S. M., *Avtoklavn. Metody Pererabotki Mineral'n. Syr'ya, Akad. Nauk SSSR, Kol'sk. Filial, Inst. Khim. i Tekhnol. Redkikh Elementov i Mineral'n. Syr'ya* 16 (1964); *Chem. Abs.* **62**: 4594.
29. Bondin, S. M., *Avtoklavn. Metody Pererabotki Mineral'n. Syr'ya, Akad. Nauk SSSR, Kol'sk. Filial, Inst. Khim, i Tekhnol. Redkikh Elementov i Mineral'n. Syr'ra* 5 (1964); *Chem. Abs.* **62**: 1315.
30. Bridgstock, G., Elkin, E. M., and Forbes, S. S., *Trans. CIM* **63**, 523 (1960).
31. Buchanan, S. C., U.S. Patent 3,090,671, May 21, 1963.
32. Bueker, E. L., and Kofron, V. K., U.S. Patent 2,860,954, Nov. 18, 1958; *Chem. Abs.* **53**: 5614.
33. Buketov, E. A., *et al.*, *Tsvet Metal.* **41** (9), 72 (1968); *Chem. Abs.* **69**: 98573.
34. Buketov, E. A., *et al.*, *Izv. Akad. Nauk Kaz. SSR, Ser. Khim. Nauk* **15** (3), 40 (1965); *Chem. Abs.* **64**, 10476.
35. Buketov, E. A., *et al.*, *Tsvet. Metal.* **39** (12), 50 (1966); *Chem. Abs.* **66**: 9170/1, 97623.
36. Buketov, E. A., Ugorets, M. Z., and Baikenov, Kh., Russ. Patent 193,076, Aug. 29, 1968; *Chem. Abs.* **70**; 13066.
37. Buketov, E. A., and Ugorets, M. Z., *Izv. Akad. Nauk Kaz. SSR, Ser. Khim.* **2** (22), 47 (1962).
38. Buketov, E. A., Ugorets, M. Z., and Alpysbaev, R., *Izv. Akad. Nauk Kaz. SSR, Ser. Tekhn. i Khim. Nauk* (3), 34 (1963); *Chem. Abs.* **62**: 11411.
39. Buketov, E. A., Ugorets, M. Z., and Moiseevich, O. Yu., *Tr. Inst. Met. i Obogashch., Akad. Nauk Kaz. SSR* **11**, 168 (1964); *Chem. Abs.* **62**: 8792.
40. Buketov, E. A., Ugorets, M. Z., and Moiseevich, O. Yu., *Tr. Inst. Met. i Obogashch., Akad. Nauk Kaz. SSR* **9**, 136 (1964); *Chem. Abs.* **62**: 9846.
41. Buketov, E. A., and Ugorets, M. Z., Canadian Patent 787,318, June 11, 1968.
42. Bunatyan, E. G., *Nauchn. Tr. Nauchn.—Issled. Gorn.—Met. Inst. Sov. Nar. Khoz. Arm. SSR* **1**, 71 (1960).
43. Canadian Copper Refiners Ltd., Brit. Patent 603,133, June 9, 1948.
44. Chizhikov, D. M., and Schastlivyi, V. P., *Selenium and Selenides (Selen i Selenidy)*, Moscow: *Nauka*, 1964.
45. Cronan, C. S., *Chem. Eng.* **58** (7), 60 (1958).
46. Darbinyan, M. V., and Gaibakyan, D. S., *Izv. Akad. Nauk Arm. SSR, Khim. Nauki* **16** (5), 443 (1963); *Chem. Abs.* **60**: 9884.
47. Darbinyan, M. V., and Kapantsyan, E. E., *Izv. Akad. Nauk Arm. SSR, Khim.* **18**, No. 3, 18 (1965).
48. Davydov, E. V., and Koz'min, Yu. A., *Tsvet. Metal.* **41** (2), 71 (1968).
49. Deutsche Akademie der Wissenschaften zu Berlin, Belg. Patent 652,746, Dec. 31, 1964; *Chem. Abs.* **64**: 9065.
50. Devyatykh, G. G., and Yushin, A. S., *Russ. J. Phys. Chem.* (Eng. transl.) **38**, 517 (1964).

51. Diev, N. P., Olesova, A. I., and Pimenov, I. V., *Trudy Inst. Met., Akad. Nauk SSSR, Ural. Filial.* **2**, 125 (1958); *Chem. Abs.* **54**: 22231.
52. Divers, E., and Shimose, M., *Chem. News* **51**, 199 (1885); ref. 153.
53. Djingheuzian, L. E., *Extr. Refining Rarer Metals, Proc. Symposium*, London, 72 (1956).
54. Doucot, R. (to Compagnie des Freins et Signaux Westinghouse) Brit. Patent 664,030 (1952); U.S. Patent 2,654,700 (1953).
55. Dutko, J., Czech. Patent 125,591, Dec. 15, 1967; *Chem. Abs.* **69**: 69110.
56. Eckart, F., *Ann. Physik* **14**, 233 (1954).
57. Eckart, F., and Berg, H., *Naturwissenschaften* **45**, 335 (1958).
58. Efendiev, G. Kh., and Alekperov, R. A., *Dokl. Akad. Nauk Azerb. SSR* **18**, No. 5, 15 (1962); *Chem. Abs.* **57**: 12098.
59. Elkin, E. M., and Tremblay, P. R., U.S. Patent 3,127,244, Mar. 31, 1964.
60. Elkin, E. M., Tremblay, P. R., and Desy, P., "Pressure Leaching of Copper Refinery Slimes," paper presented at the 18th IUPAC Congress, Montreal, August 6–12, 1961.
61. Emicke, K., personal communication.
62. Erdenbaeva, M. I., and Usenova, Z., *Izv. Akad. Nauk Kaz. SSR, Ser. Khim.* **2**, 42 (1962).
63. Erdenbaeva, M. I., and Usenova, Z. M., *Zavodsk. Lab.* **30** (10), 1190 (1964); *Chem. Abs.* **62**: 13841.
64. Erdenbaeva, M. I., and Usenova, Z. M., *Izv. Akad. Nauk Kaz. SSR, Ser. Khim. Nauk* **14** (1), 46 (1964); *Chem. Abs.* **62**: 9785.
65. Esirkegenov, G. M., Raimbekov, E. S., and Smagulova, A., *Sb. Statei Aspir. Soiskatelei, Min. Vyssh. Sredn. Spets. Obrazov. Kaz. SSR, Met. Obogashch.* **2**, 62 (1966); *Chem. Abs.* **69**: 28906.
66. Evlash, Yu. N., Russ. Patent 133,467, Nov. 25, 1960; *Chem. Abs.* **55**: 10828.
67. Feller-Kniepmeier, M., and Pawlek, F., *Metall*, **1**, 36 (1963).
68. Filippov, A. A., and Smirnov, V. I., *Izvest. Vysshikh Ucheb. Zavedenii, Tsvetnaya Met.* **3**, No. 6, 55 (1960); *Chem. Abs.* **55**: 18222.
69. Filippova, N. A., *et al.*, *Zavodskaya Lab.* **26**, 401 (1960); *Chem. Abs.* **54**: 15071.
70. Fischer, W., and Bastius, H., *Neue Huette* **9**, 434 (1964); *Chem. Abs.* **61**: 11328.
71. Flemming, W., *Chim. et ind. (Paris)* **81**, 878 (1959); *Chem. Abs.* **53**: 19624.
72. Fuller, C. S., *Ultrapurification of Semiconductor Materials*, eds. Brooks, M. S., Kennedy, J. K., The MacMillan Company, New York 1962, pp. 3–24.
73. Gaibakyan, D. S., and Darbinyan, M. V., *Izv. Akad. Nauk Arm. SSR., Khim. Nauki* **16** (3), 211 (1963); *Chem. Abs.* **59**: 14863.
74. Gaibakyan, D. S., and Darbinyan, M. V., *Izv. Akad. Nauk Arm. SSR, Khim. Nauki* **17** (5), 501 (1964); *Chem. Abs.* **62**: 11128.
75. Galimov, M. D., *et al.*, *Tsvet. Metal.* **34**, (12), 61 (1961); *Chem. Abs.* **57**: 3116.
76. Garay, L., and Demény, A., *Kohászati Lapok* **15** (93), 529 (1960); *Chem. Abs.* **55**: 9802.
77. Gardner, D., U.S. Patent 2,414,295, Jan. 14, 1947; *Chem. Abs.* **41**: 2216.
78. Garkun, V. K., and Kuzin, I. A., *Zh. Prikl. Khim.* **40** (6), 1252 (1967); *Chem. Abs.* **67**: 10565.
79. Gerlach, J., *et al.*, *Metall* **17**, 1103 (1963); *Chem. Abs.* **60**: 3769.
80. Getskin, L. S., and Leksin, V. N., *Tsvet. Metal.* **5**, 54 (1964).
81. Getskin, L. S., and Margulis, E. V., *Tsvet. Metal.* **36** (7), 56 (1963); *Chem. Abs.* **59**: 11052.
82. Getskin, L. S., Remizov, Yu. S., and Margulis, E. V., *Zhur. Priklad. Khim.* **34**, 1430 (1961); *Chem. Abs.* **55**: 23239.

83. Getskin, L. S., and Savraev, V. P., *Zhur. Priklad. Khim.* **34**, 2398 (1961); *Chem. Abs.* **56**: 8366.
84. Giesekus, H., and Schnoering, H., Ger. Patent 1,170,912, May 27, 1964; *Chem. Abs.* **61**: 7980.
85. Gladyshev, V. P., and Kozlovskii, M. T., *Izv. Vysshikh Uchebn. Zavedenii, Khim. i. Khim. Tekhnol.* **6** (5), 724 (1963); *Chem. Abs.* **60**: 8640.
86. Glukhov, V. P., Sitnikova, T. G., and Fedotov, I. A., *Tsvet. Metal.* **34**, (1), 83 (1961); *Chem. Abs.* **55**: 21500.
87. Glukhov, V. P., Sitnikova, T. G., and Ferberg, M. B., *Tsvet. Metal.* **36** (3), 83 (1963); *Chem. Abs.* **59**: 9613.
88. Godat, E. A., Jr., and Quintanilla, J. B., U.S. Patent 3,008,806; *Chem. Abs.* **56**: 11225.
89. van Goetsenhoven, F., *Mededel. Vlaam. Chem. Ver.* **21**, 117 (1959); *Chem. Abs.* **54**: 11400.
90. Goldstein, J., *Rev. Chim. (Bucharest)* **14** (3), 164 (1963); *Chem. Abs.* **59**: 12425.
91. Goldstein, J., and Laday, I., Rom. Patent 44,582, Dec. 9, 1966; *Chem. Abs.* **67**: 10409.
92. Greĭver, T. N., Russ. Patent 112,634, Aug. 15, 1958; *Chem. Abs.* **53**: 2559.
93. Greĭver, T. N., *Sbornik Tekh. Infrom. Proekt. i Nauch.—Issledovatel. Inst. Gipronikel* (4–5), 103 (1958); *Chem. Abs.* **54**: 21670.
94. Greĭver, T. N., Russ. Patent 126,875, Mar. 10, 1960; *Chem. Abs.* **54**: 18303.
95. Greĭver, T. N., *Tsvet. Metal.* **6**, 27 (1965).
96. Greĭver, T. N., *Rudobobiv Met. (Sofia)* **23** (1), 56 (1968); *Chem. Abs.* **69**: 79377.
97. Greĭver, T. N., and Burnazyan, A. S., *Nauchn. Tr. Nauchn.—Issled. Gorn.— Met. Inst., Sov. Nar. Khoz. Arm. SSR* **1**, 107 (1960); *Chem. Abs.* **57**: 16221.
98. Hamada, K., *Bull. Chem. Soc. Japan* **34**, 596, 600 (1961); *Chem. Abs.* **55**: 25435.
99. Hamada, K., *Kogyo Kagaku Zasshi* **64**, 1350 (1961).
100. Handwerk, E. C., and Monson, W. T., U.S. Patent 2,948,591, Aug. 9, 1960.
101. Hart, J. L., U.S. Patent 3,178,257, April 13, 1965; *Chem. Abs.* **62**: 15803.
102. Hennig, Y., and Pawlek, F., *Z. Erzbergbau u. Metallhüttenw.* **13**, 205 (1960).
103. Hitaka, T., *et al.*, Japanese Patent 1267, March 11, 1959; *Chem. Abs.* **53**: 22795.
104. Hobin, M. A., U.S. Patent 2,889,206, June 2, 1959; *Chem. Abs.* **53**: 16493.
105. Hoffman, J. E. and Cusanelli, D. C., German Patent 1,201,069, Sept. 15, 1965; *Chem. Abs.* **64**: 331.
106. Hollander, M. L., and Lebedeff, Y. E., U.S. Patent 2,834,652, May 13, 1958.
107. Hugot, C., *Am. Chim.—Phys.* **21**, (7), 34 (1900); ref. 153.
108. Hukki, R. T., and Runolinna, U., *Trans. Am. Inst. Mining Met. Engrs.* **187**, 1131 (1950).
109. Iguchi, A., *Chem. Soc. Japan* **31**, 748 (1958); *Chem. Abs.* **53**: 11085.
110. Ionescu, I., *et al.*, Belgian Patent 663,078, Aug. 17, 1965; *Chem. Abs.* **65**: 3393.
111. Ionescu, I., *et al.*, French Patent 1,443,989, July 1, 1966; *Chem. Abs.* **66**: 6388.
112. Isakova, R. A., *et al.*, *Tsvet. Metal*, **37** (4), 55 (1964); *Chem. Abs.* **61**: 9191.
113. Isakova, R. A., and Nesterov, V. N., *Trudy Inst. Met. i Obogashcheniya, Akad. Nauk Kazakh. SSR* **3**, 124 (1960); *Chem. Abs.* **55**: 16374.
114. Isakova, R. A., and Nesterov, V. N., *Primenenie Vakuuma v Met., Akad. Nauk SSSR, Inst. Met., Tr. Tret'ego Soveshch.* 158 (1963); *Chem. Abs.* **59**: 14933.
115. Isakova, R. A., *et al.*, *Tr. Inst. Met. Obogashch., Akad. Nauk Kaz. SSR Inst. Met. A.A. Baikov* 13 (1967); *Chem. Abs.* **70**: 49045.
116. Isakova, R. A., Nesterov, V. N., and Tseft, A. L., *Tr. Inst. Met. i Obogashch., Akad. Nauk Kaz. SSR* **4**, 8 (1962); *Chem. Abs.* **58**: 3117.

117. Ishihara, T., *Kyushu Kozan Gakkai-shi* **28**, 519 (1960); *Chem. Abs.* **55**: 13783.
118. Ishihara, T., and Sakata, H., *Ibid.*, 581; *Chem. Abs.* **55**: 13784a.
119. Ishihara, T., *Ibid.*, **29**, 22 (1961); *Chem. Abs.* **55**: 13784b.
120. Ishihara, T., Sakata, H., and Otsuka, Y., *Kyushu Kozan Gakkai-shi* **29**, 66 (1961); *Chem. Abs.* **55**: 27800.
121. Ivashentsev, Ya. I., *Tr. Tomskogo Gos. Univ.*, *Ser. Khim.* **170**, 143 (1964); *Chem. Abs.* **63**: 1501.
122. Ivashentsev, Ya. I., *et al.*, *Zh. Prikl. Khim.* (*Leningrad*) **41** (7), 1479 (1968); *Chem. Abs.* **69**: 98572.
123. Janjua, M. B. I., Yannopoulos, J. C., and Cooper, W. C., "The Corrosive Action of Selenium Towards Various Materials in the Temperature Range 300 to 700°C," in *Corrosion by Liquid Metals*, A.I.M.E. Symposium, Plenum Publishing Corporation, 1970, pp. 339–359.
124. Kabanova, L. M., and Teplyakov, B. V., *Tsvet. Metal.* **37** (7), 75 (1964); *Chem. Abs.* **62**: 1353.
125. Karaev, Z. Sh., and Efendiev, I. K., *Azerb. Khim. Zh.* (5), 119 (1961); *Chem. Abs.* **59**: 7177.
126. Kerhart, J., and Kucera, L., Czech. Patent 101,899, Dec. 15, 1961; *Chem. Abs.* **58**: 7633.
127. Kindyakov, P. S., and Katlinskii, V. M., *Trudy Moskov. Inst. Tonkoi Khim. Tekhnol.* **7**, 149 (1958); *Chem. Abs.* **54**; 21669.
128. Kindyakov, P. S., and Safonov, V. V., *Izv. Vysshikh Uchebn. Zavedenii, Tsvet. Metal.* **5**, (1), 107 (1962); *Chem. Abs.* **57**: 3115.
129. Klebanov, G. S., and Ostapkevich, N. A., *Zh. Prikl. Khim.*, **33**, 1957 (1960).
130. Knyazeva, R. N., Chernova, G. N., and Zhukovskaya, G. B., *Izv. Vyssh. Ucheb. Zaved., Khim. Khim. Tekhnol.* **9** (6), 869 (1966); *Chem. Abs.* **66**: 11086, 119318.
131. Koch, W., German Patent 973,254 (1959).
132. Koeppel, W. E., and Schellinger, A. K., *Extr. Refining Rarer Metals, Proc. Symposium*, London, 1956, 96–106.
133. Kiozumi, I., and Tenigishi, Y., Japanese Patent 1617, Mar. 18, 1961; *Chem. Abs.* **55**: 27816.
134. Koropp, K. H., and Clugston, E. J., "The Bisulphate Fusion Process for Copper Refinery Slimes Treatment," paper presented at Annual Meeting of the A.I.M.E., February 20–23, 1956.
135. Kostenko, B. N., *et al.*, *Khim. Tekhnol.*, *Respub. Mezhvedom. Nauch.—Tekh. Sb.* **7**, 116 (1967); *Chem. Abs.* **69**: 15257.
136. Kozhitov, L. V., and Vanyukov, A. V., *Zh. Prikl. Khim.* (*Leningrad*) **42** (8), 1764 (1969); *Chem. Abs.* **71**: 105806.
137. Koz'min, Y. A., *Sbornik Nauch. Trudov Vsesoyuz. Nauch.—Issledovatel. Gornomet. Inst. Tsvet. Metal.* **6**, 363 (1960); *Chem. Abs.* **56**: 5704.
138. Koz'min, Yu. A., Davydov, E. V., and Ponomarev, V. D., *Tr. Inst. Met. Obogashch., Akad. Nauk Kaz. SSR* **29**, 9 (1969); *Chem. Abs.* **71**: 93131.
139. Koz'min, Yu. A., and Kupchenko, M. M., Russian Patent 110,675, June 25, 1958; *Chem. Abs.* **52**: 19870.
140. Krebs, H., British Patent 644,743 (1950).
141. Kucera, L., and Kerhart, J., Czech. Patent 101,898, Dec. 15, 1961; *Chem. Abs.* **59**: 7218.
142. Kudryavtsev, A. A., *Izv. Vysshikh. Ucheb. Zavedenii, Khim. i Khim. Tekhnol.* **3** (1), 151 (1960); *Chem. Abs.* **54**: 15166.

143. Kudryavtsev, A. A., et al., "Separation of Sulphur, Selenium and Tellurium," Khimiko-Tekhnologicheskiy Institut. Trudy (35), 1961. Khimiya i Tekhnologiya Neorganicheskikh Veshchestv, p. 111.

144. Kudryavtsev, A. A., and Klushina, T. B., Tr. Mosk. Khim.—Tekhnol. Inst. (35), 116 (1961); Chem. Abs. 57: 3073.

145. Kuzin, I. A., and Garkun, V. K., Zh. Prikl. Khim. 40 (9), 1902 (1967); Chem. Abs. 68: 2317, 23891.

146. Lebedeff, Y. E., Stone, J. R., and von Stein, P. W., U.S. Patent 2,775,509, Dec. 25, 1956.

147. Lederer, M., and Kertes, S., Anal. Chim. Acta 15, 226 (1956); Chem. Abs. 51: 7238.

148. Loeschau, S., German Patent (East) 11,893, July 26, 1956.

149. Loeschau, S., Z. Erzbergbau u. Metallhüttenw. 12, 21 (1959); Chem. Abs. 53: 6838.

150. Loeschau, S., Freiberger Forschungsh. B34, 101 (1959); Chem. Abs. 54: 3109.

151. Loeschau, S., German Patent (East) 20,124, Oct. 19, 1960; Chem. Abs. 55: 23954.

152. Loeschau, S., Freiberger Forschungsh. B60, 7 (1961).

153. Lumbroso, H., Nouveau Traité de Chimie Minérale, Vol. XIII, ed. Pascal, P., Masson et Cie, Paris, 1961, pp. 1651–1912.

154. Malyshev, V. P., et al., Tsvet. Metal. 41 (3), 30 (1968).

155. McGee, T., Lynch, J., and Boswell, G. G. J., Talanta 15 (12), 1435 (1968); Chem. Abs. 70: 41331.

156. Mellgren, S., Opie, W. R., and Coffin, L. D., U.S. Patent 3,288,561, Nov. 29, 1966.

157. Mellgren, S., Opie, W. R., and Coffin, L. D., Canadian Patent 769,651, Oct. 17, 1967.

158. Mellgren, S., and Taubenblat, P. W., "Identification of Compounds in Copper Electrolytic Refinery Slimes," paper presented at Annual Meeting of A.I.M.E., February 1963.

159. Metzner, G., Thèse, Univ. de Paris (1898); ref. 153.

160. Mikhailovskaya, F. R., Russian Patent 145,555, Mar. 21, 1962; Chem. Abs. 57: 12272.

161. Miyamoto, M., Bunseki Kagaku 10, 211 (1961); Chem. Abs. 57: 2844.

162. Morrison, B. H., Met. Soc. Conf. 24, 227 (1963).

163. Mostecký, J., et al., Chem. průmysl 11, 2 (1961); Chem. Abs. 55: 11775.

164. Navarro, N. A., Spanish Patent 207,021, Mar. 23, 1954; Chem. Abs. 49: 10828.

165. Nielsen, S., and Heritage, R. J., J. Electrochem. Soc. 106, 39 (1959).

166. Nijland, L. M., Philips Research Reports 9, 267 (1954).

167. Nisel'son, L. A., Pustil'nik, A. I., and Soshnikova, L. A., Izv. Akad. Nauk SSSR, Otd. Tekhn. Nauk, Met. i Gorn. Delo (2), 79 (1963); Chem. Abs. 59: 2464.

168. Oberbacher, B., and Schlier, O., German Patent 1,035,113, July 31, 1958; Chem. Abs. 54: 17823.

169. Ogienko, A. S., et al., Tsvet. Metal. 42 (9), 414 (1969); Chem. Abs. 72: 23659.

170. Okunev, A. I., Byull. Tsvetnoi Met. 21, 23 (1957); Chem. Abs. 54: 10714.

171. Olesova, A. I., Kochnev, M. I., and Pimenov, I. V., Tr. Ural. Nauch.—Issled. Proekt. Inst. Medn. Prom. (8), 305 (1965); Chem. Abs. 66: 21210.

172. Orlov, A. M., Borbat, V. F., and Ferberg, M. B., Tsvet. Metal. 36 (3), 81 (1963).

173. Pacauskas, E., Janickis, J., and Rinkeviciene, E., Liet. TSR Mokslu Akad. Darb., Ser. B, (3), 45 (1966); Chem. Abs. 66: 8536, 91287.
174. Pestov, S. S., and Sablin, P. M., Tsvet. Metal. 38 (10), 88 (1965).
175. Petersen, E., Z. Physik. Chem. 8, 601 (1891); ref. 153.
176. Phillips, A. J., Lebedeff, Y. E., and Crandall, J. R., U.S. Patent 2,413,374, Dec. 31, 1946; Chem. Abs. 41: 1196.
177. Plaksin, I. N., Suvorovskaya, N. A., and Astaf'eva, A. V., J. Applied Chem. (USSR) 19, 668 (1946)
178. Pleteneva, N. B., Tsvet. Metal. 34 (1961); Eng. trans. 2, 57.
179. Pleteneva, N. B., Smirnov, M. P., and Yukhtanov, D. M., Bull. Ts. II N TsM, 11 (1960); ref. 178.
180. Polivyannyi, I. R., and Milyutina, N. A., Tr. Inst. Met. i Obogashch., Akad. Nauk Kaz. SSR 6, 64 (1963); Chem. Abs. 60; 215.
181. Polukarov, A. N., Buketov, E. A., and Makhmetov, M., Tr. Khim.—Met. Inst. Akad. Nauk Kaz. SSR 4, 45 (1967); Chem. Abs. 69: 38005.
182. Polukarov, A. N., et al., Tr. Khim.—Met. Inst., Akad. Nauk Kaz. SSR 4, 60 (1967); Chem. Abs. 69: 38019.
183. Polukarov, A. N., and Smirnov, V. I., Trudy Ural. Politekh. Inst. im. S. M. Kirova 98, 24 (1960); Chem. Abs. 55: 25658.
184. Ponomarev, V. D., Buketov, E. A., and Kononenko, G. A., Izvest. Vysshikh Ucheb. Zavedenii, Tsvetnaya Met. 2, (6) 85 (1959).
185. Ponomareva, E. I., et al., Trudy Inst. Met. i Obogashch., Akad., Nauk Kaz. SSR 1, 76 (1959); Chem. Abs. 54: 15152.
186. Porter, C. B., Labbe, A. L. Jr., and Pike, K. N., U.S. Patent 2,863,731, Dec. 9, 1958; Chem. Abs. 53: 5613.
187. Pourbaix, M., Atlas of Electrochemical Equilibria in Aqueous Solutions, Pergamon Press, Inc., New York (1966).
188. Prater, J. D., et al., U.S. Patent 3,130,012, Apr. 21, 1964.
189. Rajcic, D., Milojkovic, D., and Simic, M., Tehnika (Belgrade) 22 (5), 805 (1967) Chem. Abs. 69: 45391.
190. Rawling, J. R., and Toguri, J. M., Can. J. Chem. 44, 451 (1966).
191. Reusser, R. E., and Hart, J. L., U.S. Patents 3,239,306 and 3,239,307, Mar. 8, 1966.
192. Reuther, W., and Knaul, W., Neue Huette 7 (3), 154 (1962).
193. Reznyakov, A. A., Nesterov, V. N., and Isakova, R. A., Tr. Inst. Met. Obogashch., Akad., Nauk Kaz. SSR Inst. Met. A. A. Baikov 26, 27 (1967); Chem. Abs. 70: 49043.
194. Roberts, W. B., U.S. Patent 2,737,348, Mar. 6, 1956; Chem. Abs. 50: 8435.
195. Roseman, R., Neptune, R. W., and Allan, B. W., U.S. Patent 2,616,791, Nov. 4, 1952; Chem. Abs. 47: 1907.
196. Rowland, J. F., Mines Branch Investigation Report IR 62–95, Department of Mines and Technical Surveys, Ottawa, Canada, October 11, 1962.
197. Ruprych, M., Czech. Patent 122,333, Mar. 15, 1967; Chem. Abs. 69: 37552.
198. Samuelson, O., IVA 17, 5 (1946).
199. Satin, Ya. I., Prom. Armenii, Sov. Nar. Khoz. Arm. SSR, Tekhn.—Ekon. Byul. 5 (5), 37 (1962); Chem. Abs. 59: 8421.
200. Sato, T., and Kaneko, H., Technol. Repts. Tohoku Univ. 16, (2) (1952).
201. Sawaya, T., Japanese Patent 18, Jan. 11, 1950; Chem Abs. 46: 8603.
202. Sazhin, N. P., Zh. Vses. Khim. Obshchest 13 (5), 499 (1968); Chem. Abs. 70: 13037.

203. Schloen, J. H., and Elkin, E. M., *J. Metals* **188**, 764 (1950).
204. Scholten, J. J. F., *Chem. Weekblad* **51**, 583 (1955); *Chem. Abs.* **49**: 13852.
205. Schumann, H., German Patent (East) 32,681, Dec. 5, 1964; *Chem. Abs.* **63**: 9506.
206. Selezneva, N. A., *Izv. Akad. Nauk Kaz. SSR, Ser. Khim.* **16** (4), 31 (1966); *Chem. Abs.* **66**: 51739.
207. Shebunin, V. S., *et al.*, Russian Patent 117,141, Jan. 29, 1959; *Chem. Abs.* **53**: 20722.
208. Simons, J. H., *J. Am. Chem. Soc.* **52**, 3483 (1930).
209. Sindeeva, N. D., *Mezhdunarod. Geol. Kongress, 21–ya (Dvadstat Pervaya) Sessiya, Doklady Sovet. Geologov, Problema* **1**, 129 (1960).
210. Skowerski, M., *Przemisl Chem.* **35**, 423 (1956), *Chem. Abs.* **53**: 3621.
211. Soshnikova, L. A., and Ezernitskaya, M. E., *Sb. Tr. Gos. Nauchn.-Issled. Inst. Tsvetn. Metal.* **19**, 340 (1962); *Chem. Abs.* **60**: 1357.
212. Soshnikova, L. A., and Matveeva, Z. I., *Tsvet. Metal.* **36** (11), 62 (1963).
213. Soshnikova, L. A., and Paikin, V. A., *Tsvet. Metal.* **42** (5), 76 (1969); *Chem. Abs.* **71**: 51701.
214. Soshnikova, L. A., *et al.*, Russian Patent 223,790, Feb. 3, 1969; *Chem. Abs.* **71**: 14678.
215. Sosnovskii, G. N., *Tr. Altaisk. Gorno-Met. Nauchn.—Issled. Inst., Akad. Nauk Kazakhsk. SSR* **11**, 60 (1961); *Chem. Abs.* **56**: 13873.
216. Sugawara, M., and Kimura, K., Japanese Patent 5222, June 30, 1956; *Chem. Abs.* **52**: 10522.
217. Takahashi, I., Japanese Patent 181,066, Dec. 2, 1949; *Chem. Abs.* **46**: 2249.
218. Takei, T., Aoyama, K., and Okada, H., *Repts. Sci. Research Inst. (Japan)* **26**, 234 (1950); *Chem. Abs.* **46**: 866.
219. Tamura, T., *et al.*, Japanese Patent 7052, Oct. 3, 1955; *Chem. Abs.* **51**: 17532.
220. Tamura, T., *et al.*, Japanese Patent 2455, Apr. 4, 1956; *Chem. Abs.* **51**: 8555.
221. Tamura, T., *Nippon Kogyo Kaishi* **74**, 846 (1958).
222. Terasaki, Y., and Sasakawa, S., Japanese Patent 2560, July 9, 1952; *Chem. Abs.* **47**: 6331.
223. Thomsen, *Thermochemische Untersuchungen* **2**, 313 (1882); ref. 153.
224. Threlfall, R., *Chem. News* **95**, 207 (1907); ref. 153.
225. Timofeeva, V. K., *Sb. Nauch. Tr., Gos. Nauch.—Issled. Inst. Tsvet. Metal.* **27**, 22 (1967); *Chem. Abs.* **69**: 45369.
226. Tishchenko, A. A., *Sb. Nauchn. Tr., Ural'sk. Politekhn. Inst.* No. 134, 27 (1963); *Chem. Abs.* **61**: 3955.
227. Tishchenko, A. A., and Smirnov, V. I., *Izv. Vysshikh Uchebn. Zavedenii, Tsvetn. Met.* **5**, (3), 49 (1962); *Chem. Abs.* **57**: 12167.
228. Tishchenko, A. A., and Smirnov, V. I., *Dokl. Akad. Nauk SSSR* **145**, 863 (1962); *Chem. Abs.* **58**: 2171.
229. Tishchenko, A. A., and Smirnov, V. I., *Zh. Prikl. Khim.* **36** (11), 2363 (1963).
230. Tougarinoff, B., personal communication.
231. Tseft, A. L., Azerbaeva, R. G., and Adilova, A. A., *Vestn. Akad. Nauk Kaz. SSR* **19** (9), 58 (1963); *Chem. Abs.* **60**: 4865.
232. Tseft, A. L., Isakova, R. A., and Nesterov, V. N., Russian Patent 149,765, Sept. 14, 1962; *Chem. Abs.* **58**: 6488.
233. Tseft, A. L., and Rusina, L. D., *Trudy Vostochno-Sibir. Filiala, Akad. Nauk S.S.S.R.* **25**, 64 (1960); *Chem. Abs.* **55**: 9200.
234. Tserekova, A. M., and Giganov, G. P., *Sb. Nauch. Tr. Vses. Nauch. Issled. Gorno-Met. Inst. Tsvet. Metal.* **12**, 32 (1968); *Chem. Abs.* **70**: 23547.

235. Tsuji, Y., Japanese Patent 14,754 (1960); *Chem. Abs.* **56**: 8371.
236. Tsuzuki, R., *Nippon Kinzoku Gakkaishi* **25**, 721 (1961); *Chem. Abs.* **60**: 14149.
237. Tsvetkov, Yu. V., Basieva, I. Ya., and Chizhikov, D. M., *Primenenie Vakuuma v Met.*, *Akad. Nauk SSSR, Inst. Met., Tr. Tret'ego Soveshch.* 174 (1963); *Chem. Abs.* **59**: 12466.
238. Tuwiner, S. B., U.S. Patent 2,981,603, Apr. 25, 1961.
239. Ueda, Y., and Okamoto, K., *Kyushu Kozan Gakkaishi* **26**, 105 (1958); *Chem. Abs.* **53**: 1022.
240. Usenova, Z. M., Mamonova, G. F., and Erdenbaeva, M. I., *Zh. Neorgan. Khim.* **9** (7), 1547 (1964); *Chem. Abs.* **61**: 11614.
241. Ustyugov, G. P., Vigdorovich, E. N., and Bezobrazov, E. G., *Izv. Akad. Nauk SSSR, Neorg. Mater.* **5** (2), 363 (1969); *Chem. Abs.* **70**: 91225.
242. Vaaler, L. E., U.S. Patent 2,835,558, May 20, 1958.
243. Vaaler, L. E., British Patent 814,852, June 10, 1959; *Chem. Abs.* **53**: 22796.
244. Valeva, A., Tatarskii, A., and Gusev, G., *Tezka Prom.* **8** (12), 27 (1959); through *Chem. Abs.* **60**: 11641.
245. Vanyukov, A. V., *et al.*, *Izv. Akad. Nauk SSSR, Neorg. Mater.* **5** (1), 3 (1969); *Chem. Abs.* **70**: 91349.
246. Veale, C. R., *Analyst* **85**, 130 (1960); *Chem. Abs.* **54**: 18191.
247. Vetrenko, E. A., Diev, N. P., and Olesova, A. I., *Trudy Inst. Met., Akad. Nauk S.S.S.R., Ural Filial* **2**, 141 (1958); *Chem. Abs.* **55**: 6318.
248. Vinogradova, M. A., *Tsvet. Metal.* **32**, (6), 39 (1959); *Chem. Abs.* **53**: 19743.
249. Vocel, J., Czech. Patent 103,163, Mar. 15, 1962; *Chem. Abs.* **58**: 12204.
250. Voight, A., and Krebs, H., German Patent 800,860 (1950).
251. Voilokova, V. V., and Speranskaya, E. F., *Zh. Neorg. Khim.* **12** (11), 3035 (1967); *Chem. Abs.* **68**: 4402, 45480.
252. Vol'fkovich, S. I., *et al.*, *Zh. Prikl. Khim.* **36** (6), 1169 (1963); *Chem. Abs.* **59**: 12383.
253. Von Stein, P., U.S. Patent 3,419,355, Dec. 31, 1968; *Chem. Abs.* **70**: 49072.
254. Von Stein, P., and Lebedeff, Y. E., U.S. Patent 2,809,885, Oct. 15, 1957.
255. Wagenmann, R., and Wehle, G., German Patent 1,244,741, July 20, 1967; *Chem. Abs.* **67**: 7884.
256. Wagenmann, R., Scherz, E., and Puchta, B., German Patent (East) 27,336, Jan. 28, 1964; *Chem. Abs.* **61**: 12977.
257. Wagenmann, R., and Wehle, G., German Patent (East) 44,085, Dec. 23, 1965; *Chem. Abs.* **64**: 19032, 19033.
258. Wallden, S. J., and Edenwall, I. A. O., Swedish Patent 153,085, Jan. 17, 1956; *Chem. Abs.* **50**: 17360.
259. Ward, R. G., and Hoar, T. P., *J. Inst. Metals* **90**, 6 (1961–62).
260. Weissleder, E., German Patent 1,266,739, Apr. 25, 1968; *Chem. Abs.* **69**: 60385.
261. Yannopoulos, J. C., Unpublished work.
262. Yannopoulos, J. C., and Cooper, W. C., "Corrosive Action of Selenium at High Temperatures," paper presented at the 14th Canadian Chemical Engineering Conference, Hamilton, Ont., Oct. 26–28, 1964.
263. Yannopoulos, J. C., and Themelis, N. J., *Can. J. Chem. Eng.* **42** (5), 219 (1964).
264. Yoshino, Y., *J. Chem. Soc. Japan* **71**, 577 (1950); *Chem. Abs.* **45**: 6538.
265. Yost, D. M., and Kircher, C. E., *J. Am. Chem. Soc.*, **52**, 4680 (1930).
266. Yukhtanov, D. M., *Byull. Tsvetnoi Met.* **9**, 29 (1957); *Chem. Abs.* **54**: 12509.
267. Zarechanskaya, V. V., and Speranskaya, E. F., *Sb. Statei Aspir. Soiskatelei, Min. Vyssh. Sredn. Spets. Obrazov. Kaz. SSR, Khim. Khim. Tekhnol.* (6), 96 (1967); *Chem. Abs.* **70**: 83582.

268. Zelenov, V. I., *Sb. Materialov po Gorn. Delu, Obogashch. i Met. Tsentr. Nauchn.* —*Issled. Gorno-Razved. Inst.* **6**, 64 (1961); *Chem. Abs.* **57**: 16223.
269. Zelyanskaya, A. I., Bykov, I. E., and Gorshkova, L. S., *Trudy Inst. Met., Ural. Filial. Akad. Nauk SSSR* **1**, 151 (1957); *Chem. Abs.* **53**: 6729.
270. Zhuravleva, M. G., and Chufarov, G. I., *Zhur. Priklad. Khim.* **24**, 28 (1951); *Chem. Abs.* **46**: 8492.
271. Bovey, H. J., and Marks, S., *South African Chemical Processing*, April–May, 28 (1969).

The following articles not referred to in the chapter will be useful to the interested reader:

Dutton, W. A., Van den Steen, A. J. and Themelis, N. J., "Recovery of Selenium from Copper Anode Slimes," *Met. Trans. AIME* **2**, 3091 (1971).
Nakano Umeo, "Recovery of Selenium from Selenium Containing Materials, such as Slurry originating from the Electrolytic Refining of Copper," Ger. Offen. 2,015,535, 15 October, 1970 (*Chem. Abst.* **74**, 66160 (1971).

3

The Structure of Selenium

W. CHARLES COOPER*
*Noranda Research Centre,
Pointe Claire, Quebec, Canada*

R. A. WESTBURY
*Department of Chemistry,
Sir George Williams University, Montreal, Canada*

INTRODUCTION

It is only in recent years that reliable and consistent data have begun to appear concerning the structural aspects of the various allotropes of selenium. The development of methods for the preparation of single crystals of trigonal selenium has resulted in a significant increase in the knowledge of this form of the element. Thin selenium layers, grown epitaxially by vapor phase deposition onto single crystal substrates, can now be readily obtained, adding considerably to our knowledge of thin film structures. The characterization of the vapor itself appears to be well in hand.

However, the structural properties of the amorphous forms of selenium are not well characterized, and real, concrete evidence of their properties is difficult to obtain. Their structural properties can only be inferred from physical changes of the

* Present Address: c/o United Nations Development Programme, Casilla 197-D, Santiago, Chile.

allotropes themselves, and often these changes can be interpreted in a number of separate, but apparently convincing ways. This is an inevitable consequence of the fact that the amorphous allotropic forms are in fact amorphous, and as such defy interpretation by the normal physical methods of structural determination.

In this chapter, we shall attempt to review the numerous allotropes of selenium. As solid allotropic forms, only the following can be accepted as being positively identified at this time: red amorphous, black amorphous, vitreous, α-monoclinic, β-monoclinic, and trigonal. The expression " glassy selenium " is regarded as a synonym for " vitreous selenium."

THE VAPOR PHASE

By its very nature, the vapor phase is the simplest phase of selenium that can be described. Also, the vapor pressure of liquid selenium is important, both in its own right, and commercially, and a principal method of measuring the vapor pressure, the Knudsen technique, requires a knowledge of the composition of the effusing vapor. Furthermore, in the commercial manufacture both of selenium rectifiers and of the selenium drums used in xerography, the films of selenium are laid down via vapor phase deposition. It is obvious then that the composition of selenium vapor at any temperature should be known as accurately as possible, and this need has resulted in the publication of a number of recent papers about selenium vapor.

A variety of experimental techniques have been used for the determination of the composition of selenium vapor. These methods include electron diffraction,[130] vapor density,[93,150,153] magnetic susceptibility,[13,14,128] torsion effusion studies,[133,188] a Knudsen cell coupled with a radioactive isotope of selenium,[116,117] an electrochemical Knudsen cell,[152] spectroscopy in the visible and UV regions,[19,22,23] and mass spectrometry.[11,60]

Until the advent of modern high resolution mass spectrometers, the estimation of the composition of selenium vapor with any degree of exactitude was difficult and uncertain, and a number of early estimates of the composition of the vapor[13,14,129,133,140,141,142,150,186] are now only of historical interest. Perhaps the first significant work on the structure of selenium vapor is that of Illarionov and Lapina,[93] who from their results of the vapor density of selenium in the temperature range 550–900°C, at low pressures (max 1 mm Hg), suggested that the vapor contains, under those conditions, Se_2, Se_4, Se_6, and Se_8 molecules. The following equilibrium constants for the dissociation of higher molecules to Se_2 were obtained:

(a) Se_4 $\rightleftharpoons$ $2Se_2$ $\log K_4 \text{ (mm)} = 11.31 - 8.020 \times 10^3/T$

(b) Se_6 $\rightleftharpoons$ $3Se_2$ $\log K_6 \text{ (mm}^2) = 19.23 - 13.702 \times 10^3/T$

(c) $Se_8 \rightleftharpoons 4Se_2$ $\log K_8$ (mm^3) $= 29.41 - 20.520 \times 10^3/T$

where $K_4 = P_{Se_2}^2/P_{Se_4}$, $K_6 = P_{Se_2}^3/P_{Se_6}$, etc

Kuliev and Shakhtakhtinskii[116] studied the saturated vapor pressure of selenium in the temperature range 86–200°C. It was stated that the selenium vapor consisted of Se_6 molecules, within the temperature range studied.

Goldfinger and Jeunehomme[70] investigated the evaporation of "grey" selenium in order to obtain data for further work on the evaporation of selenides. From experiments in the temperature region 450–510°K, they observed mass spectrometer peaks corresponding to Se_n^+ ($n = 1$ to 10), but they reported that the intensities of Se_9^+ and Se_{10}^+ were very much lower than those of the lighter peaks. It was suggested that most, or perhaps all, species from Se_2 to Se_8 are present as parent molecules. However, they concluded only that the molecular species Se_8, and perhaps Se_7, are present "in a concentration of the order of 10%."

As a preliminary to work on binary selenide systems, Brebrick[19] analyzed the absorption spectrum of selenium vapor between 1900 Å and 20,000 Å (2.0μ), at five temperatures in the range 330–860°C, and for total pressures between 7×10^{-5} and 1.0 atm. Predominantly, Brebrick was concerned with an analysis of the same visible and UV vibronic bands as had been studied previously by several authors,[160] and many of his data were obtained in this region of the spectrum. For absorption at 2100 Å and 3405 Å, his analysis suggested that the equivalents of two molecular species, Se_2 and Se_n, were involved. At 860°C, Se_n was identified as Se_4, as Se_5 at 700°C and 500°C, and as $Se_{4.6}$ at 400°C. Brebrick, accepting the work of Illarionov and Lapina,[93] i.e., their identification of Se_2, Se_4, Se_6, and Se_8 as the major species in selenium vapor, interpreted his results to indicate that Se_4 and Se_6 are the predominant absorbers at 2100 Å, "although their proportions must always be such that optically it appears that only a single species besides Se_2 is present between 2100 and 2700 Å." In his 1965 paper, Brebrick's results were detailed graphically. In a more recent paper,[21] his results are available in tabular form, and are compared with more recent publications.

Berkowitz and Chupka[11] published a mass spectrometric investigation of the vapor of what they claimed to be trigonal (hexagonal) selenium and α-monoclinic selenium, although their experimental details leave some doubt that their starting materials were actually the pure allotropic forms that they expected. Trigonal selenium was prepared by vacuum distillation of commercial-grade selenium powder, retaining only the middle portion; and α-monoclinic selenium was prepared by extraction of quenched selenium (i.e., vitreous selenium) with CS_2, followed by slow evaporation of the CS_2 solution. However, there is no evidence in the paper that trigonal selenium was in fact obtained by the vaccum distillation of this "selenium powder," itself an underscribed allotropic modification; some form of x-ray pattern of the distilled selenium would have been a distinct asset. In addition, the evaporation of a saturated selenium solution of CS_2 appears to produce both α- and β- forms of monoclinic selenium, which then have to be separated by other means.

Bearing these reservations in mind, the paper is of considerable interest in the present study. The investigators, using mass spectrometric analysis, reported the vapor species of trigonal selenium effusing from a Knudsen cell at 544°K (although at this temperature the selenium must have been liquid), and of trigonal selenium and α-monoclinic selenium, by surface evaporation at considerably lower temperatures. They found the differences in the two surface evaporation spectra to be small, and from this they concluded "that the two spectra are similar enough to be due to the same antecedent." In conclusion, the workers suggested that the rate of conversion of the α-monoclinic selenium to trigonal selenium was so fast at the temperature of the experiment that the surface evaporation from α-monoclinic selenium could not be measured. Evaporation of selenium from a Knudsen cell at 544°C (i.e., liquid selenium) produced results that were sufficiently different from the free evaporation results to suggest that Se_6 is a favored species in the free evaporation of selenium.

In effect, Berkowitz and Chupka reported mass spectrometric results, under both equilibrium and nonequilibrium conditions, for evaporating liquid selenium, as a function of temperature, and for subliming trigonal selenium, at a temperature below the melting point.

These researchers also published the following expressions for the equilibrium constant, as a function of temperature, for various selenium reactions (temperatures are in °K):

$$2Se_3 \rightleftharpoons 3Se_2 \quad \log Kp \, (\text{atm}) \quad = 8.17_7 - \frac{5,625}{T}$$

$$Se_6 \rightleftharpoons 3Se_2 \quad \log Kp \, (\text{atm}^2) \quad = 14.9_6 - \frac{14,200}{T}$$

$$5Se_6 \rightleftharpoons 6Se_5 \quad \log Kp \, (\text{atm}) \quad = 18.4_4 - \frac{12,100}{T}$$

$$6Se_7 \rightleftharpoons 7Se_6 \quad \log Kp \, (\text{atm}) \quad = 3.74 - \frac{3,090}{T}$$

$$3Se_8 \rightleftharpoons 4Se_6 \quad \log Kp \, (\text{atm}) \quad = 5.60 - \frac{3,600}{T}$$

$$6Se(s) \rightleftharpoons Se_6 \quad \log Kp \, (\text{atm}^{-5}) = 9.40 - \frac{7,370}{T}$$

$$6Se(l) \rightleftharpoons Se_6 \quad \log Kp \, (\text{atm}^{-5}) = 3.30 - \frac{4,360}{T}$$

Discussing Brebrick's 1968 paper, Berkowitz and Chupka[12] acknowledged a discrepancy between their calculated equilibrium constants, as detailed here, and the partial pressures of Se_2, as presented in another table of their earlier work.[11]   A corrected series of partial pressures of Se_2 and a corrected graph were published showing the composition of saturated selenium vapor as derived from equilibrium constant measurements.   The correction of this graph was necessary, even though the equilibrium constant-temperature expressions did not change, because the entire graph is based upon the knowledge of an accurate value for the partial pressure of Se_2 as a function of temperature.   With this correction, Berkowitz and Chupka's values for the partial pressure of Se_2 are in good agreement with those reported by Brebrick[20]. Berkowitz and Chupka point out, however, that the experimental error in their results is such that close agreement cannot be regarded as strong support for Brebrick's specific values.

One of the most important conclusions of Berkowitz and Chupka concerns the importance of Se_5 as a predominant species in the vapor.   Figure 9 of their 1966 paper, as corrected in their 1968 paper, is reproduced here (Figure 3-1) and the importance of Se_5 as a contributor to the composition of the vapor phase is obvious.

Another mass spectrometric study of selenium vapor, as evolved from subliming trigonal selenium in the temperature range 102–187°C, appeared in 1966[60].   Fujisaki and co-workers also attempted to study the vapor of what they described as red monoclinic selenium, but their experimental details are such that they were obviously using vitreous, and hence amorphous, selenium.

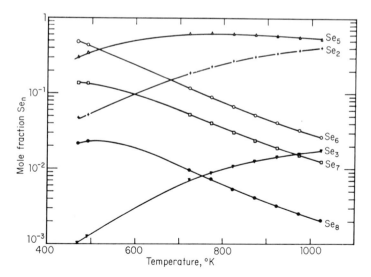

Figure 3-1   Composition of saturated selenium vapor, as derived from equilibrium constant measurements.[12]

In any case, their starting allotrope was converted to trigonal selenium within the time duration of their experiments, so this problem is of minor importance and hence the paper is concerned only with the sublimation of trigonal selenium.

The mass spectrum was scanned up to $m/e \simeq 1000$, using electrons of 40 V energy. Ion currents, due to the species Se_n^+, where $n = 1$ to 10, were found. The ion currents attributed to Se_9^+ and Se_{10}^+ were very small. Fujisaki et al. compared their results with those obtained by Goldfinger and Jeunehomme[70], under slightly different conditions (Goldfinger and Jeunehomme's results were obtained at 200°C, using 70 V electrons, while the results of the Fujisaki group were obtained at 175°C, using 40 V electrons), and for the majority of the ions, there is reasonable agreement.

Fujisaki and associates found Se^+, Se_2^+, Se_3^+, and Se_4^+ ions in the mass spectrum, and for the most part these ions were attributed to electron impact fragments of heavier species, probably by decomposition of a heavier molecule on the mass spectrometer filament, although they imply the possibility of small concentrations of Se and Se_2 in the vapor. However, they concluded that selenium vapor, at 175°C (448°K) is predominantly composed of Se_5, Se_6, Se_7, and Se_8 molecules. Their results, in tabular form, are presented in Table 3-1.

Fujisaki stated that the relative abundance, as quoted in Table 3-1, for Se_5 will be an upper limit, while those of Se_6, Se_7, and Se_8 will be lower limits. In spite of this cautiousness, their data do appear to be consistent with those of Berkowitz and Chupka (for a temperature of 448°K) as represented in Figure 3-1. The mole fraction of each component of the vapor, as calculated from Fujisaki's data, is not in exact agreement with the values to be expected on the basis of a slight extrapolation of Berkowitz and Chupka's data, but certainly

TABLE 3-1   COMPOSITION OF SELENIUM VAPOR[60]

Species	Relative Abundance at 175°C	% in Vapor[a] at 175°C	Mole Fraction[b]
Se_5	50.6	≤ 29.1	0.337
Se_6	100.0	≥ 57.5	0.555
Se_7	19.7	≥ 11.4	0.093
Se_8	3.5	≥ 2.01	0.015
Se_9	0.02	0.012	0.00006

[a] Symbols prefacing data in this column are those stated by Fujisaki et al.[60]

[b] As calculated by the authors of this chapter, assuming the "equal sign" in all cases.

the corresponding values are of the same order of magnitude, and we consider even this approximate agreement to be cross-corroborating.

Fujisaki and co-workers in an analysis of the mechanism of the sublimation, suggested that the subliming molecules are cyclic, not diradical chains. This reasoning was based on the assumption that sublimation of a cyclic molecule would be more energetically favorable than the formation of diradical chain molecules.

In a study of the thermodynamics of selenium vapor, Rau[153] measured the average molecular weight of selenium vapor, using either a quartz spiral manometer, or a technique which he developed involving three interconnected high temperature ovens, each held at a different temperature. From his data, Rau calculated the average number of atoms in the selenium molecule in the vapor state. This increased from 4 at $1000°K$ to about 5.5 at $714°K$.

In the analysis of his data, Rau assumed the presence in the vapor of Se_2, Se_4, Se_6, and Se_8, and derived dissociation constants for these species, dissociating to Se_2, on the basis of thermodynamic reasoning. His theoretically derived dissociation constants are:

(a) $Se_4 \rightleftharpoons 2Se_2$ $K_4 = P_{Se_2}^2/P_{Se_4}$
$$\log K_4 \text{ (mm)} = 11.526 - 8175/T$$

(b) $Se_6 \rightleftharpoons 3Se_2$ $K_6 = P_{Se_2}^3/P_{Se_6}$
$$\log K_6 \text{ (mm}^2\text{)} = 24.523 - 14{,}541/T - 1.510 \log T$$

(c) $Se_8 \rightleftharpoons 4Se_2$ $K_8 = P_{Se_2}^4/P_{Se_8}$
$$\log K_8 \text{ (mm}^3\text{)} = 42.024 - 23{,}573/T - 3.020 \log T$$

Using these equations, Rau was able to calculate the partial pressure of Se_2, as a function of pressure and temperature. However, both Brebrick, and Berkowitz and Chupka, have questioned the validity of Rau's results. Brebrick[21] has suggested that Rau's calculated values of the partial pressure of Se_2 are erroneous, and Berkowitz and Chupka[12] have questioned the assumptions used by Rau in the derivation of his dissociation constants.

Yamdagni and Porter[188] carried out mass spectrometric and torsion effusion studies of the evaporation of liquid selenium. The mass spectrum of the selenium vapor was scanned using electrons of 75 V energy, and most of the work was done at $600°K$. Ion currents due to the species Se_n^+, where $n = 1$ to 8, were measured; they reported that the molecular species Se_9^+ and Se_{10}^+ were too low in intensity for meaningful comparisons.

At the temperature in question, $600°K$ $(327°C)$, Yamdagni and Porter concluded that the species Se_2, Se_5, Se_6, Se_7, and Se_8 are present in the vapor, while the species Se, Se_3, and Se_4 were measured in their mass spectrometer only by reason of fragmentation processes. However, these investigators did not proceed further with their data, in that they did not calculate mole fractions in the vapor, since it was felt that the lack of knowledge concerning the relative ionization cross sections of the various molecular species precluded the possibility of an accurate and reliable vapor composition being obtained.[146] It was also realized that at least some of the Se_2^+ was a contribution from electron impact fragmentation.

A further paper in 1968,[106] reported a study of the sublimation of the trigonal and α-monoclinic forms of crystalline selenium, as well as the sublimation of the vitreous and vapor-deposited forms of amorphous selenium. The experimenter, Knox, used a technique developed to study the surfaces of solid samples in which vaporization of the substance is caused by excitation of the sample with a laser beam, operating at 6943 Å. Subsequent analysis of the vapor produced by this excitation was by a time-of-flight mass spectrometer.

Knox reported that excitation of any of the allotropic forms of selenium that were studied produced both neutral and ionic molecular species. The most predominant neutral molecular species in the vapor was Se_2, while the most predominant ionic molecular species was Se_5^+. It was suggested that the charged species arose by photon-solid interactions, while the neutral species which sublimed off the surface of the sample arose because of thermal heating of the sample by the laser pulse. Regardless of the mechanism involved, it will be noted that Se_2 and Se_5 are the two molecular species that would be expected to be predominant in selenium vapor at high temperature on the basis of the work of Berkowitz and Chupka[12].

Primarily, Knox was concerned with the slight differences in the relative intensities of the various molecular species vaporized from the surfaces of the different allotropic forms of selenium; in that sense the nature of the vapor composition above each sample was significant. However, the overall composition of the selenium vapor itself was only of secondary importance to Knox. Nevertheless, he does appear to have produced further support for the work of Berkowitz and Chupka[12], but this corroboration can not be quantitative, as it is difficult to estimate the temperature of the vapor as it sublimed from the solid sample.

From a consideration of possible mechanisms, Knox accepted the reasoning of Fujisaki et al.[60], and suggested that the Se_5, Se_6, Se_7, and Se_8 molecular species existing in the vapor were probably ring structures, while the smaller molecular species were chains, probably resulting from the fragmentation of ring structures.

A recent publication by Saure[165] has discussed the composition of subliming (trigonal) selenium vapor, in the temperature range 155–200°C. As Saure pointed out, only the vapor of the trigonal modification could be examined, because at about 110°C, every other species converted to the trigonal phase within a short time. In this work, the selenium vapor was ionized using a field ion source, and subsequent analysis of the vapor was by a single-focusing mass spectrometer. In the temperature range studied, Saure found the vapor to be composed of Se_2, Se_5, Se_6, Se_7, and Se_8 molecules. The temperature of the oven, from which the trigonal selenium sublimed, was about 160°C (433°K). Significantly, the relative peak intensities reported by Saure appear to agree, at least qualitatively, with what might be expected on the basis of the work of Berkowitz and Chupka. From other work, Saure concluded that the Se_5 molecular species was in the form of a chain.

It would appear that the composition of selenium vapor, as a function of temperature, is not yet completely determined. The data published by Berkowitz and Chupka[12] are extremely interesting, and where an overlap in temperature permits comparison with the work of Fujisaki et al.[60], the agreement appears to be significant. If the work of Berkowitz and Chupka could be more thoroughly corroborated, then the composition of the vapor would be considered understood.

Dissociation Constants for Species in the Vapor State

Scattered throughout the literature are expressions for the dissociation constants of various species in the vapor. For the convenience of the reader, these various expressions are assembled and retabulated here, with no attempt at critical evaluation. The expressions tabulated are those of Illarionov and Lapina[93] (I and L), Berkowitz and Chupka[11] (B and C), and Rau[153](R). For comparison, all pressures are assumed to be in atm. For the expressions of Illarionov and Lapina, the equivalent expressions of Brebrick[19] are used. The expressions derived by Rau[153] have been recalculated:

(a) $Se_4 \rightleftharpoons 2Se_2$ $\qquad K_4 = P_{Se_2}^2/P_{Se_4}$ atm
(I and L) $\qquad \log K_4$ (atm) $= 8.429 - 8.020 \times 10^3/T$
(R) $\qquad \log K_4$ (atm) $= 8.645 - 8.175 \times 10^3/T$
(B and C) $\qquad$ Not available

(b) $Se_6 \rightleftharpoons 3Se_2$ $\qquad K_6 = P_{Se_2}^3/P_{Se_6}$ atm^2
(I and L) $\qquad \log K_6$ (atm^2) $= 13.468 - 13.702 \times 10^3/T$
(R) $\qquad \log K_6$ (atm^2) $= 18.761 - 14.541 \times 10^3/T - 1.510 \log T$
(B and C) $\qquad \log K_6$ (atm^2) $= 14.96 - 14.200 \times 10^3/T$

(c) $Se_8 \rightleftharpoons 4Se_2$ $\qquad K_8 = P_{Se_2}^2/P_{Se_8}$
(I and L) $\qquad \log K_8$ (atm^3) $= 20.768 - 20.520 \times 10^3/T$
(R) $\qquad \log K_8$ (atm^3) $= 33.382 - 23.573 \times 10^3/T - 3.200 \log T$
(B and C) $\qquad$ As calculated by the present authors, $\log K_8$ (atm^3) $= 21.81 - 20,100/T$

Where comparison is possible, the relative agreement between Illarionov and Lapina[93] and Berkowitz and Chupka[11] is striking, in spite of the fact that Illarionov and Lapina did not find odd-numbered molecules in the vapor, while the work of Berkowitz and Chupka would seem to indicate that odd-numbered molecules are important at the temperatures in question.   Nevertheless, there are still discrepancies which should be resolved by further work.

LIQUID SELENIUM

To a large extent, the modern concept of liquid selenium began with the work of Briegleb.[24, 25]   In his experiments, Briegleb rapidly quenched molten selenium in liquid air, and then measured the solubility of the resulting vitreous selenium in carbon disulfide.   His results appeared to indicate that the vitreous selenium was heterogeneous, since part of the sample went into solution extremely quickly, while the remainder of the sample took a considerably longer time to dissolve.   From Briegleb's original work[24], Table 3-2 is reproduced, together with some auxiliary calculations by the authors of this chapter.   In considering these data, it must be noted that, for the separate samples held at the different temperatures, there was little difference in the overall solubilities; there was simply a difference in the rate of solubility of the separate fractions.

Briegleb explained his results on the basis that the quenched vitreous selenium was composed of two separate molecular species, designated by him as Se' and Se".   It was suggested that an equilibrium existed between these

TABLE 3-2   SOLUBILITY IN CARBON DISULFIDE OF
SELENIUM QUENCHED IN LIQUID AIR
FROM DIFFERENT TEMPERATURES[24]

Temperature[a] °C	% Se Sample Going[b] Quickly into Solution	Standard[c] Deviation
120	55 (53)	7.8
220	41 (41)	5.1
330	30 (28)	3.8
400	23 (23)	5.4
500	18 (18)	2.5
600	15 (15)	2.9
650	14 (13)	1.9

[a] This represents the temperature to which the selenium sample was heated before being quenched in liquid air.
[b] The averages quoted are a consequence of 9 replicate experiments. Numbers in brackets represent averages as calculated by the present authors from the raw data.
[c] Standard deviations have been calculated by the authors of this chapter.

two species, and Briegleb reasoned that because of the rapid, and presumably thorough quenching, the liquid selenium would have little time to adjust any dynamic equilibrium before the molecules were frozen into position. Hence the relative amounts of the species present in the quenched vitreous selenium could be regarded as the amounts initially present in the heated liquid selenium.

Since publication, Briegleb's data, at any given temperature, have come to be interpreted as representing the percentage of small rings present in liquid selenium at that temperature. This interpretation is based on the assumption that small ring molecules will go quickly into solution, while long-chain polymeric molecules will go into solution over a considerably longer period of time. This interpretation may well be correct, but nevertheless, the percentages quoted are difficult to accept without reservation. The experimental scatter of Briegleb's raw data alone is such that a careful reinvestigation of his work, taking all suitable precautions, would be in order.

Urazovskii and Luft[182] suggested that they had obtained results which tended to support the ideas expressed by Briegleb, but in retrospect their work does not appear to be substantive.

De Boer[40] reviewed the properties and structures of the vitreous and metallic forms of selenium. He appears to have been the first to suggest the presence of chain molecules in liquid selenium by analogy with liquid sulphur, and because of its abnormally high viscosity. He gave no data concerning the possible length of these chains, but he did suggest that they were " of a comparatively long lifetime."

Also in 1943, Richter[155] published the first of a series of papers on the structure of selenium. He could only conclude, on the basis of both x-ray and electron diffraction that liquid selenium was amorphous, and essentially he corroborated some earlier work by Das Gupta and Das[37, 38]

Krebs and Morsch[112] published a paper concerned with the viscosity of liquid selenium, both pure and with iodine as a deliberately added impurity. In their experiments, the iodine concentration ranged between 0 and 13.3 atomic %, and the workers found that as the iodine concentration increased, the viscosity of the liquid selenium, and its freezing point, decreased. In an attempt to calculate an average chain length in liquid selenium, Krebs and Morsch derived a simple theoretical analysis based upon some pioneering work by Flory.[57]

Flory had published a careful and considered paper showing an apparent linear relationship between viscosity and average chain length for a series of synthetic molten polymers with known molecular weights. Flory's equation was:

$$\log \eta = A + C\sqrt{Z}$$

where A and C were constants, C being independent of temperature, η was the viscosity of the molten polymer, and Z was the average chain length. To calculate an average chain length, Krebs modified this formula, although, as was pointed out later by Gimblett[66], the constant A in the equation was inadvertently omitted. From their theoretical analysis, and assuming that the effect of added iodine on the molten selenium was solely to serve as a chain breaker, Krebs and Morsch were able to estimate that at 200°C, the average chain was 120 atoms long, while at 300°C the average chain was only 85 atoms long. In their conclusion, Krebs and Morsch suggested that liquid selenium consisted of rings of selenium atoms, rather than chains of selenium atoms. It must be noted, however, that this was suggested simply because the experimenters felt that the presence of selenium chains, with their free radical ends, was not plausible. This idea appears to have been advanced only as a tentative hypothesis, and the fact remains that the main contribution of this paper was to suggest chain lengths at 200°C and at 300°C.

A later study by Krebs[110] also was concerned, in part, with the structure of liquid selenium. In this further work, Krebs, basing his reasoning on an approximation formula published by Powell et al.,[149] estimated a chain length for liquid selenium at 220°C of approximately 500 atoms. This estimate, however, appears to be based upon an incorrect value of the viscosity of liquid selenium at 220°C. Furthermore, Gimblett[66] has pointed out that the equation proposed by Powell et al., and used by Krebs, was derived for low molecular weight polymers only, and might well not be valid for the range in which Krebs was interested.

In retrospect, the assumptions used by Krebs and Morsch[112] and by Krebs[110] appear to be somewhat arbitrary and tenuous, and the chain lengths suggested for liquid selenium may well have to be discounted.

Nevertheless, the work of Krebs is significant because he appears to have been the first to suggest a dynamic equilibrium between rings and chains in liquid selenium. Krebs pictured the smaller rings in liquid selenium as continually breaking because of thermal dissociation, and combining with other broken rings to form chains in the melt. Ultimately, due to the coiling of the chains, rings would again be formed by the attack of a free radical chain end upon another part of its own chain. At the time, Krebs felt that this constant making and breaking of rings would have the effect of reducing the number of free radical chains to a minimum; while this idea is probably not accepted today, the entire concept of a dynamic state of flux in the melt is nevertheless present in Kreb's suggestions.

In one of a series of papers on the physics of selenium, Richter and co-workers[158] published some interference diagrams for liquid selenium, obtained by x-ray diffraction. For liquid selenium, at 230°C and at 430°C, and for vitreous selenium at room temperature, the graphical data produced by these scientists are strikingly similar, and hence they suggested that vitreous selenium and liquid selenium must have the same atomic arrangement. This is a reasonable supposition, since the quenching of liquid selenium occurs so quickly that any atomic arrangement possessed by the liquid must be "frozen in," in the vitreous state, especially if the final temperature is below the glass transition temperature. However, in this paper, no suggestions were advanced concerning the structure of liquid selenium.

It will be realized that, while this particular section is concerned solely with a survey of the structure of liquid selenium, it is impossible to be divorced completely from what was evolving at the time from concomitant studies upon the various solid modifications. Very often the same investigators were involved in studying both liquid and solid selenium, and consequently many of the ideas and suggestions appeared in papers on different aspects of selenium chemistry without apparent genesis. For example, Krebs and Schultze-Gebhardt,[113] in discussing the structure of vitreous selenium, interpreted their results on the basis of what had been published earlier pertaining to liquid selenium.[110]

A thesis by Buschert[30] has appeared, which is concerned in part with the structure of liquid selenium. Buschert used an x-ray reflection technique to investigate the nature of liquid selenium, and his raw data were ultimately resolved into radial distribution curves using a digital computer. From these radial distribution curves, it was calculated that for liquid selenium at either 235° or 310°C, there were two atoms in the first coordination sphere of a typical selenium atom, each separated from the central atom by a distance of 2.36 Å. Buschert's uncorrected results actually indicated that at this distance 2.3 atoms were involved in the first coordination sphere. After appropriate corrections, this decreased to a coordination number of about 2.1; Buschert indicated that his work was accurate only to ± 10–15%. Hence it would appear from these data that the apparent coordination number of a typical selenium atom in liquid selenium, at the temperatures studied, was 2; it may also be noted that a coordination number of 2, in a substance having only one constituent, implies a chain structure. Buschert, in fact, specifically states that, on the basis of his work, liquid selenium must have a chain structure. He admitted, however, in the thesis, that the technique of x-ray reflection and Fourier analysis cannot differentiate between an open-ended chain structure and a closed ring structure.

In a long paper, Richter and Herre[157] discussed the structure of solid amorphous selenium and liquid selenium over the temperature range -180–430°C. The paper is a review and an elaboration of work reported by Grimminger et al.,[78] Frohnmeyer et al.,[59] and Richter and Herre,[156] a discussion of experimental techniques, and a theoretical analysis of the results. The experimental method involved x-ray scattering and Fourier analysis of the data to yield radial distribution curves.

For liquid selenium, radial distribution curves were presented from two separate temperature regions: 270–300°C, and 400–430°C. In the lower temperature region, the observed atomic distance between two selenium atoms was 2.32 Å; in the region 400–430°C, the corresponding value was 2.33 Å. Both are in good agreement with the value of 2.36 Å obtained by Buschert[30]. Furthermore, the coordination number of the first coordination sphere in both temperature regions for which data were quoted was 2, again in agreement with Buschert's work.

Richter and Herre observed that their radial distribution curves for liquid selenium were virtually identical with those obtained from solid amorphous selenium (apparently referring to vitreous selenium). Hence, they concluded that the structures of liquid and solid amorphous selenium were identical. As a result, some of the structural concepts advanced for liquid selenium were transferred from the concurrent study of solid amorphous selenium. On this basis, Richter and Herre suggested that liquid selenium was composed of chains "with an extensively ordered parallel arrangement," a "quite considerable portion" of Se_6 rings, and small crystalline regions having the usual trigonal selenium lattice.

The actual molecular species present in liquid selenium may well not be Se_6, but Richter and Herre have supported the data obtained by Buschert, in regards to coordination number and interatomic distance between selenium atoms. Also, the researchers are in agreement with the concept of chains in liquid selenium.

Since 1958, there have been some significant papers concerning the structure of liquid selenium. But very often the publications have been of an on-going nature, and it would perhaps not be wise to discuss these papers chronologically. Rather, they will be discussed under a series of subheadings, but always with the aim of attempting to clarify the literature concerning the structure of liquid selenium.

Magnetic Susceptibility and Electron Paramagnetic Resonance

Some methods that would appear to offer some hope in elucidating the structure of liquid selenium are magnetic susceptibility and its companion, electron paramagnetic resonance. Some earlier work on liquid selenium had been reported.[29,104,159,180] But the first attempt at a quantitative elucidation of the structure of liquid selenium appears to have been that of Massen et al.,[128] who applied to selenium the techniques which had been used previously in a study of liquid sulfur.[147,148]

Massen and his associates determined the magnetic susceptibility of liquid selenium over a range of temperatures, using the Faraday technique. Their results were presented graphically, but by interpolation we have obtained the data presented in Table 3-3. They calculated their results on the assumption

TABLE 3-3   NUMBER-AVERAGE CHAIN LENGTH IN
LIQUID SELENIUM AS A FUNCTION OF
TEMPERATURE[128]

Temperature $^\circ C$	Number-Average Chain Length of Liquid Selenium
496	8.2×10^3
560	3.1×10^3
636	1.1×10^3
727	4.3×10^2
838	1.6×10^2

that the free radical ends of the polymer chains were the only source of para-magnetism, and they also assumed that the diamagnetism of the sample was independent of temperature. Furthermore, to derive these number-average chain length values, the group had to resort to a quantity representing the " weight fraction," ϕ, of polymers in a liquid sample; i.e., that fraction of the polymer in the form of chains. The values of ϕ, as a function of temperature, were obtained from a paper by Gee[65] but ultimately the values were derived from an analysis of the work of Briegleb[24]

The main problem in determining polymer chain lengths by magnetic susceptibility is that the paramagnetic contribution—the contribution from which the chain length is calculated—is present as one component of the total magnetic susceptibility, the other component being the diamagnetism of the sample. In fact, Poulis and his co-workers stated that the contribution attributed to paramagnetism was on the average only about 1 % of the total magnetic susceptibility. But the investigators did not feel that this was the limiting error in their experiments, and they suggest the error in the para-magnetism—and therefore in the mean chain length—to be "about a few per cent." The standard deviation of their results was about 7 %.

Perhaps a more accurate technique for the estimation of number-average chain lengths is electron paramagnetic resonance; liquid selenium is only beginning to receive much needed attention, and papers that can be related to the structure of liquid selenium are few in number. Stiles and co-workers[175] have investigated the *esr* spectrum of pure (99.9 %) vitreous selenium, in the temperature range 77–512°K. Throughout this range, which of course includes liquid selenium, a "comparatively narrow" *esr* absorption was observed, at a "g" value of 2.00. The scientists attributed this signal, by analogy with the results of *esr* work on metallic tellurium,[39] to either free electrons or free holes in the conduction band. When the selenium sample was melted in the cavity, a very broad *esr* absorption band also appeared at lower field strength. In their analysis, Stiles *et al.* were not able to attribute a specific signal, in liquid selenium, to chains of selenium atoms; and this result is somewhat surprising. It is conceivable, however, that using selenium of the purity quoted, the sample still contained sufficient impurities to negate any signal that might have been ascribed to chains with free radical ends.

Recently, a preliminary communication has appeared,[109] detailing the results of *esr* measurements on the polymerization of liquid selenium, in a temperature interval extending up to 700°C. The preliminary results reported do not appear to help in clarifying the structure of liquid selenium, but the authors state that further work is in progress. They do, however, indicate one caution that is becoming more and more evident as further *esr* work is done on selenium.[1, 33] Specifically, oxygen in liquid selenium may well act as an impurity, and tend to concentrate at the free radical endings of the chains. Thus the *esr* absorption spectrum due to the free radical chain ends might well be altered, and this could lead to an incorrect theoretical analysis of the spectrum. Future work involving *esr* absorption obviously will have to be undertaken only on the purest, most highly-outgassed selenium samples possible.

Viscosity

The viscosity of liquid selenium has been studied a number of times, but the papers that are really germane to the structure of liquid selenium are relatively few in number. Khalilov[100] reported the viscosity of molten selenium, in the temperature range 220–360°C. His experimental results compared favorably with the considerably earlier data of Dobrinski and Wesolowski.[42]

Khalilov, using his own data, and a theoretical technique developed by him previously,[101] calculated the average number of atoms comprising a polymeric unit in liquid selenium. His data for liquid selenium are reproduced as Table 3-4. These data give number-average chain lengths which seem

TABLE 3-4   AVERAGE CHAIN LENGTH IN
LIQUID SELENIUM AS A
FUNCTION OF TEMPERATURE[101]

Temperature °C	Average Chain Length in Liquid Selenium
220	7.5
260	6.1
300	5.3
330	4.7
360	4.4

remarkably small and do not appear at all consistent with other work in this field. It is conceivable that these numbers were meant by Khalilov to represent the average length of a straight segment in a randomly-kinked polymeric chain; however, Khalilov did not at all discuss the nature of his calculated values in his paper.

Shirai and co-workers[168] reported on a study of the viscosity and molecular weight of liquid selenium. The viscosities were measured, using 99.99% pure selenium, over the relatively narrow temperature interval 220–240°C. The experimenters also measured, over the same temperature range, the viscosity of liquid selenium doped with chlorine and their data indicated that the viscosity of the selenium was lowered as the concentration of dopant was increased. It will be recalled that Krebs and Morsch[112] found that iodine, as a dopant, had the same effect. Some typical data, from Shirai's paper, are reproduced in Table 3-5.

Shirai found that the data at 230°C could be fitted to a linear equation, and by extrapolation to a concentration of 0.0 wt% dopant, i.e., pure selenium he was able to quote a number-average chain length of 7.2×10^3 atoms in liquid selenium at 230°C. For the data quoted in Table 3-5, the number-average chain length was calculated from the concentration of chlorine present as dopant. It was assumed in the calculations that the chlorine acted solely as a chain terminator and further that the sample itself contained only chains. These two assumptions led to a simple mathematical expression:

$$\text{Chain length in sample} = \frac{2 \times \text{Atomic weight chlorine} \times (100 - X)}{\text{Atomic weight selenium} \times X}$$

where X is the concentration of chlorine in wt%. Thus, within the framework of these assumptions, the calculation of a chain length is a simple procedure.

TABLE 3-5 EFFECT OF CHLORINE AS DOPANT ON THE NUMBER-AVERAGE CHAIN LENGTH IN SELENIUM AT 230°C[168]

Concentration of Chlorine wt %	Calculated Number-Average Chain Length
0.0395	2275
0.0449	2001
0.0671	1339
0.0705	1274
0.0902	996
0.169	531
0.271	331
0.281	319

Harrison[82] studied the effect of pressure (up to 4 kbars) on the viscosity of liquid selenium. The experiments were made using a rolling-ball viscometer within a pressure bomb, and the equipment, by its very nature, required calibration to obtain the geometric constant of the inclined tube-rolling ball apparatus. To obtain this constant, experiments were made at each temperature using the earlier viscosity data on liquid selenium, published by Dobinski and Wesolowski[42].

Harrison was predominantly interested in observing, by changes in viscosity, any effect that pressure might have on the ring-chain equilibrium. By analogy with work by Feher and Hellwig[54] on sulfur, Harrison was able to anticipate that the difference in viscosity of liquid selenium, between 1 bar (atmospheric pressure) and 3850 bars, should represent an increase of some 6%.* Harrison's data, however, showed a viscosity increase of over 260% between atmospheric pressure and 3850 bars, and he concluded that "pressure appears to have little effect on the ring-chain equilibrium" in liquid selenium. However, he did not suggest a mechanism to which this large increase in viscosity, with pressure, could be attributed.

Harrison used the viscosity data determined at atmospheric pressure,[42] coupled with a theoretical treatment proposed by Gee,[64] to calculate what he referred to as "number-average chain lengths" in liquid selenium as a function of temperature. Actually, he calculated what should be referred to as the "number-average degree of polymerization," which is the chain length of the polymer, but expressed in units of Se_8.

The results of these calculations were presented graphically, but by interpolation, the data are listed in tabular form in Table 3-6.

Harrison also calculated number-average degrees of polymerization according to the ring-chain theory put forward by Eisenberg and Tobolsky[52].

TABLE 3-6   CALCULATED NUMBER-AVERAGE CHAIN LENGTHS
IN LIQUID SELENIUM[82]

Temperature °C	Number-Average Degree of Polymerization	Number-Average Chain Length
220	800	6.4×10^3
260	360	2.9×10^3
300	180	1.4×10^3
350	100	8.0×10^2
400	65	5.2×10^2
450	49	3.9×10^2

* This figure followed from a thermodynamic argument based upon the effect of pressure on the equilibrium constants of the ring-chain theory, as advanced by Eisenberg and Tobolsky.[52]

In the temperature range 220–260°C, excellent agreement was noted between the two separate methods of calculation, but above 300°C the values calculated by the two methods began to diverge, and Harrison suggested that this divergence might be due to the assumption, in Gee's theory, that "pure monomer selenium," a hypothetical liquid, had the same temperature dependence of viscosity as that found for pure monomeric sulfur.

Insofar as the structure of liquid selenium is concerned, Harrison's contribution was to publish a number-average degree of polymerization (and hence, by simple mathematics, a number-average chain length) for liquid selenium, as a function of temperature. These calculations were done using a theory advanced by Gee,[64] and viscosity data for liquid selenium, published by Dobinski and Wesolowski.[42]

The technique of measuring the viscosity of liquid selenium at various temperatures as a function of dopant concentration, assuming that the sole function of the dopant (at low concentration) is to act as a chain terminator, was further applied by Keezer and Bailey.[99] These workers chose thallium as the doping agent, because it has a low vapor pressure, it does not form highly volatile selenides, and the selenium-thallium phase diagram is known. The viscosities were measured in the range 500–600°K (227–327°C), with thallium concentrations varying from 20 to 10^4 ppm. However, the data obtained from samples of selenium containing more than 300 ppm thallium were not used to calculate chain lengths because the selenium-thallium phase diagram exhibits an immiscibility gap beyond about 300 ppm thallium.

By extrapolation of the isotherms of the variation of viscosity with thallium concentration to the viscosity of pure selenium, Keezer and Bailey suggested the relationship:

$$\eta = 1.57 \times 10^{-7} Z^{1.45}$$

where η was the viscosity of the liquid selenium, and Z was the number-average chain length. This expression assumed that the melt contained only chains, i.e., that all the rings assumed present in liquid selenium were opened under the influence of the thallium.

With this expression, and accepting the assumption inherent in its formulation, a number-average chain length of about 400,000 atoms can be estimated for liquid selenium at its freezing point. However, taking account of the percentage of rings believed to be present in liquid selenium (from Briegleb's work), Keezer and Bailey suggested a number-average chain length, in liquid selenium at its freezing point, of about 200,000 atoms.

Keezer and Bailey did not tabulate their experimentally obtained viscosity data for pure liquid selenium as a function of temperature. Consequently, presentation of their results of number-average chain length versus temperature in tabular form is not possible. This was an unfortunate omission on their part, because it is extremely difficult to extract sufficient significant information from a small graph. Although Keezer and Bailey compared their viscosity data for pure liquid selenium graphically with the viscosity data published by Shirai et al.,[168] they did not compare it with the more extensive data published by Dobinski and Wesolowski.[42]

In another paper, Hamada et al.[81] presented data on the viscosities of liquid selenium and partially chlorinated liquid selenium in the temperature range 235–305°C. As with their 1963 paper, they found that the viscosity of liquid selenium decreased as the concentration of dopant (chlorine) increased. In this paper, however, they observed that the decrease of viscosity with chlorine concentration was linear only above a specific concentration of dopant. At 235°C, this concentration was 0.0125 wt% chlorine. They suggested that this concentration corresponded to the situation where all the radicals at both ends of the long-chain molecules of selenium were completely terminated by chlorine atoms. As they state:

"The radicals vanish more and more with an increase in the chlorination until 0.0125%, and the average chain length of the chlorinated selenium molecule will not be changed."

Hamada and his co-workers did not use this figure of 0.0125% to calculate a number-average chain length, employing the simple mathematical expression developed previously.[168] Our calculation shows that this chlorine concentration of 0.0125% at 235°C, corresponds to a number-average chain length of 7.2×10^3 atoms. This is an interesting development since it agrees precisely with the calculated value of number-average chain length in liquid selenium at 230°C., as published by Shirai et al., in 1963. It is, of course, quite conceivable that the agreement is purely fortuitous, but nevertheless the concordance is a pleasant surprise.

Theoretical Analysis

In the last decade, several papers have been concerned with a theoretical analysis of amorphous selenium, either in the vitreous state or the liquid state. The initial paper concerned with selenium appears to have been that of Eisenberg and Tobolsky,[52] who applied to selenium a theory previously developed for sulfur.[51] In actual fact, the theory published by Eisenberg and Tobolsky had made use of an earlier theoretical analysis of sulfur by Gee,[64] and the theoretical treatment has been referred to, by Keezer and Bailey[99] as the Gee-Eisenberg-Tobolsky (GET) model.

The theoretical analysis is based upon a model which relates the number-average degree of polymerization in selenium to the enthalpies and entropies involved in the equilibrium reactions between Se_8 rings, Se_8 diradicals, and

$(Se_8)_n$ chains. To evaluate the number-average degree of polymerization Eisenberg and Tobolsky ultimately found it necessary to accept as valid the experimental data published by Briegleb.[24]   His data were interpreted on the basis that the amount of selenium going quickly into solution with CS_2 could be used as a measure of the amount of Se_8 rings present in the quenched samples. From this semi-quantitative analysis, the authors graphically presented values of the number-average degrees of polymerization for liquid selenium as a function of temperature.   These data are presented as Table 3-7.

TABLE 3-7   CALCULATED NUMBER-AVERAGE CHAIN LENGTHS IN LIQUID SELENIUM[52, 53]

Temperature °C	Number-Average Degree of Polymerization	Number-Average Chain Length
227	1000	8.0×10^3
270	450	3.6×10^3
364	100	8.0×10^2
375	90	7.2×10^2
500	20	1.6×10^2

In a later paper, Eisenberg[48] applied Gee's theory to Dobinski and Wesolowski's viscosity data for liquid selenium, and suggested that these number-average degrees of polymerization, as tabulated in Table 3-7, were too high, by a factor of 50%.   However, this suggestion has been discounted by Eisenberg[50] on the basis of recent work on the viscosity of liquid sulfur[49] where it was found that the theory of Gee, which did not take into account the possibility that long polymeric chains might be subject to entanglement, could yield erroneous results under conditions where such entanglement was possible.

As a consequence, the original published values of number-average chain length, enumerated in Table 3-7, are felt to be reasonably valid estimations.[50] Nevertheless, throughout their several publications, Eisenberg and Tobolsky, and Eisenberg, made it obvious that the theory they proposed is at best a semiquantitative analysis of the situation in liquid selenium, and that perhaps an order of magnitude, as regards the number-average degree of polymerization, is the best that can be achieved by this method of approach.   It is clear that further theoretical work on the structure of liquid selenium is necessary.

Summary

It is apparent that there are present in the literature numerous and diverse suggestions concerning the number-average chain length in liquid selenium.

Briegleb's work is generally considered to have yielded a measure of the amount of small rings present in liquid selenium, although there might be serious reservations about the actual data itself.   Therefore, most work is now directed towards the elucidation of chain lengths.   To attempt a reasonable summary of the present situation, Figure 3-2 has been prepared, and this figure appears to represent the extent to which the structure of liquid selenium has been elucidated up to the present time.   Notice that number-average chain lengths, and not number-average degrees of polymerization, are plotted in the figure.   Considering Figure 3-2, the disagreements present in the litera-

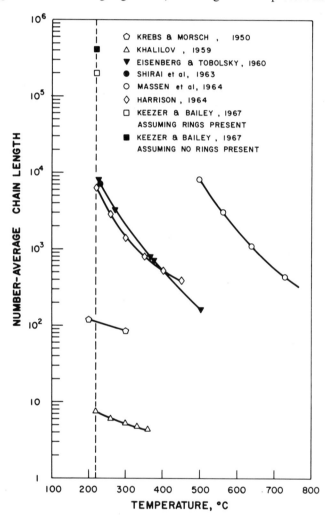

Figure 3-2 Number-average chain length in liquid selenium as a function of temperature: a comparison of data by various investigators.

ture concerning number-average chain lengths are at once obvious. Perhaps one of the more serious and pressing experimental problems is the considerable disagreement between chain lengths calculated on the basis of chlorine doping, and those calculated on the basis of thallium doping.

THE AMORPHOUS ALLOTROPES OF SOLID SELENIUM

All the noncrystalline forms of selenium are considered under the general classification of amorphous selenium. However, the literature is complex, because the term amorphous selenium can refer to red amorphous (produced in several distinct ways), black amorphous, and vitreous selenium. The structures of these separate allotropic forms have not been adequately described, and some degree of uncertainty exists in the literature. To add to the complexity of the situation, some authors refer to ground vitreous selenium as black amorphous selenium.

The red and black modifications of amorphous selenium appear to be directly related to one another, and consequently these two forms will be discussed together, while vitreous selenium will be treated separately. The section concludes with a discussion on the preparation and structure of thin amorphous vapor-deposited films.

Red Amorphous Selenium and Black Amorphous Selenium

Until recently, red amorphous selenium was produced primarily by chemical reduction from aqueous solution. Commonly, the procedure involves the reduction of an aqueous solution of selenious acid with hydrazine hydrate:

$$H_2SeO_3 + N_2H_4 \longrightarrow Se\downarrow + 3H_2O + N_2$$

sulphur dioxide:

$$H_2SeO_3 + 2SO_2 + H_2O \longrightarrow Se\downarrow + 2H_2SO_4$$

or other suitable reducing agent,[144] and the solid selenium settles out of solution as a deep-red flocculent precipitate. It has been reported[41] that under conditions where hydrazine hydrate was the reducing agent, the colloidal selenium particles were contaminated with an adsorbed layer of the reducing agent. Prolonged electrodialysis served to remove the adsorbed layer, but it will be realized that the adsorption of hydrazine hydrate by the selenium is only an example of a phenomenon exhibited by all freshly-precipitated substances which tend to be efficient adsorbers of impurities.

For the preparation of red amorphous selenium by chemical precipitation, the technique published by Gattow and Heinrich[63] can be recommended. Solid sublimed selenium dioxide is dissolved in aqueous hydrochloric acid, ensuring that only selenious acid is present in aqueous solution. The solution, with appropriate provision for cooling, is then placed in the dark, and with constant stirring, the acid is reduced with hydrazine hydrate. The reduction step is done very slowly to prevent the heat of the reaction from raising the temperature of the solution above 20°C.

The precipitate is digested over a period of 24 h, after which time the product is filtered, with care being taken to keep the powder cool, and washed free of chloride ions. At a temperature of less than 20°C, and in the dark, the powder is then dried over P_2O_5 in vacuum. The drying process is continued until a sample, heated up to 150°C, shows no loss in weight. The powdered red amorphous selenium is stored in the dark, to avoid any possibility of photolytic transformations.

Gatow and Heinrich gave no indication of the purity of the red amorphous selenium prepared in this manner, but the experimental conditions are such that no extraneous nonvolatile substances would have come into contact with the product, except for those present in the original aqueous solution of selenious acid.

As an alternative to the method just cited, the preparation of red amorphous selenium can be accomplished by the rapid condensation of selenium vapor.[154] In a recent paper,[69] selenium vapor was passed through a quartz labyrinth at 1000°C, then directly into the vapor of boiling nitrogen, where it was reported to condense into the red amorphous allotropic modification. To ensure that no high temperature oxidation would occur, the nitrogen was kept as free as possible of oxygen. As was pointed out by Gobrecht et al.,[69] nitrogen has never been observed to influence the physical properties of selenium.

Gobrecht et al. reported that the red amorphous selenium powder was prepared in this manner, in preference to a chemical preparation, to avoid any possibility of memory effects.[145] Furthermore, it was felt that in a chemical preparation, impurity atoms, such as oxygen, might be incorporated into the selenium molecules and would influence the physical properties of the pure substance. During the preparation of red amorphous selenium by a chemical reduction technique, the removal of any possible memory effects that the selenium dioxide might conceivably have retained from its manufacture should be accomplished, but the possibility of the adsorption of foreign ions into the chemically precipitated powder is a very valid point.

There is some evidence[9] to indicate that such absorption of contaminants, in trace amounts, can occur with relative ease. However, to our knowledge, the similarities and differences, if any, in the physical properties of red amorphous selenium, as prepared by chemical reduction or by condensation of high temperature vapor, have not as yet been determined. It should be remarked that differences in the physical properties of the separate red powders might well exist, since the modes of preparation are so markedly different.

Regardless of the method used to prepare red amorphous selenium powder, the importance of the effect of temperature upon the powder should be appreciated. There is much evidence[9, 61, 63, 69, 119] to suggest that red amorphous selenium powder is unstable at, or slightly above, room temperature. Hence, to ensure reproducibility of the product, the red powder should be prepared at as low a temperature as is feasible. If the preparation is by chemical reduction in aqueous medium, this would imply that the temperature of the solution should be maintained at 0°C, or if possible, even lower.[158] If the technique used is the condensation of high-temperature vapor, the criterion of a low temperature would be satisified by the boiling of nitrogen vapor. In either case, to ensure the prolonged stability of the red amorphous selenium, storage of the powder at 0°C, or lower and in the dark would not be remiss.

There are conflicting reports concerning the stability of red amorphous selenium powder upon storage. It has been stated[102] that the powder is stable for at least three to five years, if kept "cool and dry," while another report[80] has suggested that the powder slowly crystallizes on aging, even at 273°K. Recently it has been found[9] that red amorphous selenium powder, precipitated by chemical reduction, dried, and stored, at 0°C, has been stable over at least a two year period.

Several studies indicate that red amorphous selenium is truly amorphous.[61, 63, 179] Gattow and Heinrich investigated red amorphous selenium at temperatures as low as −160°C. and they found no evidence of crystallinity in x-ray diffraction photographs. An earlier work[158] suggested that some crystallinity was present in red amorphous selenium, on the basis of x-ray diffraction studies, but it is conceivable that this crystallinity was in fact induced by the long (70 to 80 h) exposure times employed.

There is however little evidence that might be used to suggest a definite structure for red amorphous selenium. Some structural parameters for the chemically precipitated modification have been published.[158] It was reported that the Se—Se bond distance in the powder was, on the average, 2.33 Å, and the coordination number of a typical selenium atom was about 2.4; also neither the reducing agent nor, within limits, the temperature of precipitation significantly affected these values.

It will be appreciated that a coordination number of 2 suggests that the selenium molecule has a chain-like structure. But as has been indicated previously, a radial distribution diagram cannot be used as a means of distinguishing between a ring structure and a chain structure.

On the basis of these structural data, and a series of assumed models, Richter et al.,[158] suggested that chemically precipitated red amorphous selenium was composed of long spiral chains.   It was also suggested that the arrangement of the spiral chains in the red amorphous selenium was such that the powder must have had a considerable degree of order.

Red amorphous selenium is readily soluble in carbon disulfide, and it was suggested some time ago[24, 25] that the allotrope might well be composed of ring molecules.   This suggestion has been accepted by several workers in the field.[61, 63, 69, 111]   Also, it has been suggested[63] that red amorphous selenium forms chains during chemical precipitation, and that during subsequent digestion many of these chains evolve into rings.   This suggestion implies that the chemically precipitated powder, after digestion, is a mixture of chains and (predominantly) rings.   Stabilization of the precipitated powder might well occur in this manner, but there are no experimental data available that can be used to confirm or refute this suggestion.   Again, there appears to be no evidence that suggests a definite structure for red amorphous selenium, but the powder is obviously a fruitful source of further work.

Occasionally, it is stated that red amorphous selenium is equivalent to ground vitreous selenium, and the red color is due only to the fact that the red amorphous selenium is in a finer state of subdivision.[38, 158]   There is however some evidence that may discredit this idea.   From heat of combustion data, it can be calculated[62] that, for the reactions:

$$\text{red amorphous selenium} \longrightarrow \text{trigonal selenium}$$

$$\text{and vitreous selenium} \longrightarrow \text{trigonal selenium}$$

the enthalpies of transition, at $298.2°K$, are -3.0 ± 0.4 kcal/mole, and -1.2 ± 0.4 kcal/mole, respectively.   These quoted values differ sufficiently to suggest that vitreous selenium and red amorphous selenium are not the same allotropic modification in differing states of aggregation.   There are also some data available from x-ray spectroscopy.[97]   From measurements on the K absorption band, it was found that vitreous selenium and red amorphous selenium differed by 10 eV in the appearance of the K absorption discontinuity.   This fact seemed to indicate that the two amorphous modifications were not equivalent in terms of electron distribution.

When red amorphous selenium is slowly heated, an endothermic heat effect occurs within the sample, and the powder turns black.[9, 61, 63, 69]   The resulting black powder is the selenium modification that is commonly referred to as black amorphous selenium, and both the color change and the endothermic heat effect are irreversible.[63]

Some data on this transition are available.   The onset of the transition was reported to occur at $37°C$,[61] and the heat of transition was suggested to be 0.1 ± 0.1 kcal/g-atom.[63]   More recent work, by an indirect means, has indicated a heat of transition of 0.09 kcal/g-atom.[62]   A transition temperature of 30°C has been reported.[69]   A recent study of the transition suggests a value of 0.107 ± 0.009 kcal/g-atom, with a transition temperature of $53.0 \pm 0.5°C$.[9]

Apart from the data quoted, little is known about black amorphous selenium.   The suggestion has been made[63] that black amorphous selenium not be considered as a separate allotropic modification, but rather it should be regarded as a "transition type" (Ubergangstyp) between red amorphous selenium and trigonal selenium.   However, the red amorphous to black amorphous transition appears to be well-defined, and characterized thermodynamically.   Under these circumstances, it seems logical at the present time to consider black amorphous selenium as a separate allotropic modification.

As a precaution, it should be clearly noted that pulverized vitreous selenium and black amorphous selenium are possibly two completely different and distinct entities.   The term black amorphous selenium should not be used interchangeably for these two different modifications, and a clear distinction must be made between black amorphous selenium and finely-ground vitreous selenium in the literature.

There is little evidence available concerning the structure of black amorphous selenium.   From x-ray diffraction experiments, it is known that black selenium powder is amorphous.[9,61,63]   Also it has been suggested[61,63] that the endothermic transition observed might be due to a cleavage of ring molecules; but this suggestion can only be tentative, and contingent upon an assumed structure for red amorphous selenium.   In actual fact, the structure of black amorphous selenium is not known, and the situation would appear to require considerable further work.

Vitreous Selenium

Vitreous or glassy selenium is prepared by the quenching of molten selenium, but the actual quench procedure employed depends, to a certain extent, upon the ultimate use that is to be made of the vitreous material.

If the investigation is such that the vitreous selenium can be used in the form of shot, then the liquid can be quenched rapidly by simply pouring the molten selenium into ice water, or liquid nitrogen.   The dynamic equilibrium present in the molten liquid can probably be maintained in the shot most efficiently by this quench procedure, and the sample should exhibit little or no tendency to initiate crystallization as it is cooled.

The experimental technique may however require the preparation of a disk, or cast slab, of vitreous selenium, and to prepare such a disk is very much a laboratory art.   If the liquid is cast into a cold mould that can serve as an efficient heat sink, the selenium will be quenched quickly, but the resultant thermal shock may well fracture the sample. This latter possibility will be minimized if the mould is allowed to cool slowly to room temperature, but such a procedure may possibly yield some unknown measure of crystallinity within the cast slab. It is also conceivable that the dynamic equilibrium existing in the liquid might be significantly altered during the cooling process. As a consequence, the preparation of a disk of vitreous selenium requires a certain degree of ingenuity.

In one apparently successful technique,[32, 73] the molten selenium was quenched rapidly, and the vitreous material was then pressed, under vacuum, into disks. Graham and Chang pressed the quenched material in a steel die at 135 atm and 56°C for 20 minutes, to produce disks 9.5 mm in diameter and 2–5 mm thick. Chang and Romo adopted a similar procedure, to obtain disks 2.5 cm in diameter and 0.25 cm thick. Graham and Chang reported that pressure variations between 15–135 atm resulted in insignificant changes in the physical and mechanical properties of the pellets, but it is conceivable that the technique used to form a disk might well induce the onset of crystallization within the sample. Kaplow et al.,[96] also prepared samples for x-ray diffraction studies by pressing. Subsequent to quenching, the vitreous selenium was ground to 325 mesh and pressed in a die, to form briquettes $1 \times \frac{1}{2} \times \frac{1}{8}$ in. thick. It was reported that there was no x-ray evidence of crystallinity in the resultant sample.

Regardless of the quench procedure employed, the temperature of the molten selenium prior to quenching should be reported, since there is evidence to suggest[32, 73] that some of the physical properties (density, elastic constants, etc.) of the quenched selenium are dependent upon the melt temperature. It should also be noted that the accepted practice at the present time is to outgas the liquid selenium before it is quenched to remove dissolved oxygen from the system.[184]

The classical method for the investigation of the nature of an amorphous material is by x-ray diffractometry, and the subsequent generation of a radial distribution function. The function obtained experimentally is then compared with a similar function derived on the basis of an appropriate structural model, and conclusions are drawn accordingly. As an amorphous material of prime commercial importance, vitreous selenium has been the subject of numerous such investigations.

In exploratory work it was reported that vitreous selenium is truly amorphous, both by x-ray diffraction[26, 87, 151] and by transmission or reflection of electrons experiments.[155]

Krebs and Schultze-Gebhardt[114] suggested on the basis of their results that vitreous selenium is composed solely of rings, of both high and low molecular weight. Furthermore, these researchers reported a coordination number of 2.1 for a typical selenium atom in the vitreous material. This value of 2.1

has been corroborated by a number of investigators[88,115,157] and a theoretical coordination number of 2.07 for vitreous selenium has been calculated.[91] Richter and Herre[157] reviewed a number of earlier papers, and elaborated upon the structure of vitreous selenium. Their work was done by the interpretation of radial distribution functions, and the vitreous selenium samples were prepared either by slow or rapid quenching of the molten selenium. Richter and Herre suggested that vitreous selenium was a mixture of Se_6 rings and long chains of selenium atoms. The suggestion was also made that the structure of the vitreous material was very dependent upon the thermal history of the amorphous sample.

The existence of Se_6 molecules in vitreous selenium has been criticized by Lucovsky,[120] who implied that the structural model used in the original analysis was questionable.

The last few years have seen the publication of some significant papers concerned with the preparation of radial distribution functions from diffraction data. Chang and Romo[32] did not suggest a possible structure for vitreous selenium. However, they did report that quenching from different melt temperatures (700°K and 775°K) produced vitreous selenium of varying intermolecular structures, and this was evidenced by small differences in the corresponding atomic radial distribution curves.

A neutron diffraction study, on a cast vitreous rod of selenium, has been reported.[88] A careful analysis of the data was made, and it was concluded that:

"amorphous selenium can be pictured as a usual liquid-like atomic assemblage with short-range order, but within which remain the chains of crystalline selenium, now in random orientation, but with sufficient remaining interchain regularity to account for a considerable part of the atomic distribution function."

Thus, Henninger and his associates have suggested that vitreous selenium is composed predominantly of long chains, in random orientations.

However, diametrically opposite conclusions were reached by Kaplow et al.,[96] who studied the structure of vitreous selenium by obtaining x-ray diffraction data, at 25°C, and −196°C, for both vitreous and trigonal selenium. The subsequent radial distribution function for vitreous selenium was matched with models which involved perturbations of the atom positions in the trigonal, α-monoclinic, and β-monoclinic crystalline forms. These perturbations were chosen by a Monte Carlo procedure which allowed only those perturbations which improved the fit to the experimental distribution function. Kaplow et al. found that relatively small static displacements, of the order of 0.20 Å, were sufficient to convert the monoclinic ring structures to the observed vitreous form. However, much larger perturbations, of the order of 0.7 Å, were required to convert the trigonal (chain) structure into a form which would give a suitable amorphous radial distribution function.

Taking into account only their diffraction data, Kaplow *et al.* could not make a choice between a structure of vitreous selenium consisting entirely of slightly perturbed rings, or one of considerably distorted chains. When their data were coupled with those of Lucovsky *et al.* (*vide infra*), Kaplow and his colleagues concluded that vitreous selenium was composed predominantly of slightly distorted Se_8 rings, although it was suggested that occasionally within the vitreous material a ring had opened to yield a deformed chain.

Some spectroscopic work also has been carried out for the purpose of elucidating the structure of vitreous selenium. Vasko[183] measured the infrared spectra of amorphous and liquid selenium and compared these spectra with the corresponding spectra of amorphous, rhombic, and liquid sulphur. The spectra of both selenium and sulfur in the liquid state, in the temperature range from 250° to 400°C, showed the same qualitative similarity as in the solid state. Furthermore, the spectrum of liquid selenium, in the same temperature range, was identical with the spectrum of amorphous selenium, with only a small shift of the bands towards longer wavelengths.

On the strength of these observations, Vasko suggested that the dominant bands of amorphous selenium in the middle infrared region are associated with the vibrations of Se_8 rings. But, apart from this conclusion, Vasko was unable to suggest a structure for the vitreous allotrope.

A further spectral study of the structure of vitreous selenium, prepared as a cast sample, has been reported.[120,122,123] From a comparison of the infrared spectra of trigonal and α-monoclinic selenium, and cast vitreous selenium, Lucovsky concluded that Se_8 molecules were present in vitreous selenium. Indeed, the entire absorption spectrum of vitreous selenium, with the exception of a band at 135 cm^{-1} and a weak shoulder at 230 cm^{-1}, could be attributed to vibrational absorption bands of the Se_8 molecular species. The two extra absorptions were suggested to be due to the presence of selenium chains in the vitreous material.

However, Lucovsky was unable to make a quantitative estimate of the relative populations of the ring and chain components in vitreous selenium. It was stated that a great deal of additional work would be necessary before any reliable assessment could be attempted. Lucovsky and his colleagues indicated that such an assessment might be possible from Raman spectroscopy.

Some exploratory work on the Raman spectrum of vitreous selenium has been reported,[137] and the results indicate the presence of both rings and chains in vitreous selenium.   Apart from this, however, definitive conclusions could not be drawn from the data.

There are some indications in the literature that small micelles of crystallinity may exist within the overall amorphous phase of vitreous selenium.[68,87,118]   Such regions of short-range order might be expected, if selenium is considered as an inorganic high polymer.[45,71,167]   But it is also possible that these crystallites are the result of heterogeneous nucleation of the vitreous sample.   This mode of crystallization can occur if small heterogeneous impurities are present in the selenium sample: during quenching, these impurity particles can act as nuclei, inducing a small amount of crystallinity within the vitreous material.

Thin Amorphous Films of Selenium

An extensive literature exists on the preparation and properties of thin amorphous films of selenium.   Commonly, thin films of selenium are prepared by deposition *in vacuo* onto a suitable substrate.   Numerous substrates have been used (e.g., Formvar, glass, tellurium, aluminium, and alkali halides), and usually the substrate is at room temperature, although for a specific purpose the substrate may be either heated or cooled.   Once the film is deposited onto the substrate, it may be used as such, or else the substrate may be stripped away to leave an unsupported selenium film.   A decision to strip the substrate from the film will usually be based upon the ultimate use of the film.†   Descriptions of the numerous techniques involved in preparing a thin film are available.[67,74,77,86,89,161,185]

Deposition of the selenium by a sublimation process is not the only means that can be used to prepare a thin film of amorphous selenium.   Kitaev and Fofanov[103] have prepared thin layers of either amorphous or crystalline selenium by chemical decomposition of a dilute Na_2SeSO_3 solution.   It was reported that above $30°C$, "grey crystalline" selenium was formed, while at $10-15°C$, amorphous selenium was obtained, although the exact allotropic modification of the selenium precipitated out is not specified.   Thin layers of amorphous selenium have also been prepared by electrodeposition onto a nickel cathode.[72]

Light transmitted by a thin layer of vacuum-deposited selenium appears deep red in color, although this characteristic is by no means indigenous only to evaporated selenium.   A thin layer of polycrystalline trigonal selenium is also red by transmitted light, as is a thin section of a single crystal of the trigonal allotrope, and a layer of vitreous selenium quenched from the melt and

† From the point of view of structure determinations, most films are examined by x-ray spectrophotometric techniques, or by optical or electron microscopy.

pressed between glass plates.   Nevertheless, as a consequence of the color of the vacuum deposited selenium, at one time it was assumed that deposition *in vacuo* produced red amorphous selenium.   However, the modern concept is that vacuum deposition produces vitreous selenium.

There appears to be no question that a thin layer of selenium, vacuum-deposited onto a substrate at either room temperature or below, is amorphous in nature.[4, 78, 107, 157]   The study by Koehler *et al.* reported that the electron diffraction pattern, taken at a low beam intensity, indicated the thin selenium film to be amorphous, while increasing the intensity of the beam caused crystal growth.   This type of behavior is quite common when thin films of selenium are being examined.   This is another indication that the technique used to study the amorphous allotropes of selenium might well be influencing the results obtained.

From radial distribution functions, the coordination number of a thin amorphous film of selenium, prepared by vapor deposition, has been estimated to be 2.1,[157] and this value has been supported by other workers.[6, 88]

Analysis of the structure of a thin amorphous film has tended to parallel concomitant work on cast vitreous selenium, and much of the present knowledge of thin selenium films has been derived from work on vitreous selenium, and vice versa.

Thus, Richter and Herre[157] suggested that a thin selenium film, laid down on a room temperature substrate, consisted of regularly oriented chains of selenium atoms and puckered Se_6 ring molecules.   It was implied that small crystallites of trigonal selenium were also present in the amorphous film, but it is conceivable that the x-ray technique may itself have induced crystallization of the sample during exposure.   For a film deposited *in vacuo* onto a substrate maintained at $-180°C$, it was suggested that the film predominantly consisted of a layered structure of Se_6 puckered ring molecules. Furthermore, Richter and Herre indicated that a thin film of amorphous selenium, prepared by sublimation onto a substrate at room temperature, is structurally equivalent, in terms of components, to a sample of vitreous selenium prepared by quenching from the melt.

The structure of a thin layer of amorphous selenium, vacuum-deposited onto a substrate of collodion, at room temperature, has been discussed by Andrievskii *et al.*[5]   As a consequence of electron diffraction studies, Andrievskii proposed that amorphous selenium, in a thin film, could exist in either of two separate amorphous forms.   It was stated that one amorphous modification existed at about 20°C, and was composed of selenium ring molecules. The second form existed at about 70°C, and was composed of selenium chain molecules.   Andrievskii reported some instability in the amorphous film, and it was suggested that the ring molecules in the thin film opened to yield the second amorphous modification, as the temperature of the film was raised to 70°C.   It is quite conceivable that the instability in the thin film (thickness 1,000 Å) was in fact the onset of crystallization.   Nevertheless, the conclusion that a thin film, at 20°C, is composed of selenium ring molecules, seems in retrospect, to be valid.

As indicated in the discussion of vitreous selenium, recent work on the structure of amorphous selenium has tended to be somewhat contradictory. Henninger et al.[88] studied thin films of amorphous selenium by x-ray diffraction; it will be recalled that studies on a cast rod of vitreous selenium by neutron diffraction were also reported by these investigators.   In accordance with their results for cast vitreous selenium, Henninger et al. concluded that a thin film of amorphous selenium consisted of chains of selenium atoms in random orientation.   It should be noted, however, that other workers suspect[96] that the thin amorphous selenium film used by Henninger et al. in the x-ray diffraction studies may have been partially crystallized.

Kaplow et al.[96] studied thin amorphous films of selenium in a manner exactly equivalent to their study of a briquette of ground vitreous selenium. They specifically state that data obtained from the briquette, and data obtained from a thin amorphous film of selenium laid down onto a room temperature substrate resulted in almost identical scattering curves, despite the two completely different modes of preparation.   As a consequence, Kaplow et al. obviously considered a thin amorphous film of selenium to be structurally equivalent to a "massive" sample of vitreous selenium.   Following the analysis outlined in the section on vitreous selenium, Kaplow et al., inferred that a thin amorphous selenium film is composed predominantly of Se_8 rings.

It will be appreciated that the conclusion arrived at by Kaplow et al. depends upon the work of Lucovsky et al.[120,122,123] who also studied both cast and evaporated samples of selenium.   Furthermore, it should be noted that Lucovsky[120] specifically stated that the samples yielded results essentially independent of the method of preparation.

Thus, structural analysis, to date, suggests that a thin film of amorphous selenium is composed predominantly of Se_8 rings.   If this analysis is correct, it implies either that Se_8 ring molecules are a favored species in the evaporation of selenium, or that considerable rearrangement of the molecular species occurs on the substrate, as the amorphous sample is deposited.

THE CRYSTALLINE ALLOTROPES OF SELENIUM

Elucidation of the structures of the three principal crystalline allotropes of selenium, namely α-monoclinic, β-monoclinic, and trigonal, selenium, is well documented.   This is in direct contrast to the amorphous solid allotropes of selenium, the structures of which are still in doubt.

α-Monoclinic and β-Monoclinic Selenium

The preparation of α-monoclinic and β-monoclinic selenium can be accomplished readily by evaporation of a saturated solution of selenium. Commonly, the solvent is carbon disulfide, and often vitreous selenium is used in the preparation of the saturated solution.

Muthmann[138] appears to have been the first investigator to report that two distinctly different crystalline modifications of selenium are formed during evaporation of a saturated solution of selenium in carbon disulfide. His observations were later confirmed by x-ray crystallographic analysis.[105] McCullough[131] reported that if a saturated solution of carbon disulfide is slowly evaporated, crystals of α-monoclinic selenium are precipitated out of solution, while if the saturated solution is rapidly evaporated, crystals of β-monoclinic selenium are formed.

As a precaution, the precipitated monoclinic selenium should be examined microscopically to ensure that only one allotrope has precipitated.

Several recent papers have discussed the growth of α-monoclinic and β-monoclinic selenium crystals from solution.[2, 92, 136]

Moody and Himes[136] examined the solubility of monoclinic selenium in several halogenated hydrocarbon solvent systems, as well as in carbon disulfide. It was reported that the most promising solvent appeared to be methylene iodide, and the experimenters suggested that some rewarding crystal-growth work might be accomplished by precipitation from this solvent, despite the fact that methylene iodide freezes at about 6°C and is unstable at elevated temperatures.

On the basis of their solubility data, Moody and Himes further suggested that the temperature coefficient of solubility of monoclinic selenium in carbon disulfide is such that monoclinic crystals of selenium might well be grown by a circulating or transport method. Some results using a circulating method have been reported.[2] A constantly circulating system was constructed, in which amorphous selenium was brought into solution at 30°C and subsequently precipitated out as monoclinic selenium, at 15°C. Because of the temperature gradient, the solution was constantly circulating by convection.

However, Abdullaev et al. gave no indication of the allotropic purity of the precipitated monoclinic selenium, and it would appear that both allotropic modifications were precipitated from solution. But the technique seems promising, and it is conceivable that, with a suitable change of conditions, only one allotropic modification of monoclinic selenium would precipitate from the cooling solution. A thorough study of the system might well be useful.

Iizima et al.[92] have grown monoclinic platelets by evaporating carbon disulfide solutions saturated with selenium on a glass substrate at room tem-

perature.   Subsequent microscopic examination showed that only crystals of
α-monoclinic selenium were precipitated; and it was reported that no crystals
of β-monoclinic selenium were ever found using this technique.   For α-
monoclinic selenium, the angle β is 117°6′ and for β-monoclinic selenium, the
angle β is 115°42′.   As Iizima et al. pointed out, a high-quality microscope,
equipped with a cross-hair ocular and a graduated rotating stage, should be
more than adequate as a means by which α- and β-monoclinic selenium may
be distinguished.   Iizima et al. further suggest that the two allotropes might
be more readily identified if the angles between the boundary lines on a
principal face rather than the angles between faces are measured.

The structure of α-monoclinic selenium has been well described.[27, 79, 105, 181]

The basic unit of α-monoclinic selenium is an Se_8 ring molecule puckered
in the shape of a crown; four such molecules form a unit cell.   In the latest
report,[181] the x-ray diffractometer data were collected from a many-faceted
polyhedron of α-monoclinic selenium, and subsequently refined to a reliability
of 9.4 % by least squares analysis using anisotropic temperature factors.

The results of Burbank[27] and Unger and Cherin[181] are compared in Table
3-8, and a diagram of the interatomic distances in α-monoclinic selenium, as
reported by Unger and Cherin, is shown here as Figure 3-3.

TABLE 3-8   STRUCTURAL PARAMETERS OF α-MONOCLINIC SELENIUM

	Space Group	a Å	b Å	c Å	B
Burbank (1951)	$p2_1/n$	9.05	9.07	11.61	90°46′
Unger and Cherin (1969)	$p2_1/n$	9.054	9.083	11.601	90.81°

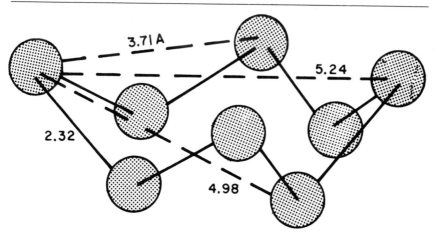

Figure 3-3   Interatomic distances in α-monoclinic selenium.[181]

The structure of β-monoclinic selenium also has been described.[28,105,124,181] The study by Burbank[28] suggested that the structure of β-monoclinic differed from α-monoclinic selenium in that the Se_8 ring molecule of β-monoclinic selenium was not completely closed. This suggestion, however, was later challenged by Marsh *et al.*[124] who reanalyzed Burbank's data, and suggested that β-monoclinic had a structure similar to that of α-monoclinic selenium, in that the β-allotrope was also composed of Se_8 ring molecules puckered into the shape of a crown. This reanalysis of Burbank's work by Marsh and his colleagues appears to be definitive, since their proposal for the structure of β-monoclinic selenium has been confirmed.[181] Starting with the assumption that β-monoclinic selenium consists of a closed Se_8 ring molecule, Unger and Cherin reported that their automatic diffractometer data could be refined to a reliability of 9%, using a least squares analysis with isotropic temperature factors. However, if the open ring structure suggested by Burbank was assumed, the least squares refinement would not converge, and the reliability remained above 30%. A diagram illustrating interatomic distances in β-monoclinic selenium is included here in Figure 3-4.

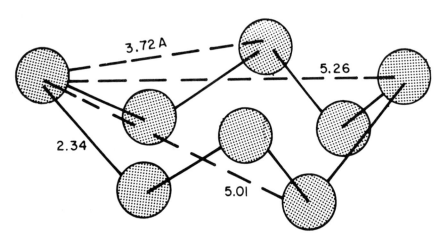

Figure 3-4 Interatomic distances in β-monoclinic selenium.[181]

In summary, the structures of α-monoclinic and β-monoclinic selenium are, in each case, that of an Se_8 closed ring molecule. The differences between the two allotropes are a consequence only of different stacking patterns within the respective unit cells in the crystalline phase. Unger and Cherin have detailed the two different stacking patterns, and their diagram is shown in Figure 3-5.

α-MONOCLINIC Se   β- MONOCLINIC Se

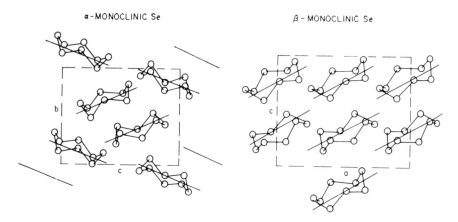

Figure 3-5   Stacking patterns of α- and β-monoclinic selenium.[181]

Trigonal Selenium

At all temperatures below the melting point, trigonal selenium appears to be the thermodynamically stable form of the element, and all other allotropic modifications revert monotropically to this form.

There are many ways of inducing crystallization to the trigonal phase. A popular method is simply to heat any other allotrope for several hours, at a temperature below the melting point of selenium. However, it is pointless to enumerate all the methods presently available, since a complete section of this chapter is concerned with the aspects of crystallization.

The structure of trigonal selenium has been investigated several times, and the allotrope is well characterized. Workers in the field have included Slattery,[170,171] Bradley,[18] Tanaka[179] and Straumanis[176]. The space group is $p3_121$ or $p3_221$[181] and, from x-ray powder diffraction data,[178] the lattice parameters are $a = 4.3662$ Å and $c = 4.9536$ Å. The trigonal lattice consists of three-fold spirals of selenium atoms running parallel to the c-axis direction. The atoms of each individual chain are held together by covalent bonding, and the chains are held together by van der Waal's forces.

A recent study[34,181] has reported some thermal and position parameters for trigonal selenium, determined by an x-ray diffraction technique on a single crystal of trigonal selenium. Two-dimensional $hk0$ data were collected using the stationary crystal–stationary counter technique. Least squares refinement began with Bradley's[18] parameters, and the data were refined to a reliability of 3.2%. A diagram of the unit cell of trigonal selenium, as outlined by Unger and Cherin, is reproduced in Figure 3-6.

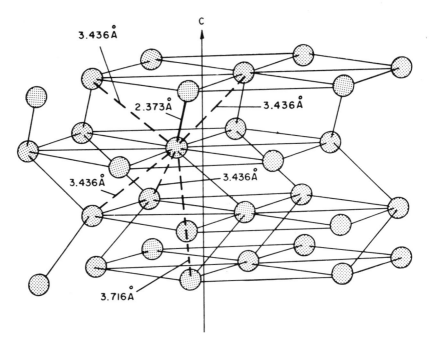

Figure 3-6   Structural parameters of trigonal selenium.[181]

Miscellaneous Crystalline Allotropes

There are several possible crystalline modifications described in the literature, but none have been confirmed by independent work.

Unger and Cherin claim, from Weissenberg measurements, to have found a crystal with a morphology very similar to β-monoclinic selenium, but yet not identical with a crystal of β-monoclinic selenium. This new form of crystal was subsequently called β'-monoclinic selenium. It was reported that the β'-monoclinic selenium crystal required a b^*-axis one-third the size of the b^* of β-monoclinic selenium in order to index all the reflections. The cell derived from the measurements was base-centered. However, since no symmetry was observed, it was concluded that the unit cell was triclinic, and reduced dimensions were published. Unger and Cherin concluded that this crystal represented a new form of selenium. But to the best of our knowledge, the existence of the new modification has not been confirmed.

It has been reported[4] that two hitherto unknown modifications, called α-cubic selenium and β-cubic selenium, may exist in thin selenium films. The modification called α-cubic selenium was described as having a simple cubic cell, with a lattice constant of $a = 2.970$ Å. The existence of the β-cubic modification was questioned[166] on the grounds that the copper grid used as the support in the objective stage of the electron microscope may have interracted with the selenium film to produce Cu_2Se. A further study,[5] done in the absence of the copper-net object holder, still gave evidence of the presence of the β-cubic modification. It was reported that the modification had a face-centered cubic unit cell, and the lattice constant was quoted as 6.04 ± 0.01 Å.

In contrast to these results, Griffiths and Sang[77] found no evidence of any modification of selenium, other than the known monoclinic or trigonal allotropes, when thin amorphous films of selenium were crystallized by heating or electron bombardment. Therefore, they stated that the reflections which led Andrievskii and Nabitovich[5] to conclude that β-cubic selenium was present in a thin film of amorphous selenium were spurious and probably due to double diffraction effects.

CRYSTALLIZATION OF SELENIUM

The crystallization behavior of amorphous selenium is of considerable importance in the electrical and photoelectrical applications of the element. Although the interest in monoclinic selenium has been increasing, the crystallization of selenium to the trigonal (hexagonal) form is the major consideration.

The morphology and growth of trigonal selenium crystals reflect the polymeric nature of selenium melts and bear a definite analogy to the behavior of organic polymers.[76]   In the crystallization process, the complex ring-chain equilibrium in vitreous amorphous selenium undergoes transformation to the orderly array of selenium chains in the displaced screw plane structure of trigonal selenium. The crystallization of these supercooled melts proceeds via the nucleation and growth of spherulites. These spherulites, the concentration of which depends upon a number of factors such as selenium purity, temperature, thermal pretreatment, and electromagnetic radiation, display in most instances a lamellar growth behavior. Thin platelike lamellae of almost constant width radiate from a common center. The concentric-ringed structure in Figure 3-7[76] has been shown to be the result of lamellae twisting cooperatively along the radial direction during growth, so that there is an in-phase progressive 180° rotation of the molecular chain axes along the radius.

In this section, the important factors affecting selenium crystallization are discussed together with a consideration of the kinetics of crystallization and the various methods which have been employed to grow single crystals of both monoclinic and trigonal selenium.

Figure 3-7   Fracture section of spherulite grown at 80°C in sample of vacuum distilled high purity selenium.[76]

Factors Affecting the Crystallization Behavior of Selenium

Thermal History. Since the crystallization of selenium proceeds by hetero-geneous nucleation and growth of nuclei, the density of nuclei is an important factor in determining the rate of crystallization. Janjua, Toguri, and Dutrizac[95] have shown that although the spherulitic growth rate is insensitive to thermal history, the nucleation rate is strongly dependent on the selenium holding temperature. By heating selenium to a sufficiently high temperature (around 500°C) prior to quenching and crystallization, a solubilization of nucleation centers can be effected and a more reproducible nucleation behavior realized for the particular selenium and substrate in question.

Even heating to a higher temperature does not necessarily destroy the thermal history of the selenium, since as shown by Piebst and Ammon,[145] the effect of thermal pretreatment was not destroyed even after the selenium was evaporated onto a substrate held at 100–145°C. These observations were confirmed by Eckart and Vogel[47] in differential thermal analysis studies of the crystallization behavior of ultrapure selenium distilled at different temperatures. Four different types of curves were obtained, depending on the pretreatment of the selenium and the distillation temperature. The crystallization point and melting point of the same initial selenium distilled at different temperatures shifted uniformly to lower temperatures with in-creasing distillation temperature.

The failure of many investigators to specify the thermal pretreatment of their samples as well as the purity of the selenium renders a comparison of results from different selenium crystallization studies very difficult, and may account in part for the seemingly contradictory data which are often reported.

Temperature. The crystallization behavior of selenium is strongly tempera-ture dependent. Even at ordinary temperatures, vitreous amorphous seleni-um undergoes a slow transformation to the trigonal form which is the most stable modification. Yamamori[187] noted that vitreous and red amorphous selenium tended to crystallize imperfectly at room temperature after 2–3 weeks. The crystallization at this temperature was mainly due to the trans-formation to trigonal selenium and not to the monoclinic form. X-ray diffraction studies gave a barely perceptible indication of monoclinic selenium. These observations have been corroborated by other investigators.[36] In the case of selenium samples heated nearer the melting point, Yamamori noted the coexistence of monoclinic and trigonal selenium.

The rate of formation of nuclei is practically zero in the temperature range 18–22°C[139] and is maximum around 125°C in the case of high purity selenium. For selenium doped with 1000 ppm of chlorine, the maximum nucleation rate was observed at a temperature of 105°C.[56] The optimum temperature for crystal growth is close to the melting point, with an optimum temperature of 210°C being noted by Mekhtieva *et al.*[132]   The effect of temperature on the rate of crystallization, as noted from x-ray diffraction studies, is indicated by the fact that 5 h were required for maximum crystallization at 70°C and 2–2.5 minutes at 217°C.

The structure of selenium films prepared by vacuum evaporation is strongly dependent on the substrate temperature.   For example, Griffiths and Sang[77] in selenium deposition studes on (0001) tellurium in the temperature range 40–140°C observed the epitaxial growth of selenium at 60–135°C with excellent single crystal growth occurring between 100–135°C.   The structure of the selenium film at 40°C appeared to be completely amorphous.   The temperature dependence of the nucleation and growth processes is considered further under the kinetics of selenium crystallization.

Impurities.   There are a number of elements which have a marked effect on the crystallization of selenium, but there is much uncertainty in the literature as to the precise manner in which they influence the crystallization behavior. The impurities of principal concern include oxygen, sulfur, tellurium, arsenic, thallium, and the halogens.   These impurities may be present or are added intentionally to enhance certain of the physical properties of selenium for applications in rectifiers, photocells, and electrophotography.   Since even trace impurities can serve as nucleation sites, it is important in crystallization studies that the purity of the selenium be indicated.   The effects of impurities, apart from providing nucleation sites, are generally considered to be: (1) promoting the cleavage of selenium rings, (2) terminating the free selenium chain ends, and (3) forming cross-links between selenium chains.

Since oxygen is an impurity in selenium which is not commonly determined, it is of interest to note that *esr* studies at x-band frequencies on amorphous and trigonal selenium between 300–77°K have been carried out to determine the nature of the selenium chain ends and the possibility of oxygen serving as a chain terminator.[33,162]   Although there exists the possibility of interaction between neighboring chains, the presence of a sharp line with $g = 2.00$ $\pm$ 0.0006 was interpreted as chain termination by an oxygen impurity.[33]

The addition of arsenic to selenium up to ~100 ppm has the very marked effect of retarding the rate of crystallization of selenium.[94]   Griffiths and Fitton[76] have pointed out that in the crystallization of selenium, it is the polymer chains and not the rings which play the primary role, because it is the free chain ends which attach to the lamella of a growing spherulite. Therefore, it is expected that the arsenic atoms, acting trivalently would tie up the free selenium chain ends and cross-link with other selenium chains, thereby reducing the crystallization rate.   This reduction in rate would increase until all the chain ends are saturated, a situation which appears to occur at the 100 ppm arsenic level. This corresponds to approximately 10^4 selenium atoms per arsenic atom. Taking account of the fact that each selenium chain has two free ends to be saturated, and that arsenic is trivalent it can be calculated that there are 7×10^3 atoms per chain for selenium annealed at 300°C and quenched to room temperature. This value agrees within the limits of experimental error with the values of 3×10^3 by Eisenberg and Tobolsky[52] and 7×10^4 by Keezer and Bailey,[99] obtained by means of viscosity measurements.

In contrast to arsenic, the effect of chlorine, bromine, and iodine additions on the crystallization behavior of selenium is quite complex. Bromine, unlike chlorine and iodine, produces an immediate, yet modest, increase in the crystallization rate, reaching a maximum around 150 ppm of bromine.[94] The effect of a bromine addition to selenium is apparently minimal, since Mamedov, Nurieva, and Gadkieva[127] observed no crystallization kinetics changes upon addition of small amounts of bromine, with a minimum activation energy of crystallization occurring at a bromine content of from 25 to 50 ppm. The activation energy increased with increasing bromine content but did not reach the value for pure selenium.

According to Fitton and Janjua,[56] chlorine doping of selenium yields a nucleation rate curve having a maximum at 100 ppm and a minimum at 300 ppm of chlorine. These changes correspond to the termination of the existing selenium chain ends and the onset of a decrease in polymerization.

Small amounts of iodine (up to 55 ppm) have marked inhibitory effect on the crystallization of selenium, whereas at the 85 ppm level, the crystallization rate approximates that for pure selenium.[94] According to Sokolov[172] and Aliyev and Akhundova,[3] the iodine atoms penetrate the selenium lattice with the formation of intercrystalline layers. Such layers can act as effective barriers to the transport of selenium to the growth front and retard the crystallization of selenium. At higher iodine concentrations, a recombination of iodine atoms into molecules can take place with the resultant loss of considerable quantities of iodine from the selenium, even at room temperature.[169]

Substrate. In the selenium crystallization process on certain substrates the spherulites develop a quite different morphology from that shown in Figure 3-7. Although the spherulites developed in contact with silica have a normal concentric ring structure, those in contact with Pyrex or normal soda glass have been found to grow at rates at least a factor of 10 faster than in the bulk. Their morphology is shown in Figure 3-8.[76] Similar structures have been found in thin films of selenium. The diffraction pattern for such spherulites was of a composite nature, made up of three orientations of a single primary pattern, each corresponding to an orientation of the hexagonal $(11\bar{2}2)$ plane.

In studies on thin (approximately 1000 Å) films of selenium on carbon and

20μ

Figure 3-8 Type II dendritic form of selenium spherulite grown in contact with soda glass at 180°C.[76]

glass (presumably soda glass), Boltov and Belan[17] observed the formation of sheaf-like crystals on the carbon substrate. The axis of the (0001) zone of these crystals was normal or parallel to the crystal plane. Spherical crystals with a highly pronounced rosette or angular structure were also observed. The sheaf-like spherulites have also been observed by other investigators in bulk selenium samples in the temperature range 100–200°C as well as in thin films.[76]

According to Boltov and Belan, the morphology of the selenium crystals on glass substrates depends strongly on the surface condition of the glass. Fourie and Van der Walt,[58] in crystallization studies on selenium layers of freshly cleaved mica surfaces, found only spherulites, whereas on mica surface exposed to moisture and on glass and rock salt faces, rosette-shaped crystals were formed. Formation of the latter on mica is considered to be due to the presence of alkali ions.

In the crystallization of selenium layers 8 to 10μ in thickness on aluminum, glass, mica, and sodium chloride substrates, Nabitovich et al.[139] claimed that the selenium in contact with the substrate was trigonal in form and was covered with selenium which was initially amorphous. Selenium crystallization was considered to start from the substrate and to depend little on the nature of the substrate itself.

The foregoing conclusion by Nabitovich et al. does not agree with the extensive work of Griffiths,[74, 76, 77, 86] in which the substrate was found to have a profound effect on crystal growth kinetics and morphology. In studies on the epitaxial growth of trigonal selenium on numerous single crystal faces, Griffiths[74, 86] observed epitaxial growth on (100) NaI, (100) KCl, and (110) KBr. However, selenium deposited on the (100) cleavage face of MgO at 90°C resulted in a film having a structure which was almost completely amorphous. This observation, which contrasts markedly with the structure of films and other substrates, indicates that MgO exercises a strong inhibitory effect on selenium crystallization. The deposition of selenium onto the (10$\bar{1}$0) and (0001) faces of tellurium resulted in single crystal films of selenium when the substrate was in the temperature range 100 to 135°C.[77]

Light, Electron, and Nuclear Irradiation. Numerous studies have shown that light, electron, and nuclear irradiation have a definite effect on the crystallization behavior of selenium. The action of light is believed to result in the creation of electron-hole pairs in vitreous selenium, with the growth rate of selenium crystallites controlled by the movement of holes to the crystal boundary, the crystallites acting as a sink for the holes.[43] According to Paribok-Aleksandrovich,[143] under the action of a light flux L at 0–6000 Lux and a temperature of 80–120°C, the rate of crystallization of selenium follows the approximate relationship $V = V_0 + KL$, where $K = $ a temperature dependent coefficient, and V_0 = the rate of crystallization in darkness at a given temperature.   Apparently only light with an energy in excess of 2.2 eV (the dissociation energy of the Se—Se bond), is effective in bringing about the crystallization of selenium.

Fitton and Griffiths[55] noted that the crystallization of amorphous selenium films can be induced by electron beam irradiation.   Trigonal selenium crystals were obtained which had a structure comprising radially growing segments with (10$\bar{1}$0), (11$\bar{2}$0), (11$\bar{2}$2), or (0003) planes parallel to the film and the c-axis normal to the growth direction.   It was possible to induce preferential growth in any direction simply by moving the electron beam in the desired direction.   The average width of lamellae so produced was approximately 2.5μ.

In work similar to that of Fitton and Griffiths, Belan and Bolotov[8] induced crystallization by the electron beam irradiation of selenium films deposited on freshly cleaved mica.   At low electron beam intensity (beam current: 10 microamps), a growth rate of 1–3μ per minute resulted in single crystals having a size as large as 10μ.   At higher intensities, a splitting of crystals occurred during the growth process.

In studies on the effect of ^{60}Co gamma rays (500–650 r/sec) on the general tion and growth of nuclei in the crystallization of amorphous selenium, Starodubtsev[173, 174] found an empirical relationship to apply: $N = aD^{4.5\pm0.5}$, where N = number of nuclei formed in vitreous selenium during dosage of high level radiation D.   Nucleation occurs by the union of mobile segments of selenium which are formed following the cleavage of selenium rings by radiation, with the rate of segment production dependent on radiation intensity.   Starodubtsev and Mikhaelyan[173] noted that the rate of growth of nuclei was constant during crystallization and did not change significantly after radiation.   This observation is contrary to that of Mikhaelyan[134] who observed that the rate of growth was influenced by the dosage of ^{60}Co gamma radiation.   This effect was accounted for on the basis of the activation of diffusion processes in selenium subjected to gamma radiation.

Solvents and Chemical Vapors.   The effect of solvents and chemical vapors upon the crystallization of selenium either to the monoclinic or the trigonal form has been known for some time.   As early as 1900, Saunders[164] compiled a list of reagents which promoted the crystallization of selenium.‡   Actually,

‡ Solvents which produce red monoclinic crystallization on heating with selenium: ethyl alcohol, benzene, thiophene, toluene, benzonitrile, propyl aldehyde, amyl nitrite, ethyl acetate, isobutyric acid, acetophenone, nitrotoluene, bromonitrobenzene, m-dinitrobenzene, dimethylaniline, acetone, propylene bromide, ethyl iodide, monoethyl aniline, chloroform, phenyl hydrazine, benzylamine, diphenyl methane.   Solvents which produce trigonal crystallization on heating with selenium: benzyl cyanide, benzimido butyl ester, quinoline, aniline, pyridine, piperidine, triethylamine, hexamethylene amine, p-β-anisaldoxime.

the best means of producing monoclinic selenium is by crystallization from a solution of amorphous selenium in carbon disulfide. In certain applications of selenium, chemicals have been used to activate the selenium. In other applications where it is of importance to maintain selenium in the amorphous modification, the action of chemicals upon selenium is an important factor in determining the environment in which the selenium can be used.

The crystallization of thin films of selenium has been studied in the presence of water vapor and organic solvent vapors such as carbon disulfide, isopropyl acetate, and trichlorethylene at room temperature.[35]   In the case of carbon disulfide, in which solvent selenium is soluble, trigonal and monoclinic selenium were present in a 1:1 ratio. The crystallization was in the form of three-dimensional crystallites with no preferred orientation. For isopropyl acetate in which selenium is slightly soluble, the crystallization proceeded to the trigonal form with the formation of a small amount of monoclinic selenium. The crystal morphology included three-dimensional crystallites, filamentary particles, lamellar platelets, and spherulites. There was no preferred orientation. In the case of trichlorethylene, in which selenium is insoluble, trigonal selenium only was present in the crystallization product with a preferred orientation in the direction (100). Three-dimensional, prism-shaped crystallites were predominant. In the case of water vapor, in which selenium is insoluble, the trigonal modification only was present with no preferred orientation. The crystal morphology was characterized by lamellar platelets and filamentary particles.

The morphology and growth characteristics of selenium crystallites induced by chemical vapors exhibit a definite similarity to those evidenced in selenium grown from the vapor. It appears that selenium molecules exhibit a high surface mobility as regards the formation of different types of crystallites in the presence of chemical vapors.

According to Modrany,[135] crystalline selenium in the trigonal form was obtained when a layer of amorphous selenium was heated at 110–140°C for 1–4 hours and then immersed for 15–24 h in pyridine, hydroquinone solution, or aniline. In this study the effect of the organic compound in orienting the selenium is not clear. Activation of selenium by contact with organic compounds is a technique which has been recognized for some time.

The crystallization of red amorphous selenium to the trigonal form can be effected readily at room temperature by contacting the selenium with a solution containing selenium ions in formal −2 and +4 oxidation states.[10]

THE KINETICS OF SELENIUM CRYSTALLIZATION

As has already been indicated, numerous factors affect the crystallization behavior of selenium to the trigonal form, not the least of which are temperature and the impurities or dopants.   Various approaches to the study of the kinetics of selenium crystallization will now be considered with particular reference to these important parameters.

The volume change attendant upon phase transformation has been applied to rate studies of both organic and inorganic polymer crystallization using Avrami's equation[7] in which:

$$(1 - \alpha) = \exp(-Kt^n)$$

where α = the fraction of the crystalline phase, t = time, K = the temperature dependent rate constant, and n = the crystal geometry parameter.   In a dilatometric study of the crystallization of pure amorphous selenium, n varied between 3.7 and 4.2 irrespective of temperature, indicating that hetero-geneous nucleation and three-dimensional growth of spherulites is the mode of crystallization.[94]   As the crystallization proceeded, n decreased towards a value of 3.   The departure from linearity became more pronounced when the crystallization temperature was decreased from 115 to 85°C.   This behavior. which appears at large degrees of crystallinity, has been noted in a number of organic polymers and is considered to represent typical homopolymer crysta-llization.[85]

In a similar study, Dzhalilov[44] reported neither the purity of the selenium nor the amount of supercooling prior to annealing.   Dzhalilov's results led to a lower crystallization rate; for example, at 100°C approximately 47 minutes was required to recrystallize 50% of the selenium, while in the present study, 27 minutes was sufficient to obtain 50% crystallization.

The activation energy calculated from the dilatometric data gave a value of 25 kcal/mole.   These results compare favorably with the value of 26 kcal/mole obtained by Mamedov and Nurieva[126] from x-ray diffraction studies on the crystallization process.

Normally, selenium crystallization proceeds via the formation of spherulites having a lamellar structure with the width of the lamellae much less than the length of the extended selenium chain.   This accommodation of selenium chains is effected by the repeated folding of the chains, giving rise to spheru-lites having a series of concentric rings.

Since the kinetics of selenium crystallization are related to the rate of forma-tion of nuclei on folded selenium chains and the rate of transport of selenium atoms to the growth face, consideration must be given to these two processes to obtain a better understanding of the crystallization behavior and kinetics. Since impurities or dopants in the selenium can saturate the free selenium chain ends and form cross-links with neighboring chains, it is evident that their presence can have a pronounced effect on the rate of crystallization. It should be noted that the chain ends are of importance in determining the crystal morphology, and the critical nucleus surface energies as well as the adherence of the molecule forming the nucleus to the existing molecular surface.

As has been indicated, the thermal history of the sample and temperature are important factors in determining the nucleation behavior of selenium. Thus, the spherulite nucleation rate decreases markedly with the length of time the selenium is held in the molten state at a definite temperature prior to crystallization, indicating a progressive decrease with time in the number of available nucleation sites.   At the crystallization temperature, the number of nuclei shows a time dependency, with the maximum number of nuclei appearing after about 20 minutes in the case of selenium containing 100 ppm of chlorine crystallized at 100°C.[56]   The same type of behavior has been observed for pure selenium.

As pointed out by Fitton and Janjua,[56] the crystal growth of selenium is controlled by the nucleation process which gives rise to a chain-folded crystal. The significance of the nucleation rate, as opposed to the spherulite growth rate, is clearly demonstrated upon doping selenium with impurities such as arsenic and chlorine (Figures 3-9 and 3-10).[56]   In the case of chlorine-doped selenium, the large changes in the nucleation rate in the region of 100–300

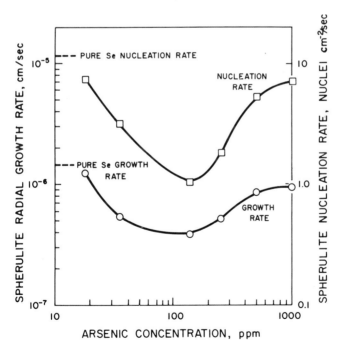

Figure 3-9   Effect of arsenic on spherulite nucleation rate and growth rate in selenium at 100°C.[56]

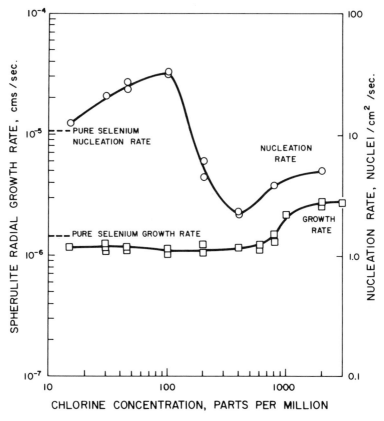

Figure 3-10   Effect of chlorine on spherulite nucleation rate and growth rate in selenium at 100°C.[56]

ppm of chlorine are associated the presence of chlorine on the ends of the selenium chains and a decrease in the extent of selenium polymerization.

Doping the selenium with chlorine produces a decrease in the activation energy for molecular transport across the phase boundary in selenium as determined from growth rate curves, *viz.*, 17.6 kcal/mole for pure selenium and 11.1 kcal/mole for selenium doped with 1000 ppm of chlorine.[56]   This decrease in the activation energy of selenium transport leads to an increase in the nucleation rate and the growth rate as shown in Figure 3-10.[56]

Arsenic, which behaves as a trivalent impurity in selenium, reacts not only with the free chain ends but also forms cross-links between selenium chains. This behavior accounts for the marked increase in the viscosity and the glass transition temperature of selenium.   As indicated in Figure 3-9,[56] both the nucleation and the growth rates decrease with increasing arsenic content up to around 100 ppm, at which concentration it is presumed that the chains are saturated. According to Fitton and Janjua, the increase in the nucleation and growth rates for selenium containing arsenic above 100 ppm is related to changes in the energies required for the formation of critically sized nuclei. This follows from Hoffman and Lauritzen's equation for the growth rate of polymers.[90]

Thallium doping leads to an increased growth rate as well as an increased activation energy (21 kcal/mole for selenium containing 1000 ppm Tl). This behavior of thallium is considered to be due to the formation of Tl_2Se_3, following the saturation of the selenium free chain ends which occurs around 300 ppm for monovalent dopants. The presence of Tl_2Se_3, which cannot be accommodated in the selenium lattice, acts as a barrier to the transport of selenium molecules to the growing crystal face and so increases the energy requirements for this process.[56]

Since the spherulites of trigonal selenium are characterized by a series of lamellae which undergo a repetitive 180° twisting to accommodate the selenium chains, it is interesting to examine the effect of impurities on the lamellar growth process. The temperature dependence of the mean lamella widths for pure selenium and selenium doped with 300 ppm of thallium and 300 ppm of arsenic gave activation energies of 12, 28, and 24 kcal/mole, respectively.[56] In the case of thallium-doped selenium, as the thallium concentration increases, an abrupt change in the mean lamella width occurs at around 300 ppm of thallium in marked contrast to arsenic-doped selenium. The widths of the selenium lamellae appear to be determined by a selenium chain diffusion process, with the energy required corresponding to that for the self-diffusion of trigonal selenium. The validity of this mechanism is underlined by the close agreement between the activation energy of 12 kcal/mole for selenium and the selenium self-diffusion activation energy of 11.8 kcal/mole as determined by Boltacks and Plachenov[16].

THE GROWTH OF SELENIUM SINGLE CRYSTALS

Trigonal Selenium

Many of the properties of a semiconductor such as selenium can be studied properly only when the material is in single crystal form. However, it is only recently that investigators have succeeded in producing single crystals of trigonal selenium of suitable size and form for meaningful studies. Unfortunately, the bulk single crystals prepared by heavily doping the selenium with elements such as thallium or chlorine contain significant amounts of the dopant. In addition, the bulk single crystals produced by crystallization in a controlled temperature gradient under pressures of 5–6 kbars leave something to be desired in terms of their perfection and purity. However, the single crystal films of selenium prepared by epitaxial growth on tellurium afford considerable promise for the study of a number of the fundamental properties of selenium. Such films have already been applied in the study of the asymmetric conduction of selenium and in the preparation of a single crystal selenium rectifier.[31, 75] As pointed out by Kolb,[108] single crystals of trigonal selenium are of interest for use as optical modulators, and parametric oscillators and amplifiers in the 10.6μ region.

Small single crystals of trigonal selenium have been prepared by both vaporization and solution methods. By evaporating selenium under a vacuum of 10^{-6} torr at a controlled temperature between 220–260°C and condensing the selenium vapor on a cold finger, Eckart[46] was able to obtain various crystal shapes including needles and platelets, depending upon the experimental conditions. Relatively large prismatic crystals having dimensions of 3 × 5 × 15 mm were obtained at an evaporation temperature of 225–240°C and a condensation temperature of 170–190°C within a period of 7 to 10 days. The ($10\bar{1}0$) planes were the boundaries of the crystals with the basal (0001) plane being rounded.

Well-formed single crystals of trigonal selenium 3–4 mm long × 1 mm thick can be formed by the slow cooling (1°C per day) from 50 to 25°C of a selenium-saturated 1.5N Na₂S solution.[108] The crystals which formed on a Teflon support rod attached to the solution stirring shaft exhibited major growth along the c-axis. Crystals perfectly hexagonal in cross section had (100) prism faces and rhombohedral faces of the (313) form. The crystals were found to contain < 50 ppm silicon resulting from an attack of the Pyrex crystallizing vessel by the Na₂S. A preliminary assay showed a sulfur content of 170 ppm.

By means of a melt temperature differential technique based on the use of two materials such as naphthalene and methyl salicylate which have boiling points slightly lower and slightly higher, respectively, than the freezing point of selenium, Stubb[177] reported the growth of single crystals of trigonal selenium.

The crystallization of selenium from the melt under ordinary conditions is a slow and difficult process. The breakdown of the complex ring-chain equilibrium which characterizes the selenium melt so as to realize ring cleavage and the rearrangement of selenium atoms into the orderly array of the trigonal lattice can be assisted greatly by drastically altering the viscosity of the melt. This is done by the use of high pressures or selective doping of the selenium. The first major advance in the growth of sizable single crystals of trigonal selenium was made by Harrison and Tiller,[83] who crystallized selenium by controlled solidification under a pressure of 5 kbars in an especially designed bomb. At this pressure, single crystals of up to 1 cm in diameter by 10 cm long were grown at 10^{-5} cm/sec by spontaneous nucleation when the tip of the vial containing the selenium formed a cone of angle of 10° or less. Unfortunately, with this technique, the position of the c-axis of the crystal relative to the long axis of the sample varied randomly from one crystal to another.

By using a seed crystal of trigonal selenium having a definite orientation, Harrison[84] was able to overcome the problem of random orientation and to increase the yield of acceptable crystals. The crystal quality was improved further by annealing the specimens during gradual depressurization to reduce residual stresses. Defects in the crystal were found to be associated primarily with these stresses or with gas bubbles resulting from the dissolution of argon in liquid selenium. Crystals were seed grown in both the c-axis (0001) and the a-axis (11$\bar{2}$0) directions but not in the (10$\bar{1}$0) direction.

According to Harrison and Tiller,[83] the nucleation and growth of trigonal selenium from the melt at high pressures involves not only the effect of higher freezing temperatures on the liquid-solid interface kinetics, but also the effect of higher temperatures on the association of selenium atoms in the liquid.

Keezer and Bailey[99] have shown that at the melting point of selenium about 50% of the atoms are present in a ring structure with the balance present in the form of chains, the average length of which is between 10^4 and 10^6 atoms per chain. By introducing impurities such as chlorine and thallium into the selenium, the viscosity of liquid selenium is greatly reduced indicating a reduction in the average chain length. The monovalent impurity atoms bond to the free chain ends. With complete occupation of the chain ends by such atoms, the viscosity should decrease markedly. Also, this process can be expected to shift the ring-chain equilibrium in favor of selenium chains. This simplification of the structure of liquid selenium greatly facilitates the crystallization of selenium from the melt. In such impurity-doped melts, the growth of single crystals may be governed by the diffusion of the dopant from the solid-liquid interface.[98]

Using melts of thallium-doped selenium, Keezer[98] was able to grow single crystals of trigonal selenium (albeit strongly contaminated with dopant) by vapor-liquid-solid (VLS), travelling solvent (TS), Czochralski, and gradient freeze techniques. As shown in Figure 3-11,[98] by careful utilization of the selenium–thallium phase diagram, it is possible to achieve a crystallization of selenium from a supersaturated selenium–thallium solution in the temperature range 172–218°C. In this procedure, known as the travelling solvent technique, the selenium crystal grows in contact with a thallium-rich selenium layer, the density of which is 5.8 g/cm^3, compared with 4 g/cm^3 for the selenium-rich liquid. Heavily etched quartz tubes, up to 25 mm in diameter and 25 cm long (with the initial 8 cm tapered to about 1 mm) were used for crystal

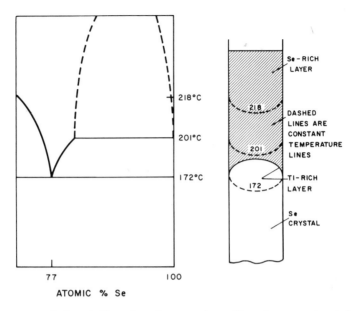

Figure 3-11   Selenium–thallium phase diagram and travelling solvent growth technique.[98]

growth.   Growth rates as high as 2.5 mm/h were realized with growth always in the $(11\bar{2}0)$ direction.

Single crystals of trigonal selenium with a $(11\bar{2}0)$ growth direction were obtained by the Czochralski technique from melts of thallium-(1–5%), chlorine-, and bromine-doped selenium.[98]   Crystals of centimeter size were obtained at growth rates as high as 2.5 mm/h for the thallium-doped melts and 0.25 mm/h for the halogen-doped melts.   Keezer's crystals, grown by various techniques utilizing thallium and chlorine doping, were found to contain from 200 to 400 ppm of residual impurities.

In consideration of the fact that selenium is used principally in thin film form in its electronic applications and in view of the imperfections and impurities associated with crystals which are grown under high pressure or from heavily-doped melts or from solution, it is significant that single crystal films of trigonal selenium have been grown epitaxially.

Griffiths and Sang[77] obtained good single crystals by depositing selenium onto cleaved $(10\bar{1}0)$ tellurium between 95–135°C.   In addition, using as a substrate the polished and annealed surface of (0001) tellurium, they were able to realize excellent single crystal growth between 100–135°C.   The films in question were deposited at a rate on the order of 1000 Å minute to give film thicknesses up to around 10μ. The heavier layers of selenium were found to be very fragile and subject to cracking on the substrate if the substrate temperature was lowered too quickly after deposition. The removal of these films from substrates for structural examination by electron diffraction and transmission electron microscopy was effected by dissolution of the tellurium in a 20% aqueous solution of nitric acid containing ferric ions to inhibit the attack of the selenium. The films themselves were strengthened with nitrocellulose or silicone rubber prior to substrate dissolution.

Although Griffiths and Sang did not study the defect structure of the single crystal films in any detail, in electron diffraction patterns from selenium they did observe reflections at d-spacings which did not correspond to either the monoclinic or trigonal selenium modifications. These were found to be due to double diffraction effects or to the diffusion of impurities into the film. Similar observations by other investigators have led to erroneous claims for the existence of other modifications of selenium.[4]

Monoclinic Selenium

Although monoclinic selenium has been found to occur under certain conditions in the course of the crystallization of selenium to the trigonal form, crystals of monoclinic selenium of any size have only been realized by solution techniques. Upon relatively slow cooling of selenium-saturated solutions in carbon disulfide, which had been maintained at high temperatures, Moody and Himes[136] were able to grow rather large crystals of α-monoclinic selenium. The best results were obtained from a carbon disulfide solution of selenium in a bomb maintained at about 175°C (20 atm) for about 12 h, and subjected to an initial cooling rate of around 30°C per hour. The resulting monoclinic crystals were either diamond or hexagonal shaped platelets with the largest crystal measuring 1 cm on an edge and about 0.5 mm thick.

Single crystals of monoclinic selenium in the order of 1.5–2 mm can be produced conveniently at room temperature by the diffusion of selenium solutions in carbon disulfide or methylene iodide into toluene gels of styrene-maleic anhydride copolymers.[15]

By circulating a selenium-saturated solution of carbon disulfide in a crystallization vessel having two temperature regions, one at 30°C and a cooler region at 15°C, Abdullaev et al.[2] observed the growth of α- and β-monoclinic selenium single crystals from the supersaturated solution in the cooler section. X-ray and microscopic examination revealed the existence of thin platelet monocrystals of α- and β-monoclinic selenium having dimensions of $1 \times 1 \times 0.3$ mm. It was observed that there was no conversion of α- to

β-monoclinic selenium.   However, the monocrystals of both α- and β-monoclinic selenium did undergo conversion into polycrystalline trigonal selenium at room temperature.   The conversion of the α-monoclinic crystals to the trigonal form at this temperature took place at a lower rate than the β-monoclinic crystals.

REFERENCES

1. Abdullaev, G. B., Ibragimov, N. I., Mamedov, Sh. V., and Zhuvarly, T. Ch., *Phys. Stat. Sol.* **16**, K113 (1966).
2. Abdullaev, G. B., Asadov, Y. G., and Mamedov, K. P., *The Physics of Selenium and Tellurium*, ed. W. Charles Cooper, Pergamon Press, Inc., Oxford, 1969, p. 179.
3. Aliev, M. I. and Akhundova, S. A., *Referativny Zh.*, *Metallurgiya* **6**, 224 (1960).
4. Andrievskii, A. I., Nabitovich, I. D., and Kripyakeivch, P. I., *Dokl. Akad. Nauk SSSR* **124**, 321 (1959); Engl. transl. *Soviet Physics-Doklady* **124**, 16 (1959).
5. Andrievskii, A. I., and Nabitovich, I. D., 1960, *Kristallografiya* **5**, 465 (1960); Engl. transl. *Soviet Physics-Crystallography* **5**, 442 (1960).
6. Andrievskii, A. I., Nabitovich, I. D., and Voloshchuk, Ya. V., *Kristallografiya* **5**, 369 (1960); Engl. transl. *Soviet Physics-Crystallography* **5**, 349 (1960).
7. Avrami, M., *J. Chem. Phys.* **7**, 1103 (1939).
8. Belan, S. A., and Bolotov, I. E., *Fiz. Metal. Metalloved.* **26**, 1114 (1968).
9. Bennett, K. G., Thesis, Sir George Williams University, Montreal (1970).
10. Berg, A., Pappas, A., and Haissinsky, M., *J. Chim. Phys.* **47**, 382 (1950).
11. Berkowitz, J. and Chupka, W. A., *J. Chem. Phys.* **45**, 4289 (1966).
12. Berkowitz, J. and Chupka, W. A., *J. Chem. Phys.* **48**, 5743 (1968).
13. Bhatnager, S. S., Lessheim, H., and Khanna, M. L., *Nature* **140**, 152 (1937).
14. Bhatnager, S. S., Lessheim, H., and Khanna, M. L., *Proc. Indian Acad. Sci.* **6A**, 155 (1937).
15. Blank, Z., Brenner, W., and Okamoto, Y., *J. Cryst. Growth* **34**, 372 (1968).
16. Boltacks, B. I., *Diffusion in Element Semiconductors*, Academic Press, Inc., New York, 1963.
17. Boltov, I. E., and Belan, S. A., *Sb. Nauch. Tr. Sverdlovsk. Fil. Mosk. Inst. Nar. Khoz.* **3**, 25 (1967).
18. Bradley, A. J., *Phil. Mag.* **48**, [6] 47 (1924).
19. Brebrick, R. F., *J. Chem. Phys.* **43**, 3031 (1965).
20. Brebrick, R. F., *J. Chem. Phys.* **43**, 3846 (1965).
21. Brebrick, R. F., *J. Chem. Phys.* **48**, 5741 (1968).
22. Brebrick, R. F., and Strauss, A. J., *J. Chem. Phys.* **40**, 3230 (1964).
23. Brebrick, R. F., and Strauss, A. J., *J. Phys. Chem. Solids* **26**, 989 (1965).
24. Briegleb, S., *Z. physik. Chem.* **A144**, 321 (1929).
25. Briegleb, S., *Z. physik. Chem.* **A144**, 340 (1929).
26. Briegleb, S., *Naturwissenschaften* **17**, 51 (1929).
27. Burbank, R. D., *Acta Cryst.* **4**, 140 (1951).
28. Burbank, R. D., *Acta Cryst.* **5**, 236 (1952).
29. Busch, G. and Vogt, O., *Helv. Phys. Acta* **30**, 224 (1957).

30. Buschert, R. C., Thesis, Purdue University, Lafayette, Indiana (1957); *Diss. Abs.* **17**, 1571 (1957).
31. Champness, C. H., Griffiths, C. H. and Sang, H., *The Physics of Selenium and Tellurium*, ed. W. Charles Cooper, Pergamon Press, Inc., Oxford, 1969, p. 349.
32. Chang, R., and Romo, P., *Acta Cryst.* **23**, 700 (1967).
33. Chen, I., *J. Chem. Phys.* **45**, 3536 (1966).
34. Cherin, P. and Unger, P., *Inorg. Chem.* **6**, 1589 (1967).
35. Chiang, Y. S. and Johnson, J. K., *J. Appl. Phys.* **38**, 1647 (1967).
36. Cooper, W. C. and Griffiths, C. H., Canadian Copper Refiners Ltd., Montreal East, Quebec, unpublished work.
37. Das Gupta, K., and Das, S. R., *Indian J. Phys.* **15**, 401 (1941).
38. Das Gupta, K., Das, S. R., and Ray, B. B., *Indian J. Phys.* **15**, 389 (1941).
39. Datars, W. R., Fischer, G., and Eastman, P. C., *Can. J. Phys.* **41**, 178 (1963).
40. de Boer, F., *Rec. trav. chim.* **62**, 151 (1943).
41. de Brouchere, L., Watillon, A., and van Grunderbuck, F., *Nature* **178**, 589 (1956).
42. Dobinski, S., and Wesolowski, J., *Bull. Intern. Acad. Polon. Sci., Classe Sci. Math. Nat.* **7**; through *Chem. Abs.* **31**, 6072 (1937).
43. Dresner, J., and Stringfellow, G. B., *J. Phys. Chem. Solids* **29**, 303 (1968).
44. Dzhalilov, S. U., *Russ. J. Phys. Chem.* **39**, 137b (1965).
45. Ebert, H., Glashütte, **75**, 105 (1948).
46. Eckart, F., *Recent Advances in Selenium Physics*, ed. ESTC, Pergamon Press, Inc., Oxford, 1965, p. 85.
47. Eckart, F. and Vogel, C. H., *Z. Naturforschg.* **22**, 709 (1967).
48. Eisenberg, A., *Polymer Letters* **B1**, 33 (1963).
49. Eisenberg, A., 1969, *Macromolecules* **2**, 44 (1969).
50. Eisenberg, A., personal communication, 1971.
51. Eisenberg, A., and Tobolsky, A. V., *J. Am. Chem. Soc.* **81**, 780 (1959).
52. Eisenberg, A., and Tobolsky, A. V., *J. Polymer Sci.* **46**, 19 (1960).
53. Eisenberg, A., and Tobolsky, A. V., *J. Polymer Sci.* **61**, 483 (1962).
54. Feher, F., and Hellwig, E., *Silicium, Schwefel, Phosphate Colloq. Munster* 1955, p. 95.
55. Fitton, B., and Griffiths, C. H., *J. Appl. Phys.* **39**, 3663 (1968).
56. Fitton, B., and Janjua, M. B. I., Noranda Research Centre, Pointe Claire, Quebec, unpublished work.
57. Flory, P. J., *J. Am. Chem. Soc.* **62**, 1057 (1940).
58. Fourie, D. J., and van der Walt, C. M., *Z. Physik.* **169**, 326 (1962).
59. Frohnmeyer, G., Richter, H., and Schmelzer, G., *Z. Metallkunde* **46**, 689 (1955).
60. Fujisaki, H., Westmore, J. B., and Tickner, A. W., *Can. J. Chem.* **44**, 3063 (1966).
61. Gattow, G., *Z. anorg. allgem. Chem.* **317**, 245 (1962).
62. Gattow, G. and Draeger, M., *Z. anorg. allgem. Chem.* **343**, 55 (1966).
63. Gattow, G. and Heinrich, G., *Z. anorg. allgem. Chem.* **331**, 275 (1964).
64. Gee, G., *Trans. Faraday Soc.* **48**, 515 (1952).
65. Gee, G., *Chem. Soc. Specl. Publ.* **15**, 67 (1961).
66. Gimblett, F. G. R., *Inorganic Polymer Chemistry*, Butterworths, London, (1963).
67. Givens, M. P., *Rev. Sci. Inst.* **25**, 1130 (1954).
68. Glocker, R. and Hendus, H., *Z. Elektrochem.* **48**, 327 (1942).

69. Gobrecht, H., Willers, G., and Wobig, D., *J. Phys. Chem. Solids* **31**, 2145 (1970).
70. Goldfinger, P. and Jeunehomme, M., *Adv. Mass. Spec.* **1**, 534 (1959).
71. Gordon, M., *High Polymers*, 2nd. ed., Addison-Wesley Publishing Co., Inc., Reading, Mass., 1963.
72. Graham, A. K., Pinkerton, H. L., and Boyd, H. J., *J. Electrochem. Soc.* **106**, 651 (1959).
73. Graham, L. J. and Chang, R., *J. Appl. Phys.* **36**, 2983 (1965).
74. Griffiths, C. H., *Recent Advances in Selenium Physics*, ed. ESTC, Pergamon Press, Inc., Oxford, 1965, p. 97.
75. Griffiths, C. H., Champness, C. H., and Sang, H., Canadian Patent 799,644, Nov. 19, 1968.
76. Griffiths, C. H. and Fitton, B., *The Physics of Selenium and Tellurium*, ed. W. Charles Cooper, Pergamon Press, Inc., Oxford, 1969.
77. Griffiths, C. H., and Sang, H., *The Physics of Selenium and Tellurium*, ed. W. Charles Cooper, Pergamon Press, Inc., Oxford, 1969, p. 135.
78. Grimminger, H., Grüninger, H. I., and Richter, H., *Naturwissenschaften* **42**, 256 (1955).
79. Halla, F., Bosch, F. X., and Mehl, E., *Z. physik. Chem.* **B11**, 455 (1931).
80. Hamada, K., Sasaki, H., and Negita, H., *J. Sci. Hiroshima Univ.*, Ser. A-II **29**, 33; *Chem. Abs.* **63**, 17609d (1965).
81. Hamada, S., Yoshida, N., and Shirai, T., *Bull. Chem. Soc. Japan* **42**, 1025 (1969).
82. Harrison, D. E., *J. Chem. Phys.* **41**, 844, 1964; *cf.* also Harrison, D. E., *Recent Advances in Selenium Physics*, ed. ESTC, Pergamon Press, Inc., Oxford, 1965.
83. Harrison, D. E., and Tiller, W. A., *J. Appl. Phys.* **36**, 1680 (1965).
84. Harrison, J. D., *J. Appl. Phys.* **39**, 3672 (1968).
85. Hay, J. N., *J, Polymer Sci.* **A3**, 433 (1965).
86. Heavens, O. S., and Griffiths, C. H., *Acta Cryst.* **18**, 532 (1965).
87. Hendus, H., *Z. Physik.* **119**, 265 (1942).
88. Henninger, E. H., Buschert, R. C., and Heaton, L., *J. Chem. Phys.* **46**, 586 (1967).
89. Hilsch, R., *Non-Crystalline Solids*, ed. V. D. Frechette, John Wiley & Sons, Inc., New York, 1960, p. 348.
90. Hoffman, J. D., and Lauritzen, J. I., Jr., *J. Res. Natl. Bur. Stds.* **65A**, 297 (1961).
91. Hosemann, R., Lemm, K., and Krebs, H., *Z. physik. Chem. N. F.* **41**, 121 (1964).
92. Iizima, S., Taynai, J., and Nicolet, M. A., *The Physics of Selenium and Tellurium*, ed. W. Charles Cooper, Pergamon Press, Inc., Oxford, 1969, p. 199.
93. Illarionov, V. V., and Lapina, L. M., *Dokl. Akad. Nauk SSSR* **114**, 1021; Engl. transl. *Soviet Physics-Doklady* **114**, 615 (1957).
94. Janjua, M. B. I., Toguri, J. M., and Cooper, W. C., *Can. J. Phys.* **49**, 475 (1971).
95. Janjua, M. B. I., Toguri, J. M., and Dutrizac, J. E., *Can. J. Phys.* **46**, 447 (1968).
96. Kaplow, R., Rowe, T. A., and Averbach, B. L., *Phys. Rev.* **168**, 1068 (1968).
97. Karal'nik, S. M., Nesenyuk, A. P., and Dobrovol'skii, V. D., *Ukr. Fiz. Zh.* **10**, 668 (1965).
98. Keezer, R. C., *The Physics of Selenium and Tellurium*, ed. W. Charles Cooper, Pergamon Press, Inc., Oxford, 1969, p. 103.

99. Keezer, R. C., and Bailey, M. V., *Mat. Res. Bull.* **2**, 185 (1967).
100. Khalilov, Kh. M., *Izv. Akad. Nauk Azerb. SSR, Ser. Fiz. Mat. i Tekh. Nauk* **6**, 67 (1959).
101. Khalilov, Kh. M., *Izv. Akad. Nauk Azerb. SSR, Ser. Fiz. Tekh. i. Khim. Nauk* **4**, 43 (1959).
102. Kharlamov, I. P., and Golubeva, G. F., *Tr. Tsentral. Nauk-Issledov. Lab.* **4**, 147 (1956); through *Chem. Abs.* **56**, 1256d (1962).
103. Kitaev, G. A., and Fofanov, G. M., *Zh. Prikl. Khim.* (Leningrad) **43**, 1694 (1970); through *Chem. Abs.* **73**, 124168n (1970).
104. Klemm, W., Spitzer, H., and Niermann, H., *Angew. Chem.* **72**, 985 (1960).
105. Klug, H. P., *Z. Krist.* **88**, 128 (1934).
106. Knox, B. E., *Mat. Res. Bull.* **3**, 329 (1968).
107. Koehler, W. F., Odencrantz, F. K., and White, W. C., *J. Opt. Soc. Am.* **49**, 109 (1959).
108. Kolb, E. D., *The Physics of Selenium and Tellurium*, ed. W. Charles Cooper, Pergamon Press, Inc., Oxford, 1969, p. 155.
109. Koningsberger, D. C., and de Neef, T., 1970, *Chem Phys. Lett.* **4**, 615 (1970).
110. Krebs, H., 1951, *Z. anorg. allgem. Chem.* **265**, 156 (1951).
111. Krebs, H., *The Physics of Selenium and Tellurium*, ed. W. Charles Cooper, Pergamon Press, Inc., Oxford, 1969, p. 347.
112. Krebs, H., and Morsch, W., *Z. anorg. allgem. Chem.* **263**, 305 (1950).
113. Krebs, H., and Schultze-Gebhardt, F., *Naturwissenschaften* **41**, 474 (1954).
114. Krebs, H., and Schultze-Gebhardt, F., *Acta Cryst.* **8**, 412 (1955).
115. Krebs, H., and Steffen, R., *Z. anorg. allgem. Chem.* **327**, 224 (1964).
116. Kuliev, A. A., and Shakhtakhtinskii, M. G., *Dokl. Akad. Nauk Azerb. SSR* **14**, 831 (1958).
117. Kuliev, A. A., and Shakhtakhtinskii, M. G., *Dokl. Akad. Nauk SSSR* **120**, 1284 (1958).
118. Lark-Horovitz, K., and Miller, E. P., *Phys. Rev.* **51**, 380 (1936).
119. Lockenvitz, A. E., and Ribe, K. H., *Phys. Rev.* **85**, 746 (1952).
120. Lucovsky, G., *The Physics of Selenium and Tellurium*, ed. W. Charles Cooper, Pergamon Press, Inc., Oxford, 1969, p. 255.
121. Lucovsky, G., *The Physics of Selenium and Tellurium*, ed. W. Charles Cooper, Pergamon Press, Inc., Oxford, 1969, p. 276.
122. Lucovsky, G., Keezer, R. C., and Burstein, E., *Solid State Comm.* **5**, 439 (1967).
123. Lucovsky, G., Mooradian, A., Taylor, W., Wright, G. B., and Keezer, R. C., *Solid State Comm.* **5**, 113 (1967).
124. Marsh, R. E., Pauling, L., and McCullough, J. D., *Acta Cryst.* **6**, 71 (1953).
125. Mamedov, K. P., and Nurieva, Z. D., *Izv. Akad. Nauk Azerb. SSR, Ser. Fiz. Mat. i Tekh. Nauk* **2**, 47 (1962).
126. Mamedov, K. P., and Nurieva, Z. D., *Kristallografiya* **9**, 271 (1964).
127. Mamedov, K. P., Nurieva, Z. D., and Gadkieva, E. A., *Izv. Akad. Nauk Azerb. SSR, Ser. Fiz. Tekh. i Mat. Nauk* **4**, 117 (1965).
128. Massen, C. H., Weijts, A. G. L. M., and Poulis, J. A., *Trans. Faraday Soc.* **60**, 317 (1964).
129. Maxwell, L. R., and Mosley, V. M., *Phys. Rev.* **55**, 238 (1938).
130. Maxwell, L. R., and Mosley, V. M., *Phys. Rev.* **57**, 21 (1940).
131. McCullough, J. D., Thesis, Cal. Inst. of Technology (1936).
132. Mekhtieva, S. I., Abdinov, D. Sh., and Aliev, G. M., *Zh. Fiz. Khim.* **42**, 243 (1968).

133. Metzger, F., *Helv. Phys. Acta* **16**, 323 (1943).
134. Mikhaelyan, V. M., *Izv. Akad. Nauk Uz. SSR, Ser. Fiz. Mat. Nauk* **7**, 97 (1963).
135. Modřany, Č. K. D., German Patent 1,075,237, Feb. 11, 1960.
136. Moody, J. W., and Himes, R. C., *Mat. Res. Bull.* **2**, 523 (1967).
137. Mooradian, A., and Wright, G. B., *The Physics of Selenium and Tellurium*, ed. W. Charles Cooper, Pergamon Press, Inc., Oxford, 1969, p. 269.
138. Muthmann, W., *Z. Krist.* **17**, 336 (1890).
139. Nabitovich, I. D., Stetsin, Ya. I., and Voloshchuk, Ya. V., *Zh. Nauchn. i Prikl. Fotogr. i Kinematogr.* **11**, 27 (1966).
140. Neumann, K., and Lichtenberg, E., *Z. physik. Chem.* **A184**, 89 (1939).
141. Niwa, K., and Sibata, Z., *J. Chem. Soc. Japan* **61**, 667 (1940).
142. Niwa, K., and Sibata, Z., *J. Fac. Sci., Hokkaido Imp. Univ., Japan*, Ser. III **3**, 53 (1940).
143. Paribok-Aleksandrovich, I. A., *Fiz. Tverd. Tela* **11**, 2017 (1969).
144. Pascal, P., ed., *Nouveau Traite de Chimie Minerale*, Vol. XIII, Masson et Cie, Paris, Part 2, 1961, p. 1670.
145. Peibst, H., and Ammon, G., *Phys. Stat. Sol.* **10**, 227 (1965).
146. Porter, R. F., personal communication, 1970.
147. Poulis, J. A., and Derbyshire, W., *Trans. Faraday Soc.* **59**, 559 (1963).
148. Poulis, J. A., Massen, C. H., and van der Leeden, P., *Trans. Faraday Soc.* **58**, 474 (1962).
149. Powell, R. E., Roseveare, W. E., and Eyring, H., *Ind. Eng. Chem.* **33**, 430 (1941).
150. Preuner, G., and Brockmöller, I., *Z. physik. Chem.* **81**, 129 (1912).
151. Prins, J. A., and Dekeyser, W., *Physica* **4**, 900 (1937).
152. Ratchford, R. J., and Rickert, H., *Z. Elektrochem., Ber. Bunsenges. physik. Chem.* **66**, 497 (1962).
153. Rau. H., *Ber. Bunsenges. physik. Chem.* **71**, 711 (1967).
154. Remy, H., *Treatise on Inorganic Chemistry*, transl., J. S. Anderson, ed. J. Kleinberg, Vol. I, Elsevier Publishing Co., New York, 1956, p. 742.
155. Richter, H., *Z. Physik.* **44**, 456 (1943).
156. Richter, H., and Herre, F., *Naturwissenschaften* **44**, 31 (1957).
157. Richter, H., and Herre, F., *Z. Naturforschg.* **13a**, 874 (1958).
158. Richter, H., Kulcke, W., and Specht, H., *Z. Naturforschg.* **7a**, 511 (1952).
159. Risi, M., and Yuan, S., *Helv. Phys. Acta* **33**, 1002 (1960).
160. Rosen, B., *Physica* **6**, 205 (1939).
161. Rouard, P., and Bousquet, P., *Opt. Acta* **16**, 675 (1969).
162. Sampath, P. I., *J. Chem. Phys.* **45**, 3519 (1966).
163. Satô, T., and Kaneko, H., *Technol. Repts. Tohoku Univ.* **14**, 45 (1950).
164. Saunders, A. P., *J. Phys. Chem.* **4**, 423 (1900).
165. Saure, H., Thesis, Technical University of Berlin (1969).
166. Semiletov, S. A., *Kristallografiya* **4**, 629; Engl. transl. *Soviet Physics-Crystallography* **4**, 588 (1959).
167. Sharples, A., *Introduction to Polymer Crystallization*, Arnold, London, 1966.
168. Shirai, T., Hamada, S., and Kobayashi, K., *Nippon Kagaku Zasshi* **84**, 968 (1963).
169. Sidyakin, V. G., *Izv. Vysshikh Ucheb. Zaved. Fiz.* **2**, 25 (1962).
170. Slattery, M. K., *Phys. Rev.* **21**, 378 (1923).
171. Slattery, M. K., *Phys. Rev.* **25**, 333 (1925).

172. Sokolov, B. V., *Uch. Zap. Vologdsk. Gos. Ped. Inst.* **23**, 171 (1958).
173. Starodubtsev, S. V., and Mikhaelyan, V. M., *Radiats. Effekty v Tverd. Telakh, Akad. Nauk Uz. SSR. Inst. Yadern. Fiz.* **32**; (1963); *Chem. Abs.* **60**, 1183g (1964).
174. Starodubtsev. S. V., Pugacheva, T. S., Mikhaelyan, V. M., and Lenchenko, V. M., *Dokl. Akad. Nauk SSSR* **150**, 1091; *Chem. Abs.* **59**, 9492c (1963).
175. Stiles, D. A., Tyerman, W. J. R., Strausz, O. P., and Gunning, H. E., *Can. J. Chem.* **44**, 1677 (1966).
176. Straumanis, M., *Z. Krist.* **102**, 432 (1940).
177. Stubb, T., *Recent Advances in Selenium Physics*, ed. ESTC, Pergamon Press, Inc., Oxford, 1965, p. 53.
178. Swanson, H. E., Gilfrich, N. T., and Ugrinic, G. M., *U.S. Nat. Bur. Stds.* (Circ. No 539) **5**, 24 (1955).
179. Tanaka, K., *Mem. Coll. Sci. Kyoto Imp. Univ.* **A17**, 59 (1934).
180. Tobisawa, S., *Bull. Chem. Soc. Japan* **33**, 889 (1960).
181. Unger, P. and Cherin, P., *The Physics of Selenium and Tellurium*, ed. W. Charles Cooper, Pergamon Press, Inc., Oxford, 1969, 1969, p. 223.
182. Urazovskii, S. S. and Luft, B. D., *J. Phys. Chem.* (USSR) **22**, 409; *Chem. Abs.* **42**, 7140e (1948).
183. Vasko, A., *The Physics of Selenium and Tellurium*, ed. W. Charles Cooper, Pergamon Press, Oxford, 1969, p. 241.
184. Vasko, A., *The Physics of Selenium and Tellurium*, ed. W. Charles Cooper, Pergamon Press, Inc., Oxford, 1969, p. 254.
185. Wainfan, N., and Parratt, L. G., *J. Appl. Phys.* **31**, 1331 (1960).
186. Wartenburg, H. V., *Z. anorg. allgem. Chem.* **56**, 329 (1908).
187. Yamamori, S., *Nippon Kinzoku Gakkai-Shi* **B15**, 274 (1951).
188. Yamdagni, R., and Porter, R. F., *J. Electrochem. Soc.* **115**, 601 (1968).

4

The Interaction with Light of Phonons in Selenium*

RICHARD ZALLEN/GERALD LUCOVSKY
Xerox Research Laboratories, Rochester, N.Y.

INTRODUCTION

This article reviews those optical properties of selenium which are determined by the atomic vibrations of the crystal or glass lattice. While we have collected for discussion some of the most important experimental and theoretical results, our emphasis will be on the qualitative features and the physical significance of the data, rather than on the numerical values of the parameters presented.

We will discuss the three principal allotropic forms of solid selenium: two crystalline modifications, trigonal Se and α-monoclinic Se; and the important amorphous modification as well. The pertinent experiments are, primarily, measurements of infrared absorption and of Raman scattering. Because we are concerned with those lattice waves (phonons) which interact with incident light waves (photons), which possess small wave-

* See also Chapter 5.

vectors ($\mathbf{q} \sim \omega/c$), the phonons involved are those near the center of the Brillouin zone ($\mathbf{q} \approx 0$, the vibrational "long waves").[2]

For any solid, a knowledge of the vibrations of the atoms about their equilibrium lattice positions is of great interest because of the insight gained into the interatomic bonding interactions, which are responsible for the very existence of the solid and for the particular structure it assumes. For selenium, however, the lattice optical properties are of special added significance due to the occurrence of reststrahlen, one-phonon absorption bands in the far-infrared which are produced by the strong, direct interaction of light with lattice fundamentals. This interaction, which is customarily associated with ionic crystals and is widely (and mistakenly) assumed to be absent in elemental solids, occurs in all three forms of Se. Trigonal Se, the most extensively studied of the three, is very special since it is, as we shall see, *the simplest reststrahlen-displaying elemental crystal.*

Because of its unique position with respect to the photon-phonon interaction and the fact that the available experimental information is most complete for this material, trigonal Se is reviewed in some detail. As a prerequisite for the interpretation of the experiments, the symmetry and the consequent optical selection rules are presented in the next section. The coupled photon-phonon spectra of trigonal Se are discussed later with particular emphasis on the responsible underlying mechanism of dynamic charge. Higher order phonon-involving optical interactions in this crystal, particularly Raman scattering and two-phonon infrared absorption, are then reviewed.

The covalent coordination of selenium is twofold, and as a consequence of this low coordination, it forms molecular crystals rather than network crystals. In trigonal Se, the molecular unit is an extended chain; in monoclinic Se, an eight-membered ring. The recent work on the two crystal forms has permitted an analysis of the spectra of the vitreous form in terms of vibrations of these two types of molecular structural units. The spectra of the glass are found to be dominated by features ascribable to Se_8 molecules, providing direct evidence for the presence of this molecular species in significant concentrations in amorphous Se. The lattice spectra of monoclinic Se are discussed elsewhere as well as the spectra and structure of amorphous Se.

SYMMETRY AND SELECTION RULES FOR TRIGONAL Se

The three-atom unit cell of trigonal Se and Te is shown in Figure 4-1. The crystal structure[36] consists of helical coils ("chains") which wind around axes parallel to the crystalline c-axis, which is a triad screw axis. The helices contain three atoms per turn and are in hexagonal array (close packing for rods). The space group is $P3_121$ (D_3^4) for the right-handed crystal of Figure

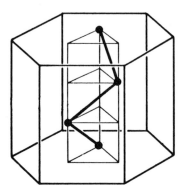

Figure 4-1   Unit cell for the crystal structure of trigonal Se, slightly schematized to clearly reveal the chain structure.

4-1.   For the enantiomorphic left-handed form, the space group is $P3_2 21$, or D_3^6. This is the same symmetry which applies to quartz and to cinnabar (trigonal HgS^{39}); Se and Te are monatomic analogues of those crystals.   This is the least complex crystal structure based on a helix.   For example, these are the simplest crystals (elemental crystals, triatomic unit cell) which exhibit circular dichroism and optical activity.

Se and Te are molecular crystals with the molecular unit, the helical chain, infinitely extended in one dimension.   Viewed in this way, they are clearly prototype polymers.   The intermolecular bonding between chains is much weaker than the molecular bonding within each chain; the ratio between the smallest chain-chain and intrachain bond lengths is 1.49 for Se and 1.31 for Te.   The weak interchain bonding is markedly manifest in the cleavage behavior; the crystals cleave only in planes containing the c-axis (no broken chains).   To our knowledge, these are the only elemental crystals with three atoms in the primitive unit cell.   As will be discussed further on, this makes them unique with respect to the interaction with light of their lattice vibrations.

The six operations of the factor group for D_3^4 crystals are 1, 3_1, $3_1 \times 3_1$, 2, $3_1 \times 2$, and $2 \times 3_1$, where:

1 denotes the identity
3_1 denotes a threefold rotation about the c-axis, followed by
   a $(1/3)c$ translation parallel to the axis
2 denotes a twofold rotation about an axis perpendicular to c

This group has been discussed by several researchers.[1,17,25]   It is simply isomorphic to the familiar point group $3m$ (C_{3v}) of the equilateral triangle and possesses three irreducible representations (I.R.'s): A_1, A_2, and E. A_1 and A_2 are one-dimensional symmetric and antisymmetric representations with respect to the twofold rotations, and E is a two-dimensional representation.

Notations other than A_1, A_2, and E which have also been used for the Γ-point ($\mathbf{q} = 0$) I.R.'s are Γ_1, Γ_1', $\Gamma_2{}^{17}$ and Γ_1, Γ_2, Γ_3.[1]   The character table for the group is contained in Table 4-1.

The symmetries of the long wavelength phonons are determined by Γ, the factor-group representation generated by the displacements of the atoms in the unit cell.   For trigonal Se, Γ is 9-dimensional and is, of course, reducible. The characters for Γ are shown in the fourth row of Table 4-1, and its reduction into I.R.'s is given in the fourth column: $\Gamma = A_1 + 2A_2 + 3E$.   Also listed is the three-dimensional polar-vector representation P.   The three zone-center acoustical phonons (rigid translations of the crystal) comprise such a representation, so that the I.R.'s of the optical phonons are contained in $\Gamma - P = A_1 + A_2 + 2E$.   Thus, there are four optical-mode eigen-frequencies (ignoring for the moment the transverse-longitudinal splitting of the infrared-active vibrations) corresponding to two nondegenerate A modes ($A_1 + A_2$), and two doubly-degenerate pairs of E modes.   The eigenvectors for the nondegenerate optical vibrations are determined by symmetry.   The A_1 mode is a breathing mode of the helical chains, the A_2 mode is a chain-rotational mode (to be discussed in another section).

Symmetry-determined selection rules governing the interaction of these vibrations with light are presented in the last two columns of Table 4-1.   For one-phonon optical absorption, the mode must induce a first-order electric moment p, so that it necessarily transforms as a vector.   The infrared-active modes are therefore those of symmetry types common to $\Gamma - P$ and P; the I.R.'s contained in this intersection are A_2 and E.   The A_2-symmetry optical

TABLE 4-1   SYMMETRY ANALYSIS AND SELECTION RULES FOR THE ZONE-CENTER PHONONS IN TRIGONAL Se

Representation		Class Characters 1 3_z 2_x			Zone-Center Modes All Acoust. Opt. Γ P $\Gamma - P$			Selection Rules Infrared	Raman
Irred reps (I.R.)	A_1	1	1	1	1	0	1	—	$\alpha_{zz}, \alpha_{xx} + \alpha_{yy}$
	A_2	1	1	−1	2	1	1	p_z	—
	E	2	−1	0	3	1	2	p_z, p_y	$\alpha_{xy}, \alpha_{yz}, \alpha_{zx},$
									$\alpha_{xx} - \alpha_{yy}$
Unit-cell rep.	Γ	9	0	−1					
Vector rep	P	3	0	−1					

Two-phonon combinations

$A_1 \times A_1 = A_2 \times A_2 = A_1,$	Raman
$A_1 \times A_2 = A_2,$	IR (z);
$A_1 \times E = A_2 \times E = E,$	IR (x, y) and Raman;
$E \times E = A_1 + A_2 + E,$	IR (x, y, z) and Raman.

mode interacts with light polarized parallel to the c-axis (c-axis = the z axis of Table 4-1); the E modes interact with light polarized perpendicular to c. The nature of these interactions is explored in the following section.

The symmetric A_1 vibration, which does not interact with light in first-order and is therefore absent from the fundamental infrared spectrum, can be acquired in Raman scattering. A Raman-active vibration induces a first-order modulation in the dielectric polarizability tensor at optical frequencies, so that it must transform as a component of a symmetric second-rank tensor. The Raman selection rules for D_3 symmetry are given in the last column of Table 4-1. The E modes, as well as the A_1, are permitted in Raman scattering. The presence of vibrations entitled to appear in both the Raman and infrared spectra, the E modes, is permitted by the absence of inversion symmetry. For completeness, the lower part of Table 4-1 shows two-phonon selection rules for combination bands produced by pairs of zone-center phonons. Both Raman scattering spectra and two-phonon absorption spectra for trigonal Se will be discussed.

THE PHOTON–PHONON INTERACTION IN TRIGONAL Se

Reststrahlen in Elemental Crystals

The phenomenon of reststrahlen, the intense absorption of light by a crystal in the far infrared which is produced by strong interaction with lattice vibration fundamentals (single phonons), is one of the most familiar effects in the spectroscopy of solids. Reststrahlen bands are normally associated with ionic crystals, NaCl being the classic prototype.[2] In fact, the oscillating electric dipole moment set up by the lattice vibrations, which gives rise to the strong photon-phonon interaction, is often interpreted as an entirely ionic effect. This, however, is not true. If it were, then crystals composed of a single atomic species, elemental crystals such as Se, could not exhibit reststrahlen because of the absence of ionic charge. The widespread incorrect idea[3,23] that an elemental crystal cannot exhibit one-phonon optical absorption has been reenforced by the circumstance that for the familiar Group IV diamond-structure semiconductors (C, Si, Ge, α-Sn), the zone-center optical phonons are indeed infrared-inactive; they do not couple to light. The explanation of this inactivity is, however, based on a symmetry consideration and is *not* because these are homopolar covalent solids. In general, lattice vibrations in elemental crystals *can* possess a first order electric moment, and thus exhibit reststrahlen, by the mechanism of displacement-induced charge redistribution (*dynamic charge*[7,10,11]).

All crystalline compounds possess infrared-active lattice fundamentals, but not all elemental crystals do. The reason for the nonoccurrence of reststrahlen in diamond-structure crystals, as first stated by Lax and Burstein[18] in 1955, is discussed in a recent article by Zallen[37] in which the following generalization is derived:

A necessary and sufficient condition for the existence of reststrahlen in an elemental crystal is a structure with at least three atoms in the primitive unit cell, $s \geqslant 3$.

For Ge and Si $s = 2$, so that the optical modes are infrared-inactive. The crystal structure of trigonal Se and Te (Figure 4-1), with $s = 3$, is the simplest structure known to occur which satisfies this minimum complexity condition for a reststrahlen-displaying elemental crystal. In other words, among elemental crystals, Ge and Si are structurally too simple to display the phenomenon under consideration; Se and Te have just the minimum amount of structural complexity ($s = 3$) necessary to exhibit it.

Trigonal Se displays pronounced reststrahlen bands. These form the subject of the following two sections. The underlying mechanism for the lattice-wave–electromagnetic-wave interaction and the experimentally-observed spectra are also reviewed. The photon-phonon interaction in Se is of particular special significance because, as mentioned previously, this material shares with Te the distinction of being structurally *the simplest known reststrahlen-displaying elemental crystal.*

Dynamic Charge in Trigonal Se

For one-phonon optical absorption, a zone-center ($\mathbf{q} \approx 0$) vibration must exhibit a first-order moment, a macroscopic polarization $\mathbf{p}$ linear in the atomic displacements $\mathbf{u}$. The mechanism for the first-order moment in an elemental crystal is schematically illustrated in a one-dimensional model in Figure 4-2d; we introduce it with a prerequisite discussion of a one-dimensional representation of NaCl in Figure 4-2a and Figure 4-2b.

Figure 4-2a shows NaCl in the rigid ion model. In this model the effective charge e^* connecting $\mathbf{u}$ and $\mathbf{p}$ is just e, the static ionic charge residing on and moving with each ion. However, it is well known that for the alkali halides, e^*/e is appreciably smaller than 1 because of the deformability of the halogen ions. The effects of dynamic charge, the deformability of ions under atomic displacements, is illustrated in Figure 4-2b–Figure 4-2d by means of the shell model, in which the ions are resolved into rigid oppositely-charged cores and shells. The shell model was originally introduced by Dick and Overhauser[12] in order to elucidate, in a simple manner, the role of dynamic charge in the alkali halides in producing the deviation of e^*/e from 1, the rigid-ion value for these crystals. How this comes about is indicated in Figure 4-2b. Recently, Chen and Zallen[10,38] applied the shell model to Se

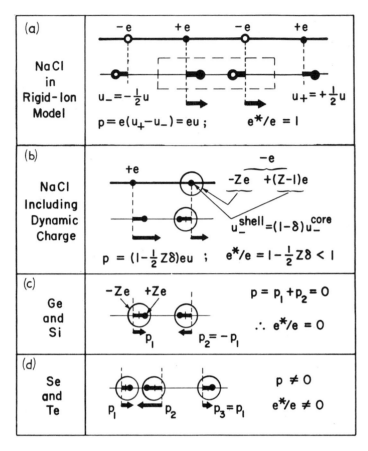

Figure 4-2   One-dimensional representations of the mechanisms for the vibration-induced first-order electric moment responsible for reststrahlen: (a) rigid-ion model for a diatomic ionic crystal; (b) effect of dynamic charge for NaCl; (c) dynamic charge in Ge and Si; and (d) dynamic charge in Se and Te.

and Te in order to elucidate the manner in which charge deformation produces the observed deviation of e^*/e from *zero*, the rigid-ion value for elemental crystals.   The dynamic charge in ionic crystals plays the role of a correction; in elemental crystals it is the whole story.

The basic feature of the dynamic-charge mechanism for Se is shown in Figure 4-2d.   For Se and Te, and for Ge and Si (shown in Figure 4-2c) as well, dynamic charge produces an induced electric dipole at each atomic site.   For Ge and Si, as for all elemental crystals with $s = 2$, the two induced dipoles within the unit cell are required by symmetry to be equal and opposite.[18,37] There is no net unit-cell electric moment, hence no reststrahlen.   For Se and Te there is no corresponding constraint.   The net moment does not vanish and these crystals exhibit strong reststrahlen bands, with values of e^*/e on the order of unity.   We shall consider here in more detail the dynamic charge associated with the simplest of the infrared-active vibrations of trigonal Se.

From Table 4-1 we see that the A_2-symmetry zone-center optical phonon in Se is responsible for the reststrahlen band observed with light polarized parallel to the trigonal c-axis.   The eigenvector for this nondegenerate vibration is determined by symmetry alone; it is a chain-rotational vibration in which intrachain bond lengths and angles are preserved.   The first-order electric moment for this mode is illustrated in Figure 4-3, following the shell-model treatment of Chen and Zallen.   The macroscopic polarization resides

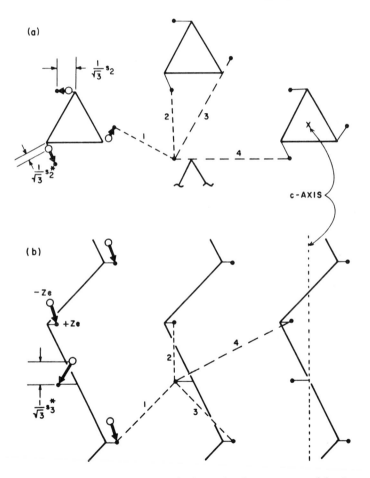

Figure 4-3   Shell-model picture for the dynamic charge electric moment, p_z, of the A_2-symmetry optical mode in trigonal Se: (a) top view, viewed along the c-axis; (b) side view, viewed $\perp c$. Solid dots represent displaced cores; open circles represent displaced shells.[10]

with the axial shell displacements (indicated by s_3^* in the figure) induced by the chain-twisting atomic displacements (indicated by s_2). Figure 4-3 accounts for the remarkable circumstance that a vibration with all of the atomic displacements perpendicular to c interacts with radiation polarized parallel to c. For this mode both the dynamic charge and the restoring force depend upon the presence of intermolecular interactions between chains, and vanish in the limit of isolated chains.

The Reststrahlen and Polariton Spectra

The far-infrared reflectivity spectra of trigonal Se have been measured by Lucovsky et al.[21] and by Geick et al.,[15] with essentially similar results. The reststrahlen spectra for the two independent polarizations are shown in Figure 4-4. The high-frequency E mode, infrared-active for $E \perp c$, is off scale to the right of the range shown; it is much weaker than the other bands and is definitively determined in transmission measurements. The infrared spectra were fitted with a set of classical oscillators, each contributing a term of the form $s\bar{v}_t^2/(\bar{v}_t^2 - \bar{v}^2 + i\gamma\bar{v}_t\bar{v})$ to the complex dielectric constant $\varepsilon(\bar{v})$. For a given oscillator, $\bar{v}_t$ is the oscillator frequency (transverse frequency in wavenumber units), s is the oscillator strength (contribution of the absorption band to the static dielectric constant), and γ is the dimensionless damping parameter. The oscillator parameters for the A_2 and E modes are listed in Table 4-2. Also included in this table, for completeness, is the A_1 eigenfrequency obtained from the Raman-scattering experiments to be discussed in

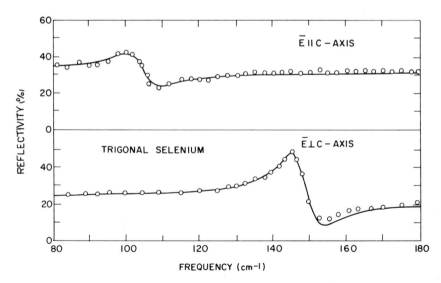

Figure 4-4   Reststrahlen spectra for trigonal Se.[21]

TABLE 4-2   ZONE-CENTER OPTICAL PHONON FREQUENCIES IN TRIGONAL Se[a]

Symmetry	Spectrum	$\bar{\nu}$ in cm^{-1}	s	γ
A_2	IR($\parallel c$)	102	0.85	0.07
E	IR($\perp c$), Raman	142	0.73	0.05
E	IR($\perp c$), Raman	231	0.02	0.04
A_1	Raman	237	—	—

[a] Refs. 15, 21, 22, and 24.

the next section.   The lattice (infrared) contribution to the static dielectric constant $\varepsilon(0)$ is of the order of 1, and it is smaller than the electronic (ultraviolet) contribution, of order 10, for this crystal.   For the isomorphic cousin from the next row of the periodic table, trigonal Te, the two contributions to $\varepsilon(0)$ are comparable: the lattice contribution is ≈ 10, the electronic interband contribution is ≈ 20.   The reason for the stronger reststrahlen absorption in Te is the higher atomic number and consequently larger polarizability under atomic displacement.

The strong photon-phonon interaction in the reststrahlen region means that at these frequencies the true elementary excitations within the crystal are coupled electromagnetic-lattice waves.   The experimental results of Figure 4-4 and Table 4-2 enable us to determine the dispersion curves for the polaritons,[6,16] the coupled photon-phonons.   In Figure 4-5a is the polariton dispersion

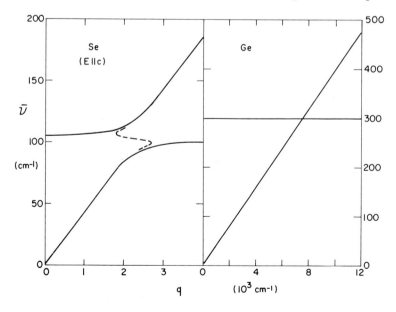

Figure 4-5   Polariton dispersion curves for Se and Ge.

curve, $\bar{v}(\mathbf{q})$, for the A_2 mode in Se; for comparison, the corresponding situation for Ge is shown in Figure 4-5b. In this figure, frequency is plotted against the real part of the complex propagation vector of the coupled waves. The range of $\mathbf{q}$ shown is of the order of photon wave-vectors ($\mathbf{q} \sim n\omega/c$, where n is the refractive index and ω an infrared frequency) and is small on the full scale of phonon wave-vectors ($\mathbf{q} \sim \pi/a$, where a is a lattice constant).

The dashed curve of Figure 4-5a is the experimentally-defined polariton dispersion curve; it corresponds to the oscillator parameters of Table 4-2 for $E \parallel c$, along with a high-frequency dielectric constant of 12 for this polarization.[15,21] Letting the damping constant go to zero, we obtain the solid curve shown in this figure, which displays the classical coupled-wave form.[16] The $q = 0$ intercept of the zero-damping curve locates the longitudinal-optical (LO) frequency; the $\mathbf{q} \to \infty$ asymptote determines the transverse-optical (TO) frequency.

Figure 4-5 conveys the essential difference between Se and Ge in respect to the interaction of their crystal lattices with light. For Ge, the dispersion curve consists of two intersecting straight lines: the horizontal line at the optical-phonon frequency corresponding to the pure lattice waves, and the line (of slope $1/2\pi n$) through the origin corresponding to the undisturbed light waves. The lattice waves and light waves are uncoupled, each is indifferent to the presence of the other. For Se, the two waves are coupled and the intersection between the lines is removed. The dispersion curve for Se is qualitatively similar to that of, say, NaCl. The decrease in slope between high and low frequencies of the mostly electromagnetic portions reflects the increase in dielectric constant and refractive index on passing through the reststrahlen band.

In Table 4-3, the longitudinal and transverse frequencies of the infrared-active optical phonons are compared. Also listed are the macroscopic effective charges[5] of these modes, defined by $e^* = \pi^{1/2} c m_a^{1/2} n_a^{-1/2} s^{1/2} \bar{v}$ where m_a is the atomic mass, n_a the number of atoms per unit volume, and s and $\bar{v}$

TABLE 4-3   TO AND LO FREQUENCIES, AND EFFECTIVE CHARGES OF THE INFRARED-ACTIVE PHONONS[a]

Mode	Transverse Frequency $\bar{v}_t$ in cm^{-1}	Longitudinal Frequency $\bar{v}_\ell$ in cm^{-1}	Effective Charge e^*/e	Electron-Phonon Coupling Constant
A_2	102	106	0.6	0.2
E	142	149	0.8	0.3
E	231	231	0.2	0.01

[a] Refs. 15, 21, and 22.

are the oscillator parameters of Table 4-2.   These values of e^* are only about
a factor of 2 smaller than the corresponding quantities for the alkali halides,
demonstrating clearly that dynamic charge is no small effect.

We close our discussion of the photon-phonon interaction in trigonal Se
by briefly mentioning another important "ionic" phenomenon that occurs
in this crystal as a direct consequence of infrared activity: the polar-mode
scattering of mobile charge carriers.   Charged particles traversing the solid
interact with the LO phonons via the macroscopic electric field associated
with these lattice waves.   The strength of the interaction is characterized by
a dimensionless quantity, the electron-phonon coupling constant,[40] which is
derivable from the infrared data.   The coupling constants for the three LO
phonons in Se are listed in the last column of Table 4-3; they are smaller than
values typical of the alkali halides (≈ 5) and slightly larger than values typical
of Group III–V semiconductors (≈ 0.1).   Polar-mode scattering has significant influence on the electrical transport properties of trigonal Se, contributing
extensively to the limitation on carrier mobilities.[21]

HIGHER ORDER OPTICAL INTERACTIONS IN TRIGONAL Se

Higher Order Phonon-Involving Optical Processes

The one-phonon optical absorption process discussed in the preceding section
is the strongest and the most direct interaction between light and the crystal
lattice.   It is a first order process involving a single photon and a single
phonon.   There are a variety of weaker higher order processes coupling the
lattice with electromagnetic radiation which involve the cooperation of additional elementary excitations.   In this section we shall review the available
experimental data on the higher order optical properties of trigonal Se which
involve at least one phonon.   The mechanisms for the interactions will be
touched upon only briefly since, unlike the situation with respect to reststrahlen, trigonal Se does not occupy a unique role involving these interactions.

In Figure 4-6, we present a set of simple diagrams intended to introduce
and illustrate the processes discussed.   The direct one-phonon absorption of
a far-infrared photon is represented in Figure 4-6a; Figure 4-6b–Figure 4-6e
show the relevant multi-particle processes.   Figure 4-6b indicates a two-
phonon combination band at higher frequencies in the infrared; the annihila-
tion of a photon is accompanied by the creation of two optical phonons.
Electronic excitations and high-energy photons (i.e., visible or near-infrared)
participate in the remaining three processes portrayed.   A Raman scattering
event is depicted in Figure 4-6c, with an incident photon virtually absorbed

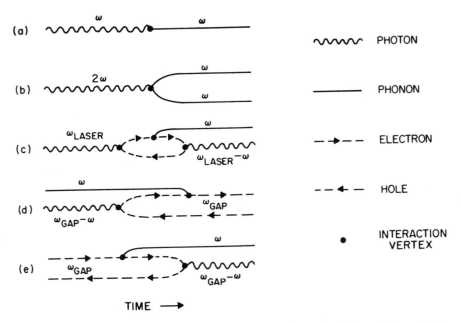

Figure 4-6 Diagrammatic representations of elementary phonon-involving optical processes: (a) one-phonon absorption (reststrahlen); (b) two-phonon absorption; (c) Raman scattering, showing a Stokes event (phonon emission); (d) phonon-assisted electronic absorption; and (e) luminescence event with emission of both photon and phonon.

and re-emitted at lower energy due to phonon emission.   Figure 4-6d depicts the phonon-assisted optical creation of an electron-hole pair, while Figure 4-6e indicates the inverse process observed in luminescence.

Raman-scattering spectra for trigonal Se are reviewed in the next section, two-phonon absorption and phonon-assisted electronic optical processes are dealt with later.

Raman Scattering

Employing a YAG : Nd^{3+} laser operating at a wavelength of 1μ, which falls within the transparent region of trigonal Se, Mooradian and Wright[24] obtained very clean scattering spectra for this crystal.   A survey spectrum (unpolarized light, polycrystalline sample) exhibiting both Stokes (phonon emission) and anti-Stokes (phonon absorption) lines is shown in Figure 4-7. Clearly indicated are the three Raman-active fundamentals: the A_1 mode and the two E modes.   The small TO-LO splittings of the E modes were not resolved.

The selection rules given in the last column of Table 4-1 were used to confirm the symmetry assignments.   Using a single crystal and polarized incident

TRIGONAL SELENIUM

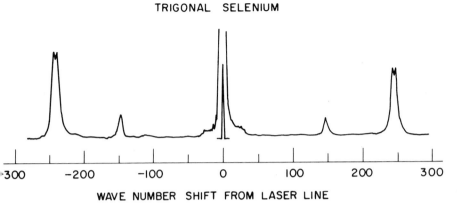

WAVE NUMBER SHIFT FROM LASER LINE

Figure 4-7   Raman scattering from trigonal Se.[24]

and scattered radiation, the A_1 line was cleanly separated from the two E lines.

Mooradian and Wright were also able to observe many weak emission bands and, in addition, determined the spectra at liquid helium temperature. One of the weak bands was identified as the Raman-forbidden A_2 fundamental, apparently weakly turned on by crystalline imperfection.   The others were ascribed to scattering processes involving a pair of phonons; their frequencies are listed in Table 4-4 along with the two-phonon absorption bands to be discussed.

TABLE 4-4   SECOND ORDER SPECTRA OF TRIGONAL Se[a]

$\bar{\nu}$ in cm^{-1}	Experimental Observation		Assignment
75	—	Raman	—
102	—	Raman	A_2
130	Infrared (‖)	—	$A_1 - A_2$
183	—	Raman	—
206	—	Raman	$2A_2$
250	Infrared ($\perp$)	—	$A_2 + E$
275	Infrared (‖, $\perp$)	Raman	$2E$
302	Infrared ($\perp$)	—	$3A_2$ (?)
328	Infrared (‖)	—	$A_1 + A_2$
345	—	Raman	—
365	Infrared (‖, $\perp$)	—	$E + E$
438	—	Raman	$2E$
451	Infrared ($\perp$)	—	$A_1 + E$
455	—	Raman	$2A_1$
475	Infrared (‖, $\perp$)	—	$2E$

[a] Refs. 22 and 24.

No theoretical treatment of Raman-scattering intensities in trigonal Se has yet been attempted. It is worth noting that the simple harmonic approximation shell model, employed by Chen and Zallen[10,38] for their treatment of reststrahlen in Se, is inadequate for a discussion of Raman scattering since it contains no cross term of electric field and atomic displacement. Such a bilinear term, which is equivalent to a modulation of the optical polarizability by the vibrations of the lattice, is essential to the character of the scattering mechanism. A more complicated model than the shell model would be required for a calculation of Raman intensities.

Two-Phonon Absorption

The two-phonon absorption process schematically indicated in Figure 4-6b occurs via the presence of a second order electric moment, an electric moment quadratic in the atomic displacements ($p \sim u^2$). Two different phonons may take part in this process, which can also involve phonons away from $q = 0$. Ge and Si, which have no first order moment, nevertheless exhibit weak two-phonon infrared absorption because of the existence of a second order moment.[18]   Se and Te, in addition to the strong one-phonon bands produced by their first-order moments, also exhibit much weaker two-phonon bands associated with their small second order moments.

Infrared transmission measurements at photon energies above the fundamental reststrahlen region have been interpreted by Lucovsky et al.[22] in terms of two-phonon combination bands. Their observed frequencies, along with the analogous results from the weak second-order Raman spectra, are listed in Table 4-4. The suggested assignments are consistent with the direct-product selection rules for zone-center phonons given in Table 4-1.

Phonon-Assisted Electronic Optical Processes

The absorption edge at the electronic interband threshold in trigonal Se has been investigated by Roberts et al.[27] and interpreted in terms of two characteristic phonon energies. For one polarization of incident light, they observed an Urbach-rule edge with an effective phonon energy of 155 cm^{-1}; for the other polarization, they observed an indirect edge involving the cooperation of a phonon of frequency 230 cm^{-1}.

Queisser and Stuke[26] have observed a set of sharp emission lines in the luminescence spectrum of Se at low temperatures. A basic triplet of lines is replicated at lower energies, with the phonon replicas yielding a frequency of 235 cm^{-1} for the emitted phonons.

Because of the electron-phonon interaction involved in these two primarily electronic processes (Figures 4-6d and 4-6e), we would anticipate the participation of LO phonons. The lattice frequencies reported are quite close to the two E-mode LO-phonon frequencies given in Table 4-2.

THE SPECTRA OF MONOCLINIC Se

Structure*

Se in the α-monoclinic form is a molecular crystal made up of eight-membered ring molecules. The space group of the crystal is $P2_1/n$ (C_{2h}^5) and there are 4 molecules, 32 atoms, in the unit cell.[36] The point group of the molecule is D_{4d}. The bonding forces within the Se_8 molecule are covalent, whereas the forces holding the molecules in the monoclinic lattice are of the weaker van der Waals type. In this sense, α-monoclinic Se (which we will refer to as α-Se, for convenience) is very similar to orthorhombic sulfur. Orthorhombic S is a molecular crystal composed of S_8 molecules which have the same point-group symmetry as the Se_8 molecule.

Large single crystals of orthorhombic S are relatively easy to grow and are stable under normal ambient conditions. On the other hand, the growth of large single crystals of α-Se is at best difficult, and this structural modification of Se is metastable. Any external force, e.g., temperature, pressure, or organic vapors, tends to promote a conversion to the thermodynamically-stable trigonal modification. As a result of this stability and growth problem, α-Se has not been studied as extensively as has been orthorhombic S. For orthorhombic S, it has been well-established that the observed infrared-active[9,29] and Raman-active[32] fundamental vibrational modes are essentially those of the S_8 molecule. The unit cell of orthorhombic S contains sixteen S_8 molecules; however, there are only two types of molecules, those lying parallel to the (110) plane and those lying perpendicular to that plane.[35] The occurrence of two molecular types manifests itself in a weak crystal-field splitting of the molecular vibrational modes. The splitting is approximately 10 cm^{-1} for two of the infrared-active modes[9] and 4–5 cm^{-1} for the Raman-active modes.[32] The characteristic molecular vibrational frequencies[29] vary from about 100 cm^{-1} to 500 cm^{-1}. The ratio of the crystal field splitting to the average molecular vibrational frequency provides a measure of the ratio of the inter- and intra-molecular bonding forces. By analogy with orthorhombic S, we expect a similar relationship between the spectra of the α-Se crystal and the Se_8 molecule.

* See also Chapter 3.

Infrared Absorption

Figure 4-8 shows the transmission spectrum of a 0.2 mm thick platelet of α-Se in the region of the fundamental absorption bands, 75 cm^{-1} to 300 cm^{-1}, as measured by Lucovsky.[19]   The transmission was measured with a nitrogen-purged grating spectrometer using unpolarized light and operating in a single beam mode.   Three strong absorption bands were observed.   Each band displays an indication of doublet splitting, with the splitting most clearly resolved in the lowest frequency band.   The absorption spectrum shown here for α-Se is similar to that reported by Chantry et al.[9] for orthorhombic S. The sulfur work was done on pellets pressed from powdered orthorhombic S. Chantry et al. also studied the absorption spectrum of α-Se; however, the absorption bands they report coincide with those reported for trigonal Se,[22] indicating that their α-Se crystals underwent a thermal conversion to trigonal Se during the course of their measurements.

The absorption spectrum of the S_8 molecule has been studied in CS_2 solution.[29]   The fundamental bands are essentially the same as those reported for the crystalline material with, of course, the absence of the crystal field splitting.   Lucovsky studied the absorption spectrum of the Se_8 molecule in CS_2 but only observed one strong absorption band at 254 cm^{-1}, rather than

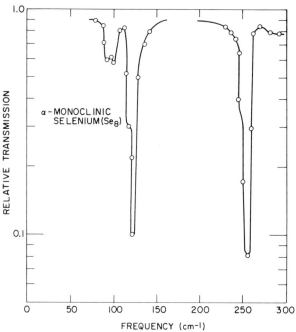

Figure 4-8   Infrared absorption of α-monoclinic Se.[19]

three as had been reported for the α-Se crystal.   He also studied the second order spectrum of α-Se using unpolarized light; those results will not be discussed here.

Raman Scattering

The Raman scattering of α-Se has been reported at 300°K by Lucovsky *et al.*[22] and at 300°K and 4°K by Mooradian and Wright.[24]   The low temperature spectrum of Mooradian and Wright is shown in Figure 4-9, along with their results for trigonal and amorphous Se.   The spectrum of α-Se sharpens considerably as the temperature is decreased from 300°K to 4°K.   Table 4-5 summarizes the vibrational frequencies of the fundamentals as obtained from the experiments performed at 300°K.[22]   The symmetry assignments are

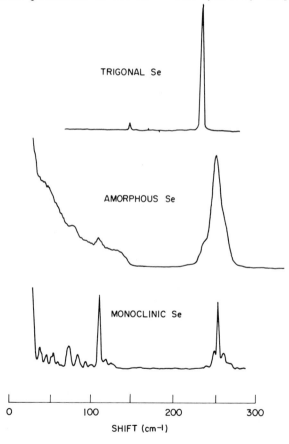

Figure 4-9   Comparison of the Raman spectra of trigonal, α-monoclinic, and amorphous Se.[24]

TABLE 4-5 SYMMETRY ASSIGNMENTS FOR THE
OBSERVED FIRST ORDER INFRARED AND
RAMAN SPECTRA OF Se_8 IN α-MONOCLINIC
Se BASED ON THE ASSIGNMENTS FOR S_8 IN
ORTHORHOMBIC S[a]

$\bar{\nu}$ in cm^{-1}	Spectrum	Assignment
50	Raman	E_2
84	Raman	E_2
92	Infrared	E_1
97	Infrared	E_1
114	Raman	A_1
116	Infrared	B_2
122	Infrared	B_2
128	Raman	E_3
239	Raman	E_3
249	Raman	A_1
254	Infrared	E_1
254	Raman	E_2

[a] Refs. 22 and 24.

based on a comparison with S_8 and are discussed in the next section.   Note here that the Raman-active fundamentals of α-Se are not resolved into doublets as are the corresponding modes of S_8 in the orthorhombic S crystal.

Vibrational Assignments

Comparisons of the infrared and Raman spectra of α-Se with the corresponding spectra of orthorhombic S and the S_8 molecule suggest that the α-Se results can be discussed in terms of vibrational modes of the Se_8 molecule.

The Se_8 molecule (and also the S_8 molecule) is a puckered eight-membered ring having D_{4d} symmetry.   There are eleven fundamental vibrational frequencies: $2A_1 + 3E_2 + 2E_3$ (Raman active), $B_2 + 2E_1$ (infrared active), and B_1 (inactive).   The assignments[22] in Table 4-5 are based on a comparison of the α-Se lines with those of the S_8 molecule.   The ratio of the vibrational frequencies of S_8 and Se_8, $\bar{\nu}(S_8)/\bar{\nu}(Se_8)$, varies between 1.72 and 2.01 with an average value of 1.9.   On the basis of the ratio of the Se and S masses, a scaling factor of 1.58 is expected.   The reported average value of 1.9 indicates a change in the effective molecular force constants that is consistent with the greater polarizability of the Se atom.   The frequency of the inactive B_1 mode of Se_8 is estimated using the average empirical scale factor of 1.9, and it is also consistent with an assignment of the bands observed in the second order infrared absorption spectrum.

For two of the modes, the infrared measurements indicate a doublet splitting of the same relative order as that reported for the corresponding vibrational modes of orthorhombic S.   That is, according to the scaling factor of 1.9, the mode splittings in orthorhombic S are $10 \, \text{cm}^{-1}$ and $12 \, \text{cm}^{-1}$, respectively, whereas the corresponding splittings in α-monoclinic Se are $5 \, \text{cm}^{-1}$ and $6 \, \text{cm}^{-1}$.   A splitting of the Raman-active modes was not reported for α-Se; by comparison with the splittings reported for orthorhombic S, a splitting of at most 2 to $3 \, \text{cm}^{-1}$ would be anticipated.

THE SPECTRA AND STRUCTURE OF AMORPHOUS Se*

Introduction

Prior to the recent infrared[22] and Raman[24] spectroscopy studies on amorphous selenium, a structural model for this material, based for the most part on indirect evidence, had been developed.   The vitreous solid was viewed as a mixture of ring molecules and chain polymers.   By analogy with the bonding forces of the crystalline modifications, the bonding of atoms within the rings and chains was considered to be covalent, and that between these units was considered to be of the weaker van der Waals type.   The presence of long (more than 10^3 atoms) polymer chains was inferred from the high viscosity of Se at its melting point,[14] while the presence of rings was inferred from the partial solubility of amorphous Se in CS_2.[4]   When applied to liquid Se, the polymerization theory of Eisenberg and Tobolsky[13] gave an estimate of the atomic population in Se_8 rings at the melting point of approximately 50% (see Chapter 3).

The infrared absorption spectrum of amorphous Se has been studied quite extensively.[8,31]   However, prior to the recent optical experiments of Lucovsky *et al.*, the fundamental ring and chain vibrational frequencies were not known, so that an interpretation of the spectrum in terms of structural units was not possible.   Nevertheless, Srb and Vasko[31] commented on similarities between the absorption spectra of orthorhombic S and amorphous Se, implying that the major features of each were due to the ring molecules.

Infrared Absorption

Figure 4-10 shows the absorption spectrum of amorphous Se in the spectral range from $50 \, \text{cm}^{-1}$ to $650 \, \text{cm}^{-1}$.[22]   Not included in the figure is a very weak absorption band at $744 \, \text{cm}^{-1}$.   The spectrum in Figure 4-10 is for a cast sample 0.2 mm thick.   The spectra of evaporated samples, quenched at several substrate temperatures, were dominated by the same absorption bands with the relative absorption strengths being essentially independent of

---

* See also Chapter 3.

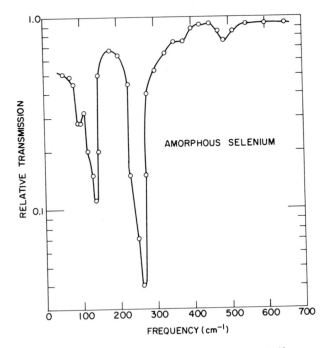

Figure 4-10   Infrared absorption of amorphous Se.[19]

the details of preparation.   The dominant bands occur at 95 cm^{-1}, 135 cm^{-1}, and 254 cm^{-1}.   Weaker structure is evident as two shoulders, at 120 cm^{-1} and 230 cm^{-1}, as a broad band from 280 cm^{-1} to 380 cm^{-1}, and as two weak bands at 490 cm^{-1} and 744 cm^{-1}.

Studies of the temperature dependence[22,30] of the absorption indicate that the bands at 95 cm^{-1}, 135 cm^{-1}, and 254 cm^{-1} are fundamentals, while the bands at 490 cm^{-1} and 744 cm^{-1} are second and third order combination bands, respectively.

Raman Scattering

The Raman spectrum[24] of amorphous Se has been measured at 300°K and 4°K; no sharpening of the spectrum was observed at the lower temperature. The results for 4°K are included in Figure 4-9.   The Raman scattering is dominated by a very strong line at 250 cm^{-1} and a strong line at 112 cm^{-1}. Weaker structure occurs as two bands at 50 cm^{-1} and 80 cm^{-1}, and two shoulders at 135 cm^{-1} and 235 cm^{-1}.   The background scattering increases with decreasing wavenumber from about 120 cm^{-1}, and is characteristic of scattering from noncrystalline solids.

The Molecular Structure of Amorphous Se

We shall first discuss the infrared absorption by developing comparisons between the spectrum of amorphous Se and those of trigonal and α-monoclinic Se. We shall then do the same for the Raman scattering. Finally, we shall summarize the infrared and Raman assignments showing that the dominant features in the reported spectra are due to vibrational modes of the Se_8 molecule.

The absorption spectrum of amorphous Se in the infrared bears a closer resemblance to that of α-Se than it does to that of trigonal Se. Figure 4-11 shows a superposition of the absorption of amorphous Se and α-Se in the region of the fundamental modes. The amorphous Se bands at 95 cm^{-1} and 254 cm^{-1} line up with the E_1 fundamentals of the Se_8 molecule as they are

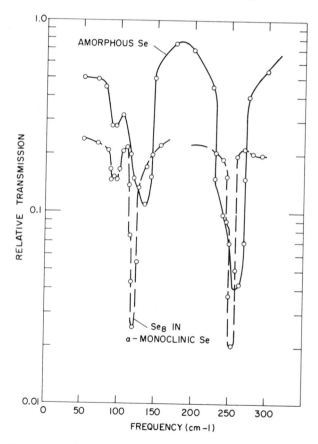

Figure 4-11 Comparison of the infrared spectrum of amorphous Se with that of Se_8 in α-monoclinic Se.[19]

reported for α-Se. The shoulder at 120 cm^{-1} in the amorphous material lines up with the third fundamental of Se_8, the B_2 mode. The bands in amorphous Se are considerably broader than those of the crystalline material, and further, they show no doublet splitting. The only strong band in amorphous Se that cannot be correlated with a fundamental in α-Se is the one at 135 cm^{-1}. This frequency is close to the trigonal Se E mode at 144 cm^{-1}; the occurrence of Raman scattering at about 138 cm^{-1} in amorphous Se prompts us to assign these spectral features near 135–140 cm^{-1} to the chain component of the amorphous solid.

The weaker features in absorption, the broad band from 280 cm^{-1} to 380 cm^{-1} and the bands at 490 cm^{-1} and 744 cm^{-1}, are also coincident with absorption bands reported for Se_8 in α-Se. In addition, the absorption in the vicinity of 490 cm^{-1} in α-Se is second order, as is the amorphous Se band. A shoulder at 230 cm^{-1} on the 254 cm^{-1} fundamental may be related to chain absorption. However, the second order spectrum of α-Se also shows structure near this energy, so that this last assignment is at best speculative.

Figure 4-9 displays an instructive comparison of the low temperature Raman spectra of amorphous Se, trigonal Se, and α-Se.[24] The dominant spectral features of the amorphous Se spectrum, at 250 cm^{-1} and 112 cm^{-1}, line up with the A_1 modes of Se_8 as they occur in α-Se. The weaker, but still resolvable, peaks at 50 cm^{-1} and 80 cm^{-1} in the amorphous Se spectrum can be attributed to E_1 vibrations of the Se_8 molecule, again by comparison with the scattering from α-Se. Based on a comparison with the scattering from trigonal Se, the shoulders in the amorphous Se spectrum at 138 cm^{-1} and 235 cm^{-1} are attributed to scattering from the polymeric chain species.

Table 4-6 summarizes the assignment of the fundamental modes of amor-

TABLE 4-6 FIRST-ORDER SPECTRA FOR AMORPHOUS
Se IN TERMS OF RING AND CHAIN
VIBRATIONS[a]

$\bar{\nu}$ in cm^{-1}	Spectrum	Assignment
50	Raman	Se_8, E_2
80	Raman	Se_8, E_2
95	Infrared	Se_8, E_1
112	Raman	Se_8, A_1
120	Infrared	Se_8, B_2
135	Infrared	Chain, E
138	Raman	Chain, E
235	Raman	Chain, A_1, E
250	Raman	Se_8, A_1, E_2
254	Infrared	Se_8, E_1

[a] Ref. 22.

phous Se in terms of vibrations of the rings and chains of the crystalline modifications. Both the infrared and Raman spectra of amorphous Se are dominated by modes of the Se_8 molecule. Although Srb and Vasko[31] implied this in their comparison of the absorption of orthorhombic S and amorphous Se, the work reported here, which is based on studies of both the infrared and Raman spectra of the glass and of the two crystal forms with their different molecular units, constitutes the first definitive evidence for the presence of Se_8 molecules in amorphous Se. It would be tempting to try to estimate the relative atomic populations in the ring and chain components on the basis of the relative infrared absorption strengths and Raman scattering intensities. This is, at best, difficult because quantitative values for the absorption and scattering cross sections are not available. However, studies of the infrared absorption and Raman scattering from alloys of Se with As, S, and Te,[28] and analyses of equilibria[33,34] in the Se-S and Se-Te systems do give some insight into this question. Estimates of the atomic fraction in the Se_8 rings range from 30% to 50%; the average value of 40% being in good agreement with studies of the preferential solubility of Se_8 in CS_2. Since the average chain length is estimated to be on the order of 10^4 atoms, the ratio of rings to chains is on the order of $10^3 : 1$.

The analysis of the infrared and Raman spectra of α-monoclinic Se indicates, as in the case of S_8 and the orthorhombic S crystal, that the bonding forces within the Se_8 molecule are considerably stronger than the intermolecular forces of the lattice. Therefore, it is expected that the vibrational frequencies of the Se_8 molecule would be relatively insensitive to the matrix in which the molecules are placed. This argument provides the basis for our interpretation of the features of the amorphous Se spectra that we associate with the Se_8 molecule. It has also been shown that the nonphotoconductive absorption in amorphous Se, in the region between 1.7 eV and 2.3 eV, can be attributed to an exciton of the Se_8 molecule.[20] The association of vibrational modes assigned to the polymer chain component of amorphous Se with modes of the trigonal Se crystal is somewhat more difficult. In trigonal Se, the axes of the helical chains are parallel to each other and the chains form an ordered hexagonal array. In amorphous Se, the chain axes are most probably not linear and the chain orientation, with respect to other chains and rings, is assumed to be random. Three of the optical modes of trigonal Se are associated with largely intrachain atomic motions: the two E modes and the A_1 mode. The restoring force and the infrared absorption strength of the A_2 mode are due to interactions with neighboring chains,[10] as discussed previously and illustrated in Figure 4-3. While the infrared and Raman spectra of amorphous Se show structure that can be attributed to the E and A_1 modes of trigonal Se, there is no evidence of the A_2 mode. This is consistent with the structural model presented here for amorphous Se.

Note Added in Proof :   This review was written in July, 1970.   Naturally our understanding of lattice–light interactions in the solid forms of selenium has deepened since that time.   More recent developments can be found in the following two articles:

R. Zallen, In: *Proc. of the Enrico Fermi Summer School on Lattice Dynamics and Intermolecular Forces, Varenna, 1972,* S. Califano, ed., Academic Press, New York, 1973;

G. Lucovsky, In: *Proc. of the Fifth Intl. Conf. on Amorphous and Liquid Semiconductors, Garmisch-Partenkirchen, 1973,* J. Stuke, ed.

REFERENCES

1. Asendorf, R. H., *J. Chem. Phys.* **27**, 11 (1957).
2. Born, M., and Huang, K., *Dynamical Theory of Crystal Lattices*, Oxford University Press, Inc., London, 1954.
3. Born, M., and Huang, K., *ibid.*, p. 87, lines 12 and 13.
4. Briegleb, G., *Z. Phys. Chem.* **A144**, 321 (1929).
5. Burstein, E., *Lattice Dynamics*, ed., R. F. Wallis, Pergamon Press, Inc., London, 1965, p. 315.
6. Burstein, E., *Comments Solid State Phys.* **1**, 202 (1969).
7. Burstein, E., Brodsky, M. H., and Lucovsky, G., *Intern. J. Quantum Chem.* **1S**, 759 (1967).
8. Caldwell, R. S., and Fan, H. Y., *Phys. Rev.* **114**, 664 (1959).
9. Chantry, G. W., Anderson, A., and Gebbie, H. A., *Spectrochimica Acta* **20**, 1223 (1964).
10. Chen, I., and Zallen, R., *Phys. Rev.* **173**, 833 (1968).
11. Cochran, W., *Nature* **191**, 60 (1961).
12. Dick, B. G., and Overhauser, A. W., *Phys. Rev.* **112**, 90 (1958).
13. Eisenberg, A., and Tobolsky, A., *J. Polymer Sci.* **46**, 19 (1960).
14. Gee, G., *Trans. Faraday Soc.* **48**, 515 (1952).
15. Geick, R., Schroder, U., and Stuke, J., *Phys. Status Solidi* **24**, 99 (1967).
16. Huang, K., *Proc. Roy. Soc. (London)* **A208**, 352 (1951).
17. Hulin, M., *Ann. Phys. (Paris)* **8**, 647 (1963).
18. Lax, M. and Burstein, E., *Phys. Rev.* **97**, 39 (1955); Lax, M., *Phys. Rev. Letters* **1**, 131 (1958).
19. Lucovsky, G., in *Physics of Selenium and Tellurium*, ed., W. C. Cooper, Pergamon Press, Inc., Oxford, 1969, p. 255.
20. Lucovsky, G., *Proc. of Tenth Intl. Semiconductor Conf.*, 1970, USAEC, Oak Ridge, p. 799.
21. Lucovsky, G., Keezer, R. C., and Burstein, E., *Solid State Commun.* **5**, 439 (1967).
22. Lucovsky, G., Mooradian, A., Taylor, W., Wright, G. B., and Keezer, R. C., *Solid State Commun.* **5**, 113 (1967).
23. Martin, D. H., *Advan. Phys.* **14**, 39 (1965), NB p. 50.
24. Mooradian, A., and Wright, G. B., in *Physics of Selenium and Tellurium*, ed., W. C. Cooper, Pergamon Press, Inc., Oxford, 1969, p. 269.
25. Nussbaum, A., in *Solid State Physics*, Vol. 18, ed., F. Seitz and D. Turnbull, Academic Press, Inc., New York, 1966, p. 225.

26. Queisser, H. J., and Stuke, J., *Solid State Commun.* **5**, 75 (1967); Queisser, H. J., in *Physics of Selenium and Tellurium*, ed., W. C. Cooper, Pergamon Press, Inc., Oxford, 1969, p. 289.
27. Roberts, G. G., Tutihasi, S., and Keezer, R. C., *Phys. Rev.* **166**, 637 (1967).
28. Schottmiller, J., Tabak, M. D., Lucovsky, G., and Ward, A. T., *J. Non-Crystalline Solids* **4**, 161 (1970).
29. Scott, D. W., McCullough, J. P., and Druse, F. H., *J. Mol. Spectrosc.* **13**, 313 (1964).
30. Siemsen, K. J., *J. Phys. Chem. Solids* **30**, 1897 (1969).
31. Srb, I., and Vasco, A., *Czech. J. Phys.* **B13**, 827 (1963).
32. Ward, A. T., *J. Phys. Chem.* **72**, 744 (1968).
33. Ward, A. T., *J. Phys. Chem.*, **74**, 4110 (1970).
34. Ward, A. T., and Myers, M. B., *J. Phys. Chem.* **73**, 1374 (1969).
35. Warren, B. E., and Burwell, J. T., *J. Chem. Phys.* **3**, 6 (1935).
36. Wyckoff, R. W. G., *Crystal Structures* Vol. 1, Interscience Publishers, New York, 1963.
37. Zallen, R., *Phys. Rev.* **173**, 824 (1968).
38. Zallen, R., and Chen, I., in *Proc. of the Ninth Intl. Conference on the Physics of Semiconductors, Moscow,* "Nauka," Leningrad, 1968, p. 1036.
39. Zallen, R., Lucovsky, G., Taylor, W., Pinczuk, A., and Burstein, E., *Phys. Rev.* **B1**, 4058 (1970).
40. Ziman, J. M., *Electrons and Phonons*, Oxford University Press, Inc., Oxford, 1960, p. 210.

5

Optical and Electrical Properties of Selenium*

Dr. JOSEF STUKE

Professor, Physikalisches Institut der Universitat, Marburg, Germany

INTRODUCTION

In the physics of semiconductors, the element selenium plays a peculiar role, partly because of its influence on the history of this branch of physics and partly because of its properties. Almost exactly 100 years ago, photoconductivity was detected on trigonal selenium by Smith.[1] This modification was termed metallic, because the conduction was found to be electrical. Selenium is one of those materials which was first categorized as a semiconductor due to its small electronic conductivity compared with metals. After the detection of this property, an appreciable number of investigations were carried out and soon, new technical applications resulted. The first photocells were made from trigonal selenium.[2] Later, "selenium-cell" and "photoconductive device" became synonymous. During the study of the

* See also Chapter 4.

behavior of different contact materials, the rectifying effect was observed on trigonal selenium by Braun.[3] The first technically reliable selenium rectifiers were developed by Presser[4] and these soon replaced most other types of power diodes. During the same period selenium photovoltaic cells were invented[5] and developed commercially.[6] In spite of the technical importance and the great number of investigations, however, the understanding of the electronic properties of trigonal selenium was poor. Even less was known about α- and β-monoclinic modifications in 1945 when the first real progress in the knowledge of semiconductors occurred. Following the invention of the transistor in 1948, the semiconductors, germanium and silicon, and a little later the III–V compounds were of such scientific and technical interest that the importance of selenium diminished considerably. This situation parallels the strong decrease of interest in crystal diodes at the end of the nineteenth century when vacuum diodes were found to be more reliable and were commercially preferred. Because the electronic properties of isotropic semiconductors are now fairly well understood, the anisotropic materials, like selenium and tellurium are being studied in order to gain a more general understanding of semiconductors. Therefore, during the last decade, particularly during the last 3–4 years, the amount of research on selenium has grown appreciably. This is probably due to the fact that trigonal selenium is exceptionally promising for the study of anisotropic materials, since its chain lattice has one of the strongest anisotropies of all semiconductors. Its structure is also sufficiently simple so that band structure calculations can be managed without major difficulties. Since selenium can be regarded as intermediate between the isotropic semiconductors and the more complicated organic compounds, its study is also necessary for an understanding of the electronic properties of organic substances. The bonding strength between the chains of trigonal selenium is about an order of magnitude less than the bonding within the chains. Therefore this semiconductor has, at least to some degree, a molecular character and can be regarded as an inorganic polymer. Moreover, the position of selenium in group VI of the periodic table, between sulfur and tellurium, makes it interesting. Sulfur is a semiconductor (or insulator) because of its tendency to form ring structures and to behave as a pronouncedly molecular material. Tellurium, on the other hand has a trigonal chain lattice except for its metallic modification at high pressure. In contrast to selenium, its bonding strength between atoms of different chains is not much less than that between nearest neighbors within the chains. Selenium bridges the gap between tellurium with its essentially normal electronic properties and sulfur, with its molecular modifications. Besides the trigonal crystal, it has the relatively stable α- and β-monoclinic forms with their Se_8 rings, so that it can be used as a simple model substance for more complicated organic materials of the aromatic type.

Despite its unusual mechanical and thermal behavior, the electronic properties of the ring and chain lattices of selenium are of particular interest. Monoclinic selenium has a very low electrical conductivity and is a photoconductor, though its photoconductivity is relatively small. The conductivity of trigonal crystals is larger, but it is still low compared with that of many other semiconductors because the density of free carriers is normally very small. One of the most intriguing problems of trigonal selenium, however, is that the Hall mobility of the carriers is much less than that of typical semiconductors and that it increases exponentially with temperature. The activation energy has been found to be 0.1–0.2 eV. This result is quite different from that found for tellurium, which has a relatively high mobility with a temperature dependence $\mu \sim T^{-3/2}$. An explanation of the anomalous behavior of selenium is one of the major problems concerned with its electrical properties. Other interesting features of trigonal selenium are its large photoconductivity and the extremely slow decay which follows the removal of illumination as well as the strong anisotropy of the optical constants for the polarizations parallel and perpendicular to the chain direction. Moreover, trigonal selenium is optically active. The rotation of the polarization can be correlated with the rotatory sense of the screw-like chains. Since its lattice has three coplanar polar axes and no inversion center, it is, like tellurium, piezoelectric and it has optically active lattice vibrations which are revealed by fairly strong reststrahlen bands in the infrared.

This unique combination of properties is the major reason for the growing interest in this semiconductor during recent years. Another reason is that due to the work of Harrison and Tiller[7] and particularly of Keezer et al.,[8] relatively large crystals of trigonal selenium of a quality sufficient for most investigations became available. Then, sensitive experimental methods were developed so that additional properties could now be measured, e.g., luminescence and magnetoresistance. The use of modulation techniques yielded more detailed information about the electronic band structure, so that trigonal selenium now attracts considerable attention.

We shall concern ourselves here with those electrical and optical properties which are fairly well understood. Data which are still problematic shall be briefly discussed, but not all aspects of the existing lack of clarity can be treated. The most important steps which led to the present knowledge are reported in order to give further insight into the problems. The properties of single crystals of trigonal selenium are emphasized because this modification has been studied much more intensively than the others. Also, these data can, to some extent, be interpreted theoretically. Studies on polycrystalline material, in spite of a large number of publications, are treated only sketchily. Only the most important results and data from research done on the properties of selenium rectifiers are mentioned. The section dealing with the two monoclinic forms is also relatively brief. The reason is that our present knowledge of the properties of the monoclinic crystals is limited compared with that of trigonal selenium.

TRIGONAL SELENIUM

Band Structure

Crystal Structure. The lattice of trigonal selenium consists of screw-like spiral chains (Figure 5-1a) which are arranged in hexagonal symmetry. Therefore, this semiconductor has often been referred to as hexagonal selenium. This is, however, inaccurate, because the chains have only a threefold

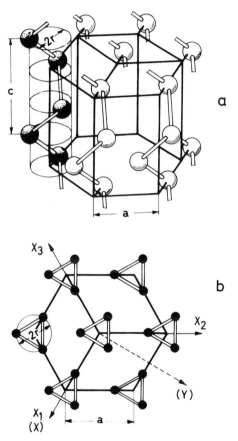

Figure 5-1 (a) Lattice of trigonal selenium. (b) Configuration of the chains (X_1, X_2, and X_3 are coplanar polar axes.)

screw axis (Figure 5-1b). Each atom within a chain has two nearest neighbors, the bond angle being 105.5°. The chains can be right-hand screws (Figure 5-1) or left-hand screws. Trigonal selenium has three coplanar axes and no inversion center:

$$X_1 = (2\bar{1}\bar{1}0), \; X_2 = (\bar{1}2\bar{1}0), \; \text{and} \; X_3 = (\bar{1}\bar{1}20)$$

The y-directions perpendicular to these axes, however, have no polar character.

The distance, c, between equivalent positions of atoms within the chains is 4.95 Å at room temperature, and the distance, a, between the chains is 4.34 Å. The ratio, c/a, is 1.14. Since, for a single turn of the screw, three steps are needed, the unit cell of trigonal selenium contains three atoms. Its lattice belongs to the space group $P3_121$ or $P3_221$ (Schoenflies: D_3), and its space group is D_3^4 for right-hand and D_3^6 for left-hand rotation of the chains.

More details about the lattice parameters at room temperature are given in Figure 5-2 and Table 5-1. The distance, d_1, between two neighboring atoms within the chains is 2.32 Å; it is appreciably smaller than d_2 (3.46 Å) the distance to the four next nearest neighbors which are located in three surrounding chains. The distance to the equivalent atoms in all six adjacent chains $d_3 = a = 4.34$ Å is almost twice d_1. The distance, d_4, to the next nearest neighbors within the chains is of minor importance, because the bonding is covalent and the main contribution is related to the neighboring atoms. Grosse[9] evaluated this bonding energy (Table 5-1) and found a value of 0.94 eV for each of the bonds whereas the values for the adjacent chains are an order of magnitude smaller. Therefore, the overlap of the orbitals is large between the neighboring chain atoms, but much smaller between atoms of different chains. The weak interaction between different chains is, however, probably not only of the van der Waals type. For isomorphic tellurium, the bonding energy between the chains is not much less than the bonding energy within the chains. This relatively strong interaction between atoms of different chains points to an overlap of orbitals. For selenium, this overlap is appreciably smaller, but it should also be of some importance for the bonding between the chains.

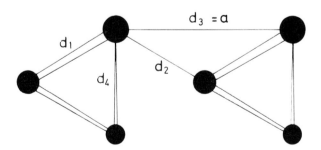

Figure 5-2 Lattice parameters of trigonal selenium.

TABLE 5-1 LATTICE PARAMETERS OF TRIGONAL SELENIUM[9]

Lattice-Parameter	Coordination Number	Distance Å	Binding Energy per Bond eV
d_1	2	2.32	0.94
d_2	4	3.46	0.08
$d_3 = a$	6	4.34	0.02
d_4	2	3.68	0.06

Reciprocal Lattice and Brillouin Zone. The reciprocal lattice of trigonal selenium is hexagonal, because it is created by the three hexagonal Bravais lattices related to the three atoms in the unit cell. Its configuration is shown in Figure 5-3b and the basic parameters of the normal lattice are shown in Figure 5-3a. The length of the reciprocal lattice vectors b_1 and b_2 in the x-y plane is $2/\sqrt{3} \cdot 2\pi/a$, and that of b_3 in the chain direction z is $2\pi/c$. The

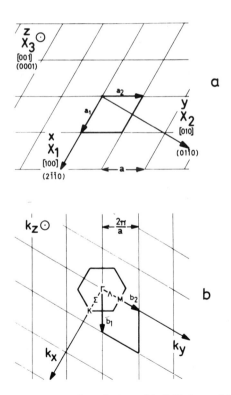

Figure 5-3 (a) Coordinates of the direct lattice. (b) X-Y plane of the reciprocal with the Brillouin zone.

x-y plane of the Brillouin zone is drawn in Figure 5-3b. The three-dimensional zone with points and axes of high symmetry can be seen in Figure 5-4. Their coordinates are found in Table 5-2. The points of the reciprocal lattice have only trigonal symmetry. Therefore, H and H' at the corners of the Brillouin zone are, strictly speaking, not equivalent. However, their energy schemes are degenerate by time reversal so that it is not necessary to distinguish between them, except that the energetic positions of H_4 and H_5

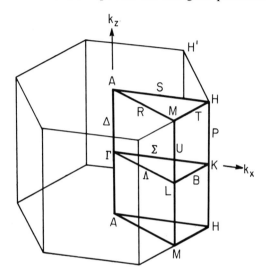

Figure 5-4 Brillouin zone for trigonal selenium with irreducible wedge.

TABLE 5-2 POINTS OF HIGH SYMMETRY IN THE BRILLOUIN ZONE

| | Coordinate in the Brillouin Zone | | |
Symmetry	k_x	k_y	k_z
Γ	0	0	0
K	$\dfrac{4}{3}\dfrac{\pi}{a}$	0	0
L	0	$\dfrac{2}{\sqrt{3}}\dfrac{\pi}{a}$	0
$A(Z)$	0	0	$\dfrac{\pi}{c}$
H	$\dfrac{4}{3}\dfrac{\pi}{a}$	0	$\dfrac{\pi}{c}$
M	0	$\dfrac{2}{\sqrt{3}}\dfrac{\pi}{a}$	$\dfrac{\pi}{c}$

are reversed.   The wedge with the corners AMH/AMH in Figure 5-4 is the irreducible part of the Brillouin zone, because the entire zone can be generated from it by applying the six point group operations plus inversion.   This corresponds to time reversal.

Band Structure Calculations.   The calculation of the band structure of semiconductors together with experimental investigations proved very valuable in obtaining information about the main features of their electronic structures. For isotropic materials like germanium, such calculations were successful relatively early.   The initial research on strongly anisotropic selenium, however, had only qualitative or, at most, a semiquantitative character.   Gaspar[10] discussed the electronic states of the trigonal lattice in terms of the point group of one selenium atom and its two nearest neighbors.   The first realistic calculation is that of Reitz,[11] who like Gaspar, used the tight-binding method (LCAO).   He found that the optically interesting part of the energy spectrum consists of bands which have mainly p-character.   Each band is split into triplets due to the three atoms in the unit cell (Figure 5-5).   Two of them are degenerate at the highly symmetrical points Γ and A, so the bands are actually only doublets.   The six valence electrons of each atom fill the two lower

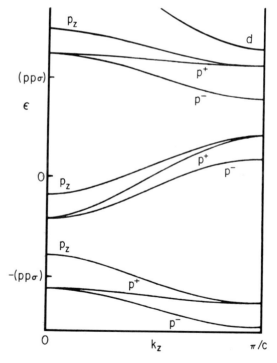

Figure 5-5   Structure of the 9 p-bands along the Δ-axis after Reitz.[11]   The position of the lowest d-band has not been calculated.

triplets, which are the valence bands, whereas the third empty triplet is the conduction band. These basic features have not been changed by more recent calculations. An interesting result of the calculation of Reitz, which was restricted to single chains and, therefore, to the Δ-axis of the Brillouin zone, is that the first direct gap is not located at the center of the zone. This is in contrast with the isotropic semiconductors such as germanium, silicon and the III–V compounds which possess a tetrahedral short range order. It is remarkable that this result is obtained even, if one neglects any interaction between the chains.

A quantitative tight-binding calculation has been performed by Olechna and Knox.[12]. Here the interaction between the chains is not taken into account, so that the band structure is obtained only for the direction parallel to the c-axis. The authors used Hartree-Fock wave functions and included hybridization of the $4p$ bands with the s-bands. The potential is a superposition of three atomic potentials and an additional exchange potential, which is the only adjustable parameter to bring the band gap into agreement with the experimental value. As in the procedure of Behrens[13] and Reitz,[11] the estimation was at first made for a fictional structure with a valence angle of 90°. A transition to the real angle of 105.5° was then performed by a perturbation method. The resulting band structure is shown in Figure 5-6. It is valid only for the c-direction, because the interaction between the chains has been ignored. The gap is, as reported by Reitz, not found in the center of the Brillouin zone, but at point A. Further approximations with a modified LCAO method have been performed by Chen and Das[14] and by Knox and Olechna.[15]

The value of these results, however, is limited, because the calculations made by Treusch and Sandrock[16] and particularly Sandrock[17] proved that it is important to take the interaction between the chains into account. This has been done for tellurium for the first time by Hulin[18] who also used the LCAO method, but in the process included the interaction between the next nearest neighbors in different chains. The quantitative agreement of his results with the experimental data on the size of the gap and the effective masses was not good. The important new result, however, was that the band gap was found at the edge of the Brillouin zone on the P-axis near point H.

The first calculation for trigonal selenium along an entirely closed path of the Brillouin zone was performed by Treusch and Sandrock[16] for the points Γ–Z–H–K, where Z corresponds to A in Figure 5-4, and along the axes Δ–S–P–Σ. They used the Green's-function method (KKR-method). Here the solution $\Psi(r)$ of the Schrödinger equation was obtained by solving an integral equation, which is equivalent to the Schrödinger equation, namely:

$$\Psi K^{(r)} = \int G_K(r', r) \; V(r') \; \Psi(r') \; dr'$$

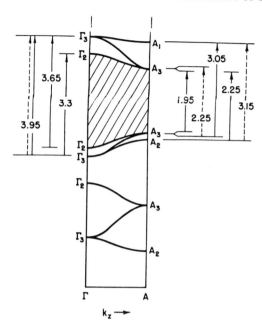

Figure 5-6   Band model along the Δ-axis of Olechna and Knox.[12]   The dotted and the solid arrows correspond to optical transitions for light polarized parallel and perpendicular to the c-axis, respectively.

where $G_K(r', r)$ is the Green's function related to this problem.[16]   The solution of this equation by a variational method gives a secular determinant, which is a function of E and $\mathbf{k}$.   For each value of $\mathbf{k}$, one obtains the corresponding value of E from the roots of the determinant.   The potential was a muffin tin potential consisting of a spherically symmetrical potential $V(r)$ within spheres $(r < \rho = d_{1/2})$ and a constant potential V_0 outside these spheres. V_0 was the only parameter used to fit the band structure to the experimental data.   The result of this calculation can be seen in Figure 5-7. The most important feature of this band structure is the location of the gap near point H at the corner of the Brillouin zone as was already found for tellurium by Hulin.   The best possible fit gave a value of 1 eV for the energy gap.   Since the experimentally found value of the band gap amounts to about 2 eV, the calculated gap and also the higher energy distances are too small.   The structure along the Δ-axis has a qualitative resemblance to the results of the tight-binding calculations (Figure 5-6).   Quantitatively there are, however, some differences, since the bands are more flat here.   They are also relatively flat along the S-axis from $A(Z)$ to H.   This is particularly the case for the lowest conduction band, S_1.   The curvature of the bands is, on the other hand, appreciably stronger from H to K along the p-axis and

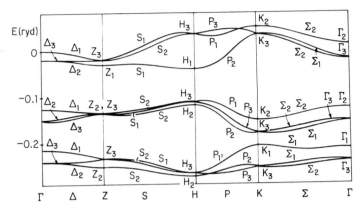

Figure 5-7   Band structure calculated by Treusch and Sandrock.[16]   Point Z corresponds to A in Figure 5-4.

also from K to Γ along the Σ-axis.   This difference of the curvature around H leads to a strong anisotropy of the effective masses for the upper valence band and the lowest conduction band.   Since the curvature perpendicular to the c-axis $[H-A(Z)]$ is less than parallel to the chain direction $(H-K)$, $m_\perp$ is larger than $m_\parallel$.   For the uppermost valence band there are found: $m_\perp = 4m_0$ and $m_\parallel = m_0$, the ratio thus being $m_\perp/m_\parallel = 4$ which seems to be in relatively good accordance with experimental values.   Another interesting result of the calculation of Treusch and Sandrock is that the uncommonly strong pressure dependence of the energy gap of trigonal selenium and its anomalous sign are quite reasonable.   By the application of hydrostatic pressure, it is mainly the distance $a = d_3$ between the chains that is diminished.   The interaction between the chains obviously is significantly involved in this phenomenon.

In order to describe the anisotropic bonding of trigonal selenium, it is questionable to use a muffin tin potential, because the muffin tin spheres fill only 24% of the unit cell for selenium.   The great advantage of this approximation, however, is that the influence of the lattice structure and of the potential can be separated mathematically and this facilitates a quantitative estimation of the influence of spin-orbit coupling.

By the inclusion of spin-orbit interaction, the more general features of the band structure are not changed.   The effect of the spin consists mainly in splitting the energy levels and in changing the selection rules for the optical transitions.   Without spin, all branches of the bands can be doubly occupied by electrons of opposite spin.   The different branches correspond to the irreducible representations of the simple groups.   The inclusion of the spin leads to new branches which can only be occupied by electrons of one spin direction.   They transform according to the representations of the double group originating from the single group by the additional symmetry operation which inverts the spin direction. The relation between the bands with and without spin is shown in Figure 5-8. The conditions at the band gap near point H, particularly for the valence band, are appreciably changed. The degeneracy of H_3 in the spinless case (Figure 5-8a) is lifted now and a splitting into the three terms H_5, H_4 and H_6 results (Figure 5-8b). Moreover, from the nondegenerate H_2-term, a lower H_6-valence band is created. The non-degenerate H_1-conduction band is transferred into a degenerate H_6-band. Also, the H_6 valence bands are degenerate at point H. For the uppermost valence band, the degeneracy is lifted, however, so that the optical properties at the absorption edge are determined by transitions from the nondegenerate H_5- and H_4-terms into the double H_6-conduction band. The significant difference between the band structure with and without spin, therefore, is that in the first case, the uppermost valence band is degenerate while the conduction band is not. With spin, the situation is just the reverse.

In order to compare these results with experimental data, it is of importance that for the optical transitions between the different levels, strong selection rules are valid, as can be seen from Figure 5-9 and Table 5-3. The lowest energy transition is possible only for light polarized perpendicular to the

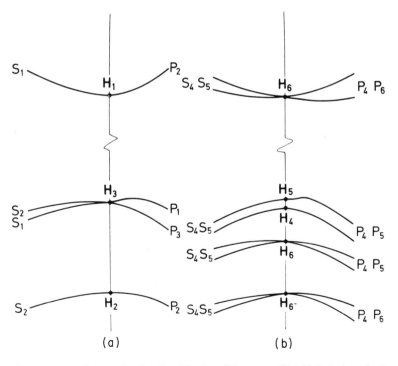

Figure 5-8   Bands around point H: (a) in the spinless case; (b) with inclusion of spin.

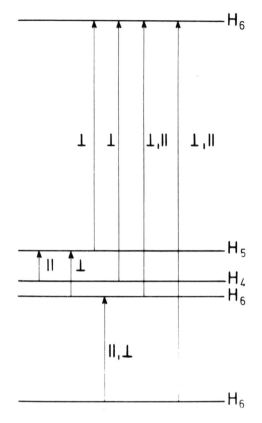

Figure 5-9 Selection rules for optical transitions at point H.

TABLE 5-3   ALLOWED ELECTRICAL DIPOLE TRANSITIONS[16]

Point H (also valid for Γ, Z, K)

	Without Spin				With Spin		
	H_1	H_2	H_3		H_4	H_5	H_6
H_1	0	$\parallel$	$\perp$	H_4	0	$\parallel$	$\perp$
H_2	$\parallel$	0	$\perp$	H_5	$\parallel$	0	$\perp$
H_3	$\perp$	$\perp$	$\parallel, \perp$	H_6	$\perp$	$\perp$	$\parallel, \perp$

		Axis P (also valid for Δ)					
	Without Spin				With Spin		
	P_1	P_2	P_3		P_4	P_5	P_6
P_1	$\parallel$	$\perp$	$\perp$	P_4	$\parallel$	$\perp$	$\perp$
P_2	$\perp$	$\parallel$	$\perp$	P_5	$\perp$	$\parallel$	$\perp$
P_3	$\perp$	$\perp$	$\parallel$	P_6	$\perp$	$\perp$	$\parallel$

c-axis.   This is true if the gap is located exactly at *H* or if it is a little outside of *H* on the *P*-axis $(P_1 \rightarrow P_2)$.   A more detailed discussion of this behavior shall be carried out later in connection with the treatment of the optical properties at the absorption edge.

A quantitative study of the influence of the spin on the band structure has been performed by Kramer and Thomas[19] with a relativistic KKR-method. In Figure 5-10, the resulting energy levels are shown for point *H*.   *A* is the spinless case; the inclusion of spin-orbit interaction leads to *B*.   With the additional inclusion of mass, velocity, and a Darwin term (relativistic term), *C* is obtained.   The spectra *A*, *B*, and *C* are calculated with the potential used by Treusch and Sandrock.   For the spectrum *D*, the constant potential

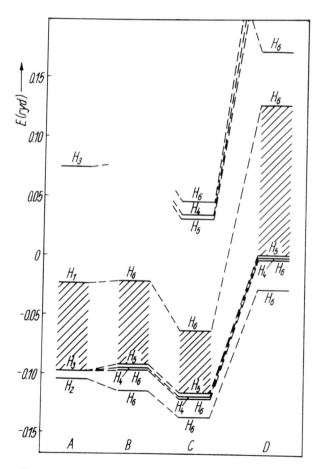

Figure 5-10   Band energies at point $H(A, B, C, D$: see text).

V_0 was made more negative in order to bring the absolute value of the forbidden band width into accordance with the experimental value of about 2 eV. An interesting feature of this alternative is that about 0.4 eV above the minimum of the conduction band, an additional level H_6 occurs. The most important result of the calculation, however, is the evaluation of the spin-orbit splittings. For the H_4–H_5 bands, it amounts to 0.04 ± 0.01 eV. The upper H_6-term originating from H_3 in the spinless case seems to be close to the H_4-level, whereas the distance to the lower H_6-term is about a factor of 10 larger than the splitting of the H_4–H_5 levels. This level, therefore, should have a distance of about 0.4 eV from the edge of the valence band. The accuracy of the H_4–H_5 splitting obtained with the KKR-method is expected to be fairly good since the wave functions are approximated very well in the neighborhood of the atomic sites, where the muffin tin potential is appropriate.

The most comprehensive and realistic band structure computation of trigonal selenium has been performed by Sandrock[17] using a pseudopotential method. He not only calculated the band structure of the whole irreducible wedge of the Brillouin zone, but also the distribution of the oscillator strength for a mesh of 48 points of this wedge. From a combination of both quantities, the optical spectra were then directly calculated. The pseudopotential form factors were taken from the values obtained for ZnSe by Cohen and Bergstresser[20] and adjusted for larger distances to a model potential after Animalu and Heine.[21] Spin-orbit splitting has not been taken into account. The result of the analysis can be seen in Figure 5-11 where a and b show the energy bands in the plane $\mathbf{k}_z = 0$ and $\mathbf{k}_z = \pi/c$, respectively. An important result is that in the first plane, the gap is generally much broader than in the plane $\mathbf{k}_z = \pi/c$ at the top of the Brillouin zone. Moreover, it is significant to note that for the vertical axes (Figure 5-11c), the gap decreases along the P-axis from K to H and from L to M in a similar way. The gap at M obviously is slightly smaller than at H. Thus, the smallest separation between the valence and the conduction band is found in the neighborhood of H, but it seems to occur on the T-axis between H and M or on the P-axis. Since halfway between H and M the selection rules for optical transitions are still approximately the same as for H, this result is in accordance with the polarization dependence of the absorption edge. Comparing the energy bands with the results of the KKR-calculation, one finds qualitative similarities. However, all band widths and interband energy differences have nearly doubled, so that the energy of the smallest gap is in better accordance with the experimental data. This band model indicates, for the first time, that below the direct edge there are indirect transitions along the T-axis and also between M and A.

In cubic semiconductors, it is a reasonable approximation to assume the interband oscillator strength M to be independent of the wave vector $\mathbf{k}$.

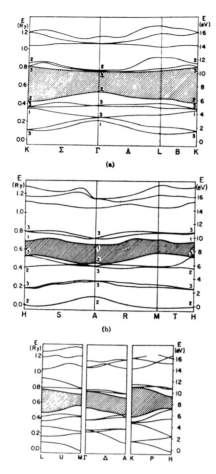

Figure 5-11   Bandmodel of Sandrock:[17] (a) points and axes with $k_z = 0$, (b) points and axes with $k_z = \pi/c$, (c) connecting vertical axes; the numbers denote the representations.

However, this is not possible for trigonal selenium, because, due to the selection rules, either the oscillator strength $M_{\parallel}$ or $M_{\perp}$ vanishes at certain symmetrical points. Therefore, Sandrock calculated, in addition to the energy bands, the size of the oscillator strength over the irreducible wedge. It turns out to vary quite rapidly between different points of the Brillouin zone. This must be taken into account when a comparison with the assignments of the optical spectra of some maxima are made with respect to certain transitions between energy bands. A detailed comparison of the optical spectra calculated by Sandrock with the experimental results shall be discussed in the following section on the optical constants of trigonal selenium.

Optical Properties

Brief Historical Review. The first optical measurements on single crystals of trigonal selenium were made in 1917 by Skinner[22] with the result that the absorption index k has a small maximum at about 2.1 eV for polarization of light perpendicular to the c-axis ($E \perp c$) which disappears for $E\|c$. The spectrum near this maximum and its temperature dependence have been measured by Gilleo[23] and by Stuke[24] for polycrystalline thin films. Its polarization dependence has been studied in single crystals in more detail by Gobrecht and Tausend.[25] The latter investigators found that the slope of the absorption edge is appreciably different for both polarizations. The curves cross because the value of the absorption coefficient K is, at high energy, larger for $E \perp c$ than for $E\|c$, whereas in the tail, at lower energies, it is reversed. Here also, the refractive index $n_{\parallel}$ turns out to be greater than $n_{\perp}$ by a factor of 1.3. Gobrecht and Tausend also studied the optical properties beyond the absorption edge in the near infrared and found some polarization dependent absorption bands which had partly been observed by Caldwell and Fan.[26] These bands later proved to be due to lattice oscillations. The influence of impurities on the spectrum in the near infrared has been investigated by Eckart and Henrion.[27] From the structure of the absorption edge these workers came to the conclusion that the edge of trigonal selenium should be due to indirect transitions for both polarizations.

The first optical measurements of the fundamental absorption on single crystals of trigonal selenium over a larger energy range have been carried out by Stuke and Keller.[28] In the reflectivity spectrum, pronounced maxima were found at 1.95 eV for $E \perp c$ and 3.1 eV for $E\|c$: these proved to be due to excitons, after the measurements of Tutihasi and Chen[29] at 20° K. These researchers also carried out the first Kramers-Kronig analysis yielding the spectra of ε_1 and ε_2, the real and the imaginary part of the dielectric constant, respectively, for the energy region up to 6 eV. The spectra show a strong dependence on the polarization, not only for the absolute values of ε_1 and ε_2, but also for the location of some maxima.   Reflectivity measurements up to 9 eV made by Mohler, Stuke and Zimmerer[30] indicate that the spectrum of trigonal selenium has a second absorption edge at about 6.5 eV, created by transitions from the lower triplet of the valence bands into the lowest conduction band.   Also in the region above 6.5 eV, they found that the spectra had a strong polarization dependence.   The investigations of Leiga[31] verified these findings.

The temperature of the absorption edge has also been studied in detail. Henrion[32] found that for both polarizations, K decreases exponentially and the slope of the curves decreases with increasing temperature approximately after the Urbach rule.   Roberts, Tutihasi, and Keezer[33] obtained the same result for perpendicular polarization and correlated it to the broadening of the exciton at the edge and to direct transitions.   In the curve for $E \| c$, kinks are observed at low temperature indicating that for this polarization indirect transitions have to be assumed.   Recently, Fischer[34] studied this structure in more detail at low temperature.   He was able to interpret not only the pronounced structure in the absorption curve, but also the numerous peaks of the luminescence spectrum by indirect excitons, taking into account the known energy spectrum of the photons.

Optical Constants in the Fundamental Absorption.   If the reflectivity, R, is known over the entire energy range, the spectra of ε_1, the real and imaginary part of the dielectric constant can be evaluated by a Kramer-Kronig analysis. For trigonal selenium with its strongly anisotropic properties, this has to be done for both polarizations of light, $E \| c$ and $E \perp c$.   Its reflectivity spectra are known up to an energy of about 14 eV, which is sufficiently high for obtaining reasonable accuracy using KK-analysis.   In Figure 5-12, reflectivity curves are shown which are a combination of the results of Tutihasi and Chen[29] at 20°K for energies up to 6eV, Mohler, Stuke and Zimmerer[30] at 90°K up to 9 eV, and those of Leiga[31] measured at about 300°K up to 14 eV.   The absolute values have been fitted together in the region where the results overlap.   This somewhat questionable procedure has been chosen in order to obtain single curves covering the whole energy range.   It may be justified, because the temperature mainly influences the sharpness of the structures, but does not change much the overall absolute value of the reflectivity.   Therefore, these curves give a fairly good picture of the occurring structure and its polarization dependence.   The corresponding ε_1 and ε_2 spectra are shown in Figure 5-13.   Since the imaginary part of the dielectric constant ε_2 is directly related to the joint density of states of the band-to-band direct transitions and to the oscillator-strength, the ε_2-spectrum deserves the most attention.   The structure observed in the spectrum of this quantity is

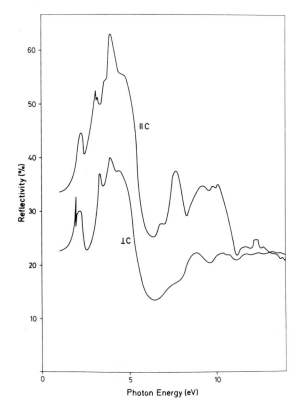

Figure 5-12   Reflectivity spectra of trigonal selenium for polarization of light parallel and perpendicular to the c-axis.[29,30,31]

tabulated in Table 5-4 which contains the peak positions and their polarization dependence.   The spectra for both polarizations are clearly divided into two main parts.   The first is the energy region between the absorption edge near 2 eV and the minimum near 6.5 eV with a number of pronounced maxima for both polarizations.   The absolute value of ε_2 is considerably larger in this region for $E\|c$ than for $E \perp c$.   The second is the energy region above 6.5 eV where a second group of maxima is observed which are also polarization dependent.

For the assignment of these maxima to the band structure of trigonal selenium, the calculation of the optical constants by Sandrock is of particular importance because the general features of the experimental spectra are fairly well reproduced by his results.   Only the position of the maxima are not exactly the same (Figure 5-14).   This difference can easily be understood because the calculated band structure does not have that numerical accuracy which is needed for a more exact agreement.   In the following discussion,

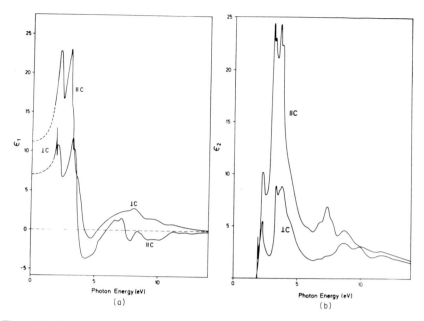

Figure 5-13   Spectra of (a) the real and (b) the imaginary part of the dielectric constant for both polarizations of light.

TABLE 5-4   ENERGY AND POLARIZATION OF OPTICAL TRANSITIONS

Peak Energy	Polarization	Character	Assignment
1.97	$\perp$	Exciton	H_5-H_6
2.01	$\perp$	Exciton	H_4-H_6
2.20	$\perp, \parallel$	Exciton	H_6-H_6
2.28	$\perp$		
2.33	$\parallel$		
3.09	$\parallel$	Exciton	
3.19	$\parallel$		
3.34	$\perp$		
3.54	$\parallel$		
3.62	$\perp$		
3.76	$\parallel$		
3.82	$\perp$		
4.30	$\perp$		
4.50	$\parallel$		
6.80	$\parallel, \perp$		Lower valence band triplet—conduction band
7.4	$\parallel$		Upper valence band— higher conduction band
8.6	$\perp$		
8.8	$\parallel$		
10.1	$\perp$		
11.6	$\parallel$		

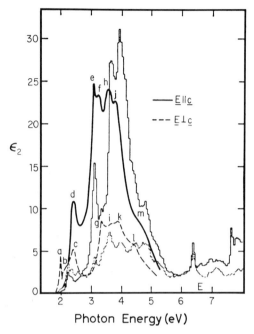

Figure 5-14 Comparison of experimental ε_2 curves[29] and calculated histograms[17] for $E \parallel c$ (solid) and $E \perp c$ (dashed).

therefore, the assignment follows mainly the proposal of Sandrock.   In those cases where the influence of spin-orbit interaction is relevant, the band models of Treusch and Sandrock and of Kramer and Thomas are taken into account. Moreover, the experimental results of modulation techniques as well as pressure and temperature dependences of some structures are considered.

The first sharp peak at 1.97 eV and also the next small one at 2.01 eV are due to excitons originating from the uppermost valence bands, H_5 and H_4, split by spin-orbit interaction (Figure 5-10).   Electroreflectance measurements by Weiser and Stuke[35] indicate that the distance between both peaks amounts to about 40 MeV in good agreement with the calculated value.[19] Both excitons are observed only for $E \perp c$ in accordance with the selection rules of Table 5-3 for point H or a region near to it.   Also, the sharp peak at 3.1 eV which occurs only for $E \parallel c$ is assigned to an exciton because of its strong temperature dependence.   Furthermore, a weak excitonic structure is observed at 2.20 eV by electroreflectance[35] and piezoreflectance.[36]   Since Sandrock has only considered direct band-to-band transitions in the one electron approximation, these excitons do not occur in his results.   The same is true for indirect transitions.

Besides the excitons, a number of further maxima are observed for both polarizations which are interpreted by band-band transitions. Pronounced maxima occur at 2.28 eV for $E \perp c$ and at 2.33 eV for $E \| c$. With modulation techniques, both of these structures are found at a little higher energy. The energy difference, however, is again about 50 MeV. Whereas most experimenters assign these relatively broad maxima to point H of the Brillouin zone, like the excitonic structure, the calculations of Sandrock indicate that they are not due to a critical point, but are created by a relatively large region of the Brillouin zone. According to his results, transitions near point H and also along the Δ-axis are not responsible for these maxima. The only safe conclusion which can be drawn about the origin of these peaks is that they are created by transitions from the highest valence band to the lowest conduction band. The main contribution obviously comes from the part of the Brillouin zone with $\pi/2c \leq k_z \leq \pi/c$. Also, the experimental results of the modulation techniques confirm that the peaks are not created by transitions at critical points. Although they are fairly intense in the ε_2-spectrum, they are observed only very weakly in the electro-reflectance and piezoreflectance spectra. This also indicates that a broader region of the Brillouin zone is involved here.

The exciton at 3.09 eV as well as the maxima at 3.19 eV for $E \| c$ and at 3.34 eV for $E \perp c$ have been attributed by Tutihasi and Chen and also by Weiser and Stuke to transitions at point $A(Z)$. The main reason for this assignment is that the selection rules of this symmetry are fulfilled. Moreover the temperature shift of these peaks has a different sign than that of the exciton near 2 eV (Figure 5-28), indicating that bands with a different temperature shift are involved. Sandrock comes, however, to a different conclusion. According to his computations, these structures are related to transitions from a lower valence band to the lowest conduction band at, or near, point H. Here too, the selection rules seem to be fulfilled.

The maxima between 3.54 and 3.82 eV for both polarizations were assigned by Tutihasi and Chen to Γ, the center of the Brillouin zone, whereas the shoulder at 4.30 eV for $E \perp c$ and at 4.5 eV for $E \| c$ were attributed to point K. In the band structure of Sandrock, the peaks below 4 eV are not assigned to transitions well-localized in the zone. The shoulder above 4 eV seems to be associated with the central part of the Brillouin zone. That is a notable result because the low energy structure is, to a large extent, arising from the regions near the hexagonal zone faces.

Of particular interest is the spectrum of the optical constants at higher energy; that is, at energies above the minimum which is observed for both polarizations at 6.5 eV in the ε_2-curves (Figure 5-13). This minimum is obviously created by a second absorption edge., which has been ascribed by Mohler, Stuke, and Zimmerer to transitions from the lower triplet of the valence band into the lowest conduction band. As can be seen from Figure

5-7, a gap between both valence band triplets exists, which leads to the strong decay of the reflectivity and of ε_2 above 4 eV and the minimum near 6 eV. The band structure of Sandrock gives basically the same result. At the energy where these transitions from the lower valence band triplet set in, a sharp peak is obtained in the calculated ε_2-spectrum for both polarizations (Figure 5-15). In the experimental curve, however, this structure is revealed only by a small maximum at 6.8 eV. The reduced size of this peak may be due to the fact that it falls into the relatively broad minimum at 6.5 eV. A strongly polarized peak occurs for $E\|c$ at 7.5 eV. Mohler, Stuke and Zimmerer attributed this pronounced maximum to transitions along the Δ-axis from the lower valence band triplet into the conduction band, because the selection rule is fulfilled here. The interpretation of Sandrock, however, claims that it is probably related to transitions from the uppermost valence band into higher conduction bands. Transitions of this type set in at approximately the same energy over the entire $k_z = 0$ region of the Brillouin zone with a large oscillator strength. Thereby, a third absorption edge is created at 7.5 eV, so that trigonal selenium has the interesting feature of three pronounced edges, namely at about 2, 6.5 and 7.6 eV. The transitions

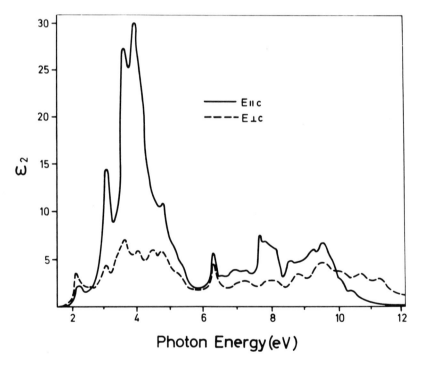

Figure 5-15   Calculated ε_2-spectrum for both polarization of light.[17]

from the upper valence band triplet into the lowest conduction band cover the energy region between 2 and 6 eV. According to Sandrock,[17] the transitions from the lower valence band triplet into the lowest conduction band start at 6.8 eV and carry up to about 10 eV; thus, they cover a similar energy range. The transitions from the upper valence band triplet into the higher conduction bands start at 7.5 eV and predominate at higher energies. It is remarkable that the experimentally found crossing of the curves for both polarizations (Figure 5-13) is also revealed by the calculated curves, indicating that Sandrock's band structure is indeed very realistic. Because of the superposition of the second and third absorption processes in the energy region above the minimum, a more detailed interpretation of all peaks is very difficult and has not been successful so far.

The Absorption Edge. Besides the fundamental absorption spectrum, the properties of the absorption edge near 2 eV have been studied in detail. It is now generally agreed that the smallest direct gap is located near point H at the edge of the Brillouin zone. According to the band model of Sandrock below this edge indirect transitions along the T-axis and also between A and M are to be expected. The following experimental results indicate that the smallest band gap of trigonal selenium is in contrast to tellurium, where a direct gap is found.

Temperature Dependence of the Absorption Edge. The influence of temperature on the absorption edge has been studied in detail by Henrion.[32] He found, for both polarizations of light, a behavior which can approximately be described by the Urbach rule (Figure 5-16). The absorption coefficient increases exponentially with energy and the slope of the curves are larger for $E \perp c$ than for $E \| c$. An interesting feature of the curves is their crossing due to their different slopes. With decreasing temperature, the slopes of both curves increase appreciably. It is for temperatures above $100°K$ that the curves approximate proportionality to $1/T$, as is to be expected according to the Urbach rule. At lower temperatures, however, the variation is considerably smaller.

Measurements down to smaller values of the absorption coefficient by Roberts, Tutihasi, and Keezer[33] showed that the temperature dependence of the edge for $E \perp c$ can be described by the Urbach rule (Figure 5-17a). These workers interpret the exponential variation of K by the long wavelength tail of the exciton near 2 eV. Due to the interaction with phonons, the exciton is strongly attenuated and broadened and the temperature dependence of the slope results. For $E \| c$, however, at low temperatures, kinks in the absorption curve are found (Figure 5-17b) which are typical of indirect transitions. From this structure, the value of the indirect gap was estimated to

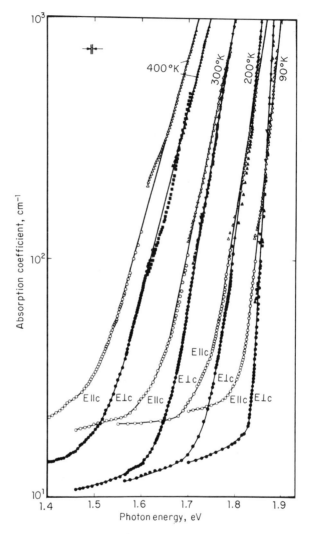

Figure 5-16 Temperature dependence of the absorption edges for $E \parallel c$ and $E \perp c$.[32]

lie near 1.8 eV. This result has been improved by the recent investigations of Fischer.[34] His measurements at 4°K have clearly resolved the phonon structure of the indirect transitions (Figure 5-18). The details of this curve can be interpreted by the assumption of indirect excitons, taking into account the phonon energies known from infrared and Raman spectroscopy. Moreover, with the same phonon energies, the pronounced structure of the luminescence spectrum can be understood (Figure 5-23). This result confirms the

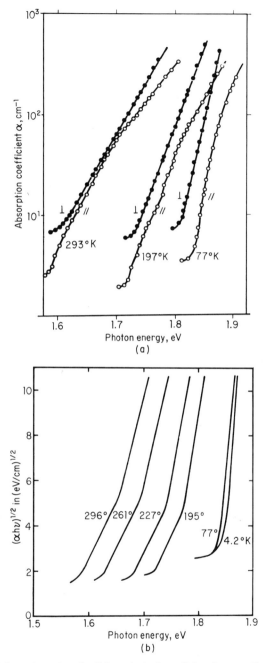

Figure 5-17 (a) Absorption edges for light polarized parallel and perpendicular to the c-axis. (b) Square root of the absorption coefficient as a function of energy for $E \parallel c$.[33]

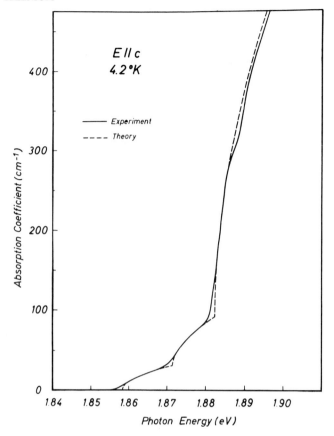

Figure 5-18 Structure of the indirect absorption edge at 4°K for $E \parallel c$.[34] Solid line: experimental results; dashed line: calculated curve.

idea that the smallest gap of trigonal selenium is indirect. The exact band gap value, however, cannot be obtained because the binding energy of the exciton involved is still unknown. Very probably, E_g lies in the vicinity of 1.85 eV. Obviously, the optical transitions across the indirect edge are not strongly polarized. The phonon structure is found not only for $E \parallel c$, but also for the long wavelength tail of the $E \perp c$-curve (see Figure 5-25b). Because of the proximity of the indirect and the direct edges, the two branches of the curve related to both edges cannot be clearly distinguished at high temperatures.

The work of Fischer indicates that the indirect gap has the same anomalous temperature dependence as the direct one. With increasing temperature, the size of the gap increases and has a temperature coefficient:

$$\alpha = \frac{dE_g}{dT} = 1.6 \times 10^{-4}\ \text{eV}/°\text{K}$$

A similar positive value of α is, according to the temperature dependence of electroreflectance, also found for the direct gap (see Figure 5-28). This similarity indicates that the energy bands involved in both gaps are located in the same part of the Brillouin zone. In fact, following the Sandrock model, the direct gap is found near H, the indirect one not far from H along the T-axis $(H\text{-}M)$ and also between M and A. Since these regions are located in the upper face of the Brillouin zone with $\mathbf{k}_z \approx \pi/c$, the positive temperature coefficient seems to be related to that part of the zone.

Pressure Dependence of the Absorption Edge. By applying hydrostatic pressure to trigonal crystals, the energy gap becomes narrower as can be seen from the results of Krisciunas, Mikalkevicius and Shileika[37] shown in Figure 5-19. The pressure coefficient $\beta = dE_g/dp$ therefore, is negative, and it amounts to $\beta = -2.8 \times 10^{-5}$ eV/atm. This value is obtained from the extrapolations of the straight lines in Figure 5-19 to an absorption

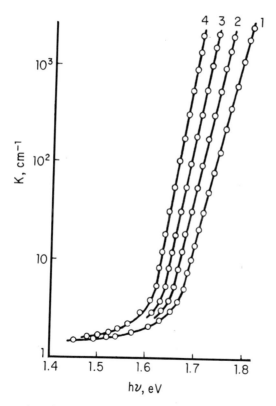

Figure 5-19   Pressure dependence of the absorption edge for $E \perp c$.[37] (1) 1 atm; (2) 1640 atm; (3) 3060 atm; (4) 4430 atm.

coefficient of 10^5 cm^{-1}. It is identical, within experimental error, for both polarizations of light. The negative sign of β is in qualitative agreement with the positive value of the temperature coefficient, since an increase of pressure corresponds to a decrease of temperature. It is interesting to note that by the application of hydrostatic pressure, not only is the absorption edge shifted to lower energy, but the slope of the edge increases in proportion to the pressure.

For uniaxial deformation, a different result has been obtained by Krisciunas and Mikalkevicius.[38]   Here the edge is shifted parallel (Figure 5-20),

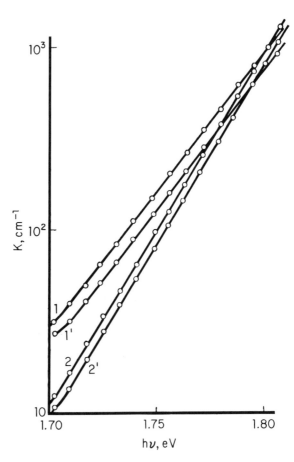

Figure 5-20   Shift of absorption edge of trigonal selenium by uniaxial stress perpendicular to the c-axis: (1 and 2) curves for $E \parallel c$ and $E \perp c$, respectively, under normal conditions; (1′ and 2′) under a stress of 120 atm.[38]

and the pressure coefficient depends on the orientation of the pressure as well as on the polarization of the light:

$$\beta_\perp^\| = -2.4 \times 10^{-5} \quad \text{and} \quad \beta_\perp^\perp = -1.1 \times 10^{-5} \text{ eV/atm}$$

where the lower indices indicates the orientation of the deformation and the upper indices give the polarization of the light relative to the c-axis. For uniaxial pressure parallel to the c-axis, the corresponding values are:

$$\beta_\|^\| = +(0.6-0.7) \times 10^5 \text{ eV/atm}$$

which is equal to $\beta_\perp^\|$, within experimental error. The positive value of $\beta_\|$ corresponds to the negative expansion coefficient along the c-axis. Detailed information about the pressure variation of the direct gap has been obtained recently by Tuomi,[36] using a combination of modulation technique and pressure. For uniaxial compression parallel to the c-axis, the pressure coefficient of the exciton at the direct gap is: $\beta_\|^\perp = +2.8 \times 10^{-5}$ eV/atm. Uniaxial pressure perpendicular to the c-axis is found to be $\beta_\perp^\perp = -6.7 \times 10^5$ V/atm. If one calculates the hydrostatic coefficient β_h with the relation: $\beta_h = \beta_\|^\perp + 2\beta_\perp^\perp$, one obtains: $\beta_h = -11 \times 10^{-4}$ eV/atm, which is about a factor of five larger than the experimental coefficient (Figure 5-19). This discrepancy is not yet understood. It may be partly due to a strong non-linearity of the pressure dependence in the case of compression perpendicular to the c-axis.

The behavior of the absorption edge of trigonal selenium can be summarized in the following way. The smallest band gap is indirect with an energy near 1.85 eV. Optical transitions seem to be possible for $E\|c$ as well as for $E \perp c$. The direct gap amounts to about 2.05 eV, the difference being 0.2 eV. These transitions are allowed only for $E \perp c$, and an exciton is created for this polarization. The absorption edge, therefore, is for $E \perp c$, a combination of the long wavelength tail of the low energy exciton and of indirect transitions. For $E\|c$, the edge seems to be due to forbidden transitions at the direct edge and the phonon-activated indirect transitions at a little lower energy.

Electronic Absorption below the Absorption Edge. At an energy lower than that of the indirect gap, a great variety of absorption maxima are observed. These are, however, mostly due to absorption processes where phonons are involved. It has been reported by Kessler and Sutter[39] that as in the inter-valence band transitions observed for trigonal tellurium at 0.11 eV, p-bands were observed in trigonal selenium near 0.3 eV. This result has, however, been questioned by other investigators.[40] From the calculations of Thomas and Kramer as well as by the modulated measurements, the energy distance between both upper-most valence bands turns out, for trigonal selenium, to be about 40 MeV.   For tellurium, a value of 110 MeV is obtained in accordance with the experimentally found position of the intervalence band transitions for this material.   Therefore, the corresponding optical absorption for selenium is not to be expected near 0.3 eV, but at a much lower energy. So far, intervalence band transitions have not been observed experimentally by absorption measurements, but by the photon emittance measurements discussed in the following section.

Emission of Photons: Luminescence.   For a detailed knowledge of the band structure near the band gap, the absorption measurements at the absorption edge and the investigation of photon emission are of interest.   The luminescence of trigonal selenium is weak compared with that of isomorphic tellurium.   As a result, it was found by Queisser and Stuke[41] at 20°K, but only by excitation with a He/Ne-laser.   More detailed measurements by Fischer, Queisser, and Zetsche[42] and by Kühler[43] at a lower temperature recently revealed a number of significant results.

Whereas for tellurium a single maximum is obtained at the band gap, the spectrum of trigonal selenium is much more complicated (Figure 5-21).   It consists mainly of six pronounced maxima near 1.82 eV and some weak

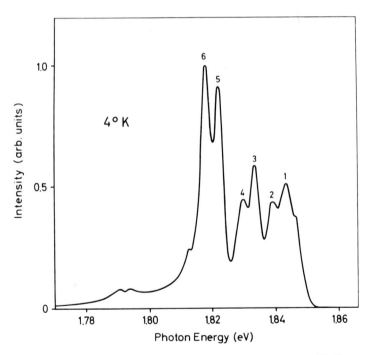

Figure 5-21   Luminescence spectrum of trigonal selenium at 4°K.[43]

phonon replica at lower energies.   The energy distances between the peaks
6–5, 4–5, and between 2–1 are approximately equal and amount to about
5 MeV.   Another remarkable feature is that the relative heights of the peaks
6, 4, and 2 and that of the maxima 5, 3, and 1 vary with temperature (Figure
5-22).   At 2°K the group 6–4–2 predominates, whereas at 20°K only the peaks
5 and 3 are obtained.   The emission peaks 6–4–2 have, therefore, a much
stronger temperature dependence than the other group.   Moreover, the
dependence of both groups on the crystal quality is different.   Whereas peak
6 can be observed over the whole surface of the crystals, peak 5 is only found
at some spots and it is very sensitive to slight plastic deformation and to the
laser intensity.   Since lattice defects are probably created by the laser light,
the dependence of peak 5 on the excitation time can also be ascribed to the
quality of the crystals.

    The partition of the spectrum into two groups of peaks with approximately
equal energy distances is, of course, of great theoretical importance.   Fischer[34]
found an interesting correlation between the structure of the indirect absorption edge and the luminescence spectrum (Figure 5-23).   If one assumes that

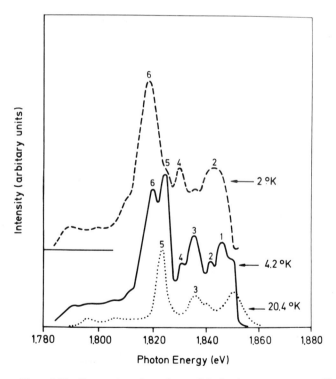

Figure 5-22   Temperature dependence of the luminescence spectrum.[43]

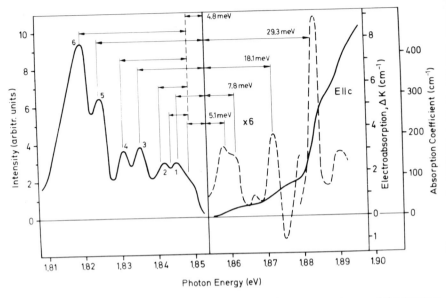

Figure 5-23   Correlation between the maxima of the luminescence spectrum and the structure of the indirect absorption edge.[34]   The dashed curve is the corresponding electroabsorption spectrum.

an indirect exciton exists with an energy of 1.853 eV, the energy distance of the structure of the absorption curve as well as that of the maxima 5–3–1 from this energy can be directly related to phonon energies known from infrared absorption and from Raman measurements.   This correlation is even more pronounced if, instead of the absorption curve, the electroabsorption spectrum is plotted (dashed curve).   The luminescence group 6–4–2 obviously is created by the same phonons, but it originates 4.8 MeV below the position of the indirect exciton.   The equal distance between the peaks of both groups, therefore, is plausible.   Fischer has explained this behavior by the following model.   The group 5–3–1 is attributed to indirect transitions from the exciton state into the uppermost valence band, whereas the group 6–4–2, with lower energy, is assigned to transitions from the exciton level into an acceptor state 4.8 MeV above the valence band.   This relatively simple model explains the energy positions of the several peaks very well. The sensitivity of the group 5–3–1 to the crystal quality, however, is difficult to understand.   Therefore, Kühler[43] proposed a model where the 6–4–2 group is interpreted in terms of transitions from the conduction band into an acceptor state 29.0 MeV above the valence band.   The group 5–3–1 is assigned to transitions from a donor state 24 MeV below the conduction band into the valence band.   The sensitivity of the 5–3–1 group to lattice defects can now be understood.   It is due to a change in the concentration of donors, while the distribution of the acceptors is assumed to be more homogeneous over the crystal.   The relative shift of 4.8 MeV, on the other hand, is only a differential effect of the donor and acceptor depths and is of minor importance.   It may be possible to explain this feature by the additional assumption that the absorption edge is caused by an indirect exciton with a binding energy of the same value as the donor depth.   This model, however, is also questionable because the exciton should be revealed in the luminescence spectrum.   Further investigations are necessary to solve these problems.

Photoemittance in the Infrared.   Emission of photons has also been studied in the infrared by Kautsch[44] and by Blätte and Kautsch[45].   With this method, trigonal crystals are heated to about 100°C and the thermal photon emission at this temperature is compared with the emission of a black body at 30°C.   The results of these difficult measurements are shown in Figure 5-24. In the energy region between 45 and 225 MeV, a large number of maxima are found between 50 and 100 MeV which can be interpreted by phonon processes.

The particularly strong structure somewhat above 50 MeV is assigned to two phonon processes involving the optical phonons E and A_2 with energies 29.0 and 29.5 MeV.   The strongly polarized peak at 48.4 MeV, however, seems to be electronic in nature.   It is tentatively assigned to transitions from the upper valence band H_5 into the upper H_6-term.   For this transition the experimentally found polarization is in line with the selection rule (Figure 5-9).   The transition from H_5 to H_4, which has a smaller energy, probably

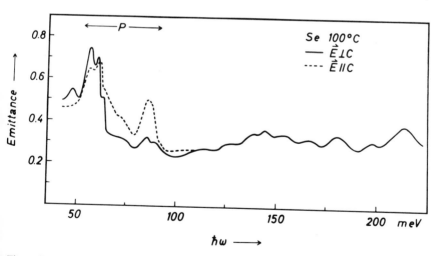

Figure 5-24   Spectra of the thermal photoemittance in the infrared for both polarizations of light.[45]

lies below the investigated energy range.   The origin of the numerous peaks above 100 MeV is not yet clear.   They seem to be due, in part, to electronic processes, because their size is markedly different for various crystals.   This points to the influence of local electronic levels with a distance of about 0.1 eV from the valence band.   About the character of these levels, nothing is presently known.

Modulation Techniques: Electroabsorption.   The   Franz-Keldysh   effect   has been observed for trigonal selenium by Stuke and Weiser[43] in the temperature range between 80 and 200°K and has recently been studied by Fischer[34] at low temperatures.   The change in the absorption coefficient, ΔK, due to the electric field, F, is found to be approximately proportional to F^2 for both polarizations of light and for both orientations of the electric field (Figure 5-25a).   For $F\|c$, the shift in the absorption edge to longer wavelengths is

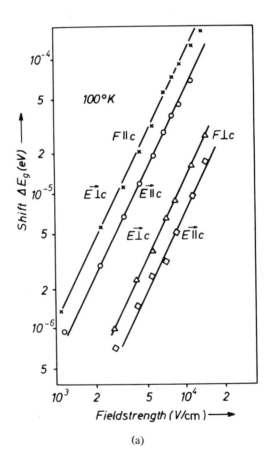

(a)

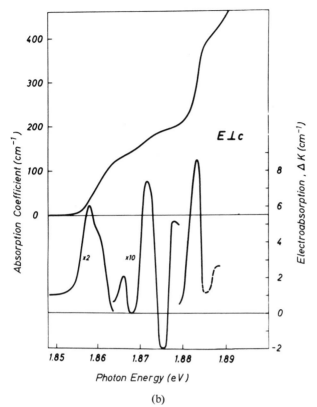

(b)

Figure 5-25   Electroabsorption of trigonal selenium: (a) field dependence for both orientations of the electric field F, and both polarizations of light;[46] (b) energy dependence compared with the structure of the indirect absorption edge.[34]

larger than for $F \perp c$.   This behavior is attributed to the anisotropy of the effective masses involved in the absorption processes.   The mass values given by Stuke and Weiser, however, are probably inaccurate because the formula of Franz[47] used for the evaluation had been derived from a simple model where only direct transitions were considered.   For the indirect transitions at the absorption edge of trigonal selenium, this relation is inadequate.

In the recent measurements of Fischer, the structure of the indirect edge reveals a number of pronounced maxima in the electroabsorption spectra (Figure 5-23 and Figure 5-25b).   This evidence indicates conclusively that the smallest band gap of trigonal selenium is indirect, and it facilitates the evaluation of the phonon energies involved.   The strongest peak is due to the optical phonon with an energy of 29.3 MeV.   The peak with 18.1 MeV distance from 1.853 eV is also created by an optical phonon.   The energies of 5.1 and of 7.8 MeV are ascribed to accoustical phonons.[48]

Modulation Techniques: Electroreflectance. Electroreflectance measurements are of great value for a detailed analysis of the band structure since an electro-reflectance response is sharper and more pronounced than the corresponding maximum of the reflectivity, which is mostly covered by background reflec-tivity. For trigonal selenium, the first electroreflectance measurements were done by Chen[49] who used an electrolytic method. The results, however, were inconsistent with the results of Gobrecht et al.,[50] Weiser and Stuke[35], and Weiser.[51] The latter researchers measured transverse electroreflectance where the field is applied between Au electrodes evaporated on the surface of the crystals. This method has the advantage that the effect can easily be studied for the two polarizations of light and also for different directions of the electric field. Due to the free crystal surface, there are also no restrictions on temperature or the energy range.

The results of such measurements at 90°K are shown in Figure 5-26 where, for the sake of comparison the reflectivity spectra are also plotted. Perhaps the most remarkable feature of the electroreflectance spectra of trigonal selenium is the fact that many of the pronounced maxima of the reflectivity between 2 and 4 eV do not appear in electroreflectance. This can be explained using the band model of Sandrock by transitions in wider parts of the Brillouin zone instead of at the critical points. For a more detailed picture, recent

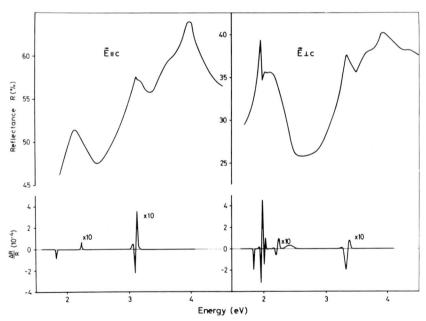

Figure 5-26 Electroreflectance $\Delta R/R$ of trigonal selenium for $E \parallel c$ and $E \perp c$ in comparison with reflectivity R.[35]

measurements at low temperatures by Weiser[51] are shown in Figure 5-27. Because indirect transitions are not observed in electroreflectance, the spectrum starts with the direct edge.   Here three relatively sharp structures are observed which are related to excitons.   The first exciton at helium temperature has an energy of 1.949 eV, and it is assigned to the transition from the uppermost valence band H_5 into the conduction band H_6 (Figure 5-9).   The second exciton found at 1.991 eV is attributed to a transition from the next valence band H_4.   This structure is slightly overlapped by the first exciton.

While these transitions are only observed for $E \perp c$, a further peak is found at 2.002 eV for parallel polarization.   For $E \perp c$, this weak structure cannot be resolved because it is covered by the two other transitions.   It is assigned to an exciton correlated to the transition from the third valence band H_6.   According to Kramer and Thomas[19] this level is close to the H_4-band.   All these sub-bands H_5, H_4, and H_6 are split by spin-orbit interactions from the valence band H_3.   From the electroreflectance data at low temperature, the splitting amounts to 42 MeV between H_5 and H_4 and to 11 MeV between H_4 and H_6.   These values are in good agreement with the theoretical results.[19]   The splitting of 53 MeV between the H_5 and the upper H_6 band is a little higher than that found in the photoemittance

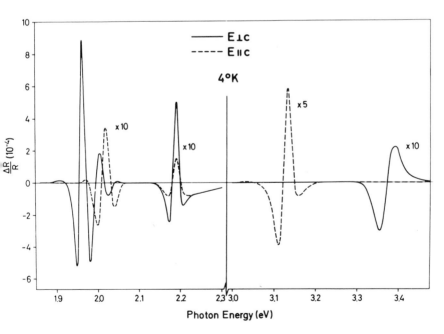

Figure 5-27   Electroreflectance spectra for $E \perp c$ (solid line) and for $E \parallel c$ (dashed curve) at 4°K.[51]

spectrum by Blätte and Kautsch.[45]   This small deviation may be due to the different temperatures used in the electroreflectance (4°K) and in the photo-emittance measurements (about 370°K).

A further structure is observed at 2.175 eV which is for $E \perp c$ about three times larger than for $E \| c$.   The strong temperature dependence of the amplitude suggests an exciton.   It is attributed to a transition from the lower H_6-valence band to the H_6-conduction band which is allowed for both polarizations (Table 5-3).   A weak and relatively broad maximum at 2.4 eV (Figure 5-26) is not polarized and corresponds to the reflectivity maximum in this energy region.   Sandrock claims, it is not due to transitions at a critical point.   It is observed only weakly in the electroreflectance and also in the piezoreflectance measurements discussed later.

At 3.1 eV, an exciton is found only for $E \| c$, which is assigned by Sandrock to a transition from a lower valence band at or near H.   For the spinless case, the transition $H_2 \to H_1$ is only allowed for parallel polarization (Table 5-3).   However, this selection rule is not valid in the double group representation: the corresponding transition should not be polarized.   As a result, the 3.1 eV exciton probably has to be attributed to another point of the Brillouin zone.   That is also suggested by the temperature shift of this structure which is opposite to the shift of the peaks at the fundamental edge (Figure 5-28).   Mostly the exciton at 3.1 eV is assigned to point $A(Z)$ on top of the Brillouin zone where, from symmetry arguments, the first transition only is allowed for a parallel polarization.   The next transition from the lower valence band $A_3(Z_3)$ to the $A_1(Z_1)$ conduction band, however, is allowed for $E \perp c$ and probably corresponds to the electroreflectance structure at 3.35 eV and to the reflectivity maximum in the same energy region.

The temperature shift in some of the electroreflectance peaks is shown in Figure 5.28.   For the excitons at the fundamental edge, the temperature coefficient is positive.   Below 100°K, it amounts to about 3.5×10^{-4} eV/°K, but it decreases with increasing temperature.   The positive sign of the coefficient is in accordance with the pressure coefficient of the absorption edge if one assumes that decreasing temperature corresponds to increasing hydrostatic pressure.   A similar behavior is found for the transition at 2.18 eV which, like the other excitons with a positive temperature shift, is also attributed to point H.   The sign of the temperature coefficient of the peaks at 3.1 eV and 3.35 eV, however, is negative.   This favours their assignment to a different point of the Brillouin zone.

Also of importance is the influence of the orientation of the electric field, F, on the electroreflectance signal.   The height of the first exciton structure is appreciably lower for $F \perp c$ than for $F \| c$ as can be seen from Figure 5-29.   This anisotropy can be ascribed to the anisotropy, of the effective masses.   Since, for interband transitions, the signal is expected to increase with a

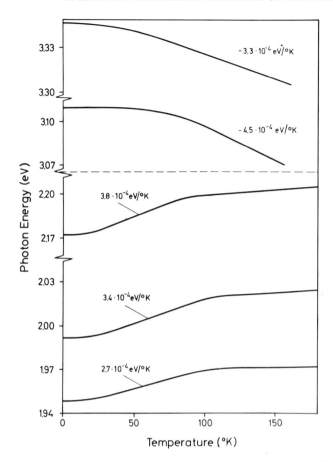

Figure 5-28   Temperature dependence of the electroreflectance structures.[35]

decrease in mass, the experimentally found anisotropy indicates that $m_\perp^x$ is appreciably larger than $m_\parallel^x$. A more accurate determination of the mass ratio, however, is not possible because the observed electroreflectance signal is due to exciton transitions. A theoretical basis does not presently exist for this effect. The dependence of the other excitons on the field orientation is not known accurately because the signal has been too small for a detailed analysis. The energy positions of the exciton spectra are not shifted with electric field strength, indicating that the change of the reflectivity is mainly due to a quenching of the excitons by the field.

Modulation techniques: Piezoreflectance. The change of the reflectivity of trigonal crystals by uniaxial stress has been investigated by Tuomi.[36]   He

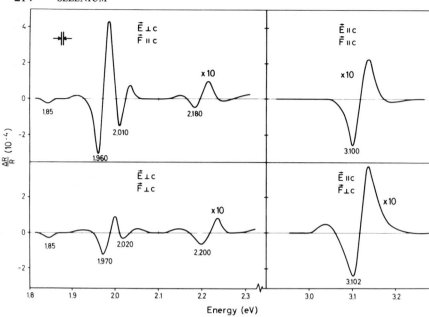

Figure 5-29 Electroreflectance at 90°K for both polarizations of light and for both orientations of the electric field.[35]

not only studied the influence of stress on the energy position of the peaks in the reflectivity spectrum by a combination of wavelength modulation and stress, but also the stress modulated reflectance spectra.   The result of the stress modulation for stress parallel to the c-axis can be seen in Figure 5-30, where the spectra for both polarizations of light are plotted.   They have a great similarity to the corresponding electroreflectance spectra (Figure 5-26). The excitons near the direct gap reveal the three pronounced peaks a, b, and d only for $E \perp c$, whereas the strong reflectivity maxima at about 2.2 eV lead to the weak, broad structure, e, which is observed for both polarizations. The strong peak at lower energy is not due to piezoreflectance, but to piezo-absorption, since this light is passing through the crystal and is reflected at the lower surface.

Of particular importance for the interpretation of the reflectivity spectra is a knowledge of the pressure coefficient, β, of the different maxima.   The measurements of Tuomi give, for the coefficients of the three excitons a, b, and d, near the band edge, the values shown in Table 5.5.   Since pressure parallel to the c-axis corresponds to an increase of temperature, the positive value of $\beta_{\parallel}$ is in accordance with the positive temperature coefficient.   It is notable that $\beta_{\parallel}^{\perp}$ is, for the exciton d, a factor of 2 larger than the coefficients for the excitons a and b.   Since the exciton d is due to an H_6-H_6 transition

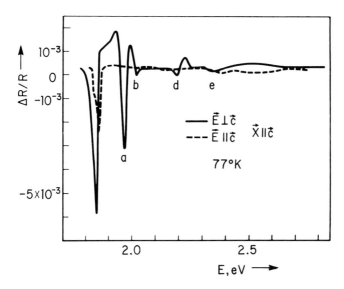

Figure 5-30   Stress modulated reflectance spectra of trigonal selenium at 77°K.   The direction of stress is parallel to the c-axis.[36]

TABLE 5-5   PRESSURE COEFFICIENT FOR DIFFERENT OPTICAL TRANSITIONS

Peak	Transition	$\beta^{\perp}_{\parallel}/10^{-5}$ eV/atm	$\beta^{\perp}_{\perp}/10^{-5}$ eV/atm
a	H_4–H_6	$+2.5 \pm 0.4$	-6.7 ± 0.5
b	H_5–H_6	$+2.8 \pm 1$	-7.9 ± 1
d	H_6–H_6	$+5.4 \pm 0.5$	

but a and b are related to H_5, $H_4 \rightarrow H_6$ transitions, the pressure shift of the lower H_6 valence band is appreciably larger than that of both upper valence bands.   For pressure applied perpendicular to the c-axis, a large negative value of the coefficient is obtained, which is approximately the same for a and b, but is still unknown for d.   For this orientation of pressure, the shift of the excitons is nonlinear and the value of $\beta^{\perp}_{\perp}$ decreases appreciably with pressure.

Optical Activity.   The spiral chains of the lattice of trigonal selenium lead to a strong activity which has been detected by Henrion and Eckart.[52]   Dextro- and levorotatory crystals are found corresponding to the different character of the chains.   For tellurium, it has been shown by Grosse[9] that dextrorotation is related to left-hand chains and, vice versa.   The different signs are not a result of a peculiar mechanism, but of a definition only.   A right-hand chain gives a right-hand rotation of the polarization plane looking in the direction of the light propagation.   The sign of the optical activity, however, is defined by the rotation looking in the reverse direction.   For tellurium, therefore, a simple relation between the rotation of the chains and the rotation of the polarization exists.   For selenium, this correlation has not been proved experimentally, but the great similarity in the structures of both elements makes it highly probable.   The absolute value of the rotatory power of trigonal selenium is relatively large compared with that of other optically active materials.   Its wavelength dependence, shown in Figure 5-31, is characterized by the typical increase with increasing energy at the absorption edge.   If one normalizes the energy to the absorption edge, a comparison with tellurium is possible.   The rotatory power of selenium is found to be larger, by a factor of about 2, than that of tellurium.   This difference is probably due to the weaker bonding between the selenium chains compared with tellurium because a strong interaction between the chains diminishes the screw character of the lattice.   The sign of the rotation can be understood in the following way.   A linearly polarized wave can be assumed to be a

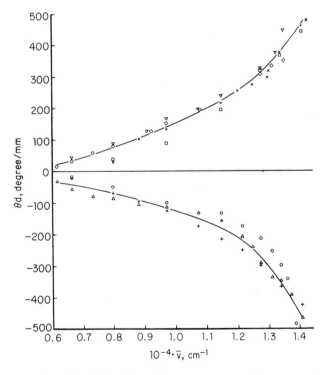

Figure 5-31 Optical activity of trigonal selenium as a function of the wave number $\bar{\nu} = 1/\lambda$.[52]

superposition of two circularly polarized waves.   For the wave with the same rotation as the chain, the interaction between wave and chain is expected to be stronger than for the reverse case.   For a chain with right-hand rotation, this stronger interaction leads to a higher value of n_R compared with n_L, the refractive indices of the right-hand and left-hand circular polarized waves, respectively.   Since the optical activity is given by:

$$\theta = \frac{\pi d}{\lambda} (n_R - n_L)$$

where θ is the rotation angle and d the thickness of the sample.   The rotation angle is positive which, according to the definition, gives a levorotatory material.

Electrical Properties

Electrical Conductivity.   The electrical conductivity, σ, of trigonal selenium single crystals is rather small, its value at room temperature lying between 10^{-6} and 10^{-5} Ω^{-1} cm^{-1}.   Besides the fact that the conductivity cannot be enhanced appreciably by the addition of impurities but only by the creation of lattice defects, σ has an additional number of unusual properties.   The current-voltage characteristic is ohmic only at low voltages and it becomes superohmic at high electric fields.   In the latter region, therefore, a conductivity is, strictly speaking, not defined.   In order to be able to compare the behavior in both regions, a voltage dependent conductivity is used.   In addition to this irregular influence of voltage, trigonal single crystals show a pronounced frequency dependence of the conductivity particularly at low temperatures.   Since the frequency and the voltage dependence of σ are strongly correlated with its temperature dependence which can be influenced by illumination and by plastic deformation, these properties shall be treated together.

Dependence on Temperature, Voltage, and Frequency.   The voltage dependence of the conductivity of trigonal single crystals has been known since the early investigations of Müller,[53] Plessner,[54] and Henckels.[55]   A more detailed study has been performed by Gobrecht and Hamisch,[56] who also investigated the influence of voltage pulses applied perpendicular to the direction of the current.   The combination of temperature and voltage dependence has been studied by Polke, Stuke, and Vinaricky,[57] Bakirov and Dzhalilov,[58] and by Graeffe and Heleskivi.[59]

According to the investigations of Stuke,[60] plastic deformation of the crystals significantly changes the voltage and temperature dependence of σ indicating that lattice defects play an important role in these phenomena. A typical result for the dark conductivity is shown in Figure 5-36. Before plastic deformation, the conductivity at room temperature is ohmic up to a voltage of 10 V, corresponding to a field strength $F = 10^3$ V/cm. For higher fields, σ becomes superohmic. By lowering the temperature, the transition between the ohmic and the superohmic parts is shifted to smaller field strengths, the superohmic part having approximately a quadratic current-voltage dependence at 93°K. In the temperature dependence of σ (Figure 5-32), the two curves are observed to be voltage independent at high temperatures, with an activation energy, $\Delta\varepsilon$, of about 0.2 eV as determined from the relation:

$$\sigma = \sigma_0 \exp\left(\frac{-\Delta\varepsilon}{kT}\right) \tag{1}$$

The curves are strongly voltage dependent at low temperatures. The slope of the curve increases with voltage and differs appreciably for various crystals. For one crystal, Polke, Stuke, and Vinaricky[57] found, at high fields, a voltage independent activation energy of 0.035 eV.

The approximately quadratic dependence on voltage of some crystals at low temperature and high field strength suggests space charge limited currents due to injection of majority carriers from the electrodes. The basic equation for the current density, j, in the trap-free case is:

$$j = \frac{9}{8} \varepsilon\varepsilon_0 \mu \frac{V^2}{L^3} \tag{2}$$

where μ is the mobility of the carriers, V the applied voltage, and L the electrode distance. An important feature of this relation and a check for occurrence of space charge limited currents is that the exponent of L is higher than that of V. This is also true in more complicated cases when traps are involved. For selenium crystals, however, it is found experimentally that $j \sim V^2/L^2$. Moreover, the threshold voltage of the superohmic part is by far too small for injection of holes with a concentration comparable to 10^{14} cm^{-3}. This is obtained by thermoelectric power and the Hall effect. These results indicate that the simple theory of space charge limited currents cannot be applied here. The behavior can, however, be explained by the assumption of thin depletion layers within the crystals[60] as is shown in the discussion of the conduction mechanism.

The influence of plastic deformation on the voltage and temperature dependence indicates that the internal depletion layers are connected with lattice defects.[60] In the ohmic region at high temperature and low voltage, the conductivity is enhanced by deformation. For crystals with a low conductivity before deformation, an increase of approximately 2 orders of magnitude can be obtained.   Subsequent annealing at relatively low temperature reduces the conductivity almost completely to its initial value (Figure 5-32).   Since the activation energy is changed only slightly this leads to an almost parallel shift of the curves.

The behavior of the voltage dependent part at low temperature, however, is quite different (Figure 5-32).   Here the conductivity decreases by plastic deformation and the onset of the voltage dependence is shifted to lower temperature and higher voltage.   By annealing, the conductivity increases somewhat, but it does reach again the value it had before deformation.   After strong plastic deformation, the reversible change is especially small.

These results indicate that two kinds of lattice defects are responsible for the change of σ.   The first one enhances the conductivity and is readily annealed.   From thermoelectric power measurements (see Figure 5-39) the increase of conductivity is due to an enhancement of the hole concentration. For this reason, the first kind of lattice defect is ascribed to vacancies which act as acceptors.   The energy for the creation of these defects can be determined from an isochrone annealing process, where the conductivity is

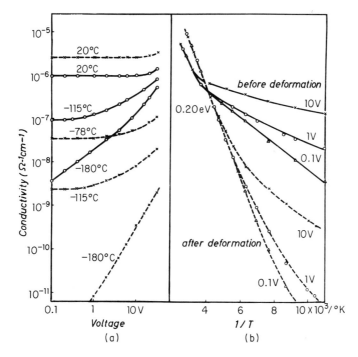

Figure 5-32   Voltage and temperature dependence of the conductivity of a trigonal selenium crystal before and after plastic deformation. Electrode distance $L = 1 \times 10^{-2}$ cm, current $\parallel c$-axis.[60]

measured while the temperature rises with a constant rate of about 10°/min (Figure 5-33).   The method of Balarin and Zetsche[61] yields a value of 0.6 eV for the energy of the vacancies.   Self-diffusion measurements in selenium by Brätter and Gobrecht[62] give for the vacancy migration enthalpy a value of about 0.5 eV along the *c*-axis and 0.7 eV perpendicular to this direction.

Since plastic deformation also creates a large number of additional dislocations, the decrease of σ at low temperature is believed due to these lattice defects.   Then it can be understood that the changes in this branch of the curves are reversed only to a limited extent by annealing because dislocations do not disappear as easily as vacancies.   If one assumes that the network of dislocations or of clusters of dislocations creates a network of depletion layers, the shift of the voltage dependent part to lower temperature and higher voltage can also be understood.   By an increase of the dislocation

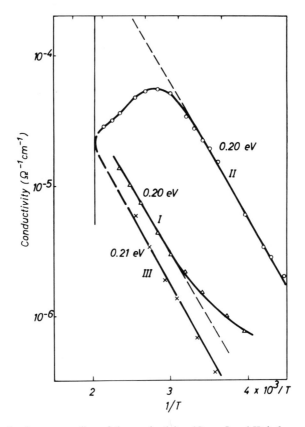

Figure 5-33   Isochrone annealing of the conductivity: (Curve I and II) before and after plastic deformation; (Curve III) after the annealing process.[60]

density, the number, N, of the depletion layers is enhanced. Thus, for a constant external voltage, the effective field strength in a single layer decreases. For the onset of the voltage dependence, therefore, a higher external voltage has to be applied.

This model implies that crystals with a low concentration of lattice defects should, also at low temperature, display a very limited voltage dependence of the conductivity. The results of Stuke[60] show this tendency. Some crystals grown from the vapor phase had, at liquid nitrogen temperature, an ohmic characteristic up to relatively high electric fields and a weak temperature dependence of σ. After plastic deformation, however, these crystals have a similar behavior as that shown in Figure 5-32. Thus, the temperature and voltage dependences of σ are not typical for the trigonal crystal itself but are created by lattice defects.

The frequency dependence of σ is, like its voltage dependence, more pronounced at low temperature than at high temperatures. As reported by Salo, Stubb and Suosara,[63] it is larger in the ohmic part of the curve than in the voltage dependent branch (Figure 5-34). At $78°K$, the conductivity rises with increasing frequency proportional to $\omega^{0.8}$ between 10^2 and 10^4 Hz for an ac-field strength of about 0.5 V/cm. By superposition of a high dc-field, the conductivity at low frequencies is noticably enhanced as is the dc-conductivity. Simultaneously, the frequency dependence strongly decreases and with rising dc-voltage, the onset of the frequency dependence is shifted to higher frequencies. At room temperature, the increase of σ with ω is much less pronounced. Therefore, the activation energy of the conductivity is considerably diminished with rising frequency. This behavior is in agreement with the result of Henckels and Maczuk[64] who found for $\Delta\varepsilon$, at a frequency of 200 MHz, a value of about 0.01 eV. This is more than a factor of 10 smaller than that obtained by dc measurements on the same crystal. At a frequency of 24 GHz, the temperature coefficient of σ seems to undergo a reversal in sign because σ decreases with increasing temperature.[65] A similar result has also been obtained for polycrystalline selenium.[66]

In this context, the influence of illumination on the magnitude of the activation energy, $\Delta\varepsilon$, is of interest. This can be studied by a stepwise-measured, thermally-stimulated current curve (TSC), which is shown in Figure 5-35. For this measurement, the crystal is at first illuminated at about $100°K$ for a few minutes. Then, after about 15 minutes in the dark, the conductivity is measured as a function of temperature at a constant heating rate. In contrast to the procedure of a normal TSC measurement (curve G), the temperature is raised stepwise. The various steps of the TSC curve are approximately reversible in the low temperature region. A simple curve is found for crystals which, due to strong plastic deformation, have a constant value of $\Delta\varepsilon$ over several orders of magnitude of σ. The activation energy steadily increases

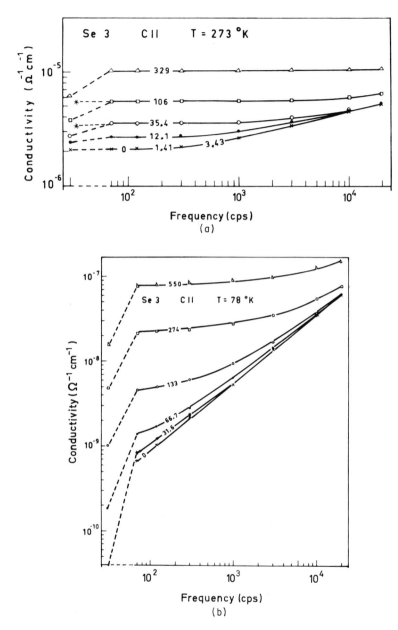

Figure 5-34   Frequency dependence of the conductivity: (a) 273°K, (b) 78°K. The parameter of the curves is the dc-electric field strength (V/cm).   The ac-field is about 0.5 V/cm.[63]

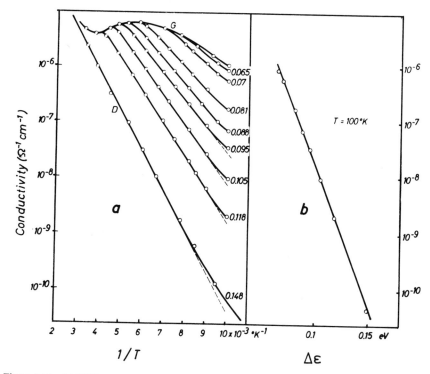

Figure 5-35 (a) TSC-curves measured stepwise. (b) Relation between the conductivity and the activation energy at constant temperature.[60]

from 0.065 eV after illumination to the dark value of 0.15 eV. The extrapolations of the straight lines to $1/T = 0$ leads to the same conductivity σ_0. These curves can be described by the equation (1):

$$\sigma = \sigma_0 \exp\left(\frac{-\Delta\varepsilon}{kT}\right)$$

with $\Delta\varepsilon$ as the variable parameter. Figure 5-35 shows the dependence of $\Delta\varepsilon$ on the corresponding values of the conductivity at 100°K. This curve is in good agreement with the relation $\Delta\varepsilon = kT \ln(\sigma_0/\sigma)$ for $T = 100°$K. Thus the increase in the conductivity due to illumination is only the result of a decrease in the activation energy.

The influence of voltage, frequency, and illumination on the temperature dependence of the conductivity can be summarized qualitatively in the following way. These three quantities enhance the conductivity in connection with a decrease of the activation energy. An enhancement of σ by one quantity diminishes the influence of the other ones. The frequency dependence at 80°K is only small when a high dc-voltage is applied simultaneously; with illumination, the voltage dependence at low temperature strongly decreases. The same seems to be true for the frequency dependence. Obviously, the magnitude of the activation energy is not only responsible for the temperature variation of the conductivity, but in addition determines the variation of σ by voltage, frequency, and illumination in the sense that the smaller $\Delta\varepsilon$ is, the less they influence the conductivity.

Anisotropy of the Conductivity. Since trigonal selenium is a crystal having a strong anisotropy, an orientation dependence of the electrical properties is to be expected. In fact, the conductivity parallel to the c-axis, $\sigma_{\parallel}$, is larger than $\sigma_{\perp}$, the conductivity perpendicular to the chains. Of particular interest is the influence of temperature, illumination, and plastic deformation on the ratio $\sigma_{\parallel}/\sigma_{\perp}$, which is shown in Figure 5-36. Before deformation,

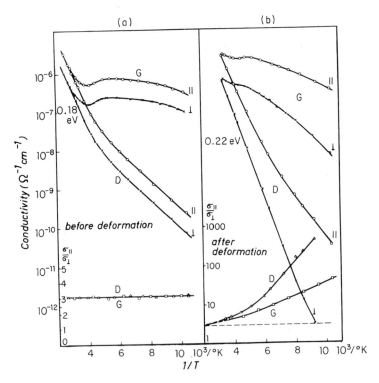

Figure 5-36 Anisotropy of the conductivity of a trigonal crystal before and after plastic deformation: $D =$ dark conductivity, $G = T$SC-curves. The ratio $\sigma_{\parallel}/\sigma_{\perp}$ is plotted linearly in (a) and logarithmically in (b).[60]

$\sigma_{\parallel}/\sigma_{\perp}$ is independent of temperature. It has the value 3 for the dark conductivity (curve D) as well as for the TSC-curve G. Since the carrier density cannot be anisotropic, the conductivity ratio equals the mobility ratio of the holes $b = \mu_{\parallel}/\mu_{\perp}$ which, therefore, does not depend on temperature. This property is similar to that found for tellurium, where the mobility ratio is 2 over a wide temperature range. The smaller absolute value of b for tellurium corresponds to the smaller anisotropy of the bonding compared with selenium. This approximate equality of both elements is of importance for the discussion of the transport properties of trigonal selenium, because it indicates that the conduction mechanism of trigonal selenium and tellurium cannot be completely different.

Strong plastic deformation, perpendicular to the c-axis, changes the mobility ratio considerably as in shown in Figure 5-36, where $\sigma_{\parallel}/\sigma_{\perp}$ is plotted logarithmically. Since the influence of deformation on the conductivity is more pronounced for $F \perp c$ than for $F \parallel c$, the ratio $\sigma_{\parallel}/\sigma_{\perp}$ becomes much larger, particularly at low temperatures where values greater than 100 are obtained. For the TSC curve, the ratio $\sigma_{\parallel}/\sigma_{\perp}$ is smaller, but with rising temperature the difference diminishes and both curves approach the value 3. This phenomenon obviously is a result of the different influence of the deformation on the two orientations of current. It can be seen from Figure 5-37 that the ratio σ_a/σ_b of the conductivities after and before elastic deformation is quite different for both directions. The increase of σ_a/σ_b at high temperature depends only slightly on the orientation of the current because it is due to an increase in the hole concentration by the creation of vacancies. The decrease of σ_a/σ_b in the voltage dependent region at low temperature, however, is strongly orientation dependent in the dark (curve D). The ratio for currents oriented perpendicular to the chains is more than 2 orders of magnitude larger than for parallel orientation. With illumination, on the other hand, the influence of deformation is strongly diminished (curve G), particularly when the current is oriented perpendicular to the c-axis. This strong influence of plastic deformation on σ_a/σ_b in the low temperature, voltage dependent branch of the curves can be explained by the orientation of the dislocations. Since, in trigonal selenium, edge dislocations are oriented in the direction of the chain, the strongly different change of the conductivity for the two field orientations can be understood if one assumes that the conductivity is determined mainly by potential barriers connected with these dislocations. This is discussed in more detail in the section on the conduction mechanism.

Pressure Dependence. The conductivity of trigonal selenium crystals varies greatly with pressure. The highest pressure was applied by Balchan and Drickamer,[67] who found a steady increase of the conductivity up to 130 katm where selenium becomes metallic. This metallic phase was

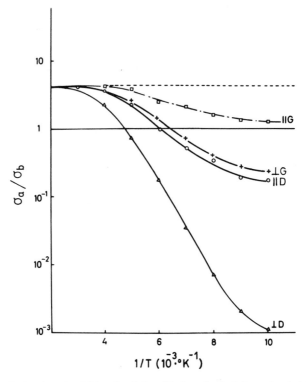

Figure 5-37   Ratio of conductivities after (σ_a) and before plastic deformation (σ_b) for orientation of the current $\parallel$ and $\perp$ to the c-axis: $D =$ dark conductivity, and $G =$ TSC-curves.[48]

shown to be superconducting by Wittig[68] with a transition temperature of about 6°K at 130 katm.   Measurements under hydrostatic pressure have been performed by Kozyrev,[69] Krisciunas and Daukantaite,[70] and more recently by Becker, Fuhs, and Stuke.[71]   These investigations found that the conductivity increases exponentially with hydrostatic pressure.   The measurements of Kozyrev, shown in Figure 5-38, give for the hydrostatic pressure coefficient $\gamma_h = d \ln \sigma/dp$ at room temperature the value $\gamma_h = + 1.6 \times 10^{-4}$ atm^{-1} which decreases with rising temperature.   The magnitude of γ_h and also its temperature dependence differ appreciably from one crystal to another.   From the temperature dependence of the conductivity at various pressures, it can be concluded that the activation energy, $\Delta\varepsilon$, in the ohmic region decreases almost linearly with hydrostatic pressure giving rise to an exponential increase of the conductivity.

Krisciunas and Daukantaite[70] found the interesting result that the pressure coefficient at room temperature can be significantly reduced by illumination.   The decrease of γ_h by light is more pronounced for crystals with a

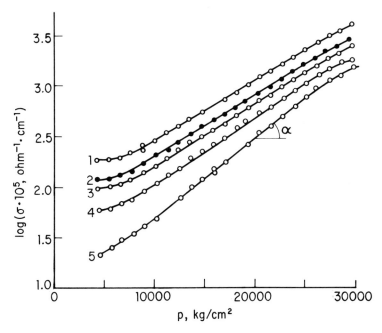

Figure 5-38 Dependence of the conductivity of a trigonal monocrystal on hydrostatic pressure for different temperatures: (1) 398°K, (2) 370°K, (3) 351°K, (4) 323°K, (5) 294°K.[69]

low dark conductivity, σ_D, Since γ_h also decreases with a rising σ_D for various crystals, there exists a correlation between the absolute value of σ and the magnitude of the pressure coefficient.   Moreover, these researchers found that γ_h is independent of the current orientation with respect to the c-axis.   In spite of the range in the conductivity ratio, $\sigma_\parallel/\sigma_\perp$, between 4 and 12, γ_h had, for the field direction parallel to the chains, the same value as for perpendicular orientation.   This behavior is similar to that of tellurium, where the ratio $\sigma_\parallel/\sigma_\perp$ has also been found to be independent of hydrostatic pressure.[71]   This is a rather surprising result since the application of hydrostatic pressure mainly decreases the distance between the chains while the alteration of the chain length is very small.   This is of particular interest for the discussion of the conduction mechanism, becaues it indicates that the mobility ratio, $\mu_\parallel/\mu_\perp$, is not influenced by the change of the lattice parameters with pressure.   From the experiments of Dolezalek and Spear,[72] the mobility obtained by acoustoelectric current saturation is not changed by hydrostatic pressure up to 5000 atm. Therefore, not only the mobility ratio, $\mu_\parallel/\mu_\perp$, but also the hole mobility (μ_p) itself is independent of pressure.   Since the activation energy, $\Delta\varepsilon$, is lowered by hydrostatic pressure, this quantity is not directly connected with the microscopic mobility, μ_p.

Recently Becker, Fuhs, and Stuke[71] studied the influence of weak uniaxial pressure parallel to the c-axis. Uniaxial stress in this direction leads to a similar elastic deformation as hydrostatic pressure, namely, elongation $\|c$ and compression $\perp c$. Because of this equivalence, different signs of the pressure coefficients, γ_h and $\gamma_\|$, are to be expected. This, however, is only found for crystals with a low activation energy. Crystals with high $\Delta\varepsilon$ values have, particularly at low temperature, a positive $\gamma_\|$. An increase in the conductivity at low temperature by the application of a high electric field or by illumination, which in both cases is accompanied by a decrease of $\Delta\varepsilon$, leads to a lowering of the positive value of $\gamma_\|$ or even to a negative value. This property, together with the fact that negative values of $\gamma_\|$ are also found for tellurium both in the extrinsic and intrinsic regions, indicates that a negative value is to be expected when the influence of the potential barriers is small. The large positive value of γ for crystals with high activation energy decreases in a manner roughly proportional to $1/T$. Obviously, $\Delta\varepsilon$ is lowered by the application of uniaxial pressure $\|c$ in contrast to the change of the band gap, which increases under the same conditions.[38] Since uniaxial stress $\|c$ is equivalent to hydrostatic pressure, this result disagrees with that of Kozyrev, who found a decrease of $\Delta\varepsilon$ under hydrostatic pressure. The reasons for this discrepancy and also for the different influence of uniaxial pressure on E_g and on $\Delta\varepsilon$ are not yet understood.

Thermoelectric Power. The thermoelectric power, θ gives information about the type and the density of the carriers. Investigations on trigonal selenium crystals have so far yielded only positive values of θ, indicating that holes are predominant in the transport mechanism. The temperature dependence shown in Figure 5-39 which has been found by Stuke and Wendt[73] can be described by the equation:

$$\theta = \frac{k}{e}\left(\ln\frac{N_v}{p} + r\right) \tag{3}$$

with a temperature independent value of the hole concentration, p. The small increase of θ with rising temperature is due to the temperature dependence of N_v, the effective density of states in the valence band which is proportional to $T^{3/2}$. The hole concentration, therefore, is temperature independent at least down to a temperature of about $100°K$. If one assumes scattering of acoustical phonons ($r = 2$) and an effective mass $m^x = m_0$, one obtains for p values between 10^{13} and 10^{15} cm^{-3}. This range in the carrier concentrations is not obtained by the addition of impurities, but is mainly due to a different density of lattice defects. The various curves of θ and of σ in Figure 5-39 were all measured on the same crystal following plastic deformation and subsequent annealing. The decrease of θ by deformation

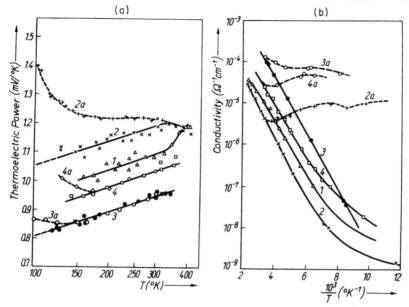

Figure 5-39 (a) Thermoelectric power and (b) conductivity of a trigonal selenium crystal in the dark (solid) and after illumination (dashed): (1) untreated, (2) after annealing, (3) after plastic deformation, (4) after annealing.[73]

accompanied by the increase in σ (curve 3) implies that the parallel shift of the conductivity curve is due to an enhancement of the temperature independent hole density. The energy distance of the acceptor levels from the valence band is about 0.01 eV or less, since they are completely ionized down to 100°K. Annealing shifts the conductivity to lower and the thermoelectric power to higher values. This correlation between θ and σ is shown in Figure 5-40 for a great number of additional measurements. The slope of the curve is k/e, as is to be expected from equation (3) in the case of constant carrier concentration. This is a relevant finding, indicating that the exponential temperature dependence of σ is due to that of the mobility:

$$\mu = \mu_0 \exp\left(\frac{-\Delta\varepsilon}{kT}\right) \qquad (4)$$

The effect of illumination on the thermoelectric power is also of great interest. In spite of an enhancement of the conductivity at low temperature by several orders of magnitude, θ is almost unchanged for curve 3a. This surprising result suggests that it is not additionally created carriers but a strong enhancement in the mobility that is the main reason for the increase in σ. As has been shown in Figure 5-35, the large change in σ is due to a lowering of the activation energy, $\Delta\varepsilon$, of the mobility according to equation (4).

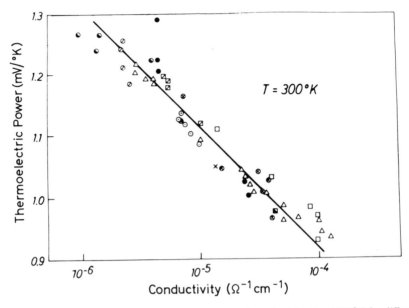

Figure 5-40   Relation between thermoelectric power and conductivity ∥ c at 300°K for different selenium crystals.[73]

For crystals with a low hole density, the thermoelectric power has a higher value after illumination which decreases with rising temperature (curve 2a). This phenomenon cannot be explained in terms of an enhancement of the density of carriers since a higher density of holes, electrons, or both would lead to a decrease in θ. The character of the temperature dependence implies a different mechanism, namely, phonon-drag. In this case, the thermoelectric power, θ, is a superposition of the usual electronic part θ_e and of θ_{ph}, the phonon drag contribution: $\theta = \theta_e + \theta_{ph}$. Therefore θ_{ph} is obtained by subtracting the extrapolated values of θ_e at low temperatures from the measured thermoelectric power. θ. This leads to the $T^{-2.5}$ dependence of θ_{ph} shown in Figure 5-41a. Similar results have also been obtained by Abdullaev, Dzhalilov, and Aliev[74] and by Tiainen and Tunkelo[75]. It can be seen from Figure 5-41b that the temperature dependence of θ_{ph} is similar to that of tellurium and also to that of germanium, silicon and diamond. This similarity leads to the conclusion that the temperature dependence of the hole mobility, μ_p, of selenium is close to that of these elements. According to Herring, both quantities are related by the formula:[76]

$$\theta_{ph} = K \frac{h^4}{k^4} \frac{v_s^5 \rho}{\mu_p m^x T^5} \tag{5}$$

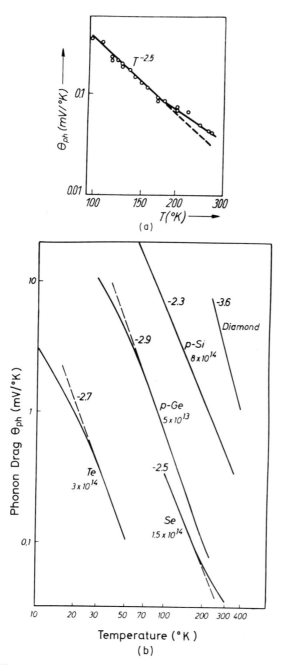

Figure 5-41 (a) Temperature dependence of the phonon drag of trigonal selenium. (b) Comparison with other semiconductors (figure denotes hole concentrations).[73]

where v_s is the velocity of sound, ρ the density, m^x the effective mass, and μ_p the hole mobility due to the interaction with phonons. K is a dimensionless constant which has to be calculated empirically. The temperature variation of θ_{ph} is mainly determined by the terms μ_p and T^5 in the denominator, since the other quantities are only slightly temperature dependent. For scattering by acoustical phonons, $\mu_p \sim T^{-3/2}$. It follows that a dependence $\mu_{ph} \sim T^{-3.5}$ would be expected. This is observed only for diamond, whereas for most other semiconductors, the exponent is found between 2 and 3. Trigonal selenium, with an exponent of 2.5, fits well into this group of semiconductors with a normal lattice mobility.

Galvanomagnetic Properties.   Galvanomagnetic measurements are of particular importance for the investigation of the transport mechanism of semiconductors. In the case of weak magnetic fields ($\mu B \ll 1$), the galvanomagnetic effects can be described by the equations:

$$j_i = \sigma_{ik} F_k + \sigma_{ikl} F_k B_l + \sigma_{iklm} F_k B_l B_m$$
$$F_i = \rho_{ik} j_k + \rho_{ikl} j_k B_l + \rho_{iklm} j_k B_l B_m \tag{6}$$

where σ_{ik} and ρ_{ik} are the mutually reciprocal tensors of conductivity and specific resistance without application of magnetic field, σ_{ikl} and ρ_{ikl} are the tensors of the Hall effect, σ_{iklm} is the magnetoconductivity, and ρ_{iklm} the magnetoresistance tensor. Since trigonal selenium belongs to the D_3 group, its conductivity tensor has 12 independent nonvanishing components. When the coordinate system is arranged as in Figure 5-1 ($x = 1$, $y = 2$, $z = 3$) these can be written as follows:

$$\sigma_{11} = \sigma_{22}, \sigma_{33}$$
$$\sigma_{123}, \sigma_{231} = \sigma_{312}$$
$$\sigma_{1111}, \sigma_{1122} = \sigma_{2211}, \sigma_{1133} = \sigma_{2233}, \sigma_{1123}$$
$$\sigma_{3333}, \sigma_{3311} = \sigma_{3322}$$
$$\sigma_{1313} = \sigma_{2323}, \sigma_{2311} = \sigma_{3112}$$

For ρ_{ik}, ρ_{ikl}, and ρ_{iklm}, a corresponding table exists. Mostly for the Hall effect the abbreviation is used: $R_1 = \rho_{231} = \rho_{312}$ and $R_3 = \rho_{123}$. Thus, R_1 is the Hall coefficient, when B is perpendicular to the c-axis and R_3 is the coefficient for the parallel orientation of B.

The Hall Effect.   Measurements of the Hall coefficient of trigonal monocrystals have been made by Plessner[54] and more recently by Heleskivi, Stubb, and Suntola.[77] Because of the small magnitude of the effect, reliable results have only been obtained for crystals with a high conductivity. The Hall coefficient turns out to be approximately temperature independent giving a hole density of about 10^{14} cm^{-3}, which is in relatively good agreement with the value obtained from thermoelectric power measurements. The combination with the conductivity ($R\sigma$) gives a value of about 0.15 cm^2/Vs for the Hall mobility at room temperature which increases exponentially with temperature according to equation (4) with the activation energy, $\Delta\varepsilon$, between 0.15 and 0.2 eV. Therefore, measurements of both the Hall effect and the thermoelectric power reveal the existence of an approximately temperature independent hole concentration and an exponentially increasing Hall mobility.

Magnetoconductivity. A completely different behavior of the mobility, however, has been found by looking at the magnetoresistance and magnetoconductivity. The latter quantity is advantageous in the case of selenium, because the measurements must be made on samples in the form of thin crystal platelets in order to obtain sufficiently low levels of resistance and noise. Mell and Stuke[78] found that the relative decrease of the conductivity, $-\Delta\sigma/\sigma$, is proportional to the square of the magnetic induction, B, which is derived from the theory for small magnetic fields. The slope of the curves is very much dependent on the temperature and on the orientation of B. With rising temperature, it decreases and approaches a dependence on T^{-3} (Figure 5-42). When the electrical field is parallel to the chain direction, the transverse effect $-\sigma_{3311}/\sigma_{33}$ is appreciably larger than the longitudinal one, $-\sigma_{3333}/\sigma_{33}$. A different pattern is found when E is perpendicular to the

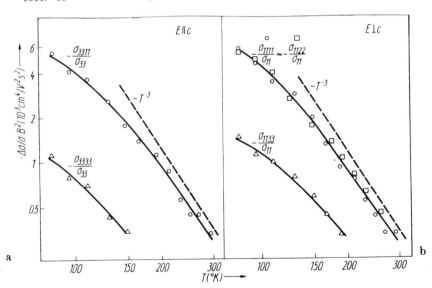

Figure 5-42 Temperature dependence of the magnetoconductivity of trigonal selenium: (a) current $\parallel c$, (b) current $\perp c$.[78]

c-axis. Then the longitudinal component $-\sigma_{1111}/\sigma_{11}$ is as large as the greater of the two transverse effects.

Since the conductivity of trigonal selenium depends only on the magnitude of the applied electrical field and on illumination particularly at low temperatures, it is of interest to study the effect of these quantities on the magnetoconductivity. The research indicates that the magnetoconductivity remains almost unchanged at constant temperature with variation in the electrical field or illumination. The same variation in σ with increasing temperature leads to a strong decrease in $-\Delta\sigma/\sigma$. Therefore, the magnetoconductivity does not depend on the σ value, but only on the temperature and the direction of B and F.

The most important result of these measurements is the approximate T^{-3} dependence of the magnetoconductivity at high temperature. Since $-\Delta\sigma/\sigma B^2$ is proportional to μ^2, the mobility varies as $T^{-3/2}$. This corresponds to scattering by acoustical phonons. The weaker temperature dependence at low temperature may be due to an additional scattering at impurities. A rough estimate of μ is possible from the relation: $-\sigma_{3311}/\sigma_{33} = \mu^2 B^2$. One obtains, at 83°K, a mobility of 80 cm²/Vs, and at room temperatures $\mu_p \approx$ 20 cm²/Vs. These values are a factor of about 100 larger than the Hall mobility. Besides this large difference in the absolute value, the different temperature dependences of these two types of mobilities are especially relevant. The $T^{-3/2}$ variation in μ_p indicates that the lattice mobility of trigonal selenium is determined mainly by phonon scattering. A detailed analysis of the various components of magnetoconductivity gives a more accurate value of μ_3, the mobility parallel to the c-axis (Figure 5-43). According to these results;

$$\mu_3 = \mu_{\parallel} = 28 \text{ cm}^2/\text{Vs}$$

at room temperature. Moreover the mobility ratio $b = \mu_3/\mu_1 = \mu_{\parallel}/\mu_{\perp}$ is obtained; near room temperature, $b = 3.5$ and it rises slightly with decreasing temperature. These values are in good agreement with those obtained from the measurements of the acoustoelectrical current saturation, which are described in the following section.

Piezoelectric Properties: Piezoelectric Effect. A remarkable feature of the trigonal selenium lattice is its large piezoelectric effect, which was detected in 1956 by Gobrecht, Hamisch, and Tausend.[79] A permanent electric dipole moment does not occur under the application of hydrostatic pressure, because the three polar axes remain coplanar. Only by uniaxial deformation perpendicular to the c-axis can a dipole moment be observed. The direction of the deformation must be such that two of these polar axes are eliminated. That is the case when the deformation is parallel to one x-direction (Figure

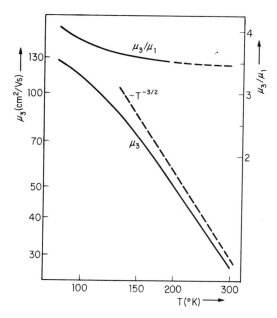

Figure 5-43   Temperature dependence of the mobility μ_3 in the chain direction and of the mobility ratio μ_3/μ_1, both derived from magnetoconductivity.[78]

5-1b).   The piezoelectric constant d_{ij} is defined by the relation: $p_i = d_{ij}\tau_j$, where p_i is the electrical polarization and τ_i the stress, which is correlated to the strain by $\varepsilon_i = c_{ij}\tau_j$, where c_{ij} is the elastic stiffness constant.   The tensor d_{ij} is of the form:

$$
\begin{matrix}
d_{11} & (-d_{11}) & 0 & d_{14} & 0 & 0 \\
0 & 0 & 0 & 0 & (-d_{14}) & (-2d_{11}) \\
0 & 0 & 0 & 0 & 0 & 0
\end{matrix}
$$

So far only d_{11} is known, for which Gobrecht, Hamisch, and Tausend give the value of 6.5×10^{-11} m/V.   It is only little larger than that of tellurium where $d_{11} = 5.5 \times 10^{-11}$ m/V, but about a factor of 30 larger than the piezoelectric constant of quartz.   The strong dipole moment responsible for this large effect originates from the deformation of the electron configuration in the chains.   Therefore, there must also exist piezoelectric lattice modes which can strongly interact with the carriers.

Piezoelectric Properties: Acoustoelectric Effect.   The phenomenon of acoustoelectric current saturation which is often observed in piezoelectric semiconductors permits the direct measurement of the drift mobility and is as such of particular interest for trigonal selenium.   Below saturation, the current rise is proportional to the applied field.   This is due to the increase in the drift velocity of the holes: $v_d = \mu_d F$.   When v_d reaches the velocity of sound, v_s, the carriers strongly couple to the piezoelectric lattice modes.   For fields where the drift velocity would exceed the velocity of sound, the additional energy acquired by the carriers from the field is passed on to the lattice modes and leads to their amplification.   Hence v_d does not exceed v_s and a saturation of the current results (Figure 5-44).   If the value of v_s is known, the measured saturation, F_{sat}, directly gives the drift mobility, $\mu_d = v_s/F_{sat}$.

Mort[80] applied this method to melt-grown trigonal monocrystals at different temperatures and for both orientations of the current with respect to the c-axis (Figure 5-45).   At high temperature, the experimental points are approaching a $T^{-3/2}$ curve (dotted line).   For temperatures below 250°K, the drift mobility passes through a broad maximum.   The decrease at low temperature has two equally reasonable interpretations.   This is due either

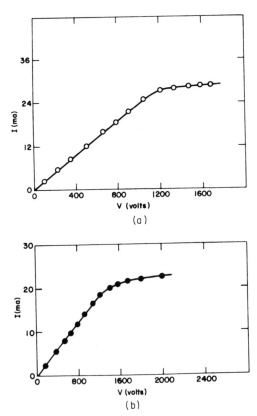

Figure 5-44 Acoustoelectric current saturation for trigonal selenium: (a) current $\parallel$ c at 300°K, (b) current $\perp$ c at 77°K.[80]

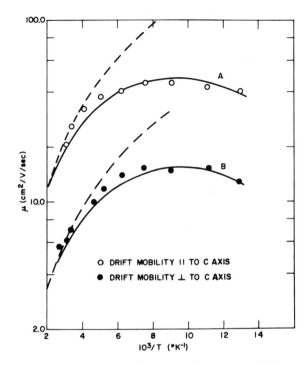

Figure 5-45   Temperature dependence on the drift mobility μ.   The dashed curves represent $\mu \sim T^{-3/2}$ normalized to the respective measured room temperature values.[80]

to a trap-controlled mobility involving traps 0.01 eV above the valence band with a density of about 10^{16} cm^{-3} or to scattering at charged impurities with a concentration of about 10^{17} cm^{-3}.

These results are in good agreement with the mobility obtained by magneto-resistance.   Not only is the same temperature dependence at high temperature observed, but also the absolute values at room temperature and the mobility ratio are very similar.   Since both methods are completely different, the good agreement indicates that at high temperature the true lattice mobility, μ_p, is approximately proportional to $T^{-3/2}$.   This rules out any hopping mechanism over this temperature range.   Because of the same temperature dependence for both orientations, it can be concluded that the transport mechanism is not qualitatively different when the holes are moving along or across the chains.

Influence of Impurities and Lattice Defects.   The strong enhancement of the conductivity by plastic deformation, connected with a decrease of the thermo-electric power indicates that the changes in σ and θ are due to the creation of shallow acceptors.   The result that these acceptors disappear by annealing at relatively low temperatures points to the existence of vacancies or double vacancies.   Since, in trigonal selenium, a vacancy has two broken bonds from the adjacent chain ends, it is possible that two acceptor states are created.   The first one is shallow, but the second must be much deeper due to the Coulomb repulsion.   In fact, the change in the photoelectric properties caused by plastic deformation and annealing suggests that a vacancy in trigonal selenium is a double acceptor.   The shallow state creates the hole density whereas the deep level acts as a recombination center and is responsible for impurity photoconductivity and photovoltage.   The conclusion that both levels belong to one center can be drawn from the fact that the impurity photovoltage is strictly proportional to the dark hole concentration (see Figure 5-56b).

There have been many attempts to increase the conductivity of trigonal monocrystals by the addition of impurity atoms, but thus far with virtually no success.   Whereas isomorphic tellurium can be easily doped with antimony up to a hole density of about 3×10^{18} cm^{-3}, selenium crystals obtained from a melt with an antimony content of about 0.1 % have practically the same conductivity as pure crystals.   The reason for this is probably that the antimony is rejected during the crystallization process.   In a manner similar to arsenic, the antimony atoms form, in the selenium melt, a star-like configuration by bonding three selenium chain ends together.   This arrangement is the reason for the strong enhancement of the viscosity of a selenium melt by the addition of arsenic or antimony.   Since it is relatively stable, this structure is not built in at crystallization from the melt.

Also of interest is the influence of oxygen.   Annealing of trigonal crystals over an extended time period in an oxygen atmosphere somewhat enhances the conductivity.   This change, however, is quite small compared with that obtained by plastic deformation.   No conclusive results have yet been reported which indicate that oxygen is an acceptor in trigonal crystals.   From a chemical point of view, this could be the case when oxygen is dissolved interstitially.   This may be possible because of the small size of the atom and the relatively loose packing of the trigonal lattice, particularly at or near dislocations.   On the other hand, oxygen possibly occupies a lattice site forming a Se—O—Se complex.   Concerning the doping nature of this configuration, however, nothing is known, so that one can only speculate about the influence of oxygen on the electrical properties of trigonal monocrystals.

The addition of halogens does not result in donor behavior as is to be expected when these atoms are located at lattice sites.   The reason for this negative effect, which is also observed for tellurium, is not yet understood but it seems to be due to a rejection of these atoms by the chains.   In the melt, halogen atoms chemically saturate the chain ends and thereby shorten the otherwise relatively long chains thus favoring crystallization. For example, chlorine enhances the crystallization rate and improves the quality of the crystals. Due to this secondary effect, it may also have some influence on the conductivity of the crystals.

Polycrystalline selenium shows a quite different behavior. Here the conductivity can be enhanced appreciably by the addition of halogens up to values of 10^{-3} Ω^{-1} cm^{-1} at room temperature. These are about 2 orders of magnitude larger than in the case of monocrystals.[81,82] However, this change is not due to a bulk effect but to the conditions at the surface of the crystallites.[83] Only at the surface or interstitially does chlorine have acceptor character. During the crystallization process, most of the chlorine atoms are shifted to the surface of the crystallites. There results a high density of surface acceptors which leads to an accumulation between layers and thereby to a conduction along the network of surfaces within the polycrystalline sample.

An opposite effect on polycrystalline selenium is noted in the case of thallium.[84,85] By the addition of atoms of this element, the conductivity strongly decreases and reaches values which are lower than those observed for most monocrystals. Thallium obviously compensates for the positive influence of chlorine either in a purely chemical manner or electronically.[86] This can be seen from Figure 5-46 where a strong decrease in σ occurs when the concentration of thallium is equal to that of chlorine.[87] Since, by the addition of thallium, conductivity values are reached which are considerably lower than those usually found for monocrystals, the thallium atoms at the surface apparently not only compensate for the chlorine acceptors at the surface but also for the acceptors within the crystallites. The influence of some other metals, particularly of Cu and of the alkaline metals,[84], is similar.

In addition to these metals and the halogens, oxygen also has some influence on the conductivity of polycrystalline selenium,[88,89,90], The conductivity of deoxygenated selenium is appreciably lower than that of material which contains some oxygen. As in the case of chlorine, the increase in conductivity probably involves a surface effect, because the oxygen atoms are expected to be acceptors at internal surfaces or at dislocations.

Conduction Mechanism. The basic problem with the transport mechanism in trigonal selenium is the discrepancy between the Hall mobility $\mu_H = \mu_0$ $\exp(-\Delta\varepsilon/kT)$ and the mobility $\mu_p \sim T^{-3/2}$ obtained in a number of experiments. The exponential temperature dependence of μ_H points to an activated transport mechanism, i.e., to a hopping process, whereas $\mu_p \sim T^{-3/2}$ indicates that the mobility is determined by phonon scattering. In the following, it shall be shown that this discrepancy can be solved by the use of a barrier model.

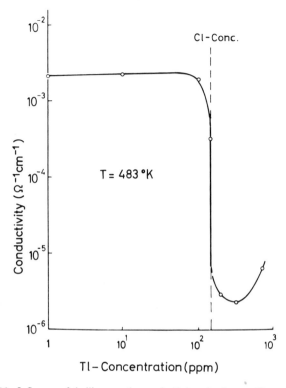

Figure 5-46   Influence of thallium on the conductivity of polycrystalline selenium.[87]

The main idea of this model is that σ is determined by a strongly inhomogeneous distribution of the hole density in the trigonal single crystals (Figure 5-47a).   Within a network of thin depletion layers, the otherwise constant hole density, p_0 is thought to be reduced to a much lower value, p.   The photoelectric properties discussed later show that this reduction is probably due to a strongly inhomogeneous compensation of the acceptors by deep donors, D, which act as electron traps.   The influence of plastic deformation on the temperature and voltage dependence indicates that the depletion layers are correlated to lattice defects.   Before discussing this in more detail, it shall be first shown that such a model can solve the mobility discrepancy.

The hole concentration in the depletion layers, p, is related to the temperature independent density, p_0, in the uncompensated crystal regions by the relation:

$$p = p_0 \exp\left(\frac{-\Delta\varepsilon}{kT}\right) \tag{7}$$

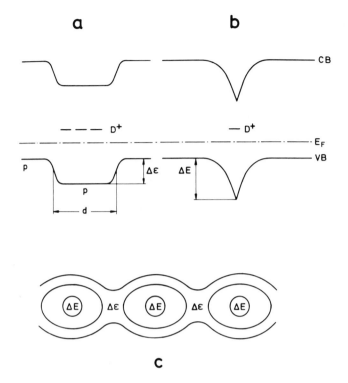

Figure 5-47 Barrier model of trigonal selenium: (a) deep donors, D, are inhomogeneously distributed, (b) donors are restricted to dislocations, (c) curves of equal potential created by the space charge surrounding adjacent dislocations. The acceptors responsible for the hole concentration, p_0, are not shown.

where $\Delta\varepsilon$ is the height of the potential barriers created by the strongly different carrier densities in both regions. As this difference depends on the degree of compensation, the ratio p/p_0 and consequently $\Delta\varepsilon$, can be greatly altered by changing the state of compensation. This happens either by partial neutralization of the donor states when optically excited electrons are trapped or by injection of additional holes into the depletion layers when a high electric field is applied. This simple model presents the important concept that an increase in the hole density within the depletion layers is necessarily connected with a decrease in the activation energy, $\Delta\varepsilon$.

The conductivity for such an inhomogeneous carrier distribution is mainly determined by the depletion layers and can be described by the relation:

$$\sigma = fpe\mu_p \tag{8}$$

μ_p is the hole mobility within the layers which is assumed to be the same as in the uncompensated parts of the crystals and f is a geometric factor arising from the details of the layers, i.e., their number, thickness, and orientation with regard to the applied field. The combination of (7) and (8) leads to

$$\sigma = f p_0 \, e \mu_p \exp\left(-\frac{\Delta\varepsilon}{kT}\right) \tag{9}$$

It is now assumed that the Hall effect and thermoelectric power are determined by the hole density, p_0, within the large crystal regions, whereas the conductivity is due to the depletion layers with their much lower hole density, p. The first assumption is valid when the thickness of the depletion layers, d, is small compared with d_0, the thickness of the region with hole density p_0. Moreover, it is necessary that $d/d_0 \gg p/p_0$, for the thin depletion layers to determine the conductivity. Since the Hall coefficient is $R = 1/p_0 e$ (the factor $3\pi/8$ is neglected for simplicity), the product $R\sigma$ gives for the Hall mobility the relation:

$$\mu_H = f \mu_p \exp\left(-\frac{\Delta\varepsilon}{kT}\right) \tag{10}$$

The relatively weak temperature dependence of $\mu_p^{3/2} \sim T$ is not observed in the exponential function for $\Delta\varepsilon$ values of about 0.1 eV. Thus, the exponential temperature dependence of μ_H is a consequence of the fact that the combination of R or θ with σ involves different hole concentrations. Hence this mobility is only of a formal nature and has nothing to do with the transport mechanism itself.

Most of the peculiar properties of trigonal single crystals can qualitatively be explained by this model. The strong voltage dependence of the conductivity at relatively low field strengths is plausible, because the field within the depletion layers is much larger than the field measured from the applied voltage and the electrode distance. The increase of σ with voltage is attributed to the injection of majority carriers from the adjacent regions with the hole density, p_0. This injection becomes more pronounced as the hole concentration, p, in the depletion layers decreases. Since p, according to formula (7), is rapidly reduced by lowering the temperature, the onset of the voltage dependence is shifted to lower voltages. The deviations from the simple theory of space charge limited currents give considerable support to the concept of an inhomogeneous distribution of the hole concentration and of the field strength within the crystals. In this case, the voltage dependence sets in at relatively low voltages when the injected carrier density reaches the much smaller value p in the depletion layers where, in addition, the field strength is much higher. The variation of the voltage dependent part of the conductivity with V^2/L^2 instead of V^2/L^3 can be understood, because the injection occurs into the N depletion layers which contribute little to the electrode distance $L = N(d_0 + d)$, for $d \ll d_0$.[93]

The curious influence of illumination on the Hall mobility can also be explained. Electrons activated to the conduction band are trapped by the deep jointed donors, D^+, thereby diminishing the degree of compensation. As a result, the activation energy is lowered and the hole density is enhanced. Each etep in the TSC curves of Figure 5-35 is characterized by a constant degree of compensation, that is, a constant density of neutral donors, D^x, and thus a constant value of $\Delta\varepsilon$. If all ionized donors, D^+, could be occupied by electrons, then $\Delta\varepsilon$ should vanish. Despite the large energetic distance of the electron traps, D^+, from the conduction band, this state cannot be reached completely even at high illumination intensity and low temperature because of the hopping recombination to be discussed later. However, a decrease of $\Delta\varepsilon$ by about an order of magnitude is possible, leading to the variety of curves in Figure 5-35. The behavior of the thermoelectric power indicates that the influence of illumination on the hole density, p_0, is negligible. The Hall mobility $\mu_H = R\sigma = \sigma/p_0 e$, therefore, strongly depends on illumination due to the variation of σ. This dependence, however, has no special physical meaning because the true hole mobility μ_p remains unchanged. Only the inadequate combination of thermoelectric power and conductivity leads to the formal change in the mobility by illumination.

The frequency dependence of the conductivity is, in this model, a result of the short circuiting of the depletion layers at high frequency. For a more detailed explanation, the hole concentration, p, of these layers has to be substituted by a spectrum of concentrations, p_i. The layers with a conductivity σ_i are short circuited for $v > v_i = \sigma_i/\varepsilon\varepsilon_0$, where v_i is the reciprocal of their dielectric relaxation time. At lower frequencies, their contribution to the resistance is $R_i \sim d_i/\sigma_i$. Here d_i is the total thickness of all layers with conductivity σ_i in the current direction. Stubb et al.,[63] calculated d_i for layers with σ_i values between 10^{-10} and $10^{-2}\ \Omega^{-1}\ cm^{-1}$. The remarkable result of these calculations is that approximately 80% of the crystal has a conductivity equal to the value measured at 24 GHz, which is assumed to represent the conductivity of the undisturbed regions with a hole concentration p_0. That gives for the geometry factor f in equation (8) a value of about 5 which is in fairly good agreement with values determined from μ_p in equation (10).

The influence of a high dc voltage on the frequency dependence is plausible because at high voltages the injection of holes into the depletion layers occurs and this diminishes the influence of the barriers. Because of this, the frequency dependence of σ is significantly lowered. There exists, unfortunately, no detailed investigation on the influence of illumination on the frequency dependence. Preliminary results indicate that the frequency dependence is considerably decreased by the application of light as would be predicted. Since, at high frequencies, the influence of the barriers is largely eliminated, the absolute value of σ is orders of magnitude higher than the dc conductivity and the activation energy is considerably reduced. Hence, a combination of the high frequency conductivity with R or θ leads to a mobility of the same order of magnitude as the hole mobility.

The value of μ_p that is obtained from magnetoconductivity measurements is also in accordance with the model. The relative change of conductivity is the same in the depletion layers and in the uncompensated regions, if the mobility in both regions is equal. Recently, Lanyon[91] has shown that the mean free path of the holes in trigonal selenium is of normal size. This can be concluded from the backward characteristic of selenium rectifiers which is interpreted by an avalanche which results from impact ionization. The onset of the multiplication process leads to a mean free path of about 200 Å. This value is at least an order of magnitude smaller than the thickness of the depletion layers, d, for which a magnitude of about 0.3 μm is obtained under the assumption that the hole density in the undisturbed regions amounts to 2×10^{15} cm^{-3}.[60] Thus, holes moving through a depletion layer are scattered more than 10 times, yielding a normal influence of the magnetic field.

Results which are at variance with those of the barrier model have been presented. Measurements of the acoustoelectrical current saturation suggest that the correct hole mobility is obtained from the relation $\mu_d = v_s/F_{sat}$ where F_{sat} is equal to the saturation voltage divided by the electrode distance. This points to a homogeneous field distribution during the short voltage pulses used in these experiments. The current saturation is observed following an ohmic part at lower voltage. Unfortunately, Mort[80] gives no conductivity values for this region. Assuming a hole density of 10^{14} cm^{-3}, the measured room temperature mobility of about 30 cm^2/Vs yields a conductivity value of about 5×10^{-4} Ω^{-1} cm^{-1} below the saturation. It is about 2 orders of magnitude larger than the usual dc conductivity. Similar high conductivity values are only obtained at high frequencies, where the influence of the barriers is substantially lessened. The same is true for measurements with short voltage pulses. The reason for this behavior is not yet understood, so that further experiments are necessary to solve this intriguing problem.

The important conclusion of these considerations, therefore, is that the discrepancy between Hall mobility and the mobility obtained by other experiments is not a real one and that most of the unusual properties of the trigonal crystals can be explained qualitatively by the barrier model. A quantitative description of the model, however, is more difficult. The most important problem involves the nature of the barriers. The influence of plastic deformation on the temperature and the voltage dependence of the conductivity implicates that they are closely related to lattice defects. The relatively small influence of annealing suggests that dislocations are involved. The network of dislocations leads to a network of potential barriers if the dislocations are surrounded by relatively thick negative space charge tubes when these tubes overlap.   The negative space charge is assumed to arise from the compensation of the holes by donors located at the core of the dislocation or close to it.   Little is known about the nature of these donors.   The photoconductive properties, particularly the spectra of optical quenching, indicate that various crystals contain at least two different kinds of donors.   This points to the presence of impurity atoms, possibly interstitial metal atoms attracted by the dislocations.   It may also be possible, however, that a dislocation itself creates the donor states, e.g., by the local change of density.[92]

The potential distribution connected with the space charge tube of a dislocation is shown schematically in Figure 5-47b.   A hole moving across the core of the dislocation has to overcome the whole potential barrier.   For holes passing between two dislocations, however, the barrier is lower, because then the height of the barrier is determined by the degree of overlap (Figure 5-47).   Midway between two neighboring dislocations the effective barrier height $\Delta\varepsilon$ is only part of the potential barrier ΔE in the core: $\Delta\varepsilon = k\Delta E$. Tuomi,[93] who has discussed this in more detail, has shown that k has values between 0.1 and 0.2.   The same result was obtained by Stöckmann[94] who gives for k a value of about 0.1.   With a hole concentration $p_0 = 2 \times 10^{15}$ cm^{-3}, the number of donors lined up in the dislocation is about 6×10^6 cm^{-1} and the distance between the dislocations is about 1×10^{-5} cm.   Lotthammer[95] could explain the dark as well as the photoconductivity behavior semiquantitatively.   The relatively small distance between the dislocations (which we assume indicates that the presence of a network of small angle boundaries) is the probable reason for the potential barriers.   In accordance with this picture, etch pit patterns often reveal a lineage of dislocations even for the best vapor-grown crystals as can be seen from Figure 5-48a.   After plastic deformation, the dislocation density is so high that the small distance needed for the creation of barriers is easily obtained, since the distribution of the dislocations is not homogeneous (Figure 5-48b).

According to its temperature dependence, the hole mobility, μ_p, is determined by phonon scattering.   Since trigonal selenium has lattice modes which are strong and optically active[96] and because it is also piezoelectric, the question arises as to what type of scattering gives rise to this dependence. Lucovsky, Keezer, and Burstein[97] calculated the effective charge and the polaron coupling constant, α, of the optical lattice modes, as is shown in Table 5-6.   The polaron mass, $m_{polaron} = (1 + \alpha/6)m^x$, is almost the same as the effective mass m^x.   Nevertheless, the scattering of carriers by polar modes can be of importance as is shown by the behavior of Group III–V compounds which have coupling constants comparable to those calculated for selenium.

Lucovsky et al.[97] estimated the absolute value of μ_p for optical and acoustical

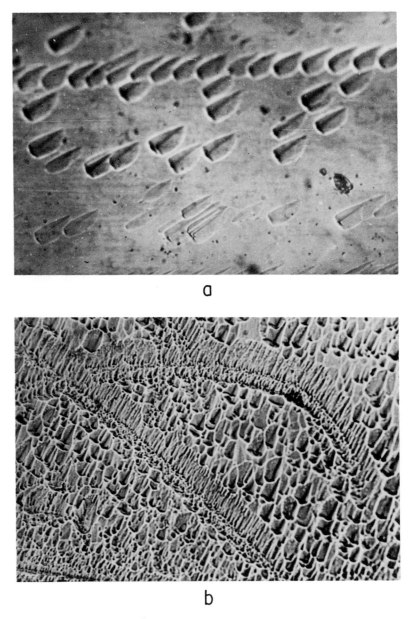

Figure 5-48 (a) Etch pits on (10$\bar{1}$0) face of vapor-grown trigonal selenium crystal. (b) Etch pits on melt-grown crystal after plastic deformation.

TABLE 5-6   EFFECTIVE CHARGE AND POLARON
COUPLING CONSTANT OF OPTICAL
LATTICE MODES[97]

Mode	Effective Charge $e*/e$	Polaron Coupling Constant α
A_2	0.7	0.18
E	1.2	0.26
E	0.15	0.005

phonon scattering.   Assuming an effective mass $m^x = 2m_0$, the Howard-Sondheimer formula yields for the mobility due to optical phonons: $\mu_p \approx$ 50 cm^2/Vs at room temperature.   Using an average value of the elastic constant and a value of 5 eV for the deformation potential, the application of the Bardeen-Shockley equation gives $\mu_p \approx 20$ cm^2/Vs for the mobility due to scattering by acoustical phonons.   Since the experimentally found value lies near 30 cm^2/Vs, both scattering mechanisms seem to be of importance at room temperature.   The observed current saturation indicates that piezoelectric scattering has also to be considered.   That has recently been discussed by Suosara.[98]   For a more detailed interpretation of the scattering mechanism, particularly for the evaluation of the effective masses, further investigations are still necessary.

Relevant to the discussion of the conduction mechanism is the finding that the mobility ratio $\mu_\parallel/\mu_\perp \approx 3.5$ and that it does not depend on temperature. It is similar to the equivalent data for tellurium where $\mu_\parallel/\mu_\perp \approx 2$ and is also approximately temperature independent.   Since $\mu_\parallel/\mu_\perp = (\tau_\parallel/\tau_\perp)(m_\perp^x/m_\parallel^x)$, this anisotropy can either be due to an anisotropy of the effective mass, of the relaxation time, or of both.   For tellurium, the mass ratio is known by cyclotron resonance to be $m_\parallel^x/m_\perp^x = 2.4$,[9] thus giving a larger mass for the orientation parallel to the c-axis than perpendicular to it.   This mass ratio, therefore, is overcompensated by a reverse ratio of the relaxation times.   For selenium, however, the mass parallel to the c-axis is probably smaller than in the orthogonal direction.   This can be inferred from the orientation dependence of the electroreflectance signal which is found to be for $F\|c$ about a factor of 3.5 larger than for $F \perp c$.   Since the height of the signal becomes larger with decreasing effective mass, this result gives qualitative evidence for $m_\parallel^x < m_\perp^x$. Unfortunately, a detailed theory about the influence of an anisotropic mass on the size of electroreflectance and electroabsorption does not yet exist. As a consequence, no quantitative values for the mass and the relaxation time ratios can presently be evaluated from $\mu_\parallel/\mu_\perp$.   Still another unsolved problem has to do with the direct experimental evidence of the potential barriers within the crystals. The solution of this crucial problem is very difficult, because the thickness of the barriers is of the order of magnitude of 0.1 μm. Therefore, even photoelectric investigations with a tiny light spot encounter great difficulties, since the photovoltage which is generated has a different sign on either side of the barrier. With the use of a much larger light spot, it averages out to an immeasurably small value. Moreover, the irregular network of the barriers restricts such investigations to very thin crystals or to light absorbed only in a thin surface region where the conditions may be altered by the crystal surface itself. Possibly, the potential distribution over a crystal with an applied voltage can be elucidated by the use of a scanning electron microscope or similar techniques. However, no conclusive results have been obtained so far.

Photoconductivity

The photoconductivity of trigonal selenium plays an unusual role in the history of photoconductivity. Since its detection 100 years ago, a large number of investigations have been performed on polycrystalline material as well as on monocrystals. The experimental results, particularly the extremely slow decay of the photocurrent after illumination and its weak dependence on the light intensity have not been understood until recently. In the following section, the most important features of these numerous investigations shall be briefly mentioned, but only the more recent results on single crystals will be discussed in detail.

Spectral Sensitivity. Since trigonal selenium is a *p*-type semiconductor, the photoconductivity is due to the creation of an additional hole concentration (Δp) by illumination. This can be done by excitation of electrons from the valence into the conduction band with subsequent trapping of the minority carriers to give a sufficiently high lifetime to the majority carriers and thus, a large value of the photocurrent. Moreover, the electrons can be excited directly into local levels of the forbidden band. These different processes are observed in the spectral sensitivity curve since the first one is related to intrinsic absorption, whereas the second one is determined by the energetic position of the local levels in the band gap.

For trigonal selenium crystals, the spectral sensitivity of the photoconductivity has a pronounced maximum near 1.7 eV (Figure 5-49) which is due to band-to-band transitions. In some cases, a further flat maximum at about 2 eV is found.[99] According to the measurements of Prosser[100] shown in Figure 5-50, the relative heights of both maxima are strongly temperature dependent, the maximum near 1.7 eV predominating at low temperatures. This maximum also depends on the polarization of light, since for $E \perp c$ it

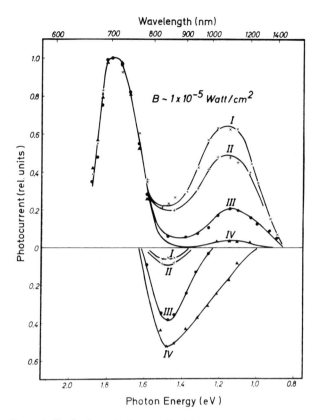

Figure 5-49 Spectral distribution of photoconductivity and optical quenching for trigonal selenium: (I) after strong plastic deformation; (II–IV) after annealing.[60]

is a little higher and sharper than for $E\|c$.   This is in accordance with the absorption characteristics, because the maximum is located at an energy where the slope of the absorption edge is larger for $E \perp c$ than for $E\|c$. It can be seen from Figure 5-17 that it lies at an energy where the absorption edge reveals no structure.   Therefore, the absolute value of the absorption coefficient, K, must be the reason for it.   The maximum at 2 eV for $E \perp c$, however, corresponds to a maximum of K.   But since it is also found with the same height for $E\|c$ where K has no maximum, this may be fortuitous. The temperature shift of the 1.7 eV maximum corresponds to the temperature dependence of the absorption edge, because with increasing temperature the absorption coefficient is enhanced.   Thereby, the value of K where the maximum occurs is shifted to lower energies.

The maximum at 1.7 eV can be correlated with the conductivity change in the depletion layers and correspondingly to a decrease in $\Delta\varepsilon$.   This can

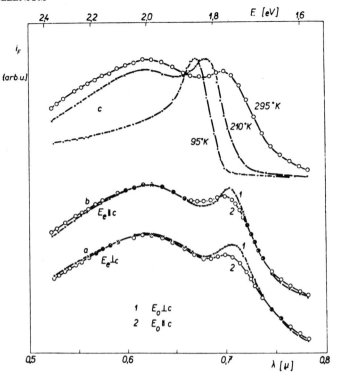

Figure 5-50 Dependence of the spectral distribution on temperature, on the orientation of the electric field, and on the polarization of light (not calculated per unit of incident light).[100]

directly be seen in Figure 5-35 where the photoconductivity due to light of 1.7 eV has been for the most part frozen in following illumination. In the TSC curve measured stepwise, the conductivity at a constant temperature is reduced in a stepwise manner and the activation energy is enhanced.

The maximum occurs when the penetration depth of the light is of the same order of magnitude as the thickness of the crystal. It is, therefore, assigned to excitation of carriers in the bulk of the crystal. The rapid decrease in the spectral sensitivity at higher energy is assigned to the transition from bulk excitation to surface excitation due to the quickly rising absorption coefficient. For high photon energies, the excitation is restricted to a very thin surface layer, giving rise to a high density of excitation processes and thus to a large recombination rate. The maximum at 1.7 eV occurs at an energy where both the penetration depth and the quantum efficiency of the light are sufficiently high so that as many electron traps as possible can be occupied.

Impurity Photoconduction and Photovoltage. The spectral sensitivity curve of trigonal Se crystals is appreciably changed in the low energy region beyond the absorption edge by plastic deformation, as has been shown by Stuke (Figure 5-49). Impurity photoconductivity is observed with a maximum at 1.2 eV, the relative height of which depends on the degree of deformation and subsequent annealing. By annealing, it disappears almost completely in much the same manner as the additional dark conductivity obtained by plastic deformation. Therefore, a complex correlation between the dark conductivity at room temperature and the relative height of the impurity maximum at 1.2 eV exists. When the size of the 1.2 eV maximum is reduced by annealing, its effective height can be further diminished by optical quenching in the same energy region. The latter effect is the reason that a simple relation to the dark conductivity is not obtained.

These difficulties can be avoided if one measures the photocurrent of a photovoltaic contact without application of an external voltage. This short circuit current is proportional to the light intensity. For small intensities, the simple relation is also valid for the photovoltage. By studying the dependence of the impurity photovoltage on plastic deformation and subsequent annealing, Geisler[101] found that the relative height of the photovoltage (Figure 5-51a) is strictly proportional to the dark hole concentration (Figure 5-51b). This discovery implies that the corresponding energy levels located near the middle of the band gap and the shallow acceptors belong to the same lattice defects, probably vacancies, as pointed out before.

There are a few results which suggest an interaction between the vacancies created by plastic deformation and impurity atoms. Lotthammer[95] found that crystals with some content of thallium, after annealing, still retained a relatively high impurity photoconduction. According to the photovoltaic measurements of Geisler,[101] some trigonal crystals reveal, after deformation, a second impurity photovoltage maximum at 1.6 eV which does not disappear by annealing in contrast with the 1.2 eV maximum. Possibly this second maximum is due to diffusion of interstitial impurity atoms to vacancies. Thereby, a relatively stable configuration can be formed which is not much changed by further annealing.

Quenching of Photoconductivity. Besides the increase of the conductivity by light with photon energy smaller than the band gap, optical quenching also occurs in this energy region. This has been reported by Stuke[60] (Figure 5-49) and has recently been studied in more detail by Fuhs and Stuke[102] and by Lotthammer.[95] In such a measurement, the crystal is illuminated by light which excites intrinsic photoconduction and by additional light of lower photon energy. Optical quenching of the photocurrent can then occur because, due to optical emptying of minority carrier traps, i.e., electron traps, the recombination is greatly enhanced. This process is of particular concern because it provides direct information about the properties of the electron

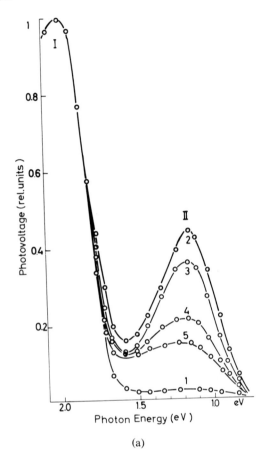

(a)

traps. Usually the long wavelength threshold of the optical quenching spectra is taken for the trap depth. The maximum is of minor importance because it is mainly due to the onset of a positive effect of the band-to-band transitions.

Fuhs and Stuke and also Lotthammer (Figure 5-52) found two types of crystals with different quenching behavior. The first one contains mainly traps with a depth of about 0.8 eV, because optical quenching starts at this energy. The second type reveals an additional threshold at about 0.65 eV, indicating that these crystals contain electron traps of somewhat smaller depth.

Further information about the levels involved is obtained from the temperature dependence of optical quenching because optical quenching must come to an end when the electron traps are thermally emptied. As can be seen in Figure 5-53, where the optical quenching rate $Q = -\Delta I_{ph}/I_{ph}$ defined by

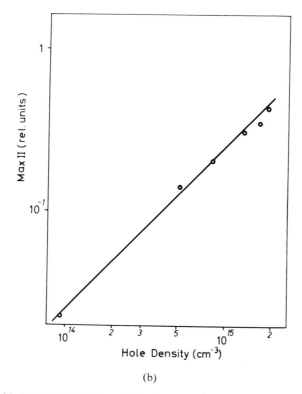

(b)

Figure 5-51 (a) Spectral distribution of the photovoltage: (1) before plastic deformation, (2) after plastic deformation, (3–5) after annealing. The curves are normalized at 2 eV. (b) Dependence of relative height of impurity photovoltage on dark hole concentration.[101]

the relative decrease of the photocurrent due to the additional light is plotted against the temperature, two electron traps are active.   Near 90°K for curve d, Q decreases significantly with rising temperature because the shallow traps start to empty by thermal activation.   They are mostly empty near 120°K so that at higher temperatures only the deep traps are involved.   The kink at about 180°K signifies that here thermal emptying of the deep traps sets in. For crystals containing only these traps, the strong decrease between 90–120°K is not observed (curve C).   Hence, up to a temperature of about 180°K, the population of the traps stays approximately unchanged by thermal activation.

Basically, the same results are obtained by thermally stimulated current measurements (TSC curves) on different crystals.   For these investigations, the electron traps are filled by illumination at a sufficiently low temperature. Then the light is switched off and the photocurrent decays slowly with time. For trigonal selenium, this decay is extremely slow so that a great part of the

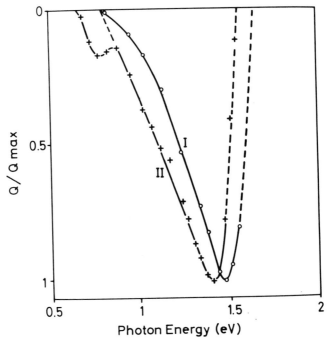

Figure 5-52   Energy dependence of optical quenching for two types of trigonal crystals. Curves normalized to maximum.[95]

photocurrent can be frozen in. When the temperature rises with an approximately constant velocity, the electrons are steadily activated into the conduction band and recombine. Since at low temperature this process predominates for the shallow traps, crystals with 2 trap levels have 2 maxima in the TSC curve (Figure 5-54) whereas for crystals with 1 kind of trap, only 1 relative broad maximum is found (Figure 5-54).

Approximate values of the thermal activation energies of traps can often be evaluated from TSC curves. The methods normally used for such calculations, however, deal with traps for majority carriers. Therefore, in the case of trigonal selenium, only a rough estimate is obtained when, for instance, the formula of Randall and Wilkins,[103] $E_t = 25 k T_{max}$, is used. Following this relation, the 2 peaks in the TSC curve correspond to activation energies of 0.25 and 0.35 eV. Still smaller values were quoted by Kolomiets and Khodosevitsch[104] for polycrystalline material, viz., 3 trapping levels at 0.12, 0.16, and 0.26 eV. For single crystals, Bakirov and Dzhalilov[105] also found 3 traps levels at 0.1, 0.14 and 0.17 eV, but these were assigned to hole traps. Henisch and Engineer[106] reported 2 electron traps of 0.45 and 0.54 eV distance from the conduction band from TSC curves measured at a relatively rapid tempera-

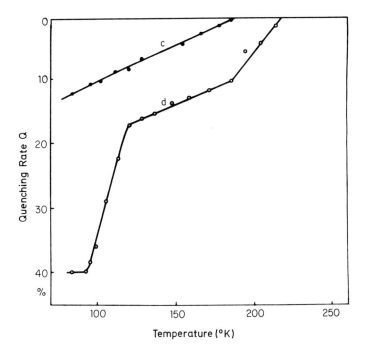

Figure 5-53 Temperature dependence of the quenching rate, Q: c corresponds to crystals type I and d to type II in Figure 5-52.

ture rise of approximately 14°/sec. These values are somewhat nearer to those found optically. The large variety of values obtained by different authors indicates that the evaluation of the trap depth from TSC curves is highly problematic. This is probably due in part to the different quality of the crystals used in these investigations. The lack of a sufficient theory for this process which, in the case of trigonal selenium, is particularly complicated by potential barriers and impurity recombination, will be discussed later.

The conclusion that barriers are involved can be drawn from the observation that optical quenching is appreciably diminished by application of the relatively low external electric field of about 10^3 V/cm (Figure 5-55). Even for a trap depth of about 0.2 eV, this field is about 2 orders of magnitude too low for the emptying of the traps by impact ionization. This is readily seen from the fact that the field strength should amount 10^5 V/cm if the holes take up 0.2 eV from the external field on a mean free path of 10^{-6} cm. Also for another process, the lowering of the thermal activation energy (similar to the Schottky effect), the field of 10^3 V/cm is by far too small, indicating that the effective field is much larger within those crystal regions which are necessary for both conductivity and photoconductivity.

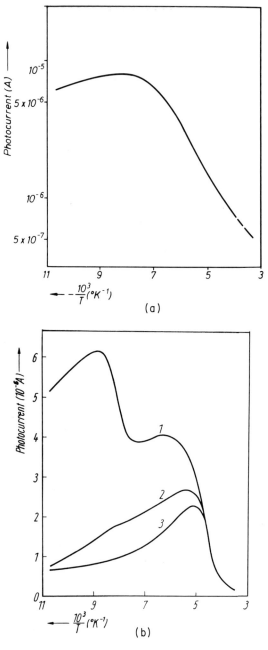

Figure 5-54 Thermally stimulated currents: (a) for crystals of type I, (b) for crystals of type II; (2) after optical quenching at 90°K, and (3) after field quenching at 90°K.[102]

256

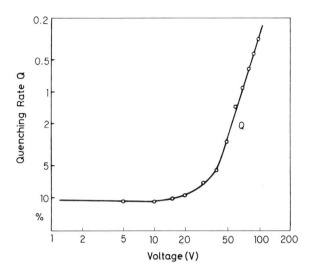

Figure 5-55   Dependence of the optical quenching rate, Q, on the applied voltage.[95]

Since these regions are assigned to lattice defects, plastic deformation should also play a role in optical quenching.   In fact, it has been found by Lotthammer that a relation between the dark conductivity created by plastic deformation and subsequent annealing and the quenching rate, Q, exists. Before deformation, Q is very small.   It rises with increasing conductivity, passes through a maximum, and then decreases quite sharply because the positive effect of the additional light predominates at high conductivity. This behavior suggests that not only the positive effect, but also optical quenching is closely related to the second acceptor state near the middle of the bandgap.   The different magnitude of Q for various crystals is also of interest.   For carefully handled vapor-grown crystals, Q is very small; following strong plastic deformation and subsequent annealing, a relatively low value of Q, compared with that of melt-grown crystals, remains.   For the latter crystals, Q does not possess a large magnitude before deformation, but the increase due to deformation can be reversed by annealing.   Different characteristics among various types of crystals are found, especially in the spectra of electron traps.   This variation is attributed to differing concentrations of these traps.   The influence of two types of energy levels on the magnitude of Q is in accordance with the fact that for a large value of Q not only the density of traps, but also that of recombination centers has to be efficiently high.

The quenching rate, Q, is plotted as a function of the intensity, B_z, of the additional illumination in Figure 5-56.   At low intensities, Q is proportional

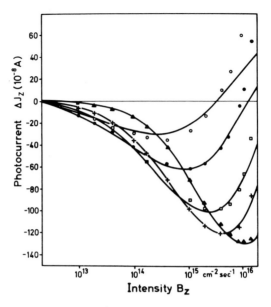

Figure 5-56 Dependence of the change ΔI_z of the photocurrent on the additional light intensity B_z. Points: experimental values; curves: calculated.[95]

to B_z. With increasing B_z, it passes through a maximum and finally reversed to a positive effect at high values of the illumination intensity where impurity photoconduction predominates. The latter process has a stronger illumination dependence than the quenching, so that it takes over for large values of B_z. This can be understood when one realizes that the positive effect is caused by the activation of electrons to the conduction band by a two photon process via energy levels near the middle of the band gap. For such processes a B_z^2-relation is to be expected and has, in fact, been found by Lotthammer in the high illumination range. The proportionality of Q to B_z, on the other hand, is due to the emptying of electron traps by photons. That the positive effect is created by a two photon process can also be seen from the fact that optical quenching is also observed when a steady illumination of light with a photon energy of 1.2 eV (the maximum of the impurity photoconductivity), is taken. Since the quenching is due to the emptying of electron traps, the photons of 1.2 eV obviously can also fill these traps.

Recombination Mechanism. Information, particularly about the character of the recombination mechanism, can be obtained by measurements of the rise and decay of the photocurrent and by studying its dependence on the light intensity. The main feature of the illumination dependence is the extremely slow increase in the photocurrent with light intensity. Fuhs and Stuke[102] have described their results by a logarithmic relation:

$$I_{ph} = c \ln\left(\frac{B}{B_0} + 1\right) \tag{11}$$

where $B_0 = 6 \times 10^{-10}\,\text{W/cm}^2$ and c decreases with rising temperature approximately proportional to $1/T$. Hemilä and Tuomi,[107] Tuomi,[93] and also Lotthammer,[95] on the other hand, found that their data were better described by a power law:

$$I_{ph} \sim \left(\frac{B}{B_0}\right)^n \tag{12}$$

where n is 0.1–0.2. Because of the small value of the exponent, it is difficult to distinguish clearly between the two relations. The illumination dependence is different from that of most other photoconductive materials for which the exponent n is usually between 0.5 and 1. This magnitude can be fairly well understood by a normal recombination mechanism involving traps for minority carriers as well as for majority carriers. The small value of n observed for trigonal selenium, however, suggests a different type of recombination process.

The steady state photocurrent is determined by the equilibrium between the generation rate of free carriers, g, and their recombination rate, which leads to the relation $\Delta p = g\tau_{eff}$, where Δp is the density of additional holes and τ_{eff} their effective lifetime. The small increase in Δp with light intensity $B \sim g$ indicates that τ_{eff} must decrease markedly when the illumination is enhanced. Concurrent with the weak intensity dependence, a very slow decay is to be expected. This feature has already been observed by the early decay studies on single crystals. It is for this reason that the photoconductive selenium cell lost its technical importance after the discovery of materials with smaller time constants. Corresponding to the power law found for the illumination dependence, Tuomi[93] described the decay by the relation:

$$I_{ph}(t) = I_0 \left(\frac{t}{t_0} + 1\right)^{-n} \tag{13}$$

where n has the same value of about 0.1–0.2 as in formula (12). The most comprehensive investigation of the decay has been performed by Fuhs and Stuke[102]. At low temperatures, according to this work, the relative photocurrent drops linearly when the time after illumination is plotted logarithmically (Figure 5-57). The decay therefore follows the relation:

$$I_{ph}(t) = I_0 \left[1 - \alpha \ln\left(\frac{t}{t_0} + 1\right)\right] \tag{14}$$

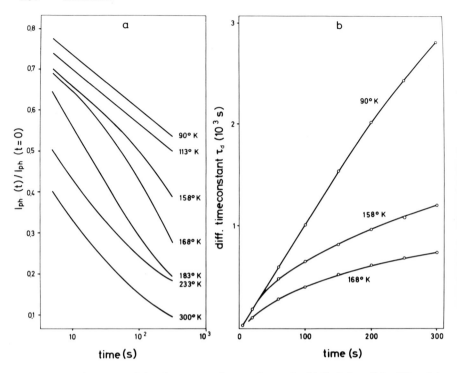

Figure 5-57 (a) Decay of the photocurrent for type I crystals. (b) Variation of the differential lifetime, τ_d, with time.[102]

A striking feature is the small slope of this curve at 90°K. The photocurrent drops only by about 10% when the time increases by an order of magnitude. The same small gradient is found at 113°K. Above 140°K, however, a second decay process becomes involved which, in this plot, leads to a different decay law with a steeper decay curve. At higher temperature this steeper decay predominates over the whole time interval investigated (5–500s). Obviously there are two competing recombination processes: process I has a very slow, approximately logarithmic decay. The slope, α, is independent of temperature, indicating that no thermal activation is involved in the process. The decay curve due to process II, however, is considerably steeper and non-logarithmic. Moreover, its decay rate is considerably enhanced with rising temperature.

The properties shown in Figure 5-57 are observed for crystals with only one kind of electron trap. Crystals which have deep as well as shallow traps reveal the transition between processes I and II a second time (Figure 5-58). For these crystals at 90°K, the photocurrent decays by process I, but at 100°K, process II, with its faster decay rate, predominates. However, unlike the

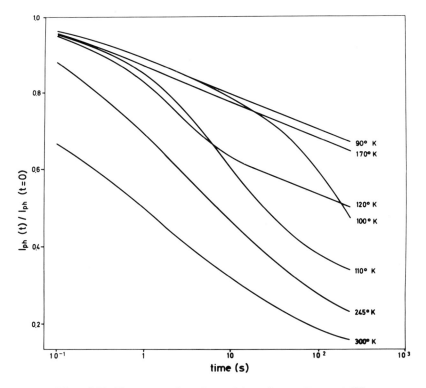

Figure 5-58   Temperature dependence of decay for type II crystals.[102]

behavior shown in Figure 5-57, at higher temperature process II is followed by process I, which again prevails at 170°K.   Finally, above 200°K the decay is determined by process II.

For the discussion of such complicated decay processes, it is useful introduce a differential lifetime, τ_d, which is defined by the relation:

$$\frac{dI_{\mathrm{ph}}}{dt} = -\frac{I_{\mathrm{ph}}}{\tau_d} \tag{15}$$

The simplest type of decay, the exponential one, is characterized by τ_d being independent of time.   The very slow logarithmic decay, however, is due to an extremely large increase of τ_d during the decay (Figure 5-57b).   The most important feature of process I, therefore, is that τ_d increases by orders of magnitude while the free hole concentration changes only slightly.   In the temperature range where process II predominates, the value of τ_d rises much more slowly, thus yielding a steeper decay curve.   At low temperature, only process I is observed during the investigated time interval.   The transition I–II is continuously shifted to smaller values with rising temperature.

Apparently, process II is related to the thermal emptying of electron traps. It is to be expected that it can also be obtained at 90°K by the application of quenching light during the decay, which in fact has been observed (Figure 5-59). The relatively simple relation between the decrease of the photocurrent Δ_1 after a constant decay time and the intensity of the quenching light B_z (Figure 5-59) is also in accordance with this model, because the trap emptying processes should be approximately proportional to B_z. From the fact that the enhancement of the optical quenching by an electric field has shown that the electron traps can already be emptied by the relatively low external field strength of about 10^3 V/cm, a strong influence of the field on the decay is to be expected. Figure 5-60 shows that at approximately the same field strength, the decay rate is considerably enhanced by the electric field. In order to prevent heating, short voltage pulses of 5 μs duration and a repetition time of 1 ms, were used. The quantity, Δ_f, defined as the difference from the low voltage curve reveals a relatively simple relation to the field strength as can be seen from Figure 5-60, where Δ_f is plotted logarithmically against the reciprocal height of the voltage pulses. Straight lines are obtained which, for crystals with two kinds of electron traps, have at 90°K a smaller slope for low field strength. At 140°K, where the shallow traps have already been thermally emptied, only the large slope is observed. This is characteristic of the deeper traps. Further important information is

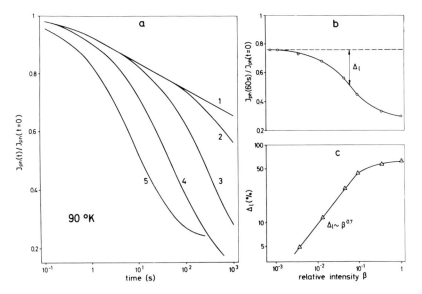

Figure 5-59 Decay of the photocurrent at 90°K under the influence of quenching light with different relative intensities β:$^{10^2}$ (1) $\beta = 0$, (2) $\beta = 1.2 \times 10^{-3}$, (3) $\beta = 1.3 \times 10^{-2}$, (4) $\beta = 9.5 \times 10^{-2}$, (5) $\beta = 1$.

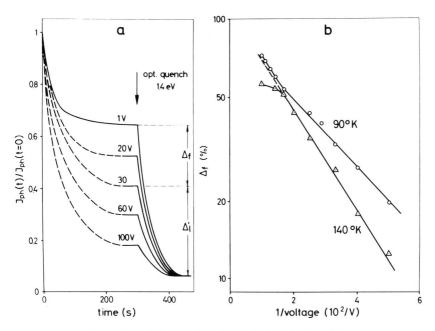

Figure 5-60   Influence of an electric field on the decay.[102]

obtained if after the field quenching, optical quenching is applied during the decay. This leads to the same final value of the photocurrent (Figure 5-60) for all curves. This result indicates conclusively that the same electron traps are involved in field and optical quenching. Trap emptying by an electric field also occurs in polycrystalline selenium. Under certain conditions, this can lead to oscillations of the photocurrent as measured by Hasler and Polke.[108] These oscillations have very low frequencies. Only recently such instabilities were found in selenium single crystals by Kühler.[43]

The interpretation of the extremely slow decay and the weak dependence of the photocurrent on the light intensity related to process I requires an approximately exponential increase of the recombination rate with the hole concentration. Since it is extremely difficult, if at all possible, to explain this phenomenon by a usual recombination model with several types of traps and recombination centers, Fuhs and Stuke assigned process I to a different recombination mechanism, namely to hopping recombination. As is shown schematically in Figure 5-61, this mechanism is characterized by a transition of the trapped electrons directly from the traps, T, into the recombination centers, R, by tunneling between these two levels. This hopping recombination is the predominant recombination process when thermal trap emptying can be neglected, specifically, at low temperatures and for large trap depths.

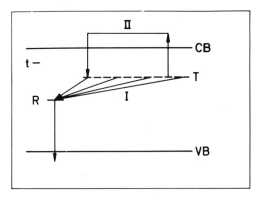

Figure 5-61   Hopping recombination model.[102]

The probability, w, of such a hop strongly depends on the distance, r, of the trap from the nearest recombination center. Since the overlap of wavefunctions is the most important factor for the hopping probability, it can be expected that $w = w_1 \exp - (r/r_0)$ with a constant r_0 of the same order of magnitude as the Bohr radius. The time constant $\tau = 1/w$ therefore increases exponentially with r. For a simple geometrical configuration the number of traps ΔN_T with a distance between $r + \Delta r$ is given by $\Delta N_T = 4\pi r^2 N_T \Delta r$. Then ΔN_T increases with r^2, but the hopping probability decreases exponentially with r, thus overcompensating for the increase in ΔN_T. Therefore, the hopping probability from traps with a small distance to the nearest recombination centre, R, is orders of magnitude greater than for large values of r. This mechanism, which is schematized in Figure 5-61 using arrows of different length, easily explains the main features of process I, including its temperature independence.

Process II, on the other hand, is obviously due to excitation of electrons from the traps into the conduction band and the subsequent recombination via the recombination center. Particularly at low temperatures, this kind of recombination occurs following numerous retrapping steps until finally, the electron recombines via R. That can take place when the electron is in the band as well as in a trap located very near to R. Then w is high enough so that $\tau = 1/w$ becomes comparable to the recombination lifetime, τ_R, and both processes cannot be distinguished by decay measurements. The decay due to process II, therefore, is more rapid as the excitation of the trapped electrons into the conduction band becomes more intense. This can be caused either by phonons, by photons, or by an electric field. It can approximately be described by the hyperbolic relation, $I_{ph}(t) = I_0(t/t_0 + 1)^{-n}$, with $n \approx 0.3$. This value is still small compared to 0.5 which is usually found for the decay of other photoconductors. Multiple trapping is the reason for this behavior.   Hole traps have to be considered too, because they can slow down the decay rate when they are in thermal equilibrium with the valence band.

The recombination model of Fuhs and Stuke has recently been refined by Stöckmann[94] who assumed that the relevant recombination processes occur inside the potential barriers.   The location of the electron traps is restricted to the dislocations while the recombination centers are statistically distributed. As is shown in Figure 5-62, the levels, R, are near the center of the barrier below the Fermi level.   Because these centers are occupied by electrons, they cannot act as recombination centers.   For the hopping recombination, only the unoccupied centers which are at a larger distance from the traps in the core of the dislocations are of importance.   In this model, therefore, the height of the barriers and the shortest hopping distance, r, are automatically related.   Since illumination diminishes the barrier height, the value of r also decreases and the hopping probability due to its exponential dependence on r increases rapidly.   Assuming reasonable values for the trap density in the dislocation and the concentration of the recombination centers, Lotthammer[95] has been able to describe a number of properties of the optical quenching and the impurity photoconduction fairly well, particularly the transition from quenching to the positive effect with increasing illumination.   Therefore, this model is adequate for the interpretation of the peculiar photoelectric properties of trigonal selenium crystals.

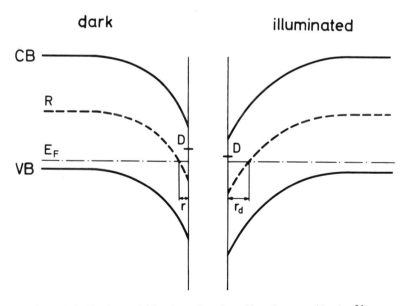

Figure 5-62   Barrier model for the explanation of hopping recombination.[94]

The transient photoconductivity measurements of Mort,[109] on the other hand, were explained by an homogeneous model, with a normal recombination mechanism. These measurements, with short light pulses of 1.5 μs duration and an integrated light flux of 2×10^{10} photons/cm^2, are of particular interest. Since the additional carrier density is far less than the dark density of majority carriers, the dominant trapping which is experimentally observed is that of the minority carriers. Because of space charge neutrality, an equal additional density of free majority carriers is maintained which thus acts as a probe of the minority carrier trapping kinetics.

Figure 5-63a shows the decay of the photocurrent following the light pulse at room temperature. This decay is nonexponential and the pronounced tail is interpreted as arising from the multiple trapping of minority carriers. Clearly, it is due to process II, which, according to the argument of the preceding section, predominates at room temperature. If, however, sufficiently strong dc ambient light is also applied so as to keep the traps filled, the recom-

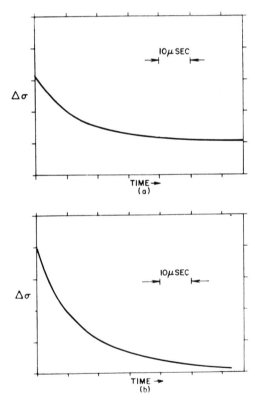

Figure 5-63   Photoconductive decay at room temperature: (a) without dc-ambient light, (b) with dc-ambient light.[109]

bination mainly occurs without minority carrier trapping and an exponential decay is obtained (Figure 5-63b). The recombination (minority carrier) lifetime, τ_R, evaluated from such measurements was found to range from 10 to 34 μs. The investigation of the decay for different orientations of the applied electric field and various temperatures indicates that τ_R is isotropic and, within experimental accuracy, independent of temperature from 125–350°K.

Additional information is gained from the initial amplitude of the photocurrent pulses. Starting with a high dc light level, the amplitude is high and remains initially constant with decreasing intensity of the ambient light. Further reduction of the ambient light then leads to a decrease in the amplitude. When it has fallen to 0.8 of its value at the highest dc light level, any further decrease of the dc light level leaves the initial amplitude unchanged.

When all the traps are filled by the ambient light, the initial current pulse amplitude is proportional to $\Delta ne(\mu_n + \mu_p)$ where Δn is the additional density of electron hole pairs. As the ambient light intensity is reduced, the density of filled minority carrier traps decreases and the free minority carrier time before trapping, τ_t, also decreases. If τ_t approaches a value smaller than the response time of the apparatus, the initial current pulse begins to drop, because those electrons which are trapped before detection do not contribute to the photoconductivity. Therefore, the initial current amplitude is, at low dc light intensities, proportional to $\Delta n\mu_p e$ only, and the ratio of both limiting values is proportional to $\mu_n + \mu_p/\mu_p$. Because, for trigonal selenium, the value of this ratio is found to be 1.25, the electron mobility μ_n can be estimated $\mu_n = \mu_p/4$. Although this may represent only a lower limit, it is nevertheless, of considerable importance, because it is the first value of μ_n quoted in the literature. Obviously, μ_n is of the same order of magnitude as the hole mobility, probably even somewhat smaller than μ_p. This is in accordance with band structure calculations and also with recent results for the temperature dependence of the thermoelectric power of Se–Te mixed crystals.[110] The sign reversal of θ to negative values at the onset of intrinsic conduction is not detected for mixtures in the selenium-rich part of the system.

With a knowledge of both the minority carrier lifetime and the mobility, the minority carrier diffusion length can be calculated. Along the c-axis, the diffusion length in the crystals studied by Mort is found to lie between 4.5–8 μm. These values, which are probably only lower limits, agree within an order of magnitude with estimates obtained in other experiments. Billington and Ehrenberg[111] concluded from a study of photovoltaic effects that electrons have a diffusion length of 8 μm. Gobrecht et al.,[112] using the travelling light spot method, could only place an upper limit of 30 μm.

The first direct experimental evidence that the electrons are mobile in trigonal selenium has also been reported by Mort who utilized a Hornbeck-Haynes bridge experiment (Figure 5-64). With the right hand side of the

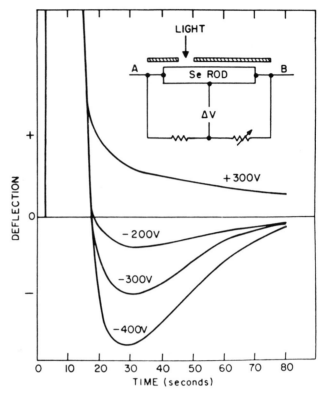

Figure 5-64 Result of Hornbeck-Haynes bridge experiment. Negative deflection represents high conductivity in the right arm. The voltages are those of point *A* with respect to *B*. Curves measured at room temperature after equal exposure.[109]

crystal masked and the bridge balanced, light was shone for a few seconds on the other half. This produced a large imbalance in the bridge due to the increased conductivity in that arm. Several seconds after the light was removed, the deflection of the detector went to the opposite side, indicating that the masked half of the bridge now had the higher conductivity. The reversal is only observed if the direction of the drift field is such that electrons can drift towards the right arm of the bridge. This furnishes direct evidence that the electrons initially trapped in the left arm of the bridge drift, under the influence of an applied field, to the right arm and become trapped there. In order to maintain space charge neutrality, an excess of free majority carriers must now exist in the right arm, yielding a higher conductivity. The return of the bridge to equilibrium is relatively slow, indicating that the minority carriers are trapped in deep levels.

An analysis of the nonexponential decay in Figure 5-63 using the Hornbeck-Haynes theory for multiple trapping shows that the experimental results can be explained by the existence of two trapping levels. At high temperature (380°K), the shallower trap empties in a time which is relatively short compared with the response time of the recorder. The observed decay is then due to the deep level alone. The analysis yields a distance of 0.70 eV from the conduction band for this trap, a capture cross section $S \approx 10^{-16}$ cm^2, and a trapping lifetime $\tau_t = 6 \times 10^{-7}$ s. The density of the unfilled deep traps amounts at 380°K to 2×10^{15} cm^{-3}, which is of the same order of magnitude as the acceptor density. The trap depth found with this method is in fairly good agreement with the value of eV obtained for the deep traps from the threshold of optical quenching.

Further significant results of these investigations is the evaluation of the hole mobility, its temperature dependence, and the mobility ratio $\mu_\parallel/\mu_\perp$ from the steady state photoconductive gain, G. If minority carrier trapping can be neglected due to a sufficiently high dc light level, then $G = \tau_R \mu^x F/d$ where $\mu^x = \mu_n + \mu_p$. Knowing τ_R from decay measurements, the photoconductive mobility μ^x can be evaluated. Though the recombination lifetime was found to be isotropic, the gain is quite anisotropic, namely by a factor of 2.5 larger along the c-axis than perpendicular to it. This anisotropy must, therefore, be due to the mobility ratio in agreement with other measurements. The average values of μ^x were 40 ± 8 cm^2/Vs along the c-axis and 17 ± 4 cm^2/Vs perpendicular to this direction. With the ratio $\mu_n/\mu_p = 1/4$, one predicts the lower limits for μ_p to be 35 cm^2/Vs parallel to the c-axis and 10 cm^2/Vs in the orthogonal direction. The agreement of these values with those obtained by the completely different methods of magnetoconductivity and current saturation is considered very good.

The temperature dependence of the hole mobility obtained from that of the gain is also in good accordance with the temperature variation of μ_p. If, aside from the light pulse, the sample is in darkness, the mobility determining the pulsed photoconductive gain is predominantly the majority carrier drift mobility, μ_p. Figure 5-65 shows the relative pulsed photoconductive gain $G = $ const. μ_p measured with current flow perpendicular to the c-axis as a function of the reciprocal temperature. For comparison, the temperature dependence of μ_p perpendicular to the c-axis determined from the acousto-electric current saturation is shown, the mobility being normalized to unity at room temperature. It is evident that the two curves are strikingly similar. The transient photoconductivity results, therefore are in good agreement with both the absolute value of the hole drift mobility and its temperature dependence.

The interpretation of these results in terms of the barrier model seems to be more difficult. This is particularly the case for the stepwise increase of the initial amplitude of the photocurrent pulse by about 25% with the addition of an intense dc illumination, which is assigned by Mort to a complete filling

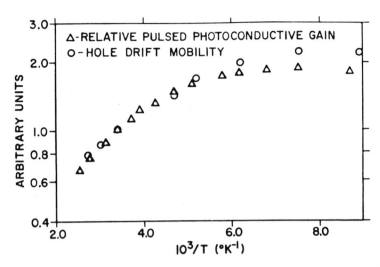

Figure 5-65 Temperature dependence of relative photoconductive gain in comparison with the hole drift mobility determined from acoustoelectric current saturation.[109]

of the electron traps. This interpretation, however, is difficult to reconcile with the temperature dependences of the steady state photocurrent and of the optical quenching rate. Also the TSC curves show that at room temperature and above, the thermal emptying of the deep electron traps is fairly extensive so that a complete filling does not occur. Therefore, it seems necessary to find an explanation of the transient photoconductivity results which does not involve complete filling of all traps. In a barrier model which includes hopping recombination, the filling of these traps is strongly hindered because those traps which are located at a short distance from a recombination center bring about a very rapid recombination. By dc illumination, only traps at farther distances can be filled. Discussing this in more detail, it can be seen that a consideration of the sum of potential barriers and hopping recombination can qualitatively explain Mort's results. For a complete understanding, however, further experimental and theoretical investigations are necessary.

Selenium Rectifiers and Photovoltaic Cells

The technical importance of the electrical and optical properties of trigonal selenium is restricted primarily to its application to devices with nonlinear current-voltage characteristics. Whereas most commercial rectifiers and photovoltaic cells are now made from monocrystalline germanium, silicon, and some Group III–V compounds, the commerical selenium rectifiers and photovoltaic cells are manufactured from polycrystalline layers of trigonal selenium.   Only recently Champness, Griffiths, and Sang[113] succeeded in producing rectifiers with thin monocrystalline layers.   These devices, however, probably have only restricted technical significance, because their manufacture is much more complicated than that of the polycrystalline rectifiers without any particular advantage in their electrical performance.   Nevertheless, the monocrystalline rectifier will certainly be of importance for a better understanding of the properties of the polycrystalline devices.

Polycrystalline Rectifiers.   The structure of a modern selenium rectifier is shown in Figure 5-66.   It consists of a trigonal selenium layer having a

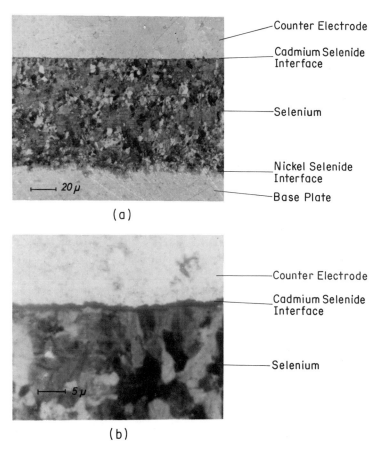

Figure 5-66   (a) Structure of a commercial polycrystalline selenium rectifier. (b) Cadmium selenide interface between counter electrode and selenium[114]

thickness of about 50–60 μm which has been evaporated on a metal plate. The latter is commonly made from an aluminum alloy for reasons of low weight. For the counter electrode, a cadmium alloy having a low melting point is used, so that it can be sprayed onto the selenium layer.

The contact between the aluminum plate and the selenium is made ohmic by a thin interface of nickel selenide. This compound is formed by vacuum or electrolytic deposition of nickel on the aluminum plate before evaporation of the selenium layer. The interface, having a thickness of less than 1 μm, gives a good ohmic contact not only for polycrystalline but also for monocrystalline selenium down to low temperatures. A barrier layer, on the other hand, exists between the counter electrode and the selenium. Here a thin cadmium selenide interface is formed between the cadmium compound and the selenium (Figure 5-66b). As has been shown by Poganski,[114] the existence of this cadmium selenide layer is of fundamental importance for the function of selenium rectifiers and photovoltaic cells.

The current voltage characteristic of a commercial selenium rectifier is shown in Figure 5-67. In the forward direction (Figure 5-67a), the current rises approximately exponentially with voltage, V_f, up to about V_d, the diffusion voltage, whereas for higher forward voltages the curve becomes linear corresponding to the ohmic resistance of the whole configuration. In the

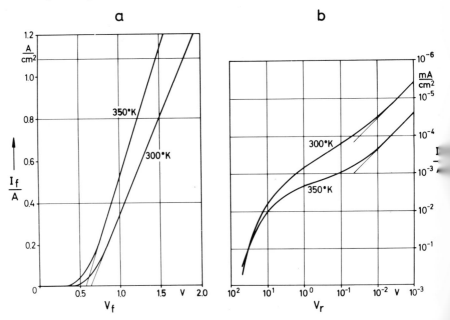

Figure 5-67 Temperature dependence of the current voltage characteristic: (a) forward direction, (b) reverse direction.[114]

reverse direction, the current is several orders of magnitude smaller as can be seen from Figure 5-67b. At low reverse voltages, V_r, the current rises linearly with voltage. Between 10^{-2} and about 10 V the rise is less than ohmic and finally, above 10 V, the current starts to increase approximately exponentially. This breakdown is, for selenium rectifiers, less sharp than for silicon devices, where the steep rise of the reverse current occurs in the region of a few volts. The cross-over of the curves for different temperatures signifies that the breakdown of selenium rectifiers is the result of carrier multiplication by impact ionization. This has been recently confirmed by the more detailed investigations of Lanyon,[91] who has shown that carriers created by light are multiplied in the same manner as thermally excited holes. From the onset of this process at about 10 V with the corresponding field strength in the blocking layer, the mean free path of the holes, λ_p, is calculated to be about 200 Å. The crossover is due to the fact that the mean free path decreases with increasing temperature so that at high temperatures a larger voltage is needed for multiplication. This decrease of λ_p with rising temperature is in qualitative agreement with the temperature dependence of the hole mobility, indicating that the normal transport mechanism is also found in the barrier layer for polycrystalline selenium, at least at high field strengths.

Whereas λ_p and thus μ_p decrease with rising temperature, the ohmic conductivity increases (Figure 5-67). Since the ohmic resistance is mainly due to the selenium layer, the discrepancy between the temperature dependences of the hole mobility, μ_p, and that of the conductivity, σ, obviously exists for the polycrystalline material. The σ value of the polycrystalline layer is enhanced by the addition of a halogen. Chlorine, which is most commonly used, leads to a conductivity of about $5 \times 10^{-3} \ \Omega^{-1} \ cm^{-1}$, which is about 2 orders of magnitude larger than that of single crystals.

For the interpretation of the current-voltage characteristic of rectifiers, particularly the voltage dependence of the barrier layer, the capacitance, C_b, is noteworthy, since it gives direct experimental evidence for the thickness of this layer and of the space charge in it. C_b has, for $V = 0$, the high value of about $30 \ nF/cm^2$. It passes in the forward direction to a maximum below V_D and falls off steeply to low values in the ohmic part of the characteristic, because there the barrier layer has vanished and the capacity is determined by the whole selenium layer. In the reverse direction, C_b decreases strongly for low reverse voltages, V_r, due to the enhancement of the barrier layer thickness in this region (Figure 5-68). With rising V_r, the C_b curve flattens off and finally reaches values of about $3 \ pF/cm^2$, indicating that the increase of d_b slows down in the high voltage region.

It is now generally accepted that rectification in a selenium rectifier does not arise from a metal semiconductor junction but from a heterojunction which consists of a thin cadmium selenide layer and selenium. Whereas

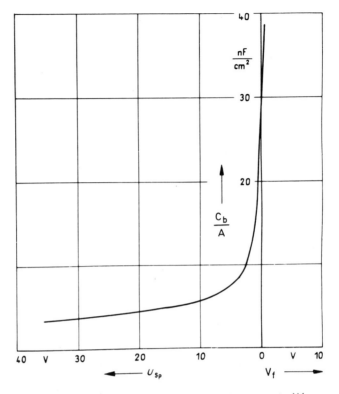

Figure 5-68   Voltage dependence of barrier layer capacity.[114]

selenium is always found to be *p*-type, cadmium selenide is a material which is generally *n*-type.   Only under extreme conditions does it seem possible to make the latter *p*-type.   As a consequence, the creation of the cadmium selenide layer at the contact of selenium and the cadmium alloy automatically leads to a *p-n* heterojunction.   The thin nickel selenide layer, on the other hand, which is producing the ohmic contact, is a *p*-type semiconductor.

Single Crystal Rectifiers.   The fabrication of monocrystalline rectifiers is somewhat different from the manufacture of the polycrystalline devices. Basically, it is done as follows.   As substrates, slices of monocrystalline tellurium of about 3 mm thickness were used by Champness, Griffiths, and Sang.[113]   These slices were cut in a (1010) or (0001) plane from a bulk crystal, polished, chemically cleaned, and finally annealed in order to remove surface damage.   Since tellurium is in the extrinsic *p*-type conduction region, like selenium, it provides the ohmic contact.   Onto the tellurium substrate, at a temperature near 100°C, selenium was evaporated at a rate between 0.2

and 0.5 μm/min.   Under these conditions, a single crystalline layer grows epitaxially on the tellurium, its thickness depending on the evaporation time. Mainly, values between 5 and 15 μm were used.   For the counter electrode, pure cadmium was evaporatated as a small spot on the selenium layer. Following this fabrication, short circuits were often found to exist between the cadmium and the tellurium due to pinholes in the selenium layer.   These had to be burned out by short current pulses.   Afterwards, the rectifier was formed in the same way as the polycrystalline devices by passing current through the rectifier unit in the reverse direction for a number of hours.

The most noticeable feature of the current-voltage characteristic of a single crystal rectifier is the curvature in the forward direction.   This is clearly shown in Figure 5-69 where it is seen that beyond the " knee " at about 0.5 V the forward characteristic of a technical unit is a straight line, whereas it is strongly curved for the single crystal rectifier.   In this region, the current density has an approximately quadratic dependence on voltage which suggests

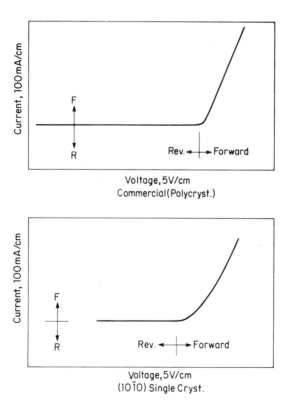

Figure 5-69   Comparison of the shapes of the dynamic current voltage characteristic of a typical commercial selenium rectifier and a single crystal unit.[113]

a space charge limited current behavior.   Below V_D, the current rises exponentially with the voltage, as is seen in the log-linear scale of Figure 5-70.
For one of these curves, the exponential extends over about five decades of
current density.   The resistance in the region above 0.8 V is, for the selenium
layer which contains 200 ppm chlorine, about 2 orders of magnitude lower
than that which is chlorine-free.   This increase in the current, however, is
much less marked at higher chlorine contents.   Since the addition of chlorine
results in no major change in the conductivity of monocrystals grown from
the melt or from the vapor phase, the influence of this element on the conductivity of these monocrystalline layers is surprising.   It is similar in characteristics to polycrystalline material, indicating that for the thin layers, the influence
of impurities seems to differ from that in monocrystals.

The absolute value of the forward current strongly depends on the thickness
of the selenium layer (Figure 5-71), indicating that it is this thickness rather
than that of the cadmium selenide layer which determines the forward resistance.   The reverse characteristic, however, is essentially the same for both
values of d_{Se}.   Moreover, it is remarkable that the breakdown of these single
crystal rectifiers occurs in the same voltage region as for commercial devices.
This result shows that once again it is an avalanche that is responsible for
the rapid increase in the reverse current and that the field strength within the
barrier layer is approximately the same as for polycrystalline devices.

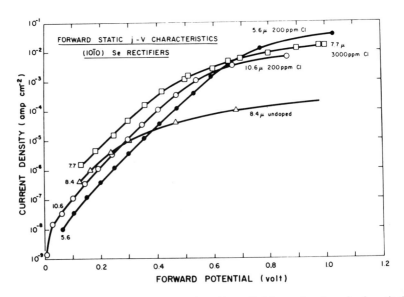

Figure 5-70 Variation of static current density with applied forward voltage for four single
crystal units.[113]

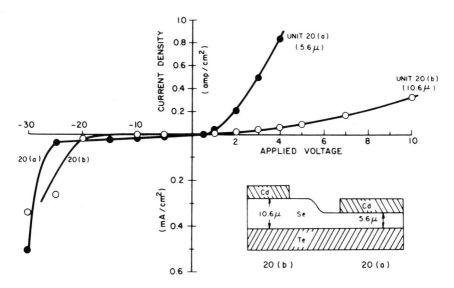

Figure 5-71 Current voltage characteristic of two single crystal rectifiers with different film thicknesses mounted on the same tellurium substrate.[113]

The barrier layer capacitance, C_b, of the single crystal units shows essentially the same behavior as commercial rectifiers. With rising reverse voltage, it decreases quickly as can be seen from Figure 5-72, where $1/C^2$ is plotted versus the reverse voltage. In this plot one obtains straight lines up to a voltage of 2 V. Their slope is correlated easily to the acceptor density as is pointed out in the following section.

Theory of the Selenium Rectifier. For a rectifier, the relation between current density, j, and applied voltage, V, both taken as positive in the forward direction can be expressed in the form:

$$j = j_0 \left\{ \exp\left[\frac{eV - jR}{\alpha k T} \right] - 1 \right\} \tag{16}$$

where R is the bulk specific area resistance of the material constituting the rectifiers and T is the absolute temperature. The quantity, α, is a coefficient equal to unity for a Schottky metal semiconductor contact or a *p-n* junction with negligible recombination in the junction region. For a symmetrical heterojunction, Dolega[115] has shown that $\alpha = 2$ and that the quantity j_0 is given by:

$$j_0 = \sigma \left(\frac{\pi e n}{2\varepsilon} \right)^{1/2} (V_D - V)^{\frac{1}{2}} \exp\left(-\frac{eV_D}{2kT} \right) \tag{17}$$

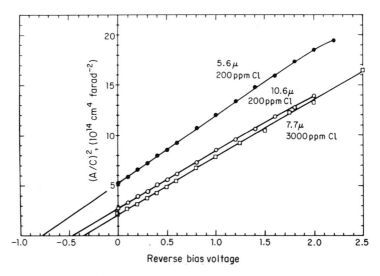

Figure 5-72   Plot of $(A/C)^2$ against reverse voltage for three single crystal units. A is the rectifier area, C the capacitance.[113]

where σ is the conductivity, n the density of carriers in the depletion layer, and ε is the dielectric constant which is assumed to be identical for the n- and p-type materials.   A reverse saturation current, such as exists in a p-n junction, does not arise here because the high recombination rate in a heterojunction prevents the injection of minority carriers from one semiconductor into the other.   The reverse differential resistance $R_{\text{diff}} = dV/dj_r$, where $j_r = -j$ and $V_r = -V$ for a symmetrical heterojunction is obtained from equations (16) and (17) by differentiation, neglecting the term jR which is small.   For $V_r \ll 2kT/e$, this leads to the equation:

$$\frac{R_{\text{diff}}}{R_0} = \frac{e}{kT}[V_D(V_D + V_r)]^{\frac{1}{2}} \tag{18}$$

where

$$R_0 = \frac{2kT}{\sigma e}\left(\frac{2\varepsilon}{\pi e n V_D}\right)^{\frac{1}{2}} \exp\left(\frac{eV_D}{2kT}\right)$$

In a symmetrical heterojunction, R_{diff} rises in a manner approximately proportional to the square root of V_r.   In an unsymmetrical heterojunction, however, the differential resistance increases to a maximum value.

The relation between the capacity, C_b, and the applied reverse voltage for a heterojunction with a uniform distribution of the acceptor density, N_A, in the p-type semiconductor and of donor concentration, N_D, in the n-type material is, according to Dolega:[116]

$$\frac{1}{C^2} = \frac{8\pi}{e} \left(\frac{1}{\varepsilon_n N_D} + \frac{1}{\varepsilon_p N_A} \right) (V_D + V_r) \tag{19}$$

For a symmetrical heterojunction, $N_A = N_D = N$ and $\varepsilon_n = \varepsilon_p = \varepsilon$, which leads to the equation:

$$C^{\frac{1}{2}} = \frac{16\pi}{e\varepsilon N} (V_D + V_r) \tag{20}$$

The acceptor or donor density, N, can thus easily be obtained from the slope of the C_b curves in the $1/C_b^2$-V plot.

The experimental results of polycrystalline rectifiers as well as of single crystal units confirm the assumption that the selenium rectifier contains a cadmium-selenide heterojunction. The slopes, measured in the exponential region of Figure 5-70, give $\alpha =$ values near 2 as is to be expected from equation (17). It is, however, not a symmetrical heterojunction since N_A, μ_p, and ε_p of the selenium layer are different from N_D, μ_n, and ε_n of cadmium selenide. The thickness of the cadmium selenide layer must be very small, because the forward resistance decreases markedly with lowering d_{Se}. The carrier concentrations are deduced from the plot to be about 3×10^{16} cm^{-3} and are remarkably independent of the chlorine content. The chlorine, on the other hand, causes the forward resistance to diminish, so that the mobility derived from both quantities seems to increase with chlorine content from about 10^{-3} to 10^{-1} cm^2/Vs. This is similar to the enhancement of the apparent mobility under the influence of light and is therefore of a formal nature only. By the addition of chlorine, the influence of the barriers is lessened as can be seen from the smaller activation energy of the temperature dependence.

As a result of the quadratic current-voltage dependence, the forward current density of single crystal units becomes larger than that of polycrystalline devices in the voltage range from 2 to 10 V. The fact that the quadratic dependence is also found if the cadmium counter electrode is replaced by an evaporated tellurium layer yielding a Te–Se–Te configuration (Figure 5-73), suggests that this dependence is due to space charge limited currents in the single crystal selenium layer. Without any traps this current amounts to

$$j_{\text{SCLC}} = \frac{9}{8} \varepsilon \varepsilon_0 \mu \frac{V^2}{d^3} \tag{2}$$

Application of this equation gives mobility values between 10 and 100 cm^2/Vs which is consistent with the forward characteristic of the rectifiers in the

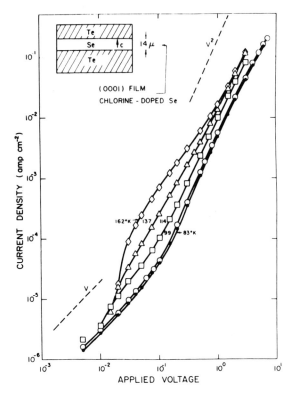

Figure 5-73 Dependence of current density on applied voltage at different temperatures for a Te–Se–Te structure.[113]

quadratic region.   A further notable result of the measurements on a Te–Se–Te structure is seen in Figure 5-73 for the curve at 162°K.   There, following ohmic behavior at low voltages, a superohmic region is observed which then reverts to approximately ohmic behavior.   A similar pattern is to be expected if the barriers vanish at high applied voltages by the injection of majority carriers.   The relatively low field strength of about 100 V/cm, however, implies that in this case it is probably due to the existence of majority carrier traps which, judging from the small temperature dependence of j_{SCLC} seem to be fairly shallow.

Photovoltaic Cells.   Photovoltaic cells are basically rectifiers with a transparent counter electrode.   Most selenium cells currently in use employ cadmium oxide as the counter electrode.   This compound is particularly suited for this purpose because it is always of an *n*-type and has an electron density up to about 10^{20} cm^{-3}.   It can be easily obtained by cathode sputtering in an atmosphere containing some oxygen.   The transparency of the thin cadmium oxide layer is almost undiminished by this high density of free carriers in the region of spectral sensitivity of the selenium photovoltaic cell.   On the contrary, the absorption edge of cadmium oxide is shifted to somewhat higher energies due to the Burstein effect.   Thus, the transparency in the high energy region of the spectral sensitivity curve is enhanced.

The construction of a modern selenium photovoltaic cell is shown schematically in Figure 5-74.   On a base plate of iron or brass, polycrystalline layer

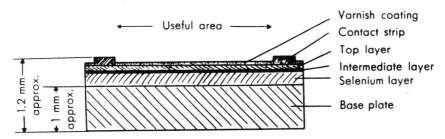

Figure 5-74   Schematical structure of a commercial selenium photovoltaic cell.

of selenium is deposited by vacuum evaporation.   Its thickness is about 100 μm.   The layer of cadmium oxide over the selenium has a thickness of about 1 μm.   In order to contact it, a ring of a low melting alloy (Cd–Bi–Sn) is sprayed onto it.   Between cadmium oxide and selenium, a very thin intermediate layer of cadmium selenide is formed.   This is involved in the function of the photovoltaic cell, because the depletion layer is located at the CdSe–Se interface.   Therefore, the selenium photovoltaic cell is CdSe–Se heterojunction illuminated through the transparent cadmium oxide electrode.

The current-voltage characteristic of a selenium photovoltaic cell with and without illumination is shown schematically in Figure 5-75.   The rectifier characteristic of the nonilluminated cell is shifted to negative values of the current.   Thus, the additional current, j_{ph}, flows in a reverse direction.   In the forward direction, j_{ph} decreases rapidly with rising voltage and reverses sign in the ohmic portion of the characteristic because the bulk resistivity of the selenium layer is diminished.   For the application of the photovoltaic cell, the lower right quarter of the coordinate system is relevant.   With no external voltage, the short circuit current, j_{pho}, is obtained assuming that the bulk resistivity, R_{Se}, of the cell and the resistance, R_c, of the circuit are zero. The j_{pho} is strictly proportional to the light intensity.   This is also approximately valid for $R = R_{Se} + R_c \ll R_b$, where R_b is the resistance of the barrier layer.   With increasing R, the operating point is shifted along the characteristic from A to B.   Therefore, the photocurrent decreases particularly at high

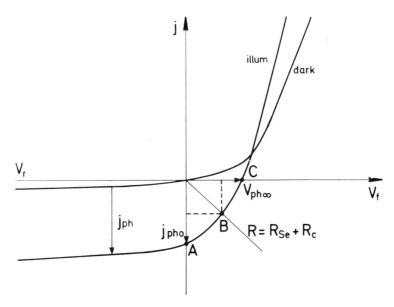

Figure 5-75 Current voltage characteristic of a photovoltaic cell in the dark and under illumination: (j_{pho}) short circuit current, ($V_{ph\infty}$) open circuit voltage, (*B*) operating point for a resistance *R*.

light intensity (Figure 5-76). The photovoltage, V_{ph}, on the other hand, increases so that when $R = \infty$, the open circuit voltage $V_{ph\infty}$ is obtained, which increases logarithmically with illumination (Figure 5-76b). This dependence and the linear characteristic of the photocurrent are of technical importance. The $j_{ph} \sim B$ relation is mainly used for scientific purposes or for commerical flux meters whereas the logarithmical dependence of V_{ph} is often applied for exposure meters and for devices where the sensitivity of the eye is involved, because the photographic process, as well as the eye has an approximately logarithmic characteristic.

Of particular interest, from a practical point-of-view, is the spectral sensitivity curve, which is seen in Figure 5-77. In contrast to the photoconductive cell, where the maximum is found near 1.7 eV, the selenium photovoltaic cell has its maximum sensitivity at about 2.3 eV. The decay at high energy is more pronounced for cells with cadmium oxide counter electrodes than for devices with a thin transparent metallic layer due to the fundamental absorption of cadmium oxide. On the low energy side, a small subsidiary maximum is often observed which is created by the thin cadmium selenide interface. Its relative size increases with the increasing thickness of this layer, since the absorption edge of cadmium selenide lies at a lower energy of trigonal selenium (Figure 5-78). The long wavelength tail of the sensitivity curve

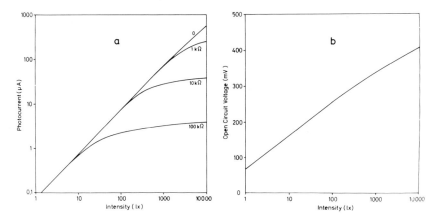

Figure 5-76 (a) Dependence of a photocurrent for different values of resistance *R*. (b) Open circuit voltage on illumination intensity.

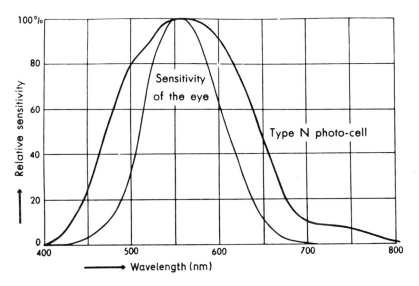

Figure 5-77 Relative spectral sensitivity for a selenium photovoltaic cell compared with the sensitivity of the human eye.

decreases exponentially like the absorption coefficient, *K*, signifying that the number of excited electron hole pairs in the selenium portion of the heterojunction is proportional to *K*. This is due to the fact that in this energy region the thickness of the depletion layer is still small compared to the penetration depth and that the recombination of excited pairs can consequently be neglected. This loss, however, becomes more important with the rising

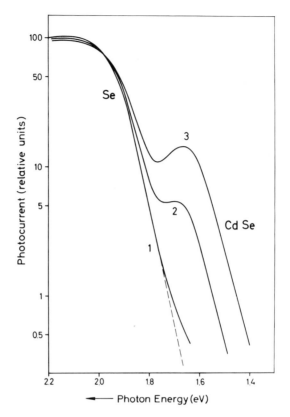

Figure 5-78 Long wavelength edge of spectral sensitivity for selenium photovoltaic cells with different thicknesses of the cadmium selenide interface.

density of the pairs and therefore the photocurrent passes through a maximum and decreases in the high energy range.   It is remarkable that cells with a relatively thick cadmium selenide interface exhibit the same pattern in the subsidiary maximum at low energy (Figure 5-78).   Here the sensitivity curve rises exponentially like the absorption coefficient of this semiconductor and passes through a maximum before the selenium sensitivity sets in.   The enhanced recombination rate appears to be the obvious reason for this behavior.   The temperature variation of the sensitivity which is seen in Figure 5-79 corresponds to that of the absorption coefficient.   Like the absorption edge of trigonal selenium, the long wavelength tail of the selenium part of the photovoltage edge shifts to lower energies with increasing temperature and simultaneously becomes steeper.   A similar temperature dependence is obtained for the cadmium selenide edge.   Moreover, the relative heights

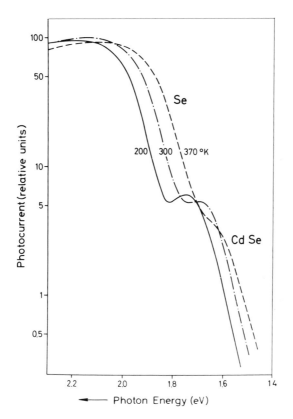

Figure 5-79 Temperature dependence of the long wavelength edge of the spectral sensitivity curve for a selenium photovoltaic cell.

of the selenium and the cadmium selenide maxima change with temperature. The ratio of the cadmium selenide maximum to the selenium maximum diminishes appreciably with decreasing temperature.   This supports the idea that the transition of the holes created in the cadmium selenide interface into the selenium is more difficult than the transition of electrons from the adjacent selenium into the cadmium selenide.   Possibly, a potential discontinuity is the reason for this if it gives rise to a potential barrier for the holes but not for the electrons.

Choyke and Patrick[117] used the long wavelength tail of a selenium photovoltaic cell for the determination of the relative absorption constant of selenium.   The cells used by these investigators had, in contrast to commercial ones, no cadmium selenide interface but a thin film of CdS between the selenium and the metallic transparent counter electrode.   The CdS forms the n-part of the heterojunction but does not have any photovoltaic response to photons of the energies used in this experiment.

In the low energy region of the absorption edge, this method is particularly advantageous in that the photocurrent arises only from fundamental absorption processes. In the case of the absorption coefficient, free carriers and imperfections also play a role. The results are interpreted in terms of indirect transitions, and the energy gap is assigned to 1.79 ± 0.01 eV at room temperature. This value is somewhat lower than that found by Fischer's luminescence and electroabsorption measurements.[34] The reasonable agreement probably is fortuitous, since the photovoltaic results were obtained for polycrystalline material. Moreover, the phonon energies necessary to explain the experimental curves depend strongly on temperature, which contradicts recent investigations of the phonon spectrum of trigonal selenium. The temperature coefficient of the indirect gap is, according to the photovoltaic results, negative above 100°K, whereas Fischer found a positive coefficient at slightly lower temperatures.

The theory of the photovoltaic cell is very similar to that of a p-n junction, except for the fact that the optical generation rate of electron hole pairs, g_{opt}, has to be taken into account. For the photocurrent of a symmetrical p-n junction, this leads to the equation:

$$j_{ph} = j \left(\exp \frac{eV}{kT} - 1 \right) - e g_{opt}(L_p + L_n) \tag{21}$$

where L_n and L_p are the diffusion lengths of the electrons and of the holes, respectively. For $V = 0$, one obtains:

$$j_{ph} = - e g_{opt}(L_p + L_n) \sim - B \tag{22}$$

For the photovoltage V_{ph}, on the other hand, a logarithmical relation between V_{ph} and B is found:

$$V_{ph} = \frac{kT}{e} \ln \left(1 + g_{opt} \frac{\tau_p}{p_n} \right) \tag{23}$$

For an asymmetrical heterojunction, these relations become somewhat more complicated since the quantities τ_p, τ_n, L_p, L_n are different in both materials. Also, the potential discontinuity at the border is of influence. However, the linear relationship between j_{ph} and B as well as the logarithmic dependence of V_{ph} on B are not affected by this complication. Only the absolute values of j_{ph} and V_{ph} are changed in going from the simplest case, the symmetrical p-n junction to the more complicated asymmetrical heterojunction.

MONOCLINIC SELENIUM

Crystal Structure*

In addition to the trigonal chain lattice structure, selenium has two modifications which consist of puckered Se_8 ring molecules: the α- and β-monoclinic selenium. Both forms crystallize from a solution of amorphous selenium in carbon disulfide.[118] Amorphous selenium is needed here since in addition to relatively long disordered chains, it contains a high percentage of disordered Se_8 rings.[119] By cooling a saturated solution of these rings, small monocrystals of monoclinic selenium are obtained, often in the form of thin platelets. The dimensions lie mostly in the range of several millimeters or less. Recently, the growth of monoclinic crystals from organic gel media has been accomplished.[120] As the β-form is more difficult to grow, most of the data in the literature are related to α-monoclinic selenium.

The main difference between both forms is due to the different packing of ring molecules in the lattices. Fairly simple is the configuration of the rings in the β-form, where all rings have the same orientation (Figure 5-80b). The smallest distance between neighboring atoms within the rings is 2.34 Å, while the shortest distance between atoms of different rings is 3.48 Å. These values are strikingly similar to those measured for the trigonal lattice (Table 5-1). Here the distance between two neighboring atoms within the chains is 2.32 Å, and the distance to the nearest atoms of neighboring chains amounts to 3.46 Å. For α-monoclinic selenium, the atomic distance within the rings is the same as for the β-form. The shortest distance between atoms of different molecules is somewhat greater (3.53 Å). The difference between both modifications, however, is that in the α-monoclinic form, half of the rings have a different orientation than those in the other half (Figure 5-80a). Thus, a configuration arises where the orientation of adjacent ring molecules varies. The unit cell of both modifications is rather complicated, since it contains, in each case, 4 molecules each consisting of 8 atoms. The high number of 32 atoms per unit cell is the reason that band structure calculations for monoclinic selenium have not yet been performed. Therefore, only experimental results which describe the energy spectrum of the monoclinic modifications of selenium are currently available. These are relatively few in number compared with the numerous investigations on the trigonal form.

* See also Chapter 3.

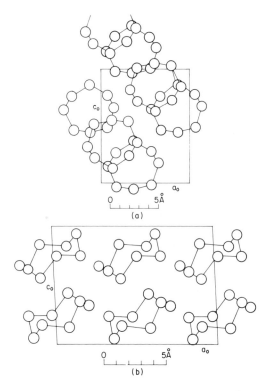

Figure 5-80   Crystal structure of monoclinic selenium: (a) α-monoclinic modification, (b) β-monoclinic modification.[122]

Optical Properties

The absorption of α-monoclinic selenium has been studied in the range of the absorption edge by Prosser[121] and by Abdullaev *et al.*[122] and within the fundamental absorption up to 4.5 eV by Caywood and Taynai.[123]. The results of these measurements are compiled in Figure 5-81. The absorption coefficient rises at the absorption edge approximately exponentially with photon energy, the edge being located at 0.2 eV higher energy than for trigonal selenium. The slope of the edge is similar to that of the trigonal modification for the polarization of light perpendicular to the *c*-axis, but it is somewhat larger than for $E\|c$. The refractive index, *n*, of monoclinic selenium at energies well below the edge is smaller than that of trigonal selenium due to the location of its edge at higher energy. According to the measurements of Prosser,[124] *n* is 2.47 at 1.2 eV. In the fundamental absorption, the absorption

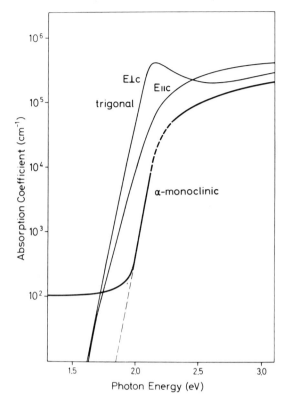

Figure 5-81   Energy dependence of the absorption coefficient of α-monoclinic selenium. For comparison, the curves of trigonal selenium are plotted for both polarizations of light.[121,122,123]

coefficient. K, rises smoothly up to values of 3×10^5 cm^{-1} at 4.5 eV.   The curve reveals no major structure except for a change in the slope at 2.85 eV. This kink is not seen in the logarithmic plot of Figure 5-81, but it is easily observed if the curve is linearly plotted.[123]   Compared with the spectrum of K for trigonal selenium, the curve of the α-monoclinic form is fairly smooth. Possibly a more detailed investigation using different methods may reveal some additional structure.

In the high energy region, the optical properties of α-monoclinic selenium are similar to those of the trigonal modification, as can be seen from the reflectance measurements of Leiga[125] shown in Figure 5-82.   The reflectivity of α-monoclinic selenium has a minimum at 6.5 eV: almost the same energy as for the trigonal modification.   At higher energies, the curve reveals a broad structured maximum near 10 eV.   Since the high energy minimum in the reflectivity curve of trigonal selenium is a result of the valence band

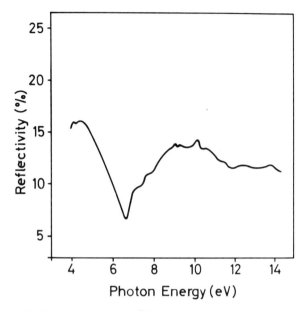

Figure 5-82   Reflectivity of α-monoclinic selenium at high photon energies.[125]

structure of this modification, particularly of the energy gap between the two upper valence band triplets, this characteristic feature also exists for the band structure of monoclinic selenium.   This is of significance from a more general point of view.   The similarity in the band structures of trigonal and monoclinic selenium in this respect is striking because the long range order is completely different.   It indicates that the short range order, particularly the almost equal distance between neighboring atoms within a chain or ring and the similar valence angles, is the reason for these features.   A more detailed knowledge of the band structure of monoclinic selenium would be of great interest, because a comparison with the band model of the trigonal modification gives information about the influence of the short- and long-range order on specific features of the structure.   This is not only of importance for a better understanding of the difference between the forms of selenium but also for a discussion of the energy spectrum of amorphous semiconductors because there, only the short-range order is retained.

In the low energy region, the curves of most monoclinic crystals have a flat tail with $K \approx 10^2$ cm^{-1}.   This relatively high value of the absorption coefficient is probably due to the fact that the crystals contain inclusions of small trigonal crystallites.   This can be concluded from the spectral distribution of the photoconductivity (Figure 5-83), where a subsidiary maximum is found at 1.7 eV: that is, approximately the energy of the maximum of the

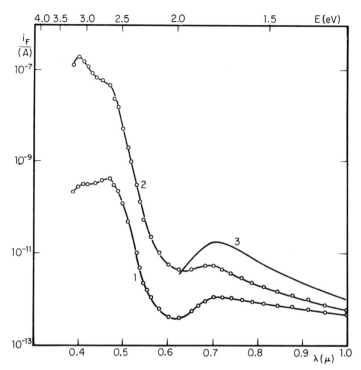

Figure 5-83 Energy dependence of photoconductivity of α-monoclinic selenium:[126] (1) not calculated per unit of incident energy, (2) calculated per unit incident energy, (3) after Gudden and Pohl.[128]

trigonal form.   Because the tail reaches down into energies well below the absorption edge of trigonal selenium, it is probably not due to absorption, but essentially to scattering of light.

Electrical Properties

The knowledge about the dark conductivity of monoclinic selenium is very limited.   This is probably caused by the low absolute value of the conductivity which amounts to about 10^{-10}–10^{-9} Ω^{-1} cm^{-1} at room temperature. Fairly detailed information from the investigation of Spear,[126] however, exists concerning the drift mobility of electrons and holes in this modification.   In these experiments, a 10 ns pulse of 40 keV electrons produces electron-hole pairs near the top electrode within a depth of a few microns below the electrode.   About 2 ms before the arrival of this excitation pulse, a pulsed electric field is applied across the specimen for a few milliseconds. Depending on the polarity, either electrons or holes are drawn out of the surface region and drift across the specimen.   From the characteristic of this current pulse, the time the carriers need to pass through the crystal, t_d, can be evaluated. Then the thickness of the crystals gives the drift velocity, v_d, and from the applied field, F, the drift mobility $\mu_d = v/F$ is obtained. This method, like the electroacoustic current saturation, has the advantage that the drift mobility is directly observed   Moreover, from the shape of the curve, the number of carriers can be derived.

The temperature dependence of the electron drift mobility is shown in Figure 5-84, where two different regions can clearly be distinguished. At low temperatures, μ_d increases exponentially with temperature and has an activation energy of 0.25 eV.   Here, for different crystals the absolute values of μ_d are appreciably different, implying that this activation energy is correla-

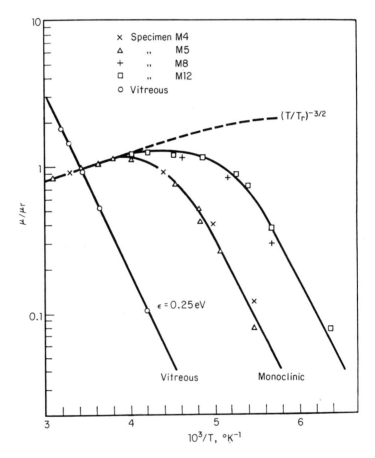

Figure 5-84 Temperature dependence of electron drift mobility in α-monoclinic selenium.[127]

ted with the trapping of the electrons, the trap density, N_t, being about 10^{14} cm^{-3}. At high temperature, however, the mobility decreases with rising temperature proportional to $T^{-3/2}$, and for various crystals the absolute values are only slightly different. In this case, therefore, the mobility is limited by lattice scattering. At room temperature, the mobility amounts to about 2 cm²/Vs, which is more than a factor of 10 less than the corresponding value in trigonal selenium. For holes, the temperature dependence of μ_d is similar to that of the electrons. However, trapping has a stronger influence on these carriers because the absolute value of μ_d is smaller than for the electrons and the $T^{-3/2}$ region is not yet reached at high temperature. Concerning the nature of the trapping centers for both carriers, not much is known. It is quite possible that self-trapping is of importance.

Of particular curiosity is the finding that at sufficiently high temperatures, monoclinic selenium, with its strongly molecular character, has a temperature dependence of μ_d similar to that of normal semiconductors. This dependence points to a normal band transport mechanism where the mobility is determined by lattice scattering. The very small absolute value, however, cannot easily be understood. As a result, Spear[127] recently ascribed the transport mechanism in monoclinic selenium to polarons. This model, however, is also beset with difficulties since the polarizability of monoclinic selenium should be much larger than that of trigonal crystals. Obviously, for a more detailed understanding of the transport mechanism of the monoclinic forms further investigations are necessary, particularly band structure calculations, in order to have an approximate value of the effective masses.

Photoconductivity

Since the early investigations of Gudden and Pohl[128], it has been well-known that monoclinic selenium possesses considerable photoconductivity. More recent measurements have been performed by Prosser[121] and by Abdullaev et al.[122] At the absorption edge, the spectral sensitivity curve rises exponentially with energy (Figure 5-83); the slope of this photoconductivity edge is approximately the same as that of the absorption coefficient (Figure 5-81). The maximum is found at relatively high values of K compared with other photoconductors, particularly when compared with trigonal selenium. According to the results of Prosser, it is located at 3.1 eV, though Abdullaev finds it at 2.8 eV. The difference between both values may be due to variations in the crystal thicknesses. The absorption coefficient in this energy region lies between 1.6 and 2 × 10^5 cm^{-1}. Thus, it is orders of magnitude higher than the corresponding value for trigonal selenium which amounts to about 10^3 cm^{-1}. This suggests that the mechanism of photoconductivity must be completely different in both forms. For amorphous selenium, however, a similarity is observed, since for this modification the maximum of the photoconductivity is found well within the fundamental absorption.[129] In addition to the photoelectrical processes, the transport mechanisms of these two modifications is similar. Probably the large concentration of disordered rings in amorphous selenium is responsible for this resemblance.

Of note is the occurrence of a subsidiary maximum at 1.7 eV. In this case the generation of the free carriers occurs in small trigonal crystallites within the monoclinic crystal. This leads, however, to an increase in the conductivity only if these carriers are injected from the crystallites into the much larger monoclinic regions which determine the resistivity of the mixture. Because, in monoclinic selenium, the mobility is larger for electrons than for holes, the injection of electrons is probably responsible for the small maximum at 1.7 eV. A similar behavior is found for amorphous selenium which is partially crystallized.

Acknowledgments   I am indebted to my coworkers H. Mell, J. Niklas, Dr. G. Weiser, and G. Zimmerer for their cooperation. In particular I am grateful to Dr. W. Fuhs for his very valuable contribution. I wish to thank Ms. G. Schultheiss for typing the manuscript.

REFERENCES

1. Smith, W., *Nature* **7**, 303 (1873); *Am. J. Sci.* (3) **5**, 301 (1873).
2. Siemens, W., *Ding. Poly. J.* **217**, 61 (1875).
3. Braun, F., *Wied. Ann.* **1**, 95 (1877).
4. Presser, E., *Funkbastler* **44**, 558 (1925).
5. Fritts, C. E., *Am. J. Sci.* **26** (3), 465 (1883).
6. Lange, B., *Phys. Z.* **31**, 139 (1930); Bergmann, L., *Phys. Z.* **32**, 286 (1931).
7. Harrison, D. E., and Tiller, W. A., *J. Appl. Phys.* **36**, 1680 (1965).
8. Keezer, R. C., and Wood, C., *Appl. Phys. Letters* **8**, 139 (1966); Keezer, R. C., and Wood, C., *Appl. Phys. Letters* **8**, 233 (1966); Keezer, R. C., Wood, C., and Moody, J. W., *Proc. Int. Conf. on Crystal Growth*, Pergamon Press, Inc., New York, 1967.
9. Grosse, P., *Die Halbleitereigenschaften des Tellur*, Springer Tracts in Modern Physics, Vol. 48.
10. Gaspar, R., *Acta Physiol. Acad. Sci. Hung.* **27**, 289 (1957).
11. Reitz, J. R., *Phys. Rev.* **105**, 1233 (1957).
12. Olechna, D. J., and Knox, R. S., *Phys. Rev.* **140**, A986 (1965).
13. Behrens, E., *Z. Phys.* **163**, 140 (1961).
14. Chen, I., and Das, T. P., *J. Chem. Phys.* **45**, 3526 (1966).
15. Knox, R. S., and Olechna, D. J., *J. Chem. Phys.* **47**, 5226 (1967).
16. Treusch, J., and Sandrock, R., *Phys. Stat. Sol.* **16**, 487 (1966).
17. Sandrock, R., *Phys. Rev.* **169**, 642 (1968).
18. Hulin, M., *J. Phys. Chem. Solids* **27**, 441 (1966).

19. Kramer, B., and Thomas, P., *Phys. Stat. Sol.* **26**, 151 (1968).
20. Cohen, M. L., and Bergstresser, T. K., *Phys. Rev.* **141**, 789 (1966).
21. Animalu, A. O. E., and Heine, V., *Phil. Mag.* **12**, 1249 (1965).
22. Skinner, C. H., *Phys. Rev.* **9**, 148 (1917).
23. Gilleo, M. A., *J. Chem. Soc.* **19**, 1291 (1951).
24. Stuke, J., *Z. Phys.* **134**, 194 (1953).
25. Gobrecht, H., and Tausend, A., *Z. Phys.* **161**, 205 (1961).
26. Caldwell, R. S., and Fan, H. Y., *Phys. Rev.* **114**, 664 (1959).
27. Eckart, F., and Henrion, W., *Phys. Stat. Sol.* **2**, 841 (1962).
28. Stuke, J., and Keller, H., *Phys. Stat. Sol.* **7**, 189 (1964).
29. Tutihasi, S., and Chen, I., *Phys. Rev.* **158**, 623 (1967).
30. Mohler, E., Stuke, J., and Zimmerer, G., *Phys. Stat. Sol.* **22**, K49 (1967).
31. Leiga, A. G., *J. Appl. Phys.* **39**, 2149 (1968).
32. Henrion, W., *Phys. Stat. Sol.* **12**, K113 (1965).
33. Roberts, G. G., Tutihasi, S., and Keezer, R. C., *Phys. Rev.* **166**, 637 (1968).
34. Fischer, R., Thesis, Univ. Frankfurt, 1969.
35. Weiser, G., and Stuke, J., *Proc. Int. Conf. Semicond.*, Moscow Vol. I, 1968, p. 228.
36. Tuomi, T. O., *Phys. Stat. Sol.* **38**, 623 (1970).
37. Krisciunas, V. Y., Mikalkevicius, M. P., and Shileika, A. Yu., *Sov. Phys.-Solid State* **7**, 2080 (1966).
38. Krisciunas, V. Y., and Mikalkevicius, M. P., *Sov. Phys.-Solid State* **8**, 2750 (1967).
39. Kessler, F. R., and Sutter, E., *Z. Phys.* **173**, 54 (1963).
40. Henrion, W., *Phys. Stat. Sol.* **14**, K51 (1966).
41. Queisser, H. J. and Stuke, J., *Sol. State Commun.* **5**, 75 (1967).
42. Fischer, R., Queisser, H. J., and Zetsche, H., *Proc. Int. Conf. Semicond.*, Moscow, Vol. I, 1968, p. 175.
43. Kühler, H., *Diplomarbeit*, Univ. Frankfurt, 1970.
44. Kautsch, H., *Diplomarbeit*, Univ. Frankfurt, 1968.
45. Blätte, M., and Kautsch, M., to be published.
46. Stuke, J., and Weiser, G., *Phys. Stat. Sol.* **17**, 343 (1966).
47. Franz, W., *Z. Naturf.* **13**, 484 (1958).
48. Stuke, J., *The Physics of Selenium and Tellurium*, ed., W. Charles Cooper, Pergamon Press, Inc., New York, 1969, p. 6.
49. Chen, J. H., *Phys. Letters* **23**, 516 (1966).
50. Gobrecht, H., *The Physics of Selenium and Tellurium*, ed., W. Charles Cooper, Pergamon Press, Inc., New York, 1969, p. 87.
51. Weiser, G., Thesis, Univ. Marburg, 1969; Weiser, G., to be published.
52. Henrion, W., and Eckart, F., *Z. Naturf.* **19a**, 1024 (1964).
53. Müller, T., *Sitzungsbericht Phys. Med. Soz. Erlangen* **70**, 7 (1938).
54. Plessner, K. W., *Proc. Phys. Soc.* **B64**, 671 (1951).
55. Henkels, H. W., *J. Appl. Phys.* **22**, 916 (1951).
56. Gobrecht, H., and Hamisch, H., *Z. Phys.* **148**, 218 (1957).
57. Polke, M., Stuke, J., and Vinaricky, E., *Phys. Stat. Sol.* **3**, 1885 (1963).
58. Bakirov, M. Ya., and Dzhalilov, N. Z., *Sov. Phys.-Sol. State* **8**, 244 (1966).
59. Graeffe, R., and Heleskivi, J., *Acta Polytechn. Scand.* **42** (1966).
60. Stuke, J., *Phys. Stat. Sol.* **6**, 441 (1964).
61. Balarin, M., and Zetzsche, A., *Phys. Stat. Sol.* **2**, 1670 (1962).
62. Brätter, M., and Gobrecht, H., *Phys. Stat. Sol.* **37**, 869 (1970).

63. Salo, T., Stubb, T., and Suosara, E., *The Physics of Selenium and Tellurium*, ed., W. Charles Cooper, Pergamon Press, Inc., New York, 1969, p. 335.
64. Henkels, H. W., and Maczuk, J., *Phys. Rev.* **91**, 1562 (1953).
65. Lilja, R., and Stubb, T., *Acta Polytechn. Scand.* **28** (1964).
66. Meilikhov, E. Z., *Sov. Phys.-Sol. State* **7**, 1407 (1965).
67. Balchan, A. S., and Drickamer, H. G., *J. Chem. Phys.* **34**, 1948 (1961).
68. Wittig, J., *Phys. Rev. Letters* **15**, 159 (1965).
69. Kozyrev, P. T., *Sov. Phys.-Sol. State* **1**, 94 (1959).
70. Krisciunas, V. Yu., and Daukantaite, O. K., *Sov. Phys.-Sol. State* **8**, 471 (1966).
71. Becker, W., Fuhs, W., and Stuke, J., to be published.
72. Dolezalek, F. K., and Spear, W. E., *J. Noncryst. Sol.* **4**, 97 (1970).
73. Stuke, J., and Wendt, K., *Phys. Stat. Sol.* **8**, 533 (1965).
74. Abdullaev, G. B., Dzhalilov, N. Z., and Aliev, G. M., *Phys. Letters* **23**, 217 (1966).
75. Tiainen, O. J. A., and Tunkelo, E., *Phys. Stat. Sol.* **36**, 567 (1969).
76. Herring, C., *Phys. Rev.* **96**, 1163 (1954).
77. Heleskivi, J., Stubb, T., and Suntola, T., *J. Appl. Phys.* **40**, 2923 (1969).
78. Mell, H., and Stuke, J., *Phys. Stat. Sol.* **24**, 183 (1967).
79. Gobrecht, H., Hamisch, H., and Tausend, A., *Z. Phys.* **148**, 209 (1957).
80. Mort, J., *Phys. Rev. Letters* **18**, 540 (1967).
81. Schweikert, H., *Z. Phys.* **128**, 47 (1950).
82. Plessner, K. W., *Proc. Phys. Soc.* London, **B64**, 681 (1951).
83. Gobrecht, H., Tausend, A., and Plümeke, P., *Z. Angew. Phys.* 496 **15**, (1963).
84. Henkels, H. W., and Maczuk, J., *J. Appl. Phys.* **25**, 1 (1954).
85. Lehovac, K., *Z. Phys.* **125**, 451 (1949).
86. Hempel, H. P., Lauckner, H., and Thiemann, H., *Z. Naturf.* **16a**, 1402 (1961).
87. Gobrecht, H., Tausend, A., and Siemsen, K., *Z. Naturf.* **18a**, 745 (1963).
88. Kozyrev, P. T., *Sov. Phys.-Solid State* **1**, 102 (1959); *Sov. Phys.-Solid State* **3**, 2716 (1962).
89. Eckart, F., *Ann. Phys.* **17**, 84 (1956); *Phys. Stat. Sol.* **2**, K23 (1962).
90. Abdullaev, G. B., Aliev, G. M., Barkinhoev, Kh. G., Askerov, Ch. M., and Larionkina, L. S., *Sov. Phys.-Solid State* **6**, 786 (1964).
91. Lanyon, H. P. D., *Phys. Stat. Sol.* **1**, 535 (1970).
92. Heine, V., *Phys. Rev.* **146**, 568 (1966).
93. Tuomi, T. O., *Acta Polytechn. Scand.* **56** (1968).
94. Stöckmann, F., to be published.
95. Lotthammer, R., Thesis, Univ. Karlsruhe, 1969.
96. Geick, R., Schröder, M., and Stuke, J., *Phys. Stat. Sol.* **24**, 99 (1967).
97. Lucovsky, G., Keezer, R. C., and Burstein, E., *Solid State Commun.* **5**, 439 (1967).
98. Suosara, E., personal communication.
99. Barnard, G. P., *The Selenium Cell, Its Properties and Applications*, Constable Company Ltd., London, 1930.
100. Prosser, V., *Proc. Int. Conf. Semicond.*, Prague, 1960, p. 993.
101. Geisler, W., *Diplomarbeit*, T. H. Karlsruhe, 1964.
102. Fuhs, W., and Stuke, J., *Phys. Stat. Sol.* **27**, 171 (1968).
103. Randall, J. T., and Wilkins, M. H. F., *Proc. Roy. Soc.* **A184**, 347 (1945).
104. Kolomiets, B. T., and Khodosevitsch, P. K., *Sov. Phys.-Solid State* **6**, 2556 (1965).

105. Bakirov, M. Ya., and Dzhalilov, N. Z., *Sov. Phys.-Solid State* **9**, 968 (1967).
106. Henisch, K. H., and Engineer, M. H., *Phys. Letters* **26A**, 188 (1968).
107. Hemilä, S. O., and Tuomi, T. O., *Annales Acad. Sci. Fennicae* Series **A**, 199 (1966).
108. Hasler, K. and Polke, M., *Phys. Stat. Sol.* **9**, 851 (1965).
109. Mort, J., *J. Appl. Phys.* **39**, 3543 (1968).
110. Beyer, W., Mell, H., and Stuke, J., to be published.
111. Billington, E. W., and Ehrenberg, W., *Proc. Phys. Soc.* London, **78**, 845 (1961).
112. Gobrecht, H., Tausend, A., and Picht, W., *Z. Naturf.* **17a**, 699 (1962).
113. Champness, C. H., Griffiths, C. H., and Sang, H., *The Physics of Selenium and Tellurium*, ed., W. Charles Cooper, Pergamon Press, Inc., New York, 1969, p. 349.
114. Poganski, S., Thesis, T.U., Berlin, 1950; *Z. Elektrochem.* **56**, 193 (1952); *Z. Physik* **134**, 469 (1953).
115. Dolega, U., *Z. Naturf.* **18a**, 653 (1963).
116. Dolega, U., *Z. Physik* **167**, 46 (1962).
117. Choyke, W. J., and Patrick, L., *Phys. Rev.* **108**, 25 (1957).
118. Moody, J. W., and Himes, R. C., *Mat. Res. Bull.* **2**, 523 (1967).
119. Lucovsky, G., *The Physics of Selenium and Tellurium*, ed., W. Charles Cooper, Pergamon Press, Inc., New York, 1969, p. 255.
120. Blank, Z., Brenner, W., and Okamoto, Y., *J. Crystal Growth* **34**, 372 (1968).
121. Prosser, V., *Czech. J. Phys.* **A10**, 35 (1960).
122. Abdullaev, G. B., Asadov, Y. G., Mamedov, K. P., *The Physics of Selenium and Tellerium*, ed., W. Charles Cooper, Pergamon Press, Inc., New York, 1969, p. 179.
123. Caywood, J. M., and Taynai, J. D., *J. Phys. Chem. Solids* **30**, 1573 (1969).
124. Prosser, V., *Czech. Phys.* **B10**, 306 (1960).
125. Leiga, A. G., *J. Opt. Soc. Am.* **58**, 1441 (1968).
126. Spear, W. E., *J. Phys. Chem. Solids* **21**, 110 (1961).
127. Spear, W. E., *Appl. Optics* (1969).
128. Gudden, B., and Pohl, R. W., *Z. Phys.* **35**, 243 (1925).
129. Hartke, J. L., Regensburger, P. J., *Phys. Rev.* **139**, 1970 (1965).

6

The Structural Aspects of Selenium Chemistry

Dr. A. W. CORDES

Department of Chemistry, University of Arkansas, Fayetteville, Arkansas

INTRODUCTION

The stereochemistry of selenium has attracted the attention of structural chemists since the earliest days of diffraction research. This interest has been sustained in large part because of the variety of molecular geometries displayed by selenium. Structures have been reported which contain selenium in all covalencies between one and six and, in cyclic structures, of four, five, six, and eight-membered rings. An excellent review of selenium stereochemistry was presented by Abrahams in 1956. For completeness, structures given in that review will be repeated here.

The review covers the literature through mid-1970, and attempts to include all of the selenium structures reported except binary and ternary metal selenide systems which have been analyzed by x-ray diffraction. All but a few of the nearly one hundred molecular structures were done by either solid state x-ray diffraction techniques

298

(almost all single crystal work) or by electron diffraction of selenium compounds in the gaseous state.   Significant figures in the structural parameters have been rounded to contain no more digits than the standard deviations warrant, and the standard deviations have been rounded to one significant nonzero digit.   Unless otherwise noted, when several independent values for the same bond distance or angle are determined in one structure, each of the experimental values is listed separately in the tables.   The estimated standard deviations are those assigned by the investigators themselves, and are given whenever they were reported.   Close attention should be paid to these standard deviations, since the x-ray and electron diffraction work span a period of forty years and tremendous changes have occurred in the quality and quantity of the data, as well as in the refinement procedures.   Fortunately, much of the relatively crude early work is least likely to be in error with respect to the stereochemistry of the selenium atom, since selenium quite often was the heaviest scatterer in the structure.   The van der Waals' radii referred to in the text are those of Pauling.[103]

Only the bond distances and angles involving selenium are reported: information on the other parts of the structure can be obtained from the original literature.   Many references are made to the literature sources in the text, but the tables provide the most efficient path to the original work.   It is not the intent of this review to afix lengthy interpretations to the structures, but merely to gather the structures together, point out obvious trends and correlations, and pass on the original interpretations of the more interesting features of their structures.

Any survey of the structural chemistry of selenium plainly shows that this area of research owes a great deal to the continuing contributions of Professor J. D. McCullough of the University of California at Los Angeles. About 20% of the structures reported in this review were done in his laboratories, and many others have been done by researchers who obtained their interest in selenium chemistry from Professor McCullough.

Elemental Selenium

Structural parameters have been reported for five well-established forms of elemental selenium.   Table 6-1 compares the bond distances and angles reported for these different forms.   The common grey allotrope, which is the only stable variety at ordinary temperature and pressure, forms trigonal (also referred to as hexagonal) crystals containing spiral chains of selenium atoms (Figure 6-1).   Two metastable forms of selenium, both occurring as monoclinic crystals (α and β), have been found to contain Se_8 molecules in the crown form (Figure 6-2) similar to elemental sulfur.   Structural studies of

TABLE 6-1   BOND DISTANCES AND ANGLES IN ELEMENTAL SELENIUM

	Se—Se (Å)	Se—Se—Se	References
Trigonal selenium (grey, spiral chains)	2.373(5)	103.1(2)°	26
α-Monoclinic selenium (red, S_8 rings)	2.34(2)[a]	105(2)°[a]	23
β-Monoclinic selenium (red, S_8 rings)	2.34(1)[a]	105.7(8)°[a]	89
Glassy selenium	2.36	104°	73
	2.32	105°	107
Gaseous Se_2	2.19(3)	—	91

[a] Average value; see text for ranges.

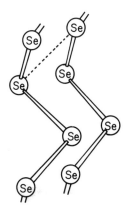

Figure 6-1   Trigonal selenium.

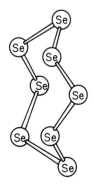

Figure 6-2   Monoclinic selenium.

the glassy form of selenium have necessarily been limited to reports of average bond distances and average numbers of neighbor atoms. The Se_2 molecule was studied by the electron diffraction of selenium gases at 750°C.

As is shown in Table 6-1, the bond distances and angles reported for solid selenium do not vary significantly from one form to another. In the trigonal crystal, as reported by Cherin and Unger,[26] only one Se—Se bond and one Se—Se—Se bond angle are needed to uniquely determine the structure. This bond angle is almost identical with that of trigonal tellurium. Perhaps the most interesting parameter of the selenium structure is the nonbonded (interchain) Se $\cdots$ Se distance of 3.436(5) Å, which is considerably shorter than twice the van der Waals' radius of 2.0 Å for selenium.

The two monoclinic (red) forms of selenium contain identical selenium rings, and both have a local (noncrystallographic) $\bar{8}2m$ symmetry within the experimental errors quoted. The range of the independently determined Se—Se bond distances for the alpha form[23] is 2.31–2.37(2) Å, and for the beta form[89], is 2.30–2.37(1) Å. The corresponding ranges for the Se—Se—Se bond angles are 102.8–109.0° and 104.2–107.1°. Intermolecular packing distances are very similar for the two forms with the shortest distances being 3.53(2) Å (alpha) and 3.48(1) Å (beta). Marsh, Pauling, and McCullough[89] have postulated a solid state mechanism for the conversion of the metastable β-monoclinic form to the stable trigonal allotrope.

X-ray scattering by glassy selenium, conducted by Krebs and his co-workers,[73] has been interpreted in terms of a model in which each selenium is separated 2.36 Å from an average of 2.07 nearest neighbor selenium atoms, and 3.72 Å from 6.0 other selenium atoms. In a similar study, Richter and Breitling[107] found the two shortest Se—Se contacts to be 2.32 Å and 3.69 Å. From considerations of the density of glassy selenium, Krebs estimated 40% of the solid is composed of rings of low molecular weight and 60% is composed of large rings which contain on the average 500 atoms. In their study of glassy selenium, Richter and Breitling found a Se · · · · Se interaction just twice the 3.69 Å distance, from which they concluded that selenium chains are essentially planar.

Evidence of a second glassy phase, as well as two new cubic crystalline phases of selenium, has been presented by Andrievskii and Russian co-workers.[6,7] The two cubic phases, observed as intermediates between the monoclinic forms and the trigonal form upon the slow heating of a thin film of selenium, were reported as a face-centered Cu lattice with a cell edge of 5.755(1) Å, and a face-centered diamond lattice with a cell edge of 6.04(1) Å.

The structures of two recently reported cations closely related to elemental selenium, S_4^{+2} and Se_8^{+2}, are covered in a later section on ring compounds.

STRUCTURES CONTAINING NONCYCLIC $X-Se-X$ LINKAGES

A significant variety of $X-Se-X$ linkages in simple molecular compounds shows the $X-Se-X$ bond angle to be relatively insensitive to the nature of the X grouping. Thus, all such bond angles, with the exception of H_2Se, have been found to lie within the 10° range of 96°–106°. Table 6-2 gives the Se—X bond distances and $X-Se-X$ bond angles for this class of structures, and the references to the original work. It should be mentioned here that a large number of $X-Se-X$ linkages are discussed in the later section on ring compounds. These are treated separately because ring closure constraints often effect the molecular geometry.

Although all of the reported Se—C distances in Table 6-2 are the same within a few standard deviations, other Se—C values will show that the shorter distances for the bonds to the aromatic carbon atoms are significant. The long Se—C distances reported by Linke and Lemmer[77,78] for selenium dicyanide were accompanied by C—N distances corresponding to values intermediate between double and triple bonds, and Se—C—N angles of 155° and

TABLE 6-2   SELENIUM PARAMETERS IN NONCYCLIC COMPOUNDS WITH
X—Se—X **LINKAGES**

Linkage	Compound	Se—X (Å)	X—Se—X	Method[a]	References
C—Se—C	Se(CH$_3$)$_2$	1.943(1)	96.2(1)°	M	12
		1.97(1)	98(10)°	E.D.	48
	Se(CF$_3$)$_2$	1.96(2)	104.4(5)°	E.D.	17
	Se(p-tolyl)$_2$	1.92(3)	106(2)°	X	14
		1.93(3)			
	Se(CN)$_2$	2.01	99°	X	77, 78
		2.08			
S—Se—S	Se(SCN)$_2$	2.21	101°	X	97
	Se(SO$_2$C$_6$H$_5$)$_2$	2.20(3)	105(3)°	X	46
	BaSe(SSO$_3$)$_2$	2.17(4)	101(3)°	X	44
H—Se—H	H$_2$Se	1.46(1)	91.0(6)°	M	69
		1.46	91.0°	V.R.	99
Cl—Se—Cl	SeCl$_2$	2.18(2)	Not given	E.D.	2
Br—Se—Br	SeBr$_2$	2.32(2)	Not given	E.D.	2
Si—Se—Si	Se(SiH$_3$)$_2$	2.273(4)	96.6(7)°	E.D.	4
Te—Se—P	Te[C$_2$H$_2$P(S)Se]$_2$	2.501(3)(Se—Te)	104.7(3)°	X	67
		2.26(1)(Se—P)			
P—Se—Hg	(HgCl$_2$(C$_6$H$_5$)$_3$PSe)$_2$	2.527(3)(Se—Hg)	98.1(1)°	X	49
		2.169(6)(Se—P)			

[a] M = Microwave, E.D. = Electron Diffraction, X = x-ray, and V.R. = Vibration-Rotation Spectr

168° (no sigmas given).   More precise values for these parameters, plus the two Se · · · · N nonbonded distances of 2.74 and 2.81 Å (the van der Waals sum is 3.5 Å), would be worth the effort of a reinvestigation of this interesting structure, since the reported parameters are a result of 2 two-dimensional refinements which gave conventional R factor values of 0.16 and 0.20.

The Se—S bond distances and the S—Se—S bond angles for the three quite different sulfur compounds in Table 6-2 are identical within experimental error, and the bond distances equal the sum of the bond radii obtained from the elemental states.   The selenium dithiocyanate molecule (Figure 6-3) is positioned on a crystallographic mirror plane of symmetry, with somewhat short Se—S distances, small SSeS/SeSC dihedral angles, but "normal" Se—S distances and S—Se—S angle.   The SeS$_4$O$_6^{-2}$ anion, reported by Foss and Tjomsland,[44] is cis in that both terminal sulfur atoms are on the same side of the S—Se—S plane.   This crystal is isomorphous with the pentathionate ion.   The orientation of the thiosulfate groups in selenium thiosulfate is shown in Figure 6-4.

The gas phase structures of selenium dichloride and selenium dibromide were determined by Akishin et al.[2] from measurements of the decomposition products of the corresponding tetrahalides.

The tellurium compound on Table 6-2 has a structure very similar to the

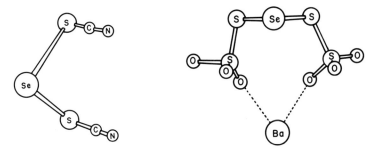

Figure 6-3 Se(SCN)$_2$.

Figure 6-4 The selenopentathionate ion in BaSeS$_4$O$_6$·2H$_2$O.

selenium analogue, which is described in the triselenide section of this chapter (see also Figure 6-10).

The structure of disilylselenide is noteworthy in that it did not reveal any bond shortening or increase of bond angle (96.6(7)°) which would have indicated $p\pi-d\pi$ bonding such as that found in the analogous disilylether (bond shortening of 0.13 Å and bond angle of 144°). In this respect the silyl selenium compound is like the corresponding disilylsulfide.

THE DISELENIDE ION AND STRUCTURES CONTAINING NONCYCLIC X—Se—Se—X LINKAGES

The seven diselenide structures which have been reported are summarized in Table 6-3. Five of the seven involve C—Se—Se—C linkages. All of the Se—C distances are within one standard deviation of the others, and the C—Se—Se angles do not appear to vary appreciably nor in any logical

BLE 6-3   SELENIUM PARAMETERS IN THE DISELENIDE ION AND NONCYCLIC COMPOUNDS WITH X—Se—Se—X LINKAGES

	Se—Se (Å)	Se—X (Å)	Se—Se—X	XSeSe/SeSeX Dihedral Angle	Method[a]	References
[CH(C$_6$H$_5$)$_2$]$_2$	2.285(5)	1.97(1)	100.1(5)°	82°	X	101
(C$_6$H$_5$)$_2$	2.29(1)	1.93(5)	105(2)°	82.0(3)°	X	88
		1.93(5)	107(2)°			
(C$_6$H$_4$Cl)$_2$	2.33(2)	1.9(1)	100(1)°	74.5°	X	74
		1.9(1)	102(1)°			
(CF$_3$)$_2$	2.34(3)	1.93(2)	103.5° (assumed)	—	E.D.	17
[P(S)(C$_2$H$_5$)$_2$]$_2$	2.33(1)	2.29(1)	106.2(5)°	104.5°	X	66
[C(NH$_2$)$_2$]$_2$Cl$_2$	2.38(1)	1.94(1)	95.5(5)°	Not given	X	27
Se$_2$	2.38(5)	—	—	—	X	41

X= x-ray, E.D. = Electron Diffraction.

manner.   The Se—Se distances average 2.34 Å, which is equal to the bonding distance in the monoclinic forms of elemental selenium, but there is a significant variance from one compound to another which does not appear to follow any simple chemical argument.   The XSeSe/SeSeX dihedral angles have been found to be 74.5 to 82° for the solid state of the diselenides bonded to carbon (see Figures 6-5, 6-6, and 6-7), and they are found to have a much higher angle of 104.5° when the selenium atoms are bonded to the second row element phosphorus (Figure 6-8).

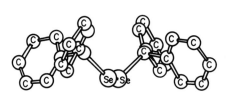

Figure 6-5 $Se_2(CH(C_6H_5)_2)_2$.

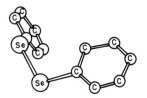

Figure 6-6 $Se_2(C_6H_5)_2$.

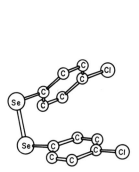

Figure 6-7 $Se_2(C_6H_4Cl)_2$.

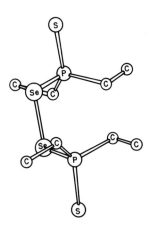

Figure 6-8 $Se_2(P(S)(C_2H_5)_2)_2$.

The structure of anhydrous sodium diselenide, done by Foppl and co-workers[41] with x-ray rotation and powder photographs, provides the only information on the diselenide ion, and the Se—Se distance reported is within one standard deviation of the Se_8 bond distances.   In the only gas phase Se_2 structure, done by Bowen,[17] the diperfluoromethyl compound was assumed to be nonplanar and *trans* for the calculation of the Se—Se and Se—C bond distances.

The diselenobisformamidinium ion was obtained by Chiesi *et al.*[27] as a product of the oxidation of selenourea.   In this ion, the end groups are

planar and hydrogen bonded to the chloride ion, and there is a crystallographic twofold rotation axis perpendicular to the Se—Se bond as shown in
Figure 6-9.

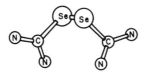

Figure 6-9   The $Se_2(C(NH_2)_2)_2^{+2}$ ion.

One additional compound containing an Se—Se bond, 1,2-diselenane, is
discussed in the section on ring compounds involving selenium.

COMPOUNDS CONTAINING NONCYCLIC X—Se—Se—Se—X LINKAGES

Two "ordinary" triselenide structures have been reported: $Se_3(CN)_2$ and
$Se_3(P(C_2H_5)_2Se)_2$.   Table 6-4 compares some of the parameters of these two
crystals.   In $Se_3(P(C_2H_5)_2Se)_2$, there is an interesting packing in which the
central selenium atom is approached by 2 selenium atoms from other molecules such that the central selenium is surrounded by a planar, almost rect-
angular arrangement of 4 selenium atoms as shown in Figure 6-10.   The

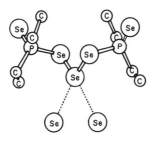

Figure 6-10 $Se_3(P(Se)(C_2H_5)_2)_2$.

**3LE 6-4   SELENIUM PARAMETERS IN COMPOUNDS WITH
TRISELENIDE LINKAGES**

	Se—Se (Å)	Se—X (Å)	Se—Se—Se	X—Se—Se	XSeSe/SeSeSe Dihedral Angles	References
CN)$_2$	2.33(3)	1.8(1)	101(2)°	95°	94°	3
P(C$_2$H$_5$)$_2$Se)$_2$	2.352(2)	2.239(7)	103.9(2)°	105.0(2)°	92.5°	68
CN)$_3$·1/2H$_2$O	2.64(1)	Not given	117.0(5)°	Not given	Not given	42
	2.70(1)					

Se · · · · Se nonbonded approach distance of 3.679(4) Å is significantly shorter than the van der Waals contact distance (4.0 Å), but is still longer than the intermolecular and interchain distances of elemental selenium.   Husebye and Helland-Madsen,[68] in discussing this structure, bring up the possibility of considering a three-centered bond for these linear Se—Se · · · · Se combinations. This will be discussed later.

In the $Se_3(CN)_2$ structure reported by Aksnes and Foss,[3] the triselenide unit was identical within experimental errors to the phosphorous compound just discussed.   However, in the phosphorus bonded triselenide, the phosphorus atoms were found to be on the same side of the plane as the selenium atoms, while in the cyanide molecule the carbon atoms are *cis* to this plane as shown in Figure 6-11.   Thus, the triselenide geometry in the latter compound, including the directions of the bonds from the terminal selenium atoms, is isostructural with a fragment of the Se_8 ring system.   The cyanide molecule has a crystallographic mirror plane bisecting the Se—Se—Se angle. The nonlinearity of the Se—C—N group (164°) was not considered significant by the investigators because of the general problem of locating the light atoms in the presence of the heavy diselenide linkage.

The inclusion of the $(SeCN)_3^-$ ion in this triselenide section is questionable, for the structure of this unit is really in a class of its own.   The general features as given by Foss and Hauge[42] in their preliminary note are shown in Figure 6-12.   The ion is a pseudohalogen analogue of the more familiar I_3^-

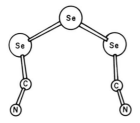

Figure 6-11 $Se_3(CN)_2$.

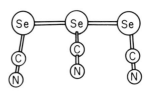

Figure 6-12   The $(SeCN)_3^-$ ion.

ion.   The two end SeCN groups are nearly coplanar and the middle SeCN unit makes a 60° angle with this plane.   The Se—Se distances of 2.64(1) and 2.70(1) Å, (*vs* 2.34 Å for elemental selenium) are considered by Foss and Hauge to be the sum of one single bond radius (1.17 Å) and one " half *p* bond " radius of 1.46 Å.   The latter value is the average of the bonding and the closest nonbonding distances in monoclinic selenium.   This approach is basically the 2 electron–3 center bonding concept of Rundle,[108] which appears to be of value in interpreting a number of selenium structures.

SELENIUM OXIDE STRUCTURES

Selenium dioxide and the tetramer of selenium trioxide have both been studied in the gas phase as well as the solid phase.

The electron diffraction work on selenium dioxide[102] was interpreted after assuming an O—Se—O angle of 120°, whereupon the Se—O distance was found to be 1.61(3) Å.   At about the same time, solid state selenium dioxide was found to consist of a polymeric system with oxygen-bridged selenium chains.[81]   The structural features of this solid are given in Table 6-5 and shown in Figure 6-13.   The packing of these SeO_2 chains gives each selenium atom a distorted octahedral environment, with oxygen distances of 1.73, 1.78, 1.78, 2.63, 3.00, and 3.00(3) Å.   The differences in bridging and nonbridging Se—O distances are not as great as might be expected.   This may be due to the lack of data in this early study which is reflected by the high standard deviations for the Se—O bond lengths.

The structure of the solid form of selenium trioxide reported by Mijlhoff[92] was a cyclic tetramer of the trioxide, shown in Figure 6-14.   The Se—O bonds to exocyclic oxygen atoms are considerably shorter (1.55(1) Å) than the Se—O cyclic bonds (1.77(1) Å).   The bond angles at the oxygen are slightly greater than 120°, as was found for the condensed dioxide, apparently implying an sp^2 hybridization for the oxygen.   The bond angles at the selenium are in general 98–108° with one notable exception: the O—Se—O angle formed by the 2 exocyclic oxygen atoms has the surprisingly high value of 128.2(9)°.   The only information from the electron diffraction of this tetramer[93] is the average Se—O distance of 1.58 Å.   A recent article by Paetzold[98] shows that the selenium oxides fit a general correlation of Se—O force constants with Se—O bond distances.

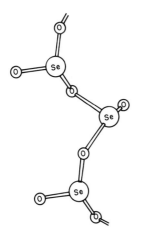

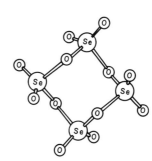

Figure 6-13 Solid state selenium dioxide.   **Figure 6-14** Solid state selenium trioxide tetramer.

TABLE 6-5 SELENIUM PARAMETERS IN THE SELENIUM OXIDES

	$Se-O_e{}^a$ (Å)	$Se-O_c$ (Å)	O_c-Se-O_c	O_e-Se-O_c	O_e-Se-O_e	$Se-O-Se$	References
SeO$_2$ (gas)	1.61(3)	—	125(5)°	98(2)°	—	—	102
SeO$_2$ (solid)	1.73(8)	1.78(3)	90.0(5)°	103.2(9)°	—	125.0(5)°	81
(SeO$_3$)$_4$ (solid)	1.54(1)	1.75(1)	98.7(9)°	103.2(9)°	128.2(9)°	123.2(9)°	92
	1.56(1)	1.80(1)		108.4(9)°			
				110.5(9)°			
(SeO$_3$)$_4$ (gas)	1.80(2) ave—	—		—		—	93

^{a} Subscript e = exocyclic; subscript c = cyclic.

OXYACIDS AND OXYANIONS OF SELENIUM

Table 6-6 shows the expected structural differences for the two oxidation states of selenium in these oxy- systems: selenic acids and the selenate ion have shorter Se—O bond distances (1.57–1.70 Å) and approximately tetrahedral angles, while the Se—O distances in selenious acids and the selenites

TABLE 6-6 SELENIUM PARAMETERS IN OXYACIDS AND OXYANIONS OF SELENIUM

	Se—O (Å)	O—Se—O	Se—X (Å)	O—Se—X	References
$_2SeO_4$	1.57, 1.57, 1.64, 1.66	110(ave)°			10
g_2SeO_4	1.61(7), 1.67(4)	106(3)° 107(3)° 108(3)° 121(4)°			32
a_2SeO_4	1.65(2)				70
$u(NH_3)_4SeO_4$	1.626(8) 1.629(8) 1.638(8) 1.647(8)	108.8(3)° 109.0(3)° 109.6(3)° 109.6(3)° 109.8(3)° 110.0(3)°			94
$nSeO_4$	1.62(2) 1.67(2) 1.66(2)	106.2(7)° 108.5(7)° 110.8(8)° 116.6(9)°			45
SeO_4	1.56(2) 1.66(5) 1.69(3)	107(1)° 107(1)° 109(2)° 120(2)°			45
SeO_4	1.64(2) 1.70(2)	108(1)° 108(1)° 107(9)°			45
$HSeO_4 \cdot H_2O$	1.56–1.70	104–114°			8
$NSeO_3$	1.67(1)		1.67(5)	112.7(6)°	40
SeO_3	1.72, 1.75, 1.76	93, 97, 112°			118
$O_2(C_6H_5)$	1.71(1), 1.76(1)	103.5(7)°	1.90(2)	99.0(9)°, 98.5(5)°	21
$O_2(C_6H_4Cl)$	1.70(5), 1.79(5)	Not given	1.85(7)	Not given	22
$SeO_3 \cdot 2H_2O$	1.72(5) 1.77(5) 1.78(5)	96.0° 98.5° 103.5°			47
$H_3(SeO_3)_2$	1.78(2) 1.78(2) 1.71(2)	100.4(5)° 102.0(6)° 104.5(5)°			113
$_3(SeO_3)_2$	1.669(4) 1.707(4) 1.730(5)	99.8(2)° 100.6(2)° 102.4(2)°			52
$eO_3 \cdot 2H_2O$	1.71				48
$SeO_3 \cdot 6H_2O$	1.69(1)	100.7(5)°			117

are longer (1.71–1.79 Å) and the O—Se—O angles are uniformly less than tetrahedral (93–104°).

Hydrogen bonding plays an important role in all of the structures containing hydrogen.   In general, no distinctions are made between terminal oxygen and protonated oxygen Se—O distances, since none of the x-ray structural determinations locate the hydrogen atoms.   In most cases, however, there are two significantly different Se—O distances which suggest the identification of the two types of bonds.

Selenic acid and selenious acid are both extensively hydrogen bonded systems.   In selenic acid, each SeO_4 tetrahedron is linked in a three-dimensional arrangement to 4 other tetrahedra (Figure 6-15), and in selenious acid, each pyramidal ion has 4 hydrogen bonds forming a double-layered structure (Figure 6-16).   In the latter structure, the selenium is in a distorted octahedral environment, with oxygen atoms at 1.72, 1.75, 1.76, 3.00, 3.08, and 3.09 Å.   Benzeneselenic acid[21] and the p-chlorobenzene analogue[22] illustrated in Figures 6-17 and 6-18 show identical molecular structure and hydrogen

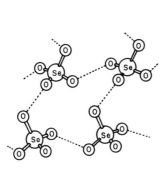

Figure 6-15   Selenic acid; the broken lines indicate hydrogen bonds.

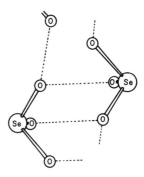

Figure 6-16   Selenious acid; the broken lines indicate hydrogen bonds.

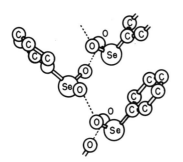

Figure 6-17   Benzeneselenic acid; the broken lines indicate hydrogen bonds.

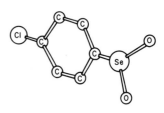

Figure 6-18 p-chlorobenzeneselenic acid.

bonding of the molecules into chains, but they have a different arrangement for packing these chain units.

The compound Ag_3NSeO_3 had been considered ionic prior to the crystal structure determination by Fawcett et al.,[40] which revealed an essentially covalent molecule with a very remarkable structure. The selenium atom is surrounded by a tetrahedron of 3 oxygen atoms and a nitrogen atom; the nitrogen, in turn, is surrounded by a tetrahedron composed of the selenium and 3 silver atoms. A crystallographic threefold rotation axis includes the nitrogen and selenium atoms, and the silver atoms are only 11.1° away from the eclipsed position with respect to the oxygen atoms (Figure 6-19). The silver ions closely approach the oxygen atoms of the next molecule (2.22(1) Å), forming a helical packing.

The orthoselenate ions in the tetraammine copper salt examined by Morosin[94] form chains with the copper atoms ($-Cu-O-Se-O-Cu-$), and a similar chain was found by Dorm[32] in mercurous selenate

$$(-Hg-Hg-O-Se-O-Hg-Hg-).$$

The considerable distortion in the selenate ion of the latter crystal is probably due to the interaction of the selenate ion with the mercury atoms, as well as the problem of locating the oxygen atoms exactly in the presence of the heavy atoms.

Interest in the structures of the alkali metal selenites, $MH_3(SeO_3)_2$, arises from their ferroelectric properties discovered by Pepinsky and Vedam in 1959. The lithium salt is ferroelectric at room temperature $(-196°–90°C)$, and the sodium salt shows a ferroelectric transformation at $-79°C$. The two salts are not isomorphous. The room temperature structure of the sodium salt[113] showed a three-dimensional network of hydrogen bonds involving every oxygen atom, which appears capable of initiating the ferroelectric state by a change in symmetry through a shift of the hydrogen positions. The room temperature structure of the potassium crystal[52] showed the structure was best understood by considering the formula to be $K_{0.5}H_{1.5}SeO_3$ (Figure 6-20).

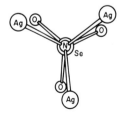

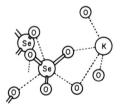

Figure 6-19 Ag_3NSeO_3.

Figure 6-20 The selenite ion in $KH_3(SeO_3)_2$; the broken lines indicate hydrogen bonds.

The three Se—O distances of 1.669(4), 1.707(4), and 1.730(5) Å correspond to the selenium bonds to oxygen atoms with 0, 1/2, and 1 hydrogen, respectively. This salt is not ferroelectric, but undergoes a phase change at −62°C which is also believed to be related to a change in the hydrogen bonding arrangement.

SELENIUM TETRAHALIDES AND ALKYL AND ARYL SELENIUM HALIDES

The structures of several diaryl selenium dihalide molecules and their tellurium analogues, reported in a series of papers by McCullough and co-workers, provided some of the first substantial evidence for the stereochemical influence of the nonbonded pair of electrons of selenium. These molecules are usually discussed on the basis of a trigonal bipyramidal geometry about the selenium, with the halogen atoms apical and the two organic groups and the lone pair of electrons in the equatorial plane (see Figure 6-21). In fact,

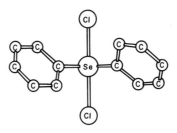

Figure 6-21 $Se(C_6H_5)_2Cl_2$.

however, the molecular data are almost exactly intermediate between the two idealized five-coordinate geometries of square pyramidal and trigonal bipyramidal, since the C—Se—C bond angles are 96°–110° (Table 6-7). The selenium-halogen bonds are about 0.2 Å longer than the single bond distances of 2.16 Å for Se—Cl and 2.36 Å for Se—Br, and thus are more comparable to the bond lengths found in the hexahaloselenates which will be discussed in a later section.

Trimethyl selenonium iodide was found by Hope[62] to be primarily ionic in the solid state, with a pyramidal $Se(CH_3)_3^+$ cation. The Se · · · · I distance of 3.776(2) Å, when compared to the van der Waals value of 4.15 Å, led Hope to suggest that the compound might be considered a weak charge-transfer complex. A similar ionic nature for triphenyl selenium chloride is suspected on the basis of the selenium chlorine separation in the solid state.[85]

The structural details of the tetrahalides of selenium are still not known. If these tetrahalides are covalent molecules, their structures would probably

TABLE 6-7   SELENIUM PARAMETERS IN SELENIUM TETRAHALIDES AND ALKYL AND ARYL SELENIUM HALIDES

	Se—Hal (Å)	Se—C (Å)	Hal—Se—Hal	Hal—Se—C	C—Se—C	References
SeBr$_2$(C$_6$H$_5$)$_2$	2.52(1)	1.91(3)	180(3)°	90°	110(10)°	82
SeCl$_2$(C$_6$H$_5$)$_2$	2.30(5)	Not given	Not given	Not given	Not given	83
SeCl$_2$(p-tolyl)$_2$	2.38(2)	1.93(3)	178(1)°	Not given	106(1)°	84
SeBr$_2$(p-tolyl)$_2$	2.55(2)	1.95(3)	177(1)°	Not given	108(1)°	84
SeCl$_2$(CH$_3$)$_2$	2.360(9)	1.94(4)	177.8(3)	87.8–91.9°	96(1)°	Cordes, Myers, and Champion (unpublished data).
	2.37(1)	1.95(4)	177.8(4)		104(2)°	
	2.42(1)	1.95(4)				
	2.43(1)	2.03(4)				
Se(CH$_3$)$_3$I	3.776(2)	1.96(2)	—	81.5(5)°	97.9(7)°	62
		1.95(2)		178.9(3)°	99.1(7)°	
SeF$_4$	1.70(3)	—	See text	—	—	18

closely resemble the dialkyl and diaryl selenium dihalides discussed previously. If the tetrahalides are ionic in the solid state, structures similar to that of trimethyl selenium iodide might be expected.   Electron diffraction of gaseous SeF_4 by Bowen[18] gave a Se—F bond distance of 1.70(3) Å, and " was consistent with " a C_{2v} symmetry for the molecule.   Numerous infrared and Raman studies of the tetrahalides, in solution and in the solid state, have given conflicting information on the molecular symmetries.[55]

SELENINYL DIHALIDES AND RELATED STRUCTURES

An alternate approach to a survey of the stereochemistry of selenium, rather than the topical approach being pursued in this review, would be to categorize the known molecular structures involving selenium according to the number of bonds formed by the selenium atom.   In such an approach, the geometry of the compounds with 2, 3, and 4 bonds per selenium atom are fairly easily rationalized.   The molecules which have been covered thus far are, in general, of this type.   However, one of the difficulties of categorization by the co-valency of selenium is presented by the sizeable group of molecules in which selenium is in a relatively high state of coordination and is found to have bonds which are intermediate in length between normal covalent bond values and nonbonded contact distances.   A number of oxychloride compounds of selenium provide a good illustration of this situation.

Selenium oxychloride, $SeOCl_2$, is a liquid at room temperature and has been one of the more popular of the nonaqueous solvents.   A good number of the solvent characteristics of $SeOCl_2$ relates to the ability of $SeOCl_2$ to act either as a Lewis acid or a Lewis base.   The structures of $SbCl_5 \cdot SeOCl_2$ (Figure 6-22) and $SnCl_4 \cdot 2SeOCl_2$ (Figure 6-23) clearly illustrate the basic character of $SeOCl_2$, in which it functions as an electron pair donor by utilizing a lone pair of electrons of the oxygen atom.   The $SeOCl_2$ molecular parameters (Table 6-8) are almost identical in these two structures.   Unfortunately, the effect of adduct formation on the structure of the $SeOCl_2$ molecule cannot be evaluated, since the uncertainties in the bond angles and

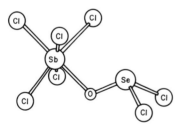

Figure 6-22 $SbCl_5 \cdot SeOCl_2$.

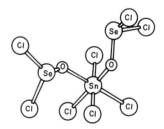

Figure 6-23 $SnCl_4 \cdot 2SeOCl_2$.

TABLE 6-8   SELENIUM PARAMETERS IN SELENINYL DIHALIDES AND RELATED STRUCTURES

	Se—O (Å)	Se—Hal (Å)	O—Se—Cl	Cl—Se—Cl	References
$SeOCl_2$(gas)	1.61	1.70	114°	106°	109
$SbCl_5 \cdot SeOCl_2$	1.66(2)[a]	2.09(1)	97.4(9)°	98.6(5)°	58
	1.71(2)	2.11(1)	99.8(9)°	99.0(5)°	
		2.13(1)	100.3(9)°		
		2.13(1)	101.8(9)°		
$SnCl_4 \cdot 2SeOCl_2$	1.73(2)	2.132(9)	98.9(6)°	97.6(3)°	57
		2.141(9)	103.3(6)°		
$N(CH_3)_4^+ \, Cl^- \cdot 5SeOCl_2$	1.59(2)[b]	2.20(1)[b]	102.9(8)[b]	95.0(4)°[b]	59
$C_9H_8NO^+SeOCl_3^-$	1.594(9)	2.234(4)	90.8(3)°	85.1(1)°	28
		2.271(4)	95.2(3)°	87.0(1)°	
		2.963(3)	102.3(3)°	91.1(1)°	
		2.992(3)	104.0(3)°	92.3(1)°	
$C_5H_2H_7^+SeOCl_3^-$	1.59(1)	2.233(5)	92.4(2)°	81.3(4)°	31
		2.251(6)	93.4(2)°	87.4(3)°	
		2.874(5)	100.1(4)°	92.9(4)°	
		2.934(5)	101.0(3)°	95.2(3)°	
$C_{10}H_{10}N_2^{+2}SeOCl_4^{-2}$	1.61(1)[a]	2.244(4)	102.1(5)°[c]	87.3–92.8(1)°	115
	1.65(1)	2.246(4)	102.2(4)		
		2.431(4)			
		2.445(4)			
		2.502(4)			
		2.525(4)			
		2.980(4)			
		2.990(4)			
$C_{10}H_9N_2^+SeO_2Cl^-$	1.62(1)	2.428(4)	100.4(3)°		116
	1.63(1)		100.5(3)°		
			106.3(4)° (O—Se—O)		
$SeOCl_2 \cdot 2C_5H_5N$	1.59(2)	2.39(1)	95.1(7)°	165.2(2)°	76
		2.57(1)	99.7(7)°		
$SeOF_2$	1.580(2)	1.727(2)	104.82(3)°	92.35(8)°	16
$SeOF_2 \cdot NbF_5$	1.60(2)	1.68(2)	100(1)°	93(2)°	35
$SeF_3^+Nb_2F_{11}^-$	—	1.63(3)	—	94.0(5)°	34
		1.67(3)			
		1.67(3)			

[a] Two formula units per asymmetric unit of the crystal.
[b] Average of five $SeOCl_2$ units.
[c] Angles to the chlorine atoms at 2.2 Å.

distances of gaseous $SeOCl_2$ are not known and the very short Se—Cl distance given for $SeOCl_2$ makes the reported parameters suspect. In the *cis* arrangement of the $SeOCl_2$ molecules in the tin complex, a tin-bonded chlorine approaches the selenium at a distance of 3.01(1) Å and two other chlorines complete the octahedron around selenium at 3.34(1) and 3.38(1) Å. Thus, $SeOCl_2$ is simultaneously a strong donor and a weak acceptor in this structure.

The ability of $SeOCl_2$ to act as a Lewis acid through expansion of the valency shell of selenium is most clearly illustrated by the structure of $SeOCl_2 \cdot 2$ pyridine, as reported by Lindquist and Nahringbauer.[76] Their results are shown in Figure 6-24. This structure also points out the difficulties in interpreting the bonding of highly coordinated selenium. The selenium atom in this adduct is surrounded by a square pyramidal arrangement consisting of 1 oxygen atom, 2 nitrogen atoms, and 2 chlorine atoms, with a lone pair of electrons presumably occupying the remaining octahedral position. In such an idealized geometry, the selenium atom of the adduct would be comparable to that of the iodine atom in IF_5. This interpretation is clouded, however, by the bond lengths involved: the 2 Se—N bonds of 2.19(2) and 2.20(2) Å are significantly longer than the Se—N bond of $C_6H_4N_2Se$ (1.83(4) Å) (see Table 6-11). This long Se—N distance suggests a dipole-dipole nature to the intermolecular bond. Moreover, the presence of the pyridine molecules has caused remarkable changes in the Se—Cl bonds; they have changed from the " normal " Se—Cl bonding distances of 2.2 or 2.3 Å to 2.39(1) and 2.57(1) Å. This bonding will be discussed later in this section.

The role of $SeOCl_2$ is different again in the case of the $N(CH_3)_4^+ Cl^- \cdot 5SeOCl_2$ structure reported by Hermodsson.[57,58] This structure may best be viewed as an example of the solvation of chloride ions by $SeOCl_2$ molecules. The structure actually consists of discrete tetramethyl ammonium ions and paired chloride ions, each pair being solvated by 10 $SeOCl_2$ molecules. These oxychloride molecules surround each chloride ion octahedrally, with 2 $SeOCl_2$ molecules being shared between the 2 adjacent octahedra as shown in Figure 6-25. The 6 selenium atoms are at distances of 2.93, 2.97, 3.05, 3.05, 3.09, and 4.02(1) Å from the chloride ion they surround, with the bridging $SeOCl_2$ molecules at 2.93 and 3.05(1) Å. That one $Cl^- \cdots Se$ distance is 4 Å, while the other 5 are 3 Å, is attributed to steric features. The immediate environment of each selenium atom consists of the oxygen and chlorine atoms of the $SeOCl_2$ at the average distances and angles shown in Table 6-8, the chloride ion at about 3.0 Å, and 1 oxygen atom from a neighboring $SeOCl_2$ unit at a comparable distance (the five distances are 2.85, 2.87, 2.89, 3.12, and 3.14(2) Å).

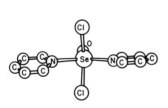

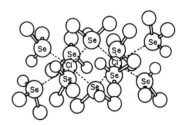

Figure 6-24 $SeOCl_2 \cdot 2C_5H_5N$.

Figure 6-25   The $(2Cl \cdot 10SeOCl_2)^{-2}$ cluster in $N(CH_3)_4Cl \cdot 5SeOCl_2$.

Similar arrays of Se—Cl distances have been found in three ionic oxy-chloride systems involving large organic cations. The structures of $C_9H_8NO^+SeOCl_3^-$, $C_5H_6N_2^+SeOCl_3^-$, and $C_{10}H_{10}N_2^{+2}SeOCl_4^{-2}$, containing the 8-hydroxyquinolinium, 2-aminopyridinium, and the 2-2'dipyridinium cations, respectively, were found to have Se—Cl distances of three different values: 2.2, 2.5, and 3.0 Å. The first of these values is considered the normal Se—Cl covalent distance; the van der Waals distance is 3.8 Å. In the 8-hydroxyquinolinium crystal, a chain-like arrangement of the oxychloride system (Figure 6-26) was formed, with each selenium atom surrounded by an oxygen at 1.594(9) Å, 2 nonbridging chlorines at 2.234(4) and 2.271(4) Å, and 2 bridging chlorines at 2.963(3) and 2.992(3) Å.

In the 2-aminopyridinium crystal, the local environment of the selenium is the same except that the chlorine bridging forms dimers instead of chains (Figure 6-27). In the dipyridinium crystal, there is 1 additional chlorine per

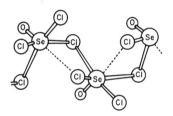

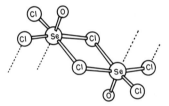

Figure 6-26 The SeOCl$_3^-$ chain in $C_9H_8NOSeOCl_3$; the broken lines indicate 3.4Å contacts.

Figure 6-27 The SeOCl$_3^-$ "dimers" in $C_5H_7N_2SeOCl_3$; the broken lines indicate 3.4 Å contacts.

selenium, and the average of the 2 noncrystallographically-equivalent di-negative units is an approximately square pyramidal structure with an oxygen atom at 1.63 Å, a chlorine atom at 2.24 Å, 2 chlorine atoms at 2.44–2.50 Å, and another at 2.99 Å. Figure 6.28 shows that the latter chlorine atom is involved in a chain-forming hydrogen bonding scheme.

All of the latter compounds in which $SeOCl_2$ behaves as an acid show molecular parameters significantly different from the antimony and tin halide adducts. When $SeOCl_2$ is an oxyelectron donor, the Se—O bond is lengthened and the Se—Cl distance is decreased (see Table 6-14).

The most intriguing question raised by the structures observed for the pyridine adduct, the tetramethyl ammonium compound, and the three structures with large organic cations, concerns the nature of those bonds which are intermediate in length between the normal covalent distance and the nonbonding contact distance. This also includes the 2.20(1) Å Se—N distances in the pyridine adduct (Se—N single bonds are usually 1.7 Å, and

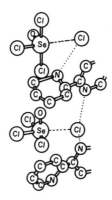

Figure 6-28   The SeOCl$_4^{-2}$ units in C$_{10}$H$_{10}$N$_2$SeOCl$_4$; the broken lines indicate 3.0 Å contacts, and the dotted lines indicate hydrogen bonds.

the van der Waals distance is 3.5 Å).   It is perhaps most convenient to consider these approaches as dipole-dipole and ion-dipole interactions.   There are two fairly general features of these structures, however, which give support to a more covalent description.   One aspect is the general tendency of the incoming groups to approach the selenium at orientations which usually give an overall octahedral coordination to the selenium, rather than an approach along the axis of the dipole.   The second common feature is that in these roughly octahedral systems of varying Se—Cl bond lengths, an interesting relationship is noticed in the Se—Cl bonds *trans* to one another.   If the normal Se—Cl distance of 2.2 is temporarily labelled a " short " bond, the bond distance of about 2.5 Å is labelled an "intermediate " bond, and the 3.0 Å distances (still very much shorter than the van der Waals value of 3.8 Å) are labelled " long" bonds, then all of the oxychloride structures mentioned which have these partial bonds (and *trans* Se—Cl bonds) have either " short-long " or "intermediate-intermediate " *trans* pairs.   This gives a relatively constant sum to the *trans* bonds in these structures, as shown in Table 6-9. This interdependence of *trans* bonding suggests a 4 electron–3 center molecular orbital scheme using the p orbital system, such as was done by Rundle[108] for the interhalogens.

The occurrence of octahedrally-oriented selenium bonds which are longer than covalent distances and significantly shorter than van der Waals sums is not limited to the selenium–chlorine system.   Such interactions have also been pointed out frequently for selenium-oxygen bonds.   The average Se—O bond is 1.6–1.7 Å (see Table 6-14), and the van der Waals Se—O distance is 3.40 Å.   There are numerous reports of Se—O distances of 2.6–3.1 Å, where these bonds " complete an octahedral environment for the selenium."   These have been noted, for example, by McCullough[81] in solid selenium dioxide, by

**TABLE 6-9   SELENIUM INTERACTIONS AT DISTANCES
BETWEEN SINGLE AND NO BOND VALUES**

$Cl \cdots Se \cdots Cl$ System ($Se \cdots Cl$ van der Waals Distance = 4.00 Å)

Structure	Se—Cl Distance (Å)	Bond (*trans*) (Å)	Sum (Å)
M_2SeCl_6	2.39(2)	2.39(2)	4.78
$SeOCl_2 \cdot 2C_5H_5N$	2.39(1)	2.57(1)	4.96
$2N(CH_3)_4Cl \cdot 10SeOCl_2$	3.091(8)	2.21(1)	5.30
	3.052(8)	2.20(1)	5.25
	2.972(8)	2.215(9)	5.187
	3.051(7)	2.219(8)	5.270
	2.298(7)	2.338(9)	5.166
$C_9H_8NO^+SeOCl_3^-$	2.963(3)	2.271(4)	5.234
	2.992(3)	2.234(4)	5.226
$C_5H_7N_2^+\,^2SeOCl_3^-$	2.874(5)	2.333(5)	5.107
	2.934(5)	2.251(6)	5.185
$C_{10}H_{10}N_2^+\,^2SeOCl_4^{-\,2}$	2.445(4)	2.525(4)	4.970
	2.431(4)	2.502(4)	4.933
	2.990(4)	2.246(4)	5.236
	2.980(4)	2.244(4)	5.234
$SnCl_4 \cdot 2SeOCl_2$	3.01(1)	2.132(8)	5.14
$SbCl_5 \cdot SeOCl_2$	3.01(1)	2.11(1)	5.12

Wells and Bailey[118] in selenious acid, by Hansen, Hazell, and Rasmussen[52] in potassium selenite, and by Hermodsson[58,59] in $N(CH_3)_4Cl \cdot 5SeOCl_2$. Se—F distances of 2.40–2.47(3) Å (the Se—F single bond value is 1.6 Å, the van der Waals value 3.35 Å) were found completing the octahedron about selenium in the SeF_3^+ ion by Edwards and Jones[34] (see the following), and the same researchers[35] found that 2.7–2.9 Å Se—F contacts complete the octahedron around selenium in $NbF_5 \cdot SeOCl_2$. As has already been mentioned, similar arguments have been used for the Se—Se interactions in $Se_3(P(Se)(C_2H_5)_2)_2$ and $K(SeCN)_3$.

The SeO_2Cl^- anion (Table 6-8) is formally equivalent to the $SeOCl_2$ molecule. The pyramidal ion has 2 Se—O bond lengths comparable to those of the acidic $SeOCl_2$ structures, and 1 elongated Se—Cl bond. This is the only reported oxychloride structure with 2 oxygen atoms per selenium, and the π bonding character of the 2 Se—O bonds apparently causes the Se—Cl bond lengthening.

Table 6-8 shows three structures related to seleninyl difluoride which have been reported. Bowater, Brown, and Burden[16] described the structure of $SeOF_2$ itself, as determined from microwave spectra. The NbF_5 adduct of $SeOF_2$[35] (Figure 6-29) was found to contain a relatively undistorted $SeOF_2$ with the selenium atom in a distorted octahedral environment caused by the approach of 3 fluorine atoms at 2.69, 2.69, and 2.88(2) Å. A similar distorted

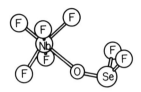

Figure 6-29 $NbF_5 \cdot SeOF_2$.

octahedral environment was found by Edwards and Jones in $SeF_3^+ Nb_2 F_{11}^-$, the cation of which is isoelectronic with $SeOF_2$. The preliminary account of the niobium compound had Se—F distances of 1.63, 1.67, 1.67, 2.40, 2.42, and 2.47(3) Å.

SELENIUM HEXAFLUORIDE AND THE HEXAHALOSELENATES

Selenium hexafluoride is the only neutral hexahalide known for selenium, and an electron diffraction study of SeF_6[39] which assumed an octahedral configuration gave an Se—F distance of 1.68(1) Å, the expected value for this linkage.

The structures of the dinegative hexahaloselenate ions have been of interest because of the consistent experimental evidence for the lack of distortion in the octahedral geometry of these anions, contrary to the bonding theories which assign a stereochemical role to the lone pair of electrons on the selenium.[111]   Both the hexahaloselenates and tellurates have been studied on this point, with the most recent x-ray investigations being made on the hexachloro- and hexabromotellurate ions.[30,56]   These latter studies provide the most significant evidence for the octahedral symmetry, since the early x-ray work was done by powder methods which were relatively insensitive to distortions and disorder in the anions.   The Se—Br bond value reported by Hoard and Dickenson[60] using powder data for the hexabromoselenate in $K_2 SeBr_6$ is 2.54 Å, and the Se—Cl bond distance in the $SeCl_6^{-2}$ ion is 2.39(2) Å.[38]   These bonds are about 0.15 Å longer than the normal single bond values, and for that reason were compared to the fractional Se—Cl bonds discussed in the previous section.

The symmetry of the selenium hexahalide system has also been studied by nuclear quadrupole resonance methods.[95]   Ammonium and potassium hexachloroselenate, and ammonium and cesium hexabromoselenate gave a single resonance line at all temperatures from room temperature to liquid nitrogen temperature, indicating equivalent halogens.   Potassium hexabromoselenate, however, did give evidence for solid state transitions, as the singlet at room temperature was split at lower temperatures.

CYCLIC SYSTEMS INVOLVING SELENIUM

A considerable number of structures have been reported in which selenium atoms are incorporated into a cyclic system.   Since the bond angles involved in a cyclic molecule are often influenced greatly by ring closure constraints, these parameters are not directly comparable to those of a " free " selenium atom and therefore are treated separately.   A notable series of these ring structures 1,4-diselenane, analogues of diselenane, and their adducts.

Structures Related to 1,4-Diselenane

The structures of 1,4-diselenane, $\overline{Se-CH_2-CH_2-Se-CH_2-CH_2}$, and related molecules have been done by single crystal x-ray techniques.   Table 6-10 compares the structural features of these compounds.

A comparison of the ten structures given in Table 6-10 shows that the Se—C bond distances do not vary significantly among the entire range of adducts. The C—Se—C bond angles, however, vary in a reasonable manner according to the size and make-up of the ring system.   Thus, the five member ring of C_4H_8Se has a C—Se—C angle of 93°, the C_4H_8OSe rings have C—Se—C angles of 94–96°, and the $C_4H_8Se_2$ rings have C—Se—C angles of about 100° There is a slight flattening of the boat conformation of the rings upon adduct formation as evidenced by the increased Se · · · Se cross-ring distance.   Among the several 1,4-diselenane adducts with iodine acids, it is apparent that the strongest interaction (shortest Se · · · I distance) takes place with elemental iodine.   The presence of the first donor-acceptor bond does not appear to affect the ability of the diselenane to form a similar bond to another iodine atom, as evidenced by the Se · · · · I distances in the doubly bonded structures.

All of these molecules and adducts have highly symmetrical structures. Many lie on symmetry elements of the lattice (1,4-diselenane itself, as shown in Figure 6-30, is located on a center of symmetry), or on noncrystallographically required elements (usually mirror planes) which are obeyed within experimental errors.   The diselenane-iodine adduct has the nonrequired 2/m symmetry illustrated in Figure 6-31.   The donor-acceptor interaction appears to be stronger in this compound than it is in the corresponding sulfur complex on the basis of a relatively shorter Se(S) · · · I distance and a longer I—I bond in the selenium compound (2.870(3) vs 2.787(3) Å).   In the disulfur compound, the iodine approach is equatorial.

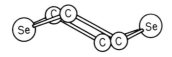

Figure 6-30   1,4-diselenane.

TABLE 6-10   SELENIUM PARAMETERS IN DISELENANE, RELATED MOLECULES, AND THEIR ADDUCTS

	Se—C (Å)	C—Se—C	C—C—Se	Se···Se (Å)	Se···X (Å)[a]	References
$C_4H_8Se_2$	1.99(4) 2.04(6)	98(3)°	108(3)° 108(4)°	3.66(2)		90
$C_4H_8Se_2 \cdot 2I_2$	1.95(2) 1.98(2)	100(2)°	115(2)°	3.75(1)	2.829(4)(a) 180.0(3) (Se—I—I)	25
$C_4H_8Se_2 \cdot C_2I_2$					3.336(7)(a) 173.1(2) (Se—I—I)	61
$C_4H_8Se_2 \cdot C_2I_4$					3.43(a) 3.40(e)	29
$C_4H_8Se_2 \cdot CHI_3$	1.89(5) 1.94(5)	100(2)°	111(3)° 111(3)°	3.70(1)	3.465(9)(a) 3.514(9)(e) 177 (I—Se—I)	13
$C_4H_8OSe \cdot ICl$	1.94(4) 2.02(4)	96(2)°			2.630(5)(a) 175.8(5) (Se—I—Cl)	72
$C_4H_8OSe \cdot I_2$	1.93(3) 1.98(3)	94(2)°			2.755(4)(a) 174.8(3) (Se—I—Cl)	87
$C_4H_8Se_2Cl_4$	1.86 1.88			3.75(1)	2.24(2) 2.24(2) ~180 (Cl—Se—Cl)	5
$C_4H_8SSeBr_2$	2.01(3) 2.02(3)	105(2)°			2.545(5) 2.548(5) 175.1(6) Br—Se—Br	11
$C_4H_8Se \cdot I_2$	1.960(2) 1.960(2)	93(2)°			2.763(5) 179.3(3) (Se—I—I)	64

[a] (a) = halogen bonded to ring in axial position; (e) = equatorial position.

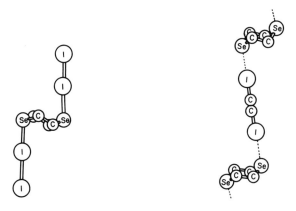

Figure 6-31 $C_4H_8Se_2 \cdot 2I_2$.   Figure 6-32 $C_4H_8Se_2 \cdot C_2I_2$.

   In the series of adducts of diselenane with C_2I_2, C_2I_4, and CHI_4, there is a steady increase in the $Se \cdots I$ distances from 3.336(7) to 3.514(9) Å.   In the C_2I_2 structure, both the diselenane and the acetylene molecule are located on crystallographic centers of symmetry such that chains are formed as shown in Figure 6-32.   Like the iodine complex, the diselenane interactions are axial and those of the sulfurs are equatorial.   The brief note on the structure of the C_2I_4 complex shows a structure very similar to the acetylene adduct.[29] Both acceptor and donor are located on centers of symmetry forming chains of molecules with linear $Se-I-C$ bonds.   Each selenium is approached in both the axial and equatorial positions by iodine atoms from different C_2I_4 molecules.   In the diselenane-iodoform adduct, each selenium is similarly approached by 2 iodine atoms, forming chains as shown in Figure 6-33.   The diselenane adducts have been reviewed, and compared to sulfur and oxygen analogues, by Hassell.[53]

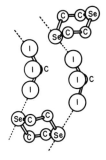

Figure 6-33 $C_4H_8Se_2 \cdot CHI_3$.

Structures of two adducts of 1,4-oxaselenane, C_4H_8OSe, have been reported: one is with iodine and the other with iodine monochloride. Both complexes have an approximate symmetry of m, with the shortest reported Se····I distances for this type of association (Table 6-10). In $C_4H_8OSe \cdot ICl$ (Figure 6-34) the Se····I separation is only 2.630(5) Å, and the I—Cl bond distance has increased from its molecular value of 2.32 Å to 2.73 Å for the complex. These parameters have led Knobler and McCullough[72] to suggest that this structure represents a step towards ion pair formation: $C_4H_8OSeI^+Cl^-$. The structure of the iodine adduct is shown in Figure 6-35. The secondary Se····I interactions which form the nonlinear chains are at a distance of 3.64(1) Å in this crystal.

The structure of the complex of the five-membered ring C_4H_8Se with iodine was determined by Hope and McCullough in 1964. The structure, as shown in Figure 6-36, has linear Se···I—I bonding at a relatively short Se—I

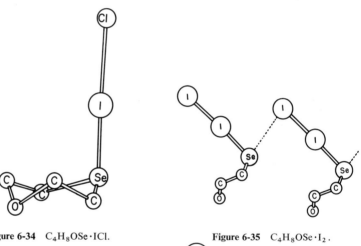

Figure 6-34 $C_4H_8OSe \cdot ICl$.

Figure 6-35 $C_4H_8OSe \cdot I_2$.

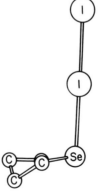

Figure 6-36 $C_4H_8Se \cdot I_2$.

separation of 2.763(5) Å, and a secondary Se $\cdots$ I interaction at 3.638(5) Å on the opposite side of the selenium atom.   There is a disorder in the positions of the 2 carbon atoms which are across the ring from the selenium atom and are 0.4 Å above and below the C—Se—C plane.   Pedersen and Hope[104] have used a broad line NMR to show that this flip-flop of the ring atoms can be quenched at liquid nitrogen temperatures.

The structures of $C_4H_8SSeBr_2$[11] and $C_4H_8Se_2Cl_4$[5] are in several respects more comparable to the diaryl selenium dihalides than to the diselenane adducts discussed previously due to the nearly linear halogen-selenium-halogen linkages and their perpendicular orientation with respect to the C—Se—C plane (Figure 6-37 through 6-39).   The bond angles of the ring in

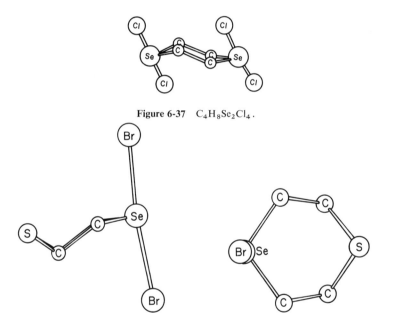

Figure 6-37 $C_4H_8Se_2Cl_4$.

Figure 6-38 $C_4H_8SSeBr_2$ ("side" view).   Figure 6-39 $C_4H_8SSeBr_2$ ("top" view).

the $C_4H_8SSeBr_2$ molecule, illustrated by Figure 6-39, reveal an interesting situation.   The sulfur ring angles have smaller values than the selenium ring angles, presumably as a result of the stronger bonding to the selenium.

Other Cyclic Systems: Five-Membered Rings

Ten structures have been reported in which selenium forms part of a five-membered ring.   Seven of these rings involve a C—Se—C linkage.   Table 6-11 summarizes the results reported for this ring system and shows that the

TABLE 6-11 SELENIUM PARAMETERS IN MOLECULES WITH FIVE-MEMBERED RINGS CONTAINING SELENIUM

		Se—C (Å)	G—Se—C	References
$C_4H_8SeI_2$		1.960(2)	94(2)°	64
$C_6H_4Se_2$		1.87(1) 1.93(2)	86.1(8)°	114
$C_{16}H_8O_2Se_2$		1.75(5)[a] 1.97(5)	80(4)°	36, 37
$C_4H_3COOHSe$		1.85(1) 1.872(8)	87.1(4)°	96
C_3H_7NSeBr		1.88(3) 1.99(3)	86(2)°	71

Formula	Structure	Bond length (Å)	Bond angle	Ref.
(top, cut off)				20
$C_{12}H_8Se$		1.885(5) 1.889(5) 1.889(5) 1.897(5)	86.6(2)° 86.7(2)°	65
		Se—N (Å)	**N—Se—N**	
$C_6H_4N_2Se$		1.83(4)	95°	80
		Se—O (Å)	**X—Se—X**	
$C_2H_4O_3Se_2$		Se—O_e 1.61(3) Se—O_c 1.82(2) Se—C 1.99(4)	O—Se—O 106(2)° C—Se—O_c 91(1)° C—Se—O_e 107(3)°	51
		Se—S (Å)	**S—Se—C**	
		2.333(3)	90°	112
		Se—C (Å)		
$C_{17}H_{12}S_2Se$		1.82(1)		

[a] Selenium to the ethylene carbon.

[b] Subscript e = exocyclic; subscript c = cyclic.

X—Se—X ring angles average less than 90°.   The Se—C bond lengths vary as a function of the hybridization of the carbon atom.

Most of these rings are nearly planar.   The structure of selenophene[20] was solved from its microwave spectrum, while the others were done by x-ray methods.   The carboxylic acid molecules shown in Figure 6-40 form nearly planar dimers in the solid state *via* hydrogen bonding.   Dibenzoselenophene[65] is very nearly planar (there is a 0.5–1.2° dihedral angle between the planes of the 2 six-membered rings), as shown in Figure 6-41, as is *trans* selenophthene (selenolo[3,2-b]selenophene) illustrated in Figure 6-42 (the Se is 0.05 Å out of the carbon plane).   In the 2-amino-1,3-selenazole ion (Figure 6-43), the carbon across the ring from the selenium atom is 0.43 Å removed from the ring plane.   The *trans* ethanediselenic anhydride structure shown in Figure 6-44[51] had the interesting background of preparation by oxidation of di-selenane and recrystallization from the aqueous solution.   The structures of piazselenole and selenoindigo are shown in Figures 6-45 and 6-46, respectively. Table 6-11 gives the bonding arrangement found by van den Hende and

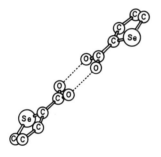

Figure 6-40 SeC$_4$H$_3$COOH dimers; the broken lines indicate hydrogen bonds.

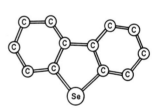

Figure 6-41 Dibenzoselenophene.

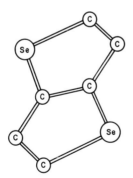

Figure 6-42 Selenophthene.

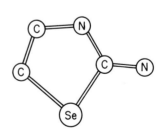

Figure 6-43 The C$_3$N$_2$H$_7$Se^{+1} cation.

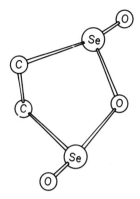

Figure 6-44   *trans*-ethanediseleninic anhydride, $C_2H_4O_3Se_2$.

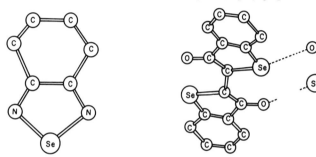

Figure 6-45 $C_6H_4N_2Se$.

Figure 6-46   Selenoindigo; the broken lines indicate chain formation.

Klingsberg[112] for a diphenyl selenothiophthene (2-(4-phenyl-SH-1,2-thia-selenol-5-ylidene)-thioacetophenone); no figure is given for lack of crystallographic coordinates). The S—S distance of 2.492(3) Å and the Se—S distance of 2.333(3) Å are interpreted to indicate that the resonance form shown in Table 6-11 is the dominant form, even though the Se—S distance is longer than the usual single bond distance of 2.21 Å. The structure of $C_4H_8Se \cdot I_2$ was discussed in the previous section (see Figure 6-36).

Other Cyclic Systems

Figure 6-47 shows the structure of nickel diethyl diselenocarbamate which has Ni—Se bond distances of 2.28(1) and 2.46(1) Å, Se—C distances of 1.84(2) and 1.97(2) Å, Ni—Se—C angles of 85(1)° and 87(1)° and a Se—N—Se angle of 81.0(1)°. The compound is planar except for the terminal methyl units, and the difference in the Ni—Se bond distances is believed to have a chemical basis.[15]

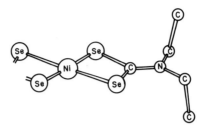

Figure 6-47 $Ni(C_5H_{10}NSe_2)_2$.

One 1,2-diselenane structure (Table 6-12) has been reported.[43]   In the 1,2-diselenane-3,6-dicarboxylic acid crystal the Se—Se bond distance is 2.32(1) Å, the Se—C distance 1.97(2) Å, and the CSeSe/SeSeC dihedral angle is 56(1)° as shown in Figure 6-48.   Thus, only the last parameter is significantly different than the corresponding parameters for noncyclic diselenides (see Table 6-3), and this change to a very small dihedral angle is certainly a constraint of ring formation.

Although chemical investigations of tetraselenium tetranitride have been hampered by its extremely explosive nature, the crystal structure has been reported by Baeringhausen, Volkmann, and Jander.[9]   The molecular structure, as shown in Figure 6-49, is similar to tetrasulfur tetranitride, and can be described as a tetrahedron of selenium atoms superimposed upon a square planar arrangement of nitrogen atoms.   The Se—N distances of 1.77, 1.78, 1.79, and 1.80(3) Å are almost exactly intermediate between the quoted Se—N single bond values of 1.86 Å and the double bond values of 1.64 Å.   The N—Se—N angles are 102 and 102(1)°, and the Se—N—Se angles are 111, 112, and 114(1)°.   Since this ring is formally unsaturated, there has been interest in the possibilities of cross-ring partial bonds in both sulfur and selenium nitride.   In the selenium compound, the Se · · · · Se cross-ring distance is 2.748(9) Å, compared to the Se—Se single bond value of 2.34 Å and the van der Waals value of 4.0 Å.

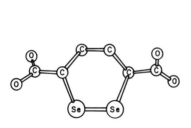

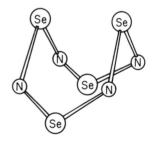

Figure   6-48   1,2-diselenane-3,6-dicarboxylic acid.

Figure 6-49   Tetraselenium tetranitride.

TABLE 6-12   SELENIUM PARAMETERS IN OTHER CYCLIC SYSTEMS CONTAINING SELENIUM

	Se—Se (Å)	Se—Se—Se			References
$Se_4^{+2}(HS_2O_7)_2^{-1}$	2.28	90°			19
$Se_8^{+2}(AlCl_4)_2^{-1}$	2.32(3)(ave)	98°			86
	2.83(1)	90°			

	Se—Ge (Å)	Ge—Se—Ge			
GeSe	2.56–2.59	96–103°			33

	Se—N (Å)	N—Se—N	Se···Se (Å)		
Se_4N_4	1.77(3)	102(1)°	2.748(9)		9
	1.78(3)	102(1)°			
	1.79(3)				
	1.80(3)				

	Se—Se (Å)	Se—C (Å)	Se—Se—C		
$C_6H_8O_4Se_2$ 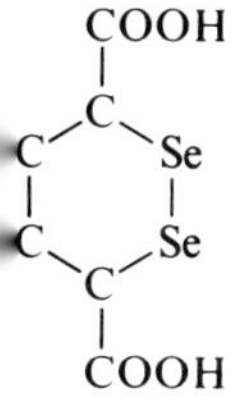	2.32(1)	1.97(2)	95.7(8)°		43

```
      COOH
       |
       C
    C´   `Se
   C|     |
    C´  ` Se
       C
       |
      COOH
```

	Ni—Se (Å)	Se—C (Å)	Ni—Se—C	Se—Ni—Se	
$Ni(C_5H_{10}NSe_2)_2$ 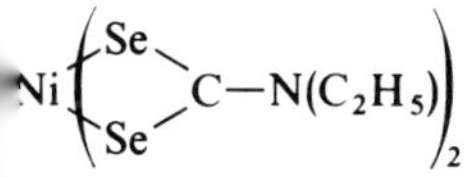	2.28(1)	1.84(1)	85(1)°	81.0(1)°	15
	2.46(1)	1.97(2)	87(1)°		

$$Ni\left(\begin{array}{c} Se \\ \diagdown \\ Se \end{array} C-N(C_2H_5)\right)_2$$

Two selenium ring structures obtained by the oxidation of elemental selenium have been reported recently. A preliminary report of the Se_4^{+2} cation by Brown, Crump, Gillespie, and Santry[19] indicated a square planar arrangement of selenium atoms with a Se—Se bond distance of 2.28 Å. The Se_8^{+2} cation, reported by McMullan, Prince, and Corbett,[86] had the interesting structure shown in Figure 6-50. The Se_8^{+2} ion has 46 valence electrons, compared to 44 for Se_4N_4 and 48 for Se_8, and it is thus very significant that the Se_8^{+2} ring has the conformation of one half of a Se_8 " crown " configuration

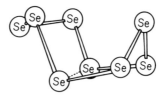

Figure 6-50   The Se_8^{+2} cation.

added to one half of the Se_4N_4 " saddle " geometry.   Discussion of Se_8^{+8} as a bicyclic unit was given on the basis of the short $(2.83(1)\ \text{Å})$ cross-ring approach noted by the dotted line in Figure 6-50.

STRUCTURES WITH SINGLY BONDED SELENIUM ATOMS

Table 6-13 gives the selenium bond distances for several structures which contain a singly bonded selenium atom.   The two similar selenourea derivative structures (Figure 6-51), done in different laboratories, gave the same Se—C bond distance.   This distance was considered by both investigators to indicate a bond intermediate between single and double bond character due to resonance forms possible for the molecules.   A shortening of the C—N bond $(1.32\ \text{Å})$ supports this reasoning.   The diphenyl crystal[63] apparently contains a weak (and rare) hydrogen bond to selenium.

TABLE 6-13   BOND DISTANCES FOR SINGLY BONDED SELENIUM

	C—Se (Å)	References
$Ni(HCON(CH_3)_2)_4(NCSe)_2$	1.71(3)	110
$Se{=}C(NHCOC_6H_5)(HCN_6H_5)$	1.82(1)	63
$Se{=}C(NHCOCH_3)(NHC_6H_5)$	1.87(3)	106
	P—Se (Å)	
$(C_6H_5)_2P(Se)OCH_3$	2.081(8)	75

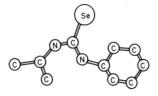

Figure 6-51 $C_9H_8N_2OSe.$

Summary of Bond Distances Involving Selenium

Table 6-14 gives the average values for the various selenium bond lengths included in this review.   The reciprocals of the individual standard deviations were used for the weighing factors to obtain the average.   Only those bond distances which had standard deviations assigned by the investigators themselves were used.

TABLE 6-14   AVERAGE BOND DISTANCE INVOLVING SELENIUM

Bond	Average Value (Å)	Number of Determinations[a]	Range of Values (Å)
Se—H	1.46	1	—
Se—F	1.672	5	1.63(3) –1.70(3)
Se—Cl (in $SeOCl_2$ as a base)	2.123	6	2.09(1) –2.141(9)
Se—Cl (in $SeOCl_2$ as an acid)	2.235	16	2.17(1) –2.251(6)
Se—Br	2.32	1	—
Se—O (selenate)	1.641	16	1.56(2) –1.70(2)
Se—O (selenite)	1.704	14	1.669(4)–1.79(5)
Se—O (in $SeOCl_2$ as a base)	1.700	3	1.66(2) –1.73(2)
Se—O (in $SeOCl_2$ as an acid)	1.609	13	1.57(2) –1.65(1)
Se—S	2.189	2	2.17(4) –2.20(3)
Se—Se	2.343	28	2.285(5)–2.38(3)
Se—Te	2.501	1	—
Se—N	1.83	1	—
Se—P	2.257	3	2.239(7)–2.29(1)
Se—C (saturated)	1.949	26	1.88(3) –2.04(6)
Se—C (aromatic)	1.890	25	1.85(7) –2.03(4)
Se—Si	2.273	1	—
Se—Hg	2.527	1	—
Se—Ni	2.28	1	—

REFERENCES

1. Abrahams, S. C., *Quart. Revs.*, **10**, 407 (1956).
2. Akishin, P. A., Spiridonov, V. P., and Mishulina, R. A., *Vestn. Mosk. Univ. Khim.*, **17**, 23 (1960).
3. Aksnes, O., and Foss, O., *Acta Chem. Scand.*, **8**, 1787 (1954).
4. Almenningen, A., Fernholt, L., and Seip, H. M., *Acta Chem. Scand.*, **22**, 51 (1968).
5. Amendola, A., Gould, E. S., and Post, B., *Inorg. Chem.* **3**, 1199 (1964).
6. Andrievskii, A. I., and Nabitovich, I. D., *Kristallografiya*, **5**, 465 (1960).
7. Andrievskii, A. I., Nabitovich, I. D., and Kripyakevich, P. I., *Doklady Akad. Nauk. SSSR*, **124**, 321 (1959).
8. Aurivillius, B., *Acta Chem. Scand.*, **18**, 2375 (1964).
9. Baernighausen, H., Volkmann, T. V., and Jander, J., *Acta Crystallogr.*, **21**, 571 (1966).

10. Bailey, M., and Wells, A. F., *J. Chem. Soc.*, 986 (1951).
11. Battelle, L., Knobler, C., and McCullough, J. D., *Inorg. Chem.* **6**, 958 (1967).
12. Beecher, J. F., *J. Mol. Spectry.*, **21**, 414 (1966).
13. Bjorvatten, T., *Acta Chem. Scand.*, **17**, 2292 (1963).
14. Blackmore, W. R., and Abrahams, S. C., *Acta Crystallogr.,*, **8**, 323 (1955).
15. Bonamico, M., and Dessy, G., *Chem. Commun.*, 1114 (1967).
16. Bowater, J. C., Brown, R. D., and Burden, F. R., *J. Mol. Spectry.*, **23**, 272 (1963).
17. Bowen, H. J. M., *Trans. Faraday Soc.*, **50**, 452 (1954).
18. Bowen, H. J. M., *Trans. Faraday Soc.*, **59**, 1241 (1963).
19. Brown, I. D., Crump, D. B., Gillespie, R. J., and Santry, D. P., *Chem. Commun.*, 853 (1968).
20. Brown, R. D., Burden, F. R., and Godfrey, P. D., *J. Mol. Spectry.* **25**, 415 (1968).
21. Bryden, J. H., and McCullough, J. D., *Acta Crystallogr.*, **7**, 833 (1954).
22. Bryden, J. H., and McCullough, J. D., *Acta Crystallogr.*, **9**, 528 (1956).
23. Burbank, R. D., *Acta Crystallogr.*, **4**, 140 (1951).
24. Chao, G. Y., and McCullough, J. D., *Acta Crystallogr.*, **13**, 727 (1960).
25. Chao, G. Y., and McCullough, J. D., *Acta Crystallogr.*, **14**, 940 (1961).
26. Cherin, P., and Unger, P., *Inorg. Chem.*, **6**, 1589 (1967).
27. Chiesi, A., Grossoni, G., Nardelli, M., and Vidoni, M. E., *Chem. Commun.*, 404 (1969).
28. Cordes, A. W., *Inorg. Chem.*, **6**, 1204 (1967).
29. Dahl, T., and Hassel, O., *Acta Chem. Scand.*, **19**, 2000 (1965).
30. Das, A. K., and Brown, I. D., *Can. J. Chem.*, **44**, 939 (1966).
31. Dill, D., and Cordes, A. W., unpublished results.
32. Dorm, E., *Acta Chem. Scand.*, **23**, 1607 (1969).
33. Dutta, S. N., and Jeffrey, G. A., *Inorg. Chem.*, **4**, 1363 (1965).
34. Edwards, A. J., and Jones, G. R., *Chem. Commun.*, 346 (1968).
35. Edwards, A. J., and Jones, G. R., *J. Chem. Soc.*, A, 2858 (1969).
36. Eller, H. v., *Bull. Soc. Chim. France*, **22**, 1429 (1955).
37. Eller, H. v., *Compt. Rend.*, **239**, 1043 (1954).
38. Engel, G., *Z. Krist.*, **90**, 341 (1935).
39. Ewing, V. C., and Sutton, L. E., *Trans. Faraday Soc.*, **59**, 1241 (1963).
40. Fawcett, J. K., Kocman, V., Nyburg, S. C., and O'Brien, R. J., *Chem. commun.*, 1198 (1969).
41. Foppl, H., Busmann, E., and Frorath, F.-K., *Z. Anorg. Allg. Chem.* **314**, 12 (1962).
42. Foss, O., and Hauge, S., *Acta Chem. Scand.*, **17**, 1807 (1963).
43. Foss, O., Johnsen, K., and Reistad, T., *Acta Chem. Scand.*, **18**, 2345 (1964).
44. Foss, O., and Tjomsland, O., *Acta Chem. Scand.*, **8**, 1701 (1954).
45. Fuess, H., and Will, G., *Z. Anorg. Allg. Chem.* **358**, 125 (1968).
46. Furberg, S., and Öyum, P., *Acta Chem. Scand.*, **8**, 42 (1954).
47. Gattow, G., *Acta Crystallogr.*, **11**, 377 (1958).
48. Gladkova, V. F., and Kondrashev, Y. D., *Kristallografiya*, **9**, 190 (1964).
49. Glasser, L. S. D., Ingram, L., King, M. G., and McQuillan, G. P., *J. Chem. Soc.*, A., 2501 (1969).
50. Goldfish, E., Hedberg, K., Marsh, R. E., and Schomaker, V., *J. Amer. Chem. Soc.*, **77**, 2948 (1955).
51. Gould, E. S., and Post, B., *J. Amer. Chem. Soc.*, **78**, 5161 (1956).

52. Hansen, F., Hazell, R. G., and Rasmussen, S. E., *Acta Chem. Scand.*, **23**, 2561 (1969).
53. Hassel, O., *Acta Chem. Scand.*, **19**, 2259 (1965).
54. Hawes, L. L., *Nature*, **198**, 1267 (1963).
55. Hayward, G. C., and Hendra, P. J., *J. Chem. Soc.*, A, 643 (1967).
56. Hazell, A. C., *Acta Chem. Scand.*, **20**, 165 (1966).
57. Hermodsson, Y., *Acta Crystallogr.*, **13**, 656 (1960).
58. Hermodsson, Y., *Acta Chem. Scand.*, **21**, 1313 (1967).
59. Hermodsson, Y., *Acta Chem. Scand.*, **21**, 1328 (1967).
60. Hoard, J. L., and Dickenson, B. N., *Z. Krist.*, **84**, 436 (1933).
61. Holmesland, O., and Roemming, C., *Acta Chem. Scand.*, **20**, 2601 (1966).
62. Hope, H., *Acta Crystallogr.*, **20**, 610 (1966).
63. Hope, H., *Acta Crystallogr.*, **18**, 259 (1965).
64. Hope, H., and McCullough, J. D., *Acta Crystallogr.*, **17**, 712 (1964).
65. Hope, H., Knobler, C., and McCullough, J. D., *Acta Crystallogr.*, **26B**, 628 (1970).
66. Husebye, S., *Acta Chem. Scand.*, **20**, 51 (1966).
67. Husebye, S., *Acta Chem. Scand.*, **23**, 1389 (1969).
68. Husebye, S., Helland-Madsen, G., *Acta Chem. Scand.*, **23**, 1398 (1969).
69. Jache, A. W., Moser, P. W., and Gordy, W., *J. Chem. Phys.*, **25**, 209 (1956).
70. Kalman, A., and Cruickshank, D. W. J., *Acta Crystallogr.*, **26B**, 436 (1970).
71. Karle, L., *Angew. Chem. Intern. Ed. Engl.*, **7**, 811 (1968).
72. Knobler, C., and McCullough, J. D., *Inorg. Chem.* **7**, 365 (1968).
73. Krebs, H., and Steffen, R., *Z. Anorg. Allg. Chem.*, **327**, 224 (1964).
74. Kruse, F. H., Marsh, R. E., and McCullough, J. D., *Acta Crystallogr.*, **10**, 201 (1957).
75. Lepicard, G., de-Saint-Giniez-Liebig, D., Laurent, A., and Rerat, C., *Acta Crystallogr.*, **25B**, 617 (1969).
76. Lindqvist, I., and Nahringbauer, G., *Acta Crystallogr.*, **12**, 638 (1959).
77. Linke, K. H., and Lemmer, F., *Z. Anorg. Allg. Chem.*, **345**, 211 (1966).
78. Linke, K. H., and Lemmer, F., *Z. Naturforsch.*, **21**, 192 (1966).
79. Lochman, F., *Nature*, **172**, 171 (1953).
80. Luzzati, V., *Acta Crystallogr.*, **4**, 193 (1951).
81. McCullough, J. D., *J. Amer. Chem. Soc.*, **59**, 789 (1937).
82. McCullough, J. D., and Hamburger, G., *J. Amer. Chem. Soc.*, **63**, 803 (1941).
83. McCullough, J. D., and Hamburger, G., *J. Amer. Chem. Soc.*, **64**, 508 (1942).
84. McCullough, J. D., and Marsh, R. E., *Acta Crystallogr.*, **3**, 41 (1950).
85. McCullough, J. D., and Marsh, R. E., *J. Amer. Chem. Soc.*, **72**, 4556 (1950).
86. McMullan, R. K., Prince, D. J., and Corbett, J. D., *Chem. Commun.*, 1438 (1969).
87. Maddox, H., and McCullough, J. D., *Inorg. Chem.*, **5**, 522 (1966).
88. Marsh, R. E., *Acta Crystallogr.*, **5**, 458 (1952).
89. Marsh, R. E., Pauling, L., and McCullough, J. D. *Acta Crystallogr.*, **6**, 71 (1953).
90. Marsh, R. E., and McCullough, J. D., *J. Amer. Chem. Soc.*, **73**, 1106 (1951).
91. Maxwell, L. R., and Moseley, V. M., *Phys. Rev.*, **57**, 21 (1940).
92. Mijlhoff, F. C., *Acta Crystallogr.*, **18**, 795 (1965).
93. Mijlhoff, F. C., dissertation, Amsterdam, (1964). (From Bowater, Brown, and Burden, 1963.)
94. Morosin, B., *Acta Crystallogr.*, **25B**, 19 (1969).

95. Nakamura, D., Ito, K., and Kubo, M., *Inorg. Chem.*, **2**, 61 (1963).
96. Nardelli, M., Fava, G., and Giraldi, G., *Acta Crystallogr.*, **15**, 737 (1962).
97. Ohlberg, S. M., and Vaughan, P. A., *J. Amer. Chem. Soc.*, **76**, 2649 (1954).
98. Paetzold, R., *Spectrochim. Acta*, **26A**, 577 (1970).
99. Palik, E. D., *J. Chem. Phys.*, **23**, 980 (1955).
100. Palik, E. D., and Oetjen, R. A., *J. Mol. Spectry.*, **1**, 223 (1957).
101. Palmer, H. T., and Palmer, R. A., *Acta Crystallogr.*, **25B**, 1090 (1969).
102. Palmer, K. J., and Elliott, N., *J. Amer. Chem. Soc.*, **60**, 1309 (1938).
103. Pauling, L., *The Nature of the Chemical Bond*, 3rd Ed., Cornell University Press, Ithaca, N.Y., 1960.
104. Pedersen, B., and Hope, H., *Acta Crystallogr.*, **19**, 473 (1965).
105. Pepinsky, R., and Vedam, K., *Phys. Rev.*, **114**, 1217 (1959).
106. Perez-Rodriguez, M., and Lopez-Castro, *Acta Crystallogr.*, **25B**, 532 (1969).
107. Richter, H., and Breitling, G., *Z. Naturforsch.*, **21a**, 1710 (1966).
108. Rundle, R. E., *Acta Crystallogr.*, **14**, 585 (1961).
109. Stevenson, D. P., and Beach, J. Y., in *Systematic Inorganic Chemistry* by Yost, D. M., and Russell, H. R., Jr., Prentice-Hall, Inc., New York, 1944, p. 306.
110. Tsintsadze, G. V., Porai-Koshits, M. A., and Antsyshkina, A. S., *J. Struct. Chem.*, **8**, 253 (1967).
111. Urch, D. S., *J. Chem. Soc.*, 5775 (1964).
112. van den Hende, J. H., and Klingsberg, E., *J. Amer. Chem. Soc.*, **88**, 5045 (1966).
113. Vijayan, M., *Acta Crystallogr.*, **24B**, 1237 (1968).
114. Villa, A. C., Nardelli, M., and Palmieri, C., *Acta Crystallogr.*, **25B**, 1374 (1969).
115. Wang, B. C., and Cordes, A. W., *Inorg. Chem.*, **9**, 1643 (1970).
116. Wang, B. C., and Cordes, A. W., unpublished results.
117. Weiss, R., Wendling, J. P., and Grandjean, D., *Acta Crystallogr.*, **20**, 563 (1966).
118. Wells, A. F., and Bailey, M., *J. Chem. Soc.*, 1282 (1949).

7

Coordination Compounds in which Selenium Functions as the Donor Atom

V. KRISHNAN/RALPH A. ZINGARO

*Department of Chemistry, Texas A&M University
College Station, Texas*

INTRODUCTION

During recent years, inorganic chemists, active in the field of coordination chemistry, have shown an increasing interest in the investigation of ligands which bear heavier donor atoms of low electronegativity. The use of metal sulfides and selenides as catalysts in the reductions of aromatic nitro groups has been discussed by Greenfield.[141] Kimura[190] has reviewed the biochemical aspects of the iron-sulfur linkage. The study of metal to sulfur bonds is becoming more widespread because of their potential importance in physiological systems. Compounds of sulfur and selenium find importance in solid state devices. Thus, cadmium sulfide and selenide are photoconductors and the sulfide and selenide of arsenic are useful as materials of construction where infrared transparency is essential. These are only a few examples of the reasons for the growing interest in the study of donor atoms of this type.

337

In a review, Livingstone[213] has compared the donor properties of sulfur, selenium and tellurium. The donor properties of sulfur and selenium resemble each other more closely than do those of oxygen. This is primarily because of their size and polarizability. The concept of d orbital expansion has been widely invoked to explain the chemical bonding in coordination compounds of sulfur and appears to be reasonably well understood. Whether reasoning of the analogous type can be applied to the case of selenium compounds has not yet been fully determined.

In the present review, a comparison is made between the selenium-bearing coordination compound and the isologous sulfur compound, wherever possible. Emphasis is given to the structural and bonding aspects of selenium compounds whenever recent literature concerned with these subjects is available.

A number of doubtful or unusual results are included because every attempt has been made to make the coverage as comprehensive as possible. In several instances, the reader has been cautioned to exercise some skepticism about the results reported. However, this has not been done whenever necessary and it is not intended that the reader should be left with the impression that every paper cited is as good as any other.

INORGANIC LIGANDS

Selenium Analogues of the Tetraoxometallates

Tetraoxo transition metal ions of the type WO_4^{-2}, MoO_4^{-2} and VO_4^{-2} are well-known. Several of the selenium analogues have been described. Thus, VSe_4^{-2},[233] $MoSe_4^{-2}$,[229] and WSe_4^{-2} [212] have been reported. The latter described the preparation of $(NH_4)_2WSe_4$ by saturating a solution of ammonium tungstate with H_2Se. The green compound is soluble in water, but insoluble in ether or benzene. When the quantity of H_2Se used was limited, a red compound which analyzed as $(NH_4)_2WO_2Se_2$ was obtained.

The crystal structure of ammonium tetraselenotungstate has been determined.[235] They report that the WSe_4^{-2} ion is tetrahedral and that the metal-selenium bonding is essentially covalent. The Se—W—Se bond angle was reported to be between 107.8–111.6°. The intermolecular Se—Se distances were reported as lying between 3.56 and 4.02 Å. Crystallographic data on the compound Cs_2MoOS_2Se have been reported.[145] The crystal is orthorhombic with $a = 9.917$ Å, $b = 7.329$ Å, and $c = 12.294$ Å. Powder data on the cesium salts of $MoOS_2Se^{-2}$ and WOS_2Se^{-2} have also been obtained.[237]

The selenium atoms in these ions can be substituted by either oxygen, sulfur, or both. Thus, the following compounds are known: olive green $(NH_4)_2MoO_2Se_2$, red $(NH_4)_2WO_2Se_2$, dark red Cs_2WOSe_3, dark red Cs_2MoOS_2Se and orange yellow Cs_2WOS_2Se. The infrared and Raman spectra of these salts have been studied.[84, 231, 232, 234, 236] The results of the infrared studies are listed in Table 7-1 and those of the electronic spectra of the anions in Table 7-2.

TABLE 7-1   METAL-SELENIUM
STRETCHING VIBRATIONS IN
TETRACHALCOGENOMETALLATES

Anion	$\nu(M-Se)$, cm^{-1}
$MoSe_4^{-2}$	340, 255
WSe_4^{-2}	310, 278
$MoO_2Se_2^{-2}$	362, 344, 285
$WOSe_3^{-2}$	326
$MoOS_2Se^{-2}$	355
WOS_2Se^{-2}	320

TABLE 7-2   PRINCIPAL VISIBLE AND
ULTRAVIOLET ABSORPTION BANDS IN
TETRACHALCOGENOMETALLATES

Anion	K
$MoSe_4^{-2}$	16,000, 17,990, 26,040
WSe_4^{-2}	18,900, 21,600, 26,500, 31,600
VSe_4^{-2}	15,620, 21,410
$MoO_2Se_2^{-2}$	22,000, 28,500, 32,000, 40,800
$WO_2Se_2^{-2}$	28,550, 34,000, 38,000
$MoOS_2Se^{-2}$	24,300, 31,000, 42,500
WOS_2Se^{-2}	29,100, 36,000, 41,500(?), 48,400

A simple molecular orbital diagram for $MoSe_4^{-2}$ and WSe_4^{-2} has been proposed[227] and is shown in Figure 7-1. The charge-transfer transitions from the nonbonding molecular orbital t_1, which is localized in the ligand, to the antibonding molecular orbital, $2e$, which is localized on metal atom, have been assigned at 17,900 and 21,600 K for $MoSe_4^{-2}$ and WSe_4^{-2}, respectively. The existence of a linear relationship between the charge-transfer bands and the ionization potential of the chalcogen atom (O, S, Se in MO_4^{-2}, MS_4^{-2} and MSe_4^{-2}) has been demonstrated. The t_1-$2e$ transition energy of MX_4^{-2} (X = O, S, Se; M = Mo, W, V) increases in the manner $VX_4^{-2} > MoX_4^{-2} > WX_4^{-2}$.[230] This suggests that the energies of the antibonding $2e$ levels increase in going from V to W. This, in turn, stabilizes the strong

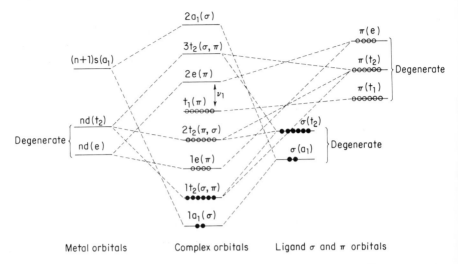

Figure 7-1   Simplified MO diagram of tetraoxo, -thio and -seleno metallate (d^0 configuration) with corresponding shift in energy levels.

π-bonding $1e$ orbitals.   The net effect is reflected in the increase in the force constants of MSe_4^{-2} (M = V, Mo, W) ions.[230, 232]   It is interesting to find that, as the electronegativity of X decreases, the force constant decreases.[232] This is reflected in the greater covalent nature of metal-selenium bonds in the tetraselenate ions.   The first band in the other compounds is attributed, with considerable certainty, to a Se → M transition.

Studies of the diffuse reflectance spectra of the compounds $Cu_3M(V)X_4$ and $Tl_3M(V)X_4$ (M = V, Nb, Ta) have been carried out.[251]   They show that the energies of the absorption edges and absorption maxima increase in the sequence V < Nb < Ta and decrease in the order O > S > Se.

Heteropoly Acids Containing Selenium

Heteropoly acids are formed by the condensation of acid anhydride molecules, most commonly the oxides of tungsten, molybdenum and vanadium, with a second acid.   The latter is considered to furnish the central atom, or central ion, of the entire complex anion.   The central atoms are represented by all groups of the periodic table, although the primary contributors are members of Groups IV, V, VI and VIII.   Some typical examples of heteropoly anions are given by the formulations: $PW_{12}O_{40}^{-3}$, $AsW_{12}O_{40}^{-3}$, $SiW_{12}O_{40}^{-4}$, $SiMo_{12}O_{40}^{-4}$, $SnMo_{12}O_{40}^{-4}$, and $TiMo_{12}O_{40}^{-4}$.

The formation of heteropoly acids and their salts containing different ratios of selenium dioxide and molybdenum trioxide was first reported in 1885.[121, 258]   The ammonium salt, $3(NH_4)_2O \cdot SeO_2 \cdot 8MoO_3 \cdot 6H_2O$, has been prepared.[277] $2(NH_4)_2O \cdot 2SeO_2 \cdot 5MoO_3 \cdot 3H_2O$ and the rubidium and barium salts and a compound of composition $(NH_4)_2O \cdot SeO_2 \cdot 6MoO_3 \cdot 4H_2O$ have also been reported.[267, 268]

Recently, a number of selenium bearing heteropoly acids have been reported in series of four papers.[269, 270, 271, 272]   Thus, a molybdoselenite having the stoichiometry, $3(NH_4)_2O \cdot 2SeO_2 \cdot 5MoO_3 \cdot 30H_2O$ was prepared in the following manner.   Selenious and molybdic acids in a molar ratio of 1 : 4 were dissolved in aqueous ammonia and refluxed for 6 h.   Following concentration, cooling and acidification to pH 3 with acetic acid, the white crystalline compound was obtained.   Silver, lead, ferric and cerous salts were also prepared.   Thermogravimetric (TGA) studies revealed changes in mass on heating which could be correlated with the loss of water and selenium dioxide.   Differential thermal analysis (DTA) was also a technique used in these studies and, in addition to heat changes associated with the loss of water, some changes were attributed to lattice transitions.

These same investigators described the preparation of phosphoselenito-molybdates,[270] $4(NH_4)_2O \cdot 1.5P_2O_5 \cdot SeO_2 \cdot 8MoO_3 \cdot 16H_2O$, as well as the lead and silver salts.   They also described[271] lead, mercuric and barium 6-molybdoselenites.   The potassium salt of this class was assigned the formulation $2K_2O \cdot SeO_2 \cdot 6MoO_3 \cdot 8H_2O$.   Infrared spectra of the various heteropoly acids reported by these workers are characterized by the following absorptions: 3360–3500 cm^{-1}, O—H stretch; 1610–1790, water bending mode; and overlapping Se—O and Mo—O stretching modes in the region of 900 cm^{-1}.

The heteropolyacid, seleno-12-tungstic acid has been described recently.[263] In order to prepare this compound, an ether extract of a strongly acidified aqueous solution containing tungstic acid and selenium dioxide in a molar ratio of 12 : 1 was crystallized *in vacuo* at 5°C.   The resultant crystals were stable at low temperatures and in the absence of light.   Seleno-12-tungstic acid was found to evolve zeolitic water at 100–270°C.   Over the temperature range ~290–320°C the remaining water was lost.   At 330–350°C, crystallization of WO_3 occurs while, above 310°C, there occurs a progressive volatilization of SeO_2.

Hydrogen Selenide

It is known that H_2S precipitates most of the heavy metals as insoluble sulfides, and that some of these sulfides are soluble in ammonium or alkali sulfides because they form complex sulfides.   This fact has been used in qualitative analysis.   The use of H_2Se for the selective precipitation of metals[314] has been investigated.   The composition and thermal behavior of these precipitates was studied.   However, because of the presumably greater toxicity and relative instability of sodium selenide, this method has not been widely adopted.

H_2Se reacts with $Pt(Ph_3P)_2$ and $Pt(Ph_3P)_3$ to form an air stable product $Pt(PPh_3)_2SeH_2$.[223] The infrared spectrum of this compound shows $v(Pt-H)$ at 2140 cm^{-1}. The NMR studies on this compound show that two possible species exist in solution. A low field peak is assigned to structure **A** and the high field peaks are assigned to structure **B**. It was

$$
\begin{array}{ccc}
Pt(PPh_3)_2 & & \overset{H_a\diagdown\quad\diagup H_a}{\underset{\underset{PPh_3\diagup\quad\diagdown PPh_3}{Pt}}{\overset{\downarrow}{Se}}} \quad + \quad \overset{\diagup H_b}{\underset{\underset{PPh_3\diagup\quad\diagdown H_c}{Pt}}{\overset{\diagdown\;\diagup}{Se\;\vdots\;PPh_3}}} \\
+ & \longrightarrow & \\
H_2Se & & \\
& & A \qquad\qquad\qquad B
\end{array}
$$

found that the coupling constant $J(^{195}Pt\text{-}H_c)$ has a high value suggesting a σ hydride complex and $J(^{195}Pt\text{-}H_b)$ is lower which suggests a weak interaction with the d_{z^2} orbital of platinum. It was noted from D_2O exchange studies that **A** and **B** do not interconvert easily at room temperature. A possible mechanism for the poisoning of the platinum surface through the abstraction of a hydrogen atom has been proposed.

Recently,[160] the oxidative addition of hydrogen selenide to Vaska's compound and to $RhCl(PPh_3)_3$ has been demonstrated. These addition compounds show an $M-H$ stretching absorption in the infrared. The carbonyl stretching frequency is raised by about 60 cm^{-1} in the addition compounds. This shift is characteristic of the product formed by oxidative addition.

Selenocyanates

The selenocyanate ion, $SeCN^-$, is an ambidentate ligand which can coordinate to a metal ion either through the nitrogen or selenium atom. It has become conventional to refer to a nitrogen bonded complex as an isoselenocyanate and to a selenium bonded complex as a selenocyanate. It is often possible to predict which atom will function as the donor toward the metal. This will depend largely upon whether the metal is classified by Ahrland, Chatt and Davis[13] as belonging to class "a" or class "b" or, in Pearson's[257] language, whether it is "hard" or "soft." It is a fact that "soft" donors bind strongly to "soft" metals and "hard" donors bind strongly to "hard" metals. There is much criticism[344] regarding the use of these terms since this classification is not clear-cut and is based largely on the polarizability of the atoms. However, for the purposes of the present discussion, it is suffice to say that the N atom in the selenocyanate group functions as a "hard" donor and Se as a "soft" donor. The classification of elements into class "a" and class "b" is based upon the underlying fact that class "a" metals belong to a system whose halide stability constants fall in the order $F > Cl > Br > I$, and the reverse is true for class "b" elements. Jorgensen[182] subdivided class "b" elements into: (1) metals in their high oxidation states, (2) metals in their low oxidation states, and (3) metals having $d^{10}s^2$ configurations.

As shall be pointed out, these considerations suffice as far as one is concerned with the neat complexes, namely, $M(NCSe)_4^{n-}$ or $M(NCSe)_6^{n-}$. However, difficulties are encountered in explaining the bonding properties in mixed complexes in which one of several ligands is a selenocyanate ion. The existence of "inorganic symbiosis" has been pointed out by Jorgensen.[182] This is concerned with the fact that in a mixed complex, all of the ligands tend to be either "soft" or "hard." There are few exceptions to this rule. In addition, the steric hindrance of the ligand, the oxidation state of the central ion and the relative donor strength of either the ligand or metal can exert a strong influence on the nature of bonding of the selenocyanate group. Golub and Skopenko[129] have reviewed the metal selenocyanates and Livingston[213] has discussed some of the aspects of thio and selenocyanate complexes. Lodinzka[214] studied the stability constants of OCN^-, NCS^-, and $NCSe^-$ complexes of various metals and compared the pK values with those of the halide complexes.

This section has been divided as follows:

(a) metal to nitrogen bonded complexes, or isoselenocyanates;
(b) metal-selenium bonded complexes, or selenocyanates;
(c) mixed complexes in which $SeCN^-$ is one of the ligands;
(d) organometallic selenocyanates.

A few examples of linkage isomerization have been reported in $SeCN^-$ complexes. These will be discussed in the appropriate sections.

Infrared and Electronic Spectra. There are three ways in which $SeCN^-$ can bond in metal complexes:

$$M-N-C-Se \qquad M-Se-C-N \qquad M-Se-C-N-M$$
$$\text{(I)} \qquad\qquad \text{(II)} \qquad\qquad\qquad \text{(III)}$$

A great deal of use has been made of infrared and electronic spectroscopic methods for the purpose of identifying the donor atom.

The crystal structure of $KSeCN$[313] shows that the $SeCN^-$ group is linear with bond distances of C—N, 1.117 Å and C—Se, 1.828 Å. The dipole

moment of the SeCN$^-$ group in organic selenocyanates[222] has been found
to be 3.48 D, which suggests a polar structure $R-\overset{+}{\underset{..}{Se}}=C=\overset{..}{N}^-$. The
infrared spectrum of KSeCN[224] exhibits fundamentals at 2070, 558 cm^{-1}
and at 424 and 416 cm^{-1}. These are attributed to C—N stretching (v_3),
C—Se stretching (v_1) and asymmetric and symmetric bending modes, v_2
[$v_2(as)$ and $v_2(s)$] of the SeCN group. The locations of these fundamentals
in various SeCN$^-$ complexes have been compared with the corresponding
frequencies in KSeCN. This serves as a useful guide in determining whether
the possible structures correspond to I, II, or III. However, one should
always bear in mind that such evidence is not conclusive. Kharitanov
et al.[186] proposed that v_3 increases in both structures I and II. In III, v_3
increases to 2130–2140 cm^{-1}, whereas v_1 decreases in structure II to 520–543
cm^{-1} and increases to 605–672 cm^{-1} in structure I, depending upon the
value of the metal-SeCN bond angle. v_3 increases in the following order
M—NCSe > MSeCN > M—SeCN—M.[329] The bending (v_2) frequency
decreases in structure II and increases in structure I. Pecile[259] found that
the integrated absorption intensity of the C—N stretching mode (v_3) of the
selenium bonded complexes is low (0.8–2.3 $\times$ 10^4 mole^{-1} cm^{-2}), but high,
(9.0–12.0 $\times$ 10^4 mole^{-1} cm^{-2}), for the isoselenocyanate complexes. These
guidelines often tend to concur with the experimental facts. However, in
mixed complexes, there exist other regular trends which will be discussed
subsequently.

Correlations derived from electronic spectral measurements depend upon
the position of the charge-transfer bands and the crystal field splitting parameter, Δ. In the case of transition metal complexes, the ligand field strength
of N bonded compounds is the same as in N bonded isothiocyanates. However, it is smaller in the Se bonded selenocyanates than in the S bonded
thiocyanates.[328, 329, 221, 47]

Barnes and Day[32] showed that the lowest allowed transitions of reducible
metal isothiocyanates are of the charge-transfer type. Reducible metal-
thiocyanate and oxidizable metal isothiocyanates exhibit only perturbed
ligand bands of the $\pi \rightarrow \pi^*$ type. Later, Schmidtke[288] made use of the
expression:

$$\sigma_{CT}(kK) = 30(\chi_L - \chi_M) + \Delta + \delta E_{sp} \qquad (1)$$

where χ_L and χ_M are Pauling electronegativities of the ligand and metal, Δ,
the ligand field parameters, and δE_{sp}, the difference in spin pairing energies
between the ground and excited states. Values calculated for the energies
of the charge-transfer bands agree well with those observed experimentally.
The optical electronegativities calculated from the charge-transfer bands are
—SCN = 2.9, —SeCN = 2.8, —NCSe = —NCS = 2.6.

N bonded Complexes or Isoselenocyanates. Most of the transition metal complexes have been isolated as anionic complexes $[M(NCSe)_6]^{n-}$ or $[M(NCSe)_4]^{n-}$. With the exception of copper(I), all of the first row transition metals form isoselenocyanates. Titanium(III) forms a dark red complex[49] which decomposes in solution. The vanadium(III) complex is reported to be N bonded.[49,306,290] The vanadium(III) complex could be easily oxidized to a blue vanadyl complex which contains N bonded $SeCN^-$.[304,290,49] The chromium(III) complex is N bonded[221,42] and the silver salt, $Ag_3[Cr(NCSe)_6] \cdot 5H_2O$ contains hexacoordinated Cr(III) with Ag—Se—C—N—Cr bridges. Successive equilibrium constants have been reported by Stancheva *et al.*[310] in Mn—NCSe—D, where D is a solvent molecule.

Hexa- and tetracoordinated N bonded derivatives of Mn(II) have been reported. The stability constants of $[Co(NCSe)_n]^{2-n}$ in nonaqueous solvents have been determined.[305] Both tetrahedral Fe(II) and octahedral Fe(III) complexes are N bonded.[278,289,47] This is not consistent with the predictions of the theory of Jorgensen.[182] Toropova[320] showed that the blue cobalt(II) complex is less stable than the corresponding isothiocyanate complex. Subsequently, it was shown that in the various solvents the strength of the Co(II)—NCSe bond increases in the order: water < methanol < acetone.[128] Pink, octahedral[109] and green-blue tetrahedral[74,245] cobalt(II) complexes are N bonded. The octahedral nickel(II) complex is N bonded, and the complex $[Ni(NCSe)_4]^{-2}$ contains bridged $SeCN^-$ groups which form an octahedral structure. Dissociation constants for $Ni(NCSe)^+$ and $Ni(NCSe)_4^{-2}$ have been reported.[303]

The stability constants of zinc(II)-isoselenocyanate complexes have been detailed[162] and it has been shown that the zinc is tetrahedrally coordinated.[110,109] Yttrium(III),[51] praeseodymium(III), and neodymium(III)[56] are also N bonded. The yellow molybdenum(III) complex is N bonded[289] and the stability constants of N bonded thorium complexes were also measured.[130]

It was shown by Forster and Goodgame[109] that the ability to form six-coordinate complexes increases in the order NCSe > NCS > NCO. The v_3 of N bonded octahedral complexes is located at a higher frequency than that in the corresponding tetrahedral complexes. The $v(M—N)$ increases in the order Ni, Co, Fe, Mn which roughly corresponds to the Irving-Williams series. The charge-transfer and ligand field spectra of N bonded isoselenocyanates are discussed in several papers.[288,290,221,109,329] It was found that charge-transfer transitions of isoselenocyanate complexes occur at frequencies lower than in the corresponding isothiocyanate complexes and the ligand field strength is comparable to that of isothiocyanate. In the spectrochemical series, it falls in the neighborhood of the value reported for H_2O and Cl^-. Recently, Salzmann and Schmidtke[279] discussed the magnetic properties of isoselenocyanate complexes and the dependence of σ- and π-bonding on the orbital reduction factor.

Se bonded Complexes, or Selenocyanates. The stability constants of $Hg(SeCN)_4^{-2}$,[320] $Cd(SeCN)_4^{-2}$,[162] $Ag(SeCN)_3^{-2}$,[126,124] and $Pb(SeCN)_6^{-4}$ [125] have been measured. In all these cases it was found that the stabilities of Se bonded selenocyanates are greater than those of corresponding S bonded thiocyanates. The following species: $CuCl(SeCN)^-$, $Cu(SeCN)_2^-$, $Cu(SeCN)_3^{-2}$, and $Cu(NCSe)_4^{-3}$ exist in solution.[126] Normal cadmium(II), mercury(II),[14,15] bismuth(III),[307] silver(I),[302,329] copper(I),[126] and lead (II)[187] selenocyanates are known. The Ag(I) compound presumably has a polymeric structure. The selenocyanates of lead, cadmium and mercury have bridging SeCN groups. All of the above mentioned elements have the d^{10} configuration and, being "soft" metals, are selenium bonded. An interesting cadmium compound has been isolated by Burmeister and Williams.[47] It has the composition $[Cd_2(NCSe)_6]^{-2}$. The following structure, based on infrared data, has been proposed (Figure 7-2). There

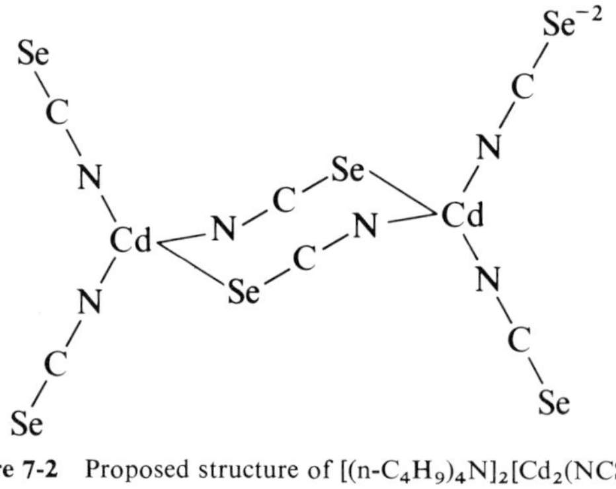

Figure 7-2   Proposed structure of $[(n-C_4H_9)_4N]_2[Cd_2(NCSe)_6]$

always exists considerable uncertainty in drawing conclusions about the structures of such complicated molecules from infrared spectra because of coupling effects and vibronic interactions. Thus, we observe the smooth transition from class "a" zinc(II) to class "b" mercury(II) through bridged selenocyanato and terminal isoselenocyanato cadmium(II).

Infrared data suggest class "b" behavior for rhodium(III), platinum(IV), gold(III), palladium(II), and platinum(II).[289,47,287,278,109] Coordination through selenium is therefore expected in these cases.

A recent x-ray study of $K_2[Pt(SeCN)_6]$[325] shows a Pt—Se distance of 2.8 Å. Platinum and palladium–selenium stretching frequencies occur in the region 220 and 240 cm^{-1}, respectively. The ligand field and charge-transfer spectra of these complexes have been studied[288] and, in all cases, the first spin-allowed transition occurs at a lower frequency in selenocyanates than in the corresponding S bonded thiocyanates. This suggests a slightly lower ligand field strength for the selenocyanates than for the thiocyanates. The charge-transfer bands of Pd(II), Pt(II) and Au(III) occur at lower wave numbers than for the corresponding thiocyanates. This suggests an optical electronegativity of 2.8 for —SeCN, compared to 2.9 for —SCN. The potassium salt, $K_2[Hg(SeCN)_4]$, is isomorphous with the corresponding thiocyanate and coordination through selenium is suggested.[15] The electronic transition at 32,800 K in $[Hg(SeCN)_4]^{-2}$ might possibly arise from a $5d \rightarrow 6s$ transition and at 39,200 K from a at_1–a_1 transition. The effective magnetic moment of $Hg[Co(NCSe)_4]$, 4.40 BM, compares favorably with the moment of the corresponding thiocyanate $Hg[Co(NCS)_4]$, 4.44 BM. This suggests the existence of Co—NCSe—Hg bridges.[245] The infrared spectra of $M[Hg(SeCN)_4]$, where M = Cd, Cu, Co, presents evidence for cross-linking of NCSe groups.[187] X-ray powder data show that $Cd[Hg(SeCN)_4]$ is not isomorphous with either $Co[Hg(SeCN)_4]$ or $Zn[Hg(SeCN)_4]$.[321] The radial distribution method suggests a tetrahedral environment for Zn(II) and Hg(II) in $Zn[Hg(SeCN)_4]$.[323, 324] Salts of the composition $M[Hg(SeCN)_3]_2$, where M = Cu or Pb, have been reported in which the SeCN acts as a bridging group.[15]

Mixed Complexes. Mixed selenocyanate complexes of vanadium(III) and oxovanadium(IV) with 1,10-phenanthroline, 2,2′-bipyridyl, pyridine, $V(bipy)_3(NCSe)_3$; $V(1,10\text{-phen})_3(NCSe)_3$; $V(py)_6(NCSe)_3$; $VO(phen)_3$ $(NCSe)_2$; and $VO(bipy)_3(NCSe)_2$ have been found.[306, 304] The NCSe$^-$ groups dissociate in solution. An N bonded isoselenocyanato group exists in $VO(py)_4(NCSe)_2$ and $VO(py)_3(NCSe)_2$. Boehland and Malitzke[37] reported an interesting compound having the composition

$$Na_3[V(NCS)_3(NCSe)_3]2CH_3CN.$$

Unfortunately, no spectral data were reported owing to its instability.

Several chromium(III) complexes which contain dissociable SeCN groups have been isolated.[321] Brusilovets et al.[42] detailed $Na[Crpy_2(NCSe)_4]\cdot$ 3dioxane, $[Crpy_3(NCSe)_3]\cdot 2py$, and $[Crpy_3(NCSe)_3]$. The latter two compounds contain N bonded groups. Yellow and red coordination isomers of $Crpy_3(NCSe)_3$ exist. Duffy and Kossel[89] isolated $[Cr(NH_3)_5(NCSe)](ClO_4)_2$ and $[Cr(NH_3)_5(NCSe)]Br_2$ and showed the presence of the isoselenocyanate group.

Octahedral Mn(II) complexes, $MnX_2(NCSe)_2$,[310] where X = 1,10-phenanthroline and 2,2′-bipyridyl, and $Mn(DMF)_4(NCSe)_4$[300, 322] where DMF is dimethylformamide, are claimed to have *trans* N bonded isoselenocyanate groups.

Grecu[140,139] prepared mixed SeCN–isonicotinic hydrazide complexes of Fe(II). A low-spin Fe(II) mixed complex, $Fe(DMG)_2(SeCN)_2$(DMG = dimethylglyoxime), is reported to have a six-coordinate structure with Se bonded selenocyanate groups along the z-axis. Baker and Bobonich[24] investigated an interesting compound, $Fe(phen)_2(SeCN)_2$, and proposed a bridged Se bonded structure at low temperatures. This was based on their observations of the variation of the magnetic moment with temperature. Later, Koenig and Madeja[193] postulated spin-state equilibria of Fe(II) in order to explain the low moments of this complex at low temperatures. The ground state of the Fe(II) complex at room temperature (298°K) is $^5T_2$ $(t_2^4 e^2)$ has a Δ value of 11,900 K and the ground state at low temperature (80°K) is $^1A_1$, (t_2^6) with Δ value of 16,300 K. The nephelauxetic ratios at 298 and 80°K were calculated to be 0.61 and 0.55. The occurrence of $v(C-N)$ in the low temperature form at 2112 and 2106 cm^{-1} is explained on the basis of the strengthening of the C—N bond as a consequence of the weakening of the Fe—NCSe bond. This decrease in the energy of the Fe—NCSe bond in $^1A_1$ ground state arises from the decreased availability of t_2 electrons to back donate into π^* orbitals of the 1,10-phenanthroline. Further, the splitting of $v(C-N)$ suggests a *cis* configuration.

Green and blue tetrahedral Co(II) complexes, $CoX_2(NCSe)_2$, where X = quinoline, triphenylphosphine oxide, and N bonded isoselenocyanate, have been reported by Cotton.[74] Interestingly, a mixed complex of the same composition, where X is pyridine, is supposed to have a bridged octahedral structure.[245] The triphenylphosphine complex with a magnetic moment of 3.4 BM is supposed to contain equal numbers of octahedrally coordinated low-spin Co(II) and tetrahedrally coordinated high-spin Co(II) ions. While such an explanation is consistent with the experimental facts, additional work seems to be in order. *Trans* octahedral structures have been proposed for $CoX_4(NCSe)_2$ where X is either pyridine or dimethylformamide. These possess N bonded isoselenocyanato groups.[300,322] Pink $[Co(NH_3)_5(NCSe)](NO_3)_2$, yellow-brown $K_3[Co(CN)_5NCSe]$, and yellow $[Co(NH_3)_4(CN)(NCSe]Cl$ complexes contain the isoselenocyanato group.[46] Recently, the light orange $[Co(NH_3)_5(NCSe)]Br_2$ complex having an isoselenocyanato group was reported.[89] An x-ray study of $[NH_4][Co(DMG)_2(SeCN)_2]\cdot 3H_2O$, shows Se bonded selenocyanates in the *trans* position.[7,45] To date, this is the only compound in which Co(II) has been shown to be selenium bonded. However, the x-ray evidence must be considered more conclusive and x-ray studies of other related molecules may reveal other similar cases.

The green $Ni(py)_2(NCSe)_2$ has a magnetic moment of 3.28 BM, and the electronic spectrum is consistent with six-coordinate, spin-free Ni(II). The bridging nature of the selenocyanate group has been demonstrated by infrared data.[246] Electronic and infrared studies show that in $Nipy_6(NCSe)_2$, $Nipy_4(NCSe)_2$ is present and the isoselenocyanate groups occupy *trans* positions.[250] The other two pyridines are present in the crystal, but outside the coordination sphere. Nelson and Shepherd[246] observed, in a series of complexes, $NiA_4(NCSe)_2$, where A = pyridine, β-picoline, or γ-picoline, that the overlap of d_π metal to p_π ligand orbitals increases in the order $N_3^- < NCO^- < NCS^- < NCSe^-$, based on the values of Δ and B (Racah's parameter). The isoselenocyanate groups occupy *trans* positions. Farago and James[104] studied the change of $v(C-N)$ of the isoselenocyanate groups with the change in the steric hindrance of the ligand in complexes of the type $NiA_2(NCSe)_2$, where A = 1,2-diaminoethane, 1,2-diaminopropane, N,N'-dimethyl-1,2-diaminoethane,   N,N'-diethyl-1,2-diaminoethane,   and *meso*-2,3-diaminobutane. Two possible isoselenocyanate structures exist:

I; $M-\overset{+}{N}\equiv C-\overset{..}{\underset{..}{Se}}\overset{-}{:}$ =and II; $\underset{M}{\diagup}\overset{..}{N}=C=\overset{..}{\underset{..}{Se}}$. It was found that the presence

of a more bulky ligand (greater steric hindrance) favors structure I. This tends to increase $v(C-N)$ and decrease the covalency of Ni—N. A recent x-ray study of $Ni(DMF)_4(NCSe)_2$, where DFM is dimethylformamide, shows that it has a *trans* octahedral structure[322] and is isomorphous with the corresponding cobalt(II) and manganese(II) complexes.[300] Reduction of the spin-orbit coupling constant below the free ion value was observed in the double salt, $[Ni(1,2-diaminoethane)_3](SeCN)_2 \cdot KSeCN$.[105] A novel five-coordinated blue Ni(II) complex with N bonded isoselenocyanate at the pyramidal position (of a square pyramid) in $[Ni\{diphenyl(o-diphenylarsinophenyl)phosphine\}_2 NCSe]ClO_4$ has been isolated.[88]

A cuprammine complex, $Cu(NH_3)_4(SeCN)_2 \cdot H_2O$ was isolated and the infrared spectrum was studied.[188,189] The complexes of the type $CuA_2(SeCN)_2$, where A is 1,2-diaminoethane, 1,2-diaminopropane, N,N'-dimethyl-1,2-diaminoethane, N,N-dimethyl-1,2-diaminoethane and, N,N'-diethyl-1,2-diaminoethane are rather unstable and the infrared data suggest either ionic or Se bonded selenocyanate groups.[104] A copper(I) salt of the composition $KCu(SeCN)Cl$ has been found.[129] $Cd(py)_2(NCSe)_2$ has a tetrahedral structure and six-coordinate $Cd(py)_4(NCSe)_2$ has a *trans* octahedral structure.[14] The mixed chloride-selenocyanate complex, $Hg(SeCN)Cl$, has a bridged (SeCN) structure. The stability constants of the mixed complexes $(HgI_2SeCN)^-$, $HgI_2(SeCN)_2^{-2}$, and $HgI_3(SeCN)^{-2}$ increase in the order listed.[77,78] An aqueous suspension of HgI_2 saturated with NaSeCN yielded colored precipitates with several metal ions and salts of the composition $Ni_2Hg_7(SeCN)_{11}I_7$ and $Co_2Hg_7(SeCN)_{11}I_7$.[77] The nature of the metal-ligand bonding in these salts appears to be an open question. Single crystal x-ray studies should prove to be most revealing. The dissociation constants of the iodide complexes of silver; $Ag(SeCN)_3I^{-3}$, $Ag(SeCN)I_3^{-3}$, and $AgI(SeCN)_2^{-2}$ have been reported.[127] A novel compound of the composition $KAg(SeCN)(SCN)$ is formed from a solution of KSCN and AgSeCN in acetone. Indium forms mixed complexes of the type $In(2,2'-bipyridyl)(NCSe)Cl_2$ and $In(py)_4(NCSe)_3$ in which N bonded isoselenocyanate groups are present.[327]

Ettore et al.,[103] from rate studies, showed that the stabilities of penta-coordinate $[PdY_2X]^{n+2}$ (where n is the charge on X, Y = o-phenylene-bis-(dimethylarsine), and X = SCN^-, N_3^-, $S_2O_3^{-2}$, and Br^-) are in the order $N_3^- < Cl^- < Br^- \approx S_2O_3^{-2} < SCN^- < SeCN^- < I$. Golden yellow $Pt(bipy)(SeCN)_2$ was reported[46] to have Se bonded selenocyanate groups. The formation of $[Pd_2I_4(SeCN)]^-$ was observed by Golub et al.[131] in the PdI_2–NaSeCN solvent system. Although a structure involving an iodine bridge and a Pd—Pd bond has been proposed (Figure 7-3), it must be viewed with some skepticism. There is really no meaningful experimental basis on which to base it. The stability constants of $[Pd_2I_4(SCN)]^-$ and $[Pd_2I_4(SeCN)]^-$ are approximately equal.

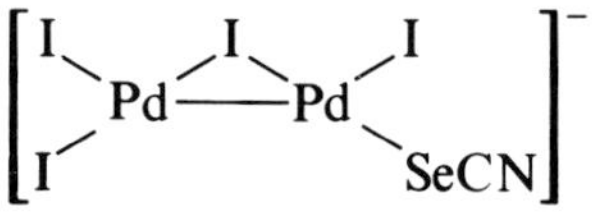

Figure 7-3   Proposed structure of $[Pd_2I_4(SeCN)]^-$

Linkage isomerism was first recognized by Werner and its existence has since been clearly demonstrated. Ambidentate ligands of the type NO_2^-, $S_2O_3^{-2}$, CN^-, SO_3^{-2}, SCN^- and $SeCN^-$ are capable of forming linkage isomers when coordinated to different donor atoms. Burmeister ably discussed these effects in two reviews.[48, 52] We will concern ourselves only with selenocyanates.

In order to induce linkage isomerism in a given complex, several factors should be taken into account. These include π-acceptor ligands, steric hindrance, etc. One group of bidentate ligands includes molecules such as ethylendiamine, 1,2-diaminobenzene, 2,2'-bipyridine, 1,10-phenanthroline, 5-nitro-1,10-phenanthroline, and 2,2',2''-tripyridine (bidentate). Another group of monodentate ligands is always found in the trans position. These include 4-methylpyridine, 4-aminopyridine, pyridine, 4-acetyl-pyridine,

N,N-diphenylthiourea, triphenylphosphine and triphenylarsine. All of these ligands have been coordinated to "soft" palladium(II)[51] and in no case was bond isomerism detected.   In all cases, only Se bonded selenocyanate compounds were formed.   Burmeister and Gysling[50] demonstrated that Pd(II) complexes bearing the non-π-donor ligand, 1,1,7,7-tetraethyl-diethylenetriamine (Et$_4$dien), and the bulky anion, [BPh$_4$]$^-$, readily isomerize in the following manner:

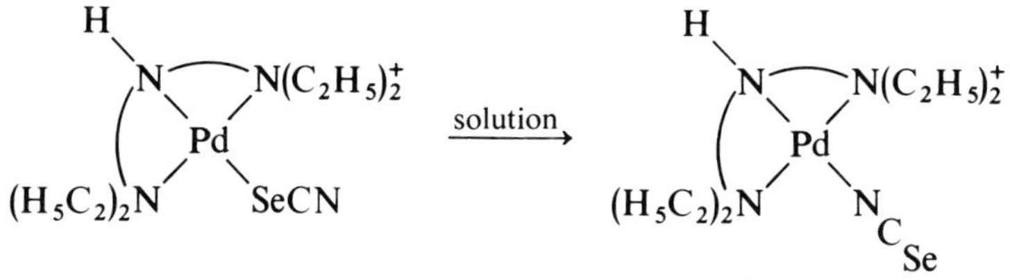

The N bonded selenocyanate is unstable in the solid state and reisomerizes to the Se bonded selenocyanate.   A dissociative intramolecular process was indicated from the rate studies.[53]   This is the first case in which a selenocyanate ion has been shown to undergo linkage isomerism. However, Burmeister and Gysling, in one of their earlier papers,[51] were able to isolate, at low temperatures, an N bonded isoselenocyanate of Pd(II) with tributylphosphine.   Unfortunately, at higher temperatures it dissociates into bridged complexes.   Thus, it is evident from Burmeister's work that neither steric hindrance nor the presence of a π-acceptor ligand can convert the "soft" Pd(II) to "hard" behavior and the resulting isoselenocyanate structure.

In another recent study, Burmeister and Lim[54] studied the kinetics of the substitution reactions of [Pd(Et$_4$dien)XCN]$^+$, where X = S or Se, with Br$^-$. When SeCN$^-$ is selenium bonded, the reaction follows an S$_{N2}$ path which involves the opening of a chelate ring.   When SeCN$^-$ is N bonded, the S$_{N2}$ mechanism is not significant.   However, a solvent-assisted ligand interchange process is operative in the case of the thiocyanate substrate.

Organometallic Selenocyanates.   Two reviews[209, 316] discuss the organo-metallic pseudohalogenides.   Here, a brief account is given concerning organometallic selenocyanates.   The compounds prepared, and their reactions are summarized in Table 7-3.   Ebsworth and Mays[93] placed NCSe$^-$ between Br$^-$ and Cl$^-$ in the "halogen interconversion series" while Thayer[317] placed SeCN$^-$ between Cl$^-$ and SCN$^-$ in the "Me$_3$Si-conversion series."

TABLE 7-3   ORGANOMETALLIC SELENOCYANATES

Compound	Preparation	Physical Properties (mp), bp(°C)
$H_3SiNCSe$	$H_3SiI + AgNCSe$	($-15.1°$), N bonded
$(CH_3)_3SiNCSe$	$(CH_3)_3SiI + AgNCSe$	($5°$), N bonded
$(CH_3)_3SiNCSe$	$(CH_3)_3SiCl + KNCSe$	175–177°, N bonded
$(CH_3)_3SiNCSe$	$(CH_3)_3SiCN + Se$	
$(C_2H_5)_3SiNCSe$	$(C_2H_5)_3SiCl + KNCSe$ or AgNCSe	195–200°/100 mm, N bonded
$(C_6H_5)_3SiNCSe$	$(C_6H_5)_3SiCl + KNCSe$	N bonded
$(C_6H_{11})_3SiNCSe$	$(C_6H_{11})_3SiCN + Se$	307–308°/250 mm, N bonded
$Me_3GeNCSe$	$Me_3GeCl + KSeCN$ or $Me_3GeNC + Se$	118–119°/232 mm (13–14°), N bonded
$Et_3GeNCSe$	$Et_3GeCl + KSeCN$	175–176°/240 mm, N bonded
$(C_4H_9)_3GeNCSe$	$(C_4H_9)_3GeI + KSeCN$	221–222°/100 mm, N bonded
$Ph_3SnNCSe$	$Ph_3SnCl + KSeCN$ or $Ph_6Sn_2 + Se(SeCN)_2$	N bonded N bonded
$Me_3PbSeCN$	$Me_4Pb + Se(SeCN)_2$	Se bonded
$Et_3PbSeCN$	$Et_4Pb + Se(SeCN)_2$ or $Et_3PbCl + KSeCN$	Se bonded
$Ph_3PbSeCN$	$Ph_4Pb + Se(SeCN)_2$ or $Ph_3PbCl + KSeCN$ or $Ph_6Pb + Se(SeCN)_2$ or $Ph_4Pb + Se(CN)_2$	Se bonded Se bonded
$Ph_2Pb(SeCN)_2$	$Ph_3Pb(SeCN) + Se(SeCN)_2$	Se bonded
$MeHgSeCN$	$Me_2Hg + Se(SeCN)_2$ or $Me_2Hg + Hg(SeCN)_2$	Se bonded
$PhHgSeCN$	$Ph_2Hg + Se(SeCN)_2$ or $Ph_2Hg + Hg(SeCN)_2$	Se bonded
$Ph_2TlSeCN$	$Ph\ Tl + Se(SeCN)_2$	Se bonded

Thayer[318] found that the chalcogenation reaction proceeds smoothly with grey selenium and he proposed the following mechanistic scheme:

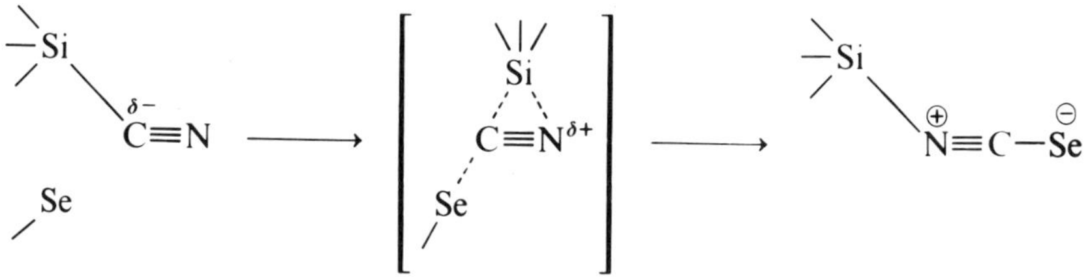

The reaction proceeds, according to this mechanism, by way of an attack by the electrophilic chalcogen atoms on the negatively polarized cyanide carbon, followed by the migration of the organosilyl group from the carbon to the nitrogen atom.

Silicon, germanium, and tin form N bonded isoselenocyanates while lead, thallium and mercury form selenium bonded selenocyanates.

Aynsley et al.,[21, 22] studied selenocyanation [$Se(SeCN)_2$] reactions as a method for cleaving the metal-metal bond in hexaphenyl dimetalloids and found that the reactivity of organometallic compounds towards selenocyanation increases with increasing atomic weight.

Few π-cyclopentadiene metal and metal carbonyl and nitrosyl selenocyanates have been studied. π-Cyclopentadienetitanium dichloride forms a greenish black N bonded isoselenocyanate[53, 57] π-$(C_5H_5)_2Ti(NCSe)_2$. The formation of $(\pi$-$C_5H_5)_2V(NCSe)_2$ was reported by Doyle and Tobias.[86] However, the bonding mode of the selenocyanate has not been determined with certainty. π-Cyclopentadienyl tungsten or molybdenum tricarbonyl hydride react with $Se(SeCN)_2$ to give dark red $C_5H_5Mo(CO)_3SeCN$ and red-orange $C_5H_5W(CO)_3SeCN$.[168] Dark brown π-$(C_5H_5)Fe(CO)_2SeCN$ is formed from the reaction of π-$C_5H_5Fe(CO)_2Cl$ with KSeCN or π-$C_5H_5Fe(CO)_2CH_2C_6H_5$ with $Se(SeCN)_2$. All these carbonyl selenocyanates are selenium bonded and the latter provides one of the few examples of an Fe(II)—Se bond.

Recently, Jennings and Wojcicki[169] reported two types of metal carbonyl selenocyanates, $[C_5H_5Fe(CO)_2(PR_3)]^+SeCN^-$ and $C_5H_5Fe(CO)(PR_3)SeCN$. The first type is ionic, as shown. The second type is prepared by the reaction of $Se(SeCN)_2$ with $C_5H_5Fe(CO)[P(C_6H_5)_3]CH_2C_6H_5$. Both the Se and N bonded isomers of the second type were prepared. At room temperature they do not undergo interconversion. At higher temperatures a deselenation reaction occurs which results in the formation of $C_5H_5Fe(CO)[PPh_3]CN$. Such a reaction was found to occur rapidly in the N bonded selenocyanate, but slowly in the Se bonded isomer.

A nitrogen bonded isoselenocyanate of chromium cyclopentadienyl nitrosyl, $C_5H_5Cr(NO)_2NCSe$, exists. Farona *et al.*[106] reported the dinuclear manganese(I) complex, $[Mn_2(CO)_6Cl_2(NCSe)_2]^{-2}$(I) and $[Mn_2(CO)_6(NCSe)_4]^{-2}$(II) in which N bonded isoselenocyanate groups are present. The similarities in the infrared spectra of I with the corresponding thiocyanate complex show the presence of bridged and terminal NCSe groups.

LIGANDS WHICH CONTAIN PHOSPHORUS-SELENIUM BONDS

Dialkyldiselenophosphates

The ligand O,O'-diethyldiselenophosphate (dsep), $(C_2H_5O)_2P(Se)Se^-$, was isolated in the form of the chromium(III) derivative, Cr(dsep)$_3$, as dark green flakes.[179]   The reaction of P_2Se_5 with alcohols yields, as the potassium salts, diethyl-, di-*n*-propyl-, di-*n*-butyl-, di-*n*-amyl-, and dicyclohexyl-diselenophosphates.[208]   They also prepared tris(diisopropylselenophosphato)chromium(III).   The Rh(III)(dsep)$_3$ was described as a dark red compound while Ir(dsep)$_3$ was obtained in the form of dark brown crystals.

Jorgensen[179] has pointed out that the arrangement of selenium atoms about the central metal atom in complexes of the type $M(Se_2P)_3$ is such that it possesses a D_3 symmetry rather than a symmetrical, octahedral O_h symmetry which would be the case for the arrangement in MSe_6.   The π-orbital on Se has much less meaning in the case of $M(Se_2P)_3$ because of the large deviations from linear symmetry.   In Figure 7-4 is shown a qualitative arrangement of the order of MO energy levels in this system.   The six σ orbitals designated as σ_{SeM} are those involved primarily in metal-selenium bonding.   The six σ_{SeP} orbitals correspond to the same symmetry types as the six σ_{SeM}.   The σ_{SeP} are placed below the σ_{SeM} levels because P is assumed to have a greater electronegativity than M.

The σ_{SeM} and σ_{SeP} orbitals utilize 12 of the 18 (4p) orbitals of selenium leaving 6 π-type orbitals.   The closely placed Se atoms in each Se$_2$P group cause these to split into 2 sets of 3 orbitals each.   The designation π_{aSeSe} refers to the higher energy antibonding combination while π_{bSeSe} refers to that of lower energy.

Electronic Spectra–Spectrochemical, Nephelauxetic Effects and Optical Electronegativties.   In the case of the tris(diethyldiselenophosphato) complexes of Cr(III), Rh(III), and Ir(III), the first spin allowed transition has been observed at some 600–1700 K below the corresponding transitions in the dithiophosphate analogues.[179,180]   This is consistent with the observation[65]

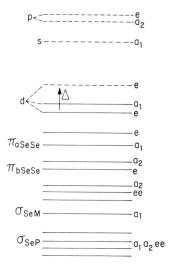

Figure 7-4   Order of MO energy levels in a system of the type $M(Se_2P)_3$.

that R_2Se falls before R_2S in the spectrochemical series.   Also, it has been reported that Co(II), in zinc selenide, has absorption bands which occur at frequencies lower than the corresponding ones in zinc sulfide.

The nephelauxetic effect, to a considerable degree, is a reflection of the polarizability of the ligand and in turn, a measure of the "covalency" of the metal-ligand bond.   This effect, which is a measure of the decrease in the electron-electron repulsion parameters in the complex compared with the corresponding terms in the gaseous ion, has been reported to be especially strong in the case of the dialkyldithiophosphates (dtp).[180]

Among the dsep complexes studied, $Crdsep_3$ was characterized by an absorption spectrum most clearly interpretable in terms of the nephelauxetic effect.   Thus, the observed separation of 3.9×10^3 K between the maxima arising from the $^4A_{2g} \rightarrow {}^4T_{2g}$ and the $^4A_{2g} \rightarrow a^4T_{1g}$ transitions corresponds to a Racah parameter $B = 340$ K which indicates $\beta_{35} = 0.37$ and the nephelauxetic function, $h = 3.0$.   In the same complex, the values of transitions to the ^2E_g, $^2T_{1g}$ and $^2T_{2g}$ levels yield a value for β_{55} of 0.77 ($\beta_{55} = 0.80$ in the $Crdtp_3$).   These observations have led to the conclusion that the nephelauxetic effect in $Crdsep_3$ on (t_{2g}, e_g) interactions is twice as large as on the (t_{2g}, t_{2g}) interactions.   This suggests that there exists a very extensive delocalization of the σ antibonding e_g subshell on the neighboring selenium atoms.

Tris(diethyldiselenophosphato) metal ion compounds display strong absorptions in the near ultraviolet region at frequencies about 3000 K below the corresponding bands in the diethyldithiophosphates. This type of experimental data has been used as a basis for the assignment of optical electronegativities of 2.6 for dsep⁻ and 2.7 for dtp⁻. Inasmuch as the optical electronegativities are not as low as should be expected from the nephelauxetic effect, selenium-containing ligands may not have electronegativities as low as one would predict on the basis of comparison with the halogens.

Jorgensen[179] has suggested that electron transfer bands are caused by transitions from the π_{aSeSe} and π_{bSeSe} groups to available subshells. Internal transitions in the partly filled shell are much stronger in the dsep complexes than in the aquo- or halide complexes. For example, they are 25 times more intense in Crdsep₃ than in $Cr(H_2O)_6^{+3}$ and 20 times stronger in Rhdsep₃ and Irdsep₃ than in the corresponding hexachloro complexes. This has been attributed[101] to the presence of forbidden Laporte transitions in the near ultraviolet which become allowed in the dsep complexes.

Krishnan and Zingaro[202] have recently prepared Kdsep, Tl(I)dsep, Pb(II)dsep₂, Sn(II)dsep₂, As(III)dsep₃, Sb(III)dsep₃, Bi(III)dsep₃, and In(III)dsep₃. The first absorption band of Kdsep in the ultraviolet region occurs at 36,630 K ($\log \varepsilon_{mol} = 3.08$) which is close to that observed for P_2Se_5 at 36,000 K. This has been ascribed to a $4p \rightarrow 5s$ Rydberg excitation in Se. It is interesting to note that in the dsep compounds of the regular elements the absorptions are very close to those observed for the bromide complexes.[102, 220, 247] Also, the absorptions observed for the dsep compounds occur at frequencies lower than those observed for the dtp compounds.[35] This could be due to the greater electronegativity of dtp. The In(III) compound shows only a single internal ligand transition. There was difficulty in the assignment of the bands because of the overlap of the internal transitions of the ligand with the $s^2 \rightarrow sp$ transitions of the central ions. In the case of Bidsep₃, bands at 22,910 K and 27,780 K may be assigned to a $^1A_{1g} \rightarrow {}^1T_{1g}$ transition by analogy with the corresponding transition in Bidtp₃.[180]

Recently,[203] the electronic spectra of Cudsep, Zndsep, Cddsep₂, Nidsep₂, Codsep₃, Rhdsep₃, Nidsep₂py₂, Kdsep, and Agdsep have been recorded. Indsep₃, which can only display internal ligand transitions, shows strong absorption bands at 35,710 K, $\log \varepsilon_{mol} = 3.80$ and 43,190 K, $\log \varepsilon_{mol} = 4.19$. Correspondingly, bands at 38,460 K, $\log \varepsilon_{mol} = 3.63$ for the Zn(II) compound; at 37,740 K, $\log \varepsilon_{mol} = 3.61$ for the Cd(II) compound, and at 40,320 K, $\log \varepsilon_{mol} = 4.21$ for the Rh(III) compound can be assigned to internal ligand transitions. In what is presumably square, coplanar Nidsep₂, a shoulder at 42,550 K, $\log \varepsilon_{mol} = 4.03$ is poorly resolved. This is very likely due to interligand conjugation.

The values of Δ, the crystal field splitting parameter, are 13,420 K in the case of Nidsep₂ and 20,200 K for Rhdsep₃. Galsbol[118] has reported a value of $\Delta = 13,600$ K for Crdsep$_3$ and a corresponding value of 14,300 K for the dtp analogue.   A comparison of the value of Δ for Nidtp$_2$, 14,500 K, has been made as well as in the case of complexes in which diselenocarbamate and dithiocarbamate function as the ligands.   The ligand field strength, based on the Δ values observed for Ni(II), Cr(III) and Rh(III) complexes, follows the order dsep < dtp < dsec < dtc.

Infrared Spectra.   Kudchadker et al.,[208] observed a pair of strong bands in the region 450–600 cm^{-1} in bis(dialkyldiselenophosphates) and the Kdsep and Crdsep$_3$.   Krishnan and Zingaro[202] also reported on the spectra of a number of dsep$^-$ salts.   All of the bands in the rock salt region could be assigned to well-known vibrations.   The region of interest, i.e., the portion of the spectrum in which P—Se vibrations were found, was from 300–600 cm^{-1}.

In the case of the potassium salt, strong absorptions at 584 and 620 cm^{-1} have been assigned to the asymmetric, v_1, and symmetric, v_2, stretching vibrations of the PSe$_2$ group, respectively.   Inasmuch as it seems reasonable to assume that both phosphorus-selenium bonds are equivalent, i.e., complete resonance of the type $P{\raise2pt\hbox{$\nwarrow$}}{\raise-2pt\hbox{$\swarrow$}}{\substack{\text{Se}\\ \ominus\\ \text{Se}}}$ exists, it is unlikely that the two frequencies arise from P—Se bonds of higher and lower bond order.

The P—Se stretching frequencies are lowered by 14–74 cm^{-1} in the Pb(II), Sn(II), As(III), Sb(III), Bi(III), and In(III) compounds relative to the frequencies observed in the K salt.   In the Bidsep$_3$, v_1 is split into a doublet while in Indsep$_3$ both v_1 and v_2 are observed as doublets.   The decreases in v_1 and v_2 have been related to the bonding of both selenium atoms to the heavier central metal ion.

One interesting feature of the spectra of the compounds mentioned above is the occurrence of a weak absorption band at, or close to, 340 cm^{-1}. The frequency is too high to be a metal-selenium stretching mode.   It may involve one of the bending or scissoring modes of P—O—C or C—C—O, or a combination of both.

The infrared spectrum of the Tl(I) compound differs from that of the other nontransition metals.   In this case, only v_1 is lowered relative to the potassium salt, and the 340 cm^{-1} band is absent.   If the Tl(I) compound had a bridged structure, this would furnish an explanation for this effect.

The infrared spectra of the Ag(I), Cu(I), Zn(II), Cd(II), Ni(II), Pd(II), Rh(III), and Cr(III) have also been examined.   The pattern, in the region of 300–600 cm^{-1}, largely parallels that just discussed for the nontransition metals.   In the case of Cd(II), v_2 is split, while in the case of Co(III), Rh(III), and Cr(III), both v_1 and v_2 appear as doublets.   The 340 cm^{-1} absorption is absent in the case of Ag(I), Cu(I), and Zn(II).

Ilina *et al.*,[166] have reported on the ligand, O,O′-diphenylthioselenophosphate, $(PhO)_2PSSe^-$. Sparingly soluble hydrates of the following metal ions were prepared: Pb(II), Ni(II), Cu(II), Zn(II), Cd(II), Hg(II), Pd(II), Pt(II), In(III), Sb(III), Bi(III), Tl(I), Ag(I), and Au(I). The brown Ni(II), mp 125°C, and the red Pd(II), mp 146°C, are diamagnetic and presumably planar. The Ni(II) complex adds 2 moles of pyridine to form a six-coordinate green adduct.

Magnetic Susceptibility Measurements. The only reported magnetic susceptibility measurements on metal dsep compounds are those of Krishnan and Zingaro[203] and these have all been recorded at room temperature. The Zn(II) and Cd(II) salts, as expected, are diamagnetic and presumably tetrahedral, although crystal structures have not yet been determined. Grey $Nidsep_2$ is diamagnetic which suggests a low-spin, coplanar complex. Red $Rhdsep_3$ is also diamagnetic and this suggests that the ligand field created by a presumably octahedral arrangement of selenium atoms is sufficiently high to stabilize the spin-paired configuration. The magnetic susceptibility of $Codsep_3$ is 1.9 BM at 298°K and 1.77 BM at 131.5°K. These values are not consistent with the diamagnetism normally encountered in the case of six-coordinate Co(III) complexes and an explanation is wanting. The determination of the crystal structure would greatly aid in explaining the value of the magnetic susceptibility.

Diselenophosphoric Diamides

Diselenophosphoric diamides, $[(RNH)_2PSe_2]^-$, have been described in only one published paper.[219] These compounds are prepared by the reaction of primary amines with P_2Se_5:

$$7RNH_2 + P_2Se_5 \longrightarrow 2(RNH)_2PSe_2^- RNH_3^+ + RNH_3^+HSe^-$$

Inasmuch as transition metal complexes of dialkyldiselenophosphates and diselenophosphinates have been studied, the study of this ligand appears to represent a logical extension of this work. Such a study should make available data which will facilitate a comparison of the binding sequences $N_2PSe_2^-$, $O_2PSe_2^-$, and $C_2PSe_2^-$.

Diselenophosphinates and Related Mixed Ligands

Diethylselenothiophosphinate salts have been described.[204] Tetraethyldiphosphinedisulfide, Na_2Se, and Se react in the molten state to yield sodium diethylselenothiophosphinate:

$$(C_2H_5)_2P(S)-P(S)(C_2H_5)_2 + Na_2Se + Se \xrightarrow[2\ h]{\sim 200°C} 2(C_2H_5)_2P(S)SeNa$$

In absolute ethanol, the same product is obtained by the reaction between diethylchlorophosphine sulfide and sodium hydroselenide. The following complexes of the type:

$$(C_2H_5)_2P \underset{S}{\overset{Se}{\diagup\diagdown}} M/n$$

were reported; bis(diethylselenothiophosphinato)zinc(II), mp 157°C; cadmium(II), mp 160°C;-nickel(II) (olive green) dec > 142°C;-lead(II)(greenish yellow), mp 132°C; and tris(diethylselenothiophosphinato)bismuth(III), mp 94°C.

Sodium diethyldiselenophosphinate has also been prepared.[205] Diethylchlorophosphine adds a mole of elemental selenium to form the phosphine selenide. The latter, with sodium hydroselenide, yields the ligand in the form of the sodium salt:

$$(C_2H_5)_2P(Se)Cl + 2NaSeH \longrightarrow (C_2H_5)_2P(Se)SeNa + NaCl + H_2Se$$

It is worth mentioning that the sodium hydroselenide was prepared by passing the stoichiometric amount of H_2Se into sodium ethoxide. Complexes prepared include those of Zn(II), mp 151°C; Cd(II), mp 158°C; lemon yellow Pb(II), mp 157–158°C; brown-red Pd(II), mp 197°C; vermillion-red Bi(III), mp 132°C; bright yellow In(III), mp 146°C, and pale yellow Tl(I), mp 129° C.

It is interesting to compare the acid strengths of the phosphinic acids which follow the sequence; $(C_2H_5)_2P(Se)SeH$, pK 2.18; $(C_2H_5)_2P(S)SeH$, pK 2.29; $(C_2H_5)_2P(S)SH$, pK 2.60; and $(C_2H_5)_2P(S)OH$, pK 4.98. The free acid, $(C_2H_5)_2P(Se)SeH$, was isolated by Kuchen and Knop[206] by exchange of the sodium salt with the acidified form of a strong cation exchanger.

Association of Phosphinato Complexes. Kuchen and Hértel[207] observed that a number of thiophosphinic and selenophosphinic acids as well as the Zn(II) and Cd(II) complexes exhibit association in benzene or chloroform which is concentration dependent. This association has been attributed to the existence of ligand bridges of the following type:

The effect is most pronounced in the thiophosphinato $[R_2P(S)O]_nM$ complexes.[207] Because this review is concerned with selenium-bearing ligands, and only modest amounts of work have been done with respect to association involving such ligands, only the principal experimental observations dealing with related nonselenium-bearing ligands are outlined here:

a) The octahedral In(III) thiophosphinato complex remains monomeric in benzene over a wide concentration range while the tetrahedral complexes of Zn(II) and Co(II) are associated.

b) In those cases where it occurs, the degree of association increases with concentration and tends to reach a limiting value that depends on the central atom and the solvent. The degree of association in various solvents decreases in the order $CCl_4 >$ benzene $> CHCl_3$. The association appears to be temperature independent up to 60°C.

c) The flow behavior of solutions of $[(C_2H_5)_2P(S)O]_2Co$ and $[(C_2H_5)_2P(S)O]_2Zn$ in benzene show no dependence of the viscosity on the shear gradient, and hence, no structural viscosity. This, together with the low temperature dependence of the association, indicates the presence of strong bonds in the associated species. This is the basis for the postulated ligand bridges.

d) The nature of the donor atoms is also important. Thus, based on the behavior of the Co(II) and Cd(II) complexes, the association decreases in the order:

$$R_2P(O)O^- > R_2P(S)O^- > R_2P(S)S^- > R_2P(Se)S^- > R_2P(Se)Se^-$$

In conclusion, it has been suggested that the association is due primarily to steric factors. The replacement of four-membered chelate rings by less strained ligand bridges leads to a more favorable steric arrangement. In a four-membered ring of the type $P\underset{Y}{\overset{X}{\diagup\diagdown}}M$, the strain has been shown to increase as the radii of the central and donor atoms decrease. However, in the octahedral complexes, the formation of four-membered rings is favored by the smaller bond angle of the metal.

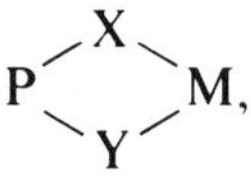

Electronic Spectra.   There appears to be a paucity of information about the electronic spectra involving the diselenophosphinates and their metal complexes. Relevant studies concerning the nonselenium-bearing phosphinates are discussed by Kuchen and Hertel.[207] However, some recent data of Galsbøl[118] are reproduced in Table 7-4. It can be seen that Galsbøl has systematically varied the chalcogen atoms in a series of phosphinato-Cr(III)

TABLE 7-4   SPECTRAL DATA OF Cr(III) COMPLEXES

	Cr(III), $B_{gas} = 918$ K					
Complexes	$^4A_{2g} \to {^4}T_{2g}$ (K)	$^4A_{2g} \to {^4}T_{1g}$ (K)	Δ (K)	B (K)	$\dfrac{\Delta}{B}$	β
$(C_2H_5)_2PSeSe^-$	12,900	17,000	12,900	391	32.9	0.43
$(C_2H_5)_2PSeS^-$	13,500	17,900	13,500	408	33.1	0.45
$(C_2H_5)_2\,PSS^-$	13,600	18,200	13,600	432	31.6	0.47
$(C_2H_5O)_2PSeSe^-$	13,800	17,700	13,800	362	38.0	0.39
$(C_2H_5O)_2PSS^-$	14,300	18,700	14,300	402	35.7	0.44
$(C_2H_5)_2NCSeSe^-$	14,800	19,100	14,800	399	37.1	0.44
$(C_2H_5)_2PSO^-$	15,600	21,900	15,600	636	24.5	0.69
$(C_2H_5)_2PSeO^-$	15,700	22,500	15,700	704	22.3	0.77
$(C_2H_5)_2NCSS^-$	15,600	20,200	15,600	433	36.1	0.47
$(C_2H_5)_2POO^-$	15,800	22,400	15,800	659	24.1	0.72
$C_2H_5SCSS^-$	15,700	19,900	15,700	383	41.0	0.42
$C_2H_5OCSS^-$	16,000	20,300	16,000	395	40.6	0.43

complexes and, based on the spectral observations, the crystal field splitting parameter, Δ, has been measured. These follow the order:

$$(C_2H_5)_2P(O)O^- > (C_2H_5)_2P(Se)O^- \cong (C_2H_5)_2P(S)O^- > (C_2H_5)_2P(S)S^- >$$
$$(C_2H_5)_2P(Se)S^- > (C_2H_5)_2P(Se)Se^-$$

Infrared Spectra. Based upon a reasonably large number of published reports, the P—S and P—Se bonds have only an extremely small π component.[301,331,67,136,332,108,161] Hence, complete resonance of the type

$R_2P\begin{smallmatrix}Se\\\\Se\end{smallmatrix}\ominus$ corresponds more closely to the true situation in these com-

pounds. A double bond formulation is not probable.

The two fundamental P—Se stretching absorptions therefore correspond to asymmetric (v_1) and symmetric (v_2) types. Thus, in the metal complexes, $[R_2PSe_2]_nM$, v_1 has been observed at 475–492 cm^{-1} and v_2 at 449–458 cm^{-1}.[207] In the sodium salt, $Et_2PSe_2Na \cdot nH_2O$, v_1 appears at 525 cm^{-1} and v_2 at 498 cm^{-1}.

Crystallographic Studies. Husebye[163,164] and Husebye and Helland-Madsen[165] determined the structures of $Se[Et_2P(S)Se]_2$, $Te[Et_2P(S)Se]_2$, and $Te[Et_2P(Se)Se]_2$.

In the case of selenium bis(diethyldiselenophosphinate), the molecules possess twofold symmetry axes passing through the central selenium atoms. In the crystal, each central selenium atom, in addition to being bonded to 2 selenium atoms in the molecule, forms close contacts with 2 selenium atoms from neighboring molecules.

The four-coordinate SeSe$_4$ arrangement is almost planar. The intermolecular Se $\cdots$ Se contacts of 3.68 Å give rise to a loose two-dimensional association of molecules. The structure is shown in Figure 7-5.

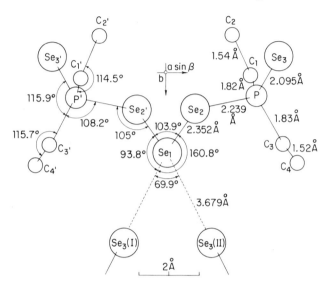

Figure 7-5 The configuration around Se$_1$ seen along the c-axis in Se[(C$_2$H$_5$)$_2$PSe$_2$]$_2$. Primed letters denote atoms in half the molecule related to the other half by a twofold symmetry axis through Se$_1$.

In the case of tellurium bis(diethylthioselenophosphate) the structure

$$\begin{array}{ccc} \text{S} & & \text{S} \\ \| & & \| \end{array}$$

consists of (C$_2$H$_5$)$_2$P—Se—Te—Se—P(C$_2$H$_5$)$_2$ molecules. The molecules have twofold symmetry with the tellurium atoms on twofold symmetry axes. One of the more significant results is the Te—Se bond length, 2.50 Å. The other bond lengths and bond angles are: Se—P = 2.26 Å, P=S = 1.93 Å, Se—Te—Se' = 100.7°, Te—Se—P = 104.7° and Se—P—S = 103.8°. The dihedral angle Se'TeSe/TeSeP is 93.1°.

There are weak intermolecular Te $\cdots$ S bonds of length 3.65 Å joining the molecules in a two-dimensional network. Each tellurium atom is involved in 2 such bonds at a S $\cdots$ Te $\cdots$ S angle of 78.1°. These bonds are roughly *trans* to the 2 Te—Se bonds so that there is a weak tendency to form a coplanar coordination around the Te atom. The results are shown in Figure 7-6.

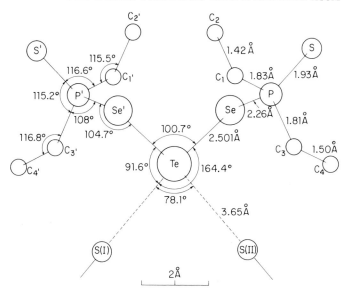

Figure 7-6 The configuration around the Te atom seen along the normal to the least squares plane through the TeSe$_2$S$_2$ group in Te[(C$_2$H$_5$)$_2$PSSe]$_2$. Primed letters denote atoms in one half of the molecule related to the other half by a twofold symmetry axis through the Te atom.

Phosphine Selenides

Bannister and Cotton[27] reported the formation of complexes between Ph$_3$PSe and Pd(II) or Sn(IV). Nicpon and Meek[248] have investigated R$_3$PSe complexes of Hg(II), R = Ph; Pd(II), R = m-tolyl; Pt(II), R = Ph; Ag(I), R = Ph; and Cd(II), R = C$_4$H$_9$. In all cases, selenium was found to be the donor atom. The 1 : 1 adducts of HgCl$_2$ with Ph$_3$PS, Ph$_3$PSe and Ph$_3$AsS are isomorphous with one another.[191]

Brown, Hill and Clifton[41] found complexes of the type MX$_5 \cdot$L where M is niobium or tantalum, X is Cl or Br and L is triphenylphosphine selenide. Because of the extreme hydrolytic susceptibility of the pentahalides, the reaction must be carried out in nonaqueous media. This study lists x-ray powder diffraction data as well as infrared spectra down to 225 cm^{-1}.

Brodie *et al.*[40] prepared complexes of trimethylphosphine selenide of the type [M(Me$_3$PSe)$_4$](ClO$_4$)$_2$ where M = Co(II), Ni(II), or Zn(II). They also prepared a corresponding series of the type M(Me$_3$PSe)$_2$X$_2$, in which X is Cl, Br, or I. Electronic spectral studies of complexes of the type [M(Me$_3$PSe)$_4$](ClO$_4$)$_2$ show that the nephelauxetic parameter, β, has a value of 0.61 for the MSe$_4$ environment. This is somewhat lower than the value reported for the corresponding sulfur ligand (0.65). The ligand field splitting parameter, Δ (3800 K) again, is slightly lower than that of the corresponding sulfide (3890 K).   In the complexes of the type ML_2X_2, Δ decreases within each of the series, ML_2Cl_2, LM_2Br_2, ML_2I_2 as L changes from the oxide to the sulfide to selenide.   The β values display considerably greater variation in the $[ML_4]^{+2}$ than in the $[ML_2X_2]$ series.   The results are explained in terms of the polarizability of the halide ion.   The effective magnetic moment of $[Co(Me_3PSe)_4]^{+2}$ is higher than in the corresponding sulfide complexes.   The complexes are reported to be tetrahedral with the cobalt ion retaining a low-field configuration.   The red-brown $Ni(Me_3PSe)_4(ClO_4)_2$ shows a charge-transfer band (Se $\rightarrow$ Ni) at 16,500 K. The reflectance spectrum is characteristic of tetrahedral Ni(II) complexes. However, the x-ray powder patterns revealed that the crystals were not isomorphous with those of the sulfide analogues.   The fundamental stretching frequency, $v(P-Se)$, which occurs at 436 cm^{-1} in the free ligand, is lowered by about 20 ± 5 cm^{-1} in the complexes.   The metal-selenium stretching vibration was observed at a lowest reported value of 202 cm^{-1} for the Zn(II)L_4^{+2} complex and at a highest value of 250 cm^{-1} for Ni(II)L_4^{+2}.

Equilibrium studies which have been concerned with the donor properties of phosphine selenides toward iodine have been reported by Condray et al.[72] Four phosphine selenides, the triphenyl, tris(1-naphthyl), diphenyl-1-naphthyl, and diphenyl-4-methylnaphthyl derivatives were investigated.   The thermodynamic values for ΔH, $\Delta G°$, and $\Delta S°$ were calculated.   These values, when compared with the corresponding ones reported for the sulfide and oxide complexes, clearly show that the donor strength toward iodine decreases in the order Se > S > 0.   This is explained in terms of the greater proximity of the $4d$ orbitals of Se to the $5p$ orbitals of iodine.   This permits a greater case of overlap and allows for $p_\pi \rightarrow d_\pi$ bonding.   This is not the case with the $3d$ orbitals of sulfur.   In addition, the importance of the sizes and polarizabilities of the donor and acceptor atoms in stabilizing the interaction are discussed.

ORGANIC LIGANDS

Neutral Ligands

Monodentate.   Dialkyl and diaryl selenides, cyclic selenides, selenoureas, selenoantipyrine and phosphine selenides (discussed previously) belong to this class of compounds in which selenium is the donor atom.

Fritzmann[112] and Fritzmann and Krinitzki[115] investigated a number of compounds of the type $Pt(R_2Se)_2Cl_2$ (R = Me, Et, Pr, isobutyl, Ph).   They exist as *cis*, *trans*, and bridged complexes.   Similar stable $Pd(R_2Se)_2X_2$, (R = Me, Et, Bu, Ph; X = Cl, Br, I) complexes were also reported.[113, 114] Jensen[170] was the first to measure the dipole moments of complexes of Me_2X,

(X = S, Se, Te) with Pt(II).   Because the moments of the *trans* isomers are largely cancelled, they are readily distinguished from the *cis* isomers which have large permanent dipole moments.

It was observed by Coates[71] that the relative strength of the coordinate bond formed between XMe_2, (X = 0, S, Se, Te) and $AlMe_3$ follows the order 0 > S > Se > Te.   Chatt and Venanzi[64] found in the bridged complexes of Pd(II), PdL_2Cl_4, (L = R_2S, R_2Se, R_2Te) that when L is an R_2S group, the complex could be more easily isolated than in the case of R_2Se or R_2Te.   They report that the stabilities fall in the order R_2S > R_2Se > R_2Te.   In the corresponding Pt(II) complexes, PtL_2Cl_4, the complexes having L = R_2Te are more easily isolated than the R_2Se derivatives.   This suggests an order of stabilities R_2S ≫ R_2Te > R_2Se.[62]   The difference in the sequence of the stabilities is explained on the basis of relative sizes of the orbitals involved in forming coordinate σ bonds.   Thus, Pd(II) and Se are comparable in size, as are Pt(II) and Te.   Subsequent studies on the vibrational spectra of coordination compounds of the type MX_2Y_2 where M = Pd(II) or Pt(II); X = Cl, Br or I and Y = Me_2S, Me_2Se or Me_2Te, have shown that the relative bond order follows the sequence M—Cl > M—Br > M—I and M—S > M—Se < M—Te.[16]   The inverse order M—Te > M—Se suggests strong π bonding between Pt and Te rather than nonbonded interactions.   Subtle effects on the internal modes of vibrations of the donor molecules, R_2S, R_2Se and R_2Te have been reported[17] in the *cis*, *trans*, and bridged halide complexes of Pd(II) and Pt(II).

Chatt *et al.*[65] showed that the value of the ligand field splitting parameter, Δ, for square planar, *trans*-[L piperidine $PtCl_2$] as observed from the first spin-allowed transition, $d_{xy} \rightarrow d_{x^2-y^2}$, is in the order Et_2S > Et_2Se > Et_2Te. In a series of complexes, *trans*-[$LamPtCl_2$] where L = Me_2S, Me_2Se, Me_2Te, R_3P, R_3Sb, $(RO)_3P$, R_3As, C_2H_4, and *am* is either R_2NH or RNH_2, it was observed[63] that the fundamental N—H stretching frequency of *am* increases in the order R_3P > R_3Sb > $(RO)_3P$ > R_3As > R_2Te > C_2H_4 > R_2Se > R_2S. This is consistent with the relative donor strength of L to Pt which is transmitted to the nitrogen atom of the *am* group and brings about an increase in the stretching frequency.   Later, Chatt and Westland,[66] from measurements of the proton chemical shifts in pyridine, attempted to distinguish inductive from mesomeric effects in complexes of the type *trans*-[$PtCl_2pyL$] where L represents alkyl phosphines, -arsines, -stibines, -sulfides, -selenides, and -tellurides.   That the d_π-d_π bonding between Pt and Se in the complex, $Pt(Me_2Se)_2Cl_2$ is less than that between Pt and P in the complex $Pt(PPh_3)_2Cl_2$ is suggested by the smaller ^{77}Se–^{195}Pt coupling constant ratio of *cis* : *trans* isomers in the respective complexes.[240]

Titanium(IV) forms a 1 : 1 adduct with Me_2Se and it may either be a five-coordinated monomer or a highly coordinated polymer.[342]

Nobel and Winfield[249] reported the formation of $WF_6 \cdot 2Et_2Se$ as a volatile, orange liquid, while a similar reaction with ether yielded $WOF_4 \cdot Me_2O$. Dimethyl selenide adducts of the boron halides have been reported[346] and the infrared spectrum of BF_3Me_2Se has been measured.[211]   Palko and Durig[255] showed, from studies of the isotopic exchange reaction:

$$^{10}BF_{3(g)} + A \cdot {}^{11}BF_3 = {}^{11}BF_3 + A \cdot {}^{10}BF_3$$

where $A = Me_2O$, Me_2S and Me_2Se, that isotopic equilibrium constants vary as $Me_2S > Me_2Se > Me_2O$, while the equilibrium constants vary inversely with the strength of boron-donor bond.

Cyclic selenides also function as donors and a large number of halogen and interhalogen addition compounds have been studied (Figure 7-7).

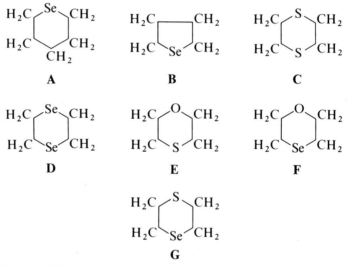

Figure 7-7   A, 1-Selenacyclohexane;   B, tetrahydroselenophene;   C, 1,4-dithiacyclohexane; D, 1,4-diselenacyclohexane; E, 1-thia-4-oxacyclohexane; F, 1-selena-4-oxacyclohexane; G, 1-selena-4-thiacyclohexane.

Baker and Fowles[23] studies the formation of selenoxan (**F**) complexes with Sn(IV) chloride and bromide, Ti(IV) chloride and Nb(V) chloride.   Infrared and NMR data suggest coordination to Se.   Sn(IV) adds 2 molecules of (**F**), while Nb(V) adds one and Ti(IV) adds either 1 or 2 moles of the same compound.   It is surprising that Ti(IV), Nb(V), and Sn(IV) behave as "soft" metals.   The suggestion has been made that when the metals are bonded to a "hard" halogen, such as fluorine, they behave as "hard" acids. However, when the substituent halogen atom is a larger, more polarizable atom, "soft" behavior is imparted to the metal. Many halogen and inter-halogen addition products of cyclic selenides (**A** through **G**) have been studied by spectrophotometric and by x-ray diffraction techniques.

Addition compounds of selenides with iodine are of the charge-transfer type[241] and show Se—I—I bonding rather than X—Se—X, as is found in R_2SeX_2 and R_2TeX_2 (X = Cl or Br). Dissociation constants of the molecular iodine complexes with **C** through **G** have been determined[242] and it was found that the complexing tendency of iodine is in the order Se > S > O, other factors being equal. The "soft" iodine chooses Se rather than S or O and chooses S rather than O. Halogen and interhalogen addition products of **G** have been prepared and their structures predicted.[243] The dissociation constants of the molecular iodine complexes of **A** and **B** have been found to be smaller than those of the corresponding sulfur analogues.[244] This suggests a greater stability for the selenium compounds. X-ray studies on molecular iodine complexes of **D**,[60] of **G**,[157] of **B**,[158, 260] of **F**,[216] and the iodine monochloride addition product of **F**[192] have been reported. The Se—I distances in these complexes are 2.829 Å (**B**), 2.755 Å (**E**), and 2.63 Å (**F**). The shortest Se—I distance is found in the iodine monochloride addition product with **F**. This suggests the formation of the selenonium salt $[C_4H_8OSeI]^+Cl^-$. X-ray studies show that in the addition compound which forms between **D** and 2 moles of iodine, there is present axial bonding from each of the Se atoms to the respective iodine molecules. The strength of the halogen-heteroatom bond is in the order O < S < Se.[149] The donor **B** coordinates a single mole of iodine. In addition to the strong, axial bonding, there exists secondary bonding to the outer iodine atom of the molecule of acceptor coordinated to the selenium atom of the next selenophene molecule. This gives rise to an arrangement of the type:

$$I_1—I_2———Se------I_1———I_2$$
$$\text{2.762 Å}\quad\text{3.64 Å}$$

In the case of **F**, which also coordinates to a single mole of iodine, the secondary bonding is weaker and the I_2—Se $\cdots I_1$ bond angle is much smaller (Figure 7-8). The Se—I_2—I_1 bond angle is almost linear, and equal to

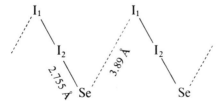

Figure 7-8 Bonding scheme in the iodine complex of 1-selena-4-oxacyclohexane. (Molecules are linked together in a helical chain about alternate 2_1 screw axes.)

174.8°. The structure of the iodine complex with **G** has a centrosymmetric structure and is randomly distorted. Both axial and equatorial bonding of iodine molecules is involved. That the iodine molecule is bound to **D** as

$\rangle Se-I_1-I_2$ and that the $I-Se$ bond is stronger than the $I-S$ bond in the

corresponding sulfur analogue has also been demonstrated.

Selenourea(Su), $(NH_2)_2CSe$, resembles thiourea in many respects. It acts as a monodentate ligand and forms complexes with transition metal ions by coordination via the selenium atom. Most metal ions can be precipitated as selenourea complexes and they are sometimes soluble in an excess of this reagent. A genuine eight-coordinate osmium(III) complex, $Os(Su)_8^{3+}$ has been isolated.[265] The precipitation of selenourea complexes is very sensitive to pH. The protonation constant of selenourea and the stability constants of Hg(II) complexes of selenourea have been detailed.[123] The behavior of Hg(II) is typically class "b" and the affinity order is $O \ll S < Se$. Also, it is surprising to find that selenourea is more basic than either urea or thiourea.

Furlani and Tarantelli[116, 315] observed that *tetrakis*- substituted selenourea complexes of Pd(II) and Pt(II) tend to form Se-coordinated species according to the following equilibrium (in the presence of excess of ligand):

$$[L_4M(II)X]^+ + X^- \rightleftharpoons [L_3M(II)X_2] + L$$

The selenoureas investigated were $RHN-C(Se)NHR'$ where R is benzhydryl and R' is phenyl, allyl or butyl.

The tendency to form five-coordinated species (C_{4v}) ML_4X, where $M = Pd(II)$ or $Pt(II)$ is greater for seleno- rather than thiourea. The frequencies of the $d-d$ transitions for the chromophores $[PdSe_4Cl]^+$, $[PtSe_4Cl]^+$ and $PdSe_2Cl_2$ are 23,200, 24,100 and 22,700 K, respectively. The positions of the charge-transfer transition, $Se \rightarrow M$ are the same, irrespective of the chromophore and slightly lower than that of sulfur, suggesting an optical electronegativity of 2.55 ± 0.1 for Se compared to 2.65 ± 0.1 for sulfur. Among the quadratic chromophores, the spectrochemical position of Se bonded selenoureas is slightly lower than that of S bonded thioureas, and quite close to that of water.

Vibrational spectra of selenourea complexes of Pd(II) and Pt(II) have been studied[151] and these complexes possess square planar symmetry. The $M-Se$ stretching frequencies in selenourea complexes occur at shorter wavelengths than is observed in the case of $M-Se(CH_3)_2$ complexes. The difference has been attributed to π bonding. The x-ray structure and infrared spectrum of $CdCl_2 \cdot 2Su$ show that it is analogous to the corresponding thiourea complex and has a tetragonal structure.[273] Domiano *et al.*[85]

showed, from x-ray powder and infrared data, that $M[SeC(NH_2)_2](NCS)_2$ compounds (M = Co(II), Ni(II), Cd(II)) are isostructural with the corresponding thiourea complexes and have a polymeric structure consisting of chains of coordinated octahedra. The metal atom is surrounded by 4 selenium atoms while the 2 N atoms of thiocyanate are *trans* with respect to the plane of Se atoms.

It has been shown recently[68] that AgCl dissolves in an excess of selenourea to form an oxidized product, the α,α'-diselenobisformamidinium cation, $[(NH_2)_2CSe]_2^{+2}$. Preliminary x-ray data of this compound showed no isostructurality with the corresponding α,α'-dithiobisformamidinium cation. The two selenourea parts of the molecule are strictly planar and the mechanism of oxidation of selenourea involves a charge rearrangement without any proton transfer.

3,4-Benzo-1,2,5-selenodiazol, forms a yellow Pd(II) complex[348] which has been used for the estimation of Pd(II).[44]

$$NH_2$$

Selenosemicarbazones, R=N—NHC=Se, where R is acetone, cyclohexanone, vanillin or salicylaldehyde behave as neutral monodentate ligands.[8] Selenosemicarbazones are stable only in alkaline or neutral media. However, *bis*(dimethylglyoximato)*bis*(selenosemicarbazone)Co(III) is exceptionally stable toward acids. This fact indicates that the bonding of selenosemicarbazone to Co(III) is via selenium. In addition, several Co(III) complexes having different R groups and a dioximato group have

been reported. Selenopyrine, complexes of Ti(IV) and Sn(IV) have been prepared. Coordination through selenium has been suggested in these complexes.[256]

It is known that dimethyl sulfoxide acts as an ambidentate ligand forming coordination compounds by donation through both oxygen and sulfur. Jensen and Krishnan[172] reported several dimethyl selenoxide complexes of transition metals and showed that the bonding invariably involves the oxygen atom of dimethyl selenoxide. Later Paetzold and Bochmann[253, 254] studied coordination compounds of dimethyl selenoxide with metal perchlorates and confirmed the earlier observation of oxygen bonding in these complexes. The position of dimethyl selenoxide in the spectrochemical series is in the order $H_2O >$ dmseo $>$ dmso; however, the difference in the nephelauxetic parameter, β, is not significant between dmso and dmseo. The coordination properties of diphenyl selenoxide (dpseo) have been similarly studied.[252]

Red $NiCl_2 \cdot 3$dpseo has an octahedral structure consisting of $[Ni(dpseo)_6]^{+2}$ and $[NiCl_4]^{-2}$ ions and has a magnetic moment of 3.40 BM. The blue $CoCl_2 \cdot 2$dpseo has a tetrahedral structure and the dark green $CuCl_2 \cdot 3$dpseo with a magnetic moment of 1.75 BM presumably has a dimeric structure. A rhodium(III) compound $RhCl_3 \cdot$dpseo which is found to be suitable for the extractive photometeric determination of Rh(III) has been reported.[349]

Bidentate ligands. Diselenides can function as chelating agents by forming bonds to metal ions via both of the selenium atoms. Fritzmann and Krinitzki,[115] and Fritzmann[114] first isolated Pt(II) and Pd(II) complexes with (IV) below:

Ph—Se—Se—Ph   (I)
diphenyl diselenide

$CH_3Se(CH_2)_2SeCH_3$ (II)
1,2-bis(methylseleno)ethane

$CH_3Se(CH_2)_3SeCH_3$ (III)
1,3-bis(methylseleno)propane

$C_2H_5Se(CH_2)_3SeC_2H_5$ (IV)
1,3-bis(ethylseleno)propane

$PhSe(CH_2)_3SePh$ (V)
1,3-bis(phenylseleno)propane

$(CH_3)_2CHSe(CH_2)_2SeCH(CH_3)_2$ (VI)
1,2-bis(isopropylseleno)ethane

$BuSe(CH_2)_2CH:CH_2$ (VII)
But-3-enyl butyl selenide

$PhSe(CH_2)_2CH:CH_2$ (VIII)
But-3-enyl phenyl selenide

$BuSe(CH_2)_3CH:CH_2$ (IX)
Butyl pent-4-enyl selenide

$PhSe(CH_2)_3CH:CH_2$ (X)
pent-4-enyl phenyl selenide

Later, Westland and Westland[342] reported the formation of a red-black Ti(IV) complex of (V). Aynsley *et al.*[20] described the synthesis of (II) and (III) and their complexes with several transition metal ions. The interesting red-brown powder, $2FeCl_3 \cdot$(III) is paramagnetic and has a magnetic moment of 5.3 BM. Complexes of (V) with MX_2 (M = Pd or Pt, X = Cl, Br, NO_3) and $HgCl_2$ have been studied.[266] Five-coordinated structures have been proposed, based on conductance measurements. It has been proposed that in these complexes, Pt(II) forms a weaker π bond to Se than does sulfur, but that the reverse is true in the case of Pd(II) complexes, as has been observed in dialkyl chalcogenide complexes of Pd(II). Greenwood and Hunter[142] have obtained molecular mass measurements which show that Pd(II) and Pt(II) complexes of (VI) exist as dimers in $CHCl_3$ but as monomers in acetone.

The dimers have a structure in which the 2 metal atoms are bridged by the selenium atoms of the 2 ligands in *trans*-configuration (Figure 7-9). The ligand (VI) acts as a bidentate chelating agent in the monomers. It is surprising to note the absence of any charge-transfer band in the spectra of the dimers while it is present in ethanol solutions. Nickel(II) complexes with (VI) which have the composition $Ni(VI)(SCN)_2$ and $Ni_2(VI)(SCN)_4$, have been reported.[143] A magnetic moment of 3.20 BM and a visible absorption band at 13,900 K $[^3A_{2g} \rightarrow {}^3T_{1g}(F)]$ for $Ni(VI)_2X_2$ suggests a six-coordinated structure. The complex, $Ni(VI)(SCN)_2$ has a square planar structure with a visible absorption band at 15,400 K. The compound, $Ni_2(VI)(SCN)_4$ is presumably polymeric in the solid state. Recently, a U(V) complex of (I) with a magnetic moment of 3.1 BM has been described.[299]

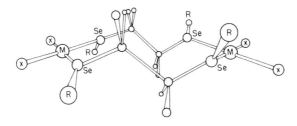

Figure 7-9   The possible structure of the dimeric complexes $M_2X_4(VI)_2$ crystallized from chloroform, $M = Pd$ or Pt, $X = Cl$ or Br.

Interesting results have been obtained by Goodall[134] in studies which deal with coordination compounds of unsaturated selenides of Pd(II) and Pt(II). The unsaturated selenides, (VII) and (VIII) form complexes with Pd(II) and Pt(II) while (IX) and (X) form complexes with Pt(II) only. A probable reason may be found in the stabilizing factor induced by the rigidity of the chelate ring containing Pt. It is considered to be more stable than the one containing Pd. The unsaturated selenides coordinate to metal ions both through Se and the olefinic double bond.

1,2-diselenocyanatoethane, $SeCN-CH_2-CH_2-NCSe$, acts as a bidentate ligand[132] by forming bonds through both of the pseudohalogen groups. Pd(II) and Pt(II) form complexes of the type $MCl_2 \cdot L$ where $M = Pd(II)$ or Pt(II) and $L = 1,2$-diselenocyanatoethane. Based on infrared studies, a planar configuration involving selenium bonding and a *trans* configuration of the ligand have been proposed. The Cl atoms occupy *cis* positions. A polymeric Co(II) complex, $[CoCl_2 \cdot L]$ and dimeric $[MCl_3L]_2$, $M = Rh(III)$ or Ir(III), $L = 1,2$-diselenocyanatoethane, have been isolated.[133] A magnetic moment of 5.01 BM and the appearance of a 19,610 K band $[^4T_1(F) \rightarrow {}^4T_1(P)]$ in the visible region for the Co(II) complex suggest octahedral coordination around Co(II). The infrared data of the Co(II) complex are indicative of the bridging of each selenocyanato group to each of the 2 Co(II) atoms. The infrared spectra of the Rh(III) and Ir(III) dimers suggest bonding to the selenium atom of the ligand, the chlorine atoms occupying *trans* positions, with the ligand retaining the *gauche* configuration. The dimers are bridged through chlorine atoms since reaction of these dimers with *p*-toluidine yields the trichloro-(1,2-diselenocyanatoethane)-*p*-toluidine-rhodium(III) monomer.

An interesting platinum complex, bis(triphenylphosphine) (carbon diselenide)platinum, has been detailed.[174]

Gould and Burlant[137] first reported the formation of 1,4-diselenane (XI) complexes of Cd(II) and Pd(II):

$$H_2C \overset{Se}{\underset{Se}{\big|\big|}} CH_2$$
$$H_2C \qquad CH_2$$

(XI)

(XII)   (XIII)

2-(2-pyridyl)benzo[b]selenophene   2-(4-pyridyl)benzo[b]selenophene

Later, Hendra and Sadasivan[150] made a detailed study of the coordination compounds of 1,4-diselenane with metal ions and their different structures. Hg(II) and Cd(II) complexes of (XI) of composition $MX_2(XI)$ have a bridged structure and the ligand retains its centrosymmetric configuration as it does in the iodine complexes. Cu(I), Ag(I), and possibly Cu(II) complexes of (XI), Cu(XI)Cl, Ag(XI)NO$_3$, and Cu(XI)Cl$_2$ contain polymeric chains of alternating 1,4-diselenane molecules having a chair configuration. The Pd(II) and Pt(II) complexes, $M(XI)_2Cl_2$, M = Pd or Pt may have either cyclic coordination of the ligand, or the ligand can have a chair configuration acting as a bridge between 2 Pd atoms.

The stated opinions concerning the structures of these compounds are based almost exclusively upon infrared spectral data. It is necessary to exercise the proper degree of caution in the interpretation of infrared spectra, especially on polycrystalline solids, as they relate to the actual bonding in compounds. X-ray studies of single crystals of these compounds would provide a proper basis upon which to base conclusions concerning their structures.

Bark and Brandon[30] investigated the possible use of (XII) and its isomer, (XIII) for the selective determination of Pd(II). Pd(II) and Pt(II) add 1 mole of (XII) and 2 moles of (XIII) while Os(IV) forms a red colored complex of the composition $(XII)_2OsCl_2 \cdot 2H_2O$. The latter is presumed to have an ionic structure. The formation of these complexes and their structures have been explained on the basis of π bonding and the ease of ring formation in the stabilization of the resultant complexes. Thus, Os(IV), (0.67 Å), which has a smaller ionic size than that of Pd(II) (1.20 Å) does not tend to form stable ring systems. Hence, the Os(IV) complex has been stated to have an ionic structure. The tendency to form a π bond depends on two effects: (a) an increase in ionic size and (b) a decrease in the electronegativity. The formation of 1:2 complexes and N bonding in (XIII) results because the two effects, (a) and (b), are almost equal. Hence, there is observed no π bond formation between Pd and Se. However, in the case of the complexes with (XII), the π bond between Pd and Se is more stable because of an apparent increase in the electronegativity of selenium. This is the result of a delocalization of electron density around selenium in the resulting completely conjugated ring system. The structure of the complex with Pd(II) is shown (Figure 7-10).

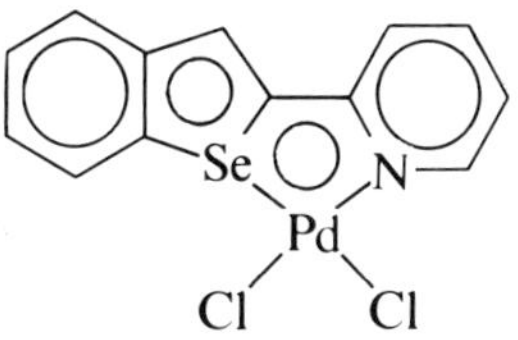

Figure 7-10   Structure of palladium complex of 2-(2-pyridyl)benzo[b]selenophene

Another class of bidentate ligands are the selenosemicarbazides. Jensen and Frederiksen[171] described the nickel compounds of 4-phenyl-selenosemicarbazide, $NH_2NHC(Se)NHPh$. Simple selenosemicarbazide, $NH_2CHC(Se)NH_2$, (XIV), forms addition compounds with Zn(II), Cu(II),[275] and Ni(II).[10] The selenosemicarbazides, as do the analogous sulfur compounds, can act as bidentate ligands by donation through both the Se and N atoms of the hydrazine group. In most instances, the complexes of (XIV) are isostructural with those of the thiosemicarbazides. Thus, the Zn(II) complex of (XIV) has a tetrahedral structure[275] and the Cu(II) complex has a ring structure:

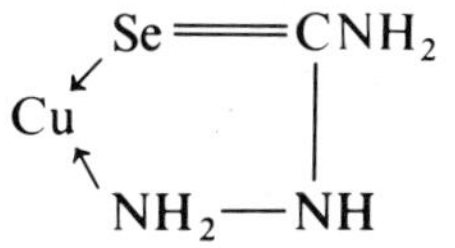

Cu(II) also forms a 1 : 2 complex which may either have a tetrahedral or an octahedral structure containing the bidentate selenosemicarbazide ligand. The addition products of (XIV) with Ni(II) salts resemble closely the analogous thiosemicarbazide complexes and have a square planar $[Ni(XIV)_2]^{+2}$ structure in the solid state. Aqueous solutions of these addition products are paramagnetic and six-coordinated complexes, $[Ni(XIV)_2]X_2$, could be isolated from the aqueous solutions. It was shown that an ammoniacal solution of the addition product precipitates a brown, diamagnetic, bis-(selenosemicarbazidato)Ni(II), with a loss of protons from each of the 2 (XIV) molecules. The protonation constant of selenosemicarbazide has been only recently determined[123] ($pK_1 = 0.80$) and shows the least basic behavior when compared with that of the corresponding sulfur and oxygen analogues. A comparison of the stability constants of complexes of (XIV) with Zn(II), Cd(II), and Hg(II) with that of the corresponding thiosemicarbazide and semicarbazide complexes reveals that Hg(II) behaves as typical class "b" metal with the order of stability being Se > S ≫ O. Cd(II) is weakly class "b" with the order being Se > S > O, and Zn(II) behaves as a class "a" metal with the order being O > S > Se. The stability constants of Hg(II) complexes of (XIV) are slightly higher than those of the selenourea complexes. This suggests different polarizabilities for Se in these ligands.

Recently it was shown[274] that the selenium analogue of "dithizone," 3-seleno-1,5-diphenylformazan (XVa and b) forms strongly colored compounds with several metal ions:

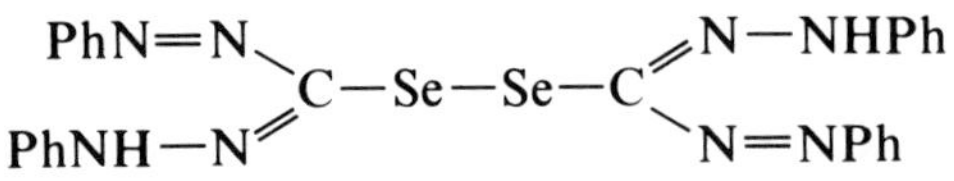

(XVI)

Interestingly enough, the oxidation product of (XV), the diselenide, (XVI) also extracts metal ions. In contrast, the corresponding sulfur compound readily undergoes auto-oxidation and reduction, to regenerate dithizone.

A new P—Se bidentate ligand, diphenyl(o-methylselenophenyl)phosphine, (XVII) and its complexes with Ni(II)[345] and Co(II)[91] have been described:

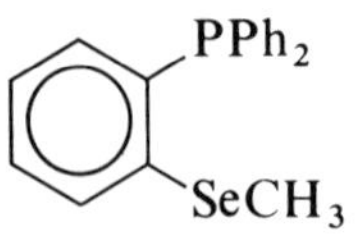

(XVII)

Nickel(II) forms four- and six-coordinate complexes, exemplified by [Ni(XVII)Cl$_2$] and [Ni(XVII)$_2$Cl$_2$], and five-coordinate complexes of the type [Ni(XVII)$_2$Br]ClO$_4$. The electronic absorption spectrum of the five-coordinate complex is consistent with a square pyramidal structure in which the bidentate ligands are on the square plane with the halide ion located at the apex of the pyramid. It was shown that highly specific steric effects are not necessary for the formation of five-coordinated Ni(II), and that it is sufficient that the ligand possess a strong π bonding tendency. It is interesting to note that intensities of the absorption bands which may be related to the covalent character of the Ni-ligand bond increase in the order S < Se < As, whereas the spectrochemical series is in the order As > S > Se. The red crystalline Co(II) complex, [Co(XVII)$_2$Br]ClO$_4$ has a low-spin, five-coordinated structure. Again, as in the case of Ni(II), Co(II) has a distorted square pyramidal structure with the halide ion at the apex. The electronic spectrum is consistent with the strong field C_{4v} symmetry and the "d" orbital energy splitting in such a symmetry is given in Figure 7-11. The charge-transfer

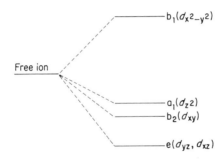

Figure 7-11 The d orbital energy level splitting in a strong ligand field of C_{4v} symmetry.

band at around 28,000 K is assigned to a $\sigma P \rightarrow b_1$ ligand-metal transition. The intensities of the d-d transitions are unusually high when compared with either tetrahedral or octahedral structures. This suggests that the ligand-metal bond has considerable covalent character. Also, in such five-coordinate compounds, the center of inversion found with C_{4v} symmetry is always absent. Again, in the Ni(II) and Co(II) complexes, the d-d bands are close to the charge-transfer bands and hence, borrow intensity from the charge-transfer bands.

Recently, Meek[218] reported an interesting demethylation reaction of coordinated (XVII) in the Ni(II) and Pd(II) complexes. Thus, when Ni(XVII)$_2$Cl$_2$ and Pd(XVII)$_2$(SCN)$_2$ are refluxed in butanol or in a butanol-dimethylformamide mixture, the demethylated complex **A** and a dimeric product **B** are formed:

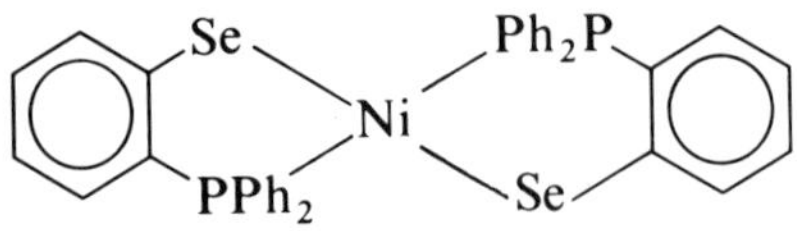

A 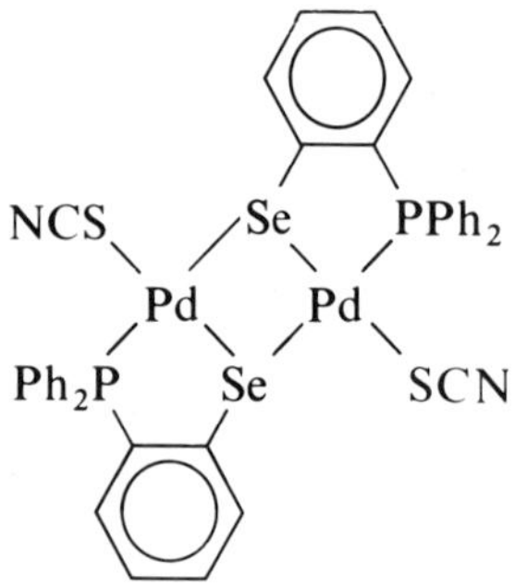   **B**

The Ni(II) complex, **A**, has a diamagnetic square-planar structure.   That structure **A** has a *trans* configuration that has been confirmed recently from x-ray crystallographic studies.[75]   A Pd(II) complex analogous to that of **A** could be prepared from $K_2Pd(NO_3)_4$.   To date, this represents the first example of an Se-demethylation reaction in a coordinated bidentate complex.

Polydentate ligands.   To date, only three selenium-containing neutral ligands have been shown to exhibit polydentate behavior.   The ligand, bis-[2-(α'-pyridyl)ethyl]selenide, bpes,

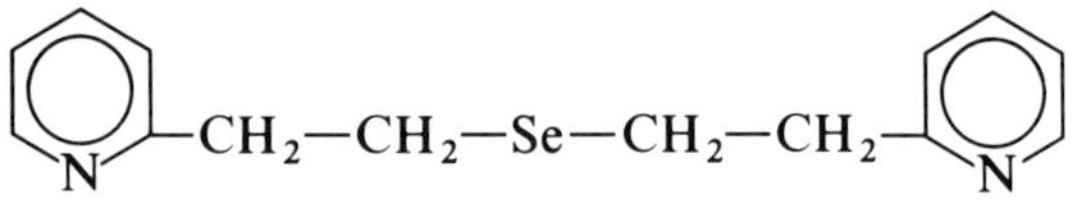

bpes

forms complexes with Cu(II) which have the composition $Cu(bpes)Cl_2$, $Cu(bpes)Br_2$, $Cu(bpes)(NO_3)_2$, $Cu(bpes)(ClO_4)_2$ and $Cu(bpes)Cl(ClO_4)$.[330] It was shown that bpes acts as a tridentate ligand by forming bonds with Cu(I) through 2 N atoms of the pyridyl group and selenium.   The Cu(bpes)Cl$_2$ behaves as a nonelectrolyte in acetonitrile while the bromide, nitrate and perchlorate complexes have greater conductivities.   A possible five-coordinate structure, Figure 7-12 for these complexes is proposed.   Again, it should be

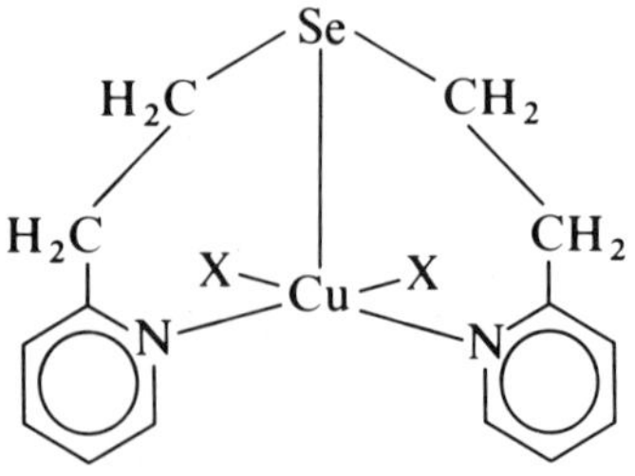

Figure 7-12   Possible structure of CuX(bpes); X = halogen.

noted that this structure is not based on meaningful structural data and it should be viewed with the proper degree of skepticism.

In the solid state, these complexes might possibly have a bridged structure such that six-coordination of Cu(II) is achieved.

Another interesting terdentate ligand, diacetyl selenosemicarbazone oxime, and its Cu(II) and Ni(II) complexes have been reported.[11, 12]

$$CH_3C{=}NOH$$
$$|$$
$$CH_3C{=}N{-}NH{-}C(Se)NH_2$$

diacetyl selenosemicarbazone oxime H_2dSeO

It was observed that Cu(II) is not reduced by this ligand and that it is iso-structural with the corresponding thio complexes. The ligand has been reported to possess a nearly planar structure and to be joined to the central metal ion through the selenium atom and the nitrogen atoms of the oxime and hydrazine groups. The coordination number of Cu(II) in these complexes may either be four, five, or six.

A tetradentate ligand tmSep and its complexes with Ni(II) have been described.[92]

$$\left[\underset{}{\bigcirc}{\text{---}}SeCH_3 \right]_3 P$$

tris(o-methylselenophenyl)phosphine(tmSep)

The Ni(II) complexes, [Ni(tmSep)X](ClO$_4$), X = Br, Cl, NCS, have a trigonal bipyramidal structure in which 3 selenium atoms are located at the equatorial positions. The energy level diagram for C_{3v} symmetry is shown in Figure 7-13.

By analogy with other trigonal bipyramidal molecules, the electronic transitions are assigned to the two $^1A_1 \rightarrow {}^1E(D)$ transitions. The spectrochemical

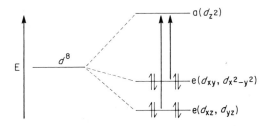

Figure 7-13 The d orbital energy level splitting in C_{3v} five-coordinate complexes.

series for different donor groups follows the order $R_3As > R_2S > R_2Se$. However, the intensities of the d-d transitions are high and show a different order, namely $As > Se > S$, as compared with the established spectrochemical series. A possible reason could involve a greater mixing of the $d_{x^2-y^2}$ and d_{xy} orbitals of the metal with the orbitals of e symmetry on the three equatorial ligands thereby making the Ni—Se bond more covalent. It has been suggested that selenoether donors are more polarized than analogous thioether donors when coordinated to metals.

Uninegative Ligands

Monodentate Ligands. Selenites, selenates, and seleninic acids, RSe(O)OH, have been investigated as ligands.[238, 308] However, in every case coordination to the metal takes place through the oxygen atom. Hence, further discussion is not included in this review.

Bidentate Ligands. Ligands of the type $[X{=}CY_2]^{-Z}$ where X is an electron withdrawing moiety, Y a sulfur or selenium atom, and Z the formal charge on the ligands, have been shown to form coordination compounds with metal ions. The section immediately following deals with two type of ligands: (1) those in which X is a dialkylamino group, and Y is a selenium atom and (2) those in which X is an alkoxy group and Y is a selenium atom:

N,N-dialkyldiselenocarbamate   O-alkyldiselenocarbonate

Dialkyldiselenocarbamates. Analogous sulfur compounds, the dialkyl-dithiocarbamates, have been extensively studied. However, the study of diselenocarbamates has been the subject of relatively recent interest.

Barnard and Woodbridge[31] prepared the first dialkyldiselenocarbamates by the reaction of a secondary amine with carbon diselenide. They also prepared Zn(II) dialkyldiselenocarbamate and Cu(II) diethyldiselenocarbamate. Jensen and Krishnan[173] extended this study to include diethyldiselenocarbamates (dsc) of Tl(I), Ni(II), Pd(II), Pr(II), Cd(II), In(III), Tl(III), Cr(III), Rh(III), nitrosylCo(II), and nitrosylFe(II). Some of these complexes have been independently prepared by Furlani *et al.*[117] Lorenz and Hoyer[215] prepared various dialkyldiselenocarbamates and some of their Cu(II) complexes. Also, Rosenbaum *et al.*,[276] in addition to the preparation of dialkyldiselenocarbamates, described the synthesis of P(III) and As(III) dialkyldiselenocarbamates.

The crystal structure of nickel diethyldiselenocarbamate was determined by Bonamico and Dessy.[38] This study revealed that Ni(dsec)$_2$ is planar, but unsymmetrical.   There are two longer Ni—Se(1) distances at 2.46 Å and two shorter Ni—Se(2) distances 2.28 Å.   This was explained in terms of the contribution of the form:

$$\text{Ni}\diagdown\underset{\text{Se(2)}}{\overset{\displaystyle \text{Se(1)}}{\diagup}}\text{C}-\text{N}\diagup$$

to the structure.   The [100] projection of the molecule is shown in Figure 7-14.

The metal diselenocarbamates closely resemble the dithiocarbamates in their physical properties.   Their electronic and infrared spectra are very similar.

The infrared spectra of dialkyldiselenocarbamate complexes have been

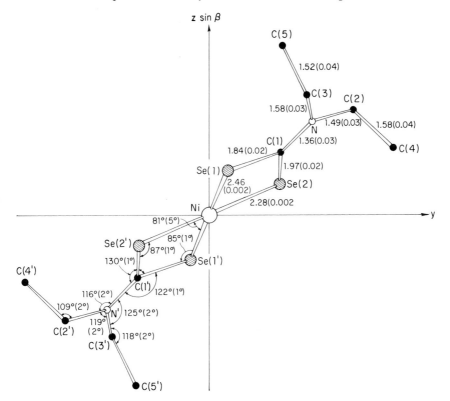

Figure 7-14   The [100] projection of Ni(dsc)$_2$ with bond lengths (in Å) and angles with standard deviations (in parentheses).

discussed recently by Jensen and Krishan.[175]   The observed infrared frequencies are consistent with a major contribution of the type $[R_2\overset{+}{N}=CSe_2^{-2}]^-$ to the structure.   The infrared stretching frequency which arises from $v(C=N)$ varies with the structures of the complexes.   Thus, the frequency at which $v(C=N)$ is observed decreases in the order planar (Ni, Cu, Pd, Pt) > tetrahedral (Cd, Zn) > octahedral (Cr, Co, Rh, In, Tl).   A similar trend is exhibited by the dithiocarbamates.[63]   Also, $v(CSe_2)$ undergoes interesting changes with changes in the stereochemistry of the complex. The observation of $v(C=N)$ at a higher frequency in the planar complexes is explained in terms of the donation of electrons from the ligand to a nonbonding orbital of the central metal atom.   This is similar to the argument proposed in the case of the dithiocarbamates.[73]   Durgaprasad et al.[90] have carried out a normal coordinate analysis of the dimethyl and diethyl-diselenocarbamate complexes of Ni(II) and assigned $v(C=Se)$ at 946 cm^{-1} and $v(Ni-Se)$ at 390 cm^{-1}, respectively.   It was later reported[175,176,178] that the experimentally observed frequencies are not quite consistent with the calculated frequencies in the case of dimethyldiselenocarbamate complex.

The electronic spectra of dsc complexes are very similar to those of the corresponding diethyldithiocarbamate (dtc) derivatives except that the differences, in wave numbers, at which corresponding bands are located occur at 2000–3000 K higher in transfer electron bands and 500–1000 K higher in the internal d^q transitions.   Furlani et al.[117] studied the electronic spectra of dsc complexes of transition metals and suggested that the position of dsc in the spectrochemical series of octahedral chromophores as Br$^-$ < dsep ⩽ Cl$^-$ < dtp < dsc < F$^-$ < urea < dtc < xanth < H$_2$O < NH$_3$.   They calculated the optical electronegativity of Se in dsc and reported a value between 2.6 and 2.7 which is essentially equal to that of S in dtc, 2.7.

Jensen et al.[177] discussed the electronic spectra of some of the dsc complexes and reported additional absorption bands.   In the case of the Cr(III) complex, they calculated the nephelauxetic ratio $\beta_{35} = 0.42$ and B = 390 K. These values are slightly larger than those of the dsep complex.   They argued that the very small spectrochemical difference between dsc and dtc is not unexpected and does not require any special explanation in terms of π back-donation between Se and the metal as proposed by Furlani et al.[117] It was pointed out that the smaller electronegativity difference between S and Se is consistent with the fact the larger electronegativity differences in the series B to F alternate in the subsequent series Al to Cl, Ga to Br and are almost negligible in Tl to At.

The Tl(I) compound is presumably an oligomer in solution and the brightly colored Tl(III) derivative is a sufficiently strong oxidizing agent that a shoulder at 20,000 K originates from an electron transfer band.

Some interesting results were reported by Cervone et al.[58] in their study of Fe(dsc)$_3$ and Fe(dsc)$_2$Cl.   The Fe(dsc)$_3$ is of a low-spin type($\mu_{eff} = 2.0–$

2.3 BM) and μ_{eff} increases with temperature. They found no evidence of either intermediate behavior (as in Fe complexes of dithiocarbamates) or an equilibrium with a high-spin form. The low-spin behavior was attributed to the balance between Δ and the spin-pairing energy. These factors, together with the strongly lower β values, tend to shift the threshold balance in favor of the low-spin type while the dithiocarbamate complexes lie very near the threshold for the magnetic cross-over $^6A_1 \leftrightarrow {}^2T_2$. The charge-transfer $t_{1u} \rightarrow 3d$ band occurs at slightly longer wavelengths compared with that of dtc complexes since the interelectronic repulsions are smaller in the selenium complex. The high quadrupole coupling constant calculated from Mössbauer spectra suggests an asymmetry in the occupancy of the valence shell as well as a geometrical asymmetry in the coordination environment. The Fe(dsc)$_2$Cl is square pyramidal and displays a strong field ground state configuration $(b_1)^2(e)^2(a_1)^1$. It has been suggested that selenium ligands appear to be more effective in stabilizing five-coordination than sulfur ligands. The esr spectrum of Cu(dsec)$_2$ has been studied.[210] In the crystalline state it has singlet of asymmetric form and in solution, has a g value of 2.022.

A quadrivalent Ni complex of dsc was reported[173] to form through the bromine oxidation of Ni(dsc)$_2$. It exhibits an intense charge-transfer band at 19,160 K. Later, it was shown by Brinkhoff et al.[39] that oxidation by bromine, or chlorine of dibutyl- or diethyldiselenocarbamates of Ni(II) yielded diamagnetic quadrivalent Ni complexes of the composition [NiL$_3$]X, where L is either diethyl- or dibutyldiselenocarbamate and X is either Cl or Br. The ligand field splitting parameter, Δ, and the Racah parameter, B, have been calculated for low-spin d^6 octahedral Ni(IV) complexes. However, Jorgensen[184] has questioned the interpretation of their spectral data. The bands reported in the region 17–20,000 K, which they attribute to d-d transitions, are of the electron transfer type Se $\rightarrow$ Ni($3d$) and are not analogous to those in octahedral Co(III) complex which arise from a $A_{1g} \rightarrow T_{1g}$ transition.

Recently, complexes of Sn(IV) with the mixed sulfur-selenium ligand, N, N-dimethylthioselenocarbamate, $(CH_3)_2NCSSe^{-1}$, were reported by Kamitani and Tanaka.[185] The Sn(IV) complexes, namely, $(CH_3)_2SnL_2$(I) and $(CH_3)_2ClSnL$(II) closely resemble the analogous dithiocarbamate complexes in their spectral properties. (I) has a distorted octrahedral trans configuration and (II), a trigonal bipyramidal one, with both (CH$_3$) groups being located in equatorial positions. The infrared spectra of these complexes indicate that the structure is probably as shown in Figure 7-15.

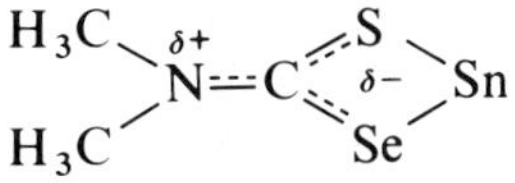

Figure 7-15   Probable structure of dimethylthioselenocarbamate complex of tin(IV).

The benzene-induced solvent shifts observed from the NMR spectra show that the CH_3 protons *trans* to Se are subject to the magnetic anistropy of benzene more than the other CH_3 protons.

The benzene molecule may interact stereospecifically with the electron-deficient nitrogen atom in order to avoid the more negative part of the molecule. As a result, it was concluded that the Sn—Se bond is more ionic than the Sn—S bond.

O-alkyldiselenocarbonates.   Rosenbaum *et al.*,[276] by the reaction of alcohols with CSe_2, prepared various O-alkyldiselenocarbonates, $ROCSe_2^-$, where R is either ethyl, *n*-propyl, isopropyl, *n*-butyl, *sec*-butyl, isoamyl, *n*-hexyl, *n*-octyl, *n*-nonyl, *n*-decyl, or *n*-dodecyl and showed that O-dodecyl-diselenocarbonate forms colored precipitates with several transition metal ions.   P(III) and As(III) form compounds with O-alkyldiselenocarbonates of the type $(ROCSe_2)_3M$ where M is either P or As.   The low-spin, square-planar complex, $Ni(EtOCSe_2)_2$, was investigated by Jensen and Krishnan[173] and it was shown that pyridine could be added to form yellow, high-spin, hexacoordinated, $Ni(EtOCSe_2)_2py_2$.   This behavior is very similar to that of the analogous selenophosphate compound.

Selenoketones.   2,4-Pentadione forms a number of complexes with several metal ions.   However, the corresponding sulfur and selenium complexes have only been recently studied.   Heath *et al.*[147,148] reported a red-brown, low-spin, square-planar complex of nickel(II), bis(4-pent-3-ene-2-selenoxo) nickel(II), $Ni(C_5H_7Se_2)_2$, from the reaction of H_2Se with nickel carbonate and 2,4-pentadione.   A tentative mechanism for this reaction has been proposed.[33]   The corresponding reaction with Co(II) chloride resulted in the formation of a tetrachlorocobaltate(II) salt of 3,5-dimethyl, 1,2-diselenolium cation:

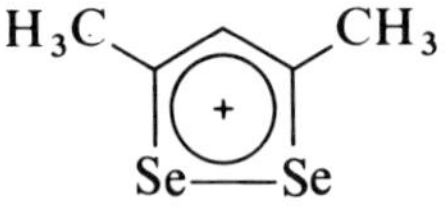

The infrared spectrum of the Ni(II) complex showed some interesting features concerning the position of the C—C, C—H in-plane, and C—Se stretching vibrations.   These results have been confirmed by normal coordinate analysis and the frequency shifts are attributed to mass differences rather than changes in the force constants or in the stereochemistry.

Quinoline-8-selenol.   Quninoline-8-selenol,

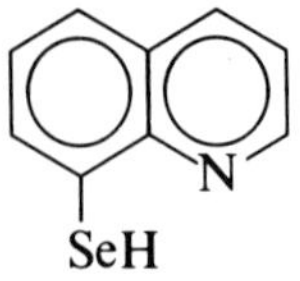

forms 1:1 complexes with zinc(II), cadmium(II), and lead(II) and their formation constants have been measured.[298]   The pK_a of the neutral ion

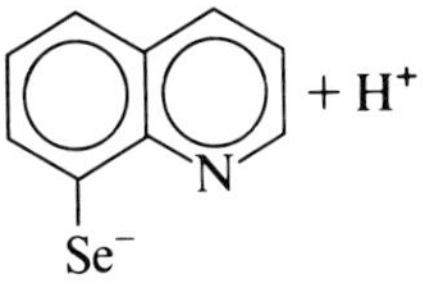

and the zwitterion

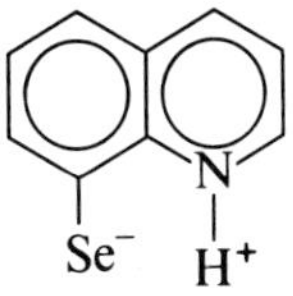

are 1.90 and 8.50, respectively.   The remarkable reduction in the pK_1 values in Cd (0.23) and Pb (0.40) suggests a neutralization of the selenium anion which results in a greater decrease in the basicity of the nitrogen atom.   The values of the formation constants were compared with those of the analogous sulfur and oxygen compounds and it was found that the values for the Zn(II) and Cd(II) chelates fall between those of the O and S analogues while the value for the Pb(II) chelate is smaller than that of the O or S analogues.   It was also observed that the sum of the protonation constants (pK_{a_1} + pK_{a_2}) of selenoxine (8.62) is less than the value for thiooxine (10.94) or oxine (15.51). The cadmium and lead complexes can be transformed to unusual protonated, MHL^{+2} species.   Selenoxine is not of any special analytical interest since it does not offer any unusual type of metal ion selectivity.   The protonation constants of selenoxine complexes show that they will form at pH values 0.6–1.0 unit lower than the thiooxine complexes, and there occurs a greater change in going from oxine to thiooxine.

Phenylselenoacetic Acids.   Pettit and co-workers[261, 262] prepared a series of substituted phenylselenoacetic acids, HL:

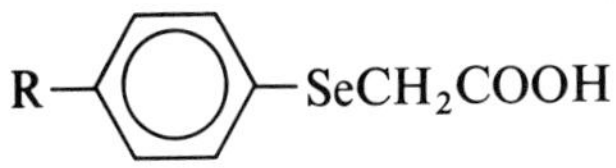

Stability constants were measured for the complexes formed with silver: AgL, AgLH$^+$ and AgL$_2^-$.   The stabilities of these complexes were compared with those of the corresponding sulfur and oxygen analogues.   The values of log K_{AgL} were 0.58, 2.83 and 3.71 for the oxygen, sulfur and selenium derivatives, respectively.   The trend in the stabilities was attributed to the increasing "softness" of the ligand.   Thus, the stability of the complex formed with the "soft" acid, silver, was correspondingly enhanced.   In this same study it was found that the differences, $\log K_{x=Se} - \log K_{x=S}$ for comparable ligands (omitting compounds with *ortho* substituents) were remarkably constant, being 0.83 ± 0.04 for the AgL and 0.65 ± 0.05 for the AgHL complexes.

The dissociation of the silver complexes,

$$\text{\Large$>$}X{-}Ag^+ \longrightarrow \text{\Large$>$}X + Ag^+ \qquad X = O, S, Se$$

was considered analogous to the acid dissociation of phenols,

$$-O{-}H \longrightarrow -O^- + H^+$$

Plots of the measured stability constants of metal complexes with ligands bearing *meta* and *para* substituents were plotted against Hammett functions for those substituents.   Good straight lines were obtained for both the protonated and unprotonated (AgHL and AgL) silver complexes of both sulfur and selenium ligands.   These investigators concluded that inductive and resonance effects which influence the phenolic proton dissociations influence the sulfur-silver and selenium-silver dissociations in exactly the same way.   They also conclude from this observation that the introduction of the concept of back-donation of electrons and d_π-d_π bonding is unnecessary since these could not be operative in the case of phenols.

Pettit, Sherrington and Whewell[262] made the generalized observation that with ligands which bear heavy donor atoms, such as selenium and sulfur, silver readily forms tetrahedral complexes.   On the other hand, with first period donors (oxygen or nitrogen), linear complexes are formed with silver. They point out the great similarities in the behavior of phenylthio- and phenyl-selenoacetic acids as donors toward silver.   Thus, a plot of $\log K_{AgL(S)}$ *vs* $\log K_{AgL(Se)}$ yields a remarkably good straight line.   On the basis of values of the thermodynamic stability constants, these workers conclude that selenium is definitely " softer " or more class " b " than sulfur.   *However, that selenium is larger and more polarizable than sulfur is a well known fact.   The need to invoke the newer and more fashionable terminology is of questionable value* (italics authors').   Pettit and co-workers also point out that this does not necessarily infer that selenium is more able to π bond via its d-orbitals.

The potential use of 2-selenophene aldoxime:

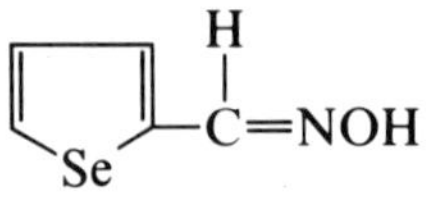

as a gravimetric reagent for Pd(II) has been described.[29]   A spectroscopic examination of a series of metal complexes of 2,5-dibenzoyl-3,4-dihydroxy-selenophen:

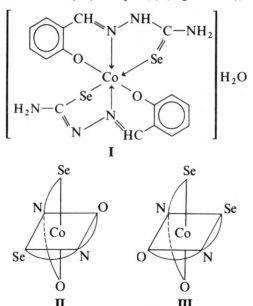

has been made.[26]   No mention of possible coordination sites was made in these studies.

Terdentate Ligands.   Ablov, Gerbeleu, and Romanov[9] have shown that salicylaldehyde–selenosemicarbazone (H_2SeSa) can be oxidized to give (salicylidineselenosemicarbazidato)$^{-1}$ (HSeSa) and (salicylidineselenosemi carbazidato)$^{-2}$ (SeSa) ions in the presence of metal ions to form a mixture of terdentate coordination compounds.   Thus, Co(III) and Cr(III) form complexes of the type:

[Co(SeSa)(HSeSa)]H_2O and [Cr(SeSa)(HSeSa)]H_2O while Fe(III) forms a pyridine adduct [Fe(SeSa)(HSeSa)]py.   These complexes are isostructural with the corresponding sulfur analogues and have a hexacoordinated structure.   In the octahedral Co(III) complex(I) (Figure 7-16), the ligand may

Figure 7-16   I: The structural formula of (salicylideneselenosemicarbazidatosalicylideneseleno-semicarbazidic)cobalt(III). II and III: Representation of enantiomorphs of complex I.

have a planar terdentate structure so that the spatial configuration of this complex can be represented as one of two (II or III) enantiomorphs. Addition of ammonia to these complexes results in the formation of ammonium salts, $(NH_4)_2[Co(SeSa)_2]$ and $(NH_4)_2[Fe(SeSa)_2]$ while the corresponding chromium complex precipitates selenium. All these complexes exhibit a behavior in their physical properties which is similar to that of the corresponding thiosemicarbazone complexes. However, the selenium compounds are found to be less stable than the sulfur compounds. The selenosemicarbazone does not replace dimethylglyoxime from the inner sphere of dioximato compounds of Co(III), unlike the salicylaldehydethiosemicarbazone complexes of Co(III).[8]

Another uni- and binegative terdentate ligand is diacetylselenosemicarbazone oxime, H_2dSeO. Neutral Cu(II) and Ni(II) complexes of this ligand have already been mentioned. It was shown that a hydrogen atom from the oxime group could be removed with the formation of a dark green crystalline Cu(II) complex, CuX(HdSeO) whose structure has been represented as:

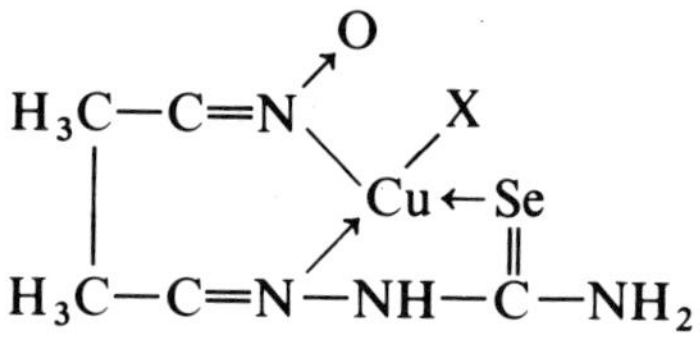
O
H3C—C=N
X
Cu←Se
H3C—C=N—NH—C—NH2

It was further observed that 2 hydrogen atoms from H_2dSeO could be removed, one from each oxime and imino group of the hydrazine with the formation of a dark brown Cu(II) complex, $Cu(dSeO)\cdot 1/2H_2O$. This complex has been represented either as a dimeric compound in which a water molecule joins the 2 planar Cu(dSeO) rings (Figure 7-17) or these groups could be packed in the crystal lattice so as to achieve a coordination number of five or six by sharing with the neighboring N or O atoms of residues.

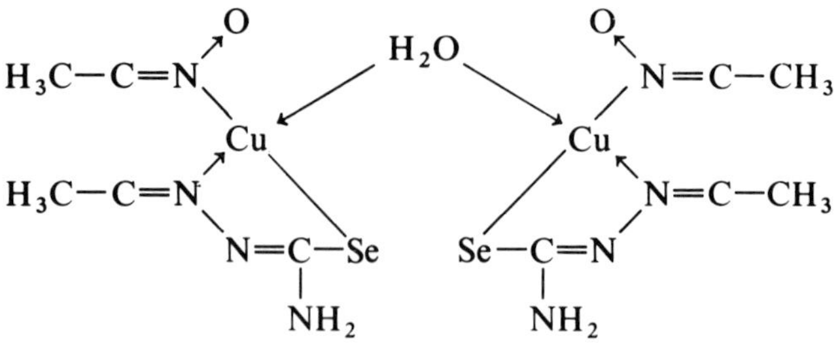
O
H2O
O
H3C—C=N
N=C—CH3
Cu
Cu
H3C—C=N
N=C—CH3
N=C—Se
Se—C=N
NH2
NH2

Figure 7-17 Schematic representation of (diacetylselenosemicarbazone oximato^{-2})copper(II), $Cu(dSeO)\cdot 1/2H_2O$.

However, the water molecule, to the best of our knowledge, has never been unequivocally shown to act as a bridging group. Hence, the structure suggested in Figure 7-17 must be viewed with some suspicion. The diacetylselenosemicarbazonemonooxime forms stable complexes with Cu(II) while salicylaldehydeselenosemicarbazone reduces Cu(II) to Cu(I).

Binegatively Charged Ligands

Bidentate Ligands.   In the previous section ligands, of the type $[X = CSe_2]^{-z}$ were discussed, in which X was either an R_2N or RO, and Z is 1. This section is concerned with ligands in which X is either Se, or $(CN)_2C$ or NCN. The 1,2-diselenolates and their coordination compounds are also discussed.

$$[CSe_3]^{-2}$$

triselenocarbonate

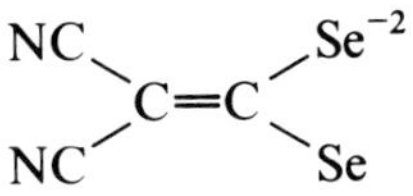

$$\underset{NC}{\overset{NC}{\diagdown}} C{=}C \underset{Se}{\overset{Se^{-2}}{\diagup}}$$

2,2-dicyanoethylene, 1,1-diselenolate

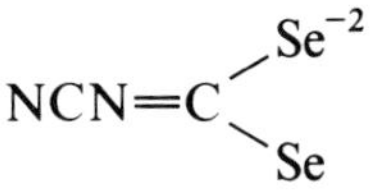

$$NCN{=}C \underset{Se}{\overset{Se^{-2}}{\diagup}}$$

cyanamidodiselenocarbonate

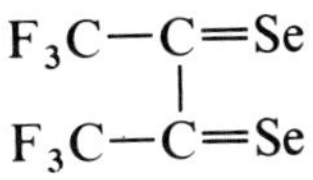

$$F_3C{-}C{=}Se$$
$$F_3C{-}\overset{|}{C}{=}Se$$

1,2-bis(trifluoromethyl)1,2-diselenolate

The triselenocarbonate ion, CSe_3^{-2} is rather unstable in solution. The barium salt has been prepared by the reaction shown below:[167, 119]

$$Ba(SeH)_2 + CSe_2 \longrightarrow BaCSe_3 + H_2Se$$

The barium salt is violet in color and its infrared spectrum shows (C—Se) asymmetric stretching frequencies at 817 and 786 cm^{-1} and a CSe_3 deformation at 420 cm^{-1}.[296] Seidel[297] reported the formation of CS_2Se^{-2}, $BaCSSe_2$, dark violet K_2CSe_3 and red K_2CS_2Se compounds and the infrared spectrum of CS_2Se^{-2}. Hoffmann-Bang and Rasmussen[156] observed that the monoselenidodithiocarbonate ion, $CSeS_2^{-2}$, reacts with mineral acids to produce, in addition to H_2Se and CS_2, small quantities of CSSe. The formation of CSSe was not, however, mentioned in Seidel's work.[297]

Gattow and Draeger[120] showed that the decomposition of H_2CSe_3 in aqueous solution is first order and calculated the dissociation constants K_1 (at 0° C) and K_2 (at 25° C) to be $6.9 \pm 4 \times 10^{-2}$ and $6.9 \pm 0.5 \times 10^{-8}$:

$$H_2CSe_3 + H_2O \rightleftharpoons HCSe_3^- + H_3O^+ \qquad K_1$$

$$HCSe_3^- + H_2O \rightleftharpoons CSe_3^{-2} + H_3O^+ \qquad K_2$$

Few thermodynamic constants have been calculated from the above work. Force constant[225] and MO calculations[226] for trichalcogenocarbonate ions have been reported. A simple molecular orbital diagram has been constructed for the trichalcogenocarbonate ion (Figure 7-18). The possible electronic transitions are listed and the observed absorption bands have been assigned. The longest wavelength absorption occurs at 46,100 K for CO_3^{-2}, 20,000 K for CS_3^{-2} and 15,800 K for CSe_3^{-2}. Hückel molecular orbital calculations on these ions are reported. The force constant calculations show that in the CO_3^{-2} ion there is a stronger p_π-p_π overlap, and the force constants of the bonds are (in mydynes/Å) 5.43 (CO_3^{-2}), 2.78 (CS_3^{-2}), 2.44 (CSe_3^{-2}).

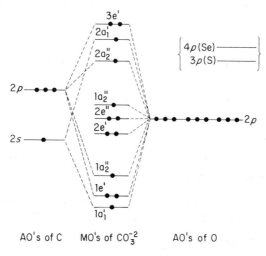

Figure 7-18   Qualitative MO diagram of trichalcogenocarbonate anion.

The 1,1- and 1,2-dithiols form an interesting series of coordination compounds. Most of these ligands stabilize planar coordination by involvement of the d_{xy}, d_{yz} and p_z valence orbitals of the metal ions. The delocalized π orbital system spans the entire network of the complex. The lack of axial perturbation in planar complexes contributes added stability resulting from the involvement of $4p_z$ orbitals in π-bonding. It was observed that the unpaired electron resides primarily within the ligand. This contributes some interesting features to the *esr* and electronic spectra of the complexes.

Few of these points are reflected in the corresponding selenium analogues and the lack of extensive study on selenium compounds precludes a better comparison with the sulfur analogues. However, some of the complexes which have been studied are the following.

1,1-diselenolates. It was shown by Jensen and Krishnan[177] that 2,2-cyano-ethylene-1,1-diselenol forms intensely colored coordination compounds with the Ni(II), Cu(II), Zn(II), Cd(II), Pt(II), Cr(III), Co(III), Rh(III), and Au(III) ions. The compounds have a stiochiometry which corresponds either to $[ML_2]^{-2}$ or $[ML_3]^{-3}$ where L is $[(NC)_2C=CSe_2]^{-2}$. A neutral, reddish-brown $(Ph_3P)_2NiL$ has also been reported. Spectral properties on these coordination compounds have been reported by the same researchers in a subsequent paper.[178] The infrared stretching frequency $v(C=C)$ in the free ligand at 1355 cm^{-1} is increased by 40 cm^{-1} in the case of Ni, Au, Zn, Cd, and Pt complexes and by about 15 cm^{-1} for the Cr, Co, and Rh complexes. This suggests different coordination numbers. However, no difference in the frequency shift was observed between planar and tetra-hedrally tetracoordinated complexes of divalent metals, like the diseleno-carbamates. A very large shift (~ 105 cm^{-1}) in $v(C=C)$ has been observed for the gold(III) complex. This has been explained in terms of the higher positive charge on the central atom which will cause an increase in the drift of electrons away from the carbon-carbon double bond and towards the selenium atoms. The electronic spectra of these complexes show intra-ligand transition L → L*, intramolecular charge transfer M → L*, and $d \leftrightarrow d$ transitions, very similar to those observed in the analogous 1,1-dithiolates.[341,138,183]

A reddish-brown, Ni(II) complex of cyanamidodiselenocarbonic acid (shown previously as L_2), has been isolated as the tetraphenyl phosphonium salt, $[Ph_4P]_2[Ni(L_2)_2]$ by Jensen and Krishnan[177]. The infrared spectrum of this compound showed $v(C\equiv N)$ and $v(C=N)$ of the free ligand at 1340 cm^{-1}. This suggests a diminished contribution of the diselenocarboxylate structure, $\bar{N}=C=N-C\begin{smallmatrix}Se\\^{-}\\Se\end{smallmatrix}$ in the complex. The electronic spectrum of this compound showed interligand, intramolecular charge-transfer, and *d-d* transitions.

1,2-diselenolates. The synthesis of novel heterocycle, bis(trifluoromethyl)-1,2-diseleneten, $F_3C-C=C-CF_3$, and its reaction with Ni(CO)$_4$ have
$$\overset{||}{Se-Se}$$
been described by Davison and Shawl.[82] A dark blue *cis*-1,2-difluoromethyl-ethylene-1,2-diselenato complex of nickel, $[(CF_3)_2C_2Se_2]_2Ni$, a light green, paramagnetic $[Ni(C_2F_6C_2Se_2)_2]^{-1}$, and orange-brown, diamagnetic $[Ni(C_2F_6C_2Se_2)_2]^{-2}$ were reported. The latter two were prepared by selective reduction at the dropping Hg electrode. The monoanion, $[Ni(C_2F_6C_2Se_2)_2]^{-1}$, showed a pronounced *g* value anisotropy $g_1 = 2.008$,

$g_2 = 2.116$, $g_3 = 2.193$. *Tris*-complexes of the type $[V(C_2F_6C_2Se_2)_3]^{-2}$ and $[V(C_2F_6C_2Se_2)_3]^{-1}$ have also been reported. It was found that $[Ni(C_2F_6C_2Se_2)_2]$ and $[V(C_2F_6C_2Se_2)_2]^{-1}$ are less stable than the corresponding sulfur complexes. The crystal structure of $[Mo(C_2F_6C_2Se_2)_3]$ has been determined[99] and it has been shown to possess a trigonal prismatic structure with distances as follows (Å): Mo—Se = 2.468, Se—C = 1.86, C—C = 1.30 and the angle, Se—Mo—Se, 83.1°.

Pierpont, Corden and Eisenberg[264] have studied coordination compounds of Ni, Cu, and Co with the mixed sulfur-selenium ligand, O-thiolatobenzeneselenolate (tbs). Only planar anions, green $[Ni(tbs)_2]^-$, blue $[Co(tbs)_2]^-$,

and green $[Cu(tbs)_2]^-$ were isolated. The polarographic half-wave potentials for the reduction of nickel, cobalt, and copper complexes, $[M(tbs)_2]^- + e^- \rightleftarrows [M(tbs)_2]^{-2}$ are found to be in good agreement with the half-wave potentials observed for the 1,2-dithiolate complexes. The *esr* spectrum of $[Ni(tbs)_2]^-$ showed pronounced g value anisotropy, similar to that observed for $[Ni(C_2F_6C_2Se_2)_2]^-$. The g tensors are little higher than the values observed for the corresponding sulfur analogues. This increase in the g tensor anisotropy has been attributed to a greater spin-orbit coupling in the case of the heavier donor atoms. The low field resonance of the unresolved ^{77}Se hyperfine splitting in the spectrum of $[Ni(tbs)_2]^-$ is suggestive of the presence of both *cis* and *trans* isomers of the anion. The electronic spectra of these complexes show close similarities to those of the analogous *bis*-dithiolate complexes.[25, 81, 343]

Tridentate Ligands. Yamasaki and Suzuki[347] and Suzuki and Yamasaki[311] studied glycolic acid derivatives of type $HOOCCH_2XCH_2COOH$, where $X = CH_2$, O, S, or Se. Both the acid dissociation constants of the ligands as well as the stability constants of the 1 : 1 complexes formed with Cu(II), Ni(II), Zn(II), Cd(II) and Pb(II) were measured. With these particular ligands, they reported the following relative affinities:

$$X =$$
$$O \quad Pb > Cu > Zn > Cd > Ni$$
$$S \quad Cu > Ni > Pb > Zn > Cd$$
$$Se \quad Cu > Pb > Ni > Cd > Zn$$

The values for the acid dissociation constants follow the predicted order, i.e., pK_1 and pK_2 for diglycolic, thiodiglycolic and selenodiglycolic acids increase in the order, $O < S < Se$. This coincides with the order of the decreasing electronegativities and increasing size of O, S, and Se atoms.

The larger stabilities measured for the complexes formed with various metal ions with the O-, S-, and Se-linked dicarboxylic acids, *cf.*, glutaric acid, was attributed to the chelate effect. The stabilities of the thio- and selenodiglycolic acid chelates were found to change in the same way that the second ionization potential varies from metal to metal. On the other hand, diglycolic acid chelates demonstrate an order of stability similar to that of the glutaric acid chelates. Because S and Se atoms are lower in electronegativity than O, this is taken as an indication that nonelectrovalent interaction makes a major contribution in the chelate formation of the S- and Se- linked acids. The results of this study show that the affinity of this group of metal ions towards Se is smaller than S, and the metals were classified into the following two groups according to Ahrland *et al.*:[13]

1) $O < S > Se$: Cu^{+2}, Ni^{+2}, Co^{+2}
2) $O > S > Se$: Pb^{+2}, Cd^{+2}, Zn^{+2}, Mn^{+2}

Suzuki, Nakano and Yamasaki[312] measured the stability constants of the chelates of *cis*-tetrahydroselenophene-2,5-dicarboxylic acid.

(TSD)

The observed order of stabilities was $Cu > Pb > Ni > Cd > Zn$. All of the TSD chelates were found to be more stable than those of selenodiglycolic acid. The greater stability of the TSD chelates was attributed to the greater pK values of this ligand.

Salicylaldehydeselenosemicarbazone may function either as a bidentate or tridentate ligand. This was discussed previously.

ORGANOMETALLIC COMPOUNDS

This section deals with organotransition and nontransition metal compounds. This group of compounds has a carbon-metal-selenium bond. The moiety containing the selenium atom may be Se, RSe- or an R_2Se- group. The organometallic compounds may be alkyl- or arylmetals, carbonyls, or cyclopentadienyl-metal derivatives. In most of these compounds selenium is bonded to the metal and in some cases the selenium, or organoselenium compound, bridges 2 metal atoms.

Carbonyls

The compounds, $[Fe(CO)_3Et_2Se]_2$[153] and $[Fe(NO)_2SeEt_2]_2$[154] have been reported and they have selenium-bridged structures.

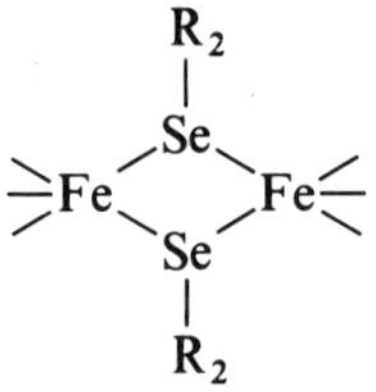

Chalcogen derivatives of iron carbonyl, $Fe_2(CO)_9X_2$ have been reported by Hieber and Gruber.[152] Compounds of the type $Fe_3(CO)_8LSe_2$ and $Fe_3(CO)_7L_2Se_2$ where L = Ph_3As, Bu_3P, or $(EtO)_3P$ have also been reported.[59] No structural data are available for these compounds. By analogy with the corresponding sulfur compounds,[79, 339] it was proposed that these compounds may have a distorted tetragonal pyramidal structure with the Fe atom located at the apex of the pyramid. Selenium-bridged complexes of iron carbonyls have been reported by Kostiner et al.[199] These include

$$[C_6F_5SeFe(CO)_3]_2, \quad [C_6H_5SeFe(CO)_3]_2,$$

$$C_6H_5SeFe_2(CO)_6SeC_6F_5, \quad C_6F_5SFe_2(CO)_6SeC_6F_5,$$

$$C_6H_5SFe_2(CO)_6SeC_6F_5 \quad \text{and} \quad C_6H_5SeFe_2(CO)_6SC_6F_5.$$

NMR and IR spectral data are given. The synthesis of compounds of the type $[Mn(CO)_4L]_2$, where L is MeSe-, EtSe-, PrSe-, CF_3Se-, C_2F_5Se-, and C_3F_7Se- have been reported.[340] They were prepared by the reaction of $Mn_2(CO)_{10}$ with the corresponding diselenides. The reaction of $Mn_2(CO)_{10}$ with $(CF_3)_2PSeCF_3$ yielded $Mn_2(CO)_8(CF_3)_2PSeCF_3$. Based on its spectral properties the following structure has been proposed (Figure 7-19).[144]

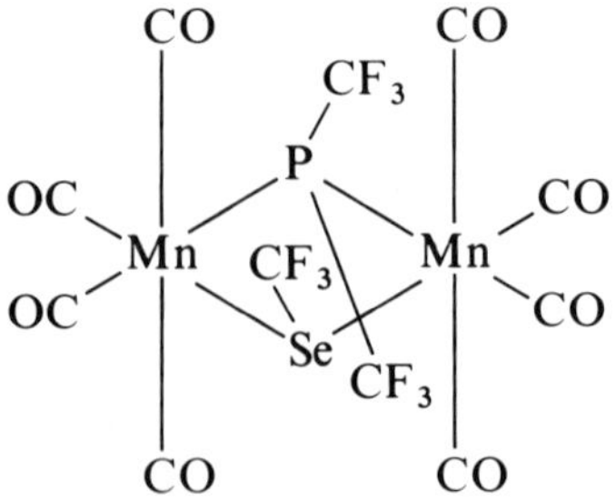

Figure 7-19   One of the limiting structures of $Mn_2(CO)_8(CF_3)_2PSeCF_3$.

It was shown[155] that rhenium pentacarbonylhalogenides, $Re(CO)_5X$, react with $R_2Se(L)$ where R = Et or Ph to produce either a disubstituted monomer, $Re(CO)_3L_2X$ or the monosubstituted dimer, $[Re(CO)_3LX]_2$. A halogen-bridged structure was proposed for the dimer. A technetium carbonyl derivative $[Tc(CO)_3SePh_2LCl]_2$ has also been reported. Abel et al.[1, 2, 3] investigated the cleavage mechanism of the Sn—Se bond in alkyl- or aryl-selenotin compounds by organometallic halides. Thus, the reaction of alkyl- or arylselenotin compounds with bromopentacarbonyl manganese or chloropentacarbonyl rhenium gave a dimer, $[M(CO)_4SeR]_2$ where M = Mn or Re, R = Me, Et, or Ph. The corresponding reaction with sulfur compounds gave a tricarbonyl derivative. This fact is explained on the basis of the initial fission of the Sn—Se bond followed by the displacement of carbonyl groups to give a tetracarbonyl derivative. The tetracarbonyl derivatives have an alkylseleno-bridged structure.

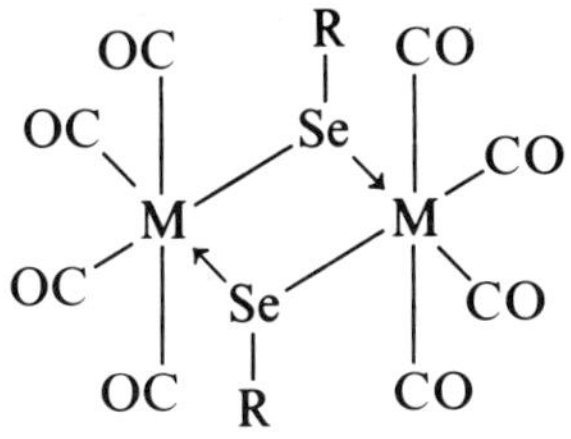

A tricarbonyl derivative, $[M(CO)_3SeR]_n$ was obtained by the pyrolysis of the tetracarbonyl derivative. The structure of the tricarbonyl derivative may involve a cluster of metal atoms which results in the formation of metal-metal bonds. Some phosphine-substituted complexes, $[Ph_3PM(CO)_3SePh]_2$, where M = Mn or Re, have been reported. The manganese and rhenium compounds have been assigned the following structures based on the number of $v(C—O)$ stretching frequencies in the infrared spectrum.

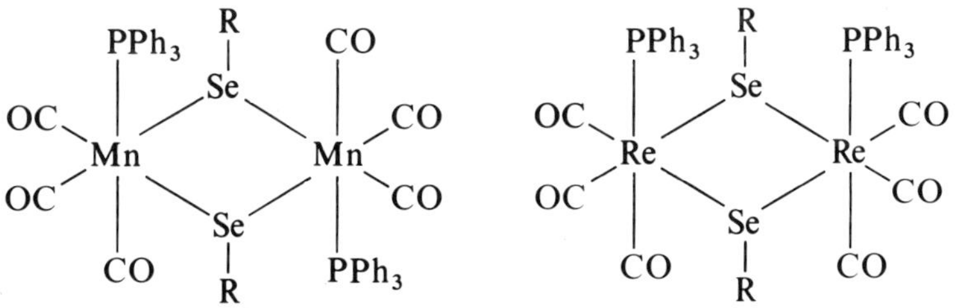

In addition to the above-mentioned compounds, some diphosphine-substituted complexes, $diphosMn_2(CO)_6(SePh)_2$ and $diphosRe(CO)_3SePh$, where diphos = 1,2-bis(diphenylphosphino)ethane have been reported.

Hsieh et al.[160] have shown that alkyl or aryl selenides undergo oxidative addition to Vaska's compound to give derivatives of the type $IrHCl(CO)$

$(PPh_3)_2SeR$. These addition compounds, as shown by IR and NMR data, have Ir—H and Ir—Se bonds. Fischer and Kiener[107] studied the reaction of metal carbene complexes, $(CO)_5MC(OMe)Me$ where M = Cr, Mo, or W with PhSeH and obtained compounds whose structure they represented as: $(CO)_3CrSe(CHMeOMe)-C_6H_5$. A dialkyl selenide of a substituted metal carbonyl, $Ph_2SeMo(CO)_3phen$, where phen = o-phenanthroline was isolated.[159]   Later, Abel and Hutson[6] reported the formation of 2,5-di-selenohexane, $C_2H_4(SeCH_3)_2$, complexes of manganese, rhenium and chromium carbonyls.   The tricarbonyl derivatives (I) were prepared from pentacarbonyl halides and diselenohexane, while the tetracarbonyl derivatives (II) were obtained by the displacement of heptadiene by selenohexane in the bicyclo-[2,2,1]-heptadienechromium tetracarbonyl:

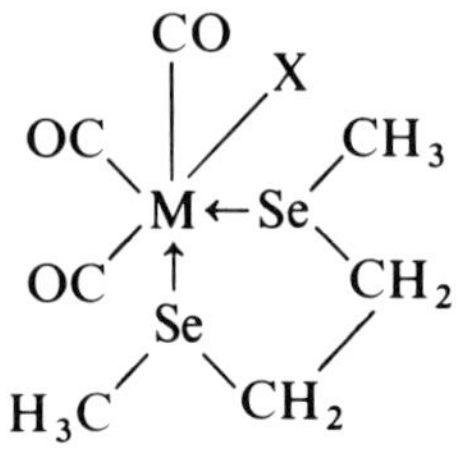

(I) 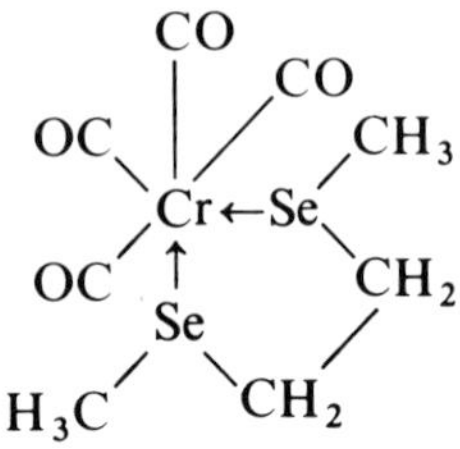   (II)

A detailed infrared analysis of the $v(C—O)$ stretching frequencies in various S, Se and Te bridged-metal carbonyl complexes has been given.[8]

Cyclopentadienyls

The replacement of halide groups in the cyclopentadienyltitanium dihalides[194] and in cyclopentadienylzirconium dichloride[197] by PhSe or SeEt groups has been demonstrated.   A bridged structure was proposed for the zirconium compound:

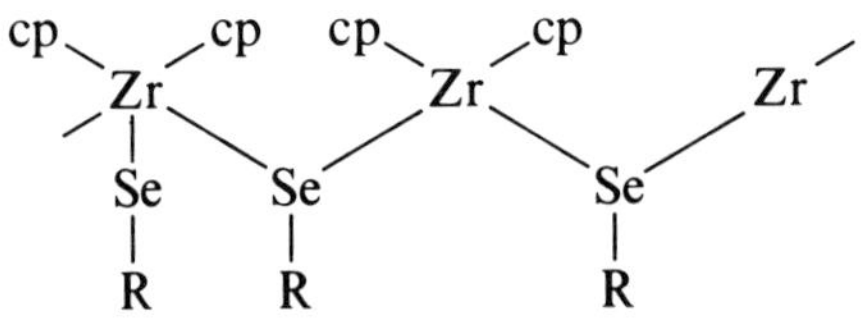

The titanium compound, $cp_2Ti(SePh)_2$, where cp = cyclopentadienyl, reacts further with molybdenum hexacarbonyl to give a selenium-bridged complex:[198]

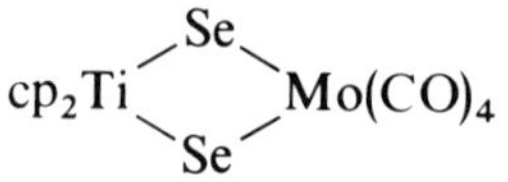

Koepf and Block[195] recently described an oxidative addition reaction of "titanocene." Thus, the dimer $[(C_5H_5)_2Ti]_2$ reacts with the diselenide, $(C_6H_5Se)_2$, to give an addition product, $(C_5H_5)_2Ti(C_6H_5Se)_2$. An interesting ring compound, $cp_2Ti(Se)_5$ has been reported by Koepf et al.[196] The NMR spectrum of this compound showed temperature dependency and the following structure has been proposed:

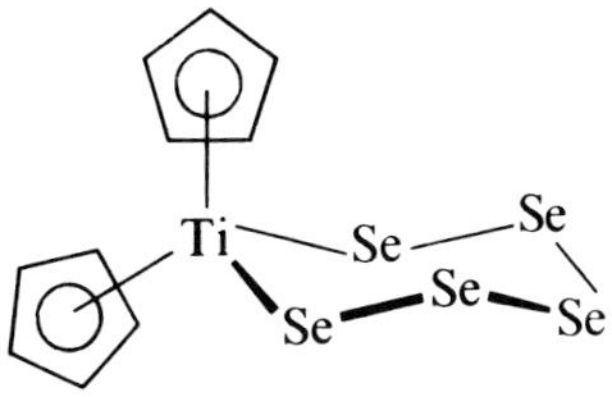

Tillay et al.[319] have shown that π-cpMo(CO)$_3$ reacts with diphenyldiselenide (Ph—Se—Se—Ph) to give both mononuclear, violet π-cpMo(CO)$_3$SePh and a completely decarbonylated red-brown $[\pi$-cpMo(SePh)$_2]_2$ dimer.

Organometallic Nontransition Metal Compounds Bonded to Selenium

This class of compounds has a carbon to metal to selenium bond. A few compounds in which the metal is not bound to carbon will also be discussed. Spinelli and Dell'Erba[309] obtained PhSeHgPh and (PhSe)$_2$Hg from the reaction of PhSeH with Ph$_2$Hg. Vyazankin et al.[337,338] synthesized several butylselenides of metals, $(BuSe)_nM$, where M = Hg, Cd, Zn, Mg, Sb, and Bi, by the reaction of BuSeH with the dialkyl metals. Diethylberyllium was found to react with diethyl diselenide and diphenyl diselenide to form Be(SeEt)$_2$ and Be(SePh)$_2$:[70]

$$Et_2Be + Et_2Se_2 \longrightarrow Et_2Se + EtBeSeEt \begin{cases} \xrightarrow{1} Be(SeEt)_2 \\ \xrightarrow{2} Et(EtSe)Be \cdot py_2 \end{cases}$$

(1) vac 50°, $= Et_2Be$ (2) in pyridine

Both diethyl ether and dipyridyl have been reported to form addition compounds with Be(SePh)$_2$. The preparation of $(CF_3Se)_2Hg$[80,100] and its reaction with mercuric halides have been studied by Clase and Ebsworth.[69] In their study, they obtained solid halides, CF_3SeHgX (X = Cl, Br, I) and measured dissociation constants for the equilibrium:

$$(CF_3Se)_2Hg + HgX_2 \rightleftharpoons 2CF_3HgSeX$$

Two boron compounds containing selenium, a triphenylselenonium salt of the triboranat-8-anion, Ph$_3$SeB$_3$Hg,[19] and 3,5-diphenyl-1,2,4,3,5-triselenadiborolan (a)[286] have been studied:

$$Ph-B \underset{Se}{\overset{Se-Se}{\diagdown}} B-Ph$$

(a)

It was shown by Masthoff et al.[217] that the C—Ca bond in Ph_3CCaCl can be cleaved by PhSeH with the formation of PhSeCaCl.   A reaction mechanism for the cleavage has been proposed.

Group IVa elements form a variety of di- and triheteroorganometallic compounds containing selenium.   Schumann and Schmidt[292] have ably discussed the various synthetic routes and reaction mechanisms leading to the formation of such compounds.   In the present review, we limit ourselves to a few salient features about these compounds which have appeared in recent years.

The bond fission of molecular selenium by nucleophilic lithiotriphenyl-metal compounds, leads to the formation of reactive intermediates Ph_3MSeLi, M = Ge, Sn and Pb.[291,292,285]   These reactive intermediates have not been isolated.   However, they undergo reactions with triorganometal halides, to form either symmetrical or unsymmetrical organometallic selenoethers, $R_3M—Se—M'R_3$, where M or M' = Ge, Sn and Pb.   The hydrolytic stability of these compounds increases as the size of M increases and as the size of the bridging chalcogen atom increases, i.e., $(R_3M)_2Te > (R_3M)_2Se > (R_3M)_2S$.   The —Ge—Se—Ge— series of compounds is an exception to this rule.

Another class of compounds which contains the hydroselenide, —SeH, group; R_3MSeH, (M = Si, Ge, Sn), has been reported.[34,333,334,335,336,337] These compounds react with triorganometallanes to give either symmetrical or unsymmetrical seleno ethers:

$$R_3GeSeH + R_3GeH \longrightarrow H_2 + R_3GeSeGeR_3$$

$$R_3GeSeH + R_3SnH \longrightarrow H_2 + R_3GeSeSnR_3$$

The strengths of the hydride bonds in reactions with sulfur, selenium, and tellurium increase in the order SnH < GeH < SiH.   In the reaction, $R_3MH + X \rightarrow R_3MXH$, where X is a chalcogen, the reactivity decreases in the order S > Se > Te.   Egorochkin et al.[97] have shown, from proton chemical shifts (—SeH), that the tendency to form hydrogen bonds decreases in the order $Et_3SiSeH > Et_3GeSeH > BuSeH$.

Compounds of the type, $(Et_3M)_2Se$ where M = Si, Ge, or Sn have been reported.[34,334]   When H_2Se is bubbled through a neutral aqueous solution of Me_3SnBr, Kreigsmann and Hoffmann[200] obtained $(Me_2SnSe)_3$ and $(MeSn)_2Se_3$.   The molecular structure of gaseous disilylselenide, $(SiH_3)_2Se$, has been determined by the electron diffraction method.[18]   The molecule has C_{2v} symmetry and an angle of 96.6° from the apex atom.   The Si—Se distance is 2.273_8 Å.   NMR studies of mixed hydrides MH_3XH, where $M = Si$, Ge and $X = S$, Se, and Te have been studied by Glidewell, Rankin and Sheldrick.[122]   Drake and Riddle[87] have shown that the Ge—P and Ge—As bonds in germylphosphines and -arsines can be cleaved by H_2Se with the formation of $(GeH_3)_2Se$.   This reaction was followed by NMR spectroscopy.   As reaction intermediates, germylhydrogenselenide and the condensation products, $(GeH_3Se)_2GeH_2$, $(GeH_3Se)_3GeH$ and $(GeH_3Se)_4Se$ have been proposed.

Some mutual transformation reactions of $(Et_3M)_2X$ where $M = Si$, Ge, or Sn and $X = S$, Se, or Te have been studied:[337]

$$(Et_3M)_2X + 2Et_3M'H \longrightarrow 2Et_3MH + (Et_3M')_2X$$

$$(Et_3M)_2Te \xrightarrow[-Te]{+Se} (Et_3M)_2Se \xrightarrow[-Se]{+S} (Et_3M)_2S$$

$$\xrightarrow[-Te]{+S}$$

The reactivities of the hydrides in the first reaction above decrease in the order $Et_3SnH > Et_3GeH > Et_3SiH$ and the reactivities of the chalcogen atom in the second reaction decrease in the order $S > Se > Te$.

Forstner and Muetterties[111] reported the formation of $(RSi)_4Se_6$ where $R = Me$, Et.   These compounds have an adamantane type structure (b) consisting of tertiary silicon atoms and secondary selenium atoms:

(b)

It was shown that selenium can be alkylated or arylated when it is heated with tetrabutyltin or tetraphenyltin.   This reaction yielded, in addition to dibutyl or diphenyl selenide, $Ph_3SnSePh$ and a trimeric dibutyltin selenide.

```
      Bu   Bu
       \  /
        Sn
     Se /  \ Se
  Bu   |    |   Bu
   \   Sn   Sn  /
    Sn      Sn
   Bu   Se    Bu

       (c)
```

```
  Me      Se      Me
   \     /  \     /
    Ge       Ge
   Me |      | Me
      Se    Se
        \  /
         Ge
        /  \
      Me    Me

       (d)
```

```
    Me   Me
     \   /
      Si
   Se    Se
      Si
     /  \
   Me    Me

     (e)
```

⇌

```
      Me   Me
       \   /
        Si
     Se    Se
   Me |      | Me
    \ Si    Si /
   Me    Se    Me

       (f)
```

The dibutyltin selenide is trimeric and has a ring structure (c), but the symmetry of this molecule is not D_{3h}.[201] An acyclic trimeric structure for dimethylgermanium selenide (d)[282] and an equilibrium between dimeric and trimeric dimethylsilicon selenide (e and f) have been proposed.[283]

The vibrational and NMR spectra of the heteroorganic compounds have been studied.[200,293,280,94–6,201,3] The bonding between chalcogen atoms and the central metal ions is explained on the basis of d_π-p_π interaction. The interaction arises from the presence of empty d orbitals on the chalcogen atoms and the unshared electron pairs of the central atoms. This results in the d orbital expansion of the chalcogen atom. It is not necessary to invoke the concept of d orbital expansion in Se and Te to explain the bonding schemes in these compounds. However, d orbital expansion in S is a generally accepted conclusion. It was observed that there is a diminution of d_π-p_π interaction between Si or Ge and chalcogens and this follows the order $O > S > Se > Te$. The infrared stretching frequencies $v(M-C)$ show a regular variation with the electronegativities of the bonded chalcogen atoms. Schmidbauer and Rudisch[280] have interpreted their NMR data on $(Me_3Ge)_2Se$ and $(Me_3Sn)_2Se$, on the basis of anisotropy or dispersion effects rather than the hyperconjugative or d_π-p_π interaction effects.

MISCELLANEOUS COMPLEXES

β-Diketones of the selenophene series have been reported to form complexes with metal ions. These complexes do not contain Se—M bonds and hence they are not considered in any detail in this review. Most of these selenophene diketonates have been tested as reagents for solvent extraction. The spectral properties of the metal complexes closely resemble those of the metal-β-diketonates.

Metal selenites and selenates have been reported. An interesting platinum selenite, $H[Pt(SeO_3)(OH)_3 \cdot H_2O]$, was reported by Derbisher.[83] The selenite group in this compound is bonded to Pt via oxygen atoms in a cyclic manner, and occupies two coordination positions.

Acknowledgments   During the preparation of this review, the authors were engaged in research projects supported by the Robert A. Welch Foundation, the United States Atomic Energy Commission, and the Research Council of Texas A&M University. We express our appreciation to them for allowing us to devote sufficient time toward the preparation of this manuscript.

REFERENCES

 1. Abel, E. W., Crosse, B. C., and Hutson, G. V., *Chem. & Ind.* 238 (1966).
 2. Abel, E. W., Crosse, B. C., and Hutson, G. V., *J. Chem. Soc.* A, 2014 (1967).
 3. Abel, E. W., and Brady, D. B., *J. Organometal. Chem.* **11**, 145 (1968).
 4. Abel, E. W., Dalton, J., Paul, I., Smith, J. G., and Stone, F. G. A., *J. Chem. Soc.* A, 1203 (1968).
 5. Abel, E. W., Atkins, A. M., Crosse, B. C., and Hutson, G. V., *J. Chem. Soc.* A, 687 (1968).
 6. Abel, E. W., and Hutson, G. V., *J. Inorg. Nucl. Chem.* **31**, 3333 (1969).
 7. Ablov, A. V., and Shaffranskii, V. V., *Zh. Neorg. Khim.* **9**, 585 (1964).
 8. Ablov, A. V., Gerbeleu, N. V., and Romanov, A. M., *Zh. Neorg. Khim.* **13**, 787 (1968).
 9. Ablov, A. V., Gerbeleu, N. V., and Romanov, A. M., *Zh. Neorg. Khim.* **13**, 3024 (1968).
10. Ablov, A. V., Gerbeleu, N. V., and Romanov, A. M., *Zh. Neorg. Khim.* **14**, 981 (1969).
11. Ablov, A. V., Gerbeleu, N. V., and Negryatse, N. Ya., *Zh. Neorg. Khim.* **14**, 986 (1969).
12. Ablov, A. V., Gerbeleu, N. Y., and Negryatse, N. Ya., *ibid.*, **15**, 119 (1970).
13. Ahrland, S., Chatt, J., and Davies, N. R., *Quart. Rev.* **12**, 265 (1958).
14. Alasaniya, R. M., Tsintsadze, G. V., and Skopenko, V. V., *Tr. Gruz. Politekh. Inst.* 28 (1968).
15. Alasaniya, R. M., Skopenko, V. V., and Tsintsadze, G. V., *Tr. Gruz. Politekh. Inst.* 21 (1967).
16. Allkins, J. R., and Hendra, P. J., *J. Chem. Soc.* A, 1325 (1967).
17. Allkins, J. R., and Hendra, P. J., *Spectrochim. Acta.* **24A**, 1305 (1968).
18. Almenningen, A., Fernholt, L., and Seip, H. M., *Acta Chem. Scand.* **22**, 51 (1968).
19. Amberger, E., and Gut, E., *Chem. Ber.*, **101**, 1200 (1968).
20. Aynsley, E. E., Greenwood, N. N., and Leach, J. B., *Chem. & Ind.* 379 (1966).
21. Aynsley, E. E., Greenwood, N. N., Hunter, G., and Sprague, H. J., *J. Chem. Soc.* 1344 (1966).
22. Aynsley, E. E., Greenwood, N. N., and Sprague, M. J., *ibid.*, 2395 (1965).

23. Baker, K. L., and Fowles, G. W. A., *J. Chem. Soc.* A 801 (1968).
24. Baker, Jr., W. A., and Bobonich, H. M., *Inorg. Chem.* 3, 1182 (1964).
25. Baker-Hawkes, M. J., Billig, E., and Gray, H. B., *J. Amer. Chem. Soc.* 88, 4870 (1966).
26. Balenovic-Solter, A., Tomaskovic, M., and Stefanac, Z., *Mikrochim. Acta.* 344 (1968).
27. Bannister, E., and Cotton, F. A., *J. Chem. Soc.* 1959 (1960).
28. Barcza, L., and Sommer, L., *Z. Anal. Chem.* 192, 304 (1963).
29. Bark, L. S., and Griffin, D., *Analyst*, 92, 162 (1967).
30. Bark, L. S., and Brandon, D., *Talanta*, 14, 759 (1967).
31. Barnard, D., and Woodbridge, D. T., *J. Chem. Soc.* 2922 (1961).
32. Barnes, J. C., and Day, P., *ibid.*, 3386 (1964).
33. Barraclough, C. G., Martin, R. L., and Stewart, I. M., *Aust. J. Chem.* 22, 891 (1969).
34. Bochkarev, M. N., Sanina, L. P., and Vyazankin, N. S., *Zh. Obshch. Khim.* 39, 135 (1969).
35. Bode, H., and Arnswald, W., *Z. Anal. Chem.* 185, 179 (1962).
36. Boehland, H., and Niemann, E., *Z. Chem.* 8, 191 (1968).
37. Boehland, H., and Malitzke, P., *Z. Anorg. Allg. Chem.* 350, 70 (1967).
38. Bonamico, M., and Dessy, G., *J. Chem. Soc.* D, 1114 (1967).
39. Brinkhoff, H. C., Cras, J. A., Steggerda, J. J., and Willemse, J., *Rec. Trav. Chim*, 88, 633 (1969).
40. Brodie, A. M., Rodley, G. A., and Wilkins, C. J., *J. Chem. Soc.* A, 2927 (1969).
41. Brown, D., Hill, J., and Clifton, E. F., *J. Less-Common Metals*, 20, 57 (1970).
42. Brusilovets, A. I., Skopenko, V. V., and Tsintsadze, G. V., *Zh. Neorg. Khim.* 14, 467 (1969).
43. Buerger, K., and Goetze, U., *J. Organometal. Chem.* 10, 380 (1967).
44. Bunting, T. G., and Meloan, C. E., *Anal. Chem.* 40, 435 (1968).
45. Burger, K., and Pinter, B., *Magy. Kem. Poly.* 73, 209 (1967).
46. Burmeister, J. L., and Al-Janabi, M. Y., *Inorg. Chem.* 4, 962 (1965).
47. Burmeister, J. L., and Williams, L. E., *ibid.*, 5, 1113 (1966).
48. Burmeister, J. L., *Coord. Chem. Rev.* 1, 205 (1966).
49. Burmeister, J. L., and Williams, L. E., *J. Inorg. Nucl. Chem.* 29, 839 (1967).
50. Burmeister, J. L., and Gysling, H. J., *J. Chem. Soc.* D, 543 (1967).
51. Burmeister, J. L., and Gysling, H. J., *Inorg. Chim. Acta.* 1, 100 (1967).
52. Burmeister, J. L., *Coord. Chem. Rev.* 3, 225 (1968).
53. Burmeister, J. L., Gysling, H. J., and Lim, J. C., *J. Amer. Chem. Soc.* 91, 44 (1969).
54. Burmeister, J. L., and Lim, J. C., *J. Chem. Soc.* D, 1154 (1969).
55. Burmeister, J. L., Deardorff, E. A., VanDyke, C. E., *Inorg. Chem.* 8, 170 (1969).
56. Burmeister, J. L., and Patterson, S. D., *Inorg. Chim. Acta*, 3, 105 (1969).
57. Burmeister, J. L., Deardorff, E. A., Jensen, A., and Christiansen, V. H., *Inorg. Chem.* 9, 58 (1970).
58. Cervone, E., Camassei, F. D., Luciani, M. L., and Furlani, C., *J. Inorg. Nucl. Chem.* 31, 1101 (1969).
59. Cetini, G., Stanghellini, P. L., Rossetti, R., and Gambino, O., *J. Organometal. Chem.* 15, 373 (1968).
60. Chao, G. Y., and McCullough, J. D., *Acta Cryst.* 14, 940 (1961).

61. Chatt, J., Duncanson, L. A., and Venanzi, L M., *J. Chem. Soc.*, 4461 (1955).
62. Chatt, J., and Venanzi, L. M., *J. Chem. Soc.* 2787 (1955).
63. Chatt, J., Duncanson, L. A., and Venanzi, L. M., *Soumen Kemishleki* **1329**, 75 (1956).
64. Chatt, J. and Venanzi, L. M., *J. Chem. Soc.* 2351 (1957).
65. Chatt, J., Gamlen, G. A., and Orgel, L. E., *ibid.*, 1047 (1959).
66. Chatt, J. and Westland, A. D., *J. Chem. Soc.* A 88 (1968).
67. Chernik, L. C., Pedley, I. B., and Skinner, H. A., *ibid.*, 1851 (1957).
68. Chiesi, A., Grossini, G., Nardelli, M., and Vidoni, M. E., *J. Chem. Soc.* D, 404 (1969).
69. Clase, H. J., and Ebsworth, E. A. V., *J. Chem. Soc.* 940 (1965).
70. Coates, G. E., and Fishwick, A. H., *J. Chem. Soc.* A, 635 (1968).
71. Coates, G. E., *J. Chem. Soc.* 2003 (1951).
72. Condray, B. R., Zingaro, R. A., and Kudchadker, M. V., *Inorg. Chim. Acta.* **2**, 309 (1968).
73. Cotton, F. A., and McCleverty, J. A., *Inorg. Chem.* **3**, 1398 (1964).
74. Cotton, F. A., Goodgame, D. M. L., Goodgame, M., and Haas, T. E., *ibid.*, **1**, 565 (1962).
75. Curran, R., Cunningham, J. A., and Eisenberg, R., *Inorg. Chem.* **9**, 2749 (1970).
76. Czakis-Sulikowska, D. M., *Roczniki Chem.* **40**, 1821.
77. Czakis-Sulikowska, M., *Chem. Anal.* (Bucharest), **10**, 1189 (1965).
78. Czakis-Sulikowska, M., *Roczniki Chem.* **39**, 1161 (1965).
79. Dahl, L. F., and Sutton, P. W., *Inorg. Chem.* **2**, 1067 (1963).
80. Dale, J. W., Eméleus, H. J., and Haszeldine, R. N., *J. Chem. Soc.* 2939 (1958).
81. Davison, A., Howe, D. V., and Shawl, E. T., *Inorg. Chem.* **6**, 458 (1967).
82. Davison, A., and Shawl, E. T., *J. Chem. Soc.* D 670 (1967).
83. Derbisher, G. V., *Zh. Neorg. Khim.* **5**, 1441 (1960).
84. Diemann, E., and Mueller, A., *Inorg. Nucl. Chem. Lett.*, **5**, 339 (1969).
85. Domiano, P., Manfredotti, A. G., Grossoni, G., Nardelli, M., and Tani, M. E. V., *Acta. Cryst.* **B25**, 591 (1969).
86. Doyle, G., and Tobias, R. S., *Inorg. Chem.* **7**, 2479 (1968).
87. Drake, J. E., and Riddle, C., *J. Chem. Soc.* A 1573 (1969).
88. Dubois, T. D., and Meek, D. W., *Inorg. Chem.* **6**, 1395 (1967).
89. Duffy, N. V., and Kossel, F. G., *Inorg. Nucl. Chem. Lett.* **5**, 579 (1969).
90. Durgaprasad, G., Sathyanarayana, D. N., and Patel, C. C., *Can. J. Chem.* **47**, 639 (1969).
91. Dyer, G. and Meek, D. W., *J. Amer. Chem. Soc.* **89**, 3983 (1967).
92. Dyer, G., and Meek, D. W., *Inorg. Chem.* **6**, 149 (1967).
93. Ebsworth, E. A. V., and Mays, M. J., *J. Chem. Soc.* 3893 (1963).
94. Egorochkin, A. N., Vyazankin, N. S., Razuvaev, G. A., Kruglaya, O. V., and Bochkarev, M. N., *Dokl. Akad. Nauk. SSSR.* **170**, 333 (1966).
95. Egorochkin, A. N., Khorshev, S. Ya., Vyazankin, N. S., Bochkarev, M. N., Kruglaya, O. A., and Semchikov, G. S., *Zh. Obshch. Khim.* **37**, 2308 (1967).
96. Egorochkin, A. N., Vyazankin, N. S., Bochkarev, M. N., and Khorshev, S. Ya., *ibid.*, **37**, 1165 (1967).
97. Egorochkin, A. N., Vyazankin, N. S., Bochkarev, M. N., Bychkov, V. T., and Burov, A. I., *ibid.*, **38**, 396 (1968).
98. Egorochkin, A. N., Khorshev, S. Ya., and Vyazankin, N. S., *Dokl. Akad. Nauk. SSSR.* **185**, 353 (1969).

99. Eisenberg, R., personal communication.
100. Emeléus, H. J., and Welcman, N., *J. Chem. Soc.* 1268 (1963).
101. Englman, R., *Mol. Phys.* **3**, 48 (1962).
102. Eppler, R. A., *Chem. Rev.*, **61**, 523 (1962).
103. Ettore, R , Peloso, A , and Dolcetti, G., *Gazz. Chim. Ital.* **97**, 968 (1967).
104. Farago, M. E., and James, J. M., *Inorg. Chem.* **4**, 1706 (1965).
105. Farago, M. E., James, J. M., and Trew, V. C. G., *J. Chem. Soc.* A 728 (1967).
106. Farona, M. F., Frazee, L. M., and Bremer, N. J., *J. Organometal. Chem.* **19**, 225 (1969).
107. Fischer, E. O., and Keiner, V., *Angew. Chem. Intern. Ed. Engl.* **6**, 961 (1967).
108. Fluck, E., *The Nuclear Magnetic Resonance and its Application in Inorganic Chemistry*, Berlin, Springer Verlag, 1963, p. 268.
109. Forster, D., and Goodgame, D. M. L., *ibid.*, **4** 1712 (1965).
110. Forster, D., and Horrocks. Jr., W. De. W., *ibid.*, **6**, 339 (1967).
111. Forstner, J. A., and Muetterties, E. L., *Inorg. Chem.* **5**, 552 (1966).
112. Fritzmann, E., *Z Anorg. Chem.* **73**, 239 (1911).
113. Fritzmann, E., *ibid.*, **133**, 119 and 133 (1924).
114. Fritzmann, E., *J. Russ. Chem. Soc.* **47**, 588 (1924).
115. Fritzmann, E., and Krinitzke, V. V., *J. Appl. Chem. USSR*, **11**, 1610 (1939).
116. Furlani, C. and Tarantelli, T., *Inorg. Nucl. Chem. Lett.* **2**, 391 (1966).
117. Furlani, C., Cervone, E., and Camassei, F. D., *Inorg. Chem.* **7**, 4265 (1968).
118. Galsbol, F., *Nordisk Kemikermode*, **13**, 54 (1968).
119. Gattow, G., and Draeger, M., *Z. Anorg. Allg. Chem.* **348**, 229 (1967).
120. Gattow, G., and Draeger, M., *ibid.*, **349**, 202 (1967).
121. Gibbs, W., *Chem. Ber.*, **18**, 1089 (1885).
122. Glidewell, C., Rankin, D. W. H., and Sheldrick, G. M., *Trans. Faraday Soc.* **65**, 1409 (1969).
123. Goddard, D. R., Lodam, B. D., Ajayi, S. O., and Campbell, M. J., *ibid.*, 506 (1969).
124. Golub, A. M., and Pomerants, G. B., *Zh. Neorg. Khim.* **4**, 769 (1959).
125. Golub, A. M., *ibid.*, **4**, 1577 (1959).
126. Golub, A. M., and Skopenko, V. V., *ibid.*, **6**, 140 (1961).
127. Golub, A. M., Skopenko, V. V., and Pomerants, G. B., *ibid.*, **10**, 344 (1965).
128. Golub, A. M., and Skopenko, V. V., *ibid.*, **7**, 1012 (1962).
129. Golub, A. M., and Skopenko, V. V., *Usp. Khim.* 2098 (1965).
130. Golub, A. M., and Kalibabchuk, V. A., *Zh. Neorg. Khim.* **12**, 2370 (1967).
131. Golub, A. M., Pomerants, G. B., and Ivanova, S. A., *ibid.*, **14**, 2826 (1969).
132. Goodall, D. C., *J. Inorg. Nucl. Chem.* **30**, 1269 (1968).
133. Goodall, D. C., *ibid.*, **30**, 2483 (1968).
134. Goodall, D. C., *J. Chem. Soc.* A, 890 (1969).
135. Goubeau, J., *Angew. Chem.* **69**, 77 (1957).
136. Goubeau, J., and Sawodny, W., *ibid.*, **78**, 565 (1966).
137. Gould E. S., and Burlant, Wm., *J. Amer. Chem. Soc.* **78**, 5825 (1956).
138. Gray, H. B., *Prog. Transition Metal Chem.* **1**, 240 (1965).
139. Grecu, I., and Neamtu, M., *Farmacia* (Bucharest) **13**, 209 (1965).
140. Grecu, I., *Omagiu. Raluca. Ripan*, 263 (1966).
141. Greenfield, H., *Trans. New York Acad. Sci.*, Ser II **31**, 1038 (1969).
142. Greenwood, N. N., and Hunter, G., *J. Chem. Soc.* A, 1520 (1967).
143. Greenwood, N. N., and Hunter, G., *ibid.*, 929 (1969).
144. Grobe, J., *Z. Anorg. Allg. Chem.* **331**, 63 (1964).

145. Guenther, J. R., and Oswald, H. R., *Z. Naturforsch.* **B24** 1481 (1969).
146. Haq, M., and Samuel, R., *Proc. Indian. Acad. Sci.* **5A**, 423 (1937).
147. Heath, G. A., Steart, I. M., and Martin, R. L., *Inorg. Nucl. Chem. Lett.*, **5**, 169 (1969).
148. Heath, G. A., Martin, R. L., and Stewart, I. M., *Aust. J. Chem.* **22**, 83 (1969).
149. Hendra, P. J., and Sadasivan, N., *Spectrochim. Acta.* **21**, 1127 (1965).
150. Hendra, P. J., and Sadasivan, N., *J. Chem. Soc.* 2063 (1965).
151. Hendra, P. J., and Jovic, Z., *Spectrochim. Acta. Part* A, **24**, 1713 (1968).
152. Hieber, W., and Gruber, J., *Z. Anorg. Allg. Chem.* **296**, 91 (1958).
153. Hieber, W., and Beck, W., *ibid.*, **305**, 275 (1960).
154. Hieber, W., and Beck, W., *ibid.*, **305**, 265 (1960).
155. Hieber, W., Opavsky, W., and Rohm, W., *Chem. Ber.* **101**, 2244 (1968).
156. Hoffman-Bang, W., and Rasmussen, B. V., *Naturwissenschaften*, **52**, 660 (1965).
157. Hope, H., and McCullough, J. D., *Acta Cryst.* **15**, 806 (1962).
158. Hope, H., and McCullough, J. D., *ibid.*, **17**, 712 (1964).
159. Houk, L. W., and Dobson, G. R., *Inorg. Chem.* **5**, 2119 (1966).
160. Hsieh, M., Krishnan, V., Zingaro, R. A., *Int. J. Sulfur Chem.*, A (1), 197 (1971).
161. Hudson, R. F., *Structure and Mechanism in Organophosphorus Chemistry*, Academic Press, Inc., London, 1965.
162. Humffray, A. A., Bond, A. M., and Forrest, J. S., *J. Electroanalyt. Chem. Interfacial Electrochem.* **15**, 657 (1967).
163. Husebye, S., *Acta Chem. Scand.* **20**, 2007 (1966).
164. Husebye, S., *ibid.*, **23**, 1389 (1969).
165. Husebye, S., and Helland-Madsen, G., *ibid.*, **23**, 1398 (1969).
166. Ilina, L. A., Zemlyanski, N. I., Larimonov, S. V., and Chernaya, N. M., *Izv. Akad. Nauk. SSSR, Ser Khim.* **1**, 198 (1969).
167. Ives, D. J. G., Pittman, R. W., and Wardlaw, W., *J. Chem. Soc.*, 1080 (1947).
168. Jennings, M. A., and Wojcicki, A., *J. Organometal. Chem.* **14**, 231 (1968).
169. Jennings, M. A., and Wojcicki, A., *Inorg. Chim. Acta.* **3**, 335 (1969).
170. Jensen, K. A., *Z. Anorg. Allg. Chem.* **225**, 97 and 115 (1935).
171. Jensen, K. A., and Frederiksen, E., *ibid.*, **230**, 33 (1936).
172. Jensen, K. A., and Krishnan, V., *Acta. Chem. Scand.* **21**, 1988 (1967).
173. Jensen, K. A., and Krishnan, V., *ibid.*, **21**, 2904 (1967).
174. Jensen, K. A., *Proc. 10th Intl. Conf. Coord. Chem.*, 259 (1968).
175. Jensen, K. A., and Krishnan, V., *Acta Chem. Scand.* **24**, 743 (1970).
176. Jensen, K. A., and Krishnan, V., *ibid.*, **24**, 1088 (1970).
177. Jensen, K. A., and Krishnan, V., and Jorgensen, C. K., *ibid.*, **24**, 1090 (1970).
178. Jensen, K. A., and Krishnan, V., *ibid.*, **24**, 1092 (1970).
179. Jorgensen, C. K., *Mol. Phys.* **5**, 485 (1962).
180. Jorgensen, C. K., *J. Inorg. Nucl. Chem.* **24**, 1571 (1962).
181. Jorgensen, C. K., *Absorption Spectra and Chemical Bonding in Complexes*, Academic Press Inc., Oxford, 1962.
182. Jorgensen, C. K., *Inorg. Chem.* **3**, 1201 (1964).
183. Jorgensen, C. K., *Inorg. Chim. Acta Rev.* **2**, 65 (1968).
184. Jorgensen, C. K., personal communication.
185. Kamitani, T., and Tanaka, T., *Inorg. Nucl. Chem. Lett.* **6**, 91 (1970).
186. Kharitanov, Iu. Ia., Tsintsadze, G. V., and Porai-Koshits, M. A., *Dokl. Akad. Nauk. SSSR.*, **160**, 1351 (1965).
187. Kharitanov, Iu. Ia., and Skopenko, V. V., *Zh. Neorg. Khim.* **10**, 1803 (1965).

188. Kharitanov, Iu. Ia., and Tsintsadze, G. V., *ibid.*, **10**, 1191 (1965).
189. Kharitanov, Iu. Ia., and Tsintsadze, G. V., *ibid.*, **10**, 35 (1965).
190. Kimura, T., *Structure and Bonding*, **5**, 1 (1968).
191. King, M. G., and McQuillan, G. P., *J. Chem. Soc.* A 898 (1967).
192. Knobler, C., and McCullough, J. D., *Inorg. Chem.* **7**, 365 (1968).
193. Koenig, E., and Madeja, K., *ibid.*, **6**, 48 (1967).
194. Koepf, H., Block, B., and Schmidt, M., *Z. Naturforsch.* **22b**, 1077 (1967).
195. Koepf, H., and Block, B., *ibid.*, **23b**, 1536 (1968).
196. Koepf, H., Block, B., and Schmidt, M., *Chem. Ber.* **101**, 272 (1968).
197. Koepf, H., *J. Organometal. Chem.*, **14**, 353 (1968).
198. Koepf, H., and Raethlein, K. H., *Angew. Chem.* **81**, 1001 (1969).
199. Kostiner, E. S., Reddy, M. L. N., Urch, D. S., and Massey, A. G., *J. Organometal. Chem.* **15**, 383 (1968).
200. Kreigsmann, H., and Hoffmann, H., *Z. Chem.* **7**, 268 (1963).
201. Kreigsmann, H., Hoffmann, H., and Geissler, H., *Z. Anorg. Allg. Chem.* **359**, 58 (1968).
202. Krishnan, V., and Zingaro, R. A., *Inorg. Chem.* **8**, 2337 (1969).
203. Krishnan, V., and Zingaro, R. A., *J. Coord. Chem.*, **1**, 1 (1971).
204. Kuchen, W., and Knop, B., *Angew. Chem.* **76**, 496 (1964).
205. Kuchen, W., and Knop, B., *ibid.*, **77**, 259 (1965).
206. Kuchen, W., and Knop, B., *Chem. Ber.* **99**, 1663 (1966).
207. Kuchen, W., and Hertel, H., *Angew. Chem. Intern. Ed. Engl.* **8**, 89 (1969).
208. Kudchadker, M. V., Zingaro, R. A., and Irgolic, K. J., *Can. J. Chem.* **48**, 1415 (1968).
209. Lappert, M. F., and Pyszora, H., *Adv. Inorg. Chem. Radiochem.* **9**, 133 (1966).
210. Larin, G. M., Solozhenkin, P. M., Kopitsya, N. I., and Kirspuu, H., *Izv. Akad. Nauk. SSSR, Ser. Khim.* 968 (1969).
211. LeCalve, J. and Lascomb, J., *Spectrochim. Acta* **24A**, 736 (1968).
212. Lenher, V., and Fruehan, A. G., *J. Amer. Chem. Soc.* **49**, 3076 (1927).
213. Livingstone, S. E., *Quart. Rev.* **19**, 386 (1965).
214. Lodinzka, A., *Roczniki Chem.* **41**, 1007 (1967).
215. Lorenz, B. and Hoyer, E., *Z. Chem.* **8**, 230 (1968).
216. Maddox, H., and McCullough, J. D., *Inorg. Chem.* **5**, 522 (1966).
217. Masthoff, R., Guenther, K., and Vieroth, Ch., *Z. Anorg. Allg. Chem.* **364**, 316 (1969).
218. Meek, D. W., *Inorg. Nucl. Chem. Lett.* **5**, 235 (1969).
219. Melton, R. G., and Zingaro, R. A., *Can. J. Chem.* **46**, 1425 (1968).
220. Merritt, Jr., C., Hershenson, H. M., and Rogers, L. B., *Anal. Chem.* **25**, 572 (1953).
221. Michelson, K., *Acta Chem. Scand.* **17**, 1811 (1963).
222. Millefiori, S., and Foffani, A., *Tetrahedron*, **22**, 803 (1966).
223. Morelli, D., Segré, A., Ugo, R., La Monica, G., Cenini, S., Conti, F., and Bonati, F., *J. Chem. Soc.* D, 524 (1967).
224. Morgan, H. W., *J. Inorg. Nucl. Chem.* **61**, 367 (1961).
225. Mueller, A., Gattow, G., and Seidel, H., *Z. Anorg. Allg. Chem.* **347**, 24 (1966).
226. Mueller, A., Seidel, H., and Rittner, W., *Spectrochim. Acta* **23A** 1619 (1967).
227. Mueller, A., Krebs, B., Glemser, O., and Diemann, E., *Z. Naturforsch.* **22b**, 1235 (1967).
228. Mueller, A., Krebs, B., and Diemann, E., *Angew. Chem. Intern. Ed. Engl.* **6**, 257 (1967).

229. Mueller, A., Krebs, B., and Diemann, E., *ibid.*, **6**, 1081 (1967).
230. Mueller, A., Glemser, O., and Diemann, E., *Z. Anal. Chem.* **241**, 136 (1968).
231. Mueller, A., and Diemann, E., *Naturwissenschaften*, **55**, 560 (1968).
232. Mueller, A., Krebs, B., Kebabcioglu, R., Stockburger, M., and Glemser, O., *Spectrochim. Acta* **24A**, 1831 (1968).
233. Mueller, A., and Diemann, E., *Z. Chem.* **8**, 197 (1968).
234. Mueller, A., and Diemann, E., *Z. Naturforsch.* **B23**, 1605 (1968).
235. Mueller, A., Krebs, B., and Beyer, H. H., *ibid.*, **B23**, 1537 (1968).
236. Mueller, A., and Diemann, E., *ibid.*, **24b**, 353 (1969).
237. Mueller, A., Diemann, E., and Heidborn, U., *ibid.*, **B24**, 1482 (1969).
238. Myasoedov, B. F., Palshin, E. S., and Molochnikova, N. P., *Zh. Anal. Khim.* **23**, 66 (1968).
239. MacDiarmid, A. G., *Quart. Rev.* **10**, 208 (1956).
240. McFarlane, *J. Chem. Soc.* D 755 (1968).
241. McCullough, J. D., and Mulvey, D., *J. Phys. Chem.* **64**, 264 (1960).
242. McCullough, J. D., and Zimmerman, I. C., *ibid.*, **65**, 885 (1961).
243. McCullough, J. D., *Inorg. Chem.* **3**, 1425 (1964).
244. McCullough, J. D., and Brunner, A., *ibid.*, **6**, 1251 (1967).
245. Nelson, S. M., *Proc. Chem. Soc.* 372 (1961).
246. Nelson, S. M., and Shepherd, T. M., *Inorg. Chem.* **4**, 813 (1965).
247. Neuman, L., and Hume, D. N., *J. Amer. Chem. Soc.* **79**, 4581 (1957).
248. Nicpon, P., and Meek, D. W., *J. Chem. Soc.* D, 398 (1966).
249. Noble, A. M., and Winfield, J. M., *Inorg. Nucl. Chem. Lett.* **4**, 339 (1968).
250. Norbury, A. H., Ryder, E. A., and Williams, R. F., *J. Chem. Soc.* A, 1439 (1967).
251. Omloo, W. P. F. A. M., and Jellinek, F., *Rec. Trav. Chim.* **88**, 1205 (1969).
252. Paetzold, R., and Vordank, P., *Z. Anorg. Allg. Chem.* **347**, 296 (1966).
253. Paetzold, R., and Bochmann, G., *Z. Chem.* **8**, 308 (1968).
254. Paetzold, R., and Bochmann, G., *Z. Anorg. Allg. Chem.* **368**, 202 (1969).
255. Palko, A. A., and Durig, J. S., *J. Chem. Phys.* **46**, 2297 (1967).
256. Panyuskin, V. T., Garnovskii, A. D., Grandberg, I. I., Osipov, O. A., Minkin, V. I., Troitskaya, V. S., and Vinokurov, V. G., *Zh. Obshch. Khim.* **38**, 1154 (1968).
257. Pearson, R. G., *J. Amer. Chem. Soc.* **85**, 3533 (1963).
258. Pechard, M. E., *Ann. Chim. Phys.* **30**, 395 (1893).
259. Pecile, C., *Inorg. Chem.* **5**, 210 (1966).
260. Pederson, B., and Hope, H., *Acta Cryst.* **19**, 473 (1965).
261. Petit, L. D., Royson, A., Sherrington, C., and Whewell, R. J., *J. Chem. Soc.* D, 1179 (1967).
262. Petit, L. D., Sherrington, C., and Whewell, R. J., *J. Chem. Soc.* A, 2204 (1968).
263. Petrini, G., Pilati, O., and Giordano, N., *Chim. Ind.* (Milan) **50**, 1002 (1968).
264. Pierpont, C. G., Corden, P. J., and Eisenberg, R., *J. Chem. Soc.* D, 401 (1969).
265. Pilipenko, A. I., and Serada, I. P., *Zh. Neorg. Khim.* **6**, 1881 (1961).
266. Pluścec, J., and Westland, A. D., *J. Chem. Soc.* 5371 (1965).
267. Prandtl, W., Von Blochin, W., and Obpacher, H., *Z. Anorg. Allg. Chem.* **93**, 70 (1915).
268. Prandtl, W., and Von Blochin, W., *ibid.*, **93**, 45 (1915).
269. Prasad, S., and Garg, V. N., *Bull. Chem. Soc. Jap.* **38**, 1533 (1965).
270. Prasad, S., and Garg, V. N., *J. Ind. Chem. Soc.* **43**, 708 (1966).
271. Prasad, S., and Garg, V. N., *Indian J. Chem.* **5**, 595 (1967).

272. Prasad, S., and Garg, V. N., *Bull. Chem. Soc. Jap.* **41**, 493 (1968).
273. Proskina, N. N., Chulskaya, S. M., Volodina, G. F., and Ablov, A. V., *Zh. Strukt. Khim.* **9**, 1095 (1968).
274. Ramakrishna, R. S., and Irving, H. M. N. H., *Chem. & Ind.* 325 (1969).
275. Romanov, A. M., Ablov, A. V., and Gerbeleu, N. V., *Zh. Neorg., Khim.* **14**, 381 (1969).
276. Rosenbaum, A., Kirchberg, H., and Leibnitz, E., *J. Prakt. Chem.* **19**, 1 (1963).
277. Rosenheim, A., and Krause, L., *Z. Anorg. Allg. Chem.* **118**, 177 (1921).
278. Sabatini, A., and Bertini, I., *Inorg. Chem.* **4**, 959 (1965).
279. Salzmann, J. J., and Schmidtke, H. H., *Inorg. Chim. Acta* **3**, 207 (1969).
280. Schmidbauer, H., and Ruidisch, I., *Inorg. Chem.* **3**, 599 (1964).
281. Schmidt, M., and Ruf, H., *Angew. Chem.* **73**, 64 (1961).
282. Schmidt, M., and Ruf, H., *J. Inorg. Nucl. Chem.* **25**, 557 (1963).
283. Schmidt, M., and Ruf, H., *Z. Anorg. Allg. Chem.* **321**, 270 (1963).
284. Schmidt, M., and Schumann, H., *Chem. Ber.* **96**, 780 (1963).
285. Schmidt, M., and Ruf, H., *ibid.*, **96**, 784 (1963).
286. Schmidt, M., Siebert, W., and Rittig, F., *ibid.*, **101**, 281 (1968).
287. Schmidtke, H. H., *J. Inorg. Nucl. Chem.* **28**, 1735 (1966).
288. Schmidtke, H. H., *Ber. Phys. Chem.* **71**, 1138 (1967).
289. Schmidtke, H. H., and Garthoff, D., *Helv. Chem. Acta* **50**, 1631 (1967).
290. Schmidtke, H. H., and Garthoff, D., *Z. Naturforsch.* **24A**, 125 (1969).
291. Schumann, H., Thom, K. F., and Schmidt, M., *J. Organometal. Chem.* **2**, 364 (1964).
292. Schumann, H., and Schmidt, M., *Angew. Chem. Intern. Ed. Engl.* **4**, 1007 (1965).
293. Schumann, H., and Schmidt, M., *J. Organometal. Chem.* **3**, 485 (1965).
294. Schumann, H., Thom, K. F., and Schmidt, M., *ibid.*, **4**, 28 (1965).
295. Schumann, H., and Schmidt, M., *ibid.*, **4**, 22 (1965).
296. Seidel, H., *Naturwissenschaften*, **51**, 257 (1965).
297. Seidel, H., *ibid.*, **52**, 539 (1965).
298. Sekido, E., Fernando, Q., and Freiser, H., *Anal. Chem.* **37**, 1556 (1965).
299. Selbin, J., Ahamed, N., and Pribble, M. J., *J. Chem. Soc.* D, 759 (1969).
300. Shevchenko, I. G., and Tsintsadze, G. V., *Zh. Neorg. Khim.* **9**, 2675 (1964).
301. Siebert, H., *Z. Anorg. Allg. Chem.* **275**, 210 (1954).
302. Skopenko, V. V., *Visn. Kiivsk. Univ.*, **4**, Ser. Astrom. Fiz. ta. Khim. **81** (1961).
303. Skopenko, V. V., and Brusilovets, A. I., *Ukr. Khim. Zh.* **30**, 24 (1964).
304. Skopenko, V. V., Ivanova, E. I., and Tsintsadze, G. V., *ibid.*, **34**, 1000 (1968).
305. Skopenko, V. V., and Brusilovets, A. I., *ibid.*, **34**, 1210 (1968).
306. Skopenko, V. V., and Ivanova, E. I., *Zh. Neorg. Khim.*, **14**, 742 (1969).
307. Skopenko, V. V., and Zumbaev, A. Zh., *Ukr. Khim. Zh.* **35**, 428 (1969).
308. Sotnikov, V. S., Alimarin, I. P., and Lomovasov, M. V., *Talanta* **8**, 588 (1961).
309. Spinelli, D., and Dell'Erba, C., *Annali di chimica*, **51**, 45 (1961).
310. Stancheva, P., Skopenko, V. V., and Tsintsadze, G. V., *Ukr. Khim. Zh.* **35**, 166 (1969).
311. Suzki, K., and Yamasaki, K., *J. Inorg. Nucl. Chem.* **28**, 473 (1966).
312. Suzki, K., Nakano, I., and Yamasaki, K., *ibid.*, **30**, 545 (1968).
313. Swank, D. D., and Willett, R. D., *Inorg. Chem.* **4**, 499 (1965).
314. Taimni, I. K., and Rakshpal, R., *Anal. Chim. Acta* **25**, 438 (1961).
315. Tarantelli, T., and Furlani, C., *J. Chem. Soc.* A, 1717 (1968).
316. Thayer, J. S., and West, R., *Advan. Organometal. Chem.* **5**, 169 (1967).

317. Thayer, J. S., *J. Organometal. Chem.* **9**, P30 (1967).
318. Thayer, J. S., *Inorg. Chem.* **7**, 2599 (1968).
319. Tillay, E. W., Shermer, E. D., and Baddley, W. H., *ibid.*, **7**, 1925 (1968).
320. Toropova, V. F., *Zh. Neorg. Khim.* **1**, 243 (1956).
321. Tsintsadze, G. V., Skopenko, V. V., and Shvelashvili, A. E., *Soobshch. Akad. Nauk. Gruz. SSSR.* **41**, 337 (1966).
322. Tsintsadze, G. V., Porai-Koshits, M. A., and Antsyshkina, A. S., *Zh. Strukt. Khim.* **8**, 296 (1967).
323. Tsintsadze, G. V., Shvelashvili, A. E., and Skopenko, V. V., *Tr. Gruz. Politekh. Inst.* **1**, 29 (1967).
324. Tsintsadze, G. V., Shvelashvili, A. E., and Skopenko, V. V., *ibid.*, **4**, 53 (1967).
325. Tsintsadze, G. V., Skopenko, V. V., and Shvelashvili, A. E., *ibid.*, **1**, 19 (1968).
326. Tsintsadze, G. V., Skopenko, V. V., and Brusilovets, A. I., *Soobshch. Akad. Nauk. Gruz. SSSR*, **50**, 109 (1968).
327. Tsintsadze, G. V., Skopenko, V. V., Mikitchenko, V. F., and Zhumbaev, A., *Dopov. Akad. Nauk. Ukr. RSR Ser B.* **31**, 130 (1969).
328. Turco, A., Pecile, C., Nicolini, M., *Proc. Chem. Soc.* 213 (1961).
329. Turco, A., Pecile, C., and Nicolini, M., *J. Chem. Soc.* 3008 (1962).
330. Uhlig, E., Brock, B., and Glaenzer, H., *Z. Anorg. Allg. Chem.* **348**, 189 (1966).
331. Van Wazer, J. R., *J. Amer. Chem. Soc.* **78**, 5709 (1956).
332. Van Wazer, J. R., Callis, C. F., Shoolery, I. N., and Jones, R. C., *ibid.*, **78**, 5715 (1956).
333. Vyazankin, N. S., Bochkarev, M. N., and Sanina, L. P., *Zh. Obshch. Khim.* **35**, 1154 (1966).
334. Vyazankin, N. S., Bochkarev, M. N., and Sanina, L. P., *ibid.*, **36**, 1961 (1966).
335. Vyazankin, N. S., Bochkarev, M. N., and Sanina, L. P., *ibid.*, **37**, 1037 (1967).
336. Vyazankin, N. S., Bochkarev, M. N., Sanina, L. P., Egorochkin, A. N., and Khorshev, S. Ya., *ibid.*, **37**, 2576 (1967).
337. Vyazankin, N. S., Bochkarev, M. N., and Sanina, L. P., *ibid.*, **38**, 414 (1968).
338. Vyazankin, N. S., Bychkov, V. T., Vostokov, I. A., and Linzina, O. V., *ibid.*, **38**, 663 (1968).
339. Wei, C. H., and Dahl, L. F., *Inorg. Chem.* **4**, 493 (1965).
340. Welcman, N., and Rot, I., *J. Chem. Soc.* 7515 (1965).
341. Werden, B., Billig, E., and Gray, H. B., *Inorg. Chem.* **5**, 78 (1966).
342. Westland, A., and Westland, L., *Can. J. Chem.* **43**, 426 (1965).
343. Williams, R., Billig, E., Waters, J. H., and Gray, H. B., *J. Amer. Chem. Soc.* **88**, 43 (1966).
344. Williams, R. J. P., and Hale, J. D., *Structure and Bonding*, **1**, 249 (1966).
345. Workman, M. O., Dyer, G., and Meek, D. W., *Inorg. Chem.* **6**, 1543 (1967).
346. Wynne, K. J., and George, J. W., *J. Amer Chem Soc* **87**, 4750 (1965).
347. Yamasaki, K., and Suzki, K., *Proc. Intl. Conf. Coord. Chem. 8th*, 357 (1964).
348. Ziegler, M., and Glemser, O., *Z. Anal. Chem.* **146**, 29 (1955).
349. Ziegler, M., and Schroeder, H., *Mikrochim. Acta* 782 (1967).

8

Organic Chemistry of Selenium

KURT J. IRGOLIC/MOHAN V. KUDCHADKER*

Department of Chemistry, Texas A&M University

INTRODUCTION

The chemistry of organic selenium compounds has developed rapidly over the past ten years especially in the area of selenocarbohydrates, selenoamino acids, selenopeptides and other selenium containing molecules of biological interest. Much attention has been directed towards selenium derivatives of carbonic acid like selenourea, selenosemicarbazide, selenocarbazide and diselenocarbamic acids. Five- and six-membered selenium-containing hetero-cyclic compounds, especially selenophene, have been investigated in great detail.

The authors were thus faced with the problem of a huge number of papers to be condensed to a manageable volume. An exhaustive treatment of organic selenium compounds covering the past 130 years would be desirable, but such a treatise would

* Present Address: Texaco Research Laboratories, Houston, Texas.

become a voluminous book in its own rights.   It was therefore decided, to exclude heterocyclic selenium compounds and all selenium derivatives of carbonic acid and further limit the coverage to the literature, which appeared between 1955 and March 1971.   The work in this area performed before 1955 has been reviewed by Rheinboldt,[435] to whom reference has been made frequently in this chapter.

An attempt was made to present the synthetic methods available in the field of organic selenium chemistry, report on their advantages and drawbacks whenever possible and describe the physical and chemical properties of organic selenium compounds.   A detailed discussion of infrared, nuclear magnetic resonance, mass spectral and ultraviolet and visible spectral data was not possible due to space limitations.   The reader will, however, find references to such investigations in the sections dealing with the properties of organic selenium compounds.   In the text, references supplying such data are marked by the supercripts *(IR), †(NMR), ^{x}(mass spectral), and ^{u}(uv-visible).

Yields, melting points and/or boiling points are given only for seleno-carbohydrates.   These compounds have, to the best of our knowledge, never been reviewed before.   All the available literature on this subject has therefore been incorporated into its own section.

All the other classes of organic selenium compounds have been discussed in a more compressed manner.   Following the equations describing a particular reaction are listed the individual compounds prepared in this way. The condensed presentation of these formulas will sometimes require a little more work from the reader in order to find the compound of interest to him than would be necessary with an unfortunately more space consuming tabular listing.   The reader, who is only interested in general reactions, can conveniently pass over these passages.   It is our hope that this chapter will serve both kinds of readers.

The authors wish to express their appreciation to the Robert A. Welch Foundation of Houston, Texas and to the Selenium-Tellurium Development Association, Inc., for financial assistance during the time this chapter was written.

SELENOLS, RSeH

Selenols, the selenium analogues of alcohols, are colorless liquids or solids which oxidize very easily to the diselenides.   Traces of these diselenides dissolved in the selenols impart a yellow color to the latter.   Selenols are very poisonous and have an extremely penetrating and bad odor.   All operations involving selenols must be carried out in a very well-ventilated hood.   Any gases evolved from a reaction apparatus should be trapped in alkali.   Selenols are more acidic than thiols and alcohols.

The selenium containing compounds which have been employed as starting materials for the synthesis of selenols are elemental selenium, sodium hydrogen selenide, hydrogen selenide, carbon diselenide, carbon oxide selenide, diorganyl diselenides, R_2Se_2, organyl selenocyanates, RSeCN, and seleninic acids, RSe(O)OH.

Selenols from Elemental Selenium

Elemental selenium reacts with organyl lithium compounds, with sodium acetylides,[73, 419] with Grignard reagents, and with trialkyl aluminum compounds.[570]

In spite of the large number or organic lithium compounds reported in the literature, only a few aromatic derivatives and no aliphatic compounds have been studied with respect to their reaction with selenium:

$$R{-}Li + Se \xrightarrow{\text{ether}} R{-}Se{-}Li \xrightarrow[H^+]{H_2O} R{-}SeH \qquad (1)$$

R: 3-bromobiphenyl, 4-bromobiphenyl, 4-bromophenoxyphenyl;[200] 2-furyl;[373] 2-benzothienyl;[510] 3,4-methylenedioxyphenyl.[135]

In most cases the selenols were not isolated but oxidized to the diselenides or converted to other more stable products.

Sodium acetylides, which are easily synthesized from the unsaturated hydrocarbons and metallic sodium in liquid ammonia, combine with selenium according to equation (2):

$$R{-}C{\equiv}C{-}Na + Se \longrightarrow R{-}C{\equiv}C{-}Se{-}Na \qquad (2)$$

R: C_6H_5;[73] CH_2: $C(CH_3)$;[419] CH_3C: C;[77,81] HC: C.[81]

Hydrolysis of the sodium selenolate would produce the selenol. In most cases, however, the selenolates are not isolated, but converted to the diorganyl selenides by reaction with an organyl halide.

The reaction between Grignard reagents and selenium is the most prolific method for the synthesis of selenols. The reaction is carried out in absolute ether under an atmosphere of dry nitrogen. After hydrolysis of the reaction mixtures with dilute acids at 0°C the selenols are isolated in yields ranging from 30–60% by steam distillation or by extraction with aqueous sodium hydroxide solution. Selenides, diselenides, and hydrogen selenide are formed as by-products. Scheme (3) outlines possible ways in which these compounds have been formed.

$$2R-MgBr + 2Se \longrightarrow 2R-Se-MgBr \longrightarrow 2R-SeH$$

$$R-Se-R$$

Se   R MgBr

$$H_2Se \xleftarrow{H_2O} Se(MgBr)_2 + R-SeSe-R \xleftarrow{O_2}$$

(3)

This method seems to be of general applicability for the preparation of aromatic selenols provided the Grignard reagents can be prepared. References to the older literature were collected by Rheinboldt.[435] This method has recently been employed for the synthesis of the following selenols, $R-C_6H_4-SeH$ (R given): H;[64, 108] 2-CH$_3$, 2-CH$_3$O;[429] 2-C$_2$H$_5$;[234] 4-Cl, 4-Br, 2-,3- and 4-CH$_3$, 4-CH$_3$O, 4-NO$_2$;[108] 4-F.[473] 2,4,6-Triphenyl-benzeneselenol was prepared similarly.[275] A detailed procedure for the synthesis of benzeneselenol was published by Foster.[161]

Only a few alkaneselenols have been prepared in this manner. The yields seem to be lower than those obtained with aromatic Grignard reagents. Recently, butaneselenol,[191] 2-methyl-2-butaneselenol (Brit. Pat., 782,887), and phenylmethaneselenol[162] were obtained in about 30% yield. For the reaction between i-amyl-, $tert$-butyl- and cyclohexylmagnesium chloride with selenium, see Rheinbolt.[435]

Zakharkin[570] reported the isolation of ethane- and 2-methylpropaneselenol in yields of 71 and 64%, respectively, from the reaction of equimolar amounts of the appropriate trialkyl aluminum compounds and selenium.

Selenols from Sodium Hydrogen Selenide, Hydrogen Selenide, Carbon Diselenide and Carbon Oxide Selenide

The alkylation of sodium hydrogen selenide to the alkaneselenols has been carried out according to equation (4) with alkyl bromides and iodides and potassium alkyl sulfates.[435] The only chloride employed thus far has been the octyl derivative:

$$NaSeH + RX \Longrightarrow R-SeH + NaX \qquad (4)$$

R: C$_8$H$_{17}$(Belg. Pat., 670,823); sec-C$_4$H$_9$, i-C$_3$H$_7$, (C$_6$H$_5$)$_2$CH, C$_6$H$_4$(CH$_3$)CH.[334]

This reaction is also applicable to the synthesis of selenoglycerins from the appropriate bromohydroxypropanes.[45] 2-Hydroseleno derivatives of heterocyclic compounds were obtained from the corresponding chlorides and sodium hydrogen selenide in aqueous medium (Belg. Pat., 670,823).

$$(X = O, S, Se)$$

$$(R = cyclohexyl, phenyl, C_5H_{11})$$

Selenopyridine and selenopurine derivatives, which can exist in a selenol form (see equation 148b) are discussed later.

Hydrogen selenide was found to add to ethylene in the presence of an MoS_2 catalyst[306] to give ethaneselenol in 13% yield:

$$CH_2=CH_2 + H_2Se \xrightarrow[52\ atm.]{200°\ 1hr} C_2H_5SeH \qquad (6)$$

Draguet[149] obtained 2-aminoethaneselenols from hydrogen selenide and aziridines:

$$R: H, CH_3, C_6H_5. \qquad (7)$$

Scheithauer[460] reported that hydrogen selenide converted cyclohexanone in dimethylformamide in presence of pyridine into cyclohexaneselenol. This compound decomposed during vacuum distillation.

Jensen[258] obtained the very explosive potassium salt of 2-nitroethaneselenol by the reaction of carbon diselenide with nitromethane in ethanol containing potassium ethoxide:

$$CSe_2 + CH_3NO_2 \xrightarrow[-80°C]{KOC_2H_5} O_2NCH_2CSe_2K \qquad (8)$$

$$O_2NCH_2CSe_2K \xrightarrow[-Se]{H_2Se} O_2NCH_2CH_2SeK$$

Tyerman[507] found that the excited selenium atoms produced by flash photolysis of COSe insert into the carbon-hydrogen bonds of ethane, propane, cyclopropane, cyclobutane, and i-butane to form the corresponding selenols.

Selenols by Reduction of Diselenides, Selenocyanates and Seleninic Acids

Since selenols are very sensitive towards oxidation, diselenides, R_2Se_2, their oxidation products, are convenient, stable starting materials for the preparation of selenols. The reduction of the diselenides can be accomplished with a variety of reagents. Fredga[168] and Golmohammadi[187] reduced diselenides of the general formula $[HOOC(CH_2)_nSe]_2$ (n = 1–7, 10) with Rongalite in aqueous ammonia. 1-Anthraquinoneselenol was the product of the reaction between the corresponding diselenide and glucose in basic aqueous ethanol.[436] Gunther[217] obtained selenols by the reduction of $[HOOC-(CH_2)_n]_2Se_2$, (n = 1–3), $[H_2NCH_2CH_2]_2Se_2$, $[(CH_3)_2N(CH_2)_n]_2Se_2$ (n = 2, 3), $[HOOCCH_2CH_2C(O)NHCH_2CH_2]_2Se_2$, 4'-phospho-Se-pantethine, selenocoenzyme A and 4-carboxy-1-thia-2-selenacyclopentane with dithiothreitol in aqueous solutions of pH 7.6.

Sodium in liquid ammonia was used for the preparation of dimethyl,[1, 38, 230] diethyl,[59,80] and di-i-propyl diselenide.[80] Sodium in methanol served as a reducing agent for the aromatic diselenides, $(R—C_6H_4)_2Se_2$ (R = 2-CH_3, 3-CH_3, 3-Cl, 2-CH_3O, 4-CH_3O).[336] Hypophosphorous acid in $6M$ hydrochloric acid (R = 8-quinolyl[472]) or in absolute ethanol or methanol [R = CH_3, $(CH_3)_2NCH_2CH_2$;[215] $(CH_3)_3N(CH_2)_2$, $(CH_3)_3N(CH_2)_3$[214]], zinc and hydrochloric acid (R = C_6H_5[520]), sodium borohydride [R = $(CH_3)_2NCH_2CH_2$;[115] $H_2NCH_2CH_2$;[287] $HOOCCH_2$, CH_3OOCCH_2;[304] 2-$C_{10}H_7$;[486] $C_6H_5CH_2$[573, 574]] and sodium benzenethiolate (R = 4-$NO_2C_6H_4$)[396] have also been employed as reducing agents for diorganyl diselenides, R_2Se_2.

Bis(trifluoromethylseleno) mercury, obtained from the diselenide and elemental mercury, treated with anhydrous hydrogen chloride gave trifluoromethaneselenol.[132] Heptafluoropropaneselenol could not be prepared in this manner.[154]

Aliphatic and aromatic selenocyanates, RSeCN, which are easily obtainable from alphatic halides or aromatic diazonium salts and KSeCN, are reduced by zinc and hydrochloric acid either in aqueous or methanolic solution (R = 4-$CH_3C_6H_4$;[192] C_6H_5;[223] 3,5-($tert$-C_4H_9)$_2$-4-HOC_6H_2;[362] C_6H_5, 4-ClC_6H_4, 4-$CH_3C_6H_4$[520]), by zinc and sulfuric acid in ethanol (4-$NO_2C_6H_4$;[523] 2- and 4-$CH_3OOCC_6H_4$[528]) or by zinc in basic medium (2-$H_2NC_6H_4$[363]). The yields in these reactions vary between 50 and 90%. Aryl selenocyanates heated with strong aqueous sodium hydroxide gave the sodium salts of the corresponding selenols and seleninic acids:

$$Ar\text{-}SeCN + 8NaOH \longrightarrow 3 \; ArSeNa + ArSeO_2Na + 2NaCN$$

$$+ \; 2NaOCN + 4H_2O \quad (9)$$

Since better methods for the preparation of selenols from selenocyanates are available, reaction (9) has not been used recently. The older literature is covered by Rheinboldt.[435]

Seleninic acids, $RSeO_2H$, can also be reduced to selenols.[436] Since seleninic acids are obtained by oxidation of diselenides or selenocyanates, their reduction to selenols is not of preparative importance.

Physical Properties and Reactions

The Raman and IR-spectra of CH_3SeH, CH_3SeD, CD_3SeH and CD_3SeD were analyzed and the rotational constants determined.[229, 230] The Se—H stretching mode gives rise to an adsorption band between 2280 and 2330 cm^{-1}.[229,230,344,473] The location of these bands was independent of concentration in benzene, dioxane, and tetrahydrofuran indicating that the selenols are not associated through hydrogen bonds.[473] The dissociation constants of hydroselenoacetic acid,[304] 2-aminoethaneselenol,[499] and 8-quinolineselenol[368] have been determined. Ultraviolet and visible spectal data were reported by Chierici,[103] Gunther,[217] Kiss[286] Mangini,[333] and Parker.[396]

The $^1H(Se—H)$ NMR resonance in selenols in which the hydroseleno group is bonded to a methylene group is found as a $1:2:1$ triplet with $J_{H-C-Se-H}$ of approximately 7 cps between 10.09 and 10.7τ, while singlets at 8.54 to 8.67τ were observed for aromatic selenols.[304, 305, 328, 329, 344, 351] Lalezari's[305] assignments for alkaneselenols are incorrect. The $^{77}Se—H$ coupling constants in CH_3SeH and C_6H_5SeH were determined by McFarlane.[329]

Selenols and selenolate anions have been employed extensively as reagents to introduce the organylseleno group into organic molecules either through reactions with organic halides, expoxides, carboxylic acid chlorides, lactones, diazonium salts, or addition to a carbon-carbon multiple bond. These reactions, especially those leading to selenocarbohydrates and selenoamino acids, are discussed in the appropriate sections.

SELENIUM COMPOUNDS OF THE GENERAL FORMULA R-Se-X

$[X = OH, Cl, Br, SCN, SeCN, NCO, NR_2, OR, OC(O)CH_3]$

Selenium compounds of the general formula RSeX can be looked upon as derivatives of selenenic acids, RSe—OH.

Selenenic Acids, R—Se—OH

Aromatic selenenic acids, especially nitro substituted derivatives, are prepared in almost quantitative yield by reduction of the corresponding seleninic acids with the calculated amount of hydrazine sulfate:[436, 441]

$$2RSeO_2H + N_2H_4 \longrightarrow 2RSeOH + N_2 + 2H_2O \qquad (10)$$

This reaction is, however, not of general applicability, since many seleninic acids are reduced to the diselenides under these conditions. Sodium hypophosphite, NaH_2PO_2, in hydrochloric acid solution of pH 0.62,[438] and ethane- and benzenethiol in benzene solution[437] can also be used as reducing agents. The hydrolysis of organyl selenium halides was used only for the synthesis of a few aromatic selenenic acids (for references see Rheinboldt[435]). The only known aliphatic compound, 2-amino-ethaneselenenic acid, was formed by oxidation of the corresponding diselenide with hydrogen peroxide.[288] 1,4-Anthraquinonylene diselenenic acid[u] was obtained from its anhydride with acetic acid by hydrolysis.[253]

Organyl Selenium Halides

The selenenyl halides, RSeX, better named organyl selenium halides, have been isolated only as the chlorides or bromides. Diselenides, organyl selenocyanates, and organyl selenium trihalides are the most common starting materials. In the reaction of diselenides with chlorine, bromine, or sulfuryl chloride, an excess of the halogenating agent must be carefully avoided to prevent trihalide formation. Excess thionyl chloride[48] does not produce trihalides and seems to be the reagent of choice for the conversion of diselenides into organyl selenium chlorides:

$$R_2Se_2 \xrightarrow[\text{SOCl}_2]{\text{Cl}_2, \text{Br}_2, \text{SO}_2\text{Cl}_2 \text{ or}} 2RSeX \qquad (11)$$

R, X: CF_3, Cl;[382, 566] CF_3, Br;[132, 566] CFH_2, Br;[567] CH_3, Cl; C_2H_5, Cl;[382] C_2F_5, Cl;[537*†] C_2H_5, Br;[59, 60] C_6H_5, Br;[64] 2-$HOOCC_6H_4$, Cl.[432]

Heptafluoropropyl selenium chloride and bromide was obtained from $(C_3F_7Se)_2Hg$ and excess halogen.[154] 2-Naphthyl[512,u] 1-anthraquinonyl selenium bromide,[239] and 1,4-anthraquinonylene di(selenium bromide)[253] were obtained similarly. Rheinboldt[439] found that bis(2-nitrophenyl) triselenide gave, with chlorine and bromine, the respective monohalides.

Organyl selenocyanates, RSeCN, are cleaved by chlorine (R = 2-$CH_3COC_6H_4$)[432] and bromine in chloroform solution and by hydrochloric acid at 60°C (R = 2-$ClCOC_6H_4$):[480]

$$RSeCN + Br_2 \longrightarrow RSeBr + BrCN \qquad (12)$$

R: CH_2COOH;[59]† 2-$CH_3COC_6H_4$;[195] 2-$C_6H_5COC_6H_4$;[441] 2,4-$(NO_2)_2C_6H_3$, 1-anthraquinonyl.[239]

Methyl phenyl selenides (I) are cleaved by bromine and thionyl chloride producing the respective selenium monohalides.

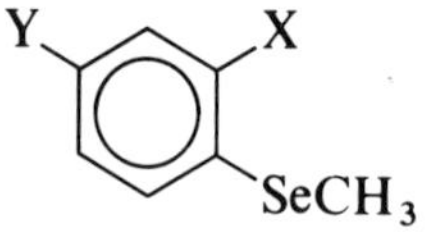

X,Y: COOH, H;[432] COOH, Br; CH_3CO, Br.[510]

(I)

Since Gosselck[192] was able to obtain 2-acetyl-4-methylphenyl selenium bromide by boiling the substituted phenyl methyl selenium dibromide in ethanol, it is likely that the above reactions also proceeded via the selenium dihalides.

The thermal dissociation of $RSeCl_3$ and $RSeBr_3$ to the monohalides and halogen can be accomplished by rapidly heating the trihalides in vacuum above the melting point. Slow heating will produce halogenated products.[435] Lawson[311] prepared in this way, 2,4-dinitrophenyl selenium chloride.

The following reactions were also employed to synthesize organyl selenium halides. Nothing is known, however, about the applicability of these reactions to other compounds:

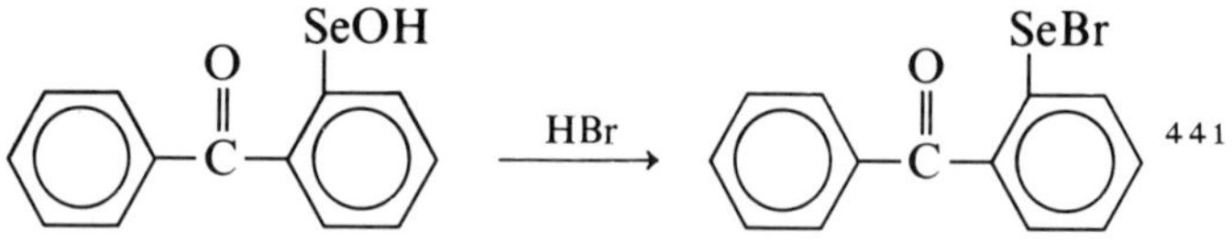

$$(C_2F_5Se)_2Hg + 2X_2 \longrightarrow 2C_2F_5SeX + HgX_2 \quad [537] \qquad X = Cl, Br$$

$$CF_3SeCl_3 + (CF_3)Se_2 \longrightarrow 3CF_3SeCl \quad [132]$$

$$(C_2H_5)_2P(S)SeR \xrightarrow{X_2} RSeX \quad [302] \qquad X = Br, Cl; R = C_2H_5, C_3H_7$$

$$CSe_2 \xrightarrow[-80° \text{ to } -30°C]{Cl_2/CCl_4} CCl_3SeCl \quad [565], (73\% \text{ yield})$$

An NMR investigation of the mixtures obtained from dimethyl diselenide and Se_2Cl_2 suggested the presence of CH_3Se_nCl (n = 1–6).[204]

Organyl Selenium Isocyanates, Thiocyanates, Selenocyanates, and Organyl Selenosulfates

Organyl selenium halides react with silver thiocyanate, silver isocyanate and potassium selenocyanate[539]* according to reaction (14):

$$R-Se-Cl \begin{cases} -AgSCN \longrightarrow RSe-SCN + AgCl\ (R = CF_3, C_3F_7) \\ -AgNCO \longrightarrow RSe-NCO + AgCl\ (R = CF_3) \\ -KSeCN \longrightarrow RSe-SeCN + KCl\ (R = CF_3) \end{cases} \quad (14)$$

The perfluoropropyl selenium isocyanate was too unstable for isolation.

Rheinboldt[439] obtained 1-anthraquinonyl and 2-nitro-4-R-phenyl selenium selenocyanates (R = H, Cl, Br, CH$_3$O, NO$_2$) in good yields from the organyl selenium bromides and potassium selenocyanate in benzene solution. The corresponding thiocyanates are also known (see Rheinboldt).[435]

Alkyl selenosulfates are obtained by alkylation of potassium selenosulfate in aqueous ethanol with alkyl halides. The compounds

$$R_3\overset{+}{N}-CH_2CH_2-SeSO_3^- \ (R = H, CH_3)$$

were isolated in 40–80% yield.[212, 287]

Esters, Anhydrides, and Amides of Selenenic Acids

Esters of selenenic acids, R—Se—OR′, cannot be obtained from a selenium halide and sodium alkoxide, since the formation of diselenides seems to prevail. The methyl esters of 2-nitrobenzene-, 2,4-dinitrobenzene-, 1-anthraquinone- and 4-hydroxyanthraquinoneselenenic acids were formed by treating the selenocyanates in methanol with silver acetate, silver oxide or copper acetate in the presence of pyridine.[241] The methyl esters treated with glacial acetic acid give R—Se—OC(O)CH$_3$ (R = 2-NO$_2$C$_6$H$_4$, 1-anthraquinonyl).[240] Silver acetate in glacial acetic acid converted 1,4-anthraquinonylene bis-(selenium bromide) into the diacetyl compound, which in absolute ethanol produced the diethyl ester of the diselenenic acid.[253] 2-Nitrophenyl selenium bromide treated with methanol and silver acetate gave the corresponding methyl ester[241] and not the compound with the CH$_3$C(O)O group bonded to selenium atom as claimed by Behagel.[49] Lawson[311] showed that 2,4-dinitrophenyl selenium trichloride reacts with ethanol in the presence of pyridine and with aniline in diethyl ether to give the selenenic ester and amide, respectively.

From reaction mixtures containing pentafluoroethyl selenium chloride and ammonia, and primary amines or secondary amines, the following compounds were isolated:[537]

$$2C_2F_5-Se-Cl + nRR'NH$$

$$\begin{array}{l} \xrightarrow[n=2]{R,R' = CH_3} C_2F_5-Se-N(CH_3)_2 \quad (Ia) \\[2ex] \xrightarrow[n=3]{R,R' = H} (C_2F_5-Se)_2NH^{*,\dagger} \\[2ex] \xrightarrow[n=2]{R = CH_3, R' = H} C_2F_5-Se-NHCH_3^{*,\dagger} \\[2ex] \xrightarrow[n=3]{R = CH_3, R' = H} (C_2F_5-Se)_2NCH_3^{*,\dagger} \end{array}$$

$$(15)$$

Emeleus[154]* synthesized the perfluoromethyl and -propyl derivatives of (Ia).

The compounds $2\text{-}NO_2\text{-}4\text{-}R\text{-}C_6H_3\text{-}Se\text{-}NHC_6H_4\text{-}R'$ were obtained similarly from the selenium bromides and amines (R, R': NO_2, H; NO_2, CH_3; H, CH_3).[440] N-Alkylanilines, however, condense with selenium halides in *para* position to the amino group (see reaction 79).

Reactions

Many of the reactions given by the compounds R—Se—X have already been discussed in this section. Organyl selenium halides serve as reagents to introduce the organylseleno group into other molecules to give symmetric and unsymmetric selenides. Selenium halides and methyl esters of selenenic acids add to carbon-carbon multiple bonds producing unsymmetric selenides. With potassium cyanide, selenocyanates are obtained.

ORGANYL SELENOCYANATES AND ISOSELENOCYANATES

Organyl Selenocyanates

Synthesis of Organyl Selenocyanates. Aliphatic and aromatic selenocyanates, R—SeCN, can be easily prepared. Potassium selenocyanate serves in almost all reactions as the source of selenium. The aliphatic compounds are formed when the alkyl bromides or chlorides are heated with KSeCN in acetone, methanol or ethanol as solvent:

$$R-X + KSeCN \longrightarrow R-SeCN + KX \quad (16)$$

$$X = Cl, Br, I$$

The following compounds were obtained in this manner: C_3H_7, $i\text{-}C_3H_7$, C_4H_9;[166] allyl;[501] C_4H_9, $i\text{-}C_5H_{11}$, C_8H_{17}, $C_{12}H_{25}$;[359] $C_{16}H_{33}$;[422] $C_{18}H_{37}$[68]

(Brit. Pat. 839,351); CH_2COOH;[59] $(CH_2)_{10}COOH$;[69] $(CH_2)_{12}COOH$;[424] $C_{12}H_{25}SC(O)(CH_2)_{10}$[69] (Brit. Pat., 839,351); $C_6H_5CH\colon CHCO_2CH_2CH_2$ (U.S Pat., 2,934,420); diphenylmethyl;[398, 501] bis (4-methylphenyl)methyl;[501] 2-phthalimidoethyl;[406] and 6-quinolymethyl.[317]

Alkyl iodides have been employed in the synthesis of ethyl and methyl selenocyanate.[37,166]

α,ω-Polymethylene dibromides combine with KSeCN to give the corresponding diselenocyanates in yields higher than 80%:

$$Br-(CH_2)_n-Br+2KSeCN \longrightarrow NCSe-(CH_2)_n-SeCN+2KCl \quad (17)$$

n:2;[190,380] 4;[85] 8;[69] 10, 12, 14;[424] 10 (Brit. Pat., 839,351).

1,3-Dibromo-1,3-dicarbethoxypropane,[63]* 1-chloro-2-benzoylamino-ethane,[211] and $[Br(CH_2)_{10}C(O)S(CH_2)_5]_2$[69] (Brit. Pat., 839,351) have been converted into the substituted selenocyanates in the same manner.

Pichat[406] prepared 2-chloroethyl selenocyanate in 63% yield from 1-bromo-2-chloroethane and KSeCN. A short note by Wille[544] describes the addition of HSeCN to the triple carbon-carbon bonds in propynal and 3-butynone:

$$\underset{\displaystyle \|}{R-\overset{\displaystyle O}{C}}-C\equiv CH + HSeCN \longrightarrow \underset{\displaystyle \|}{R-\overset{\displaystyle O}{C}}-CH=CH-SeCN$$

$$R=H, CH_3$$

Selenium diselenocyanate, $Se(SeCN)_2$, reacts with tetramethyl, tetraethyl and tetraphenyl lead to give organyl selenocyanates:[39]

$$R_4Pb + Se(SeCN)_2 \xrightarrow{CHCl_3} R_3PbSeCN + RSeCN + Se \quad (19)$$

$$R = CH_3, C_2H_5, C_6H_5$$

Haloalkyl selenium chlorides treated with AgCN or KCN eliminated AgCl or KCl and formed selenocyanates. Trifluoromethyl,[132, 539*, 566] difluoromethyl,[568] perfluoroethyl,[539] trichloromethyl,[565] and 2-carbethoxy- and 2-carbomethoxyphenyl selenocyanates[432] became available through this reaction. Agenas[11]* obtained 2-carboxyethyl selenocyanate from propiolactone and potassium selenocyanate.

Aromatic selenocyantes are generally not obtainable from halides and KSeCN unless additional substituents in the aromatic ring enchance the reactivity of the halogen atom. Thus, 4-bromoquinoline N-oxide[495] and 2,4-dinitrochlorobenzene condensed easily with KSeCN to the selenocyanate.[239, 311, 436] 2,4,5-Trinitrotoluene and phenyl selenourea gave 2,4-dinitro-5-tolyl selenocyanate. A similar reaction with 2,4,5-trinitrobromobenzene produced 2,4-dinitro-1,5-diselenocyanatobenzene.[182]   Copper(II) selenocyanate and azulene in acetonitrile gave compound (II):[370]

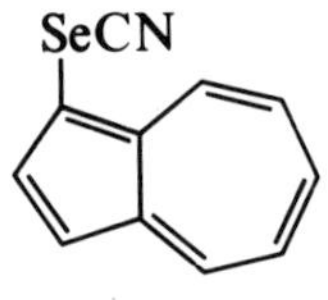

SeCN

(II)

A large number of aromatic selenocyanates has been prepared by adding an aromatic diazonium salt to an aqueous KSeCN solution at 0°C.

In recent years this method has been employed for the preparation of 2-iodophenyl,[119] 8-quinolyl,[472] 2-biphenylyl,[104] and substituted biphenylyl,[99,101,549] naphthyl,[298] substituted naphthyl,[297] 1-anthraquinonyl,[239,253] 4-hydrooxyanthraquinonyl,[241] 4-selenocyanatophenyl,[24*] and various substituted phenyl selenocyanates.[37,106,151,195,363,436,441]

Agenas has made extensive use of selenocyanogen, $(SeCN)_2$, as a reagent for introducing the selenocyanato group into aniline and N-substituted anilines,[16] pyrroles,[9] indoles,[2] and indolines.[22]   Selenocyanogen, a compound of extreme lability, was generated in methanolic solution from potassium selenocyanate and bromine below $-50°C$.   To this solution was added the stoichiometric amount of the aromatic compound.   The selenocyanate group entered *para* to the amino group in anilines and indoline and occupied the 3-position in pyrroles.   Mueller[362] has used this method earlier for the preparation of 3,5-di-*tert*-butyl-4-hydroxyphenyl selenocyanate.   The selenocyanation proceeded according to reaction (20):

$$2KSeCN + Br_2 \xrightarrow[CH_3OH]{-50°C} NCSe-SeCN + 2KBr$$

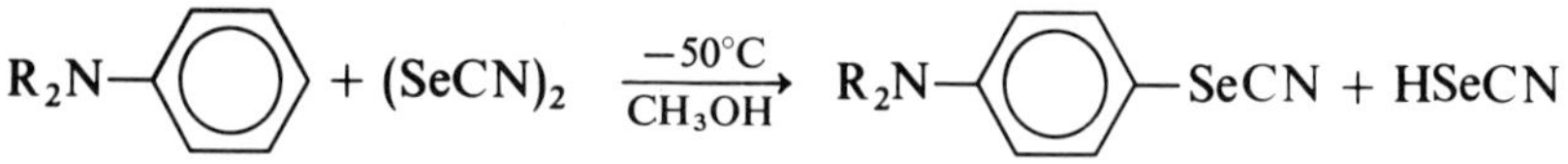

$$R_2N-\!\!\!\bigcirc\!\!\!- + (SeCN)_2 \xrightarrow[CH_3OH]{-50°C} R_2N-\!\!\!\bigcirc\!\!\!-SeCN + HSeCN$$

$$HSeCN \xrightarrow{>-50°C} Se + HCN \tag{20}$$

Agenas[13,14,15,21,22,23] investigated the mass spectral behavior of these aromatic selenocyanates and obtained NMR spectra of aminophenyl selenocyanates.[24]   The crystal and molecular structure of 1,4-diselenocyanatobenzene was determined by McDonald.[326]   The selenium atom in aromatic selenocyanates does not donate electrons to the aromatic system.[335]

Reactions of Selenocyanates.   The reaction of selenocyanates are summarized in Figure 8-1.   A more detailed discussion of these reactions can be found

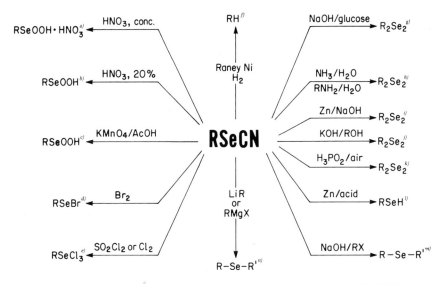

References: a) 424, 380. b) 297, 298, 101. c) 299, 300. d) 59, 195, 239, 441. e) 311, 432. f) 548. g) 359. h) 528. i) 362, 363.
j) 2, 14, 17, 22, 101, 106, 181, 211, 336, 406, 422, 436, 512, 520, 549. k) 472. l) 192, 520, 523, 528. m) 103, 151, 198, 370.
n) 206.

Figure 8-1   The reactions of selenocyanates.

in the sections dealing with the products.   The references given as footnotes
should serve as a guide to the original papers.

2-Chloroformylphenyl selenocyanate gave with benzenethiol and dimethyl-
amine the corresponding ester and amine.[432]   2-Biphenylyl selenocyanate
treated with nitric acid (d = 1.52) at low temperatures produced a mixture of
4'-nitro-2-biphenyl and 4,4'dinitro-2-biphenylyl selenocyanates.   2-Biphenyl
selenocyanate treated similarly gave only the seleninic acid.[101]   Aromatic
selenocyanates condense with selenols and thiols to produce diselenides and
selenide sulfides, respectively.[369]   Exchange reactions occur between organyl
selenocyanates and $K^{14}CN$ and $KSe^{14}CN$.[461]

Organyl Isoselenocyanates

Most of the organyl isoselenocyanates, $R-N=C=Se$, have become available only recently.   The earlier unsuccessful attempts to prepare these compounds were summarized by Rheinboldt.[435]

Most of the organyl isoselenocyanates have been prepared by refluxing
isonitriles with black elemental selenium in chloroform, acetone or petroleum
ether:

$$R-N\equiv C: + Se \longrightarrow R-NCSe \qquad (21)$$

R: CH_3, C_2H_5, C_3H_7, $i\text{-}C_3H_7$;[166] $tert\text{-}C_4H_9$;[256] C_3H_7, C_4H_9, C_5H_{11}, C_6H_{13}, cyclohexyl;[125] cyclohexyl;[398] C_nH_{2n+1} (n = 4, 6–12), $C_6H_5CH_2$, $C_6H_5CH_2CH_2$;[314] C_6H_5;[254, 125, 87, 89, 90] $2\text{-}CH_3C_6H_4$, $4\text{-}CH_3C_6H_4$;[87, 89, 90] $2\text{-}CH_3OC_6H_4$, $4\text{-}CH_3OC_6H_4$, $4\text{-}ClC_6H_4$[89, 90]

The yields in these preparations vary from 20–80% depending upon the nature of the organic group. Thayer[502] converted organometallic cyanides $R_3M\text{—}CN$ (R = CH_3, C_2H_5, C_4H_9, C_6H_{13}, C_6H_5; M = Si, Ge) into the corresponding isoselenocyanates.

Other methods which have found limited applications, employ Na_2Se and $RNCX_2$ (R = C_6H_5, X = Cl;[125] R = C_2H_5, X = Br^{244}) and C_6 $H_5\text{—}NCO$ and P_2Se_5[125] as the reagents.

While aliphatic halides and potassium selenocyanate gave the selenocyanates, $R\text{—}SeCN$, the reaction with benzoyl chloride,[148, 125, 126, 88, 89, 90, 92] 4-nitrobenzoyl chloride,[92] acetyl chloride,[92, 148] crotonyl chloride,[92] triphenyl-methyl chloride,[398, 500] cinnamoyl chloride,[92] and imidocarboxylic acid chlorides, $RC(=NR)Cl$ (R = C_6H_5, $4\text{-}ClC_6H_4$)[285] produced the isoseleno-cyanates, $R\text{—}NCSe$. Rheinboldt's report[434] that triphenylmethyl seleno-cyanate is formed from the corresponding chloride and KSeCN is incorrect.

Tarantelli[501] showed that diphenylmethyl and bis(4-methylphenyl)methyl selenocyanates isomerize to the isoselenocyanates upon heating their benzene solutions to 120°C in a sealed tube. The isomerization of allyl selenocyanate in cyclohexane at 80°C led to an equilibrium mixture containing both seleno-cyanates and isoselenocyanates.

These isomeric compounds can be distinguished through chemical means and through their ultraviolet and infrared spectra. Amines react only with isoselenocyanates to give selenoureas. The ultraviolet spectra of isoseleno-cyanates have an adsorption band in the region 250–280 mμ with log $\varepsilon \simeq 3.3$. The selenocyanates exhibit only an inflection in this region with a log ε value 0.1 to 0.2 logarithmic units lower than in the case of the iso-compounds.[501] A strong, broad, IR band between 2000 and 2200 cm^{-1} centered at about 2100 cm^{-1} with a shoulder or an additional band at 2050 cm^{-1} is charac-teristic for the isoselenocyanates. The selenocyanates have a sharp band at 2150 cm^{-1}.[16, 89]

The reaction of isoselenocyanates with ammonia[88, 90] amines,[88, 90, 92, 183, 285, 314, 501] hydrazine, and substituted hydrazines[87, 91, 125, 126, 244, 256, 257, 314, 398, 501] produced selenoureas and selenosemicarbazides according to reaction (22):

$$R\text{—}N{=}C{=}Se \begin{cases} \xrightarrow{R'NH_2} RNH\text{—}C(Se)\text{—}NHR' \\ \xrightarrow{R'NHNH_2} RNH\text{—}C(Se)\text{—}NH\text{—}NHR' \end{cases} \quad (22)$$

Franklin[166] reported that alkyl isoselenocyanates deposited selenium upon treatment with acid.   In the presence of zinc/HCl the amines were formed.

SELENINIC ACIDS, SELENONIC ACIDS AND THEIR DERIVATIVES

Seleninic Acids

Pure seleninic acids, $RSe(O)OH$, are colorless and odorless solids.   The aliphatic compounds are more soluble in water than the aromatic derivatives. In aqueous solution they are weaker than the carboxylic acids, but they can still be titrated with base using phenolphthalein as indicator.[435]   The pK values of $4\text{-}XC_6H_4SeO_2H$ (X = H, CH_3, Cl, Br) were found to be 4.70, 4.55, 4.30 and 4.28, respectively.[139]   Salts of 5-nitro-[297,298] and 8-nitro-napthaleneseleninic acid[296,298] with various metal and ammonium ions and of 1-[299] and 2-naphthaleneseleninic acid[300] with the benzylisothiuronium cation have been prepared.   Alimarin[26] suggested benzeneseleninic acid as a reagent for the gravimetric determination of scandium, and Myasoedov[366] used it for the extraction of protactinium and other elements into amyl alcohol.   Paetzold isolated the silver[381] and sodium and potassium salts[376] of ethane- and methaneseleninic acid and analyzed the infrared spectra of the alkali metal salts.   The infrared spectra of benzene- and 4-methylbenzene seleninic acid[143] have been reported.   The structure of benzeneseleninic acid was determined by Bryden.[86]   Chierici[107] investigated the UV spectral properties of aromatic seleninic acids.   The amphoteric character of seleninic acids is manifested by their formation of adducts with strong acids like HCl and HNO_3:

$$R{-}Se(O)OH + HX \longrightarrow [RSe(OH)_2]^+X^- \qquad (23)$$

Paetzold[378] found that the infrared and Raman spectral data obtained from the HCl adducts of methane- and ethaneseleninic acids are consistent with the formula of an alkyl dihydroxy selenonium chloride.   The NMR spectrum of $C_8H_{17}SeO_2H \cdot HNO_3$ is also in agreement with the proposed selenonium structure.[423]   These adducts are the products of the interaction of seleninic acids with concentrated mineral acids.[378,423]   The nitric acid adducts are frequently isolated when a diselenide or selenocyanate is oxidized with concentrated nitric acid,[8,380,423,424] but are easily converted to the free acids by treatment with water or neutralization of the mineral acid by base. According to results of an x-ray analysis, the adduct $H_2NCH_2CH_2SeOOH \cdot HCl$ must be formulated as an ammonium salt.[264]

Aliphatic and aromatic seleninic acids are prepared by oxidation of diselenides or selenocyanates with concentrated nitric acid, 30% hydrogen peroxide, potassium permanganate in acetic acid, chlorine in aqueous medium and ammonium peroxydisulfate (for 4-carboxy-1,2-diselenacyclopentane[61]). In the oxidation of aromatic selenium compounds to seleninic acids by nitric acid, nitration has been observed.[298, 300]

$$R-Se-Se-R \xrightarrow{HNO_3} 2RSe(O)OH \qquad (24)$$

R: CH_3, C_2H_5;[377, 381] CF_3;[132] C_nH_{2n+1} (n = 6, 8, 10–18, 21);[423] $HOOC(CH_2)_{10}$, $C_2H_5OOC(CH_2)_{12}$;[424] C_6H_5, $2\text{-}XC_6H_4$ (X = NO_2, SO_2CH_3, OCH_3, COOH), $4\text{-}NO_2C_6H_4$, $2\text{-}NO_2\text{-}4\text{-}XC_6H_3$ (X = NO_2, CH_3, Br, Cl, OCH_3), 1-anthraquinonyl[436] $2,4\text{-}(NO_2)_2\text{-}5\text{-}CH_3C_6H_2$.[182]

In these reactions the seleninic acids were obtained in yields up to 96%.

$$R-Se-Se-R \xrightarrow{H_2O_2} 2R-Se(O)OH \qquad (25)$$

R: C_6H_{13};[423] $H_2NCH_2CH_2$;[287] C_6H_5, $2\text{-}ClC_6H_4$, $2\text{-}CH_3OOCC_6H_4$;[436] $2\text{-}C_6H_5C(O)C_6H_4$.[441]

$$R-SeCN \xrightarrow{\text{Oxidation}} R-Se(O)OH \qquad (26)$$

(Oxiding agent); R: (HNO_3); $C_{11}H_{23}$;[423] $1,5\text{-}NO_2C_{10}H_6$, $1,8\text{-}NO_2C_{10}H_6$, $1\text{-}NO_2\text{-}2\text{-}C_{10}H_6$;[296, 297, 298] $4\text{-}(4\text{-}NO_2C_6H_4)C_6H_4$;[101] (H_2O_2); $2\text{-}NO_2\text{-}4\text{-}C_6H_5C_6H_3$;[99] $(KMnO_4)$; $1\text{-}C_{10}H_7$, $2\text{-}C_{10}H_7$.[299, 300]

Diselenocyanates, $NCSe-(CH_2)n-SeCN$, (n=2;[380] n = 10, 12, 14[424]) are transformed into the diseleninic acids upon oxidation with concentrated or fuming nitric acid. Bis(trifluoromethylseleno) mercury and concentrated nitric acid produced the mercury salt of trifluoromethaneseleninic acid.[122] Selenoacetals, $RCH(SeR')_2$ were oxidized by nitric acid to $R'SeO_2H \cdot HNO_3$.[309]

Organyl selenium trihalides can be hydrolyzed to seleninic acids, when the hydrohalic acid formed is neutralized by base. 2-Benzoylphenyl,[441] 4-(4-nitrophenyl)phenyl[101] selenium tribromide, 2-phthalimidoethyl[406] and chloromethyl selenium trichloride[388*] were thus converted into the corresponding acids. Trifluoromethyl selenium trichloride was hydrolyzed by moist air.[132] Phenyl selenium trichloride produced, with an equimolar amount of water in the presence of pyridine, the complex $[C_6H_5Se(O)Cl_2]^-C_5H_6N^+$.[382]

The isolation of the trihalides required for the synthesis of seleninic acids can be avoided. Chlorine or bromine in aqueous medium converted $ClCH_2CH_2SeCN$, $(H_2NCH_2CH_2Se)_2$, and 2-phthalimidoethyl selenocyanate into seleninic acids.[406]   Seleninic acid anhydrides, $[RSe(O)]_2O$, are hygroscopic and easily form the acids.[383]   The intramolecular anhydride of 1,2-ethanediseleninic acid, obtained through oxidation of 1,2,5,6-tetraselenacyclooctane with nitric acid, is not hydrolyzed by water.[380]   The structure[202] and the preparation[203] of the *trans* form of this anhydride from 1,4-diselenacyclohexane has been investigated.   Dibutyl, dibenzyl, didodecyl, and diphenyl diselenide treated with ozone or *tert*-butyl hydroperoxide gave seleninic acid anhydrides. 2,4-Dinitrophenyl phenyl diselenide produced the mixed seleninic anhydride.[41]   The benzene-, methane-, trifluoromethane-, and ethaneseleninic acid anhydrides were obtained when the corresponding diselenides were oxidized with NO_2 or ozone in CCl_4.[383]   The easily hydrolyzable seleninic acid esters $RSe(O)OR'$ ($R = CH_3$, C_2H_5, CF_3; $R' = CH_3$, C_2H_5) formed when the anhydrides were treated with methanol or when the silver salts of the acids were treated with an alkyl iodide,[381] or when the seleninic acid chlorides were combined with sodium methoxide in methanol.[382] Oxidation of diphenyl diselenide and phenyl selenium chloride with ozone in the presence of ethanol and ozonization of ethyl 2,4-dinitrobenzeneselenenate gave seleninic acid esters.[41]   Transesterification makes it possible to exchange the alcoholic R' groups.[381]   The cyclic anhydride (III) was the product of the reaction between SeO_2 and 2,3-dimethylbutadiene.[354]   The previously reported structure of a cyclic selenone[44] is not consistent with the IR and NMR results obtained with (III):

(III)

The acid chlorides, $RSe(O)Cl$ ($R = CF_3$, CH_3, C_2H_5;[382] C_6H_5, 2,4-$(NO_2)C_6H_3$[41]) were produced by action of ozone or NO_2 on the organyl selenium chlorides, $RSeCl$, in CCl_4 solution at $-20°C$ or by treatment of the diselenide with chlorine and subsequent oxidation with ozone.   Water in tetrahydrofuran converted the acid chlorides to alkyl dihydroxy selenonium chlorides.[382]

The reduction of aromatic seleninic acids by thiols leads to diselenides and selenenic acids, $RSeOH$.[437]   2-Nitrobenzeseleninic acids were reduced by NaH_2PO_2 only to the selenenic acids while derivatives without the nitro group were converted to diselenides.[438].

2-Biphenyl- and 4-biphenylseleninic acids gave 4'-nitro derivatives upon treatment with nitric acid.[101]

Selenonic Acids, $RSeO_3H$

Selenonic acids have been prepared by oxidizing the seleninic acids with potassium permanganate in aqueous potassium hydroxide solution:

$$R-SeO_2H \xrightarrow{\text{KMnO}_4/\text{H}_2\text{O}/\text{KOH}} R-SeO_3K \qquad (27)$$

$$R: CH_3, C_2H_5, C_3H_7;^{67} C_8H_{17}, C_{10}H_{21}.^{425}$$

The yields in these reactions are as high as 92%. The octane- and decane-selenonic acids obtained by ion exchange from the potassium salts in aqueous solution partially decomposed upon evaporation of the solvent. The benzylisothiuronium salts of these acids have been prepared.[425] Paetzold[386] discussed the SeO force constants for $CH_3SeO_3^-$ in the context of other sulfur and selenium compounds. The IR spectrum of a polystyreneselenonic acid has been published by Zundel.[582] Pichat[406] oxidized R_2Se_2 with hypochlorous acid and $RSeO_2H$ with chlorine in water (R = H_2NCH_2CH) to the unstable selenonic acid. The aromatic selenonic acids have been synthesized by oxidation of seleninic acids and diselenides with $KMnO_4$ and $KMnO_4$ together with chlorine in aqueous solution, respectively. The introduction of the selenonic acid group into aromatic hydrocarbons can be accomplished with H_2SeO_4 without solvent or in acetic acid/acetic anhydride medium, or with SeO_3 in liquid sulfur dioxide. The latter method gave benzene-4-bromobenzene- and 4-chlorobenzeselenonic acids in 70–80% yield. The same reactions can be carried out with SeO_3 in H_2SeO_4 as a solvent.[146] Benzeneselenonic acid, a very hygroscopic substance, melts at 64°. The melting point of 142°C claimed for this compound by Schmidt[462] and Doughty[147] belongs to the adduct $[C_6H_5Se(OH)_2]^+ [C_6H_5SeO_3]^-$, which was prepared by Paetzold[379] from benzeneseleninic and -selenonic acid. Agrawal[25] synthesized ^{75}Se-2-fluororeneselenonic acid in 87% yield from the hydrocarbon and $H_2^{75}SeO_4$ in acetic acid/acetic anhydride and converted it into the 7-nitro, 7-amino and 7-acetylamino compounds. The selenonic acids reported in the older literature were compiled by Rheinboldt.[435]

Selenonic acids are oxidizing agents. They are reduced by concentrated hydrochloric acid to seleninic acids and by H_2S, SO_2 and zinc and acid to selenols. Upon heating above the melting point, the selenonic acids decompose, while their salts explode on strong heating. Mother liquors from the preparation of selenonic acids should therefore not be concentrated. The mass spectra of methyl 4-chloro- and 4-bromobenzeneselenonates were reported by Rebane.[426]

ALIPHATIC AND AROMATIC SELENIUM TRIHALIDES

Most of the selenium trihalides, $RSeX_3$, which have thus far been prepared, have an aromatic group bonded to the selenium atom. The tribromides are of greater stability than the trichlorides. The trichlorides are very moisture sensitive. All the trihalides decompose on melting. The corresponding fluorine and iodine compounds are unknown. Phenyl selenium trichloride in ether treated with water and pyridine formed $[C_6H_5Se(O)Cl_2]^-C_5H_6N^+$.[382]

The condensation of selenium tetrachloride with an aromatic hydrocarbon has found only limited application for the synthesis of selenium trichlorides (see Rheinboldt[435]). When diphenylacetylene and selenium tetrachloride were combined, 2-phenyl-3-chlorobenzoselenophene was isolated. The 1,2-diphenyl-2-chloroethenyl selenium trichloride was postulated as an intermediate.[444]

The reaction of diselenides and selenocyanates with bromine, chlorine or sulfuryl chloride in chloroform or carbon tetrachloride gave almost quantitative yields of the corresponding trihalides:

$$R_2Se_2 \underset{RSeCN}{\overset{}{\Big\rangle}} \; Cl_2 \; \Big\langle \; \begin{array}{l} \longrightarrow 2RSeCl_3 \\ \longrightarrow RSeCl_3 + ClCN \end{array} \tag{28}$$

These reactions were recently employed for the synthesis of methyl,[559]*[†] pentafluoroethyl,[537] 2-benzoylphenyl,[441] 2-carbothiophenoxyphenyl,[432] 2,4-dinitrophenyl,[331] and 2-phthalimidoethyl[406] selenium trihalides. 4-(4-Nitrophenyl)phenyl selenium tribromide was synthesized similarly.[101] Trifluoromethyl selenium trichloride was obtained from bis(trifluoromethyl) selenide and chlorine.[132] Methylene diselenide (CH_2Se_2) was cleaved by sulfuryl chloride to give chloromethyl selenium trichloride and selenium tetrachloride.[388] Seleninic acids, RSeOOH, treated with concentrated hydrobromic acid gave organyl selenium bromides (see Rheinboldt[435]).

The selenium trihalides prepared before 1955 can be found in Rheinboldt's review article. Chierici[107] investigated the UV spectra of aromatic selenium trichlorides.

Organyl selenium trihalides lose halogen on heating. 2,4-Dinitrophenyl selenium trichloride showed in solution the chemical behavior of the monochloride.[311] Acidic potassium iodide solutions and zinc in an inert solvent reduce the trihalides to diselenides.

DIORGANYL DISELENIDES, R_2Se_2

Diorganyl diselenides are yellow to orange-red liquids or solids. They can be purified by distillation or by recrystallization. Their color is caused by the chromophoric diselenide group. The structure of these compounds is correctly represented by the formula, R—Se—Se—R. The branched structure $R_2Se \rightarrow Se$ claimed for bis(2-naphthyl) diselenide has been proven incorrect.[56] Unsymmetric diselenides with two different aromatic groups or an aromatic and aliphatic group in the molecule have been prepared. Unsymmetric aliphatic derivatives seem to be unknown. Most of the diselenides reported in the literature are symmetric.

Symmetric Diselenides

The symmetric diselenides can be synthesized by introducing the diselenide group into organic molecules employing inorganic selenium compounds like alkali metal diselenides or converting other organic selenium compounds into diselenides.

Diselenides from Inorganic Selenium Compounds. The alkylation of sodium and potassium diselenides with aklyl chlorides, alkyl bromides, methyl iodide, dimethyl sulfate, and diethyl sulfate produces the diselenides in yields as high as 95%. The reactions can be carried out in aqueous medium,[4,377] in aqueous ethanol,[421] and in liquid ammonia.[71,80,477] Carboxyalkyl halides are first neutralized with an alkali carbonate solution before they are treated with sodium diselenide.[186]

$$Na_2Se_2 + 2RX \longrightarrow R—Se—Se—R + 2NaX \qquad (29)$$

R: CH_3, CD_3;[205,230] CH_3;[72,377,477] C_2H_5;[59,72,80,377] C_6H_{13}, C_8H_{17};[423] C_nH_{2n+1} (n = 10–18, 21);[422] $HOOC(CH_2)_n$ (n = 3–9);[17x] $HOOC(CH_2)_7$;[187] $HOOC(CH_2)_{10}Se(CH_2)_3$;[186] $C_2H_5OOC(CH_2)_{12}$;[421] $HO(CH_2)_n$ (n = 3,4), $HOCH_2C(CH_3)_2CH_2$, $HOCH_2CH(CH_3)CH_2$, $HOCH_2CH_2CH(CH_3)$, $CH_3COO(CH_2)_5$, $CH_3COO(CH_2)_3CH$-(CH_3);[19x] 3-indolylethyl;[4,18] 2-cyanophenylmethyl;[433] diphenyl-methyl;[127] 1-anthraquinonyl.[436]

Dibromomethane and sodium diselenide gave $(CH_2Se_2)_x$. Aromatic halides combine with sodium diselenide only when activated by a nitro group (see Rheinboldt[435] for references).

Diazonium salts form diselenides upon treatment with sodium diselenide:

$$2ArN_2^+Cl^- + Na_2Se_2 \longrightarrow Ar_2Se_2 + 2N_2 + 2NaCl \qquad (30)$$

Ar: 2- and 4-carboxyphenyl;[450] 5-benzothienyl, 2-carboxy-4-bromo-phenyl;[510] 2-carboxy-n-methylphenyl (n = 3, 4, 5).[454]

A 90% yield was reported for the bis(2-carboxyphenyl) diselenide.

The sodium diselenide required in these reactions can be prepared from the elements in liquid ammonia, from selenium in basic aqueous solution containing Rongalite or from selenium in strongly alkaline solution. Gunther[216] introduced $(CH_3O)_2Mg_2Se_2 \cdot 4CH_3OH$ as a reagent for the synthesis of diselenides. It was prepared by treating magnesium, activated by heating with iodine, with selenium in dry methanol. To the resulting brown solution were added aminoethyl bromide, benzyl chloride, butyrolactone or phthalide. Bis(2-aminoethyl) (55%), dibenzyl (82%), bis(3-carboxypropyl) (66%) and bis(2-carboxyphenylmethyl) di-selenide (82%) were isolated after hydrolysis of the reaction mixture. Dibenzyl diselenide was also obtained in 79% yield from benzaldehyde and the magnesium selenide in the presence of morpholine.

Potassium selenosulfate, obtained by treating potassium sulfite solutions with selenium, react with organic halides (reaction 31):

$$RX + K_2SeSO_3 \longrightarrow RSeSO_3K + KX \qquad (31)$$

The potassium organyl selenosulfates, which can be isolated[212, 287] are, however generally oxidized to the diselenides by addition of iodine or hydrogen peroxide.[435] The organyl selenosulfates are also transformed into the diselenides upon heating in aqueous solution. By refluxing the acidified reaction mixtures until the sulfur dioxide evolution had ceased, methyl, 3-carboxypropyl[215] 3-dimethylaminopropyl,[214] 2-amino-, 2-methyl- amino-2-dimethylaminoethyl,[212] and ^{75}Se-3-carboxy-propyldiselenide[488] were prepared. Klayman[287] proposed the following mechanism for this conversion:

$$^-O_3S-SeCH_2CH_2-NH_3^+ + H_2O \longrightarrow H_3N^+-CH_2CH_2Se^- + H_2SO_4$$

$$H_3N^+-CH_2CH_2Se^- + {}^-O_3SSe-CH_2CH_2-NH_3^+ \longrightarrow$$
$$(H_3N^+-CH_2CH_2)_2Se_2 + SO_3^{-2}$$

$$H_2SO_4 + SO_3^{-2} \longrightarrow SO_2 + SO_4^{-2} + H_2O \qquad (32)$$

Elemental selenium has been employed in a few isolated cases for the preparation of diselenides. Trifluoromethyl iodide at 260–285°C[132] (15%), mercury bis(trifluoroacetate)[382, 566] (20%), silver trifluoroacetate at 280°C[122, 155, 566] (28%), silver heptafluorobutyrate and bis-(heptafluoropropylseleno) mercury[154] gave with selenium bis(perfluoroalkyl) diselenides in the yields indicated. Methyl difluoroacetate similarly produced 26.5% of bis(difluoromethyl) diselenide.[568] i-Butyl aluminum (0.075 moles) reacted with selenium (0.22 moles) at 140–150°C. Di(i-butyl) diselenide was obtained in 70% yield after hydrolysis with sulfuric acid.[570]

Bis(pentafluorophenyl) diselenide was the product of reactions at 230°C between selenium and pentafluorophenyl iodide (16% yield) and pentafluorophenyl mercuric chloride (30%).[124] Under similar conditions a 6% yield of

$[H(CF_2)_6CH_2]_2Se_2$ was obtained from selenium and the fluoroalkyl iodide.[159] Bergson[56] observed the formation of bis(2-napthyl) diselenide upon heating a mixture of selenium and the monoselenide.

Selenium dioxide and trifluoroacetic anhydride in a sealed tube at 280°C gave a low yield of $(CF_3)_2Se_2$.[155] With ethylene in acetic acid at 50psi and 100–125°C, selenium dioxide is reported to form bis(2-acetoxyethyl) diselenide among other products.[375] Indole[53] and 2-methylindole[547] produced in a reaction with selenium dioxide in benzene the corresponding bis(3-indolyl) diselenides in low yields. Sebe[470] obtained diselenides by treating tropolone and its bromo derivatives with selenium dioxide in i-pentanol.

Diselenium dihalides are not good reagents for the synthesis of diselenides.[435] Diselenium dichloride and pentafluorophenyl lithium[124] produced only a 30% yield of the diselenide. Diselenium dibromide was observed to add to tetrafluoroethylene. Bis(2-bromotetrafluoroethyl) diselenide was isolated in 11% yield.[567]

Dialkyl diselenides obtained from ketones and hydrogen selenide are discussed later.

Diselenides from Organic Selenium Compounds. The organic selenium compounds which have been converted into diselenides, are, in decreasing order of importance, selenocyantes, R—SeCN, selenols, R—SeH, unsymmetric diorganyl selenides, organyl selenium chlorides, R—SeCl, selenenic acids, R—SeOH, and seleninic acids, RSe(O)OH.

Organyl selenocyanates, easily prepared from alkyl halides as well as activated aromatic halides and potassium selenocyanate or by selenocyanation of aromatic compounds are hydrolyzed in basic and acidic aqueous alcoholic medium. The diselenides are in most cases isolated in excellent yields:

$$2R-SeCN \xrightarrow{\text{hydrolysis}} R_2Se_2 \qquad (33)$$

The exact mechanism of this reaction is not yet known. The various postulated reaction sequences have been summarized by Rheinboldt.[435]

The following diselenides have been obtained by basic hydrolysis of selenocyanates with potassium or sodium hydroxide in water, methanol or ethanol (R given): C_nH_{2n+1} (n = 10–13, 15–18, 21);[421] 2-(benzoylamino)-ethyl;[211] $HOOCCH_2$;[304] $HOOCCH_2$, $HOOCCH_2CH_2$[17x] C_4H_9, i-C_5H_{11}, C_8H_{17}, $C_{12}H_{25}$;[359] 2-phthalimidoethyl;[406] C_6H_5;[14*, 520, 436] 2- and 3-$CH_3C_6H_4$;[336] 4-$CH_3C_6H_4$;[14] 2,4-$(CH_3)_2C_6H_3$;[106] 2-$CH_3OC_6H_4$,[436, 336] 4-$CH_3OC_6H_4$;[336] 2-ClC_6H_4;[436] 3-ClC_6H_4;[336] 2-IC_6H_4;[119] 2-$HOOCC_6H_4$;[436, 433] 2-$O_2NC_6H_4$;[99, 436] 4-NO_2-C_6H_4;[436] 2-NO_2-4-XC_6H_3 (X = CH_3, CH_3O, Cl, Br);[436] 2,4-$(NO_2)_2C_6H_3$;[241] 2,4-$(NO_2)_2$-5-$CH_3C_6H_2$;[182] 4-$C_6H_5C_6H_4$;[99] 2-$C_6H_5C_6H_4$, 2- and 4-(4-$NO_2C_6H_4)C_6H_4$;[101] 2-C_6H_5-4-NO_2-C_6H_3;[549]

$2\text{-}C_{10}H_7$;[512] 1-anthraquinonyl;[239] 5-nitro-2-pyridyl;[108] 3-indolyl;[2] 5-indolyl, 5-indolinyl.[22*, x, u]

1,4-Diselenocyanatoanthraquinone[253] and 1,4-tetramethylene diselenocyanate[85] produced polymeric diselenides. Hölzle[239] employed a dilute sodium carbonate solution to hydrolyze 2,4-dinitrophenyl selenocyanate, while Rheinboldt[436] used ammonia in methanol for the 2,4-dinitrophenyl and 2-methylsulfonylphenyl derivatives. 2-Bromophenyl selenocyanate was converted to the diselenide in a reaction with 2-bromoaniline.[181] 2-Carbomethoxy and 4-carbomethoxyphenyl selenocyanates were transformed by methylamine into the corresponding diselenides with concomitant conversion of -COOR to $CONHCH_3$.[528]

The cleavage of carboxyalkyl selenocyanates is best carried out in aqueous acetic or hydrochloric acid.[436] Bis(2-carbomethoxyphenyl) diselenide[436] and bis(6-quinolylmethyl) diselenide[317] were obtained in this way. The reaction of 1-naphthyl selenocyanate with 20% nitric acid produced nitronaphthyl diselenides, dinaphthyl diselenide and seleninic acids.[298] Sekido[472] reacted 8-quinolyl selenocyanate with hydrochloric acid in the presence of H_3PO_2 and oxidized the resulting selenol to the diselenide. The overall yield was only 12%.

Selenols are easily oxidized to diselenides. It is not necessary to isolate the selenols. The reaction mixture in contact with atmospheric oxygen or treated with an oxidizing agent like iodine or hydrogen peroxide will produce the diselenide:

$$2R\text{—}SeH \xrightarrow{\text{oxidation}} R_2Se_2 + H_2O \qquad (34)$$

R: $i\text{-}C_3H_7$, $sec\text{-}C_4H_9$, $(C_6H_5)_2CH$, $C_6H_5(CH_3)CH$;[334] $HOCH_2\text{-}CH_2$;[567] $4\text{-}FC_6H_4CH_2$;[305] C_6H_5;[64] $2,4,6\text{-}(CH_3)_3C_6H_2$;[106] $3,5\text{-}(tert\text{-}C_4H_9)_2\text{-}4\text{-}HOC_6H_2$;[362] 3,4-methylenedioxyphenyl;[134] 2-naphthyl;[56] 2- and 5-benzothienyl;[510] 2-pyridyl;[340] 4-pyridyl;[529] 1-anthraquinonyl.[436]

The hydrolyzed reaction mixtures obtained from a Grignard reagent and elemental selenium, conveniently give the diselenides upon oxidation:

$$RMgX + Se \longrightarrow R\text{—}Se\text{—}MgX \xrightarrow{H_2O} R\text{—}SeH \xrightarrow{O_2} R_2Se_2 \qquad (35)$$

R: $4\text{-}ClC_6H_4$, $4\text{-}BrC_6H_4$;[367] $2,4,6\text{-}(CH_3)_3C_6H_2$;[106] $2,4,6\text{-}(C_6H_5)_3\text{-}C_6H_2$.[33, 275]

Dithienyl diselenide was prepared by treating the reaction mixture obtained from thienyl lithium and selenium with $K_3Fe(CN)_6$ (USSR Pat., 165,752).

A number of selenium compounds which upon hydrolysis give selenols, can be transformed into diselenides as outlined in the following reactions:

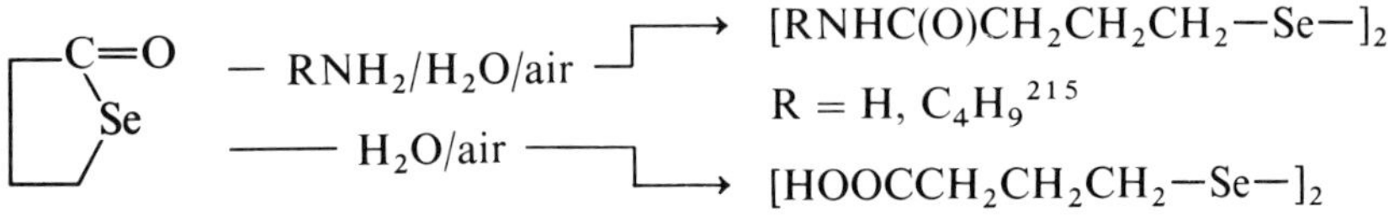

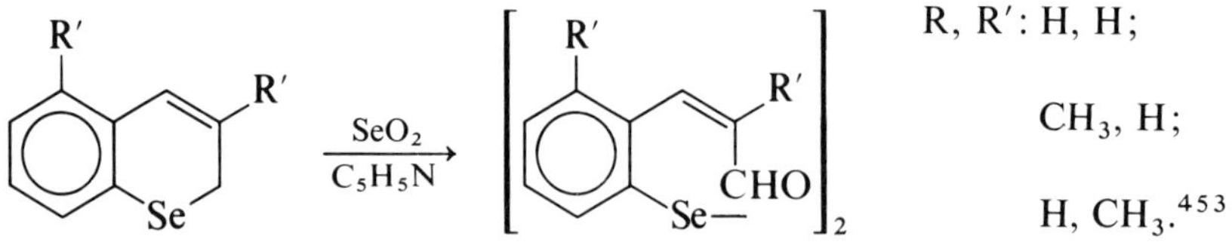

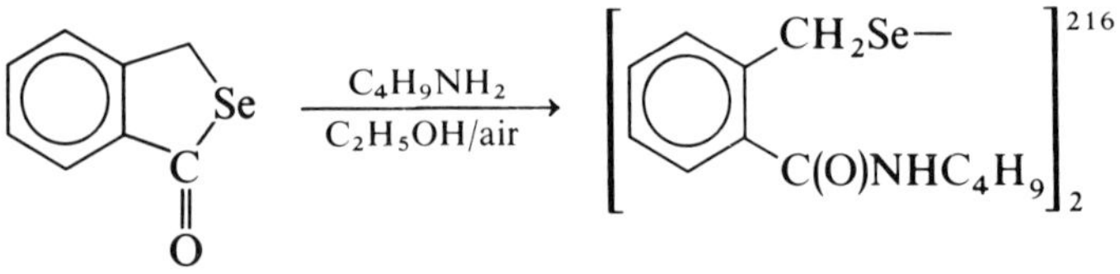

$$
\text{(cyclic } -C(=O)\text{-Se)} \quad
\begin{array}{l}
\xrightarrow{-\,RNH_2/H_2O/air} [RNHC(O)CH_2CH_2CH_2-Se-]_2 \quad R = H,\ C_4H_9{}^{215} \\
\xrightarrow{\quad H_2O/air \quad} [HOOCCH_2CH_2CH_2-Se-]_2
\end{array}
$$

$$
\xrightarrow[C_5H_5N]{SeO_2}
\left[\; \text{(benzo ring)} \begin{array}{c} R' \\ =CR' \\ Se-\ CHO \end{array} \right]_2
\qquad
\begin{array}{l}
R,\ R' :\ H,\ H; \\
CH_3,\ H; \\
H,\ CH_3.{}^{453}
\end{array}
$$

$$
\text{(isobenzo-Se-C(=O))} \xrightarrow[C_2H_5OH/air]{C_4H_9NH_2}
\left[\; \begin{array}{c} CH_2Se- \\ C(O)NHC_4H_9 \end{array} \right]_2^{216}
\qquad (36)
$$

Quantitative yields have been obtained in these reactions.

$$
\begin{array}{c} O \\ \| \\ C_6H_5C-NH-CH_2CH_2-Se- \end{array}
\begin{array}{c} O \\ \| \\ CC_6H_5 \end{array}
\xrightarrow[C_6H_5CONH(CH_2)_2SH/air]{C_4H_9NH_2/air\ or}
$$
$$
[C_6H_5CONH-CH_2CH_2-Se-]_2{}^{341} \qquad (37)
$$

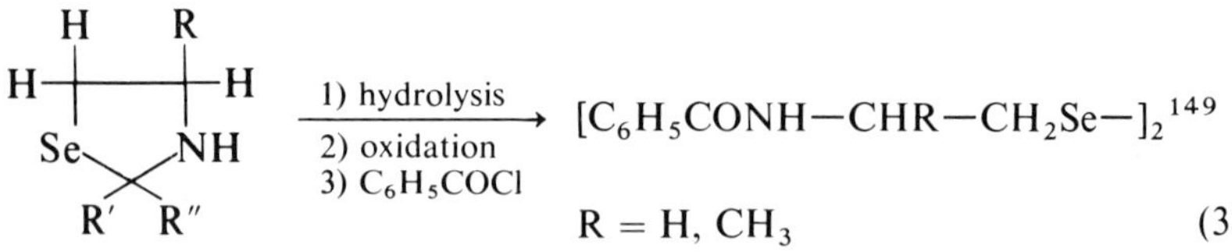

$$
\begin{array}{c} H\quad R \\ H-\!\!\underset{Se}{\overset{\,\,\,}{|}}\!\!\underset{NH}{\overset{\,\,\,}{|}}\!\!-H \\ R'\ R'' \end{array}
\begin{array}{l}
\xrightarrow{\substack{1)\ \text{hydrolysis} \\ 2)\ \text{oxidation} \\ 3)\ C_6H_5COCl}}
\end{array}
[C_6H_5CONH-CHR-CH_2Se-]_2{}^{149}
$$
$$
R = H,\ CH_3 \qquad (38)
$$

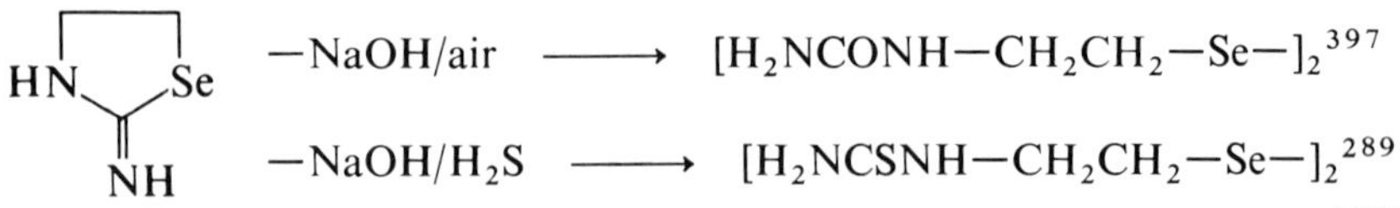

$$
\begin{array}{c} HN \\ \big\backslash \\ NH \end{array}\!\!Se
\quad \xrightarrow{-NaOH/air} \quad [H_2NCONH-CH_2CH_2-Se-]_2{}^{397}
$$
$$
\xrightarrow{-NaOH/H_2S} [H_2NCSNH-CH_2CH_2-Se-]_2{}^{289}
$$
$$
(39)
$$

Isoselenuronium salts, obtained from a halide and selenourea or a N-substituted selenourea, are hydrolyzed in basic medium to the diselenides. The isoselenuronium salts need not be isolated.

$$[R-Se-C(NH_2)(NHR)]^+X^- \xrightarrow[\text{air}]{H_2O/OH^-} R-Se-Se-R \quad (40)$$

R: CH_3, $C_6H_5CH_2$[90] 2-(17-oxoandrost-5-en-3β-yl), 2-(20-oxypregn-5-en-3β-yl) (US Pat., 3,372,173).

The compounds $(C_2F_5Se)_2Hg$[537] and $C_6H_5SeHgC_6H_5$[489] were oxidized by iodine to the corresponding diselenides.

The reduction of seleninic acids,[296, 297, 299, 300, 437, 438] selenenic acids[288, 436, 437, 441] and organyl selenium halides, $R-Se-Cl(Br)$[195, 270, 402, 565, 567] to diselenides is not of preparative importance, since all these compounds are in turn synthesized from the diselenides or the selenocyanates. The conversion of benzeneselenol by sulfuryl chloride[347] and phenylseleno magnesium bromide by halogen containing phosphorus compounds[401] to the diselenide seems to proceed via the intermediate selenium halide. The formation of a number of nitro substituted phenyl diselenides upon thermal decomposition of triselenides, R_2Se_3, has been reported.[439]

Benzyl selenides, $C_6H_5CH_2-Se-R$, are cleaved in liquid ammonia containing sodium according to reaction (41):

$$2C_6H_5CH_2-Se-R \xrightarrow[\text{2)hydrolysis/air}]{1)Na/NH_3} R-Se-Se-R \quad (41)$$

R: $CH_2C(CH_3)_2CH_2COOH$, $CH_2CH_2C(CH_3)_2CH_2COOH$, $\dot{C}H_2CH_2-C(CH_3)_2COOH$;[6] $C(CH_3)_2CH_2COOH$;[5] $CH_2C(CH_3)_2CH(OH)-COOH$;[7] $CH_2CH_2CH(OH)(CH_2)_4COOH$.[66]

Diphenyl diselenide was obtained upon heating phenyl allyl selenide in benzene or quinoline to at least 170°C.[112, 266] Hannig[227] isolated the diaryl diselenides from the reaction of $4-RC(O)C_6H_4SeCH_3$ (R=CH_3, C_5H_{11}) with bromine in pyridine/chloroform in 82% yield.

The following reactions modifying functional groups in the organic moiety of diselenides have been carried out:

$$(HOCH_2CH_2)_2Se_2 \xrightarrow{HCl} (ClCH_2CH_2)Se_2{}^{567} \quad (42)$$

$$[C_2H_5OOC(CH_2)_n]_2Se_2 \longrightarrow [HOOC(CH_2)_n]_2Se_2 \quad (43)$$

$$n = 12;^{421} \ n = 3.^{488}$$

$$(HOOCCH_2)_2Se_2 \xrightarrow{CH_2N_2} (CH_3OOCCH_2)_2Se_2{}^{304}$$

$$[RC(O)NH-(CH_2)_2]_2Se_2 \xrightarrow{\text{HCl}} [HCl \cdot H_2N(CH_2)_2]_2Se_2 \qquad (44)$$

$$R = C_6H_5;^{211} \ R = 2\text{-HOOCC}_6H_4.^{406}$$

$$2 \ \ 4\text{-CF}_3COOC_6H_4COCl + (H_2NCH_2CH_2)_2Se_2 \longrightarrow$$

$$(4\text{-CF}_3COC_6H_4CONH-CH_2CH_2)_2Se_2$$

(Ger. Offen., 1,943,787)

$$[HCl \cdot H_2N(CH_2)_2]_2Se_2 + \quad \underset{\substack{| \\ H_2C-C}}{\overset{H_2C-C\diagup\!\!\diagdown}{}} \quad O \longrightarrow$$

$$[HOOC-(CH_2)_2C(O)NHCH_2CH_2]_2Se_2{}^{217} \qquad (45)$$

$$[(CH_3)_2N-(CH_2)_{2 \text{ or } 3}-Se]_2 \xrightarrow{\text{CH}_3\text{I}}$$

$$[(CH_3)_3\overset{+}{N}-(CH_2)_{2 \text{ or } 3}-Se-]_2^+ \ 2I^- \ {}^{212, \ 214}$$

$$(2\text{-NC}-C_6H_4CH_2)_2Se_2 \xrightarrow{\text{NaOH/H}_2\text{O}} (2\text{-HOOCC}_6H_4CH_2)_2Se_2{}^{433}$$

$$(46)$$

Physical Properties. The spectral properties of diselenides have been determined by mass spectral, NMR and EPR techniques. Their infrared, visible, and ultraviolet spectra have been recorded. The leader in mass spectrometry of organic selenium compounds is Agenas. In his publications on this subject[13, 14, 15, 19, 20, 21, 23] detailed data on the mass spectral behavior of aromatic and aliphatic diselenides are available.

NMR investigations have been carried out with the following diselenides: CH_3;[405] CH_3, C_2H_5, C_3H_7;[538] $4\text{-FC}_6H_4CH_2$;[305] $HCl \cdot H_2NCH_2CH_2$;[288] 2-methyl-3-indolyl;[547] CH_3;[327] $CH_3(^{77}Se)$;[329] CH_3, C_2H_5, $C_6H_5(^1H, ^{77}Se)$;[310] $R-(Se)_n-R$ (chemical shift as a function of n);[511, 515] bis(3-acetyl-acetonyl).[144]

Windle[545, 546] obtained EPR signals from irradiated samples of $(C_4H_9)_2Se_2$, $[HOOCCH(NH_2)CH_2Se]_2$ and $(C_6H_5CH_2)_2Se$, but not from the methyl and phenyl derivatives. $[2,4,6-(C_6H_5)_3C_6H_2]_2Se_2$ did not dissociate into radicals in solution.[33] A theoretical treatment of such radicals has been carried out.[102]

A complete treatment of the IR spectra of dimethyl and bis(trideuteriomethyl) diselenide has been published.[205, 166] The IR spectra of other aliphatic diselenides[6,7,12,20,27,58,66,538] and aromatic diselenides[22, 547] have been reported.

Ultraviolet and visible spectra of many aromatic and aliphatic diselenides were recorded by Agenas,[22] Bergson,[55, 56, 58, 64, 65] Bogolyubov,[71] Chierici,[104,108] Dewar,[144] Kataeva,[275] Mautner,[340] and Wilshire.[547] Optical rotary dispersion studies have been carried out with $[HOOCCH(NH_2)-CH_2Se]_2$ and $[HOOCCH(R)Se]_2$ ($R = CH_3$, C_2H_5, C_6H_5).[145] The dipole moments of dimethyl and diethyl diselenide were determined by Mingaleva.[352] Bis(diphenylmethyl) diselenide has been subjected to an x-ray structural analysis.[393]

Unsymmetric Diselenides R—SeSe—R′

Rheinboldt[435] reported that 21 unsymmetric diselenides had been synthesized up to 1955, mainly from selenenyl halides and selenols. In the past 15 years only a few more compounds have been synthesized. Nakazaki[369] was able to condense selenocyanates with selenols:

$$R—SeCN + R'SeH \Longrightarrow R—Se—Se—R' + HCN \qquad (47)$$

$$R, R': C_6H_5, 2-C_{10}H_7; C_6H_5, 2-NO_2C_6H_4$$

McFarlane[329] mixed equimolar amounts of benzeneselenol and dimethyl diselenide and pumped off the methaneselenol formed. Methyl phenyl diselenide was isolated. It decomposed on distillation to the symmetric diselenides.

Reactions of Diselenides

Most reactions given by diselenides affect the diselenide group. Upon reduction and oxidation the selenium-selenium bond is cleaved. The products obtained are summarized in Figure 8-2. The reactions, in which the organic part of the molecule is changed without cleavage of the diselenide link, have already been discussed.

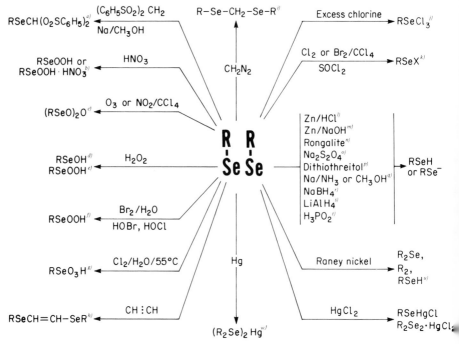

References: a) 392. b) 6, 107, 182, 298, 377, 381, 423, 424, 436, 441. c) 41, 383. d) 288. e) 287, 423, 436. f) 406. g) 406. h) 277. i) 403. j) 101, 388, 537, 559. k) 59, 60, 239, 253, 432, 382, 512, 537, 566, 567, 568. l) 367, 520. m) 191, 195, 226. n) 4, 18, 168, 186, 187, 188, 189. o) 433, 450, 510. p) 217. q) 1, 38, 60, 230, 336. r) 115, 213, 215, 217, 287, 486, 510, 573. s) 113. t) 215, 216, 472. u) 30, 53, 233, 548. v) 201, 372. w) 334, 428.

Figure 8-2   The reactions of diselenides.

DIORGANYL SELENIDE SULFIDES, R—Se—S—R

Diorganyl selenide sulfides have been prepared by reacting an organyl chalcogen halide with the sodium salt of a thiol or selenol:

$$R—SeBr + R'SNa \longrightarrow R—Se—S—R' + NaBr \qquad (48)$$

R, R': C_2H_5, C_2H_5;[59] C_6H_5, 2-$NO_2C_6H_4$.[64]

Nakazaki[369] obtained diphenyl, phenyl 2-nitrophenyl and bis(2-nitrophenyl) selenide sulfide by heating the selenocyanates with thiols. 2-Aminoethyl selenosulfate in methanolic sodium hydroxide solution at 0°C gave with decanethiol the selenide sulfide which disproportionated to the disulfide and diselenide.[289]   Such a disproportionation has also been observed by Bergson.[59] Bis(pentafluorophenyl) selenide sulfide, however, was formed by refluxing a mixture of the diselenide and disulfide in petroleum ether at 80–100°C for 18 h.[295]   Renson[432] reacted 2-carbethoxyphenyl, 2-acetylphenyl, and 2-chloroformylphenyl selenium chloride with benzenethiol and obtained 2-

carbethoxyphenyl phenyl, 2-acetylphenyl phenyl, and 2-phenylthioformyl-phenyl phenyl selenide sulfide.

Degani[140] determined the reactivity of a number of aromatic selenide sulfides towards bis(phenylsulfonyl)methyl sodium in methanol:

$$R-SeS-R' \xrightarrow{(C_6H_5SO_2)_2CHNa} (C_6H_5)_2SO_2CH-S-R' \quad (49)$$

In the series of compounds R_2S_2, R_2SeS, R_2Se_2, the disulfide is the least and the diselenide the most reactive. Diphenyl selenide sulfide added to acetylene and phenylacetylene forming $C_6H_5Se-C(R)=CH-SC_6H_5$.[277] UV data are available for diethyl and 2-nitrophenyl phenyl selenide sulfide.[59, 64] Structural and IR investigations were also carried out with the latter compound.[64]

DIORGANYL TRISELENIDES, R_2Se_3

The only known aliphatic triselenide is the diethyl derivative, which was formed in 10% yield in the reaction between ethaneselenol and $SeOCl_2$. The major product was the diselenide (for references see Rheinboldt[435]). Rheinboldt[439] prepared seven aromatic triselenides including the 1-anthraquinonyl derivative in 90% yield according to reaction (49a):

$$2X-\text{⟨O⟩}-SeSeCN + 2C_2H_5SH \longrightarrow 2HCN + Se + (C_2H_5)_2S_2 + Ar_2Se_3$$

$X = H, CH_3, Cl, Br, CH_3O, NO_2$ \quad (49a)

The 2-mehtyl- and 1,2-dimethyl-3-indolyl triselenides were obtained in about 20% yield from the indole and selenium dioxide in benzene. The UV spectra of these triselenides have been reported.[547] The diphenyl triselenide was isolated as one product of the thermal decomposition of diphenyl diselenide, and of the reaction of diphenyl diselenide and diphenyl selenide at 290°C in vacuo with elemental selenium.[479] Grant[204] showed by analyzing the proton NMR spectra of liquids obtained from dimethyl diselenide and selenium, that dimethyl polyselenides, $CH_3-Se_n-CH_3$ (n = 2–10), exist.

The other polychalcogenides, R_2Se_2S, R_2S_2Se, R_2S_3Se, and R_2Se_3S can be found in Rheinboldt's review. No work has been done in this area since 1951.

DIORGANYL SELENIDES

The large number of diorganyl selenides prepared in the last fifteen years makes it necessary to subdivide this section according to the organic groups bonded to the selenium atom. The discussion of symmetric aliphatic

selenides is followed by the treatment of unsymmetric aliphatic derivatives. Symmetric aromatic selenides, unsymmetric diaryl selenides, aryl
alkyl selenides, aryl trifluoromethyl selenides, and selenides of the type
$R—Se(CH_2)_nSe—R$ are then discussed in the order indicated.

Symmetric Dialkyl Selenides

Symmetric dialkyl selenides have been synthesized by alkylation of sodium
selenide, elemental selenium, selenium dioxide, hydrogen selenide, and
selenium tetrachloride, by reduction of dialkyl selenium dihalides, and by
modifying organic groups in dialkyl selenides.

Dialkyl selenides from inorganic selenium compounds. Symmetric dialkyl
selenides, R_2Se, can be prepared by alkylation of sodium selenide. The reaction is performed in basic aqueous medium, when the sodium selenide has
been synthesized by dissolving elemental selenium in sodium hydroxide
solution in the presence or absence of Rongalite:

$$Se \xrightarrow[\text{Rongalite}]{\text{NaOH/H}_2\text{O}} Na_2Se \xrightarrow{RX} R_2Se \qquad (50)$$

$R:CH_3, C_2H_5, C_3H_7, C_4H_9;$[231, 350] $C_5H_{11};$[350] $^{14}CH_3;$[50] $C_6H_5CH_2;$[98]
$CH_2COOH.$[158, 284]

The yields obtained have been as high as 90%, but decrease with increasing
chain length of the halide.

Sodium selenide is also formed from the elements in liquid ammonia.
Addition of a halide or a dialkyl sulfate produces the selenide:

$$2Na + Se \xrightarrow{\text{NH}_{3(l)}} Na_2Se \xrightarrow[R_2SO_4]{\text{RX or}} R_2Se \qquad (51)$$

$R: CH_3, C_2H_5;$[384] $C_2H_5, i\text{-}C_3H_7;$[80] $C_2H_5;$[71] $C_6H_5CH_2.$[477]

The selenides were isolated in yields of at least 80%.

The alkylation of elemental selenium has been carried out with trifluoromethyl iodide,[132] silver heptafluorobutyrate,[154] mercury pentafluoropro-
pionate,[537] $H(CF_2)_6CH_2I$, $H(CF_2)_{10}CH_2I$,[159] and triethyl and tri-i-butyl
aluminum.[570] The yields of fluoroalkyl selenides ranged from 10 to 40%.
The dialkyl selenides were isolated in 40–50% yields from reaction mixtures
originally containing 2 moles of selenium for each mole of trialkyl aluminum.

Selenium dioxide treated with trifluoroacetic anhydride or silver trifluoroacetate at 260°C gave bis(trifluoromethyl) selenide.[155] A selenide of unknown
structure was observed as a product of the reaction of β-pyronene with selenium dioxide[508, 509] and of 5,5-dimethyl-1,3-cyclohexanedione.[175] Propene

and i-butene reacted with SeO_2 in water at $80°C$ to form $[(CH_3)_2CH(OH)-CH_2]_2Se$ and $[CH_3CH(OH)CH_2]_2Se$, respectively (Ger. Pat., 1,105,414), while 1-octene,[498] 2-butene,[243] ethylene,[375] and methyl undecenoate[497] in acetic acid at $110°C$ yielded the selenides $[R—CH(OOCCH_3)CH(R')]_2Se$ (R, R': C_6H_{13}, H; CH_3, CH_3; H, H; $CH_3OOCC_8H_{16}$, H).

Aldehydes[79] were transformed into selenides according to reaction (52):

$$R—C(O)H + HCl_{(g)} + H_2Se$$

$$\xrightarrow[N_2]{0° \text{ to } -10°C} R—CH(Cl)—Se—CH(Cl)R \quad (52)$$

$$R=CH_3, C_2H_5$$

Landa[306] kept ethylene and hydrogen selenide for 1 h over a MoS_2 catalyst at a presssure of 52 atm and a temperature of $220°C$ and isolated diethyl selenide in 65% yield.   Hydrogen selenide added to the triple bond in phenyl carbethoxyacetylene to give bis(2-carbethoxy-1-phenyl-ethenyl) selenide.[196] Kaabak[263] electrolyzed a mixture of selenium and vinyl cyanide in $1N$ sodium sulfate and obtained bis(2-cyanoethyl) selenide.   It is likely, that hydrogen selenide, which was observed as a by-product, was generated by electrolysis and then added to the double bond.   The same selenide was obtained by subjecting vinyl cyanide and $(C_2H_5)_3Ge—SeH$ to UV radiation at $20°C$.[519]

Selenium tetrachloride adds to carbon-carbon double bonds, producing bis(2-chloroalkyl) selenium dichlorides.   When vinyl cyanide (1.5 moles) was mixed with $SeCl_4$ (0.1 mole) at room temperature, bis(2-chloro-2-cyanoethyl) selenide and not the expected selenium dichloride was obtained in 77% yield.[443]   While 1-cyanobutadiene gave with selenium dioxide in concentrated hydrochloric acid at $-15°C$, the expected selenium dichloride,[174] 2-naphthol yielded only bis(2-hydroxy-1-naphthyl) selenide.[173]   A selenide is expected to form in such reactions, when the selenium-chlorine bonds in the selenium dichlorides are weakened by a functional group and/or the unsaturated starting material is easily chlorinated by the selenium dichloride.

The addition of selenium monochloride to propene at $-40°C$ in methyl chloride gave a mixture of bis(2-chloropropyl) and bis(1-chloro-i-propyl) selenide in 43% yield.[312]

Dialkyl Selenides by Reduction of Dialkyl Selenium Dihalides.   The reduction of dialkyl selenium dihalides is of importance only when the dihalides can be synthesized from an organic compound and an inorganic selenium halide.

$$R_2SeCl_2 \xrightarrow{K_2S_2O_5/H_2O} R_2Se \quad (53)$$

Funk[173,174,175] prepared in this manner the following selenides: $[R-CH-(X)CH_2]_2Se$ (R, X: H, Br; Cl, Cl; C_6H_5, Cl; CH_3, Cl; $ClCH_2$, Cl; $NCCH:CH$, Cl), $(ClCH:CH)_2Se$, $[ClCH_2CCl(CH_3)CH_2]_2Se$, bis(2-chlorocyclopentyl) selenide, and bis(2-chlorocyclohexyl) selenide.

Modifications of the Organic Group. The reactions modifying the organic moiety in symmetric dialkyl selenides are summarized in the following equations:

$$(NC-CH_2CH_2)_2Se \xrightarrow{HCl} (HOOC-CH_2CH_2)_2Se^{263} \qquad (54)$$

$$(HOOC-CH_2CH_2)_2Se \xrightarrow{CH_3OH/HCl} (CH_3OOC-CH_2CH_2)_2Se^{158} \qquad (55)$$

$$[C_2H_5OOCCH=C(C_6H_5)]_2Se \longrightarrow [HOOCCH=C(C_6H_5)]_2Se^{196}$$

$$Se(CH_2COOH)_2 \xrightarrow{CH_3COCl} Se(CH_2CO)_2O^{284}$$

$$[R-CH_2CH(Cl)]_2Se \xrightarrow{(C_2H_5)_2NC_6H_5, 150°C} (R-CH=CH)_2Se^{79} \qquad (56)$$

The yields of alkenyl selenides in reaction (56) were 15–20%.

Other methods. Kataev[271] reacted carbomethoxyacetylene with selenourea and obtained as one of the products bis(2-carbomethoxyvinyl) selenide. Other methods, which have not been used for the synthesis of selenides during the period covered by this review, are based on the cleavage of dialkyl diselenides with sodium ethoxide followed by addition of the proper halide, and the reaction of ethers with aluminum selenide at 300–350°C. For more details concerning these reactions and a list of selenides prepared prior to 1955, the reader is referred to Rheinboldt.[435]

Physical Properties and Reactions. Mass spectral data are available for selenides.[14,17,20] The UV spectrum of diethyl selenide has been reported.[71] The infrared and Raman spectra of dimethyl selenide,[28,171,405,478] bis(trideuteriomethyl) selenide,[29] bis(pentafluoroethyl) selenide,[537] diindolyl selenide,[20] and bis-(ω-carboxyalkyl) selenides[12] have been studied. The analysis of the microwave spectrum of dimethyl selenide gave a $C-Se-C$ bond angle of 90° and a dipole moment of 1.141 D. The potential barrier for the internal rotation of the methyl groups is 1500 ± 20 cal/mole.[47] Dipole moments for dimethyl, diethyl and di-*tert*-butyl selenide were calculated by Krishnamurthy.[301] The EPR spectra of radicals generated by irradiation of dialkyl selenides with light of 2537 Å at liquid nitrogen temperature has been studied.[545,546] NMR data for dimethyl and diethyl selenide,[310, 351] bis(pentafluoroethyl) selenide,[537] diethyl selenide (^{77}Se),[310,84] dimethyl, diphenyl (^{1}H, ^{77}Se),[310] and ^{13}C-dimethyl selenide[150, 327] are available in the literature.

An electron diffraction study of dimethyl selenide has been published.[185a] The crystal and molecular structure of bis(4-methylphenyl) selenide was determined by Blackmore.[69a] Literature references to older structural work on organic selenium compounds are given in this paper.

The reaction of diorganyl selenides are summarized in Figure 8-3. Perfluoromethyl and -ethyl selenides are stable towards water, alkali, dilute acid and chlorine at room temperature. With chlorine at higher temperatures, $SeCl_4$, RSeCl, RCl, and $RSeCl_3$ are formed.[132, 537] West[540] found, that phenol is weakly hydrogen bonded to dibutyl selenide in carbon tetrachloride.

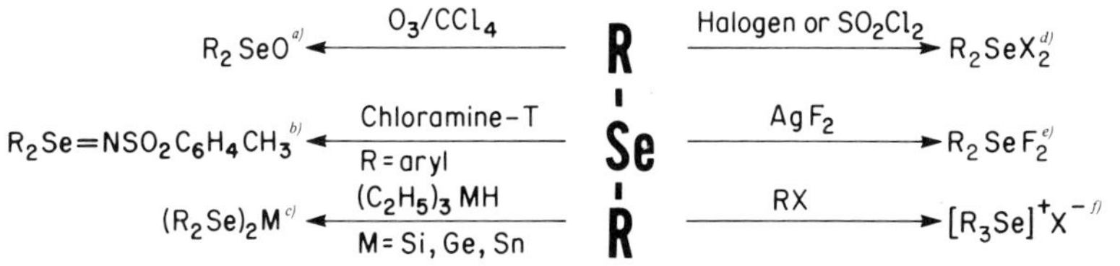

References: a) 41, 374. b) 490. c) 516, 517, 518. d) 181, 192, 206, 227, 231, 236, 308, 324, 372, 384, 385, 387, 443, 479, 556, 581.
e) 558, 560 f) 118, 137, 180, 181, 199, 231, 232, 371, 559.

Figure 8-3 The reactions of diorganyl selenides.

Unsymmetric Aliphatic Diorganyl Selenides

Unsymmetric, aliphatic diorganyl selenides have been obtained by reacting selenols or their salts or organyl selenium halides with appropriate organic compounds and by modifying organic groups in selenides.

Unsymmetric Aliphatic Selenides from Selenols. The required selenols or their salts are prepared by reducing diselenides with Rongalite in basic aqueous medium, with sodium borohydride in water,[212, 213, 214] by hydrolyzing reaction mixtures containing organylseleno magnesium halide, and by reacting sodium acetylides with selenium in liquid ammonia.

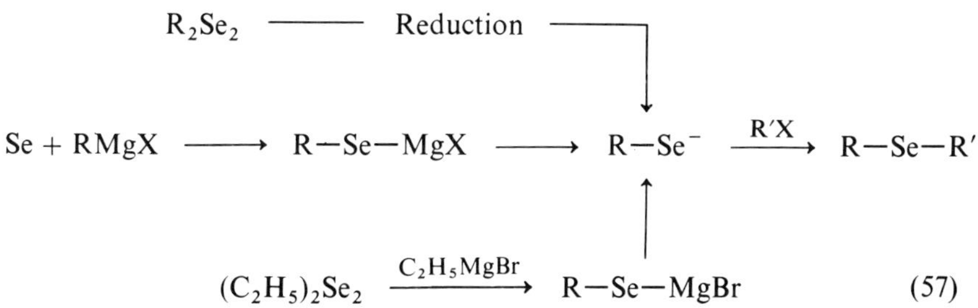

$$Se + RMgX \longrightarrow R-Se-MgX \longrightarrow R-Se^- \xrightarrow{R'X} R-Se-R'$$

$$(C_2H_5)_2Se_2 \xrightarrow{C_2H_5MgBr} R-Se-MgBr \tag{57}$$

R, R'X: CH_3, $C_6H_5C(O)CH_2Br$;[73] CH_3, CH_3COCH_2Br;[74] C_2H_5, $CH_3C(O)CH_2Cl$;[60] C_2H_5, $ClCH_2COOH$;[60*] C_2H_5, $Cl(CH_2)_3Br$;[188] C_4H_9, $CH_2:CH-(CH_2)_nBr (n = 2,3)$;[191] $(CH_3)_2NCH_2CH_2$, CH_3I;[212] $(CH_3)_3N(CH_2)_{2\ or\ 3}$, CH_3I;[212, 214] CH_2COOH, $C_8H_{17}CH(CH_3)-$ $(CH_2)_6Br$; $(CH_2)_7COOH$, $C_8H_{17}CH(CH_3)Br$;[187] $(CH_2)_nCOOH$ (n = 1-4), 3-indolyl-2-ethyl bromide;[18x, 3*] $(CH_2)_nCOOH$, $C_mH_{2m+1}Br$ (m + n = 13 or 14);[168] $(CH_2)_{10}COOH$, $C_nH_{2n+1}Br$ (n = 1, 2, 5-18, 21, 22); $(CH_2)_{10}CONH_2$, $C_{10}H_{21}Br$;[189] $(CH_2)_{10}COOH$, $CH_3CH_2CH_2Cl$ and $HOCH_2CH_2Br$;[186] $(CH_2)_{12}COOH$, CH_3I and C_2H_5Br;[421] $C_6H_5CH_2$, $H_2NCH_2CH_2CH_2Br$;[213] $C_6H_5CH_2$, $ClCH_2CH_2CH(OH)-$ $(CH_2)_4COOH$;[66] $C_6H_5CH_2$, $HN(O)CHNC(O)-\underline{C}H-CH_2CH_2-$ Br;[291] $CH_3C:C$, C_2H_5Br and i-C_3H_7I;[77,82] $CH_3C:C$, CH_3I;[74] $CH_2:CH-C:C$, CH_3X;[74,404*†] $CH_2:C(CH_3)C:C$, CH_3X and C_2H_5X;[419*†] $C_6H_5C:C$, CH_3I.[73]

Carboxyalkaneselenols, obtained by reduction of the diselenides with Ronga-lite, gave upon treatment with 1- and 2-naphthylmethyl chloride, naphthyl-methylselenoacetic, -2- and -3-propionic, -2-, -3- and -4-butyric, -2- and -3-isobutyric, -5-valeric, -6-caproic and -11-undecanoic acids.[170] Fredga[169] prepared a series of amidocarboxyalkyl alkyl selenides by alkylation of the admidocarboxyalkaneselenolates with alkyl bromides. The selenolates and their precursor compounds were not isolated.

$$Br-(CH_2)_n-CN \xrightarrow[H_2O]{Na_2Se_2} [-Se-(CH_2)_n-CN]_2 \xrightarrow[ether]{HCl/H_2O} \downarrow$$

$$[-Se-(CH_2)_nC(O)NH_2]$$

$$\downarrow \text{Rongalite } NH_3/H_2O$$

$$C_mH_{2m+1}Se-(CH_2)_nCONH_2 \xleftarrow{C_mH_{2m+1}Br} {}^-Se-(CH_2)_nCONH_2$$
$$n = 2, m = 1-6, m + n = 8$$

$$Br-(CH_2)_nCOOC_2H_5 \xrightarrow[NH_3/H_2O]{(NH_4)_2Se_2} [-Se-(CH_2)_nCOOC_2H_5] \qquad (57a)$$

$$\text{Rongalite}/NH_3/H_2O$$

$${}^-Se-(CH_2)_n-COOC_2H_5 \xrightarrow{NH_3/H_2O}$$

The preparation of other compounds of this type is described in reactions (72) and (73) later in this section. The yields in these reactions range as high as 90%. In some cases, however, much of selenide is lost during the purification process of the crude product.[3,18]

Bis(pentafluoroethylseleno) and bis(heptafluoropropylseleno) mercury treated with methyl and ethyl iodide in sealed tubes produced perfluoroalkyl alkyl selenides in good yield.[154*,537*†] The cyclic seleninic acid ester (IV) was reduced with sodium borohydride, the resulting seleno compound alkylated with benzyl chloride and the hydroxy group in the selenide (V) acetylated:[354†*]

$$\text{(IV)} \quad \xrightarrow[\text{2) C}_6\text{H}_5\text{CH}_2\text{Cl}]{\text{1) NaBH}_4} \quad \text{C}_6\text{H}_5\text{CH}_2-\text{Se}-\text{CH}_2-\overset{\overset{\displaystyle \text{CH}_3}{|}}{\text{C}}=\overset{\overset{\displaystyle \text{CH}_3}{|}}{\text{C}}-\text{CH}_2\text{OH} \quad (58)$$

(IV)   (V)

Selenols react with epoxides according to reaction (58a):

$$\text{R}-\text{SeH} + \text{R}'-\overset{\overset{\displaystyle \text{R}''}{|}}{\text{CH}}-\text{CH}_2 \quad \xrightarrow[\text{UV}]{\text{Na or}} \quad \text{R}'-\overset{\overset{\displaystyle \text{OH}}{|}}{\underset{\underset{\displaystyle \text{R}''}{|}}{\text{C}}}-\text{CH}_2-\text{Se}-\text{R} \quad (58a)$$

R, R', R'': CH_3, H, H;[215] C_3H_7, CH_3, CH_3; C_6H_5, CH_3, CH_3;[283] C_3H_7, CH_3, H; C_3H_7, $ClCH_2$, H; C_3H_7, CH_2:CH, H; C_3H_7, CH_2:CH, CH_3; C_3H_7, CH_2:CCl, H; C_6H_5, CH_3, H; C_6H_5, $ClCH_2$, H.[282*†]

Propaneselenol added to 3,4-epoxy-1-butene and 3,4-expoxy-3-methyl-1-butene mainly in the 1,4-position ($C_3H_7Se-CH_2=CR-CH_2OH$), while propane- and benzeneselenol gave with expoxyethylbenzene and 3,4-epoxy-1-pentene mixtures containing $R'CH(OH)CHR''-SeR$ and $R'CH(SeR)-CHR''OH$ ($R=C_3H_7$, C_6H_5; R', $R'':C_6H_5$, H; CH_2:CH, CH_3).[282] The yields range from 30–85%. 3-Indolyldimethylaminomethane and phenylmethaneselenol gave 3-indolylmethyl benzyl selenide in 100% yield.[20*,x,u]

Lactones were employed as starting materials for the synthesis of benzyl carboxyalkyl selenides. The reactions were carried out in dimethylformamide at a temperature of 120°C:

$$R''\diagdown C \diagup (CH_2)_n \diagdown C{=}O$$

(structure with R', R_2C, CH_2 groups forming a ring) ——$C_6H_5CH_2Se^-$——→

$$C_6H_5CH_2{-}Se{-}CH_2{-}\overset{\overset{R}{|}}{C}{-}\overset{\overset{R''}{|}}{C}{-}(CH_2)_n COOH \quad (59)$$

with lower labels R, R'

R, R′, R″, n, % yield: H, H, H, 0, 73; H, CH₃, CH₃, 0, 35; CH₃, H, H, 0, 78; CH₃, H, OH, 0, 23;[6*] H, H, CH₃, CH₃, 1, 41.[7*]

Propiolactone and the lactone of 2-carboxyphenylmethanol gave in similar reactions the corresponding benzyl selenides.[10,11*] The 3,1-selenazolidines (VI) upon methylation under nitrogen generated the selenides (VII):[149]

$$H_2C{-}C\diagup R \diagdown R' \quad \xrightarrow{CH_3X} \quad CH_3{-}Se{-}CH_2{-}\overset{\overset{R}{|}}{\underset{\underset{R'}{|}}{C}}{-}N{=}CH{-}C_6H_5 \quad (60)$$

(ring with Se, NH, CH, C_6H_5)

(VI)   (VII)

R, R′: H, H; H, CH₃; CH₃, CH₃.

R, R′: H, H; H, CH₃; CH₃, CH₃.

Unsymmetric selenides are also obtainable through addition of a selenol to triple and double carbon-carbon bonds:

$$HC{\equiv}C{-}R + R'SeH \longrightarrow R'{-}Se{-}CH{=}CH{-}R \quad (61)$$

R, R′: C_2H_5O, C_2H_5; C_2H_5S, C_2H_5;[513*] $C_6H_5C(O)$, $C_6H_5CH_2$;[196*] HC:C, C_3H_7.[153]

These addition reactions proceed in absolute ether or methanol with or without the presence of a catalytic amount of base. Elsakov[153] reported that his base catalyzed addition in methanol had produced exclusively the *cis* form, while in other solvents, mixtures of *cis* and *trans* compounds had been obtained.

The addition reactions to olefins carried out in the presence of triethylamine in ethanol or diethyl ether or with the neat reactants gave yields of 75%:

$$\begin{matrix} R \\ R^1 \end{matrix}\!\!>C{=}C<\!\!\begin{matrix} R^2 \\ R^3 \end{matrix} + C_6H_5CH_2SeH \longrightarrow C_6H_5CH_2{-}Se{-}\underset{R^1}{\overset{R}{C}}{-}CH<\!\!\begin{matrix} R^2 \\ R^3 \end{matrix} \qquad (62)$$

R, R¹, R², R³: H, H, H, COOH;[532,534] H, H, H, C(O)H;[578] H, H, H, $CH_3C(O)$;[574,577] H, H, COOH, $CH_2SCH_2C_6H_5$;[62] H, H, CH_3, C(O)H; H, CH_3, H, C(O)H; H, CH_3, CH_3, C(O)H; CH_3, CH_3, H, C(O)H; CH_3, CH_3, H, $CH_3C(O)$[577]; CH_3, CH_3, COOH, COOH.[5*]

Organyl Selenium Halides as Starting Materials for Unsymmetric Aliphatic Selenides. Trifluoromethyl selenium chloride condensed with acetone to trifluoromethyl acetonyl selenide.[566] The following selenides became available through the addition of halomethyl selenium halides to ethylene: $F_2CH{-}Se{-}CH_2CH_2Br$,[568] $Cl_3C{-}Se{-}CH_2CH_2Cl$,[565] and $CF_3{-}Se{-}CH_2CH_2Cl$.[566] The products were obtained in 100% yield, when the neat selenium halides were kept in contact with ethylene.

Phenyl selenium bromide added to triphenyl carbethoxymethylene phosphorane (VIII). The intermediate (IX) lost hydrogen bromide:[400]

$$\underset{\text{(VIII)}}{(C_6H_5)_3P{=}CH{-}COOC_2H_5} + R{-}SeBr$$
$$\downarrow \qquad (63)$$
$$\underset{\text{(IX)}}{[(C_6H_5)_3P{-}\overset{SeR}{CH}{-}COOC_2H_5]^+Br^-}$$
$$\downarrow +\text{(VIII)}$$
$$[(C_6H_5)_3P{-}CH_2COOC_2H_5]^+Br^- + (C_5H_6)_3P{=}C(SeR){-}COOC_2H_5$$

Modification of Organic Groups. Modification of the organic groups in selenides can easily be carried out without effecting the selenium-carbon bonds, when a multiple carbon-carbon bond or functional groups are present in the organic moiety. Brandsma[78] used divinyl selenide to prepare other unsaturated selenides:

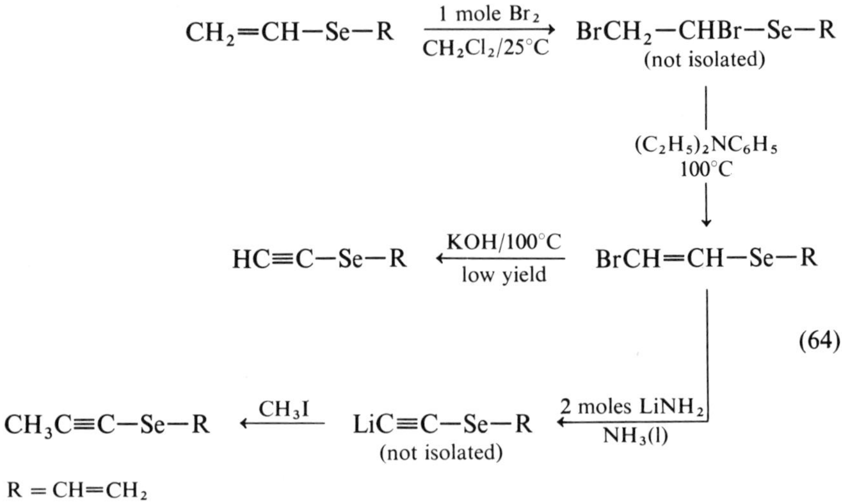

$$CH_2=CH-Se-R \xrightarrow[CH_2Cl_2/25°C]{1 \text{ mole } Br_2} \underset{\text{(not isolated)}}{BrCH_2-CHBr-Se-R}$$

$$(C_2H_5)_2NC_6H_5 \atop 100°C$$

$$HC\equiv C-Se-R \xleftarrow[\text{low yield}]{KOH/100°C} BrCH=CH-Se-R$$

(64)

$$CH_3C\equiv C-Se-R \xleftarrow{CH_3I} \underset{\text{(not isolated)}}{LiC\equiv C-Se-R} \xleftarrow[NH_3(l)]{2 \text{ moles } LiNH_2}$$

$$R = CH=CH_2$$

Bis(2-methylvinyl) selenide was cleaved with lithium in liquid ammonia. Subsequent treatment of the reaction mixture with ethyl bromide afforded 2-methylvinyl ethyl selenide.[83] Bis(ethylseleno)ethyne and sodium acetylide in liquid ammonia reacted to give a 30% yield of ethyl ethynyl selenide,[81*] which was also obtained together with phenyl ethyl selenide by mixing bis(ethylseleno)ethyne with phenyl lithium in ether.

A 1,3-dipolar addition of benzonitrile N-oxide, N-α-diphenylnitrone,[418*†] and 1,3-diphenylnitrilimine,[420*†] to the carbon-carbon double bond in the selenides R—Se—C≡C—CH=CH$_2$ produced the following selenides:

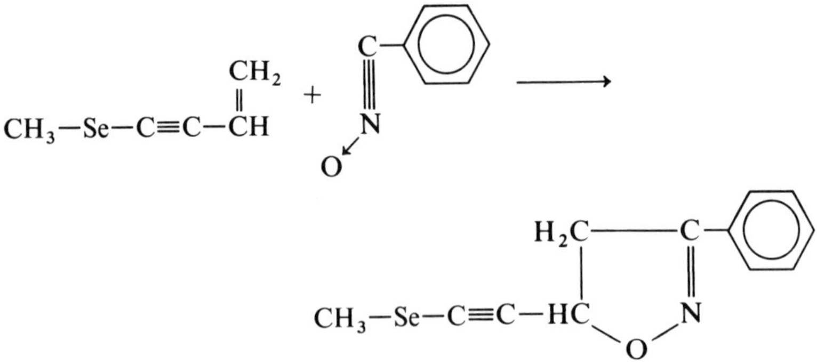

$$CH_3-Se-C\equiv C-\overset{\overset{\displaystyle CH_2}{\|}}{CH} + \underset{\overset{\displaystyle N}{\underset{\displaystyle O}{}}}{C}\text{—}\bigcirc \longrightarrow$$

$$CH_3-Se-C\equiv C-HC\underset{O}{\overset{H_2C-\!\!\!-\!\!\!-C-\bigcirc}{\diagdown \diagup N}}$$

(65)

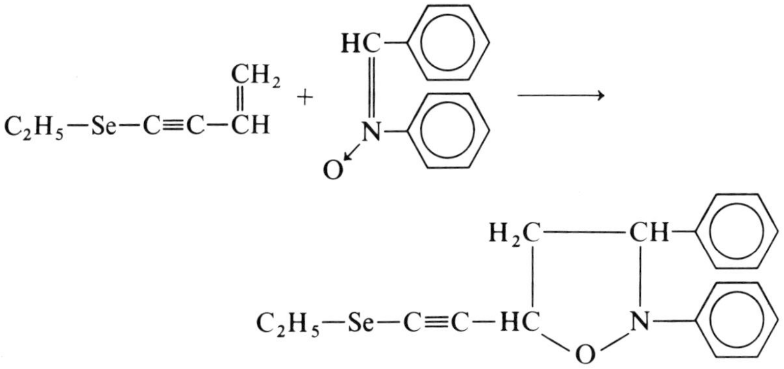

(66)

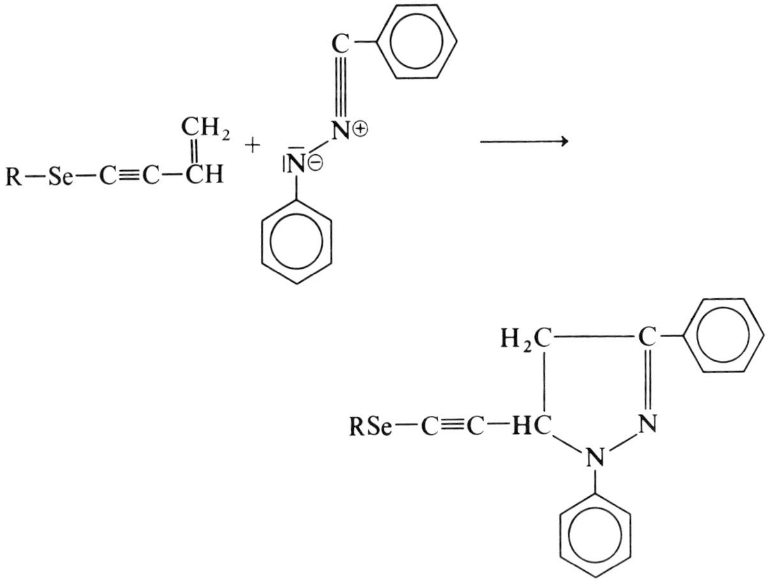

(67)

(R = CH$_3$, C$_2$H$_5$)

Benzyl γ-oxoalkyl selenides (for their preparation see reaction 62) in methanol solution added to an aqueous solution containing sodium cyanide, ammonium carbonate and chloride and ammonia gave hydantoins, starting materials for the synthesis of benzylselenoaminocarboxylic acids:[577,578]

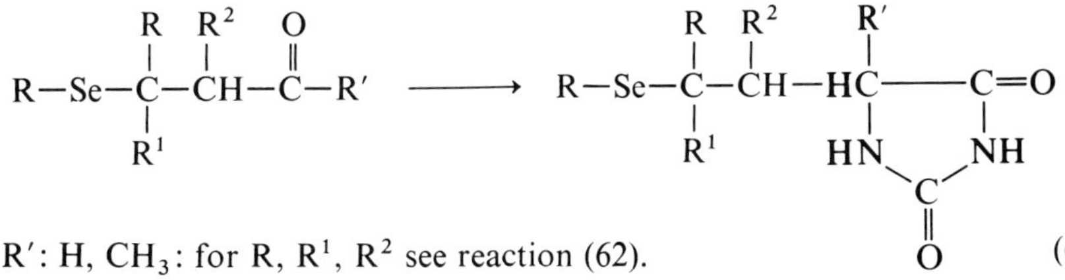

R': H, CH$_3$: for R, R^1, R^2 see reaction (62).   (68)

15-Chloro-12-selenapentadecanoic acid in an aqueous sodium carbonate solution reacted with sodium diselenide to give bis(1-carboxy-12-selena-15-pentadecyl) diselenide.[186]

Ethyl 2-ethoxyvinyl and ethyl 2-ethylthiovinyl selenides dissolved in sulfuric acid and ethanol afforded upon treatment with 2,4-dinitrophenylhydrazine the hydrazone of 2-ethylselenoacetaldehyde.[513]

Alkyl carboxyalkyl selenides have been decarboxylated, esterfied, reacted with methyl lithium and converted to the acid chlorides and amides as described in equations (69)–(73):

$$C_6H_5CH_2-Se-C(CH_3)_2CH(COOH)_2$$

$$\xrightarrow{140°C} \quad C_6H_5CH_2-Se-C(CH_3)_2CH_2COOH^5 \quad (69)$$
$$76\% \text{ yield}$$

$$C_6H_5CH_2-Se-CH_2CH_2COOH + 4\text{-}NO_2C_6H_4OH \xrightarrow{DCC}$$

$$C_6H_5CH_2-Se-CH_2CH_2COO-\langle\bigcirc\rangle-NO_2{}^{532,534}$$
$$(70)$$

$$4\text{-}FC_6H_4-Se-CH_2COOAg + ClCH_2O-\langle\bigcirc\rangle-Cl \longrightarrow$$
(with Cl, Cl substituents)

$$4\text{-}FC_6H_4-Se-CH_2-COOCH_2O-\langle\bigcirc\rangle-Cl$$
(with Cl, Cl substituents)

$$CH_3CH_2-Se-CH_2COOH \xrightarrow{CH_3Li} CH_3CH_2-Se-CH_2C(O)CH_3{}^{60}$$
$$20\% \text{ yield} \quad (71)$$

$$R-Se-(CH_2)_n-COOH \xrightarrow[\text{2) NH}_3 \text{ conc}]{\text{1) ClC(O)C(O)Cl}} R-Se-(CH_2)_nC(O)NH_2$$
$$80\% \text{ yield} \quad (72)$$

R, n: C_mH_{2m+1} (m = 1–13), 10;[189] $C_8H_{17}CH(CH_3)$, 7, $C_8H_{17}CH(CH_3)(CH_2)_6$, 1.[187]

$$CH_3CH_2-Se-CH_2COOH \xrightarrow{ClC(O)C(O)Cl} CH_3CH_2-Se-CH_2C(O)Cl$$
$$83\% \text{ yield}$$
$$CH_3CH_2-Se-CH_2C(O)NH_2$$
$$63\% \text{ yield} \quad (73)$$

The Clemmensen reduction of Se-methyl selenobenzoate gave benzyl methyl selenide in 32% yield.[345]

2,5-Dimethyl-3-iodoselenophene treated with methyl lithium or ethyl lithium in ether below $-60°C$ experienced ring cleavage with formation of $CH_3C\equiv C-CH=C(CH_3)-SeR$ ($R=CH_3$, C_2H_5) in high yields.[208,209*,†,x]

Physical Properties and Reactions. The compounds for which IR, NMR, mass spectral, and UV data are available have been marked in the text. In addition to these investigations, Simonnin[482] has carried out NMR studies with the compounds $C_2H_5-Se-C(R)=C=CH_2$ ($R=H$, CH_3). The mass spectrum of 2-carboxyphenylmethyl benzyl selenide was described.[14]

Morgan and co-workers[356,357,358] reacted selenium tetrachloride with acetylacetone. Dewar[144] on the basis of IR, NMR, and UV data suggested the following structures for the products obtained in this reaction:

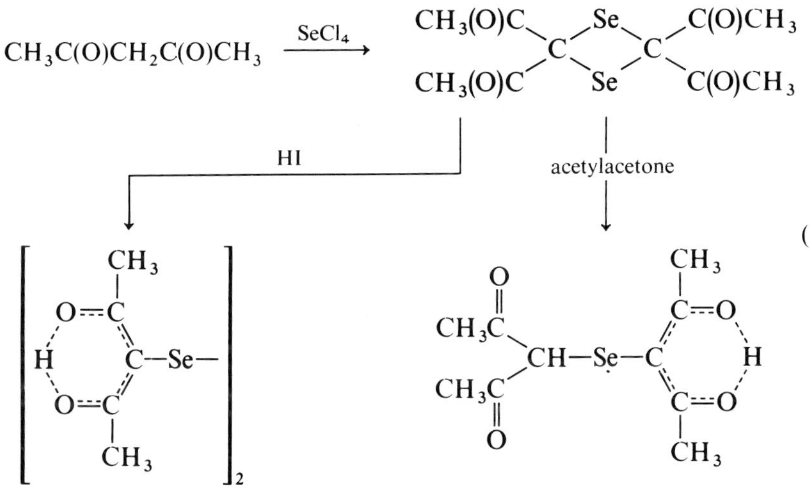

$$(74)$$

Acetylacetone in glacial acidic acid in the presence of anhydrous zinc chloride gave upon treatment with hydrogen selenide polyselenoadamantanes.[375]

The reactivity of the unsymmetric selenides is governed by the functional groups and the selenium atom. The reactions given by the functional groups have already been discussed earlier in this section. Reactions affecting the selenium atom (e.g., conversions to dihalides or selenonium compounds) have received little attention. Benzyl alkyl selenides experience selenium-carbon(alkyl) bond cleavage in liquid ammonia containing sodium. Dibenzyl diselenide is isolated on working up these reaction mixtures in the presence of air.[5,6,7,66]

Symmetric Diaryl Selenides

Symmetric diaryl selenides have received relatively little attention during the period covered by this review as compared to unsymmetric diaryl and aryl alkyl selenides.   Symmetric diaryl selenides have been obtained:

1. from aryldiazonium salts and alkali selenides or selenolates;
2. from aromatic compounds with a hydrogen atom activated by a dialkyl-amino, hydroxy, alkoxy or similar group and inorganic selenium compounds like selenium oxychloride, diselenium dichloride, selenium dioxide or hydrogen selenide, from diaryl mercury or aryl lithium compounds and selenium halides or elemental selenium, from aromatic chlorides bearing activating nitro substituents and potassium selenide, and from diraryl sulfones or diaryl tellurides and selenium;
3. through conversion of diorganyl diselenides into diaryl selenides;
4. by modification of organic groups in diaryl selenides.

Diaryl Selenides from Aryldiazonium Salts.   The reaction of diazonium salts with aqueous solutions of alkali selenides at $0°$ has been used to prepare diphenyl,[231] bis-(2-napthyl)[56u] and bis(5-benzothienyl) selenide.[510]   The decomposition of aryldiazonium tetrafluoroborates in acetone solution in the presence of zinc dust and selenium tetrachloride gave the diaryl selenides, $(RC_6H_4)_2Se$, $(R = H, 2\text{-}CH_3, 4\text{-}CH_3, 4\text{-}Cl, 4\text{-}CH_3O)$ in yields ranging from $3\text{--}33\%$.[371]   The selenides were probably formed by reduction of the intermediate diorganyl selenium dichlorides.

Diaryl Selenides from Selenium Dioxide, Elemental Selenium, Diselenium Dichloride, Selenium Tetrachloride, Selenium Oxychloride and Hydrogen Selenide.   Funk, employing selenium dioxide in hydrochloric acid solution, obtained selenides from 4-hydroxychlorobenzene, 2-hydroxycarboxybenzene and 2-methyl-4-hydroxychlorobenzene.[175]   The position of the selenium atom on the aromatic rings is not known.

Bis(3-indolyl) and bis(1-methyl-3-indolyl) selenide were the products of the interaction of selenium dioxide and the respective indoles in benzene.[53x] Wilshire[547*,†,u] found that the yields in the indole reactions reported to be as high as 22% were not reproducible and that, in the absence of benzene, only a 7% yield was obtained.   Similar reactions with 2-methyl- and 1,2-dimethyl-indole produced diselenides and triselenides.   Ziegler[581] was able to syn-thesize diphenyl selenide by refluxing benzene in the presence of selenium, selenium tetrachloride and anhydrous aluminum chloride.

Bis(pentaflurophenyl) selenide was formed by heating the aryl mercuric chloride,[124] the diaryl mercury,[93] or diaryl thallium bromide[138] with ele-mental selenium.   The same selenide was obtained in 60% yield from the aryl iodide and selenium kept in a sealed tube at 230°C[123,124] and from the aryl lithium compound and diselenium dichloride or selenium tetrachloride.[124] Heating the diaryl telluride with selenium to 320°C gave the selenide quantitatively. Selenium oxychloride heated with 2-*tert*-butyl-4-methylphenol in chloroform produced bis(2-hydroxy-3-*tert*-butyl-5-methylphenyl) selenide.[283,294*,x]

Bubbling hydrogen selenide through an aqueous solution of tropylium bromide or a suspension of 4-pyridylpyridinium chloride in pyridine gave ditropyl selenide in 95 % yield[281*,u] and bis(4-pyridyl) selenide,[261] respectively. Bis(2,4,6-trinitrophenyl) selenide was the product of the reaction between picryl chloride and potassium selenocyanate in ethanol.[182]

Conversion of Diorganyl Diselenides into Selenides. Diaryl diselenides, R_2Se_2, heated with Raney nickel (R = C_6H_5, 2-$CH_3C_6H_4$),[233] pyrolyzed (R = C_6H_5),[479] reduced with sodium borohydride (R = 5-benzothienyl),[510] or cleaved with R'Li (R = 2-thienyl) (USSR Pat., 165,753), yielded the corresponding selenides. Phenyl allyl selenide heated to 240°C produced diphenyl selenide among other products.[112,266]

Bis(2,4,6-trimethylpenyl) selenide was a by-product of the reaction between the arylseleno magnesium bromide and methyl iodide.[105]

Modification of the Organic Groups. Ivanova[245] (USSR Pat., 209,445) condensed diphenyl selenide with dimethylcarbamoyl chloride in the presence of aluminum chloride at 150°C and modified the functional group to give the selenide shown in reaction (75):

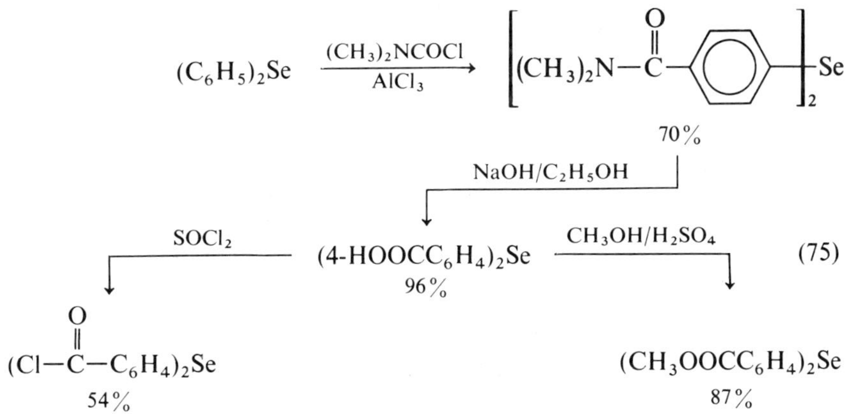

$$(C_6H_5)_2Se \xrightarrow[AlCl_3]{(CH_3)_2NCOCl} \left[(CH_3)_2N-\overset{\overset{O}{\|}}{C}-\underset{}{\bigcirc}-\right]_2 Se$$

70%

$$\Big\downarrow NaOH/C_2H_5OH$$

$$\overset{SOCl_2}{\Big\downarrow} \quad (4\text{-}HOOCC_6H_4)_2Se \xrightarrow{CH_3OH/H_2SO_4} \quad (75)$$

96%

$$(Cl-\overset{\overset{O}{\|}}{C}-C_6H_4)_2Se \qquad\qquad (CH_3OOCC_6H_4)_2Se$$

54%   87%

Diphenyl, bis(2-hydroxyphenyl), bis(2-hydroxy-3-hexadecylphenyl) and bis(2-hydroxy-3,5-R_2-phenyl) selenide were synthesized by an undisclosed method and studied with regard to their anticorrosive, antioxidative and lubricating properties.[303]

Reactions and Physical Properties of Diaryl Selenides. References to the reactions given by aromatic selenides are included in Figure 8-3. The reactions of symmetric diaryl selenides leading to unsymmetric derivatives are described subsequently. The few reactions not mentioned elsewhere are discussed below.

Diphenyl and bis(2-methylphenyl) selenide[233] treated with Raney nickel in benzene at 180°C produced diaryls, while ethoxybenzene was obtained from the corresponding selenide in refluxing benzene.[548] Bis(1-methyl-3-indolyl) selenide with Raney nickel in refluxing dioxane yielded the corresponding indole and 3,3'-diindolyl.[53] Diphenyl selenide was converted to selenanthrene by potassium amide in liquid ammonia.[464] Bis(pentafluorophenyl) selenide heated with sulfur at 230°C gave the sulfide.[124]

Krishnamurthy[301] has determined the dipole moments of bis(4-methoxyphenyl) and the isomeric ditolyl selenides in benzene solution. Polarization data are also given. Similar experiments with diphenyl selenide were carried out by LeFevre.[313] The mass spectrum of diphenyl selenide[53] and ultraviolet absorption spectra of diphenyl chalcogenides including the selenide have been reported.[286]

Unsymmetric Diaryl Selenides

Unsymmetric diaryl selenides can be prepared from organic selenium compounds, R—Se—X (X = CN, Cl, Br, SCN) through condensation with an appropriate aromatic hydrocarbon derivative, by replacing the alkyl group in alkyl aryl selenides by an aryl group and by modifying functional groups in diaryl selenides.

Unsymmetric Diaryl Selenides from Aryl Selenocyanates, Aryl Selenium Halides and Thiocyanates and Aryl Alkyl Selenides. The most important method for the synthesis of unsymmetric diaryl selenides is the reaction of a selenium containing compound Ar—Se—Y (Y = H, Br, SCN, CN, alkali metal) with an aromatic Grignard reagent, an aryl lithium compound or an aromatic compound bearing an activated halogen or hydrogen atom.

Greenberg[206] converted aryl selenocyanates into unsymmetric selenides employing aryl lithium or aryl magnesium bromide in ether solution. The 39 diaryl selenides prepared in this manner were obtained in yields of 56–88%. The best previous method, the reaction of diselenides with aryl magnesium bromide, utilized only one of the two organic groups of the diselenides. Morever, the preparation of the diselenides from the selenocyanates often proceeded with low yields. In the period covered by this review, the diselenide reaction was used only once. Phenyl 3,5-di(*tert*-butyl)-4-hydroxyphenyl selenide was obtained from diphenyl selenide and the appropriate Grignard reagent in 25% yield.[362] Greenberg's method (76) cannot be used when the aryl selenocyanate bears substituents which would react with the organometallic reagent.

$$R-C_6H_4-SeCN \quad \xrightarrow[\substack{M=Li,\,MgBr}]{M-C_6H_4R'} \quad R-C_6H_4-Se-C_6H_4-R' + MCN$$

$$(76)$$

R,R': H, 4-CH$_3$; H, 3-CH$_3$; H, 4-C$_6$H$_5$; 4-Cl, H; 4-Cl, 4-CH$_3$; 4-Cl, 3-CH$_3$; 4-Cl, 4-C$_6$H$_5$; 3-Cl, H; 3-Cl, 3-CH$_3$; 3-Cl, 4-CH$_3$; 3-Cl, 4-Cl; 3-Cl, 4-Br; 3-Cl, 4-C$_6$H$_5$; 4-Br, H; 4-Br, 4-Cl; 4-Br, 4-CH$_3$; 4-Br, 3-CH$_3$; 4-Br, 4-C$_6$H$_5$; 4-CH$_3$, 4-C$_6$H$_5$; 3-CH$_3$, 4-CH$_3$; 3-CH$_3$, 4-C$_6$H$_5$; 4-CH$_3$O, H; 4-CH$_3$O, 4-Cl; 4-CH$_3$O, 4-Br, 4-CH$_3$O, 4-CH$_3$; 4-CH$_3$O, 3-CH$_3$; 4-CH$_3$O, 4-C$_6$H$_5$; 2-C$_6$H$_5$, H; 2-C$_6$H$_5$, 4-Cl; 2-C$_6$H$_5$, 4-Br; 2-C$_6$H$_5$, 4-CH$_3$; 2-C$_6$H$_5$, 3-CH$_3$; 2-C$_6$H$_5$, 4-C$_6$H$_5$.

The selenides 2-C$_{10}$H$_7$-Se-C$_6$H$_4$-R' were similarly prepared from 2-napthyl selenocyanate (R' given): H, 4-Cl, 4-Br, 4-CH$_3$, 3-CH$_3$, 4-C$_6$H$_5$.

Aryl selenium halides reacted with aryl magnesium bromide to form diaryl selenides. Thus, Chierici[103] prepared phenyl 2-methyl-phenyl selenide from phenyl selenium bromide. Renson,[432] treating 2-chloroformylphenyl selenium chloride with phenyl cadmium, isolated 2-benzoylphenyl phenyl selenide.

Aromatic selenides are obtained in high purity and almost quantitative yield from aryl selenium bromides and aryl mercuric chloride or diaryl mercury. This method, useful for the preparation of small quantities of a selenide, has been employed for the synthesis of 2-naphthyl phenyl selenide.[512] Benzene- and 4-chlorobenzeneselenol refluxed with 2-iodobenzoic acid in aqueous potassium hydroxide solution in the presence of elemental copper gave aryl 2-carboxyphenyl selenides.[485]

Nitroaryl aryl selenides can be prepared by condensation of a nitroaryl halide with an aromatic selenolate in ethanolic solution:

$$(77)$$

Y; X; R: CH; Br; H;[103] N; Cl; H, 4-Cl, 4-Br, 2-CH$_3$, 3-CH$_3$, 4-CH$_3$, 4-CH$_3$O, 4-NO$_2$.[108]

Aniline and 4-methylaniline condense with nitroaryl selenium thiocyanates but not with chlorides and bromides in chloroform solution to give in quantitative yields, the nitroaryl 4-aminophenyl and the nitroaryl 2-amino-5-methylphenyl selenides:[440]

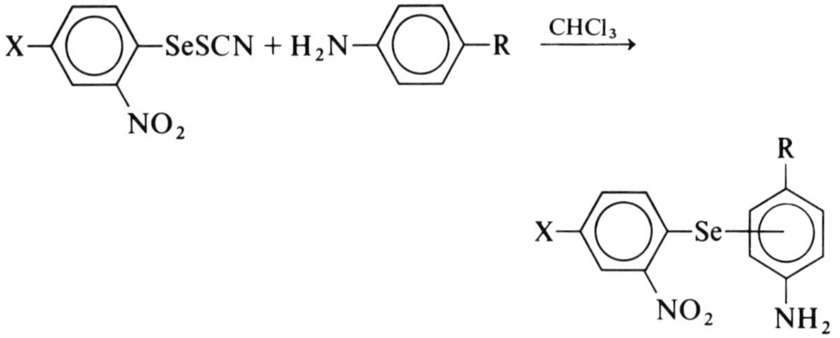

(78)

R; position of NH_2; X: H; 4; H, Cl, Br, NO_2. CH_3; 2; H, NO_2.

N,N-Dialkylanilines and N-alkylanilines react not only with nitroaryl selenium thiocyanates but also with the chlorides and bromides:[442]

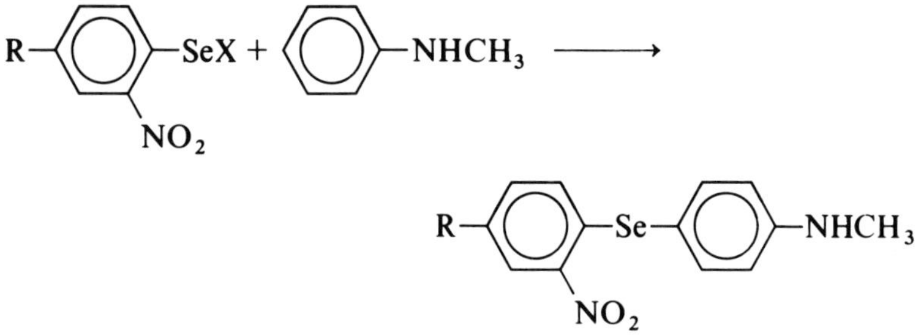

(79)

X = Cl, Br, SCN; R = H, Cl, Br, CH_3, CH_3O, NO_2.

The diaryl selenides prepared as described in equations (78) and (79) have been converted into the hydrochlorides and N-acetyl compounds. 3-Nitrophenyl 4-dimethylaminophenyl[103] and 2,4-dinitrophenyl-4-dimethylaminophenyl selenide[311] were obtained similarly.

Arcoria[34] found that phenyl alkyl selenides are dealkylated by 2,4-dinitrophenyldiazonium salts with formation of phenyl 2,4-dinitrophenyl selenide. Other diazonium salts did not react.   Marziano[336] extended this reaction to 2- and 4-chlorophenyl, 2- and 3-nitrophenyl and the isomeric methoxyphenyl and methylphenyl methyl selenides.   While similar reactions with oxygen and sulfur analogues produced coupling products, the diaryl selenides were the sole products obtained with the methyl aryl selenides.

Modification of Functional Groups in Diaryl Selenides. Unsymmetric diaryl selenides have also been prepared by modifying functional groups attached to one aromatic ring or by changing only one of the functional group in a symmetric diaryl selenide. The acetylation of some aminoaryl aryl selenides and the preparation of hydrochlorides has already been mentioned in this section. 2-Aminophenyl phenyl selenide heated with acetic anhydride gave the N,N-diacetyl compound.[103] A convenient way of synthesizing aminoaryl selenides is the reduction of the corresponding nitro compounds:

$$(80)$$

Y, R, R' (reducing agent): N, H, H; N, 4-Cl, H; N, 4-Br, H; N, 4-CH_3, H; N, 3-CH_3, H; N, 2-CH_3, H; N, 4-CH_3O, H (Sn/HCl);[108] CH, H, H (Sn/HCl);[103] CH, H, CH_3O; CH, H, C_2H_5O; CH, H, C_3H_7O; CH, 4-CH_3, C_3H_7O; CH, 4-C_2H_5, CH_3O (FeSO$_4$) [Jap. Pat., 19,630].

Sodium disulfide in ethanol reduced only one amino group in bis(4-nitrophenyl) selenide[103] to give 4-nitrophenyl 4-aminophenyl selenide.

2,4-Dinitrophenyl pentachlorophenyl selenide gave upon hydrogenation with a platinum oxide catalyst the diamino compound, which was converted to the diisocyanato derivative, a monomer for the preparation of resins, plastics, and self-extinguishing polyurethan coating systems (U.S. Pat., 3,281,477 and 3,284,502). A Russian patent (USSR Pat., 209,445) claims the synthesis of phenyl carboxyphenyl selenide through acylation of diphenyl selenide with dimethylcarbamoyl chloride in dichloroethane in the presence of aluminum chloride.

Provided that unsymmetric diaryl sulfones are easily available, their reaction with elemental selenium at 400°C can be used to prepare selenides. Chierici[108] obtained in this way 2-methylphenyl phenyl selenide.

Physical Properties and Reactions. Chierici,[103,108] Modena,[355] and Vicentini[512] reported UV spectra of unsymmetric diaryl selenides. Cecchini[98] carried out a thermal analysis of the system 2-nitrophenyl phenyl sulfide/ 2-nitrophenyl phenyl selenide. Besides the general reactions of these selenides as summarized in Figure 8-3, the formation of radicals upon treatment of phenyl 3,5-di-(*tert*-butyl)-4-hydroxyphenyl selenide with potassium hexacyanoferrate(III) or lead dioxide[362] and the study of the nucleophilic reactivity of nitrophenyl and dinitrophenyl phenyl selenides and their oxygen and sulfur analogues[46] should be mentioned.

Aromatic Aliphatic Selenides

During the past fifteen years many alkyl aryl selenides have been synthesized and their physical properties studied.   These selenides bearing suitable substituents are convenient starting materials for heterocyclic selenium compounds.   The selenium-carbon(alkyl) bond is rather weak, being cleaved, for instance, by halogenating agents before the selenium-carbon(phenyl) bond is attacked.   This section is subdivided according to the various methods available for the synthesis of alkyl aryl selenides:

1. Alkylation of aromatic selenols.
2. Mannich condensation employing benzeneselenol.
3. Alkylation of aryl selenium halides.
4. The addition of $Ar—Se—X$ ($X = H$, Cl, Br, CH_3O, CH_3CO_2) to carbon-carbon multiple bonds.
5. Alkyl aryl selenides from diaryl selenides.
6. Modification of organic groups in alkyl aryl selenides.
7. Physical properties and reactions of alkyl aryl selenides.

Alkylation of Aromatic Selenols.   The most general method for the preparation of aryl alkyl selenides is the reaction of a sodium, lithium, or zinc salt of an aromatic selenol with an alkyl halide or dialkyl sulfate.   The selenols, which are prepared as described in section I, need not be isolated.   The reaction mixture made alkaline with sodium hydroxide or sodium ethoxide can be immediately treated with the alkylating agent.   If the selenols are isolated, they are dissolved in aqueous or ethanolic sodium hydroxide before alkylation.   The yields in these reactions vary widely depending on the nature of the selenol and the alkylating agent.

$$Ar—Se^- + RX \longrightarrow Ar—Se—R + X^- \qquad (81)$$

Ar: R: C_6H_5: C_nH_{2n+1} ($n = 1$–5);[223] CH_3;[180] C_2H_5, C_3H_7,
i-C_4H_9;[177] $CH_2CH\text{:}CH_2$;[112,266] $CH_2CH\text{:}CHCH_3$;[268*]
$CH_2CH_2CH\text{:}CH_2$;[191]; $CH(CH_3)CH\text{:}CH_2$;[268]
$(CH_2)_3CH\text{:}CH_2$;[191] $CH(CH_3)CH\text{:}CHCH_3$, $CH_2CH\text{:}CHC_6H_5$,
$CH_2CH\text{:}CHCH_2C_6H_5$;[268*] $CH_2C\text{:}CH$, $CH(CH_3)C\text{:}CH$,
$CH_2C\text{:}CCH_3$;[411,112] CH_2CH_2COOH;[267,192]
$CH(CH_3)CH_2COOH$;[133] $CH_2CH(OCH_3)_2$, 2-oxocyclohexyl[353];
2-$CH_2C_6H_4NO_2$;[141] $(CH_2)_3COOH$, $(CH_2)_3COOCH_3$.[52]

4-ClC_6H_4: $CH_2CH\text{:}CH_2$, $CH_2CH\text{:}CHCH_3$,
$CH(CH_3)CH\text{:}CHCH_3$, $CH_2CH\text{:}CHC_6H_5$,

$CH_2CH\!:\!CHCH_2C_6H_5;^{269}$ $CH_2COOH.^{367}$

3-ClC_6H_4: $CH_3.^{336}$

4-BrC_6H_4: $CH_2CH\!:\!CH_2$, $CH_2CH\!:\!CHCH_3$,
$CH(CH_3)CH\!:\!CHCH_3$, $CH_2CH\!:\!CHC_6H_5;^{269}$ $CH_2COOH;^{367}$
$(CH_2)_3COOH$, $(CH_2)_3COOC_2H_5.^{52}$

2-Br_6H_4: $CH_3.^{181}$

4-$CH_3C_6H_4$: $CH_3;^{105,192}$ $C_2H_5;^{105}$ $CH_2CH\!:\!CH_2$,
$CH_2CH\!:\!CHC_6H_5$, $CH_2CH\!:\!CHCH_3$, $CH(CH_3)CH\!:\!CHCH_3$,
$CH_2CH\!:\!CHCH_2C_6H_5;^{269}$ $CH_2CH_2COOH;^{133}$ $(CH_2)_3COOH$,
$(CH_2)_3COOC_2H_5.^{52}$

3-$CH_3C_6H_4$: $CH_3;^{336}$ $CH_2CH_2COOH.^{133}$

2-$CH_3C_6H_4$: $CH_3;^{336}$ $CH_2CH_2COOH.^{133}$

4-$tert$-$C_4H_9C_6H_4$: $(CH_2)_3COOH$, $(CH_2)_3COOC_2H_5.^{52}$

4-$CH_3OC_6H_4$: $CH_3;^{336}$ $CH_2CH\!:\!CH_2$, $CH_2CH\!:\!CHC_6H_5$,
$CH_2CH\!:\!CHCH_3$, $CH_2CH\!:\!CHCH_2C_6H_5$,
$CH(CH_3)CH\!:\!CHCH_3.^{269}$

2-$CH_3OC_6H_4$: $CH_3.^{336}$

2-$(C_2H_5O)_2CHC_6H_4$: $C_6H_5CH_2.^{454}$

2-$(O)HCC_6H_4$: CH_3, CH_2COOH, $C_4H_9.^{119}$

4-$HOOCC_6H_4$: C_nH_{2n+1} (n = 1–8), $C_6H_5CH_2;^{226}$ $CH_3.^{450}$

2-$HOOCC_6H_4$: $CH_3;^{450}$ $CH_2COOH.^{433}$

2-$CH_3C(O)C_6H_4$: CH_3^{195}

2-$HOOC$-4-BrC_6H_3: CH_3, $CH_2COOH.^{510}$

4-$NO_2C_6H_4$: $CH_3.^{396u}$

3,5-$(tert$-$C_4H_9)_2$-4-HOC_6H_2: CH_3, i-C_3H_7, $tert$-$C_4H_9.^{362}$

2-$HOOCC_6H_3(R')$: CH_3; R' = H, 4-CH_3, 5-CH_3, 6-$CH_3.^{454}$

1-$C_{10}H_7$: CH_2COOH, $CH_2COOC_2H_5$, $CH_2CH_2COOH;^{51}$
$(CH_2)_3COOC_2H_5$, $(CH_2)_3COOH.^{52}$

$2\text{-}C_{10}H_7$: CH_2COOH;[51,486] $CH_2COOC_2H_5$, CH_2CH_2COOH;[51]
$(CH_2)_3COOH$, $(CH_2)_3COOC_2H_5$.[52]

8-quinolyl: CH_3.[337]

5-benzothienyl: CH_2COOH, $CH_2CH(OCH_3)_2$.[510]

2-benzothienyl: CH_2COOH.[510]

Benzeneselenol and phthalide refluxed in ethanol in the presence of sodium ethoxide gave 2-carboxybenzyl phenyl selenide in good yield.[484] Furyl and thienyl alkyl selenides were synthesized according to reaction (82):

$$ \begin{array}{ccc} R{-}Y & \xrightarrow[\text{ether}]{C_4H_9Li} & R{-}Y{-}Li & \xrightarrow[\text{2) R'X}]{\text{1) Se}} & R{-}Y{-}R' \end{array} \qquad (82) $$

Y, R, R': O, H, CH_3; O, H, $C_6H_5CH_2$; O, H, $CH_2CH{:}CH_2$;[373]
O, H, C_2H_5; O, CH_3, C_2H_5; O, CH_3, CH_2COOCH_3;[184] S, H,
C_2H_5; S, C_2H_5, CH_2COOCH_3; S, C_2H_5, C_2H_5.[184]

Grignard reagents, obtained from aromatic halides and treated with elemental selenium, afforded solutions containing arylseleno magnesium halides. An alkyl halide added to such a solution gave the aryl alkyl selenide in approximately 50% yield:

$$ R{-}\bigcirc{-}MgX \xrightarrow[\text{2) CH}_3\text{I}]{\text{1) Se}} R{-}\bigcirc{-}SeCH_3 \qquad (83) $$

R: 4-Br, 3-Br, 4-Cl, 3-Cl, 2-Cl;[181] H, 4-C_6H_5.[493]

2,4,6-Trimethylphenyl methyl,[105u] carboxymethyl, 1-carboxyethyl, 2-carboxyethyl, 2-naphthyl and 1-naphthyl carboxymethyl selenide were prepared similarly.[486]

Aryl selenocyanates are convenient starting materials for selenide syntheses. Their conversion to selenols, which are then alkylated, is accomplished in basic medium with or without the presence of a reducing agent like glucose or sodium dithionite. In this way, 2-methylphenyl,[336] 2-bromophenyl,[151] 3-nitrophenyl[103u] and 1-azulenyl methyl,[370*] 2-acetylphenyl[118] and 2-naphthyl carboxymethyl, 2-naphthyl methylcarboxymethyl,[486] and allyl, 3-phenylallyl, and 2-buten-1-yl 4-nitrophenyl selenide[269] were obtained. Aryl selenocyanates are reported to react smoothly with organyl lithium compounds with elimination of lithium cyanide producing aryl organyl selenides.[94]

Mannich Condensation with Benzeselenol. Benzeneselenol condensed with formaldehyde and benzylamine, dibenzylamine, aniline, methylaniline and ammonia in ethanol to form N,N-bis(phenylselenomethyl)benzylamine, N-phenylselenomethyldibenzylamine,   N,N-bis(phenylselenomethyl)aniline, N-methyl-N-phenylselenomethylaniline and tris(phenylselenomethyl)amine, respectively. When the reactions were carried out with aniline, methylaniline, and dimethylaniline in the presence of an equimolar amount of concentrated hydrochloric acid, the corresponding 4-aminobenzyl phenyl selenides, $4-R_2NC_6H_4CH_2-Se-C_6H_5$, were obtained.[409,410*†] Benzeneselenol produced phenyl chloromethyl selenide when treated with formaldehyde and concentrated hydrochloric acid at $-5°C$.[467]

Alkylation of Aryl Selenium Halides. Aryl selenium halides have been alkylated with Grignard reagents. Renson[432] employed organic cadmium compounds, prepared from the Grignard reagents and cadmium chloride, to synthesize 2-acylphenyl alkyl selenides from 2-chloroformylphenyl selenium chloride:

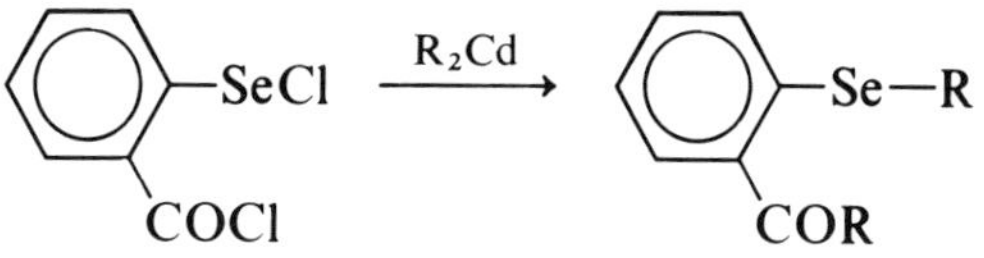

$$R = CH_3, C_2H_5, C_3H_7, C_4H_9$$

2,4-Dinitrophenyl selenium chloride condensed with acetone to give the aryl acetonyl selenide.[311] Organyl selenium bromides are reported to produce mainly diaryl diselenides.

The Addition of Ar—Se—X (X = H, Cl, Br, CH₃O, CH₃COO) to Carbon-Carbon Multiple Bonds. The addition of aromatic compounds of the type R—Se—X to multiple carbon-carbon bonds has been employed extensively for the preparation of aryl alkyl selenides.

Benzeneselenol added to substituted and unsubstituted olefins in benzene in the presence or absence of sodium alkoxide:

R, R', R″: H, H, H; CH₃, H, H;[177] H, H, COOH; H, CH₃, COOCH₃;[267] H, CH₃, COOC₂H₅; CH₃, H, COOC₂H₅;[133] C₆H₅, H, NO₂; 1-C₁₀H₇, H, NO₂ (Pol. Pat., 43,915); cis- and trans-COOH, H, COOH.[267]

The phenylseleno group attaches itself to the unsubstituted carbon atom in reactions with olefins bearing a nitro, carboxy, or carbalkoxy group. In addition reactions with propene and *iso*-butene, phenyl *iso*-propyl, and phenyl *tert*-butyl selenide were obtained, respectively. Cyclohexene gave the expected selenide.[177] Pentachlorobenzeneselenol and 1-trichloromethyl-2-nitroethene produced phenyl 1-trichloromethyl-2-nitroethyl selenide (Pol. Pat., 43,915).

Cis- and *trans*-carboxychloroethene gave the respective aryl carboxyethenyl selenides and not the saturated adducts with benzeneselenol[108,110] and 4-nitrobenzeneselenol[110] in ethanol containing sodium bicarbonate. The reactions proceeded with retention of configuration. Benzeneselenol and diketene produced Se-phenyl acetylselenoacetate.[456]

Aryl selenium acetates, $R-Se-OC(O)CH_3$, which can be generated *in situ* from selenium bromides or selenium methoxides, $R-Se-OCH_3$, and glacial acetic acid, react in glacial acetic acid with olefins according to reaction (86):

$$R-Se-\overset{\overset{O}{\|}}{O}CCH_3 + CH_2=C\overset{R'}{\underset{R''}{\diagdown}} \longrightarrow R-Se-CH_2-\overset{\overset{R'}{|}}{\underset{R''}{\underset{|}{C}}}-\overset{\overset{O}{\|}}{O}CCH_3 \quad (86)$$

R: R', R" or olefin: 2-NO$_2$C$_6$H$_4$: H, CH$_3$CO$_2$; H, C$_6$H$_5$;[240] 1-anthraquinonyl: CH$_3$, CH$_3$; H, CH$_3$CO$_2$; H, H;[240] H, C$_6$H$_5$;[252] cyclohexene;[240,252] 4-hydroxy-1-anthraquinonyl: H, CH$_3$CO$_2$;[240] 4-methoxy-1-anthraquinonyl: cyclohexene.[252]

The addition of anthraquinonyl selenium acetate to *iso*-butene gave the *iso*-butyl derivative,[240] while benzeneselenol produced the *tert*-butyl selenide.[177] When the olefin employed in the reactions with the acetates is 2,3-dihydropyran or an ethylene derviative bearing a methyl and a phenyl or two phenyl groups on the same carbon atom, the addition of the RSe$^+$ group is followed by elimination of a hydrogen ion:

$$R-Se\overset{\overset{O}{\|}}{O}CCH_3 + H_2C=C\overset{R'}{\underset{R''}{\diagdown}} \xrightarrow{-CH_3CO_2^-}$$

$$R-Se-CH_2-\overset{\oplus}{C}\overset{R'}{\underset{R''}{\diagdown}} \xrightarrow{-H^+} R-Se-CH=C\overset{R'}{\underset{R''}{\diagdown}} \quad (87)$$

R: R', R", or olefin: $2\text{-}NO_2C_6H_4$: C_6H_5, C_6H_5; $4\text{-}CH_3OC_6H_4$, $4\text{-}CH_3OC_6H_4$; $2,4\text{-}(CH_3O)_2C_6H_3$, $2,4\text{-}(CH_3O)_2C_6H_3$; $3,4\text{-}(CH_3O)_2C_6H_3$, $3,4\text{-}(CH_3O)_2C_6H_3$; 2,3-dihydropyran-5-yl.[240] $2,4\text{-}(NO_2)_2C_6H_3$: $4\text{-}(CH_3)_2NC_6H_3$, $4\text{-}(CH_3)_2NC_6H_3$.[240] 1-anthraquinonyl: CH_3, C_6H_5; 2,3-dihydropyranyl-5.[240]

These addition reactions are catalyzed by pyridine. In pyridine as solvent, 2-nitrophenyl selenium methoxide and bromide, 2,4-dinitrophenyl selenium methoxide and cyanide, and 1-anthraquinonyl selenium bromide added to 2,2-bis(dimethylaminophenyl)ethene to give aryl 2,2-bis(dimethylamino-phenyl) ethenyl selenide.[240] Michalski[349] reacted ethyl selenium chloride with cyclohexene and obtained ethyl 2-chlorocyclohexyl selenide. 2,4-Dinitro-phenyl 2-chlorocyclohexyl selenide was prepared similarly.[311]

In the reaction between benzeneselenol and butadiene and its derivatives at room temperature without solvent, a mixture of cis- and trans-1,4-adducts is obtained:

$$R-Se-X + R^1-CH=C(R^2)-C(R^3)=C\underset{R^5}{\overset{R^4}{\diagdown}} \longrightarrow$$

$$R-Se-CH(R^1)C(R^2)=C(R^3)-HC\underset{R^5}{\overset{R^4}{\diagdown}} \quad (88)$$

R, X: R^1, R^2, R^3, R^4, R^5: C_6H_5, H: H, H, H, H, H; H, CH_3, H, H, H; H, H, H, H, CH_3; CH_3, H, H, H, CH_3; H, CH_3, CH_3, H, H.[177*†] 1-anthraquinonyl, CH_3O: H, H, H, H, H.[240]

The addition of selenols to acetylenic compounds occurs when the re-agents are mixed in the neat state or dissolved in ethanol. A basic catalyst like sodium ethoxide or piperidine will make the reaction highly stereospecific in favor of the cis compounds.[196,197] In the absence of a catalyst, mixtures of cis and trans isomers are obtained. Azerbaev,[42,43] however, reports that trans addition products are predominantly formed in reactions with acetylenyl-carbinols. The conditions for these reactions vary with the acetylene and selenol employed:

$$R-SeH + R'-C\equiv C-R'' \longrightarrow R-Se-C(R')=CH(R'') \quad (89)$$

R: R', R", configuration: C_6H_5: H, C_4H_9, cis;[265] H, COOH, trans;[110] H, COOH, cis;[109,110,278] H, $COOCH_3$, trans; H, $COOCH_3$, cis; H, $OC(O)CH_3$, cis;[278] H, C_6H_5, cis;[196,265] H, $C_6H_5C(O)$; C_6H_5, $CH_3C(O)$;[196] C_6H_5, C_6H_5, cis;[265] C_6H_5, $COOC_2H_5$;[195,196*] $COOCH_3$, $COOCH_3$; C_6H_5, COOH, cis;[278] H, 6-methyl-3-pyridyl,

cis;[265] CH_3, $P(O)(OC_2H_5)_2$;[417] H, $CH_3CH(OH)$; H, CH_2OH; H,
$C_3H_7C(CH_3)OH$; H, $C_2H_5C(CH_3)OH$; H, $(CH_3)_2COH$, all *trans*.[43*†]
2-$H_2NC_6H_4$: H, COOH.[364]   4-$O_2NC_6H_4$: H, COOH, *cis*.[109]

Infrared spectroscopy and the comparison of calculated and experimental dipole moments helped in the determination of the configuration of these compounds.[278,279, 280,266]

Similar addition reactions are given by organyl selenium chlorides and bromides and by diphenyl selenide sulfide. Chierici[109,110] bubbled acetylene at −10°C through an ethyl acetate solution containing either phenyl, 3-nitrophenyl or 4-nitrophenyl selenium bromide and isolated the respective *trans*-2-bromovinyl aryl selenides. 2-Nitrophenyl selenium bromide did not react under these conditions.[109] The addition of phenyl selenium chloride and bromide to acetylene, phenylacetylene and diphenylacetylene is reported to produce the *cis*-2-halovinyl phenyl selenides,[276] while some nitrophenyl selenium halides give *trans* compounds.[274,280] The configuration of the adduct is solvent dependent. Acetylene and phenylacetylene with diphenyl selenide sulfide in a base catalyzed reaction gave *cis*-2-phenylthiovinyl phenyl selenide.[277*] In addition reactions to phenylacetylene the phenylseleno group attaches itself to the carbon atom bearing the phenyl group.

Benzeneselenol adds to 1,2-epoxy-3-R-3-butenes in the presence of sodium benzeneselenolate to produce phenyl 2-hydroxy-3-R-3-buten-1-yl selenides (R = H, Cl, CH_3), while in the absence of the catalyst, mixtures with phenyl 4-hydroxy-2-R-buten-1-yl selenide were obtained (R = H, Cl).[569]

Alkyl Aryl Selenides from Diaryl Diselenides. Diaryl diselenides and sodium added to a solution of bis(phenylsulfonyl)methane in methanol gave bis-(phenylsulfonyl)methyl aryl selenides:

$$(C_6H_5SO_2)_2CH_2 + (R{-}C_6H_4)_2Se_2$$

$$\xrightarrow[\text{Na}]{CH_3OH} (C_6H_5SO_2)_2CH{-}Se{-}C_6H_4{-}R \quad (89a)$$

R: H, 2-Cl, 4-Cl, 4-Br, 3-CH_3, 4-CH_3O, 3-NO_2, 4-NO_2.

When a nitroaryl diselenide and another aromatic diselenide not containing a nitro group was allowed to compete for the sulfonylmethane, only the nitroaryl selenide was isolated.[392]

Triethyl phosphite caused a heterolytic cleavage of the selenium-selenium bond in diphenyl diselenide with a concurrent migration of an ethyl cation to the selenolate anion. Se-Phenyl diethyl selenophosphate and ethyl phenyl selenide were obtained.[348]

Brandsma[81] prepared ethyl phenyl selenide by reacting phenyl lithium with bis(ethylseleno)ethyne in ether.   The other product was ethyl ethynyl selenide, which, with excess phenyl lithium, was also converted to phenyl ethyl selenide.

Modification of Organic Groups in Alkyl Aryl Selenides.   A large number of alkyl aryl selenides have been synthesized by modifying either the aliphatic portion of the molecule, by introducing new substituents into the aromatic ring, or by modifying substituents already present in the aryl group.

Reactions Modifying the Alphatic Group.   The following reactions modifying the aliphatic groups have been carried out with alkyl aryl selenides: double bond migration, conversion of an acetylenic to an allenic system, *cis-trans* isomerization of olefins, bromine addition to double bonds, dehydrobromination, hydrazone formation from acetals, saponification of carboxylic acid esters, esterification of carboxyl groups, and formation of anhydrides from carboxylic acids.

Isomerization of Olefinic and Acetylenic Systems.   The following isomerization proceeded in solution in the presence of an equivalent amount of sodium ethoxide employing an aliphatic alcohol as the solvent:

$$C_6H_5-Se-CH_2-CH=CH_2 \longrightarrow C_6H_5-Se-CH=CH-CH_3^{112,266}$$

$$(90)$$

The conversion, however, was quantitative only in dimethyl sulfoxide with potassium *tert*-butoxide.[266]   Similar rearrangements with substituted compounds of the type shown above have been reported by Kataev.[272,273]   A thermal rearrangement of $C_6H_5-Se-CH(CH_3)CH=CH_2$ into C_6H_5-Se $-CH_2-CH=CH-CH_3$ was also observed.[268*]

Phenyl propargyl selenides isomerize to the allenic compounds under the influence of strong base:

$$C_6H_5-Se-CH(R)C \equiv C-R' \xrightarrow{\text{base}} C_6H_5-Se-C(R)=C=CH-R'$$

$$(91)$$

R, R': H, H; H, CH$_3$;[412] H, H; CH$_3$, H.[411]

Phenyl propargyl selenide in 0.1N potassium *tert*-butoxide in *tert*-butanol heated to 85°C and added to 0.2N ethanolic sodium hydroxide containing potassium iodomercurate gave a mixture of the allenyl and methylacetylenyl phenyl selenide.[412]

cis-2-Carboxyvinyl phenyl selenide was converted to the *trans* compound by heating to 260°C.[109,110]   The *trans* derivative did not isomerize on heating.[109]   *cis*-2-Phenylvinyl phenyl selenide gave the *trans* derivative upon irradiation with a mercury-quartz lamp.[265]

Bromine Addition to Olefins and Dehydrobromination.   Chierici[109,110] showed that phenylcarboxyvinyl selenides react with bromine.   The brominated products lose hydrogen bromide and carbon dioxide, as indicated in reaction (92):

$$cis\text{- or } trans\text{-4-XC}_6\text{H}_4-\text{Se}-\text{CH}=\text{CHCOOH}$$

$$\downarrow \text{Br}_2, 0°\text{C} \quad | \quad \text{CH}_3\text{COOH}$$

$$4\text{-XC}_6\text{H}_5-\text{Se}-\text{CHBr}-\text{CHBr}-\text{COOH}$$

$$\downarrow \begin{array}{l} +\text{H}_2\text{O}, -\text{CO}_2 \\ -\text{HBr} \end{array}$$

(92)

$$4\text{-XC}_6\text{H}_4-\text{Se}-\text{C}\equiv\text{CH} \quad \xleftarrow[\;25°\text{C}\;]{\text{KOH/C}_2\text{H}_5\text{OH}} \quad cis\text{-}, trans\text{-4-XC}_6\text{H}_4-\text{Se}-\text{CH}=\text{CH}$$

$$X = H, NO_2$$

The *cis*-bromovinyl selenides are dehydrobrominated at a much faster rate than the *trans* compounds.[109]

Formation of Hydrazones.   Phenylselenoacetaldehyde dimethyl acetal,[353] 2-nitrophenylselenoacetaldehyde diacetyl acetal, 2-benzylselenobenzaldehyde and 2-methylseleno-n-R-benzaldehyde (R = H, 4-, 5- and 6-CH$_3$)[454] gave with 2,4-dinitrophenylhydrazine the corresponding hydrazones.

Saponification and Esterification.

$$\text{C}_6\text{H}_5-\text{Se}-\text{Y}-\text{COOC}_2\text{H}_5 \quad \xrightarrow[\text{saponification}]{\text{basic or acidic}} \quad \text{C}_6\text{H}_5-\text{Se}-\text{Y}-\text{COOH} \quad (93)$$

$$Y: \quad \text{CH(CH}_3)\text{CH}_2, \text{CH}_2\text{CH(CH}_3)^{133} \quad \text{C(C}_6\text{H}_5)=\text{CH}.^{195,196*}$$

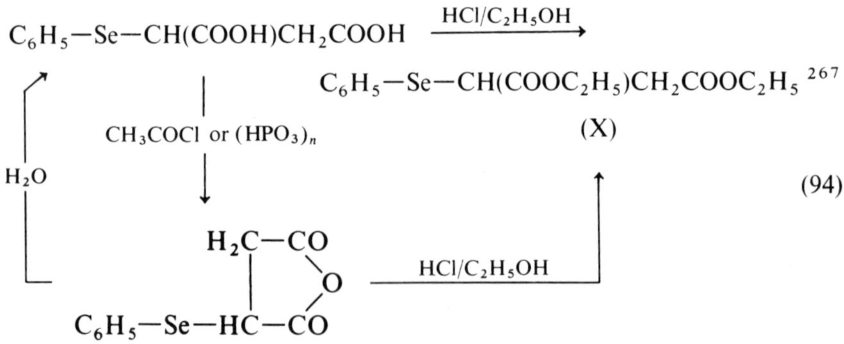

$$\text{C}_6\text{H}_5-\text{Se}-\text{CH(COOH)CH}_2\text{COOH} \quad \xrightarrow{\text{HCl/C}_2\text{H}_5\text{OH}}$$

$$\text{C}_6\text{H}_5-\text{Se}-\text{CH(COOC}_2\text{H}_5)\text{CH}_2\text{COOC}_2\text{H}_5 \quad ^{267}$$

(X)

$$\begin{array}{c} \text{H}_2\text{C}-\text{CO} \\ | \quad \quad \searrow\text{O} \\ \text{C}_6\text{H}_5-\text{Se}-\text{HC}-\text{CO} \end{array} \quad \xrightarrow{\text{HCl/C}_2\text{H}_5\text{OH}}$$

CH$_3$COCl or (HPO$_3$)$_n$

H$_2$O

(94)

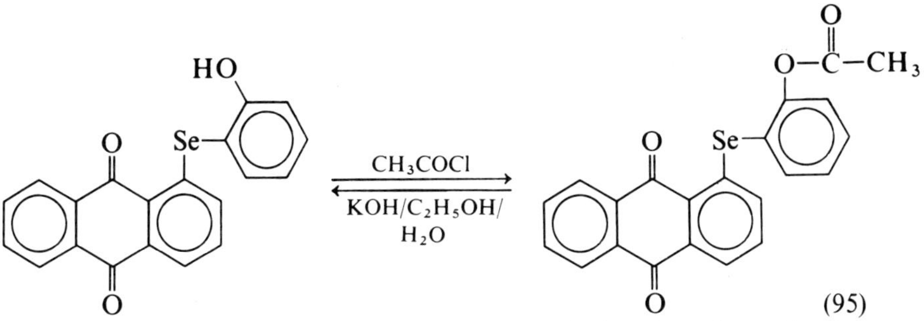

(95)

Diethyl benzylselenosuccinate (X) cannot be hydrolyzed to the acid.   Diphenyl diselenide, fumaric acid and maleic acid are formed upon saponification.[267]   When Schoellkopf[467†] treated phenyl chloromethyl selenide in the presence of potassium *tert*-butoxide with *iso*-butene, *cis*- and *trans*-2-butene, cyclohexene and 1,3-cyclohexadiene, (XI)–(XVI) were isolated in 70% yields:

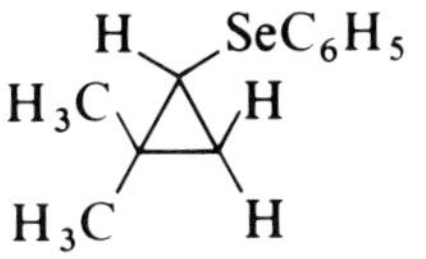

The base generates phenylselenocarbene, $C_6H_5-Se-\overset{..}{C}-H$, from the chloro methyl selenide, which then adds stereospecifically to the olefin. With cyclohexene, a mixture of *exo*-(XIV) and *endo*- (XV) compounds was obtained with the *exo*-derivative predominating.   A similar mixture was produced through 1,2-addition to 1,3-cyclohexadiene.   1,4-Addition, which would lead to a norbornene derivative, was not observed.

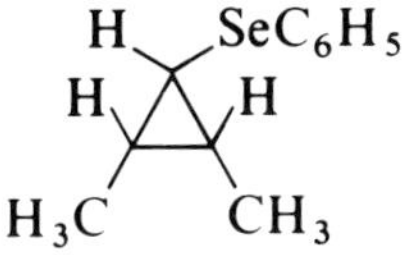

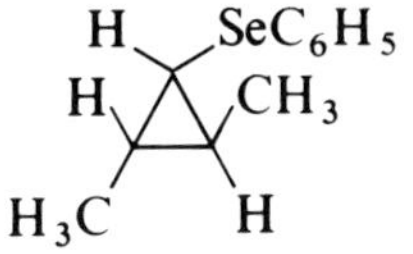

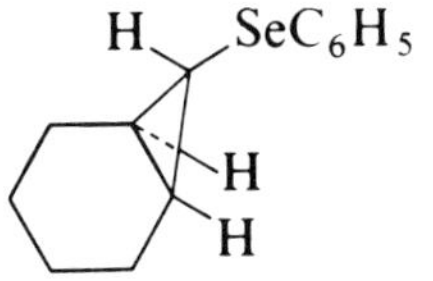

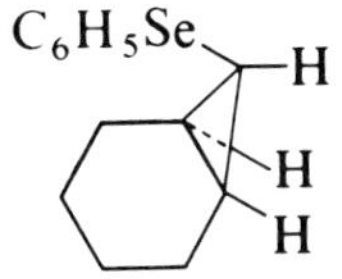

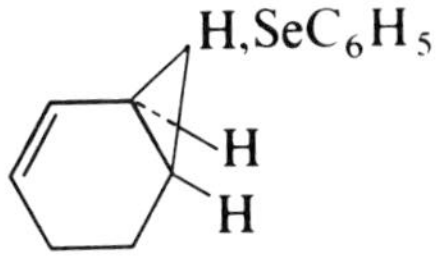

Introduction of Substituents into the Aromatic Ring.   Hannig[225,226] acylated alkyl phenyl selenides in the *para* position with carboxylic acid chlorides and aluminum chloride in chloroform solution at 0–5°C:

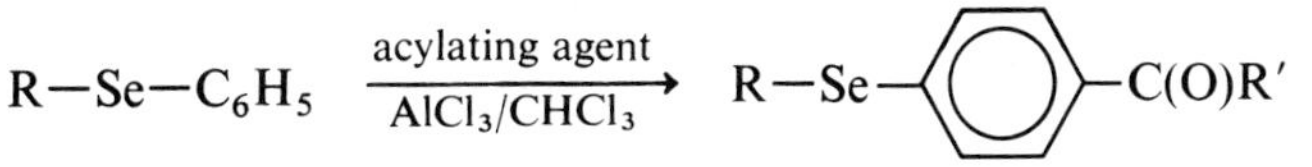

$$R-Se-C_6H_5 \xrightarrow[\text{AlCl}_3/\text{CHCl}_3]{\text{acylating agent}} R-Se-\underset{}{\bigcirc}-C(O)R' \qquad (96)$$

R; acylating agent: C_nH_{2n+1} (n = 1–4, 6), i-C_5H_{11}; $CH_3C(O)Cl$;[225] CH_3; $CH_3C(O)Cl$;[493] C_nH_{2n+1} (n = 1–8); $C_2H_5C(O)Cl$, $C_3H_7C(O)Cl$; [226] C_nH_{2n+1} (n = 1–4, 6–8), i-C_5H_{11}; succinic anhydride;[226] $C_nH_{2n+1}(n = 1-5)$; $ClCH_2CH_2CH_2C(O)Cl$.[223]

The yields in these reactions approach 80%. 2-Acetyl-5-methylphenyl methyl selenide was obtained by treating the selenide with acetyl chloride/ aluminum chloride in carbon disulfide.[192] The bromoacetyl group was introduced with bromoacetyl chloride into the *para* position of diphenylyl methyl selenide under similar conditions.[494] The phosphorus oxychloride complex of *N*-methylformanilide converted 3,4-methylene-dioxyphenyl methyl selenide into the 6-formyl compound.[136] 3-Methoxyphenyl methyl selenide coupled with the 2,4-dinitrophenyldiazonium salt in acetic acid in the *para* position to the methoxy group.[336] The main product, however, was the unsymmetric diaryl selenide. Bromination of methyl 2-thienyl selenide with a bromide/bromate mixture gave the 5-bromo derivative.[185]

Modifications of Substituents in the Aromatic Group. The reactions modifying the substituents in aromatic rings are arranged according to the substituent being modified.

—Br:

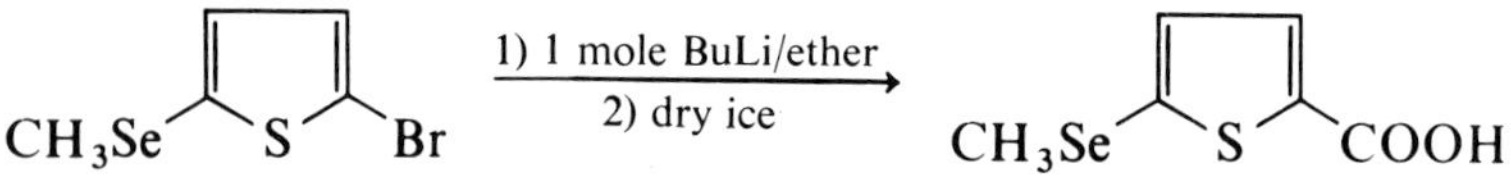

(97)

When 2 moles of butyl lithium were employed or when methyl 2-thienyl selenide was allowed to react with butyl lithium, the selenium-carbon(thienyl) bond was cleaved.[185] 2-Bromophenyl methyl selenide, treated with butyl lithium in ether, gave, after addition of phosphorus trichloride, tris(2-methyl-selenophenyl)phosphine.[151] 4-Bromophenyl methyl selenide reacted with magnesium. Subsequent addition of carbon disulfide formed 4-dithio-carboxyphenyl methyl selenide.[345]

—C(O)H:

2-Methylselenobenzaldehydes[450,454] and 2-methylseleno-4,5-methylenedi-oxybenzaldehyde[136] were converted into the 2,4-dinitrophenylhydrazone and thiosemicarbazone, respectively.

2-Methylselenobenzaldehyde and its 3-, 4-, and 5-methyl derivatives condensed with malonic, phenylacetic and cyanoacetic acids, phenylacetonitrile, and propionic acid anhydride to give cinnamic acids.[455]

$-COOH$, $-C(O)Cl$, $-C(O)NR_2$, $-C(S)SH$:

The silver salts of 4-carboxyphenyl alkyl selenides reacted with 2-chlorodiethylaminoethane hydrochloride[226] according to equation (98):

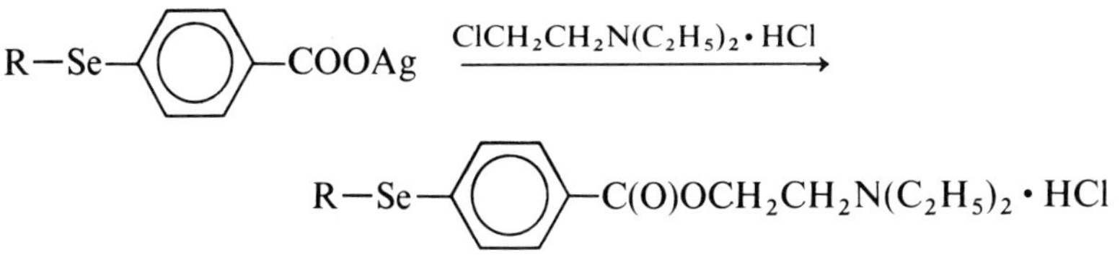

$$R-Se-\underset{}{\bigodot}-COOAg \quad \xrightarrow{ClCH_2CH_2N(C_2H_5)_2 \cdot HCl}$$

$$R-Se-\underset{}{\bigodot}-C(O)OCH_2CH_2N(C_2H_5)_2 \cdot HCl \quad (98)$$

$$R: C_nH_{2n+1}(n = 1-6), C_6H_5CH_2$$

2-Methylselenobenzoic acid,[450] 2-methylseleno-5-bromobenzoic acid,[510] 2-methylseleno-, 2-methylseleno-4-methyl, -5-methyl- and -6-methylbenzoic acid[454] formed the carboxylic acid chlorides when treated with butoxydichloromethane and zinc chloride. With thionyl chloride the selenium-methyl bond is cleaved producing chloroformylphenyl selenium chlorides.[432,510]

Lithium aluminum hydride in toluene/ether reduced 2-methylselenobenzoic acid and its 4-, 5-, and 6-monomethylderivatives to the corresponding 2-methylselenophenylmethanols, while a mixture of the aldehyde and alcohol was obtained from the methyl phenyl amide of 2-methylselenobenzoic acid.[450]

The reactions of the carboxylic acid chlorides[450] are described by reaction scheme (99).

In a similar reaction 4-methylselenobenzoic acid gave compound (XVII) ($R' = CH_3$).[450] The compounds corresponding to (XVIII), (XIX), and (XX) were also obtained from 2-methylseleno-5-bromobenzoic acid chloride.[510] Meyer[345] esterified 4-methylselenodithiobenzoic acid with methyl sulfate.

$-OH$:

3-Hydroxyphenyl and 4-hydroxyphenyl methyl selenide were converted into the corresponding methoxy compounds upon treatment with methyl sulfate.[336]

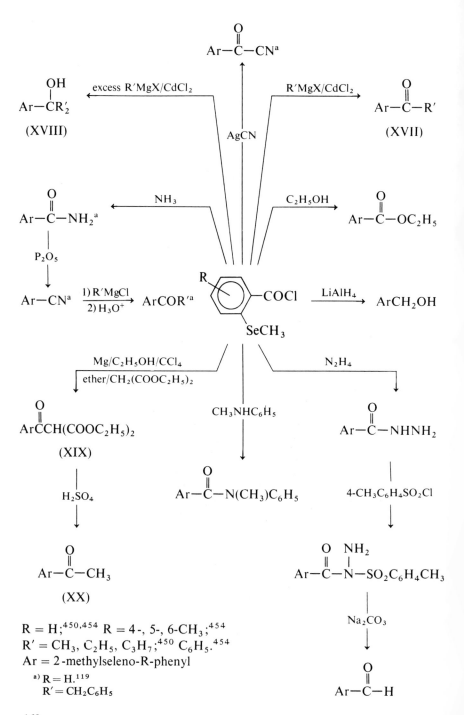

$$R = H;^{450,454} \quad R = 4\text{-},\ 5\text{-},\ 6\text{-}CH_3;^{454}$$
$$R' = CH_3,\ C_2H_5,\ C_3H_7;^{450}\ C_6H_5.^{454}$$
Ar = 2-methylseleno-R-phenyl
  [a] R = H.[119]
    R' = CH_2C_6H_5

$-NH_2$:

The above mentioned hydroxphenyl selenides were prepared by diazotization of the corresponding amines and decomposition of the diazonium salts.[336]

$-NO_2$:

The following selenides containing a nitro group were reduced to the amines: 2-nitrophenyl, 3-nitrophenyl, 4-nitrophenyl methyl (Sn/HCl);[103] 4-nitrobenzyl phenyl (Fe/CH_3COOH).[141]

$-C(O)R$:

2-Methylselenoalkylacetophenones, $2\text{-}CH_3SeC_6H_4(R)COCH_3$ (R = H, 4-CH$_3$, 5-CH$_3$, 6-CH$_3$) were reduced to the alcohols with lithium aluminum hydride.[454] With cadmium organic compounds, tertiary alcohols (R = H) were obtained.[450]

The Reformatsky reaction employing 2-methylselenobenzophenone, -acetophenone, and bromocarboxylic acid esters yielded a number of 2-methylselenocinnamic acid derivatives.[455]

2-Methylselenophenyl benzyl ketone was oxidized by selenium dioxide to $2\text{-}CH_3SeC_6H_4C(O)C(O)C_6H_5$.[119] 4-Alkylselenophenyl alkyl ketones were oxidized in 50% yield to the 4-alkylselenobenzoic acids by $2N$ nitric acid:[226]

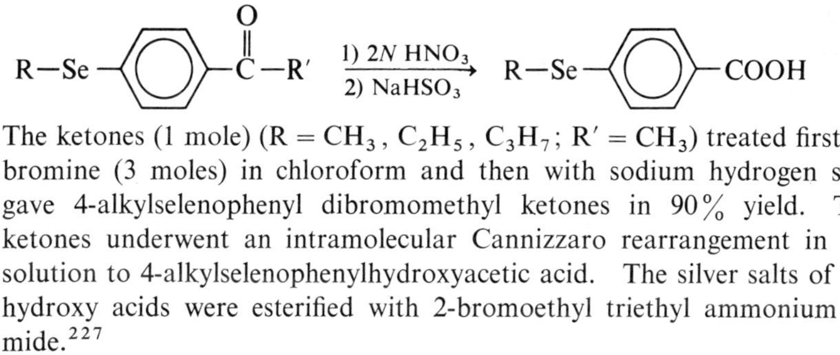

$$R-Se-\!\!\bigcirc\!\!-\overset{\overset{O}{\|}}{C}-R' \xrightarrow[\text{2) NaHSO}_3]{\text{1) 2}N \text{ HNO}_3} R-Se-\!\!\bigcirc\!\!-COOH \qquad (100)$$

The ketones (1 mole) (R = CH$_3$, C$_2$H$_5$, C$_3$H$_7$; R$'$ = CH$_3$) treated first with bromine (3 moles) in chloroform and then with sodium hydrogen sulfite gave 4-alkylselenophenyl dibromomethyl ketones in 90% yield. These ketones underwent an intramolecular Cannizzaro rearrangement in basic solution to 4-alkylselenophenylhydroxyacetic acid. The silver salts of these hydroxy acids were esterified with 2-bromoethyl triethyl ammonium bromide.[227]

The synthesis of 4-alkylselenophenyl monobromomethyl ketones from 4-acetylphenyl alkyl selenium dichlorides is described in reaction scheme (101). The reactions of the bromo ketones are included.

Supniewski[492,493] obtained compound (XXII) (R = CH$_3$) (Scheme 101) in about 50% yield from the bromination of 4-acetylphenyl methyl selenide. Hannig's method[227] gave an 80% yield. Compound (XXIIb) was obtained via the hexamethylenetetramine adduct (XXIIa). (XXIIb) was treated with dichloroacetyl chloride. The product $R'-C(O)CH_2-NH-C(O)CHCl_2$ with formaldehyde gave $R'-C(O)CH(CH_2OH)-NH-C(O)CHCl_2$, which was reduced to $R'-CH(OH)CH(CH_2OH)-NH-C(O)CHCl_2$(R$'$ = 4-CH$_3$Se$-$C$_6$H$_4$).

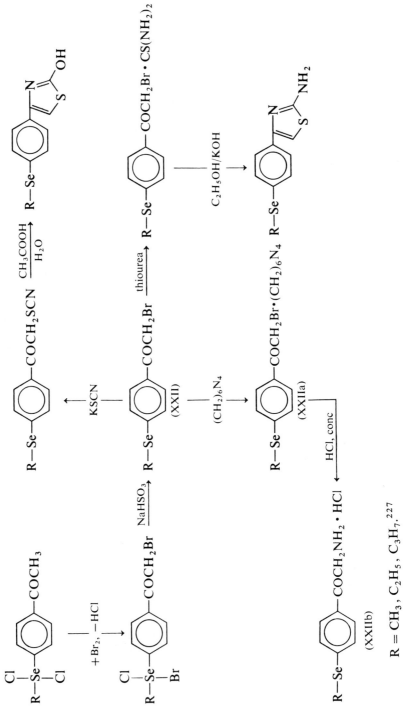

$R = CH_3, C_2H_5, C_3H_7.^{227}$

The following reactions were carried out with 4-(4-methylselenophenyl)-phenyl bromoethyl ketone (reaction 102):[494]

$$R-C(O)CH_2Br \xrightarrow{(CH_2)_6N_4} R-C(O)CH_2Br \cdot (CH_2)_6N_4 \qquad (102)$$

$$\downarrow HCl$$

$$R-C(O)CH_2NH-C(O)CH_3 \xleftarrow[CH_3COOH, CH_3COONa]{(CH_3CO)_2O} R-C(O)CH_2NH_2 \cdot HCl$$

$$\downarrow CH_2O/CH_3OH$$

$$R-C(O)CH(CH_2OH)NH-C(O)CH_3$$

$$\downarrow (i\text{-}C_3H_7O)_3Al$$

$$R-CH(OH)CH(CH_2OH)-NH-C(O)CH_3$$

$$\downarrow \begin{array}{l}1)\ HCl \\ 2)\ NH_3\end{array}$$

$$R-CH(OH)CH(CH_2OH)-NH_2$$

$$Cl_2CHCH(OH)_2 \Big| NaCN$$

$$R-CH(OH)CH(CH_2OH)-NH-C(O)CHCl_2$$

$$R = 4\text{-}(4'\text{-}CH_3SeC_6H_4)C_6H_4$$

2-Methylselenophenyl and 5-methyl-2-methylselenophenyl methyl ketone refluxed with benzaldehyde or 4-hydroxybenzaldehyde in glacial acetic acid in the presence of piperidine or in absolute ethanol in the presence of sodium ethoxide produced compounds (XXI).[195]

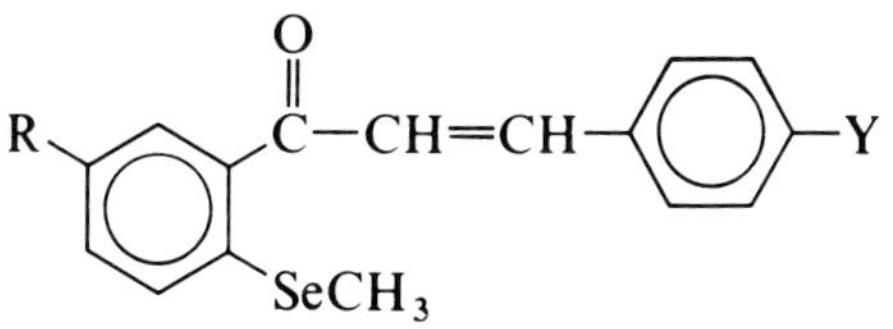

(XXI)

R, Y: H, H; CH$_3$, H; CH$_3$, OH

4-Alkylselenophenyl methyl ketones undergo a Mannich condensation upon the addition of paraformaldehyde, azacycloheptane and a drop of concentrated hydrochloric acid to a dioxane solution containing the ketone:[225]

$$R: C_nH_{2n+1} \ (n = 1\text{--}4, 6), \ i\text{-}C_5H_{11} \tag{103}$$

The reactions of 4-alkylselenophenyl 2-chloroethyl ketones are outlined by reaction (104):

$$\tag{104}$$

$R = C_nH_{2n+1}(n = 1\text{--}5)^{223}$

The condensation of alkylselenophenyl methyl ketones dissolved in diethyl ether with ethyl acetate in the presence of sodium yielded 4-methylseleno-benzoyl acetone.[228]   The diketone was used for the synthesis of heterocyclic ring systems (reaction 105):

$$R = C_nH_{2n+1} \ (n = 1\text{–}7); \ R' = 4\text{-}CH_3SeC_6H_4$$

The silver salt (XXIII) condensed with the chloroethylamine (XXIV)[226] according to reaction (106):

$$R: C_nH_{2n+1} \ (n = 1\text{–}4, \ 6\text{–}8), \ i\text{-}C_5H_{11}; \ R' = C_2H_5 \ or \ (CH_2)_5$$

$$-C(OH)(CH_2R)R':$$

The dehydration of the hydroxy compounds (XXV) lead to olefins:[450]

$$
\begin{array}{ccc}
\underset{\underset{\displaystyle (XXV)}{\overset{\displaystyle \text{Se—R''}}{\overset{\displaystyle R'}{\underset{}{\overset{}{\text{C—CH}_2\text{R}}}}}}{\overset{\displaystyle \text{OH}}{|}} & \longrightarrow & \underset{\text{Se—R''}}{\overset{\text{C=CHR}}{\underset{R'}{}}}
\end{array} \qquad (106a)
$$

R, R′, R″: H, CH$_3$, CH$_3$; CH$_3$, CH$_3$, CH$_3$; CH$_3$, C$_2$H$_5$, CH$_3$; CH$_3$, CH$_3$, C$_2$H$_5$; CH$_3$, C$_2$H$_5$, C$_2$H$_5$.

Compound (XXV) (R = R′ = H) and 2-methylselenoalkylphenylmethanol, 2-CH$_3$Se(R)C$_6$H$_3$CH$_2$OH (R = H, 4-CH$_3$, 5-CH$_3$, 6-CH$_3$) treated with phosphorus oxychloride or butoxydichloromethane produced the corresponding alkyl chlorides.[454]   The hydroxymethyl group in 4-CH$_3$SeC$_6$H$_4$CH$_2$OH was oxidized to the aldehyde group by aerating its refluxing dimethyl sulfoxide solution.[450]   The alkyl chlorides obtained above were converted into aldehydes in reactions with sodium carbonate/dimethyl sulfoxide, hexamethylenetetramine/acetic acid, or *iso*-propyl nitrate.[454]

Physical Properties and Reactions.   The compounds for which spectral data are available in the literature are marked with the proper superscripts.   In addition, the following investigations should be mentioned.   Chierici,[103,105] Modena,[355] and Kiss[286] reported the UV spectra of a number of aryl alkyl selenides.   Gasco[179] investigated the UV spectra of 1- and 2-naphthyl methyl selenides and carried out appropriate molecular orbital calculations.   The near UV vapor phase spectrum of methyl phenyl selenide was reported by Mangini.[333]

The NMR spectra of C$_6$H$_5$—Se—C(CH$_3$)=C=CH$_2$, C$_6$H$_5$—Se—CH= C=CH$_2$,[482] C$_6$H$_5$—Se—C≡CCH$_3$,[481] and C$_6$H$_5$—Se—CH$_2$C≡CH[483] were recorded and compared with the spectra of similar organometallic compounds.

References to the general reactions of alkyl aryl selenides are included in Figure 8-3.   The replacement of methyl groups by aryl groups in reactions of aryl methyl selenides with diazonium salts has already been discussed. The rate of demethylation of methyl phenyl selenide by 2,4-dinitrophenyl-diazonium chloride was found to be greater than that for the sulfide.[332] The cleavage of the selenium-methyl bond by halogenating agents has been discussed previously.   The rate of the reaction of methyl aryl selenides and sulfides with dimethyl sulfate has been investigated.   The selenides react faster than the sulfides.[193]

Alkyl aryl selenides having appropriate functional groups bonded to the aryl group and a hydrogen atom or a suitable substituent in one of the *ortho* positions to the selenium atom may form a five- or six-membered selenium containing ring system:

$$R \xrightarrow[-H_2O]{P_2O_5} R \qquad (107)$$

R, R', R": 2-CH$_3$, H, H; 3-CH$_3$, H, H; 4-CH$_3$, H, H; H, CH$_3$, H; H, H, CH$_3$

$$\xrightarrow[\text{benzene}]{POCl_3} \qquad (108)$$

X = Cl, OH

Similar heterocyclic compounds were prepared by Gosselck,[192,195] Chmutova,[112] Degani,[141] Renson[433] Ruwet[452] and Bellinger.[51,52]

Benzoselenophene and its derivatives were synthesized from 2-dimethoxyethyl phenyl selenide,[353] 2-acetylphenyl carboxymethyl selenide,[118] 2-carboxy-4-bromophenyl carboxymethyl selenide,[510] and allyl phenyl selenide.[112]

Aryl Trifluoromethyl Selenides

Trifluoromethyl selenium chloride reacts with aromatic Grignard reagents in ether at a temperature of $-50°C$ giving the alkyl aryl selenides in about 50% yield:

$$CF_3SeCl + R-C_6H_4MgBr \xrightarrow{-50°C} CF_3-Se-C_6H_4R \qquad (109)$$

R: H, 4-CH$_3$, 4-F, 3-F;[562] 4-CH$_3$O.[563]

Anilines condense directly with trifluoromethyl selenium chloride, when the ether solutions of the reagents are mixed at $-20°C$ and then kept at $+20°C$ for one hour:

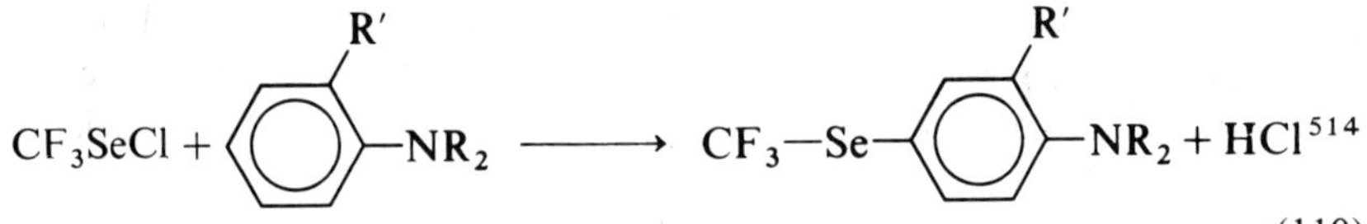

$$CF_3SeCl + \underset{\text{R}'}{\bigcirc}-NR_2 \longrightarrow CF_3-Se-\underset{\text{R}'}{\bigcirc}-NR_2 + HCl^{514}$$

(110)

R, R': H, H; CH_3, H; H, CH_3;[561] H, F; H, CO_2CH_3[514]

Phenol with an excess of trifluoromethyl selenium chloride in pyridine forms in quantitative yield 2,4,6-tris(trifluoromethylseleno)phenol.[563] The reaction modifying the aryl group without affecting the selenium-carbon bonds are presented in scheme (111). Lutskii[318] determined the dipole moments of some of these compounds and discussed the participation of the vacant d orbitals of selenium in conjugation with the aromatic ring.

Diorganyl Selenides, $R-Se-(CH_2)_n-Se-R$

Selenides with more than one selenium atom in the molecule have been synthesized from selenols and organic dihalides, by addition of diselenides to acetylene, from alkali metal acetylides, selenium, and an alkyl halide, and from formaldehyde, benzeneselenol and ammonia or amines. The following equations give more details about these reactions.

$$2R-Se^- + X-Y-X \longrightarrow R-Se-Y-Se-R + 2X^- \quad (112)$$

R, X, Y: CH_3, Br, $(CH_2)_3$; CH_3, Br, $(CH_2)_2$;[38*†] i-C_3H_7, Br, $(CH_2)_2$;[207] $C_6H_5CH_2$, Br, $CH_2CH(COOH)CH_2$;[63] $C_6H_5CH_2$, Br, $CH_2CH_2CH(CH_2)_4COOH$;[57] C_6H_5, Cl, $CH_2C\colon CCH_2$;[412] C_6H_5, Br, $(CH_2)_3$;[541] C_6H_5, Cl, 1,4-$(CH_2)_2C_6H_4$.[121]

Bis(pentafluoroethylseleno) mercury heated with diiodomethane gave bis(pentafluoroethylseleno)methane.[537] The reaction of α-chloroethers, $RCHCl(OCH_3)$, with selenols in the presence of zinc in diethyl ether gave $RCH(SeR')_2$ in yields as high as 68%.[309]

$$[R-Se(CH_2)_3]_2Se_2 \xrightarrow[\text{C}_2\text{H}_5\text{OH/NH}_3/\text{H}_2\text{O}]{\text{Rongalite}} R-Se-(CH_2)_3-Se^-$$

(113)

$$R-Se(CH_2)_3-Se-R' \xleftarrow{R'X}$$

(XXVI)

$R = HOOC(CH_2)_{10}$, $R' = C_2H_5*, C_6H_{13}*, C_{11}H_{23}, C_2H_5Se(CH_2)_3*$;
$R = H_2NC(O)(CH_2)_{10}$, $R' = C_2H_5, C_6H_{13}$.[186]

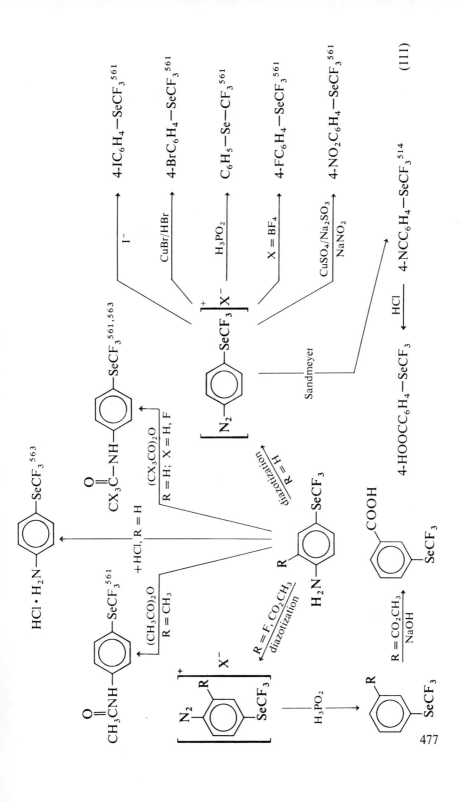

(111)

477

The compounds (XXVI) can also be synthesized from $[HOOC(CH_2)_{10}]_2Se_2$ and $R'Se(CH_2)_3Cl$ or from R'_2Se_2 and $Cl(CH_2)_3Se(CH_2)_{10}COOH$. The yields can be as high as 80%. These selenium containing carboxylic acids were converted into the amides by treatment with oxalyl chloride followed by concentrated ammonia.[186]

Brandsma[81] obtained bis(ethylseleno)acetylene in 80% yield by adding successively elemental selenium, sodium amide, elemental selenium and finally ethyl bromide to monosodium acetylide in liquid ammonia:

$$HC{\equiv}CNa \xrightarrow{\ Se\ } HC{\equiv}C{-}SeNa \xrightarrow{\ NaNH_2\ } NaC{\equiv}C{-}SeNa$$

$$\Big\downarrow Se \qquad (114)$$

$$C_2H_5{-}Se{-}C{\equiv}C{-}Se{-}C_2H_5 \xleftarrow{\ C_2H_5Br\ } NaSe{-}C{\equiv}C{-}SeNa$$

The Mannich condensation presented in reaction (115) gave amino-selenides[409]*† in good yields:

$$nC_6H_5SeH + RNH_2 + nCH_2O \xrightarrow{\ C_2H_5OH\ } (C_6H_5Se{-}CH_2)_nNR_{3-n}$$

$$(115)$$

$$n, R:\quad 2,\ C_6H_5;\ 2,\ C_6H_5CH_2;\ 3,\ H.$$

Diorganyl diselenides add to acetylenes in reactions catalyzed by alkoxides and amines to give unsaturated selenides, $RSe{-}C(R'')={CH}{-}SeR'$, with a *cis* configuration:[277]*

$$R{-}Se{-}Se{-}R' + R''C{\equiv}CH \longrightarrow RSe{-}C(R'')={CH}{-}SeR' \quad (116)$$

$$R,\ R',\ R'';\ CH_3,\ CH_3,\ H;\ CH_3,\ CH_3,\ C_6H_5;\ CH_3,\ CH_3,\ 2{-}CH_3{-}$$
5-pyridyl; $C_2H_5,\ C_2H_5,\ C_6H_5;\ C_2H_5,\ C_2H_5,\ 2{-}CH_3{-}5{-}pyridyl;$
$C_6H_5,\ C_6H_5,\ H;\ C_6H_5,\ C_6H_5,\ C_6H_5;\ C_6H_5,\ CH_3,\ H.$

4-Chloro-2,5-bis(phenylseleno)nitrobenzene was reduced to the amine by $FeSO_4$ in basic medium [Jap. Pat., 19,630(64)]. Emeleus[154] obtained $(RSe)_2CH_2$ from $(RSe)_2$ and diiodomethane upon heating to 120°C for 16 h ($R = CF_3,\ C_3F_7$). Diazomethane and diphenyl and dibenzyl diselenide gave bis(organylseleno)methane.[403] Selenocyclobutane was photochemically converted to a polymeric selenide.[129] Seebach[471] synthesized $(C_6H_5Se)_n{-}CH_{4-n}$ ($n = 2, 3, 4$) and investigated the reactions of these compounds with organic lithium compounds leading to selenium stabilized carbanions.

Se-ESTERS OF SELENOCARBOXYLIC ACIDS

Most of the Se-esters of selenocarboxylic acids have been prepared from an acid chloride and a selenol in the presence of a base to neutralize the hydrogen chloride generated in the reaction:

$$R-SeH + R'-C(O)Cl \xrightarrow{-HCl} R'C(O)-SeR \qquad (117)$$

$R = CH_3$, $R' = C_6H_5$;[345]

$R = C_4H_9$, $R' = CH_3$, $CH_3(CH:CH)_2$, $(CH_3)_2C:CH$, $2\text{-}X-C_6H_4$ $(X = H, CH_3, CH_3O, F, Cl, Br, I)$;[429]

$R = C_6H_5$, $R' = CH_3$, $(CH_3)_2C:CH$,[429] $CH_2:C(CH_3)$, $C_6H_5SeCH_2\text{-}CH(CH_3)$,[416] C_6H_5,[429,436] $2\text{-}XC_6H_4$ $(X = Cl, Br, I, CH_3O)$;[429] $(X = SCN, SCH_3)$;[432]

$R = 2\text{-}CH_3C_6H_4$, $R' = 2\text{-}XC_6H_4$ $(X = H, Br, CH_3O)$;[429]

$R = 2\text{-}CH_3OC_6H_4$, $R' = 2\text{-}XC_6H_4$ $(X = H, I, CH_3O)$, CH_3;[429]

$R' = C_6H_5CH_2CH_2$, $R = X\text{-}C_6H_4$ $(X = H, 3\text{-}Cl, 4\text{-}Cl, 4\text{-}CH_3, 4\text{-}CH_3O, 4\text{-}NO_2, 3\text{-}NO_2)$;[194*]

$R' = C_6H_5$, $R = $ androst-5-en-17-on-3β-yl, pregn-5-en-3-β-yl (U.S. Pat., 3,372,173); $X-CH_2CH_2$ $[X = (CH_3)_2N$, $CH_3(C_6H_5CO)N$, $(CH_3)_3\overset{\oplus}{N}\text{-}$,[212] C_6H_5CO,[213,338] $HOOCCH_2CH_2C(O)NH$,[217] $(CH_3)_3\text{-}NCH_2$;[214]

$R' = CH_3$, $R = $ 2-furyl,[373] $X\text{-}CH_2CH_2$ $[X = (CH_3)_3\overset{\oplus}{N}\text{-}$,[212] $HOOC\text{-}CH_2CH_2C(O)NH$,[217] $(CH_3)_3\overset{\oplus}{N}CH_2$.[214]

The thiobenzoyl compound $C_6H_5C(S)SeCH_2CH_2N(CH_3)_2$ was prepared similarly.[115]

Oxalyl chloride and cyclohexaneselenol and malonyl chloride and benzeneselenol gave the compounds $RSe-C(O)(CH_2)_nC(O)-SeR$ $(n = 1,$[580] $n = 0$[460*]).

Draguet[149] treated 3,1-selenazolidines with base and benzoyl chloride and obtained the seleno esters (XXVII):

$$(118)$$

$$C_6H_5C(O)N(R^3)CR^1R^2CH_2-Se-C(O)C_6H_5$$

$$(XXVII)$$

R^1, R^2, R^3:   H, H, H; H, CH_3, H; CH_3, CH_3, H; H, H, CH_3; H, H, C_6H_5; R, R': various organic groups.

Lozinskii[316] prepared a number of carboxylic acid selenocyanates, $X-C_6H_5-NH-N=C(Cl)C(O)SeCN$ (X = 4-CH_3, 4-Cl, 2-Br, 4-Br, 2-CH_3O) from the acid chlorides and sodium selenocyanate.

Boiko[73,74] converted ethynyl methyl selenides into Se-methyl organyl-selenoacetic acids*† through the mercuric sulfate catalyzed addition of water to the triple bond.   In the absence of the mercury salt the main product was the O-methyl ester.

2,3,-Dimethylbenzoselenazolium iodide hydrolyzed with aqueous sodium hydroxide yielded after the addition of acyl chlorides Se-[2-(N-methyl-N-acetylaminophenyl] selenobenzoates.   The acyl chlorides employed were benzoyl, 4-bromobenzoyl, 4-nitrobenzoyl and 3,5-dinitrobenzoyl chloride.[142]

A comparison of the rates of alkaline hydrolysis of 3- and 4-substituted Se-phenyl seleno-3-phenylpropionate with those of the corresponding sulfur and oxygen derivatives produced the sequence Se $\gg$ O > S, indicating that the substituent effects are transmitted in a similar manner through oxygen, sulfur, and selenium.[198]

SELENIUM CONTAINING CARBOHYDRATES

As a logical consequence of the synthesis and study of sulfur containing carbohydrates, these investigations were extended to include the analogous selenium derivatives.   Early work in this area was performed between 1917 and 1925 by Schneider and Wrede.   Since 1961, Wagner and co-workers have published extensively on the preparation and properties of selenocar-bohydrates.

Selenium derivatives of carbohydrates appear not to be found in nature. Their properties are quite different from the naturally occurring carbohydrates. Wrede[551] reported, for instance, that the selenoisotrehalose is indifferent toward emulsine, myrosine and yeast extract.   Selenium derivatives of the following D-carbohydrates have been synthesized: glucose, galactose, mannose, altrose, arabinose, xylose, and cellobiose.   The formulas of the types of compounds which have been prepared are displayed in scheme (119).   The known selenium derivatives of carbohydrates all have the D-configuration.   The prefix " D " is therefore not included in the names of the compounds treated in this section.   All of the hexose derivatives have been isolated in the pyranose form while all of the pentose derivatives have been shown to occur in the furanose form:

organyl glycosyl selenide
R = aliphatic or aromatic aglycone
(Tables 8-1 to 8-3)

nonreducing diglycosyl selenide
(Table 8-4)

nonreducing diglycosyl diselenide
(Table 8-4)

(119)

dicellobiosyl selenide
(Table 8-4)

Y = difunctional aglycone
X = NH, O or Se (Table 8-5)

bis(6-deoxyglucose) 6-(di)selenide
$n = 1, (2)$ (Table 8-6)

The aliphatic and aromatic selenoglucosides are listed in Tables 8-1 and 8-2. These compounds were synthesized using the following methods:

a) The potassium salts of selenols react with 2,3,4,6-tetra-O-acetyl-α-D-glucopyranosyl bromide according to reaction (120). Bonner[75] dissolved the selenol in ethanolic potassium hydroxide and combined this solution with the 2,3,4,6-tetra-O-acetyl-α-D-glucopyranosyl bromide. Wagner[520,523,528] first dissolved the sugar derivative and then the selenol in acetone and immediately

TABLE 8-1   PHENYL SELENOGLUCOSIDES

Substituents in the Phenyl Group	Glucoside				Tetraacetylglucoside				
	Method	mp °C	Yield, %	[α]D(°C)	mp °C	Yield %	[α]D (°C)	Method	Reference
Only H	d	107	70	−51.7(22)	99	70	−26.6(23)	a	520
Only H	d	108	83	−52.5(20)	100	87.6	−27.2(25)	a	75
Only H	d	94–96	55	+237(20)	85	16	+198(20)	a	172
4-CH₃	d	147.5[1]	—	−46.8(22)[1]	126.5	60	−33.7(23)	a	520
2-CH₃	d	126	70	−61.5(23)	100.5	65	−20.9(23)	a	520
4-Cl	d	173.5	70	−49.8(22)	128	55	−40.8(23)	a	520
4-OH	d,c	161–163	30	−50.0(20)	147	50	−45.6(20)	c	523
4-NO₂	d	127[2]	90	−90.5(19)[2]	167	40	−39.7(19)	a	523
		—	—	—	—	25	—	b	525
2-NO₂	d	196–198	95	−208(22)	152–154	25	−104(20)	b	525
2,4-(NO₂)₂	d	112–114	75	−155(20)	192–194	60	−100(20)	b	525
	d	112–114	—	−154.7(20)	192–194	60	−100.4(20)	b	522
2,4,6-(NO₂)₃	—	—	—	—	198–200	30	−74.3(20)	b	522, 525
4-NH₂	d	128–129	83	−56.9(20)	126–127	80	−49.1(19)	c	523, 524
2-NH₂	d	78–81	80	−52.8(21)	92–95	85	−23.0(21)	c	525
4-NHC(O)CH₃	d	163–164	85	−49.1(20)	144	85	−53.3(20)	c	523
4-NHC(S)OCH₃	c,d	93–97	70	−44.7(20)	171–174	85	−50.3(20)	c,d	524

4-N[C(O)CH₃]C(S)OCH₃	—	—	—	—	124–127	65	−23.7(20)	c,d	524
4-NHC(S)NH₂	d	—³	—	—	—	—	—	—	524
4-NHC(S)NHCH₃	d	109–111	75	−42.0(20)	164–167	70	−35.6(20)	c	524
4-NHC(S)NHC₆H₅	d	167–168	55	−39.8(20)	153–156	70	−32.8(20)	c	524
4-N[C(O)CH₃]C(S)NH₂	—	—	—	—	161–164	75	−30.2(20)	c,d	524
4-NHC(S)NHNH₂	d	—³	—	—	164–167	60	−40.8(20)	c	524
4-NHC(S)NHN=CHC₆H₅	d	141–144	55	−33.8(20)	164–167	80	−29.0(20)	c	524
4-NHC(S)NHN=CH[o-(OH)C₆H₄]	d	128–131	60	−20.9(25)	122–130	55	−23.1(20)	c	524
4-NCS	—	—	—	—	120–122	75	−53.9(19)	c	524
4-COOH	c	213–214	95	−59.5(20)	—	—	—	—	526
2-COOH	c	208	95	−128(20)	—	—	—	—	526
4-CONH₂	d	188–190	80	−69.3(20)	162–163	35	−36.3(17)	c,d	526
2-CONH₂	d	114–117	80	−67.7(20)	197–200	40	−33.2(20)	c,d	526
4-CONHCH₃	—	—	—	—	157–158	60	−33.9(20)	c,d	526
2-CONHCH₃	—	—	—	—	170–171	70	−21.0(20)	c,d	526
2-CON(CH₃)₂	—	—	—	—	133–134	20	−20.8(20)	c,d	526
4-COOCH₃	d	151–152	70	−71.7(20)	138–139	70	−29.4(20)	a	526
2-COOCH₃	d	91	75	−125(20)	114–115	90	−51.6(20)	a	526
4-OC(O)CH₃	—	—	—	—	127	40	−32.3(20)	a	523

1 The melting point of the monohydrate is 140°C. The optical rotation was determined with the hydrated compound.
2 The melting point of the hydrated compound is 73°C. The optical rotation was determined with the anhydrous compound.
3 Did not crystallize.

TABLE 8-2 PYRIDYL SELENOGLUCOSIDES,

Pyridyl	Substituents in Pyridine Ring	R	Method	mp °C	Yield %	$[\alpha]_D$ (°C)	Reference
2	None	H	d	76–79	34	−60.8(20)	529
2	None	Acetyl	a	95–97	45	−30.6(20)	529
4	None	H	d	102	50	−63.4(20)	529
4	None	Acetyl	a	117.5–120	57	−49.2(20)	529
2	3-nitro	Acetyl	b	127–129	48	−88.4(20)	522, 525
2	5-nitro	Acetyl	b	176–178	80	−3.6(20)	522, 525
4	3-nitro	Acetyl	b	124–126	30	−102(20)	525
2	3,5-dinitro	Acetyl	b	211.5–213.5	42	+22.7(20)	522, 525
2	3,5-dinitro	H	d	—	Very small	—	521
2	5-amino	Acetyl	c	—	—	—	522
2	5-acetylamino	Acetyl	c	134–136	—	−28.2(20)	522
2	5-acetylamino	H	d	—[a]	—	−71.8(20)	522

[a] Amorphous.

added the required amount of potassium hydroxide to neutralize the selenol (reaction 120):

$$(120)$$

(R = acetyl,   R' = organic group)

These reactions are accompanied by a Walden's inversion and thus produce the β-selenoglucosides from 2,3,4,6-tetra-O-acetyl-α-D-glucopyranosyl bromide. Substituted and unsubstituted benzeneselenols (Table 8-1) and pyridineselenols (Table 8-2) have also been employed as the selenium containing components. When 2,3,4,6-tetra-O-acetyl-α-D-galactopyranosyl bromide was permitted to react with benzeneselenol, the expected phenyl 2,3,5,6-tetra-O-acetyl-1-seleno-β-D-galactopyranoside (mp 108°C, 61 % yield, $[\alpha]_D^{21}$ −41.9°) was obtained[520] which was deacetylated to phenyl 1-seleno-β-D-galactopyranoside (mp 90°C, 35% yield, $[\alpha]_D^{24}$ − 4.2°). Frenzel and co-workers[172] obtained a mixture of phenyl 1-seleno-α (and β)-D-gluco-pyranoside when 3,4,6-tri-O-acetyl-1,2-anhydro-α-D-glucopyranose was treated with benzeneselenol (reaction 121):

(R = acetyl, R′ = phenyl)   (121)

The anomeric products were separated by TLC. The α-configuration was confirmed by the positive molecular rotation (see Table 8-1) and the NMR spectrum of this compound.

b) The second method available for the preparation of selenoglucosides involves the use of organic halides and glucose derivatives, which already have a selenium atom in the 1-position. Since hydroselenoglucosides are too sensitive towards oxidation, they are generated in the reaction mixture by basic hydrolysis of the 2-(2,3,4,6-tetra-O-acetyl-β-D-glucopyranosyl)-2-selenopseudourea hydrobromide (see Table 8-3):

(122)

The halogen atoms in the aromatic halides (R = phenyl or pyridyl, X = F or Cl) must be activated by nitro groups in o- and/or p-positions.[522,525] Ethyl and i-propyl iodide are the only aliphatic halides which have been used in this reaction.[525] Ethyl iodide was observed also to react with potassium

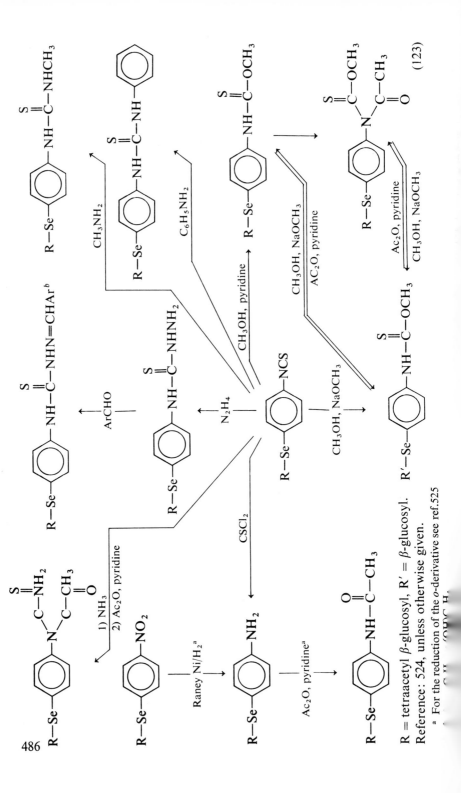

486

(123)

R = tetraacetyl β-glucosyl, R' = β-glucosyl.

Reference: 524, unless otherwise given.

[a] For the reduction of the *o*-derivative see ref.525

TABLE 8-3   SELENOGLYCOSIDES $C_nH_n(OR')_{n-2}$—Se—R^a

R	R'	n	Method	mp °C	Yield %	$[\alpha]_D$ (°C)	Reference
H	Acetyl	6	c	—	—	—	522
K	H	6	c	60–70(dec)	95	—	525
Na	H	6	d	190–200	84	+11.6(20)	292
C_2H_5	Acetyl	6	b	100–104	40	−46.4(20)	525
i-C_3H_7	Acetyl	6	b	100–102	40	—	525
—C(=NH$_2$)NH$_2$[b]	Acetyl	6	—	187–189	—	−23.2(20)	522
			—	187–189	60–70	−18.8(20)	525
—C(=NH$_2$)NH$_2$[c]	Acetyl	6	—	74–76(dec)	—	—	521
—C(=NH$_2$)NH$_2$	Acetyl	5	—	166–168(dec)	70	−105(20)	525
—C(O)C$_6$H$_5$	Acetyl	6	a	112	25	−15.2(20)	292

[a] n = 6, glucose derivative; n = 5, xylose derivative.
[b] The isoselenuronium cation is combined with a bromide ion.
[c] The isoselenuronium cation is combined with a bicarbonate ion.

glucosyl selenolate, obtained by reduction of diglucosyl diselenide with potassium in methanol, but gave the ethyl selenoglucoside in only 20% yield as compared to 40% when the isoselenuronium reaction was used.[525]

c) Reactions of the functional groups in the aromatic aglycone (benzene or pyridine derivatives) of selenoglucosides produce new selenocarbohydrates. Reaction scheme (123) and reactions (124–127) summarize these conversions.

(R' = β-glucosyl[528])   (124)

(R = tetraacetyl-β-glucosyl, R" = H, CH$_3$[528])   (125)

(R = tetraacetyl-β-glucosyl[523])   (126)

(R = tetraacetyl-β-glucosyl[522])   (127)

d) The deacetylation of the phenyl 2,3,4,6-tetra-*O*-acetyl-1-seleno-glucoside with methanol/sodium methoxide (for references see Tables 8-1 to 8-3), methanol/ammonia,[75] or methanol/hydrogen chloride[522] produced the free glucosides, which could be easily acetylated with acetic anhydride and pyridine:

$$\text{(128)}$$

(R = acetyl, R' = organic group)

The corresponding aliphatic 2,3,4,6-tetra-*O*-acetyl-1-seleno-glucosides give, upon deacetylation, syrupy, free 1-seleno-glucosides.[525] Kocourek[292] reported that benzoyl 2,3,4,6-tetra-*O*-acetyl-β-glucosyl selenolate gave sodium β-glucosyl selenolate. The deacetylation of tetraacetyl 3-nitro-, 3,5-dinitro- and 5-nitro-2-pyridyl selenoglucosides with methanol/sodium methoxide, however, were found to take place with cleavage of the selenium-carbon(pyridyl) bond,[521] while the unsubstituted 2- and 4-pyridyl derivatives gave the free glucosides.[529]

Selenocarbohydrates, in which two glycosyl groups are bonded to a selenium atom or a diselenide group, have been prepared by Wagner, Schneider and Wrede. Literature references are given in Table 8-4. Wagner[525] treated a tetra-*O*-acetylglycosylisoselenuronium bromide with potassium hydroxide in aqueous acetone or with sodium hydrogen sulfite in water. The selenol generated in this reaction was immediately oxidized to the diglycosyl diselenide by atmospheric oxygen. When the hydrolysis of the isoselenuronium salt was carried out in presence of tetra-*O*-acetyl-α-bromomonosaccharides, symmetric and unsymmetric diglycosyl selenides were isolated in very good yields. In reaction (129), the glucose derivatives have been used as examples. The other derivatives including pentosyl compounds are listed in Table 8-4.

Catalytic deacetylation with sodium methoxide in methanol[525] or methanol and ammonia[465,466,552,553] produced the free diglycosyl selenides and diselenides, which, upon treatment with acetic anhydride and pyridine, gave back the acetylated compounds.[552,553]

Schneider[465,466] and Wrede[551,552,553] were able to synthesize the acetylated and free diglucosyl selenide, digalactosyl selenide, glucosyl galactosyl

TABLE 8-4 DIGLYCOSYL SELENIDES AND DISELENIDES, R—Se$_n$—R'

R[a]	R'[a]	R″[b]	n	mp °C	Yield %	[α]$_D$ (°C)	Reference
β-glucosyl	β-glucosyl	H	1	193	—	−83.58(26)	551
							465
		H	1	193–194	95	—	525
		Acetyl	1	186–187	80	—	525
		Acetyl	1	186	45	−51.24(21)	465
							551
		H	2	173–175	95	−97.7(20)	522, 525
		H	2	—	—	−93.98(18)	552
		Acetyl	2	133	—	−133.8(18)	552
		Acetyl	2	136–8[c]	90	−141.7(20)	522, 525
β-glucosyl	β-galactosyl	H	1	—	—	−48.35(16)	553
		Acetyl	1	161	—	−30.7(16)	553
		Acetyl	1	163–165	60	−35.7(20)	525
β-galactosyl	β-galactosyl	H	1	228	100	−37.6(19)	466
		Acetyl	1	202	45	−12.5(19)	466
β-glucosyl	β-xylosyl	H	1	115–117	85	−109(20)	525
		Acetyl	1	156–158	75	−108(21)	525
β-xylosyl	β-xylosyl	H	2	192–194	95	−200(22)	525
		Acetyl	2	121–123	90	−275(22)	525
β-cellobiosyl	β-cellobiosyl	H	1	215(dec)	—	−86.35(20)	553
		Acetyl	1	252	—	−47.74(18)	553
β-cellobiosyl	β-glucosyl	H	1	∼160(dec)	—	—	553
		Acetyl	1	141	—	−40.36(17)	553

[a] All glycosyl groups have D-configuration.
[b] R″ = group or atom bonded to the hydroxylic oxygen atoms.
[c] The second modification melts at 158–159°.

selenide, and dicellobiosyl selenide by the reaction of potassium selenide or diselenide in absolute ethanol with the proper tetraacetylbromosaccharides. Glycosyl pentosyl selenides could not be obtained using this method.

Diglucosyl selenide is inactive towards emulsine, myrosine, trehalase and yeast extract. Hot sodium hydroxide solutions and moderately concentrated mineral acids attack the selenide only slowly.[465,551] Fehling's solution precipitates CuSe on boiling. Silver nitrate and mercuric oxide form Ag$_2$Se and HgSe, respectively.[465,555] The selenide is not poisonous to mice upon injection.[551]

Selenocarbohydrates of the general formula R—Se—Y—X—R', in which two glycosyl groups are connected through a difunctional aglycone, were prepared as described in equations (130–132).[523,525] The glucosyl derivatives are used as examples. The compounds obtained through these reactions, the configuration of the glycosidic carbon atoms, literature references and other pertinent data are collected in Table 8-5.

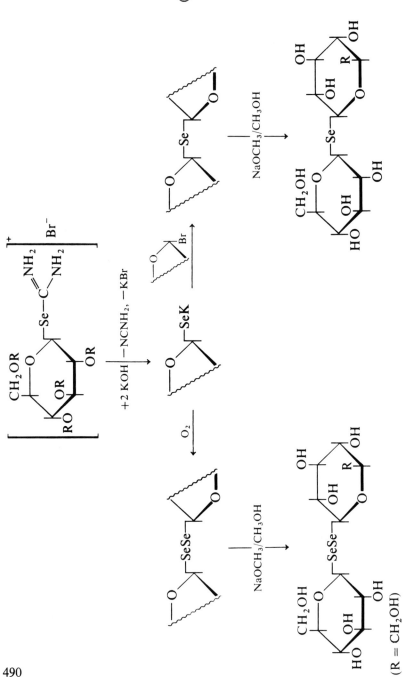

(129)

490

(R = CH₂OH)

TABLE 8-5 SELENIUM DERIVATIVES OF CARBOHYDRATES, R—Se—Y—X—R', CONTAINING TWO GLYCOSYL GROUPS CONNECTED BY A DIFUNCTIONAL AGLYCONE

R[a]	Y	X	R'	mp °C	Yield %	$[\alpha]_D$ (°C)	Reference
β-glucosyl	$-C_6H_4-N=N(\rightarrow O)-C_6H_4-$	Se	β-glucosyl	225–229(dec)	90	−80.0(20)	523
ta-β-glucosyl	$-C_6H_4-N=N(\rightarrow O)-C_6H_4-$	Se	ta-β-glucosyl	227–228(dec)	85	−47.5(20)	523
β-glucosyl	p-phenylene	O	β-glucosyl	130–132	70	−68.6(20)	523
ta-β-glucosyl	p-phenylene	O	ta-β-glucosyl	191	20	−40.8(20)	523
β-glucosyl	o-phenylene	NH	α-glucosyl	171–174(dec)	60	+35.3(23)	525
ta-β-glucosyl	o-phenylene	NH	ta-α-glycosyl	100–104	95	+107(23)	525
β-glucosyl	p-phenylene	NH	β-glucosyl	134(dec)	—	−78.7(20)	523
ta-β-glucosyl	p-phenylene	NH	ta-β-glucosyl	110–111	95[b]	−69.0(20)	523
β-glucosyl	p-phenylene	NH	β-mannosyl	192–193	95	−111.2(20)	523
ta-β-glucosyl	p-phenylene	NH	ta-β-mannosyl	188–192(dec)	95[b]	−90.6(20)	523
β-glucosyl	p-phenylene	NH	β-galactosyl	133–185(dec)	45	−66.4(20)	523
ta-β-glucosyl	p-phenylene	NH	ta-β-galactosyl	amorphous	95[b]	−39.6(20)	523
β-glucosyl	p-phenylene	NH	β-arabinosyl	134–138(dec)	—	−47.0(20)	523
ta-β-glucosyl	p-phenylene	NH	ta-β-arabinosyl	syrup	95[b]	—	523
β-glucosyl	p-phenylene	NH	β-xylosyl	125–128(	—	−46.9(20)	523
ta-β-glucosyl	p-phenylene	NH	ta-β-xylosyl	164–167dec)	95[b]	−43.0(20)	523

[a] ta = tetraacetyl.
[b] Crude product.

octa–acetyl
derivative   (R = CH$_2$OH)   (130)

From the corresponding reaction with 2-aminophenyl β-seleno-glucoside, 2-(β-glucosylseleno)-(α-glucosylamino)benzene was isolated:[525]

deacetylated
derivative   (R = acetyl)   (131)

Reaction (131), performed in the absence of bis(4-nitrophenyl) diselenide, gave only 4-aminophenyl tetracetyl-α-selenoglucoside:

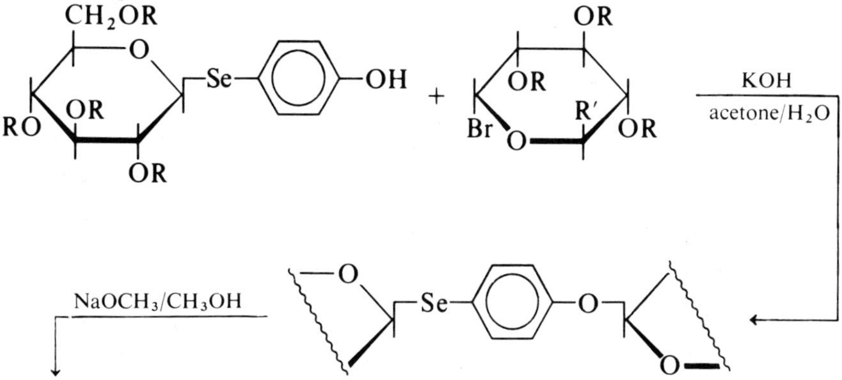

$$(R = \text{acetyl}, \ R' = CH_2OR) \tag{132}$$

deacetylated derivative

The glycosyl group bonded to the selenium atom in the N,Se-diglycosides can be cleaved by acids only with great difficulty, while the N-bonded glycosyl group is easily released under these conditions.[523]

The bis(6-deoxyglucose) 6-selenides and diselenides, which were synthesized by Wrede[554] from methyl 2,3,4-tri-O-acetyl-6-bromo-6-deoxy-β-D-glucopyranoside and K_2Se and K_2Se_2, respectively, are listed in Table 8-6. The reactants, dissolved in ethanol, were heated in a sealed tube to 130°C. Deacetylation, demethylation and acetylation of the free sugar derivative produced the other compounds found in Table 8-6. The reaction sequence is presented in reaction (133):

BLE 8-6 BIS(6-DEOXYGLUCOSE) 6-(DI)SELENIDE,

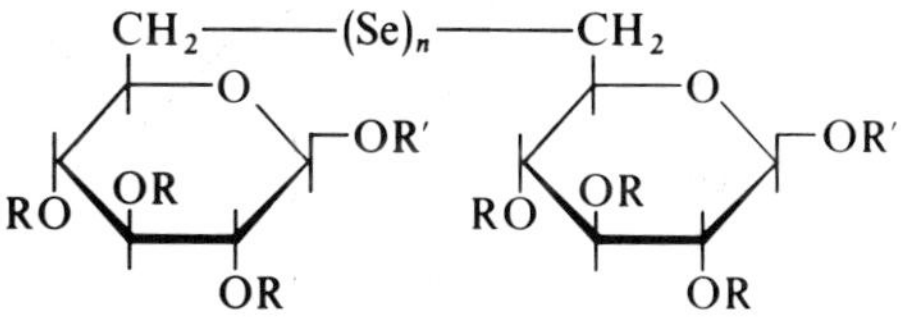

R	R'	n	mp °C	Yield %	$[\alpha]_D$ (°C)	Reference
H	H	1	200(dec)[a]	—	+70.3(14)	554
Acetyl	Acetyl	1	150–155	—	—[b]	554
H	CH$_3$	1	138	—	+14.58(14)	554
Acetyl	CH$_3$	1	179–180	39	−3.1(16)	554
H	H	2	125(dec)	—	+139.3(14)	554
Acetyl	Acetyl	2	175–179	—	—	554
Acetyl	CH$_3$	2	148	40	+49.74(17)	554
H	CH$_3$	2	96–97	—	+75.65(14)	554

[a]Sinters at 160C°.
[b]Specific rotation +40° in ethyl acetate.

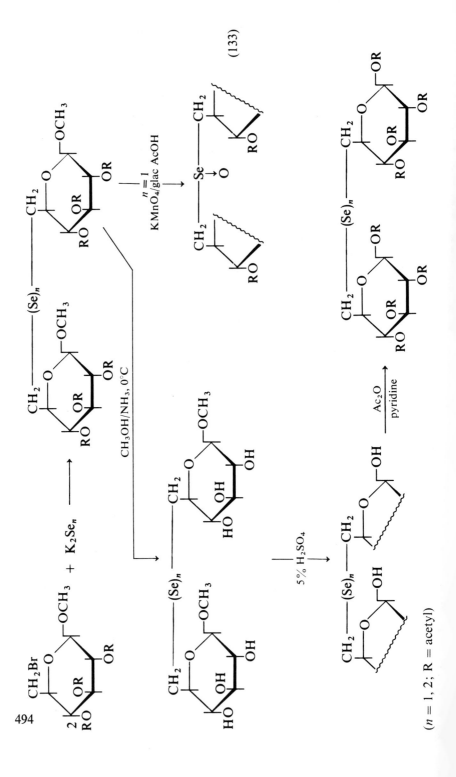

494

(133)

($n = 1, 2$; R = acetyl)

Wrede formulated all of these derivatives as glucofuranosides. Although the ring size has not been determined for these selenium derivatives, it is very likely that they crystallize as pyranosides. Potassium permanganate in glacial acetic acid oxidizes bis(1-methyl 2,3,4-tri-O-acetyl-6-deoxy-β-D-glucopyranoside) 6-selenide to the selenoxide (mp 231°C, $[\alpha]_D^{20}$ −18.9°). Deacetylation with methanol/ammonia gave a noncrystalline product which, upon treatment with acetic anhydride/pyridine, regenerated the original compound.[555]

All the selenocarbohydrates discussed above have the selenium atom bonded to the carbon atom 1 or 6. Van Es[156] prepared altrose and glucose derivatives bearing a selenocyanate group in the 3 and 2 positions, respectively, from methyl 2,3-anhydro-4,6-O-benzylidene-α-mannopyranoside, according to reaction (134):

(134)

An attempt to substitute the selenium atom for the ring oxygen atom by the reaction of 5,6-anhydro-1,2,-O-isopropylidene-α-glucofuranose or 5,6-anhydro-1,2-O-isopropylidene-α-L-idofuranose with potassium selenocyanate failed.[156]

In a series of reactions selenium derivatives of xylose[157] were prepared as shown in reaction scheme (135):

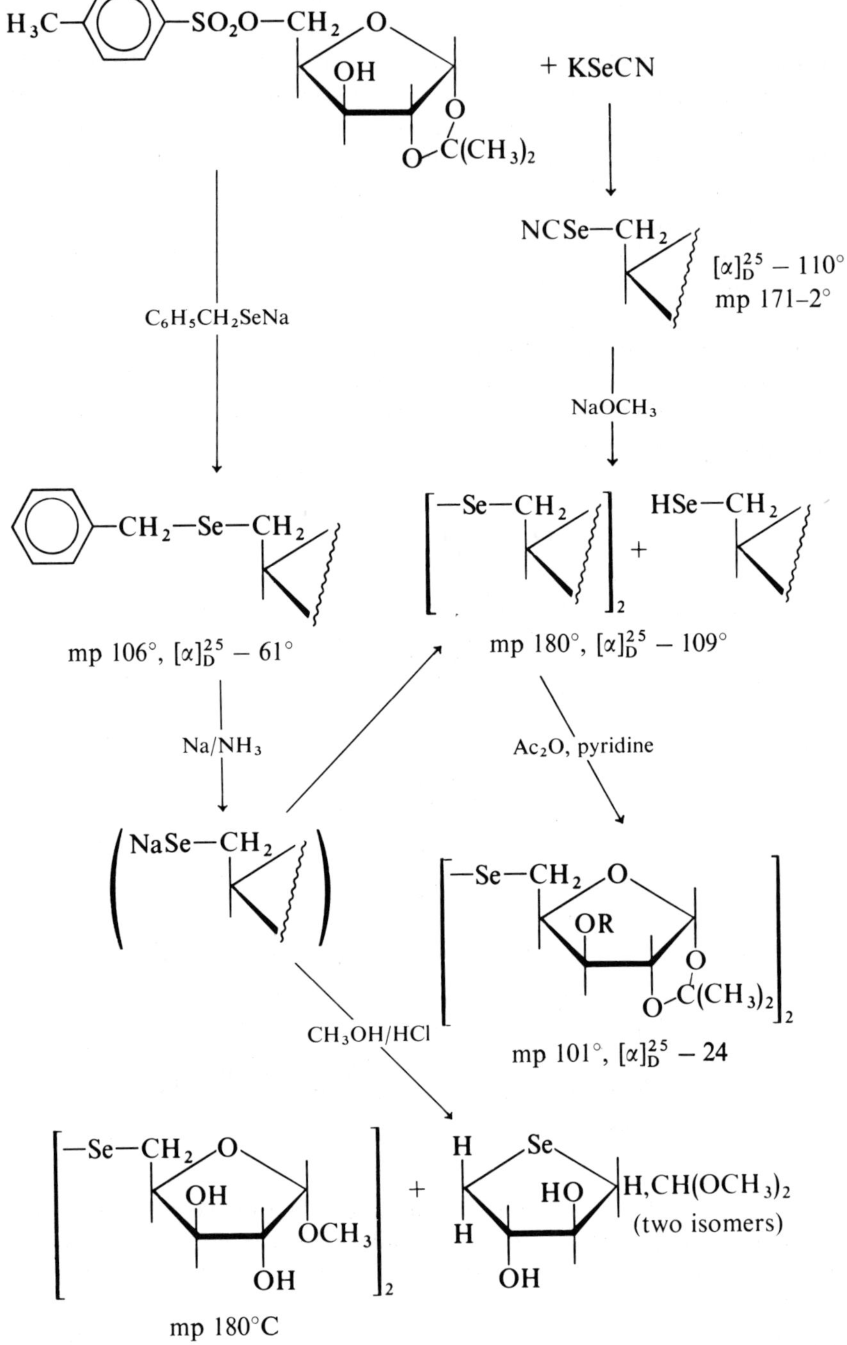

H$_3$C— (ring) —SO$_2$O—CH$_2$ O

OH

O—C(CH$_3$)$_2$

+ KSeCN

C$_6$H$_5$CH$_2$SeNa

NCSe—CH$_2$ $[\alpha]_D^{25} - 110°$
mp 171–2°

NaOCH$_3$

(phenyl)—CH$_2$—Se—CH$_2$

mp 106°, $[\alpha]_D^{25} - 61°$

$\left[-Se-CH_2 \right]_2$ + HSe—CH$_2$

mp 180°, $[\alpha]_D^{25} - 109°$

Na/NH$_3$

Ac$_2$O, pyridine

$\left(NaSe-CH_2 \right)$

$\left[-Se-CH_2 \quad O \right.$
OR
$\left. O-C(CH_3)_2 \right]_2$
mp 101°, $[\alpha]_D^{25} - 24$

CH$_3$OH/HCl

$\left[-Se-CH_2 \quad O \right.$
OH
$\left. OCH_3 \right]_2$
OH
mp 180°C

+ H Se
HO H, CH(OCH$_3$)$_2$
H   (two isomers)
OH

(R = acetyl)

(135)

496

The tetrahydroselenophene derivative in reaction (135) was obtained in two isomeric forms, probably differing with respect to the position of the dimethoxymethyl group.

Phenyl selenoglycosides can be hydrolyzed by acids and bases. Bond cleavage occurs between the selenium atom and the glycosidic carbon atom. The rates of cleavage for the Group VI glycosides increase in the sequence Se < S < O and O < S ≪ Se for acid[520,527] and base hydrolysis,[526] respectively. When a nitro group is present at the p-position in the phenyl ring, the rates are slower in acidic media and more rapid in basic media than those of the unsubstituted phenyl compounds. The rates for acid hydrolysis for nitro compounds increase in the order S < Se ≪ O, while the sequence for the reaction in basic medium is given by Se < O < S. Although the selenols, or their salts, are the primary hydrolytic products, their oxidation products, the diselenides, were generally isolated. The stability towards acid-catalyzed hydrolysis is different for the phenyl α-selenoglucoside and the phenyl β-selenoglucoside. Frenzel[172] found that the stability of the β-anomers increased in the order O < S < Se, while the α-anomers fell into the sequence O < Se < S.

Bonner[76] showed that bromine in acetic acid or carbon tetrachloride cleaved the selenium-carbon(carbohydrate) bond in phenyl 2,3,4,6-tetra-O-acetyl-1-seleno-β-glucoside, producing phenyl selenium tribromide and diphenyl diselenide.

While bis(6-deoxyglucose) 6-selenide yielded the selenoxide upon oxidation with potassium permanganate,[554] cleavage of the phenylseleno group occurred when phenyl 2,3,4,6-tetra-O-acetyl-1-seleno-α-glucoside was treated with hydrogen peroxide in glacial acetic acid.[520]

DIORGANYL SELENIUM DIHALIDES AND RELATED COMPOUNDS

Diorganyl selenium dihalides are prepared by reacting diorganyl selenides with elemental halogen, condensing hydrocarbons with selenium tetrachloride or selenium oxychloride, adding selenium tetrachloride or oxychloride to carbon-carbon multiple bonds, converting diorganyl selenoxides to dihalides, and exchanging halogen in diorganyl selenium dihalides. Diorganyl selenium dialkoxides, dicarboxylates and imines are also discussed in this section.

The Reaction of Diorganyl Selenides with Halogens

Most of the diorganyl selenium dihalides have been prepared from the diorganyl selenides. Chlorine, bromine and iodine dissolved in diethyl ether, chloroform, or carbon tetrachloride convert diorganyl selenides quantitatively into the dihalides. This reaction can be employed to purify the selenides via their dihalides, which are easily reduced back to the parent selenides:

$$R-Se-R' + X_2 \longrightarrow RR'SeX_2 \qquad (136)$$

$R = R'$, X: CH_3, Cl;[556,557*] CH_3, Br; C_2H_5, Br;[384] ClCH: CH, Br;[173] CH_3, I;[324u] C_4H_9, Br;[387] CH_2CH_2CN, Br;[263] $BrCH_2CH_2$, Cl;[174] $CH_2CH(Cl)CN$, Cl;[443] C_6H_5, Cl;[385,479,557,581] C_6H_5, Br;[308,372,557] C_6H_5, I;[324u] $Y-C_6H_4$, Br (Y = 4-Cl, 4-CH_3, 2-CH_3, 4-CH_3O);[372] $Y-C_6H_4$, I (Y = 4-Cl, 2-Cl, 4-CH_3, 2-CH_3, 2-C_6H_5, 4-CH_3O, 4-C_2H_5O).[324u]

$R = CH_3$; R', X: C_6H_5, Br and Cl;[181] 2-CH_3CO-4-$CH_3C_6H_3$, Br.[192]

$R = CF_3$, X = Cl; R': $Y-C_6H_5$ (Y = H, 4-F, 3-F, 4-CH_3);[562] (Y = 4-Br, 4-CH_3O, 4-NO_2, 4-COOH, 3-COOH, 4-$NH_2 \cdot HCl$, 4-CF_3CONH).[563]

$R = 4$-$CH_3COC_6H_4$, X = Cl, R': CH_3, C_2H_5, C_3H_7.[227]

Greenberg[206] synthesized in this way a large number of unsymmetric diaryl selenium dichlorides and dibromides (see reaction (76) for aryl groups).

Dimethyl selenium difluoride, the first example of a diorganyl selenium difluoride, was synthesized from dimethyl selenide and silver difluoride in 1,1,2-trichloro-1,2,2-trifluoroethane. The compound is a liquid decomposing above 120°C. NMR data suggest a trigonal-bipyramidal structure for the molecule.[558] Other aliphatic and aromatic diorganyl selenium difluorides were obtained similarly.[560†]

The Condensation of Selenium Tetrachloride and Oxychloride with Hydrocarbons

The selenium dichloride group can be directly introduced into properly substituted aliphatic and aromatic compounds through condensation reactions with selenium tetrachloride, selenium oxychloride, or selenium dioxide in hydrochloric acid. Methyl aryl ketones and methyl alkyl ketones react without difficulty producing bis(acylmethyl) selenium dichlorides. Ethyl aryl ketones and methyl ethyl ketones condense also, but the position of the selenium atom in the products is not known with certainty.[458,175] Propyl phenyl ketone, however, does not form a selenium dichloride (see Reinboldt[435]).

$$R-\overset{\overset{\displaystyle O}{\|}}{C}-CH_2R' \xrightarrow{\text{SeOCl}_2 \text{ or SeO}_2/\text{HCl}} (R-\overset{\overset{\displaystyle O}{\|}}{C}-HR'-)_2SeCl_2 \quad (137)$$

R, R', or compound: C_6H_5, H;[176] C_6H_5, H; C_6H_5, CH_3;
4-BrC_6H_4, H; *tert*-C_4H_9, H; cyclohexanone;[458] CH_3, H;[173] CH_3,
CH_3; C_6H_5, H.[175]

Funk performed these condensations with selenium dioxide in aqueous hydrochloric acid.   From acetone, selenium dioxide, and hydrobromic acid bis(acetonyl) selenium dibromide was obtained in low yield.[175]

For the successful synthesis of diaryl selenium dichlorides employing selenium oxychloride, the aromatic compounds must bear an alkoxy, aryloxy, hydroxy, or acetylamino group.   The selenium dichloride group then enters in *para*-position to these activating groups (see Rheinboldt,[435] for references). The direct introduction of the $SeBr_2$ group is feasible in a few cases, but does not have preparative importance.   The diorganyl selenium dibromides and diiodides are easily accessible through reduction of the selenium dichlorides to the selenides followed by their reaction with bromine or iodine.

The Addition of Selenium Tetrachloride and Oxychloride to Carbon-Carbon Multiple Bonds

The addition of selenium tetrachloride and selenium oxychloride to olefins and acetylenes gives diorganyl selenium dichlorides generally in good yields:

$$R-CH=CH_2 + SeCl_4 \longrightarrow [RCH(Cl)CH_2]_2SeCl_2 \quad (138)$$

R or Compound: C_3H_7, C_4H_9, cyclohexene.[443]

The additions, depending on the olefins employed, have been carried out without solvent at $-78°C$ or in ether at room temperature.   Acetylene treated similarly gave bis(2-chlorovinyl) selenium dichloride in only 9% yield, while phenylacetylene produced the adduct in 73% yield.[443].

Funk[173,175] employed selenium dioxide in these reactions.   The dioxide in concentrated hydrochloric or hydrobromic acid was shaken with the liquid olefins.   Gaseous olefins were bubbled through the cooled solutions:

$$R-CR'=CH_2 \xrightarrow{\text{SeO}_2/\text{HX}/\text{H}_2\text{O}} [R-CR'(X)-CH_2]_2SeX_2 \quad (139)$$

R, R', X: H, H, Cl;[173] H, H, Br;[173,174] Cl, H, Cl; CH_3CO, H, Cl;
$CH_3CH\!:\!CH$, H, Cl; $NC-CH\!:\!CH$, H, Cl; $ClCH_2$, CH_3, Cl;[174]
CH_3, H, Cl; CH_3, H, Br; $ClCH_2$, H, Cl; $ClCH_2$, H, Br; $BrCH_2$,
H, Cl; $BrCH_2$, H, Br; C_6H_5, H, Cl; C_6H_5, H, Br; cyclopentene,
Cl;[174] cyclohexene, Cl.[173,174]

The yields in these reactions are reported to be 70–80%. Acetylene treated similarly gave bis(2-bromovinyl) selenium dibromide[174] and bis(2-chlorovinyl) selenium dichloride.[173] While 1-methyl- and 1-cyanobutadiene gave the 1,2-addition products as described by reaction (139), butadiene and 1,5-hexadiene yielded 1,1,3,4-tetrachloroselenacyclopentane and 1,1,3,6-tetra-chloroselenacycloheptane, probably by intramolecular addition of the inter-mediate 2-chloroalkenyl selenium trichloride to the remaining carbon-carbon double bond.[173,174]

Diorganyl Selenium Dihalides by Halogen Exchange and from Diorganyl Selenoxides

Diorganyl selenoxides, R_2SeO, and their hydrates, $R_2Se(OH)_2$, react with hydrohalic acids in water or in an organic solvent to give diorganyl selenium dihalides. Dimethyl selenium dichloride was prepared in this manner by Paetzold.[384] The selenoxide need not be isolated. Thus, addition of con-centrated hydrochloric acid to the reaction mixture obtained from diphenyl selenide and nitric acid[231] gave diphenyl selenium dichloride.

The halogen exchange reactions carried out with diorganyl selenium di-halides are described by reaction (140):

$$R_2SeCl_2 \; \underset{\text{conc HCl}}{\overset{\text{KBr or conc HBr}}{\rightleftarrows}} \; R_2SeBr_2$$

$$KI \downarrow \qquad\qquad\qquad\qquad \downarrow KI$$

$$R_2SeI_2 \qquad\qquad (140)$$

The reactions proceed quantitatively in aqueous or ethanolic medium. Bis-(2-chloropropyl), bis(3-bromo-2-chloropropyl), bis(2-chloro-2-chloromethyl-propyl)[174] and bis(2-chloroethyl) selenium dibromide were thus obtained from the dichlorides and concentrated hydrobromic acid.

The mixed selenium dihalide, methyl 4-bromoacetylphenyl selenium bro-mide chloride was obtained when the methyl 4-acetylphenyl selenium dichloride was treated with bromine.[227]

Physical Properties and Reactions of Diorganyl Selenium Dihalides

Diorganyl selenium dihalides are crystalline solids. The structure of the dichlorides and dibromides is represented by a distorted trigonal bipyramid with axial halogen atoms.[320,321,322] The diiodides, however, are probably molecular adducts with only one selenium-iodine bond, $R_2Se—I—I$, as has been shown for the iodine adduct of 1,4-diselenacyclohexane.[323]

The thermal stability of the dihalides decreases from the chlorides to the iodides. While iodine is expelled upon heating (and slowly even at room temperature) from the diorganyl selenium diiodides, diaryl selenium dichlorides and dibromides are converted to haloaryl selenides upon heating. Dialkyl and alkyl aryl selenium dichlorides and dibromides lose alkyl halides.[192] Trifluoromethyl aryl selenium halides, however, expel halogen upon prolonged heating in solution giving trifluoromethyl aryl selenides.[562] Bis(β-oxoalkyl) selenium dichlorides were converted to the monochloroketones upon heating (see reaction (137) for compounds).[458]

The dissociation of a number of diorganyl selenium dibromides and diiodides has been studied by UV spectroscopy. The dissociation constants for the diiodides have values of 10^{-1} to 10^{-3}, depending on the nature of the organic group. The dibromides are less dissociated than the diiodides.[324,325,505]

Diorganyl selenium diiodides are insoluble in water. The dichlorides and dibromides dissolve in warm water giving a strongly acidic solution, from which the unchanged dihalides crystallize on cooling. When the hydrohalic acid was, however, neutralized by silver oxide,[384,387,562] sodium carbonate,[563] sodium hydroxide,[581] or sodium acetate,[562,563] the diorganyl selenoxides were obtained. The bis(2-chlorovinyl) and bis(2-chloroethyl) selenium dihalides hydrolyzed to the selenoxides when treated with water alone.[443] Selenones, R_2SeO_2, are formed from the dihalides and hydrogen peroxide[563] or ozone.[387]

Diaryl selenium dihalides can be transformed into triaryl selenonium salts. Diphenyl selenium dichloride was thus converted to triphenyl selenonium chloride in a reaction with benzene in the presence of aluminum chloride.[231] Aryl mercuric chlorides can also be employed.

The reduction of diorganyl selenium dihalides is easily accomplished with sodium hydrogen sulfite,[227] potassium disulfite,[174] lithium aluminum hydride,[307] zinc, sodium sulfide nonahydrate and sodium thiosulfate (see Rheinboldt[435]).

Diorganyl Selenium Dialkoxides, Dicarboxylates and Imines

Diaryl selenium dialkoxides, $R_2Se(OR')_2$, are formed when the dihalides react with sodium alkoxide in the appropriate alcohol. Dimethyl selenium dimethoxide* and diphenyl selenium dimethoxide* and diethoxide* were prepared in this manner.[385] Bis(4-methoxy-3-R-phenyl) selenium dichloride was converted to the corresponding carboxylates upon treatment with sodium acetate or benzoate in ethanol (R = H, Br[408]).

Diaryl selenium imines, $R_2Se=NR'$, were obtained from diphenyl selenium dichloride and NH_3 (reaction 141), and from diaryl selenides and chloramine-T (reaction 142):

$$2(C_6H_5)_2SeCl_2 + NH_3 \xrightarrow{\ CH_2Cl_2\ } [(C_6H_5)_2Se{=}N{=}Se(C_6H_5)_2]^+Cl^- \quad ^{32}$$

$$[C_6H_5)_2Se{=}N{=}Se(C_6H_5)_2]^+X^- \xleftarrow{\ +MX\ }$$

X = Br, N_3, IO_4, $Fe(CN)_6^{-3}/3$ \hfill (141)

$$R{-}Se{-}R' + \text{chloramine-T} \longrightarrow \underset{R'}{\overset{R}{>}}Se{=}N{-}SO_2{-}\!\!\bigcirc\!\!{-}CH_3$$

R = C_6H_5; R': C_6H_5, 4-$CH_3C_6H_4$, $C_6H_5CH_2$, 4-$NCC_6H_4CH_2$;
R = R' = X-C_6H_4 (X = 4-CH_3, 2-CH_3, 4-CH_3O, 2-CH_3O). \hfill (142)

Similar compounds have been prepared from dibenzoselenophene and 3,8-dimethyldibenzoselenophene.[236]   Dimethyl selenium dichloride formed adducts with triethylamine at low temperatures.[556]

DIORGANYL SELENOXIDES

Diorganyl selenoxides are basic substances, which can be titrated with perchloric acid in nonaqueous medium.[491]   The protonation constants (pK_a) for dimethyl and diphenyl selenoxide are 2.55 and 0.35, respectively.[290] With nitric, hydrochloric, and perchloric acid, adducts of the formula R_2Se-(OH)X are formed.   The Raman spectrum of the aqueous solution of the nitric acid adduct suggests a partial dissociation of the molecule $R_2Se(OH)$-ONO_2 into $(R_2SeOH)^+$ and NO_3^-.   The crystalline hydrogen chloride adduct is represented by the formula $R_2Se(OH)Cl$, which on dissolution in water dissociates partially.   From solutions containing 2 moles of hydrochloric acid for each mole of selenoxide, the diorganyl selenium dichloride is obtained upon evaporation of the solvent (R = CH_3).[384]   Diorganyl selenoxides also form complexes with pyridine (e.g. $(CH_3)_2SeO\cdot$ 2 pyridine),[384] dinitrogen tetroxide [e.g., $R_2SeO\cdot N_2O_4$ (R = CH_3, C_2H_5)],[384] iodine, [e.g., $(C_6H_5)_2$-$SeO\cdot I_2$],[210] and 4-nitrophenol, diphenylmethanol and tris-(4-R-phenyl)-methanol (R = H, CH_3, CH_3O, NO_2) [$(C_6H_5)_2SeO\cdot R'OH$].[457]   It has been suggested that the complexes with the hydroxy compounds are due to hydrogen bonding involving the oxygen atom in the selenoxide.

The two most important methods for the synthesis of selenoxides are the oxidation of diorganyl selenides and the hydrolysis of diorganyl selenium dihalides.   The oxidation of selenides has been accomplished with hydrogen peroxide, peroxycarboxylic acids, ozone, potassium permanganate and potassium dichromate.   Not all selenides are equally well oxidized by all of the above oxidizing agents.   In some cases, the solvent drastically influences the yields of selenoxides (see Rheinboldt[435]).   Paetzold[384] used a mixture of oxygen and nitrogen dioxide to oxidize dimethyl and diethyl selenide in dry carbon tetrachloride at $-5°C$ to the selenoxides, which were obtained as the adducts $R_2SeO \cdot N_2O_4$.   In aqueous solution, the compounds $R_2Se(OH)$-NO_3 are formed.   While the diethyl adduct loses NO_2 at room temperature, the stable dimethyl derivative begins to decompose at $70°C/13$ torr to nitrogen oxide, formaldehyde, and methaneseleninic acid.   The oxidation of the trifluoromethyl phenyl selenide with nitric acid $(d = 1.41)$ produced the selenoxide in 40% yield.[562]   Ozone passed through a carbon tetrachloride solution of benzyl phenyl,[374] dimethyl, dibutyl, didodecyl, diphenyl selenide,[41] and 1,4-bis(phenylselenomethyl)benzene[121] gave the selenoxides in high yields. Sodium metaperiodate in methanol/water or $C_6H_5I \cdot Cl_2$ in pyridine/water oxidized diorganyl selenides to selenoxides with yields of 90%.[120]

The hydrolysis of diorganyl selenium dihalides is generally carried out in aqueous medium containing sodium acetate,[562,563] sodium carbonate,[563] silver oxide, or alkali metal hydroxide.   Diethyl and dipropyl selenium dibromides, treated with silver oxide in water, lose an alkyl group, producing the alkaneseleninic acids.   The dimethyl derivative, however, gave the selenoxide.   The dialkyl selenium oxides, R_2SeO, $(R = C_nH_{2n+1}$, $n = 1$–$4)$ were, however, obtained by reaction of the selenium dibromides with silver oxide in absolute methanol at $-15°C$.[384,387*]

$$CF_3(R-C_6H_4)SeCl_2 \quad \xrightarrow{\text{H}_2\text{O/base}} \quad CF_3(R-C_6H_4)SeO \qquad (143)$$

R: H, 3-F, 4-F, 4-CH_3;[562] 4-Br, 4-CH_3O, 4-COOH, 3-COOH, 4-NH_2, 4-CF_3CO, 4-NO_2.[563]

The selenium dichlorides, $[R-CH(Cl)CH_2]_2SeCl_2$, $(R = C_3H_7$, C_4H_9, $C_6H_5)$ and bis(2-chloro-2-phenylvinyl) selenium dichloride gave the selenoxides upon treatment with water alone.[443]   The selenoxides can be obtained in these hydrolysis reactions in yields as high as 90%.   The halogen atoms in diaryl selenium dihalides are bonded with varying strength depending on the nature of the aryl group necessitating special conditions for the hydrolysis of many compounds.

The possibility of the direct introduction of the SeO group into organic molecules has not yet been investigated in detail.   Diorganyl selenium oxides have been isolated as products of the reactions of olefins with selenium dioxide.   Thus, 1,3-diphenylpropene and 1-octene in glacial acetic acid are reported to react with selenium dioxide to give bis(1,3-diphenyl-1-acetoxy-i-propyl) selenoxide[459] and bis(2-acetoxyoctyl) selenium acetate hydroxide, respectively.[498]

Diorganyl selenoxides are colorless and odorless solids, which can absorb water to form dihydroxides. Most of the dihydroxides are easily converted back to the oxides by dehydration. Alkyl aryl selenoxides, however, decompose upon heating to aliphatic aldehydes, alkyl aryl selenides and diaryl diselenides. Diaryl selenoxides liberate oxygen at temperatures above their melting points to yield selenides. Trifluoromethyl aryl selenoxides lost trifluoromethane when treated with sodium hydroxide.[563]

Selenoxides have oxidizing properties liberating iodine from potassium iodide solutions.[563] Their reduction to selenides has been accomplished with zinc, sulfites, hydrazine sulfate and hydrazine hydrate. Treatment of diphenyl selenoxide in tetrahydrofuran with potassium gave a red solution and a red precipitate, possibly containing ketyl-like compounds.[235] With strong oxidizing agents, diorganyl selenones are obtained.

The ^{19}F NMR spectrum of trifluoromethyl phenyl selenide has been determined.[564] The vibrational spectra of dialkyl selenoxides, R_2SeO (R = CH_3, C_2H_5, C_3H_7, C_4H_9) have been investigated by Paetzold.[389] A correlation of the Se=O valence force constants in dimethyl selenoxide and similar selenium compounds with those of corresponding sulfur derivatives[386] has been published.

The resolution of simple selenoxides, RR'SeO, into their optically active components has so far not been possible. Oki,[374] studying benzyl phenyl selenoxide by NMR techniques, estimated the lifetime of an enantiomer to be 4 seconds at room temperature. Cinquini,[121] however, separated the diastereoisomers of $4\text{-}C_6H_4[CH_2Se(O)C_6H_5]_2$ by fractional crystallization from methylene chloride.

The mass spectra of diphenyl and bis(4-methylphenyl) selenoxide have been investigated by Rebane.[427] The interaction of the Se(O)CF$_3$ group in aryl trifluoromethyl selenoxides with the π-electron system in the ground and excited states has been determined.[319]

DIORGANYL SELENONES

Diorganyl selenones, R_2SeO_2, are generally obtained by oxidation of diorganyl selenides or selenoxides with hydrogen peroxide, potassium permanganate, or peroxycarboxylic acids. Although dialkyl, alkyl aryl, and diaryl selenones are known, this area of organoselenium chemistry has not been explored in detail.

Dimethyl selenide is decomposed by hydrogen peroxide and potassium permanganate. Dimethyl, diethyl, dipropyl, and dibutyl selenone were, however, isolated in good yields by oxidizing the selenides or selenoxides with ozone in carbon tetrachloride solution at $-5°$C.[387,*389*] Yagupol'skii[563] oxidized trifluoromethyl aryl selenides, selenoxides and selenium dichlorides with a mixture of trifluoroacetic anhydride and 85% hydrogen peroxide at low temperature:

$$X-C_6H_4-\underset{Cl_2}{Se}-CF_3 \xrightarrow{H_2O_2/(CF_3CO)_2O} X-C_6H_4-\overset{O}{\underset{O}{\overset{\uparrow}{\underset{\downarrow}{Se}}}}-CF_3 \quad (144)$$

X: H, 3-F, 4-F, 4-Br, 4-CH₃O, 4-CH₃CONH

The yields were as high as 90%. The 3- and 4-nitrophenyl selenoxides could not be transformed into the selenones. The CF_3SeO_2 group has been found by ^{19}F NMR techniques to be a far stronger electron attracting group than CF_3SO_2.[564] The interaction of π-electron systems with the SeO_2CF_3 group has been investigated.[319]

Certain cyclic selenones and naphthyl alkyl selenones were obtained by heating haloalkaneseleninates and reacting naphthylseleninates with alkyl iodides, respectively (see Rheinboldt[435]).

The direct introduction of the SeO_2 group into benzene rings has been reported by Dostal.[146] Benzene and bromobenzene treated with selenium trioxide in liquid sulfur dioxide gave diphenyl and bis(4-bromophenyl) selenones in 4% yields. The yields increased to 20% when the reaction was carried out in anhydrous selenic acid. The main products in these reactions were the selenonic acids.

The properties and the reactions of the selenones have not been investigated in detail. They oxidize concentrated hydrochloric acid to chlorine and slowly liberate iodine from potassium iodide solutions.[563,387] Dimethyl selenone, however, gave methyl chloride and methaneseleninic acid upon treatment with hydrochloric acid.[387] Trifluoromethyl aryl selenones lose trifluoromethane when treated with sodium hydroxide, carbonate, or bicarbonate.[563] The dialkyl selenones are not hygroscopic.[387] Diphenyl selenone produced with potassium in tetrahydrofuran a brown solution and a brown precipitate.[235]

TRIORGANYL SELENONIUM COMPOUNDS

Triorganyl selenonium compounds, $[R_3Se]^+X^-$, have three organic groups bonded to the selenium atoms. Examples for all the possible variations with aliphatic and aromatic organic groups are known with the exception of compounds of the type $[ArAr'Ar''Se]^+X^-$. The anion may be F^-, Br^-, Cl^-, I^-, Br_3^-, I_3^-, $ClBr_2^-$, ClI_2^-, BrI_2^-, ClO_4^-, BF_4^-, $Cr_2O_4^{-2}$, picrate$^-$, and complex metal ions. The selenonium chlorides have a tendency to separate in noncrystalline form. Their conversion to the more insoluble iodides or picrates, which crystallize easily, helps in separating and purifying the selenonium compounds. Dialkyl selenides react exothermically with alkyl iodides. With increasing chain length of the alkyl groups in the selenides as well as in the iodides, the reactions require longer times and the yields decrease. While the compounds with small alkyl groups combine at room temperature, longer alkyl chain derivatives require heating. Alkyl bromides, bromocarboxylic acids and α-bromoketones have been employed as halide components:

$$R_2Se + R'X \longrightarrow [R_2R'Se]^+X^- \qquad (145)$$

R = R', X, % yield: CH_3, I, 91; C_2H_5, I, 54;[231] CH_3, Cl, –.[559]
R, R', X: CH_3, 4-$YC_6H_4COCH_2$ (Y = H, F, Cl, Br), Br.[232]

The selenides treated with alkyl iodides in the presence of silver perchlorate or tetrafluoroborate in 1,2-dichloroethane gave the corresponding perchlorates and tetrafluoroborates (R, % yield: X = BF_4; CH_3, 60; C_2H_5, 30; C_3H_7, 35; C_4H_9, 30; X = ClO_4; C_3H_7, 68; C_4H_9, 60).[231] Alkylation of methyl phenylethynyl selenide by triethyl oxonium fluoroborate without a solvent gave the methyl ethyl phenylethynyl selenonium salt, which was isolated as the picrate.[199*] This is the first compound of the type $[RR'R''Se]^+X^-$.

The alkyl aryl selenides are less reactive towards methyl iodide than the dialkyl selenides. The alkylation of methyl aryl selenides proceeds smoothly with dimethyl sulfate, α-bromocarboxylic acids, and α-bromoketones:

(146)

R = R' = CH_3, X = CH_3SO_4, Y = 2-Br, 3-Br, 4-Br, 2-Cl, 3-Cl, 4-Cl;[181]   Y = H,   2-NO_2,   3-NO_3,   4-NO_2;[180]   Y = 4-CH_3;[192]
R = CH_3, X = Br, Y = 2-CH_3CO; R' = CH_2COCH_3, $CH_2COC_6H_5$, CH_2COOH;   Y = 2-C(O)H;   R' = $CH_2COC_6H_5$, $CH_2CO_2CH_3$;
Y = 2-$C_6H_5C(O)$; R' = CH_2CO_2H;
R = C_4H_9, X = Br, Y = 2-C(O)H, R' = CH_2COOH.[118]

The selenonium bromides were obtained in quantitative yields when the reactants were refluxed for 2 days.[118] The formation of the butyl 2-formylphenyl carboxymethyl selenonium bromide is surprising, since ethyl phenyl selenide and bromoacetic acid in ether are reported to produce carboxymethyl phenyl selenide (see Rheinboldt[435]). 2-Formyl-4,5-methylenedioxyphenyl methyl selenide and chloroacetic acid kept at 100°C for 24 h gave compound (XXVIII), which was converted to the thiosemicarbazide.[137]

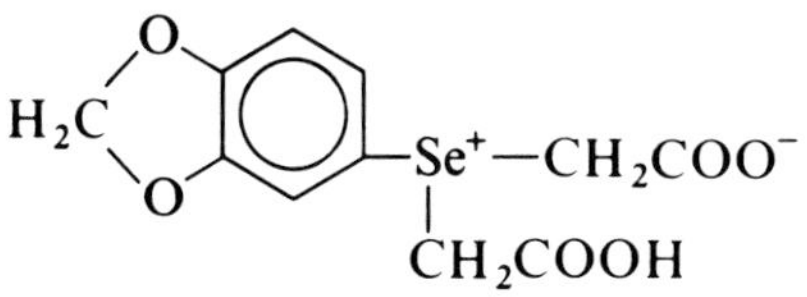

$$H_2C \underset{O}{\overset{O}{\diagup}} \text{—} Se^+ \text{—} CH_2COO^-$$
$$\underset{CH_2COOH}{|}$$

(XXVIII)

Diaryl selenides are reported to be unreactive towards alkyl halides (Rheinboldt[435]). Hashimoto,[232] however, succeeded in preparing diphenyl phenacyl and diphenyl 4-bromophenacyl selenonium bromide from diphenyl selenide and the phenacyl bromides.

Triaryl selenonium halides cannot be obtained from diaryl selenides and an aromatic halide. Friedel-Crafts reactions with diaryl selenium dichlorides, the arylation of diaryl selenium dichlorides with aryl mercuric chlorides and the reaction of certain aromatic hydroxy compounds with selenium oxychloride lead to triaryl selenonium salts. Hashimoto[231] obtained a 65% yield of triphenyl selenonium chloride from diphenyl selenium dichloride and benzene in the presence of aluminum chloride. The corresponding tetrafluoroborate was formed by heating diphenyl selenide with diphenyl iodonium tetrafluoroborate.[371]

Selenium dioxide in concentrated aqueous hydrochloric acid heated with phenol,[173] o- and p-cresol at 70°C produced triaryl selenonium chlorides. The position of the selenium atom in these compounds is not known.[175] Diphenyl selenide and diazotetraphenylcyclopentadiene kept at 140°C for 10 minutes produced in almost quantitative yield the selenonium salt (XXIX.)[315*,u]

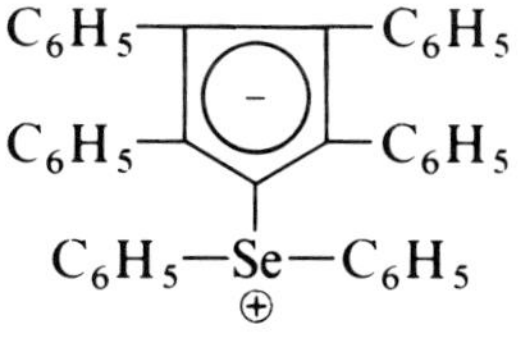

$$C_6H_5 \text{—} \overset{C_6H_5}{\underset{C_6H_5}{\diagdown}} \text{—} C_6H_5$$

$$C_6H_5 \text{—} \underset{\oplus}{Se} \text{—} C_6H_5$$

(XXIX)

Anion exchange in selenonium salts is possible using either the solubility differences between these salts or by taking advantage of the insolubility of certain metal salts. Iodides and bromides can thus be converted into the chlorides employing silver chloride. Addition of potassium iodide to selenonium chlorides will give the selenonium iodide. The anion exchange reactions carried out recently are summarized in reaction scheme (147):

$$[R_2R'Se]^+Br^{-a} \qquad\qquad [R_2R'Se]^+Br^{-b}$$

$$\begin{bmatrix} R \\ | \\ R-Se \\ | \\ R' \end{bmatrix}^+ BF_4^- \underset{AgBF_4^d}{\overset{KCl}{\rightleftarrows}} [R_2R'Se]^+Cl^{-c} \qquad (147)$$

$$[R_2R'Se]^+ClO_4^- \xleftarrow{AgClO_4^f} [R_2R'Se]^+I^{-d}$$

(diagram arrows labeled: AgBF$_4$ (down from a), KBr, AgCl (down to b/c), KI / AgBF$_4^e$, KI, AgCl)

^{a} R = C_6H_5, R' = 4-X$C_6H_4COCH_2$ (X = H, F, Cl, Br).[232]
^{b} R = R': CH_3, C_2H_5, C_6H_5.
^{c} R = R': CH_3, C_3H_7, C_4H_9.
^{d} R = R': C_6H_5.
^{e} R = R': CH_3, C_2H_5, C_3H_7, C_4H_9.
^{f} R = R': C_3H_7, C_4H_9.[231]

Treating phenacyl and 4-bromophenacyl diphenyl selenonium bromides with $HgBr_2$, $SbBr_3$, and $SnBr_4$ gave the selenonium tribromomercurates(II), tetrabromoantimonates(III), and pentabromostannates(IV).[232]   Addition of the heavier halogen to chloroform solutions of triphenyl selenonium chloride and bromide caused the formation of the interhalide salts $ClBr_2^-$, ClI_2^- and BrI_2^-.   The selenonium tribromide and triiodide was prepared similarly.[231] The basic triorganyl selenonium hydroxides are obtained from the halides and moist silver oxide.

Phenyl dimethyl selenonium methyl sulfate is nitrated by a mixture of concentrated nitric and sulfuric acid forming 91.3% *meta*, 6.1% *para*, and 2.6% *ortho* substitution products.[180]   The chlorination and bromination at room temperature catalyzed by silver salts gave *ortho*, *meta*, and *para* monohalophenyl derivatives with the *meta* substituted products predominating.   A few percent of the disubstituted compounds were also formed.   Bromination at 80°C without a catalyst produced predominantly the *para* substituted product.[181]

Aryl methyl carboxymethyl selenonium bromides are cleaved by bromide into methyl bromide and aryl carboxymethyl selenium dibromide.[435]   Aryl dimethyl selenonium salts are converted into aryl methyl selenides by treatment with sodium methoxide in methanol.[180]

All selenonium salts are thermally decomposed to selenides.   Although there are no systematic investigations available, the smaller organic group seems to be preferentially cleaved.   Certain *ortho* substituted phenyl dialkyl selenonium salts can be thermally transformed into benzoselenophenes.[118,137]

The trimethyl selenonium iodide molecule has the shape of a distorted trigonal bipyramid with two methyl groups and an electron pair in the equatorial positions and one methyl group and the iodide ion occupying the apices.[242] The signs and the magnitudes of the reduced coupling constants between carbon and selenium in trimethyl selenonium iodide was determined by heteronuclear double resonance techniques.[327]   Magnetic nonequivalence of the methyl groups attached to selenium in $[(CH_3)_2SeCH(CH_3)C_6H_5]^+Br^-$ has been observed.   The chemical shift difference was strongly solvent dependent. The inversion was slow on the NMR time scale.[330]

TETRAORGANYL SELENIUM COMPOUNDS

The reaction of triphenyl selenonium chloride with phenyl lithium gave only diphenyl selenide and biphenyl.[550]   Bis(2,2'-biphenylylene) and bis(4,4'-dimethyl-2,2'-biphenylylene) selenium are the only known tetraorganyl selenium compounds.   These crystalline compounds are of lesser stability than the corresponding tellurium derivatives.[236x]   The synthesis and the reactions of these compounds are presented in reaction scheme (148) (p. 510).

COMPOUNDS WITH A FORMAL CARBON-SELENIUM DOUBLE BOND: SELENOALDEHYDES, SELENOKETONES, SELENO-CARBOXYLIC ACID AMIDES AND RELATED COMPOUNDS

Organic selenium compounds containing a formal carbon-selenium double bond comprise the selenium analogs of aldehydes, ketones and various selenium derivatives of carboxylic acids.   The selenium derivatives of carbonic acid, selenourea, selenosemicarbazide and selenocarbazide are not included. The selenoaldehydes and selenoketones are polymeric substances, for which little is known about the molecular size with the exception of selenoformaldehyde.   Selenoacetals have also been described.

Selenoaldehydes

Selenoaldehydes have been prepared by bubbling hydrogen selenide through a solution of the aldehyde in concentrated aqueous or ethanolic hydrochloric acid (see Rheinboldt[435]).   Brandsma,[79] however, isolated bis(1-chloroethyl) and bis(1-chloropropyl) selenide after treating neat acetaldehyde and propionaldehyde with gaseous hydrogen chloride and hydrogen selenide at low temperatures.   Hende[237] obtained compound $(XXX)(Y = Se)/(Y = O)$ from

(148)

510

the aldehyde and diphorphorus pentaselenide in chlorobenzene. The molecule is planar, indicating interaction between the selenium and the sulfur
atom.

(XXX)

Selenoformaldehyde has been investigated in more detail. The latest developments in this area have been reviewed by Russo.[449] Treatment of
formaldehyde in concentrated hydrochloric acid with hydrogen selenide[96,97,360,361] and of dichloro- and dibromomethane in methanol or ethanol with
sodium selenide[413*,414,448,463] produced mixtures of trimeric,[360,361,448,463]
tetrameric,[448] pentameric,[463] and linear polymeric products, which were
separated by extraction procedures. Methylene radicals, generated from
diazomethane, gave with selenium mirrors trimeric selenoformaldehyde.[542*†]
The trimer, 1,3,5-triselenacyclohexane, $(CH_2Se)_3$, exists in the chair form.[331]

The linear polymers have a melting point of about 135°C. Upon annealing
at 125°C, a small weight decrease and an increase of the melting point to about
177°C has been observed.[96,361] The annealed polymer has a hexagonal unit
cell with the polymer chain wound into a helix. Twenty-one monomer units
in 11 turns give an identity period of 46.25 Å.[95]

The polymerization of the trimer with boron trifluoride etherate at 235°C
produced linear hexagonal polyselenoformaldehyde,[361] while the tetramer at
100°C gave an orthorhombic polymer,[448] which was also obtained from the
trimer upon gamma irradiation.[97] The orthorhombic changes to the hexagonal form at 185–190°C.[448] Above 200°C, depolymerization to the trimeric
or tetrameric form occurs.[413]

The reaction between bis(chloromethyl) ether and sodium selenide in
methanol gave 1,3,5-oxadiselenacyclohexane, which was polymerized to a
formaldehyde-selenoformaldehyde copolymer with $BF_3 \cdot (C_2H_5)_2O$ as the
initiator.[447]

A solution of acetaldehyde in ethanol saturated with hydrogen chloride
gave with hydrogen selenide trimeric selenoacetaldehyde, while the reaction
in aqueous 2N hydrochloric acid produced 1,3,5-dioxaselenacyclohexane.[130]

Selenoketones

The older literature (see Rheinboldt[435]) contains reports that dimeric selenoketones are obtained when hydrogen selenide interacts with ketones in
aqueous hydrochloric acid. Margolis[334] repeated some of the earlier

investigations and was unable to explain the properties of the products on the basis of a selenoketone formula. The pyrolysis product, the UV spectra, and the mixed melting points with compounds of known structure showed that acetone, methyl ethyl ketone, methyl phenyl ketone, and diphenyl ketone produced the corresponding diselenides with hydrogen selenide in either aqueous or ethanolic hydrochloric acid. The selenium precipitated in these reactions in the absence of air was the amount required by reaction (148a):

$$2 \ \overset{R}{\underset{R'}{>}}C{=}O + 3H_2Se \longrightarrow \overset{R}{\underset{R'}{>}}CH{-}Se{-}Se{-}HC\overset{R}{\underset{R'}{<}} + Se + 2H_2O$$

(148a)

R, R': CH_3, CH_3; CH_3, C_2H_5; CH_3, C_6H_5; C_6H_5, C_6H_5.

An x-ray analysis of the product obtained from diphenyl ketone ascertained the diselenide structure. The selenoketone and the hydrogen selenide adduct of the ketone might be intermediates in these reactions.

Scheithauer[460] obtained cyclohexaneselenol from cyclohexanone and hydrogen selenide in dimethylformamide in the presence of pyridine. These results contradict the earlier investigations. Additional experiments in this area would be desirable.

Selenocarboxylic Acid Amides

Selenocarboxylic amides, $R{-}C(Se)NH_2$, are rather unstable substances, which are rapidly discolored in light and slowly converted into polymeric materials. Their stability increases with increasing number of organic groups on the nitrogen atom.[128] For selenoamides bearing at least one hydrogen atom on the nitrogen atom, the following tautomeric and mesomeric forms must be considered:

$$R{-}C\overset{Se^-}{\underset{NH_2^+}{<}} \rightleftharpoons R{-}C\overset{Se}{\underset{NH_2}{<}} \rightleftharpoons R{-}C\overset{SeH}{\underset{NH}{<}}$$

(148b)

(XXXIII)   (XXXI)   (XXXII)

For dialkylamides, structure (XXXII) is not possible. Asinger[35] found that 4-selenoxoimidozolidines form metal salts, lending support to the presence of the imide form (XXXII). The comparison of ^{14}N NMR spectra of selenobenzamide with analogous sulfur and oxygen compounds pointed towards an increased contribution of the resonance form (XXXIII) in the heavier chalcogen derivatives.[221,222] This view is supported by Hückel calculations applied to selenoamides.[40u] Formula (XXXIII) implies hindered rotation about the C=N double bond. Schwenker[469] determined the barrier for this rotation in dimethyl chalcogenobenzamides from the temperature dependence of the ^{1}H NMR signals to be 7.5(O), 15.4(S) and 21.1(Se) kcal/mole. Jensen[260] found 19.5 $\pm$ 0.3 and 26.0 $\pm$ 0.4 kcal/mole for the Se-benzamide and Se-acetamide, respectively. Infrared investigations of the N—H stretching frequencies in chalcogenobenzamides and -furamides lead to the same conclusion. The H—N—H angles calculated from the frequencies of the symmetric and antisymmetric N—H vibrations in the primary amides increased in the sequences $0 < S < Se$ indicating a trend from sp^3 towards sp^2 hybridization of the nitrogen atom.[218] Jensen[257] carried out an extensive study of the infrared spectra of thio- and selenoamides.

Selenocarboxylic acid amides are formed when nitriles are treated with hydrogen selenide in ethanol. Not all nitriles, however, react under these conditions. Diselenomalonic acid diamide was obtained in this manner in 36% yield.[257] Refluxing amides with diphosphorus pentaselenide in benzene produced the selenoamides in yields not higher than 22% (reaction 149):

$$R—C(O)NR'(CH_3) \xrightarrow{P_2Se_5} R—C(Se)NR'(CH_3) \tag{149}$$

R, R': H, CH_3; CH_3, H; CH_3, CH_3; C_2H_5, CH_3; C_2H_5, H; C_3H_7, H;[128*] C_6H_5, CH_3.[257*]

Lactams, $\left[-(CH_2)_n—C(O)NH-\right]$ (n = 3, 4, 5, 7, 11), refluxed for 24 hours with red phosphorus and gray selenium in xylene gave very low yields of the corresponding selenolactams.[219*] The IR spectra of these selenolactams were investigated with the aim to identify the bands associated with C=Se vibrations.[220]

Asinger[35] prepared a large number of 4-selenoxoimidazolidines with substituents in the 3- and 5-positions according to reaction (150):

$$\tag{150}$$

$R = R^1 = H$; R^2, R^3: CH_3, CH_3; $(CH_2)_5$; CH_3, $CH_2CO_2C_2H_5$;
$R = R^1 = CH_3$, $R^2 = H$; R^3: CH_3, C_2H_5, C_3H_7, i-C_3H_7, C_6H_5;
$R = R^2$, $R^1 = R^3$: CH_3, C_2H_5; CH_3, i-C_3H_7; $R = R^1 = R^2 = R^3$:
CH_3, C_2H_5; R, R^1, R^2, R^3: CH_3, CH_3, $(CH_2)_5$; H, C_6H_5, CH_3,
CH_3; H, C_6H_5, $(CH_2)_5$; $(CH_2)_5$, H, C_6H_5; $(CH_2)_5$, $(CH_2)_5$.

The imidazolidine ring is cleaved by concentrated sulfuric acid and lithium
aluminum hydride.

Compounds containing the group 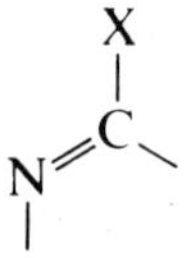 (X = Cl, Br, SCH₃) react with

hydrogen selenide,[35,54] sodium hydrogen selenide,[340,506] and selenourea[152]
producing the compounds (XXXIV)–(XXXVII).   Methylation of (XXXVIa)
with methyl iodide produced methylseleno derivatives.[54]

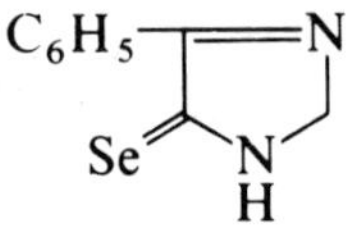

XXXIV [36]

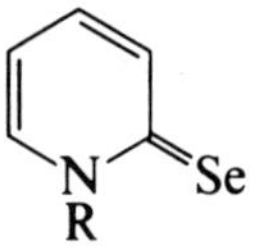

R = H, CH₃[340u]

XXXV

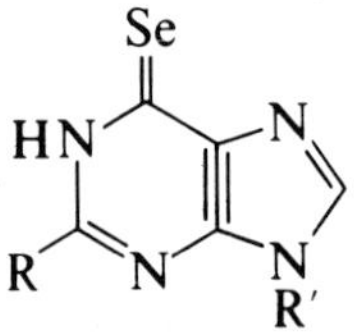

R = NH₂, R′ = COOC₂H₅[152u]
R = R′ = H[337u]
R = NH₂, R′ = H[339,342u]

XXXVI

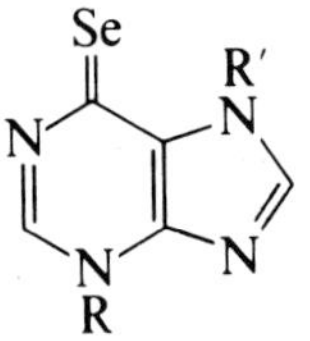

R, R′: CH₃, H; CH₃, CH₃;
C₆H₅CH₂, CH₃[54]

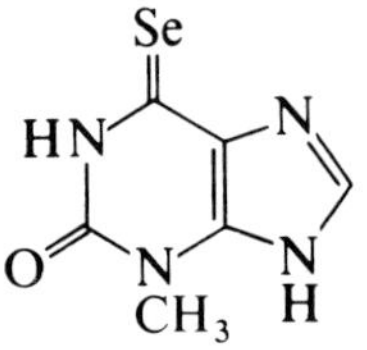

XXXVIa

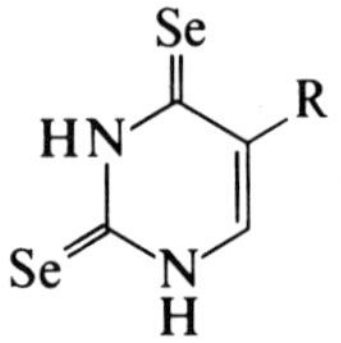

R = H[337u]
R = CH₃[342u]

XXXVII

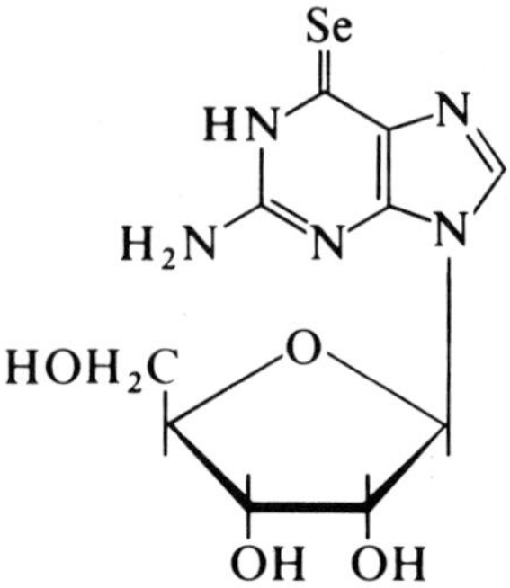

XXXVIb[506]

2-Selenopyridone (XXXV, R = H) was found to exist largely in the seleno-amide form.   It is dimeric in benzene.[300a]

Chloroform and sodium selenide in the presence of di-i-propylamine gave a 3% yield of $HC(Se)N(i\text{-}C_3H_7)_2$.[257*]   Dimethyl selenobenzamide was also synthesized from hydrogen selenide and $C_6H_5CCl_2N(CH_3)_2$.[469]

Selenoacetamide was obtained from S-butyl thioacetimide and hydrogen selenide, while O-methyl benzimide and O-methyl 4-methylbenzimide produced the O-methyl selenobenzoates.[127]   Selenoamides are alkylated by alkyl halides to give $\left[RC\underset{NR_2'}{\overset{SeCH_3}{\diagup\diagdown}} \right]^+ X^-$ [257,340]   Selenobenzamide was converted to benzonitrile in a reaction with N-isothiocyanatodiisopropylamine.[31]

Other Selenocarboxylic Acid Derivates

Cyclic selenocarboxylic acid esters were prepared according to reaction (150a):

$$X = O, S, Se$$

The dipole moments of these compounds were determined.[530]   Diseleno-carboxylic acids are reported to be formed from carbon diselenide and diorganyl zinc compounds.[255]   The reaction of CSe_2 with alkali metal cyanides produced $NC-CSe_2H$, which was isolated as the tetraphenyl arsonium salt.[415*]

SELENIUM COMPOUNDS OF BIOLOGICAL IMPORTANCE

The discovery that selenium is an essential trace element required for the life processes of animals gave impetus to the investigation of selenium analogues of biologically active, sulfur containing compounds. Seleno-aminocarboxylic acids, selenium containing peptides, selenium derivatives of pyrimidines, purines, cystamine, cysteamine, choline, pantetheine and coenzyme A have been prepared and studied.   Reviews covering selenium containing aminoacids and peptides[536] and various aspects of biologically important organic selenium compounds have been published.[251]   The biomedical and biochemical aspects are treated in two recent books.[365,445]

Organylselenoaminocarboxylic Acids

The selenium containing aminocarboxylic acids, prepared thus far, belong to one of the following three classes of compounds, using 2-aminopropionic derivatives as examples:

$$R-Se-CH_2-\underset{\underset{NH_2}{|}}{CH}-COOH \qquad\qquad R-\underset{\underset{NH_2}{|}}{CH}-\overset{\overset{O}{\|}}{C}-Se-R'$$

$$\text{(XXXVIII)} \qquad\qquad\qquad \text{(XXXIX)}$$

$$(Se)_n \begin{cases} CH_2-\underset{\underset{NH_2}{|}}{CH}-COOH \\ \\ CH_2-\underset{\underset{NH_2}{|}}{CH}-COOH \end{cases} \qquad (n = 1, 2)$$

$$\text{(XL)}$$

The preparation alkylselenoaminocarboxylic acids (XXXVIII) has been accomplished by alkylation of alkaneselenolates with haloaminocarboxylic acids or other suitable aminoacid derivatives, by alkylation of 1-carboxy-2-aminoalkaneselenolates with organic halides, by addition of phenyl-methaneselenol to unsaturated aminoacids or unsaturated aldehydes, which are then converted to aminoacids, and by basic hydrolysis of hydantoins. Unless otherwise stated, the DL-aminoacids have been used.

Alkylation of Alkaneselenolates with Haloaminocarboxylic Acids. Phenyl-methaneselenol, almost exclusively employed as the selenium containing reagent, is dissolved in ethanol. The selenolate, formed after addition of sodium, then reacts with the haloaminocarboxylic acid:

$$C_6H_5CH_2SeNa + Cl(CH_2)_n-\underset{\underset{NH_2}{|}}{CH}-COOR$$

$$\downarrow$$

$$C_6H_5CH_2Se(CH_2)_n\underset{\underset{NH_2}{|}}{CHCOOH} \qquad (151)$$

n = 1: Se-benzyl-L-selenocysteine; n = 2: Se-benzyl-L-selenohomo-cysteine. n, configuration: 1,L;[162,390] 2, L.[249]

In most of these reactions, the ester is saponified before the selenoamino-acid is isolated.   2-Benzoylamino-4-benzylselenobutyric acid was obtained similarly from ethyl 2-benzoylamino-4-chlorobutyrate.[391]   Benzeneseleno-late and methyl 3-chloro-2-aminopropionate hydrochloride produced methyl 3-phenylseleno-2-aminopropionate.[390]

Tosylate (4-methylphenylsulfonyl) groups in L-serine tosylates are replaced nucleophilically by the benzylseleno groups:[503,504,395]

$$H_3C-\underset{O}{\overset{O}{\underset{\downarrow}{\overset{\uparrow}{S}}}}-O-CH_2-\underset{\underset{R'}{\overset{|}{N}H}}{\overset{|}{C}H}-\overset{O}{\overset{\|}{C}}-OR \xrightarrow[DMF/NaOH/acetone]{C_6H_5CH_2SeH}$$

$$\underset{}{\bigcirc}-CH_2-Se-CH_2-\underset{\underset{R'}{\overset{|}{N}H}}{\overset{|}{C}H}-\overset{O}{\overset{\|}{C}}-OR$$

R' = C₆H₅CH₂OC(O):   R = C₆H₅CH₂,[503,504]   4-NO₂C₆H₄CH₂,
(C₆H₅)₂CH,   CH₃.[504]   R' = tert-C₄H₉OC(O);   R = C₆H₅CH₂,[503]
4-NO₂C₆H₄CH₂.[504]

R' = $C_6H_5CH_2OC(O)$:   R = $C_6H_5CH_2$,[503,504] $4\text{-}NO_2C_6H_4CH_2$,
$(C_6H_5)_2CH$, CH_3.[504]   R' = $tert\text{-}C_4H_9OC(O)$;   R = $C_6H_5CH_2$,[503]
$4\text{-}NO_2C_6H_4CH_2$.[504]

Plieninger[407] obtained 4-methylseleno-2-aminobutyric acid from methane-selenolate and 2-aminobutyrolactone.   The yields in these reactions range from 50–80%.   Theodoropoulos[504] reported that the method used by Painter[390,391] and Frank[162] gave inconsistent results.

Alkylation of 1-Carboxy-2-aminoalkaneselenolates.   Benzylselenoaminocar-boxylic acids, treated with sodium in liquid ammonia, release the benzyl group.   The selenolates produced can then react with an organic halide:

$$C_6H_5CH_2-Se-CHR-CHR'-\underset{\underset{NH_2}{\overset{|}{}}}{CR''}-COOH \xrightarrow{Na/NH_3}$$

$$^-Se-CHR-CHR'-\underset{\underset{NH_2}{\overset{|}{}}}{CR''}-COOH$$

$$R'''-Se-CHR-CHR'-\underset{\underset{NH_2}{\overset{|}{}}}{CR''}-COOH \xleftarrow{R'''I} \qquad (152)$$

R, R', R", R''', configuration: H, H, H, CH_3, L;[249,578*] H, H, H, CH_3, D;[578] H, H, H, CH_3, DL;[391,573] H, H, H, $^{14}CH_3$, DL; H, H, H, $^{14}C_2H_5$, DL;[394] H, H, H, C_2H_5, L;[249] H, H, CH_3, CH_3, DL; H, CH_3, H, CH_3, DL; CH_3, H, H, CH_3, DL.[576]

3-Benzylseleno-2-aminopropionic acid was cleaved similarly. The product treated with bis(chloromethyl) sulfide, gave bis(2-amino-2-carboxyethyl-selenomethyl) sulfide.[572*]

Addition of Phenylmethaneselenol to Carbon-Carbon Double Bonds.   Methyl 2-acetylaminoacrylate added benzeneselenol in an exothermic reaction in the presence of sodium methoxide, producing 3-benzylseleno-2-acetylamino-propionate in 95 % yield.[571]   The addition of the selenol to acrolein, methyl-acrolein, and crotonaldehyde in ether solution with piperidine as a catalyst, and conversion of the benzylselenoaldehyde to the amino acid through treatment with hydrocyanic acid and ammonia (Strecker's amino acid synthesis) gave 4-benzylseleno-2-amino-,[573*] 4-benzylseleno-2-methyl-2-amino-, 4-benzylseleno-3-methyl-2-amino-, and 4-benzylseleno-4-methyl-2-amino-butyric acid.[574*]

The Basic Hydrolysis of Hydantoins.   Hydantoins prepared as described by reaction (68) are hydrolyzed by aqueous sodium hydroxide at 165°C to the amino acids (reaction 153):

(153)

R, R^1, R^2, R': H, H, H, H;[291,578] H, H, H, CH_3; H, H, CH_3, H; CH_3, H, H, H; CH_3, H, CH_3, H; CH_3, CH_3, CH_3, H.[577]

The first selenoaminoacid, 3-benzylseleno-2-aminopropionic acid, was prepared by Fredga.[169]   He reacted 3-chloro-2-aminopropionic acid with potassium diselenide. The diorganyl diselenide formed was cleaved with mercury and then benzylated to produce the selenoaminoacid. Painter[390] did not obtain good results with Fredga's method. It has not been used since for the synthesis of selenoaminoacids. The reaction of benzyl chloro-methyl selenide with diethyl phthalimidosodiomalonate followed by saponi-fication and decarboxylation yielded 3-benzylseleno-2-aminoprioponic acid.[543]

Only the methods represented by reactions (151) and (152) and Fredga's method will give L-, D- or DL-selenoaminoacids depending upon the con-figuration of the starting aminoacid derivative. The method represented by reaction (153) and the addition of selenols to carbon-carbon double bonds generate the asymmetric center during the reaction and yield racemic products. Zdansky[578,579] has resolved racemic 4-benzyl-seleno-2-acetyl-aminobutyric acid and 3-benzylseleno-2-acetylaminopropionic acid with papain in the presence of aniline. Only the L-compound precipitated as the anilide. Walter[534] resolved the propionic acid derivative with hog acylase I. The separation of cysteine and selenocysteine derivatives by ion-exchange chromatography has been investigated.[535] The mass spectrum of Se-methyl selenocysteine has been reported.[496] The IR spectrum and the chemical, biochemical, and chromatographic behavior of selenomethionine has been studied.[476]

Se-Aryl Aminomonoselenocarboxylates

Se-Aryl aminomonoselenocarboxylates (XXXIX) have been synthesized by Jakubke[246,247,248] according to reactions (154)–(156):

$$R, \text{configuration}: H, -; CH_3, DL; i\text{-}C_3H_7, DL; i\text{-}C_4H_9, DL.^{246,247,248}$$

The 3-aminopropionic acid derivative was prepared similarly:

R, configuration: H, –; CH_3, L, DL; C_3H_7, L; i-C_3H_7, DL; i-C_4H_9, L; $C_6H_5CH_2$, L, DL; $C_6H_5CH_2SCH_2$, L, DL; CH_3SCH_2-CH_2, DL; $C_6H_5CH_2SCH_2CH_2$, DL; $C_6H_5CH_2OOCCH_2$, DL; $CH_3OOCCH_2CH_2$, L.[246,247]

The α-benzyl-γ-Se-phenyl N-benzyloxycarbonyl-L-selenoglutamate was prepared similarly:

$$R-\underset{\underset{O=C-OCH_2C_6H_5}{\overset{|}{NH}}}{\overset{|}{C}H}-COOH \xrightarrow[\text{2) 1-C}_{10}\text{H}_7\text{SeH}]{\text{1) THF/(C}_2\text{H}_5)_3\text{N/ClCO}_2\text{C}_2\text{H}_5} R-\underset{\underset{O=C-OCH_2C_6H_5}{\overset{|}{NH}}}{\overset{|}{C}H}-\overset{\overset{O}{\|}}{C}-Se-$$

(156)

R, configuration: H, –; CH_3, L, DL; i-C_4H_9, L; $C_6H_5CH_2$, L, DL; $CH_3OOCCH_2CH_2$, L; $C_6H_5CH_2SCH_2$, L.[248]

The 3-aminopropionic acid derivative and α-benzyl-γ-Se-1-naphthyl N-benzyloxycarbonyl-L-selenoglutamate were prepared similarly.

The benzyloxycarbonyl group in these N-protected esters is cleaved upon treatment with hydrobromic acid in glacial acetic acid. The free amino groups can be reacylated with benzoyl chloride[246,247] and acetyl chloride.[70,246,247]

Some of the N-protected selenoaminoacid Se-phenyl and Se-naphthyl esters prepared according to reaction (154) to (156) were aminolyzed with the sodium salts of aminoacids, aminoacid ester, and their hydrochlorides producing peptides.[246,247,248]

Bis(aminocarboxyalkyl) Diselenides and Selenides

The synthesis of the diselenides of type (XL) has been accomplished by cleaving the benzyl group with sodium in liquid ammonia from the benzyl seleno compounds (XXXVIII, R = $C_6H_5CH_2$) and oxidizing the selenol:

$$C_6H_5CH_2Se-(CHR)_x-\overset{\overset{\displaystyle R'}{|}}{C}H-CH-COOH$$

$$(XXXVIIIa) \quad \overset{|}{N}H_2$$

1) $NH_{3(1)}/Na$ | 2) $H_2O/Fe^{+3}/air$

$$\left[-Se-(CH)_x-\overset{\overset{\displaystyle R}{|}}{C}H-\overset{\overset{\displaystyle R'}{|}}{C}H-CH-COOH \atop \overset{|}{N}H_2 \right]_2 \quad (157)$$

(XLa)

x, R, R', configuration: 0, –, H, L;[163,390,571*] 1, H, H, DL;[391,574*] 1, H, H, DL;[487] 1, H, CH$_3$, DL; 1, CH$_3$, H, DL.[574*]

Williams[543] and Painter[390] showed that (XXXVIIIa) (x = 0, R' = H) is also cleaved by concentrated hydriodic acid, giving (XLa) in low yields. Potassium diselenide,[169] sodium hydrogen selenide,[390,391] and barium and calcium hydrogen selenide[543] combined with 3-chloro-2-aminopropionic acid and gave after oxidation (if necessary) the diselenide (XLa) (x = 0, R' = H) in yields of about 20%.

The monoselenides (XL) (n = 1) were prepared by Zdansky. 3-Benzyl-seleno-DL-2-acetylaminopropionic acid was cleaved by sodium in liquid ammonia and the product allowed to react with ethyl 2-acetylaminoacrylate. After removal of the acetyl groups, a mixture of *meso-* and DL-bis(2-carboxy-2-aminoethyl) selenide (selenolanthionine, see also Roy[446]) was isolated in 78% yield.[575*] Starting with the optically active components, the D-, L-, and *meso-*selenides were obtained.[579*] 3-Carboxy-3-aminopropyl 2-carboxy-2-aminoethyl selenide (selenocystathionine) was synthesized from 2-carboxy-3-aminopropyl benzyl selenide and β-chloralanine in a similar manner. By combining the appropriate optically active reagents, L(+)-, D(–)-, L(–)-*allo-* and D(+)-*allo-*selenocystathionines were formed.[579*]

Selenium Containing Peptides

3-Benzylseleno-2-aminopropionic acid (Se-benzylselenocysteine) has been incorporated into a large number of peptides. This is possible because the selenium-carbon bonds are not affected when the amino group is acylated with benzyl chloroformate,[162] *tert*-butyl azidoformate,[249] and acetyl chloride[70,578] or deacylated with 2N HBr in glacial acetic acid. The carboxyl groups can be esterfied yielding ethyl, methyl,[162] diphenylmethyl,[504] 4-nitrophenyl[162,250,531,534] and 4-nitrobenzyl esters.[534] The benzyl and

4-nitrobenzyl esters have been converted to the hydrazides and azides.[503,504] The N-protected Se-benzylselenocysteine has been condensed with the amino group of other aminoacids employing the dicyclohexylcarbodiimide,[162] the mixed acid anhydride ($ClCO_2C_2H_5$),[162] or the 4-nitrophenyl ester method.[162,531,532]

Frank[162] used DL- and L-Se-benzylselenocysteine as the amino as well as the carboxyl component in the preparation of a number of di- and tripeptides. The saponification of the methyl and ethyl esters of these peptides with $1N$ NaOH in acetone or dioxane at room temperature resulted in partial cleavage of the benzylseleno group. Theodoropoulos[503,504] replaced the tosylate group by the benzylseleno group, as shown in reaction (158):

$$CH_3 \underset{O}{\overset{O}{\underset{\uparrow}{\overset{\uparrow}{\diagdown}}}} S-O-CH_2-\underset{\underset{Z}{\overset{|}{NH}}}{\overset{|}{CH}}-\overset{O}{\overset{\|}{C}}-NH-CH_2-\overset{O}{\overset{\|}{C}}-OCH_2C_6H_5 \quad \boxed{C_6H_5CH_2Se^-}$$

$$\diagdown-CH_2-Se-CH_2-\underset{\underset{Z}{\overset{|}{NH}}}{\overset{|}{CH}}-\overset{O}{\overset{\|}{C}}-NH-CH_2-\overset{O}{\overset{\|}{C}}-OCH_2C_6H_5 \quad (158)$$

Z: $C_6H_5CH_2OC(O)$, *tert*-$C_4H_9OC(O)$

Among the naturally-occurring sulfur-containing peptides are glutathione and oxytocin. Selenium analogs of both compounds have been prepared. Frank[163] synthesized diselenoglutathione (XLI) in two different ways:

$$\begin{array}{c} NH_2 \\ | \\ O=C-CH_2-CH_2-CH-COOH \\ | \\ NH \quad O \\ | \quad \| \\ Se-CH_2-CH-C-NH-CH_2-COOH \\ | \\ Se-CH_2-CH-C-NH-CH_2-COOH \\ | \quad \| \\ HN \quad O \\ | \\ O=C-CH_2CH_2-CH-COOH \\ | \\ NH_2 \end{array}$$

(XLI)

Firstly, ethyl Se-benzyl-L-selenocysteinylglycinate was reacted with α-ethyl N-benzyloxycarbonyl-L-glutamate. Secondly, N,N'-bis(benzyloxycarbonyl)-L-diselenocystine was condensed with ethyl glycinate and, after deacylation of the amino group, with the appropriate glutamic acid derivative. Saponification of the acyl groups and cleavage of the benzyl groups with sodium in liquid ammonia followed by oxidation gave diselenoglutathione. The N-protected dibenzyl ester of the precursor tripeptide of diselenoglutathione was prepared by Theodoropoulos.[503,504] Janicki[250] had synthesized the tripeptide ethyl N-benzyloxycarbonyl-γ-L-glutamyl(α-ethyl ester)-Se-benzyl-DL-selenocysteinylglycinate but was unable to saponify the ester group and convert the peptide into diselenoglutathione because of decomposition.

Selenium derivatives of oxytocin and related compounds have the formula (XLII):

(XLII)

H, NH$_2$; X, Y: Se, Se; Se, S; S, Se.

Frank[164] prepared diselenoxytocin (Z = NH$_2$, X = Y = Se) beginning with the benzyloxycarbonyl-protected 4-nitrophenyl leucinate. Condensation with ethyl glycinate followed by deacylation of the dipeptide, and reaction with 4-nitrophenyl N-benzyloxycarbonylprolinate gave the tripeptide $C_6H_5CH_2OC(O)$-7-8-9-OC_2H_5, which was then converted to the glycyl amide. The nonapeptide, having benzyl groups bonded to X and Y (X, Y = Se), was then built up by adding one amino acid at a time. The same nonapeptide was obtained from three tripeptides, as shown in reaction (159):

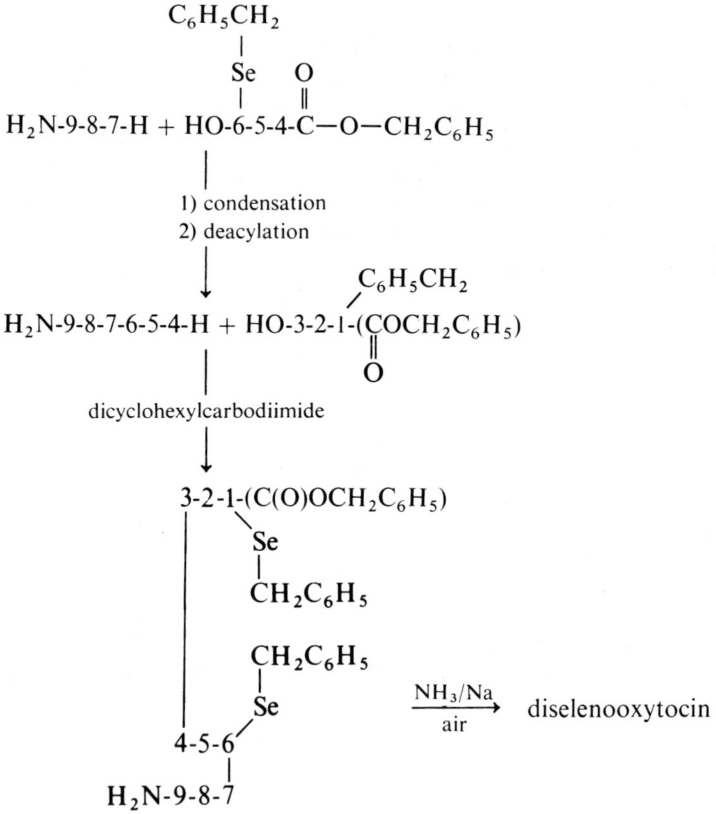

$$C_6H_5CH_2$$
$$|$$
$$Se \quad O$$
$$| \quad \parallel$$
$$H_2N\text{-}9\text{-}8\text{-}7\text{-}H + HO\text{-}6\text{-}5\text{-}4\text{-}C\text{-}O\text{-}CH_2C_6H_5$$
$$|$$

1) condensation
2) deacylation

$$\downarrow \qquad\qquad C_6H_5CH_2$$
$$\diagup$$
$$H_2N\text{-}9\text{-}8\text{-}7\text{-}6\text{-}5\text{-}4\text{-}H + HO\text{-}3\text{-}2\text{-}1\text{-}(\underset{\underset{O}{\parallel}}{C}OCH_2C_6H_5)$$
$$|$$

dicyclohexylcarbodiimide   (159)

$$\downarrow$$

$$3\text{-}2\text{-}1\text{-}(C(O)OCH_2C_6H_5)$$
$$\diagdown$$
$$Se$$
$$|$$
$$CH_2C_6H_5$$

$$CH_2C_6H_5$$
$$|$$
$$Se \qquad \xrightarrow[\text{air}]{NH_3/Na} \quad \text{diselenooxytocin}$$
$$\diagup$$
$$4\text{-}5\text{-}6$$
$$|$$
$$H_2N\text{-}9\text{-}8\text{-}7$$

Starting with the tetrapeptide $H_2N\text{-}9\text{-}8\text{-}7\text{-}6(CO_2CH_2C_6H_5)SeCH_2C_6H_5$,[503,504] employing the nitrophenyl ester method and incorporating the 3-benzyl-selenopropionyl group as component 1, gave 1-deamino-1,6-diselenooxytocin ($X = Y = Se$; $Z = H$),[532] for which crystal data were reported.[533]

6-Selenooxytocin ($X = Se$, $Y = S$, $Z = NH_2$) and 1-deamino-6-seleno-oxytocin ($X = Se$, $Y = S$, $Z = H$) were prepared similarly.[531,534,111] 1-Selenooxytocin ($X = S$, $Y = Se$, $Z = NH_2$) and 1-deamino-1-selenoxytocin ($X = S$, $Y = Se$, $Z = H$) were obtained from the octapeptide with a benzylthio group attached to component 6 and Se-benzyl-selenocysteine and 3-benzyl-selenopropionic acid, respectively.

The monoselenooxytocins and their deamino derivatives were degraded to selenocystine upon heating in 6N HCl at 100°C for 22 h.[535]

Di- and tripeptides bearing a Se-phenyl ester group at the end-carboxyl group were obtained from the Se-phenyl selenoaminocarboxylates and the N-protected amino acid or dipeptide.[246,247,70] Similar reactions were carried out with the Se-1-naphthyl esters.[248] Se-phenyl N-benzyloxy-carbonylglycylselenoglycinate was synthesized from the N-protected gly-cylglycine and benzeneselenol in tetrahydrofuran containing triethylamine and ethyl chloroformate. The tripeptides Se-phenyl glycylglycylseleno-glycinate, Se-phenyl glycylglycyl-DL-selenoalaninate, and Se-phenyl gylcylglycyl-β-selenoalaninate refluxed in pyridine gave the corresponding cyclic hexapeptides.[246]

Other Organic Selenium Compounds of Biological Importance

In addition to the selenoaminoacids and selenopeptides already discussed, selenium derivatives of cystamine, cysteamine, choline, homocholine, pyrimidine, pantetheine, coenzyme A and selenium derivatives of steroids have been prepared.

The synthesis of choline diselenide (XLIII) (n = 1) (R, R^1, R^2 = CH_3) and homocholine diselenides (n = 2) and related compounds and the di-selenides (XLIV) has been described. Compounds of the type (XLV) and (XLVI) with R = alkyl and acyl have also been discussed.

Chu[114] investigated the hydrolysis and aminolysis of benzoyl seleno-choline and its dimethylamino analogue. Shefter[474,475] determined the crystal and molecular structure of acetyl selenocholine iodide and acetyl selenocholine. The high resolution NMR spectrum of acetyl selenocholine was studied by Cushley.[131]

$$\left[\begin{array}{c} Se-(CH_2)_n-CH_2-N\underset{R^2}{\overset{R^1}{\diagup}}\overset{R}{\vert} \\ \vert \\ Se-(CH_2)_n-CH_2-N\underset{\underset{R}{\vert}R^1}{\overset{R^2}{\diagup}} \end{array} \right]^{++} \quad 2X^-$$

(XLIII)

$n = 1$; R, R^1, R^2, X: H, H, -, -; [149,211,216] H, H, H, Cl; [212,287,288, 341,343] H, H, H, picrate; [287] CH_3, H, H, Cl; [212] CH_3, CH_3, H, Cl; [212,214] CH_3, CH_3, CH_3, I. [212,214]

$n = 2$, R, R^1, R^2, X: CH_3, CH_3, H, Cl; CH_3, CH_3, CH_3, I. [214]

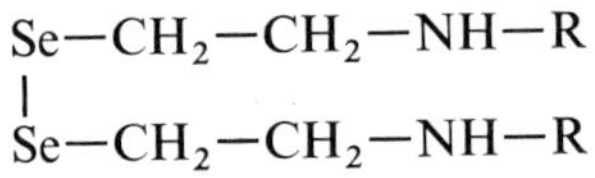

$$\begin{array}{l} Se-CH_2-CH_2-NH-R \\ | \\ Se-CH_2-CH_2-NH-R \end{array}$$

(XLIV)

R: C_6H_5CO; [211,341] H_2NCS; [289] $HOOCCH_2CH_2CO$. [217]

$$\left[(CH_3)_3\overset{\oplus}{N}-(CH_2)_n=Se-R\right]^+ X^-$$

(XLV)

$n = 2$; R, X: H, I; [214] CH_3, I; CH_3CO, I; CH_3CO, Br; C_3H_7CO, I; C_6H_5CO, I; C_6H_5CO, Br; [212] C_6H_5CS, Br. [115]

$n = 3$; R, X: H, I; CH_3, I; CH_3CO, I; C_6H_5CO, I. [214]

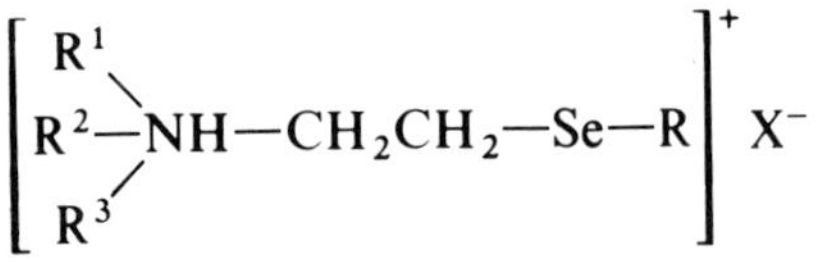

$$\left[\begin{array}{l} R^1 \\ \diagdown \\ R^2-NH-CH_2CH_2-Se-R \\ \diagup \\ R^3 \end{array}\right]^+ X^-$$

(XLVI)

R, R^1, R^2, R^3, X: CH_3, CH_3, CH_3, H, I; C_6H_5CO, CH_3, CH_3, H, I; C_6H_5CO, CH_3, C_6H_5CO, -, -; [212] C_6H_5CO, H, C_6C_5HO, -, -; [338, 341] C_6H_5CS, CH_3, CH_3, H, Cl. [115] CH_3CO, H, $HOOCCH_2CH_2CO$, -, -; C_6H_5CO, H, $HOOCCH_2CH_2CO$, -, -. [217]

The selenopyrimidine derivatives, selenocytosine, 5-methylselenocytosine 2-selenouracil, diselenouracil, diselenothymine, 2-selenothymine and seleno-purine derivatives were discussed previously.

Selenopantetheine derivatives and selenopantethine were synthesized by Gunther[211,213] as shown in reaction scheme (160):

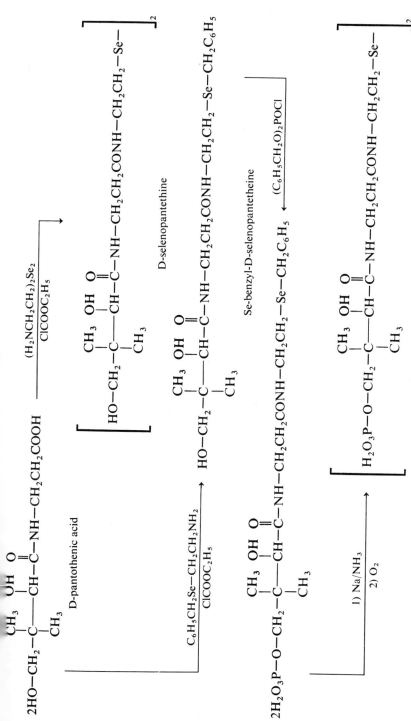

D-pantothenic acid

$(H_2NCH_2CH_2)_2Se_2$ / $ClCOOC_2H_5$

D-selenopantethine

$C_6H_5CH_2Se-CH_2CH_2NH_2$ / $ClCOOC_2H_5$

Se-benzyl-D-selenopantetheine

$(C_6H_5CH_2O)_2POCl$

1) Na/NH_3
2) O_2

4′-phospho-D-selenopantethine

527

528

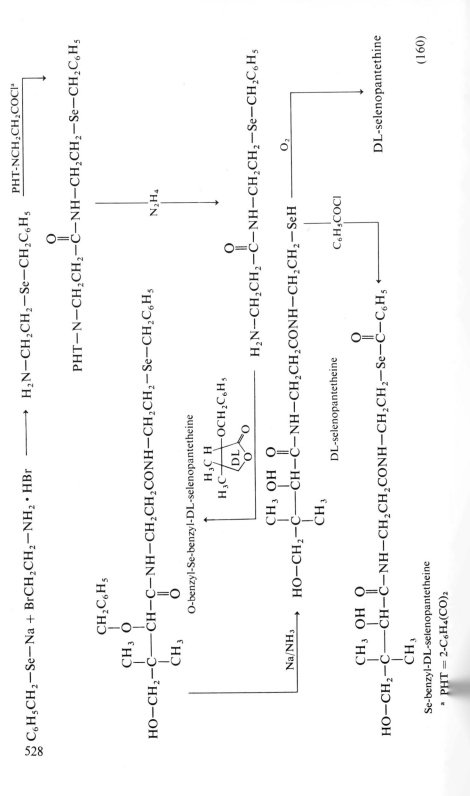

$$C_6H_5CH_2-Se-Na + BrCH_2CH_2-NH_2 \cdot HBr \longrightarrow H_2N-CH_2CH_2-Se-CH_2C_6H_5 \xrightarrow{PHT-NCH_2CH_2COCl^a}$$

$$PHT-N-CH_2CH_2-C-NH-CH_2CH_2-Se-CH_2C_6H_5 \xrightarrow{N_2H_4}$$

$$H_2N-CH_2CH_2-C-NH-CH_2CH_2-Se-CH_2C_6H_5$$

O-benzyl-Se-benzyl-DL-selenopantetheine

$$H_2N-CH_2CH_2-CH_2CONH-CH_2CH_2-SeH \xrightarrow{O_2} DL\text{-selenopantethine}$$

DL-selenopantetheine

Se-benzyl-DL-selenopantetheine

(160)

^a PHT = 2-C₆H₄(CO)₂

For the preparation of selenocoenzyme A, the primary hydroxyl group in selenopantethine must be esterified with phosphoric acid.[213] Since a direct phosphorylation of D-selenopantethine had been unsuccessful, Se-benzyl-D-selenopantetheine was phosphorylated with dibenzyl phosphoric acid chloride and the phospho derivative treated with sodium in liquid ammonia. The intermediate phosphoselenopantetheine was then oxidized to the phosphoselenopantethine, which was then converted to selenocoenzyme A, as shown in reaction scheme (161):

(161)

The isomeric selenocoenzyme A diselenides were transformed into the Se-benzoyl derivatives, which were then separated chromotagraphically. Selenonicotineamide adenine dinucleotide phosphate has been prepared and studied by Christ.[116,117]

Selenium-containing steroids were obtained from the appropriate steroid derivatives and selenium dioxide,[160,346] steroid tosylates and selenourea,[238] (US Pat 3,372,173) and from steroid methyl sulfonates and benzene-selenolate.[262]

PATENTS

Belg. Patent 670,823 (1966); *CA* **65**, 21030B.
Brit. Patent 782,887 (1957); *CA* **52**, 10192H.
Brit. Patent 839,351 (1960); *CA* **55**, 72951I.
Ger. Patent, 1,105,414 (1960); *CA* **56**, 8566D.
Ger. Patent 1,943,787 (1970); *CA* **72**, 111093P.
Jap. Patent 19,630 (1964); *CA* **62**, 10378B.
Pol. Patent 43,914 (1960); *CA* **58**, 4471C.
Pol. Patent 43,915 (1960); *CA* **60**, 4061C.
U.S. Patent 3,284,502 (1966); *CA* **67**, 64045J.
U.S. Patent 3,372,173 (1968); *CA* **69**, 77632Q.
U.S. Patent 3,281,447 (1966); *CA* **66**, 19156Q.
U.S. Patent 2,934,420 (1960); *CA* **54**, 17780C.
USSR Patent 165,752 (1964); *CA* **62**, 6185D.
USSR Patent 196,815 (1967); *CA* **68**, 95975N.
USSR Patent 209,445 (1968); *CA* **69** 76949M.

REFERENCES

1. Abel, E. W., Crosse, B. C. and Hutson, G. V., *J. Chem. Soc.*, A, 2014 (1967).
2. Agenäs, L. B., *Acta Chem. Scand.*, **17**, 268 (1963).
3. Agenäs, L. B., *Ark. Kemi*, **23**, 145 (1964).
4. Agenäs, L. B., *Ark. Kemi*, **23**, 155 (1964).
5. Agenäs, L. B., *Ark. Kemi*, **23**, 463 (1964).
6. Agenäs, L. B., *Ark. Kemi*, **24**, 415 (1965).
7 Agenäs, L. B., *Ark. Kemi*, **24**, 573 (1965).
8. Agenäs, L. B., *Acta Chem. Scand.*, **19**, 764 (1965).
9. Agenäs, L. B., *Ark. Kemi*, **28**, 145 (1967).
10. Agenäs, L. B., and Persson, B., *Acta Chem. Scand.*, **21**, 835 (1967).
11. Agenäs, L. B., and Persson, B., *Acta Chem. Scand.*, **21**, 837 (1967).
12. Agenäs, L. B., and Lindgren, B., *Ark. Kemi*, **29**, 471 (1968).
13. Agenäs, L. B., *Ark. Kemi*, **29**, 479 (1968).
14. Agenäs, L. B., *Acta Chem. Scand.*, **22**, 1763 (1968).
15. Agenäs, L. B., *Acta Chem. Scand.*, **22**, 1773 (1968).
16. Agenäs, L. B., *Ark. Kemi*, **30**, 417 (1969).
17. Agenäs, L. B., and B. Lindgren, *Ark. Kemi*, **30**, 529 (1969).
18. Agenäs, L. B., *Ark. Kemi*, **30**, 471 (1969).
19. Agenäs, L. B., *Ark. Kemi*, **30**, 497 (1969).
20. Agenäs, L. B., *Ark. Kemi*, **31**, 31 (1969).
21. Agenäs, L. B., *Ark. Kemi*, **31**, 45 (1969).
22. Agenäs, L. B., *Ark. Kemi*, **31**, 159 (1969).
23. Agenäs, L. B., *Ark. Kemi*, **30**, 549 (1969).
24. Agenäs, L. B., *Ark. Kemi*, **30**, 433 (1969).
25. Agrawal, K. C., and Ray, F. E., *Int. J. Cancer*, **2**, 257 (1967).
26. Alimarin, I. P., and Shakhova, N. V., *Zh. Anal. Chim.*, **16**, 412 (1961).
27. Allum, K. G., Creighton, J. A., Green, J. H. S., Minkoff, G. J., and Prince, L. J. S., *Spectrochim. Acta*, A, **24**, 927 (1968).
28. Allkins, J. R., and Hendra, P. J., *Spectrochim. Acta*, **22**, 2075 (1966).

29. Allkins, J. R., and Hendra, P. J., *Spectrochim. Acta, A*, **23**, 1671 (1967).
30. Anger, V., and Fischer, G., *Mikrochim. Acta*, 501 (1962).
31. Anthoni, U., Larsen, C., and Nielsen, P. H., *Acta Chem. Scand.*, **21**, 2580 (1967).
32. Appel, R., and Buechler, G., *Z. Anorg. Allgem. Chem.*, **348**, 175 (1966).
33. Arbuzov, B. A., Kataeva, L. M., Kataev, E. G., and Il'yasov, A. V., *Izv. Akad. Nauk, SSSR, Otd. Khim. Nauk*, 360 (1962).
34. Arcoria, A., and Cordella, G., *Boll. Sedute Accad. Gioenia Sci. Nat. Catania*, **3** [4], 314 (1956).
35. Asinger, F., Berding, H., and Offermanns, H., *Monatsh. Chem.*, **99**, 2072 (1968).
36. Asinger, F., Berding, H., and Offermanns, H., *Monatsh. Chem.*, **99**, 2084 (1968).
37. Aynsley, E. E., Greenwood, N. N., and Sprague, M. J., *J. Chem. Soc.*, 2395 (1965).
38. Aynsley, E. E., Greenwood, N. N., and Leach, J. B., *Chem. Ind. (London)*, 379 (1966).
39. Aynsley, E. E., Greenwood, N. N., Hunter, G., and Sprague, M. J., *J. Chem. Soc., A*, 1344 (1966).
40. Azman, A., Drofenik, M., Hadzi, M., and Lukman, B., *J. Mol. Struct.*, **1**, 181 (1968).
41. Ayrey, G., Barnard, D., and Woodbridge, D. T., *J. Chem. Soc.*, 2089 (1962).
42. Azerbaev, I. N., Asmanova, A. B., Tsoi, L. A., and Bazalitskaya, V. S., *Izv. Akad. Nauk, Kaz. SSR. Ser. Khim.*, **20**, 53 (1970).
43. Azerbaev, I. N., Asmanova, A. B., Tsoi, L. A., and Dzhamaletdinova, M. K., *Izv. Akad. Nauk, Kaz. SSR. Ser. Khim.*, **20**, 75 (1970).
44. Backer, H. J., and Strating, J., *Rec. Trav. Chim.*, **53**, 1113 (1934).
45. Baroni, A., *Atti Accad. Lincei, Classe Sci. Fis., Mat. Nat.*, **26**, 460 (1937).
46. Bartolotti, L., and Cerniani, A., *Boll. Sci. Fac. Chim. Ind. (Bologna)*, **14**, 33 (1956).
47. Beecher, J. F., *J. Mol. Spectrosc.*, **21**, 414 (1966).
48. Behagel, O., and Seibert, H., *Chem. Ber.*, **66**, 708 (1933).
49. Behagel, O., and Müller, W., *Chem. Ber.*, **68**, 1540 (1935).
50. Benes, J., and Prochazakova, V., *J. Chromatogr.*, **29**, 239 (1967).
51. Bellinger, M., and Cagniant, P., *Compt. Rend. Acad. Sci., Ser. C*, **268**, 1385 (1969).
52. Bellinger, N., Cagniant, P., and Renson, M., *Compt. Rend. Acad. Sci., Ser. C*, **269**, 532 (1969).
53. Bergman, J., *Acta Chem. Scand.*, **22**, 1883 (1968).
54. Bergmann, F., and M. Rashi, *Israel J. Chem.*, **7**, 63 (1969).
55. Bergson, G., *Ark. Kemi*, **9**, 121 (1955).
56. Bergson, G., *Ark. Kemi*, **10**, 127 (1956).
57. Bergson, G., *Acta Chem. Scand.*, **11**, 1607 (1957).
58. Bergson, G., *Ark. Kemi*, **13**, 11 (1958).
59. Bergson, G., and Nordström, G., *Ark. Kemi*, **17**, 569 (1961).
60. Bergson, G., and Delin, A. L., *Ark. Kemi*, **18**, 441 (1961).
61. Bergson, G., *Acta Chem. Scand.*, **15**, 1611 (1961).
62. Bergson, G., and Biezais, A., *Ark. Kemi*, **18**, 143 (1961).
63. Bergson, G., *Ark. Kemi*, **19**, 195 (1962).
64. Bergson, G., and Wold, S., *Ark. Kemi*, **19**, 215 (1962).

65. Bergson, G., Claeson, G., and Schotte, L., *Acta Chem. Scand.*, **16**, 1159 (1962).
66. Bergson, G., *Ark. Kemi*, **21**, 439 (1963).
67. Bird, M. L., and Challenger, F., *J. Chem. Soc.*, 527 (1942).
68. Blackman, L. C. F., and Dewar, M. J. S., *J. Chem. Soc.*, 162 (1957).
69. Blackman, L. C. F., and Dewar, M. J. S., *J. Chem. Soc.*, 171 (1957).
69a. Blackmore, W. R., and Abrahams, S. C., *Acta Crystallogr.* **8**, 323 (1955).
70. Blaha, K., Fric, I., and Jakubke, H. D., *Collect. Czech. Chem. Commun.*, **32**, 558 (1967).
71. Bogolyubov, G. M., and Shlyk, Y. N., *Zh. Obshch. Khim.*, **39**, 1759 (1969).
72. Bogolyubov, G. M., Shlyk, Y. N., and Petrov, A. A., *Zh. Obshch. Khim.*, **39**, 1804 (1969).
73. Boiko, Yu. A., Kupin, B. S., and Petrov, A. A., *Zh. Org. Khim.*, **4**, 1355 (1968).
74. Boiko, Yu. A., Kupin, B. S., and Petrov, A. A., *Zh. Org. Khim.*, **5**, 1553 (1969).
75. Bonner, W. A., and Robinson, A., *J. Amer. Chem. Soc.*, **72**, 345 (1950).
76. Bonner, W. A., and Robinson, A., *J. Amer. Chem. Soc.*, **72**, 356 (1950).
77. Brandsma, L., Wijers, H., and Arens, J. F., *Rec. Trav. Chim.*, **81**, 563 (1962).
78. Brandsma, L., and Arens, J. F., *Rec. Trav. Chim.*, **81**, 539 (1962).
79. Brandsma, L., and Arens, J. F., *Rec. Trav. Chim.*, **81**, 33 (1962).
80. Brandsma, L., and Wijers, H., *Rec. Trav. Chim.*, **82**, 68 (1963).
81. Brandsma, L., *Rec. Trav. Chim.*, **83**, 307 (1964).
82. Brandsma, L., Wijers, H. E., and Jonker, C., *Rec. Trav. Chim.*, **83**, 208 (1964).
83. Brandsma, L., and Schuijl, P. J. W., *Rec. Trav. Chim.*, **88**, 513 (1969).
84. Breuninger, V., Dreeskamp, H., and Pfisterer, G., *Ber. Bunsenges. Physik. Chem.*, **70**, 613 (1966).
85. Brown, J. R., Gillman, G. P., and George, M. H., *J. Polymer Sci.*, *Part A-1*, **5**, 903 (1967).
86. Bryden, J. H., and McCullough, J. D., *Acta Crystallogr.*, **7**, 833 (1954).
87. Bulka, E., and Ahlers, K. D , *Z Chem.*, **3**, 348 (1963).
88. Bulka, E., and Ahlers, K. D., *Z. Chem.*, **3**, 387 (1963).
89. Bulka, E., Ahlers, K. D., and Tucek, E., *Chem. Ber.*, **100**, 1367 (1967).
90. Bulka, E., Ahlers, K. D., and Tucek, E., *Chem. Ber.*, **100**, 1459 (1967).
91. Bulka, E., Ahlers, K. D., and Tucek, E., *Chem. Ber.*, **100**, 1373 (1967).
92. Bulka, E., Ehlers, D., and Storm, E., *Z. Chem.* **10**, 403 (1970).
93. Burdon J., Coe, P. L., and Fulton, M., *J. Chem. Soc.*, 2094 (1965).
94. Burlant, W. J., *Diss. Abs.*, **15**, 1170 (1955).
95. Carazzolo, G., and Valle, G., *J. Polymer Sci.*, *Part A*, **3**, 4013 (1965).
96. Carazzolo, G., Mortillaro, L., Credali, L., and Bezzi, S., *J. Polymer Sci.*, *Pt. B*, **3**, 997 (1965).
97. Carazzolo, G., and Mammi, M., *Makromol. Chem.*, **100**, 28 (1967).
98. Cecchini, M. A., and Giesbrecht, E., *J. Org. Chem.*, **21**, 1217 (1956).
99. Cerriani, A., and Passerini, R., *Boll. Sci. Fac. Chim. Ind. Bologna*, **14**, 107 (1956).
*
101. Cerriani, A., and Passerini, R., *Ann. Chim.* (Rome), **47**, 58 (1957).
102. Chen, I., and Das, T. P., *J. Chem. Phys.*, **45**, 3526 (1966).
103. Chierici, L., and Passerini, R., *Atti Accad. Nazl. Lincei, Rend., Classe Sci. Fis. Mat. e Nat.*, **15**, 69 (1953).

* There is no reference 100.

104. Chierici, L., and Passerini, R., *J. Chem. Soc.* 3249 (1954).
105. Chierici, L., Lumbroso, H., and Passerini, R., *Boll. Sci. Fac. Chim. Ind. Bologna*, **12**, 127 (1954).
106. Chierici, L., and Passerini, R., *Boll. Sci. Fac. Chim. Ind. Bologna*, **12**, 131 (1954).
107. Chierici, L., and Passerini, R., *Boll. Sci. Fac. Chim. Ind. Bologna*, **12**, 138 (1954).
108. Chierici, L., and Passerini, R., *Ricerca Sci.*, **25**, 2316 (1955).
109. Chierici, L., and Montanari, F., *Gazz. Chim. Ital.*, **86**, 1269 (1956).
110. Chierici, L., and Montanari, F., *Boll. Sci. Fac. Chim. Ind. Bologna*, **14**, 78 (1956).
111. Chiu, C. C., Schwartz, I. L., and Walter, R., *Science*, **163**, 925 (1969).
112. Chmutova, G. A., *Sb. Aspir. Rab. Kazan. Gos. Univ., Khim., Geol.*, 70 (1967).
113. Chu, S. H., Gunther, W. H. H., and Mautner, H. G., *Biochem. Prepn.*, **10**, 153 (1963).
114. Chu, S. H., and Mautner, H. G., *J. Org. Chem.*, **31**, 308 (1966).
115. Chu, S. H., and Mautner, H. G., *J. Med. Chem.*, **11**, 446 (1968).
116. Christ, W., and Coper, H., *Fed. Europ. Biochem. Soc. Lett.*, **2**, 267 (1969).
117. Christ, W., Schmidt, D., and Coper, H., *Hoppe-Seyler's Z. Physiol. Chem.*, **351**, 427 (1970).
118. Christiaens, L., and Renson, M., *Bull. Soc. Chim. Belges*, **77**, 153 (1968).
119. Christiaens, L., and Renson, M., *Bull. Soc. Chim. Belges*, **79**, 133 (1970).
120. Cinquini, M., Colonna, S., and Giovini, G., *Chem. Ind. (London)*, 1737 (1969).
121. Cinquini, M., Colonna, S., and Landini, D., *Bull. Sci. Fac. Chim. Ind. Bologna*, **27**, 207 (1969).
122. Clase, H. J., and Ebsworth, F. A. V., *J. Chem. Soc.*, 940 (1965).
123. Cohen, S. C., Reddy, M. L. N., and Massey, A. G., *Chem. Commun.*, 451 (1967).
124. Cohen, S. C., Reddy, M. L. N., and Massey, A. G., *J. Organometal. Chem.*, **11**, 563 (1968).
125. Collard-Charon, C., and Renson, M., *Bull. Soc. Chim. Belges*, **71**, 531 (1962).
126. Collard-Charon, C., Huls, R., and Renson, M., *Bull. Soc. Chim. Belges*, **71**, 541 (1962).
127. Collard-Charon, C., and Renson, M., *Bull. Soc. Chim. Belges*, **71**, 563 (1962).
128. Collard-Charon, C., and Renson, M., *Bull. Soc. Chim. Belges*, **72**, 304 (1963).
129. Cozzens, R. F., and Harvey, A. B., *J. Polymer Sci. Part A-2*, **8**, 1279 (1970).
130. Credali, L., Russo, M., Mortillaro, L., DeChecchi, C., Valle, G., and Mammi, M., *J. Chem. Soc. B*, 117 (1967).
131. Cushley, R. J., and Mautner, H. G., *Tetrahedron*, **26**, 2151 (1970).
132. Dale, J. W., Emeleus, H. J., and Haszeldine, H. R., *J. Chem. Soc.*, 2939 (1958).
133. Danze, J. M., and Renson, M., *Bull. Soc. Chim. Belges*, **75**, 169 (1966).
134. Dallacker, F., Eschelbach, F. E., and Zegers, H., *Justus Liebigs Ann. Chem.*, **689**, 16 (1965).
135. Dallacker, F., and Zegers, H., *Justus Liebigs Ann. Chem.*, **689**, 163 (1965).
136. Dallacker, F., and Eschelbach, F. E., *Justus Liebigs Ann. Chem.*, **689**, 171 (1965).
137. Dallacker, F., Kaiser, E., and Uddrich, P., *Justus Liebigs Ann. Chem.*, **689**, 179 (1965).
138. Deacon, G. B., and Parrott, J. C., *J. Organometal. Chem.*, **22**, 287 (1970).

139. DeFilippo, D., and Momicchioli, F., *Tetrahedron*, **25**, 5733 (1969).
140. Degani, J., and Tundo, A., *Ann. Chim.*, **50**, 140 (1960).
141. Degani, J., Fochi, R., and Spunta, G., *Boll. Sci. Fac. Chim. Ind. Bologna*, **23**, 151 (1965).
142. Denes, V. I., Ciurdaru, G., and Farcasan, M., *Rev. Roumaine Chim.*, **10**, 1045 (1965).
143. Detoni, S., and Hadzi, D. J., *J. Chim. Phys.*, **53**, 760 (1956).
144. Dewar, D. H., Fergusson, J. E., Hentschel, P. R., Wilkins, C. J., and Williams, P. P., *J. Chem. Soc.*, 688 (1964).
145. Djerassi, J., *Acta Chem. Scand.*, **15**, 417 (1961).
146. Dostal, K., Mosna, P., and Zak, Z., *Z. Chem.*, **6**, 153 (1966).
147. Doughty, H W., *Amer. Chem. J.*, **41**, 326 (1909).
148. Douglass, I. B., *J. Amer. Chem. Soc.*, **59**, 740 (1937).
149. Draguet, C., and Renson, M., *Bull. Soc. Chim. Belges*, **75**, 243 (1966).
150. Dreeskamp, H., and Pfisterer, G., *Mol. Phys.*, **14**, 295 (1968).
151. Dyer, G., and Meek, D. V., *Inorg. Chem.*, **6**, 149 (1967).
152. Dyer, G., and Minnier, C. E., *J. Med. Chem.*, **11**, 1232 (1968).
153. Elsakov, N. V., and Petrov, A. A., *Optika i Spectroskopiya*, **16**, 797 (1964).
154. Emeleus, H. J., and Welcman, N., *J. Chem. Soc.*, 1268 (1963).
155. Emeleus, H. J., and Dunn, M. J., *J. Inorg. Nucl. Chem.*, **27**, 752 (1965).
156. Es, T. van, *Carbohyd. Res.*, **5**, 282 (1967).
157. Es, T. van, and Whistler, R. L., *Tetrahedron*, **23**, 2849 (1967).
158. Euler, H. V., and Hasselquist, H., *Z. Physiol. Chem., Hoppe-Seyler's*, **306**, 49 (1956).
159. Faurote, P. D., and O'Rear, J. G., *J. Amer. Chem. Soc.*, **78**, 4999 (1956).
160. Florey, K., and Restivo, A. R., *J. Org. Chem.*, **22**, 406 (1957).
161. Foster, D. G., *Org. Synth.*, Coll. Vol. **III**, 771 (1965).
162. Frank, W., *Z. Physiol. Chem.*, **339**, 202 (1964).
163. Frank, W., *Z. Physiol. Chem.*, **339**, 214 (1964).
164. Frank, W., *Z. Physiol. Chem.*, **339**, 222 (1964).
165. Frankiss, S. G., *J. Mol. Struct.*, **3**, 89 (1969).
166. Franklin, W. J., and Werner, R. L., *Tetrahedron Lett.*, 3003 (1965).
167. Fredga, A., *Svensk. Kem. Tid.*, **48**, 160 (1936).
168. Fredga, A., and A. Lindgren, *Acta Chem. Scand.*, **15**, 938 (1961).
169. Fredga, A., *Acta Chem. Scand.*, **17** (Suppl. 1) S51, (1963).
170. Fredga, A., *Acta Chem. Scand.*, **24**, 1117 (1970).
171. Freeman, J. M., and Henshall, T., *J. Mol. Struct.*, **1**, 31 (1967).
172. Frenzel, H., Nuhn, P., and Wagner, G., *Arch. Pharm.*, **302**, 62 (1969).
173. Funk, H., and Weiss, W., *J. Prakt. Chem.*, **1** [4], 33 (1955).
174. Funk, H., and Papenroth, W., *J. Prakt. Chem.*, **8** [4], 256 (1959).
175. Funk, H., and Papenroth, W., *J. Prakt. Chem.*, **11** [4], 191 (1960).
176. Futekov, L., Iordanov, N., and Georgieva, A., *Natura (Plovdiv)*, **2**, 59 (1968).
177. Gabdrakhmanov, F. G., *Sb. Aspir. Rab., Kazan. Gos. Univ., Khim., Geol.*, 85 (1967).
178. Gabdrakhmanov, F. G., Samitov, Yu. Yu., and Kataev, E. G., *Zh. Obshch. Khim.*, **37**, 761 (1967).
179. Gasco, A., Dimodica, G., and Barni, E., *Ann. Chim. (Rome)*, **58**, 385 (1968).
180. Gilow, H. M., and Walker, G. L., *J. Org. Chem.*, **32**, 2580 (1967).
181. Gilow, H. M., Camp, Jr., R. B., and Clifton, E. C., *J. Org. Chem.*, **33**, 230 (1968).

182. Giua, M., and Blanco, R., *Gazz. Chim. Ital.*, **89**, 693 (1959).
183. Giudicelli, J. F., Mennin, J., and Najer, H., *Bull. Soc. Chim. Fr.*, 1099 (1968).
184. Gol'dfarb, Ya. L., and Litvinov, V. P., *Izv. Akad. Nauk SSSR, Ser. Khim.*, 2088 (1964).
185. Gol'dfarb, Ya. L., Litvinov, V. P., and Sukiasyan, A. N., *Dokl. Akad. Nauk SSSR*, **182**, 340 (1968).
185a. Goldish, E., Hedeberg, K., Marsh, R. E., and Shoemaker, V., *J. Amer. Chem. Soc.*, **77**, 2948 (1955).
186. Golmohammadi, R., *Acta Chem. Scand.*, **17**, 1779 (1963).
187. Golmohammadi, R., *Acta Chem. Scand.*, **20**, 479 (1966).
188. Golmohammadi, R., *Acta Chem. Scand.*, **20**, 563 (1966).
189. Golmohammadi, R., *Ark. Kemi*, **25**, 279 (1966).
190. Goodall, D. C., *J. Inorg. Nucl. Chem.*, **30**, 1269 (1968).
191. Goodall, D. C., *J. Chem. Soc.*, *A.*, 890 (1969).
192. Gosselck, J., *Chem. Ber.*, **91**, 2345 (1958).
193. Gosselck, J., and Barth, H., *Z. Naturforsch.*, **16b**, 280 (1961).
194. Gosselck, J., *Z. Naturforsch.*, **16b**, 404 (1961).
195. Gosselck, J., and Wolters, E., *Chem. Ber.*, **95**, 1237 (1962).
196. Gosselck, J., and Wolters, E., *Z. Naturforsch.*, **17b**, 131 (1962).
197. Gosselck, J., *Angew. Chem., Intern. Ed. Engl.*, **2**, 660 (1963).
198. Gosselck, J., Barth, H., and Beress, L., *Justus Liebigs Ann. Chem.*, **671**, 1 (1964).
199. Gosselck, J., Beress, L., Schenk, H., and Schmidt, G., *Angew. Chem. Intern. Ed. Engl.*, **4**, 1080 (1965).
200. Gould, E. S., and McCullough, J. D., *J. Amer. Chem. Soc.*, **73**, 1109 (1951).
201. Gould, E. S., and Amendola, A., *J. Amer. Chem. Soc.*, **77**, 2103 (1955).
202. Gould, E. S., and Post, B., *J. Amer. Chem. Soc.*, **78**, 5161 (1956).
203. Gould, E. S., and Burlant, W. J., *J. Amer. Chem. Soc.*, **78**, 5825 (1956).
204. Grant, D., and VanWazer, J. R., *J. Amer. Chem. Soc.*, **86**, 3012 (1964).
205. Green, W. H., and Harvey, A. B., *J. Chem. Phys.*, **49**, 3586 (1968).
206. Greenberg, B., Gould, E. S., and Burlant, W., *J. Amer. Chem. Soc.*, **78**, 4028 (1956).
207. Greenwood, N. N., and Hunter, G., *J. Chem. Soc. A*, 1520 (1967)
208. Gronowitz, S., and Frejd, T., *Acta Chem. Scand.*, **23**, 2540 (1969).
209. Gronowitz, S., and Frejd, T., *Acta Chem. Scand.*, **24**, 2656 (1970).
210. Grundnes, J., and Klaeboe, P., *Acta Chem. Scand.*, **18**, 2022 (1964).
211. Gunther, W. H. H., and Mautner, H. G., *J. Amer. Chem. Soc.*, **82**, 2762 (1960).
212. Gunther, W. H. H., and Mautner, H. G., *J. Med. Chem.*, **7**, 229 (1964).
213. Gunther, W. H. H., and Mautner, H. G., *J. Amer. Chem.*, **87**, 2708 (1965).
214. Gunther, W. H. H., and Mautner, H. G., *J. Med. Chem.*, **8**, 845 (1965).
215. Gunther, W. H. H., *J. Org. Chem.*, **31**, 1202 (1966).
216. Gunther, W. H. H., *J. Org. Chem.*, **32**, 3929 (1967).
217. Gunther, W. H. H., *J. Org. Chem.*, **32**, 3931 (1967).
218. Hadzi, D., Jan, J., Azman, A., and Oblak, S., *Rev. Roumaine Chim.*, **10**, 1163 (1965).
219. Hallam, H. E., and Jones, C. M., *J. Chem. Soc.*, *A*, 1033 (1969).
220. Hallam, H. E., and Jones, C. M., *Spectrochim. Acta, Part A*, **25**, 1791 (1969).
221. Hampson, P., and Mathias, A., *Mol. Phys.*, **13**, 361 (1967).
222. Hampson, P., and Mathias, A., *J. Chem. Soc. B*, 673 (1968).
223. Hannig, E., *Arch. Pharm.*, **296**, 441 (1963).
224. Hannig, E., *Pharmazie*, **19**, 201 (1964).

225. Hannig, E., and Ziebandt, H., *Pharm. Zentralhalle*, **104**, 301 (1965).
226. Hannig, E., and Ziebandt, H., *Pharmazie*, **22**, 626 (1967).
227. Hannig, E., and Ziebandt, H., *Pharmazie*, **23**, 552 (1968).
228. Hannig, E., *Pharmazie*, **23**, 688 (1968).
229. Harvey, A. B., and Wilson, M. K., *Inorg. Nucl. Chem. Letters*, **1**, 101 (1965).
230. Harvey, A. B., and Wilson, M. K., *J. Chem. Phys.*, **45**, 678 (1966).
231. Hashimoto, T., Sugita, M., Kitano, H., and Fukui, K., *Nippon Kagaku Zasshi*, **88**, 991 (1967).
232. Hashimoto, T., Kitano, H., and Fukui, K., *Nippon Kagaku Zasshi*, **89**, 83 (1968).
233. Hauptmann, H., and Walter, W. F., *J. Amer. Chem. Soc.*, **77**, 4929 (1955).
234. Hansch, C., and Geiger, C. F., *J. Org. Chem.*, **24**, 1025 (1959).
235. Hein, F., and Hecker, H., *Z. Naturforsch.*, **11b**, 677 (1956).
236. Hellwinkel, D., and Fahrbach, G., *Justus Liebigs Ann. Chem.*, **715**, 68 (1968).
237. Hende, J. H., and Klingsberg, E., *J. Amer. Chem. Soc.*, **88**, 5045 (1966).
238. Hiscock, S. M., Swann, D. A., and Turnbull, J. H., *Chem. Commun.*, 1310 (1970).
239. Hölzle, G., and Jenny, W., *Helv. Chim. Acta*, **41**, 356 (1958).
240. Hölzle, G., and Jenny, W., *Helv. Chim. Acta*, **41**, 593 (1958).
241. Hölzle, G., and Jenny, W., *Helv. Chim. Acta*, **41**, 331 (1958).
242. Hope, H., *Acta Crystallogr.*, **20**, 610 (1966).
243. Huguet, J. L., *Advan. Chem. Ser.*, **76**, 345 (1968).
244. Huls, R., and Renson, M., *Bull. Soc. Chim. Belges*, **65**, 684 (1965).
245. Ivanova, V. M., Nemleva, S. A., Seina, Z. N., Kaminskaya, E. G., Gitis, S. C., and Kaminskii, A. Ya., *Zh. Org. Khim.*, **3**, 146 (1967).
246. Jakubke, H. D., *Z. Chem.*, **3**, 65 (1963).
247. Jakubke, H. D., *Chem. Ber.*, **97**, 2816 (1964).
248. Jakubke, H. D., *Justus Liebigs Ann. Chem.*, **682**, 244 (1965).
249. Jakubke, H. D., Fischer, J., Jost, K., and Prudinger, J., *Collect. Czech. Chem. Commun.*, **33**, 3910 (1968).
250. Janicki, J., Scupin, J., and Zagalak, B., *Rocz. Chem.*, **36**, 353 (1962).
251. Jauregui-Adell, J., *Advan. Protein Chem.*, **21**, 387 (1966).
252. Jenny, W., *Helv. Chim. Acta*, **36**, 1278 (1953).
253. Jenny, W., *Helv. Chim. Acta*, **41**, 317 (1958).
254. Jensen, K. A., and Frederiksen, E., *Z. Anorg. Allgem. Chem.*, **230**, 31 (1936).
255. Jensen, K. A., Carlsen, J. B., Holm, A., and Nielsen, P. H., *Acta Chem. Scand.*, **17**, 550 (1963).
256. Jensen, K. A., Felbert, G., Pedersen, C. T., and Svanholm, U., *Acta Chem. Scand.*, **20**, 278 (1966).
257. Jensen, K. A., and Nielsen, P. H., *Acta Chem. Scand.*, **20**, 597 (1966).
258. Jensen, K. A., Buchardt, O., and Lohse, C., *Acta Chem. Scand.*, **21**, 2797 (1967).
259. Jensen, K. A., and Henriksen, U., *Acta Chem. Scand.*, **21**, 1991 (1967).
260. Jensen, K. A., and Sandstrom, J., *Acta Chem. Scand.*, **23**, 1911 (1969).
261. Jerchel, D., Fischer, H., and Thomas, K., *Chem. Ber.*, **89**, 2921 (1956).
262. Jones, D. N., Mundy, D., and Whitehouse, R. D., *Chem. Commun.*, 86 (1970).
263. Kaabak, L. V., Tomilov, A. P., and Varshavskii, S. L., *Zh. Vses. Khim. Obshchestva im. D. I. Mendeleeva*, **9**, 700 (1964).
264. Karle, I. L., and Estlin, J. A., *Z. Kristallogr.*, **129**, 147 (1969).

265. Kataev, E. G., and Petrov, V. N., *Zh. Obshch. Khim.*, **32**, 3699 (1962).
266. Kataev, E. G., Kataeva, L. M., and Chmutova, G. A., *Zh. Org. Khim.*, **2**, 2244 (1966).
267. Kataev, E. G., and Gabdrakhmanov, F. G., *Zh. Obshch. Khim.*, **37**, 772 (1967).
268. Kataev, E. G., Chmutova, G. A., and Yarkova, E. G., *Zh. Org. Khim.*, **3**, 2188 (1967).
269. Kataev, E. G., and Chmutova, G. A., *Zh. Org. Khim.*, **3**, 2192 (1967).
270. Kataev, E. G., Mannafov, T. G., and Kostina, G. I., *Zh. Obshch. Khim.*, **37**, 2059.
271. Kataev, E. G., Konovalova, L. K., and Yarkova, E. G., *Zh. Org. Khim.*, **5**, 621 (1969).
272. Kataev, E. G., Chmutova, G. A., Yarkova, E. G., and Plotnikova, T. G., *Zh. Org. Khim.*, **5**, 1514 (1969).
273. Kataev, E. G., Chmutova, G. A., Musina, A. A., and Yarkova, E. G., *Dokl. Akad. Nauk SSSR*, **187**, 1308 (1969).
274. Kataev, E. G., and Mannafov, T. G., *Zh. Org. Khim.*, **6**, 1959 (1970).
275. Kataeva, L. M., and Kataev, E. G., *Zh. Obshch. Khim.*, **32**, 2710 (1962).
276. Kataeva, L. M., Kataev, E. G., and Mannafov, T. G., *Zh. Strukt. Khim.*, **7**, 226 (1966).
277. Kataeva, L. M., Kataev, E. G. and Idiyatullina, D. Ya., *Zh. Strukt. Khim.*, **7**, 380 (1966).
278. Kataeva, L. M., Kataev, E. G., Titova, Z. S., and Aleksandrova, N. A., *Zh. Strukt. Khim.*, **7**, 715 (1966).
279. Kataeva, L. M., Kataev, E. G., and Mannafov, T. G., *Zh. Strukt. Khim.*, **10**, 830 (1969).
280. Kataeva, L. M., Podkovyrina, N. S., Mannafov, T. G., and Kataev, E. G., *Zh. Strukt. Khim.*, **10**, 1124 (1969).
281. Kemppainen, A. E., and Compere, E. L., Jr., *J. Chem. Eng. Data*, **11**, 588 (1966).
282. Khazemova, L. A., and Al'bitskaya, V. M., *Zh. Org. Khim.*, **5**, 1926 (1969).
283. Khazemova, L. A., and Al'bitskaya, V. M., *Zh. Org. Khim.*, **6**, 935 (1970).
284. Khilya, V. P., and Lezenko, G. A., *Zh. Org. Khim.*, **6**, 2048 (1970).
285. Kirsanova, N. A., and Derkach, G. I., *Ukr. Khim. Zh.*, **36**, 372 (1970).
286. Kiss, A. I., and Muth, B. R., *Acta Chim. Acad. Sci. Hung.*, **24**, 231 (1966).
287. Klayman, D. L., *J. Org. Chem.*, **30**, 2454 (1965).
288. Klayman, D. L., and Lown, J. W., *J. Org. Chem.*, **31**, 3396 (1966).
289. Klayman, D. L., and McIntyre, P. T., *J. Org. Chem.*, **33**, 884 (1968).
290. Klofutar, C., Krasovec, F., and Kusar, M., *Croat. Chem. Acta*, **40**, 23 (1968).
291. Klosterman, H. J., and Painter, E. P., *J. Amer. Chem. Soc.*, **69**, 2009 (1947).
292. Kocourek, J., Klenha, J., and Jiracek, V., *Chem. Ind. (London)*, 1397 (1963).
293. Korcek, S., Holotik, S., Lesko, J., and Vesely, V., *Chem. Zveste*, **23**, 281 (1969).
294. Korcek, S., Vesely, V., and Baxa, J., *Collect. Czech. Chem. Commun.*, **35**, 504 (1969).
295. Kostiner, E. S., Reddy, M. L. N., Urch, D. S., and Massay, A. G., *J. Organometal. Chem.*, **15**, 383 (1968).
296. Kozlov, V. V., *Zh. Obshch. Khim.*, **26**, 1755 (1956).
297. Kozlov, V. V., *Zh. Obshch. Khim.*, **27**, 3011 (1957).
298. Kozlov, V. V., and Suvorova, S. E., *Zh. Obshch Khim.*, **31**, 3034 (1961).
299. Kozlov, V. V., and Pronyakova, V. M., *Zh. Org. Khim.*, **1**, 493 (1965).

300. Kozlov, V. V., and Pronyakova, V. M., *Zh. Vses. Khim. Obshchest.*, **12**, 471 (1967).
300a. Krackov, M. H., Lee, C. M., and Mautner, H. G., *J. Amer. Chem. Soc.*, **87**, 892 (1965).
301. Krishnamurthy, S. S., and Soundararajan, S., *J. Organometal. Chem.*, **15**, 367 (1968).
302. Kuchen, W., and Knop, B., *Angew. Chem.*, **76**, 496 (1964).
303. Kuliev, A. M., Zul'fugarova, A. G., and Levshina, A. M., *Tr. Azerb. Gos. Univ.*, *Ser. Khim.*, 71 (1959).
304. Kurz, J. L., and Harris, J. C., *J. Org. Chem.*, **35**, 3086 (1970).
305. Lalezari, I., and Shargi, N., *Spectrochim. Acta*, **23**, 1948 (1967).
306. Landa, S., Weisser, O., and Mostecky, J., *Collect. Czech. Chem. Commun.*, **24**, 2179 (1959).
307. Langen, J. O. M. van, and Plas, T. V. D., *Research Correspondence, Suppl. to Research* (*London*), **7**, S12 (1954).
308. Lane, E. S., and Williams, C., *J. Chem. Soc.*, 1468 (1955).
309. Lapkin, I. I., Pavlova, N. N., and Pavlov, G. S., *Zh. Org. Khim.*, **6**, 71 (1970).
310. Lardon, M., *J. Amer. Chem. Soc.* **92**, 5063 (1970).
311. Lawson, D. D., and Kharasch, N., *J. Org. Chem.*, **24**, 857 (1959).
312. Lautenschlaeger, F. K., *J. Org. Chem.*, **34**, 4002 (1969).
313. LeFevre, R. J. W., and Saxby, J. D., *J. Chem. Soc.*, *B*, 1064 (1966).
314. Lipp, M., Dallacker, F., and Meier, I., *Monatsh. Chem.*, **90**, 41 (1959).
315. Lloyd, D., and Singer, M. I. C., *Chem. Commun.*, 390 (1967).
316. Lozinskii, M. O., and Pel'kis, P. S., *Zh. Org. Khim.* **1**, 1415 (1965).
317. Lugovkin, B. P., *Zh. Obshch. Khim.*, **29**, 3762 (1959).
318. Lutskii, A. E., Obukhova, E. M., Yagupol'skii, L. M., and Voloshchuk, V. G., *Zh. Strukt. Khim.*, **10**, 349 (1969).
319. Lutskii, A. E., Obukhova, E. M., Granzhan, V. A., Volchenok, S. A., Kanevskaya, Z. M., Yagupol'skii, L. M., and Voloshchuk, V. G., *Teor. Eksp. Khim.*, **5**, 614 (1969).
320. McCullough, J. D., and Hamburger, G., *J. Amer. Chem. Soc.*, **63**, 803 (1941).
321. McCullough, J. D., and Marsh, R. E., *Acta Crystallogr.*, **3**, 41 (1950).
322. McCullough, J. D., *Acta Crystallogr.*, **6**, 746 (1953).
323. McCullough, J. D., Chao, G. Y., and Zuccaro, D. E., *Acta Crystallogr.* **12**, 815 (1959).
324. McCullough, J. D., and Mulvey, D., *J. Phys. Chem.*, **64**, 264 (1960).
325. McCullough, J. D., and Zimmermann, I. C., *J. Chem. Phys.*, **64**, 1084 (1960).
326. McDonald, W. S., and Pettit, L. D., *J. Chem. Soc. A*, 2044 (1970).
327. McFarlane, W., *Mol. Phys.*, **12**, 243 (1967).
328. McFarlane, W., *Chem. Commun.*, 963 (1967).
329. McFarlane, W., *J. Chem. Soc. A.*, 670 (1969).
330. McFarlane, W., and Nash, J. A., *Chem. Commun.*, 524 (1969).
331. Mammi, M., Carazzolo, G., Valle, G., and Delpra, A., *Z. Kristallogr.*, **127**, 401 (1968).
332. Magelli, A., and Passerini, R., *Boll. Sci. Fac. Chim. Ind. Bologna*, **14**, 52 (1950).
333. Mangini, A., Trombetti, A., and Zauli, C., *J. Chem. Soc.*, *B*, 153 (1967).
334. Margolis, D. S., and Pittman, R. W., *J. Chem. Soc.*, 799 (1957).
335. Martin, D., and Brause, W. M., *Chem. Ber.* **102**, 2508 (1969).
336. Marziano, N., and Passerini, R., *Gazz. Chim. Ital.*, **94**, 1137 (1964).
337. Mautner, H. G., *J Amer. Chem. Soc.*, **78**, 5292 (1956).

338. Mautner, H. G., and Gunther, W. H. H., *J. Amer. Chem. Soc.*, **83**, 3342 (1961).
339. Mautner, H. G., and Jaffe, J. J., *Biochem. Pharmacol.*, **5**, 343 (1961).
340. Mautner, H. G., Chu, S. H., and Lee, C. M., *J. Org. Chem.*, **27**, 3671 (1962).
341. Mautner, H. G., Chu, S. H., and Gunther, W. H. H., *J. Amer. Chem. Soc.*, **85**, 3458 (1963).
342. Mautner, H. G., Chu, S. H., Jaffe, J. J., and Sartorelly, A. C., *J. Med. Chem.*, **6**, 36 (1963).
343. Mautner, H. G., *U.S. Dept. Com., Office Tech. Serv.*, **AD 418,409** (1963).
344. Merijanian, A., Zingaro, R. A., Sagan, L. S., and Irgolic, K. J., *Spectrochim. Acta*, **25A**, 1160 (1969).
345. Meyer, R., Scheithauer, S., and Kunz, D., *Chem. Ber.*, **99**, 1393 (1966).
346. Meystre, C., Frey, H., Voser, W., and Wettstein, A., *Helv. Chim. Acta*, **39**, 734 (1956).
347. Michalski, J., and Wieczorkowski, J., *Rocz. Chem.*, **28**, 233 (1954).
348. Michalski, J., and Wieczorkowski, J., *J. Chem Soc.*, 885 (1960).
349. Michalski, J., and Tulimowski, Z., *Bull. Acad. Polon. Sci., Ser. Sci. Chim.*, **14**, 217 (1966).
350. Mila, J. P., and Labarre, J. F., *Compt. Rend. Acad. Sci. Paris, Ser. C.*, **263**, 1481 (1966).
351. Mila, J. P., and Laurent, J. P., *Bull. Soc. Chim. Fr.*, 2735 (1967).
352. Mingaleva, K. S., Bogolyubov, G. M., Shlyk, Y. N., and Petrov, A. A., *Zh. Obshch. Khim.*, **39**, 2679 (1969).
353. Mitra, R. B., Rabindran, K., and Tilak, B. D., *Current Sci. (India)*, **23**, 263 (1954).
354. Mock, W. L., and McCausland, J. H., *Tetrahedron Lett.*, 391 (1968).
355. Modena, G., *Proc. Intern. Meeting Mol. Spectry.*, 4th, Bologna **2**, 483 (1959).
356. Morgan, G. T., and Drew, H. D. K., *J. Chem. Soc.*, 1456 (1920).
357. Morgan, G. T., Drew, H. D. K., and Barker, T. V., *J. Chem. Soc.*, 2432 (1922).
358. Morgan, G. T., and Drew, H. D. K., *J. Chem. Soc.*, 922 (1922).
359. Morimoto, S., *Nippon Kagaku Zasshi*, **75**, 557 (1954).
360. Mortillaro, L., Credali, L., Mammi, M., and Valle, G., *J. Chem. Soc.*, 807 (1965).
361. Mortillaro, L., Credali, L., Russo, M., and DeChecchi, C., *J. Polymer Sci., Pt B*, **3**, 581 (1965).
362. Mueller, E., Stegman, H. B., and Scheffler, K., *Justus Liebigs Ann. Chem.*, **657**, 5 (1962).
363. Mushkalo, L. K., and Sheiko, D. I., *Zh. Obshch. Khim.*, **33**, 157 (1963).
364. Mushkalo, L. K., and Sheiko, D. I., *Ukr. Khim. Zh.*, **30**, 384 (1964).
365. Muth, O. H., ed., *Selenium in Biomedicine*, Avi Publishing Co., Westport, Conn., 1966.
366. Myasoedov, B. F., Pal'shin, E. S., and Molochinikova, N. P., *Zh. Anal. Khim.*, **23**, 66 (1968).
367. Naito, T., Ina, S., Inoue, S., Hanai, H., and Katagiri, S., *Nagoya Shiritsu Daigaku Yakugakubu Kiyo*, **5**, 43 (1957).
368. Nakamura, N., and Sekido, E., *Talanta*, **17**, 515 (1970).
369. Nakazaki, M., *J. Chem. Soc. Japan, Pure Chem. Sect.*, **75**, 338 (1954).
370. Nefedov, A. V., *Zh. Obshch. Khim.*, **38**, 2191 (1968).
371. Nesmeyanov, A. N., and Makarova, L. G., *Uchenye Zapiski Moskov. Gosudarst. Univ. im. M. V. Lomonosova*, **132**, *Org. Khim.*, **7**, 109 (1950).
372. Nesmeyanov, A. N., Vinogradova, V. V., and Makarova, L. G., *Izvest. Akad. Nauk SSSR., Otdel. Khim. Nauk*, 1710 (1960).

373. Niwa, E., Aoki, H., Tanaka, H., Munakata, K., and Namiki, M., *Chem. Ber.*, **99**, 3215 (1966).
374. Oki, M., and Iwamura, H., *Tetrahedron Lett.*, 2917 (1966).
375. Olsson, D. H., *Tetrahedron Lett.*, 2035 (1966).
376. Paetzold, R., Schumann, H. D., and Simon, A., *Z. Anorg. Allgem. Chem.*, **305**, 78 (1960).
377. Paetzold, R., Schumann, H. D., and Simon, A., *Z. Anorg. Allgem. Chem.*, **305**, 88 (1960)
378. Paetzold, R., Schumann, H. D., and Simon, A., *Z. Anorg. Allgem. Chem.*, **305**, 98 (1960).
379. Paetzold, R., and Lienig, D., *Z. Chem.*, **4**, 186 (1964).
380. Paetzold, R., and Lienig, D., *Z. Anorg. Allgem. Chem.*, **335**, 289 (1965).
381. Paetzold, R., and Roensch, E., *Z. Anorg. Allgem. Chem.*, **338**, 195. (1965).
382. Paetzold, R., and Wolfram, E., *Z. Anorg. Allgem. Chem.*, **352**, 167 (1967).
383. Paetzold, R., Borek, S., and Wolfram, W., *Z. Anorg. Allgem. Chem.*, **353**, 53 (1967).
384. Paetzold, R., Lindner, U., Bochmann, G., and Reich, P., *Z. Anorg. Allgem. Chem.*, **352**, 295 (1967).
385. Paetzold, R., and Lindner, U., *Z. Anorg. Allgem. Chem.*, **350**, 295 (1967).
386. Paetzold, R., *Spectrochim. Acta. A*, **24**, 717 (1968).
387. Paetzold, R., and Bochmann, G., *Z. Anorg. Allgem. Chem.*, **360**, 293 (1968).
388. Paetzold, R., and Dieter, K., *Z. Chem.*, **10**, 269 (1970).
389. Paetzold, R., and Bochmann, G., *Spectrochim. Acta. Part A*, **26**, 391 (1970).
390. Painter, E. P., *J. Amer. Chem. Soc.*, **69**, 229 (1947).
391. Painter, E. P., *J. Amer. Chem. Soc.*, **69**, 232 (1947).
392. Pallotti, M., and Tundo, A., *Boll. Sci. Fac. Chim. Ind. Bolgona* **17**, 71 (1959).
393. Palmer, H. T., and Palmer, R. A., *Acta Crystallogr. B*, **25**, 1090 (1969).
394. Pan, F., Natori, Y., and Tarver, H., *Biochim. Biophys. Acta*, **93**, 521 (1964).
395. Pande, C. S., Rudick, J., and Walter, R., *J. Org. Chem.*, **35**, 1440 (1970).
396. Parker, A. J., and Brody, D., *J. Chem. Soc.*, 4061 (1963).
397. Pavlova, L. V., and Rachinskii, F. Yu., *Zh. Obshch. Khim.*, **35**, 493 (1965).
398. Pedersen, C. T., *Acta Chem. Scand.*, **17**, 1459 (1963).
399. Petragnani, N., and de Moura Campos, M., *Chem. Ber.*, **94**, 1759 (1961).
400. Petragnani, N., and de Moura Campos, M., *Chem. Ind.*, (*London*). 1461 (1964).
401. Petragnani, N., and de Moura Campos, M., *Chem. Ind.* (*London*), 1076 (1965).
402. Petragnani, N., and de Moura Campos, M., *Tetrahedron*, **21**, 13 (1965).
403. Petragnani, N., and Schill, G., *Chem. Ber.*, **103**, 2271 (1970).
404. Petrov, A. A., Radchenko, S. I., Mingaleva, K. S., Savich, I. G., and Lebedev, V. B., *Zh. Obshch. Khim.*, **34**, 1899 (1964).
405. Pfisterer, G., and Dreeskamp, H., *Ber. Bunsenges. Phys. Chem.*, **73**, 654 (1969).
406. Pichat, L., Herbert, M., and Thiers, M., *Tetrahedron*, **12**, 1 (1961).
407. Plieninger, H., *Chem. Ber.*, **83**, 265 (1950).
408. Poggi, A. R., and Pirisi, R., *Rend. Seminar. Fac. Sci. Univ. Cagliari*, **30**, 134 (1960).
409. Pollak, I. E., and Grillot, G. F., *J. Org. Chem.*, **31**, 3514 (1966).
410. Pollak, I. E., and Grillot, G. F., *J. Org. Chem.*, **32**, 3101 (1967).
411. Pourcelot, G., *Compt. Rend.*, **260**, 2847 (1965).
412. Pourcelot, G., and Cadiot, P., *Bull. Soc. Chim. Fr.*, 3016 (1966).
413. Price, M., and Bremer, B., *J. Polymer Sci., Part B*, **5**, 843 (1967).
414. Price, M., and Bremer, B., *J. Polymer Sci., Part B.*, **5**, 847 (1967).

415. Prokop, P., Lorentz, B., and Hoyer, E., *Z. Chem.*, **9**, 156 (1969).
416. Pudovik, A. N., and Kashevarova, E. I., *Dokl. Akad. Nauk SSSR*, **158**, 157 (1964).
417. Pudovik, A. N., Khusainova, N. G., and Galeeva, R. G., *Zh. Obshch. Khim.*, **36**, 69 (1966).
418. Radchenko, S. I., Chistokletov, V. N., and Petrov, A. A., *Zh. Obshch. Khim.*, **35**, 1735 (1965).
419. Radchenko, S. I., and Petrov, A. A., *Zh. Org. Khim.*, **1**, 2115 (1965).
420. Radchenko, S. I., Chistokletov, V. N., and Petrov, A. A., *Zh. Org. Khim.*, **1**, 51 (1965).
421. Rebane, E., *Ark. Kemi*, **20**, 205 (1962).
422. Rebane, E., *Ark. Kemi*, **25**, 363 (1966).
423. Rebane, E., *Ark. Kemi*, **26**, 345 (1967).
424. Rebane, E., *Acta Chem. Scand.*, **21**, 652 (1967).
425. Rebane, E., *Acta Chem. Scand.*, **21**, 657 (1967).
426. Rebane, E., *Acta Chem. Scand.*, **23**, 1817 (1969).
427. Rebane, E., *Acta Chem. Scand.*, **24**, 717 (1970).
428. Reddy, M. L., Wiles, M. R., and Massey, A. G., *Nature.*, **217**, 740 (1968).
429. Renson, M., and Draguet, C., *Bull. Soc. Chim. Belges*, **71**, 260 (1962).
430. Renson, M., *Bull. Soc. Chim. Belges.*, **73**, 483 (1964).
431. Renson, M., and Collienne, R., *Bull. Soc. Chim. Belges*, **73**, 491 (1964).
432. Renson, M., and Piette, J. L., *Bull. Soc. Chim. Belges*, **73**, 507 (1964).
433. Renson, M., and Pirson, P., *Bull. Soc. Chim. Belges*, **75**, 456 (1966).
434. Rheinboldt, H., and de Campos, H. V., *J. Amer. Chem. Soc.*, **72**, 2784 (1950).
435. Rheinboldt, H., in " Methoden der organischem Chemie," Vol. IX, *Schwefel-Selen- und Tellurverbindungen*, 4th Ed., E. Müller, O. Bayer, H. Meerwein, and K. Ziegler, ed., Georg Thieme (Stuttgart), 1955, p. 1337.
436. Rheinboldt, H., and Giesbrecht, E., *Chem. Ber.*, **88**, 666 (1955).
437. Rheinboldt, H., and Giesbrecht, E., *Chem. Ber.*, **88**, 1037 (1955).
438. Rheinboldt, H., and Giesbrecht, E., *Chem. Ber.*, **88**, 1947 (1955).
439. Rheinboldt, H., and Giesbrecht, E., *Chem. Ber.*, **88**, 1 (1955).
440. Rheinboldt, H., and Perrier, M., *Bull. Soc. Chim. Fr.*, 445 (1955).
441. Rheinboldt, H., and Giesbrecht, E., *Chem. Ber.*, **89**, 631 (1956).
442. Rheinboldt, H., and Perrier, M., *Bull. Soc. Chim. Fr.*, 251 (1956).
443. Riley, R. F., Flats, J., and Bengels, D., *J. Org. Chem.*, **27**, 2651 (1962).
444. Riley, R. F., Flats, J., and McIntyre, P., *J. Org. Chem.*, **28**, 1138 (1963).
445. Rosenfeld, I., and Beath, O. A., *Selenium Geobotany, Biochemistry, Toxicity and Nutrition*, Academic Press Inc., New York, 1964.
446. Roy, J., Gordon, W., Schwartz, I. L., and Walter, R., *J. Org. Chem.*, **35**, 510 (1970).
447. Russo, M., Mortillaro, L., Credali, L., and DeChecchi, C., *J. Polymer Sci., Part B*, **4**, 167 (1965).
448. Russo, M., Mortillaro, L., Credali, L., and DeChecchi, C., *J. Polymer Sci., Part A-1*, **4**, 248 (1966).
449. Russo, M., and Credali, L., *J. Macromol. Sci., Part A*, **1**, 387 (1967).
450. Ruwet, A., and Renson, M., *Bull. Soc. Chim. Belges*, **75**, 157 (1966).
451. Ruwet, A., and Renson, M., *Bull. Soc. Chim. Belges*, **77**, 465 (1968).
452. Ruwet, A., and Renson, M., *Bull. Soc. Chim. Belges*, **78**, 449 (1969).
453. Ruwet, A., Meessen, J., and Renson, M., *Bull. Soc. Chim. Belges.*, **78**, 459 (1969).

454. Ruwet, A., and Renson, M., *Bull. Soc. Chim. Belges*, **78**, 571 (1969).
455. Ruwet, A., and Renson, M., *Bull. Soc. Chim. Belges*, **79**, 61 (1970).
456. Ruwet, A., Janne, D., and Renson, M., *Bull. Soc. Chim. Belges*, **79**, 81 (1970).
457. Saffioti, W., *Anais Acad. Brasil. Cienc.*, **35**, 317 (1963).
458. Schaefer, J. P., and Sonnenberg, F., *J. Org. Chem.*, **28**, 1128 (1963).
459. Schaefer, J. P., Horvath, B., and Klein, H. P., *J. Org. Chem.*, **33**, 2647 (1968).
460. Scheithauer, M., and Mayer, R., *Z. Chem.*, **6**, 375 (1966).
461. Schiavon, G., *Ric. Sci. Rend. Ser. A*, **32**, 69 (1962).
462. Schmidt, M., and Wilhelm, I., *Chem. Ber.*, **97**, 872 (1964).
463. Schmidt, M., Blaettner, K., Kochendorfer, P., and Ruf, H., *Z. Naturforsch.*, **21b**, 622 (1966).
464. Schmitz-DuMont, O., and Ross, B., *Angew. Chem., Int. Ed. Engl.*, **6**, 1071 (1967).
465. Schneider, W., and Wrede, F., *Chem. Ber.*, **50**, 793 (1917).
466. Schneider, W., and Beuther, A., *Chem. Ber.*, **52**, 2135 (1919).
467. Schoellkopf, U., and Kueppers, H., *Tetrahedron Lett.*, 105 (1963).
468. Schuijl, P. J. W., Brandsma, L., and Arens, J. F., *Rec. Trav. Chim.*, **85**, 889 (1966).
469. Schwenker, G., and Rossway, H., *Tetrahedron Lett.*, 4237 (1967).
470. Sebe, E., and Matsumoto, S., *Science Repts. Tohoku Univ. First. Ser.*, **38**, 297 (1954).
471. Seebach, D., and Peleties, N., *Angew. Chem., Int. Ed. Engl.*, **8**, 450 (1969).
472. Sekido, E., Fernando, Q., and Freiser, H., *Anal. Chem.*, **36**, 1768 (1964).
473. Shargi, N., and Lalezari, I., *Spectrochim. Acta*, **20**, 237 (1964).
474. Shefter, E., and Kennard, O., *Science*, **153**, 1389 (1966).
475. Shefter, E., and Mautner, H. G., *Proc. Nat. Acad. Sci. U.S.*, **63**, 1253 (1969).
476. Shepherd, L., and Huber, R. E., *Can. J. Biochem.*, **47**, 877 (1969).
477. Shlyk, Y. N., Bogolyubov, G. M., and Petrov, A. A., *Zh. Obshch. Khim.*, **38**, 1199 (1968).
478. Sieberth, H., *Z. Anorg. Allgem. Chem.*, **271**, 65 (1952).
479. Silverwood, H. A., and Orchin, M., *J. Org. Chem.*, **27**, 3401 (1962).
480. Simchen, G., *Angew. Chem., Int. Ed. Engl.*, **7**, 464 (1968).
481. Simonnin, M. P., *Bull. Soc. Chim. Fr.*, 1774 (1966).
482. Simonnin, M. P., and Pourcelot, G., *C. R. Acad. Sci., Ser. C*, **262**, 1279 (1966).
483. Simonnin, M. P., *Compt. Rend.*, **275**, 1075 (1963).
484. Sindelar, K., Metysova, J., Svatek, E., and Protiva, M., *Collect. Czech. Chem. Commun.*, **34**, 2122. (1969).
485. Sindelar, K., Svatek, E., Metysova, J., Metys, J., and Protiva, M., *Collect. Czech. Chem. Commun.*, **34**, 3792 (1969).
486. Sjöberg, B., and Herdevall, S., *Acta Chem. Scand.*, **12**, 1347 (1958).
487. Skuping, J., *Rocz. Chem.*, **36**, 631 (1962).
488. Spencer, R. P., and Brody, K. R., *J. Nucl. Med.*, **9**, 621 (1968).
489. Spinelli, D., and Dell'Erba, C., *Ann. Chim. (Rome)*, **51**, 45 (1961).
490. Spinelli, D., and Monaco, G., *Boll. Sci. Fac. Chim. Ind. Bologna*, **20**, 56 (1962).
491. Stefanac, Z., and Tomaskovic, M., *Bull. Sci. Conseil Acad. RSF Jougoslavie*, **10**, 317 (1965).
492. Supniewski, J., Misztal, S., and Krupinska, J., *Bull. Acad. Polon. Sci. Cl. II*, **2**, 153 (1954).
493. Supniewski, J., Misztal, S., and Krupinska, J., *Archiv. Immunol. Terap. Doswiadczalnej*, **3**, 531 (1955).

494. Supniewski, J., Rogoz, F., and Krupinska, J. *Bull. Acad. Polon. Sci. Biol.*, **9**, 231 (1961).
495. Suzuki, Y., *Yakugaku Zasshi*, **81**, 1151 (1961).
496. Svek, H. J., and Junk, G. A., *J. Amer. Chem. Soc.*, **89**, 790 (1967).
497. Takaoka, K., and Toyama, Y., *Nippon Kagaku Zasshi*, **89**, 405 (1968).
498. Takaoka, K., and Toyama, Y., *Nippon Kagaku Zasshi*, **89**, 618 (1968).
499. Tanaka, H., Sakurai, H., and Yokoyama, A., *Chem. Pharm. Bull.*, **18**, 1015 (1970).
500. Tarantelli, T., and Pecile, C., *Ann. Chim. (Rome)*, **52**, 75 (1962).
501. Tarantelli, T., and Leones, D., *Ann. Chim. (Rome)*, **53**, 1113 (1963).
502. Thayer, J. S., *Inorg. Chem.*, **7**, 2599 (1968).
503. Theodoropoulos, D., Schwartz, I. L., and Walter, R., *Tetrahedron Letters*, 2411 (1967).
504. Theodoropoulos, D., Schwartz, I. L., and Walter, R., *Biochemistry*, **6**, 3927 (1967).
505. Tideswell, N. W., and McCullough, J., *J. Amer. Chem. Soc.*, **79**, 1031 (1957).
506. Townsend, L. B., and Milne, G. H., *J. Heterocycl. Chem.*, **7**, 753 (1970).
507. Tyerman, W. J. R., O'Callaghan, W. B., Kebarle, P., Strauss, O. P., and Gunning, H. E., *J. Amer. Chem. Soc.*, **88**, 4277 (1966).
508. Uzarewicz, A., and Zacharewicz, W., *Rocz. Chem.*, **35**, 541 (1961).
509. Uzarewicz, A., and Zacharewicz, W., *Rocz. Chem.*, **35**, 887 (1961).
510. Vafai, M., and Renson, M., *Bull. Soc. Chim. Belges*, **75**, 145 (1966).
511. Van Wazer, J. R., and Grant, D., *Am. Chem. Soc., Div. Polymer Chem., Preprints*, **5**, 621 (1964).
512. Vicentini, G., Giesbrecht, E., and Pitombo, L. R. M., *Chem. Ber.*, **92**, 40 (1959).
513. Volger, H. C., and Arens, J. F., *Rec. Trav. Chim.*, **77**, 1170 (1958).
514. Voloshchuk, V. G., Yagupol'skii, L. M., Syrova, G. P., and Bystrov, V. B. *Zh. Obshch. Khim.*, **37**, 118 (1967).
515. Voronkov, M. G., Pereferkovich, A. N., and Pestunovich, V. A., *Zh. Strukt. Khim.*, **9**, 627 (1968).
516. Vyazankin, N. S., Bochkarev, M. N., and Sanina, L. P., *Zh. Obshch. Khim.*, **36**, 166 (1966).
517. Vyazankin, N. S., Bochkarev, M. N., and Sanina, L. P., *Zh. Obshch. Khim.*, **36**, 1961 (1966).
518. Vyazankin, N. S., Bochkarev, M. N., and Sanina, L. P., *Zh. Obshch. Khim.* **37**, 1037 (1967).
519. Vyazankin, N. S., Bochkarev, M. N., and Maiorova, L. P., *Zh. Obshch. Khim.*, **39**, 468 (1969).
520. Wagner, G., and Lehmann, G., *Pharm. Zentralhalle*, **100**, 160 (1961).
521. Wagner, G., and Nuhn, P., *Z. Chem.*, **3**, 62 (1963).
522. Wagner, G., and Nuhn, P., *Z. Chem.*, **3**, 64 (1963).
523. Wagner, G., and Nuhn, P., *Arch. Pharm.*, **296**, 374 (1963).
524. Wagner, G., and Nuhn, P., *Arch. Pharm.*, **297**, 81 (1964).
525. Wagner, G., and Nuhn, P., *Arch. Pharm.*, **297**, 461 (1964).
526. Wagner, G., and Nuhn, P., *Arch. Pharm.*, **298**, 686 (1965).
527. Wagner, G., and Nuhn, P., *Arch. Pharm.*, **298**, 692 (1965).
528. Wagner, G., and Nuhn, P., *Pharm. Zentralhalle*, **104**, 328 (1965).
529. Wagner, G., and Valz, G., *Pharmazie*, **22**, 548 (1967).
530. Wallmark, I., Krackov, M. H., Chu, S. H., and Mautner, H. G., *J. Amer. Chem. Soc.*, **92**, 4447 (1970).

531. Walter, R., and du Vigneaud, V., *J. Amer. Chem. Soc.*, **87**, 4192 (1965).
532. Walter, R., and du Vigneaud, V., *J. Amer. Chem. Soc.*, **88**, 1331 (1966).
533. Walter, R., and Schwartz, I. L., *Advan. Exp. Med. Biol.* **2**, 101 (1967).
534. Walter, R., and Chan, W. Y., *J. Amer. Chem. Soc.*, **89**, 3892 (1967).
535. Walter, R., Schlesinger, H. D., and Schwartz, I. L., *Anal. Biochem.*, **27**, 231 (1969).
536. Walter, R., *American Peptide Symp., 1st.*, 467 (1968).
537. Welcman, N., and Regev, H., *J. Chem. Soc.*, 7511 (1965).
538. Welcman, N., and Rot, I., *J. Chem. Soc.*, 7515 (1965).
539. Welcman, N., and Wulf, M., *Israel J. Chem.*, **6**, 37 (1968).
540. West, R., Powell, D. L., Lee, M. K. T., and Whatley, L. S., *J. Amer. Chem. Soc.*, **86**, 3227 (1964).
541. Westland, A. D., and Westland, L., *Can. J. Chem.*, **43**, 426 (1965).
542. Williams, F. D., and Dunbar, F. X., *Chem. Commun.*, 459 (1968).
543. Williams, L. R., and Ravve, A., *J. Amer. Chem. Soc.*, **70**, 1244 (1948).
544. Wille, F., Scherl, A., Kaupp, G., and Capeller, L., *Angew. Chem.*, **74**, 753 (1962).
545. Windle, J. J., Wiersema, A. K., and Tappel, A. L., *J. Chem. Phys.* **41**, 1966 (1964).
546. Windle, J. J., Wiersema, A. K., and Tappel, A. L., *Nature*, **203**, 404 (1964).
547. Wilshire, J. F. K., *Australian J. Chem.*, **20**, 359 (1967).
548. Wiseman, G. E., and Gould, E. S., *J. Amer. Chem. Soc.*, **76**, 1706 (1954).
549. Wiseman, G. E., and Gould, E. S., *J. Amer. Chem. Soc.*, **77**, 1061 (1955).
550. Wittig, G., and Fritz, H., *Justus Liebigs Ann. Chem.*, **577**, 39 (1952).
551. Wrede, F., *Biochem. Z.*, **83**, 96 (1917).
552. Wrede, F., *Chem. Ber.*, **52**, 1765 (1919).
553. Wrede, F., *Z. Physiol. Chem.*, **112**, 1 (1920).
554. Wrede, F., *Z. Physiol. Chem.*, **115**, 284 (1921).
555. Wrede, F., and Zimmermann, W., *Z. Physiol. Chem.*, **148**, 65 (1925).
556. Wynne, K. J., and George, J. W., *Inorg. Chem.*, **4**, 256 (1965).
557. Wynne, K. J., and George, J. W., *J. Amer. Chem. Soc.*, **87**, 4750 (1965).
558. Wynne, K. J., and Puckett, J., *Chem. Commun.*, 1532 (1968).
559. Wynne, K. J., and George, J. W., *J. Amer. Chem. Soc.*, **91**, 1649 (1969).
560. Wynne, K. J., *Inorg. Chem.*, **9**, 299 (1970).
561. Yagupol'skii, L. M., and Voloshchuk, V. G., *Zh. Obshch. Khim.*, **36**, 165 (1966).
562. Yagupol'skii, L. M., and Voloshchuk, V. G., *Zh. Obshch. Khim.*, **37**, 1543 (1967).
563. Yagupol'skii, L. M., and Voloshchuk, V. G., *Zh. Obshch. Khim.*, **38**, 2509 (1968).
564. Yagupol'skii, L. M., Syrova, G. P., Voloshchuk, V. G., and Bystrov, V. F., *Zh. Obshch. Khim.*, **38**, 2591 (1968).
565. Yarovenko, N. N., Gazieva, G. B., Shemanina, V. N., and Fedorova, N. A., *Zh. Obshch. Khim.*, **29**, 940 (1959).
566. Yarovenko, N. N., Shemanina, V. N., and Gazieva, G. B., *Zh. Obshch. Khim.*, **29**, 942 (1959).
567. Yarovenko, N. N., Raksha, M. A., and Shemanina, V. N., *Zh. Obshch. Khim.*, **30**, 4069 (1960).
568. Yarovenko, N. N., and Raksha, M. A., *Zh. Obshch. Khim.*, **30**, 4064 (1960).
569. Zaitseva, G. I., and Al'bitskaya, V. M., *Zh. Org. Khim.*, **4**, 2057 (1968).

570. Zakharkin, L. I., and Gavrilenko, V. V., *Izv. Akad. Nauk SSSR.*, *Otdel Khim. Nauk*, 1391 (1960).
571. Zdansky, G., *Ark. Kemi*, **17**, 273 (1961).
572. Zdansky, G., *Ark. Kemi*, **17**, 519 (1961).
573. Zdansky, G., *Ark. Kemi*, **19**, 559 (1962).
574. Zdansky, G., *Ark. Kemi*, **21**, 211 (1963).
575. Zdanksy, G., *Ark. Kemi*, **26**, 213 (1966).
576. Zdanksy, G., *Ark. Kemi*, **27**, 447 (1967).
577. Zdansky, G., *Ark. Kemi*, **29**, 47 (1968).
578. Zdansky, G., *Ark. Kemi*, **29**, 437 (1968).
579. Zdansky, G., *Ark. Kemi*, **29**, 443 (1968).
580. Ziegler, E., and Nölken, E., *Monatsh. Chem.*, **89**, 737 (1958).
581. Ziegler, M., and Schroeder, H., *Mikrochim. Acta*, 782 (1967).
582. Zundel G., Metzger, H., and Scheuing, I., *Z. Naturforsch.*, **22b**, 127 (1967).

9

Biochemistry of Selenium[*]

HOWARD E. GANTHER

*Department of Nutritional Sciences,
University of Wisconsin,
Madison, Wisconsin 53706*

INTRODUCTION

Selenium, as an essential trace element and as a naturally occurring toxicological agent, is notable for its very high biological activity. Selenium is toxic at dietary levels comparable to the nutritional requirement for some trace elements, and sustains life in animals at dietary levels in the parts per billion range. Geological and meteorological factors determine the relative abundance of selenium in the soil in different regions of the world, giving rise to practical problems of selenium toxicity or selenium deficiency that are of considerable economic importance.

Inorganic forms of selenium present in the soil or administered to animals are readily converted to organoselenium compounds. This process may enhance or reduce the biological activity of the

* Supported in part by National Institute of Arthritis and Metabolic Diseases grant no. AM 10692 and by a grant from the Selenium-Tellurium Development Association, Incorporated.

selenium.   Certain plants of the genus *Astragalus* that are found only in highly seleniferous regions have evolved mechanisms for concentrating selenium from the soil in amounts far in excess of other plants growing side by side.   The tolerance of these plants to high concentrations of selenium is correlated with a metabolically distinct pattern of selenium assimilation into low molecular weight organoselenium compounds.   Dietary supplements of inorganic selenium compounds effectively prevent a variety of nutritional disorders, but there is general agreement that the active biological form of selenium is an organic form.   Thus a parallel can be drawn between selenium and certain other trace elements, such as iodine, where an inorganic nutrient is converted to a biologically active organic form.

The isolation and characterization of the selenium compounds produced in living organisms is made very difficult by the factors of low concentration and high lability.   The most successful efforts have been in the identification of detoxification products of excessive selenium intake.   Very little is known about the structure or mechanism of action of organoselenium compounds concerned with the essential role of selenium.

Organoselenium compounds and their biosynthesis in animals will be given the major emphasis in this chapter.   Coverage of the literature has emphasized those papers that have appeared since 1957, especially those dealing with animals.   An excellent review of the comparative biology and metabolism of selenium in plants has recently appeared.[224]   The proceedings of the symposium on selenium held at Oregon State University in 1966[160] is a valuable source of information covering many of the active areas of biological research on selenium.   A monograph by Rosenfeld and Beath[193] provides a comprehensive coverage of earlier research on selenium, including many interesting aspects of its history.

TOTAL CONTENT AND DISTRIBUTION OF SELENIUM

In considering the biochemistry of selenium there can be no parameter more fundamental than the amount of selenium present in living cells. Consideration of this point leads to certain restrictions concerning the possible biological roles of selenium.   This point will be developed more fully in a later section.

Selenium occurs in minute amounts in nearly all substances.   The amount of selenium in soil depends on the quantity of selenium in the parent material, as modified by processes during or after soil formation that add or remove selenium.   Lakin and Davidson[118] have reviewed this topic and summarized the tendency of selenium to accompany sulfur.   The form of selenium in soil, as well as the total amount, determines the amount available to plants. For example, selenite is normally taken up readily by plants, but can become associated with iron in acid soils to form a poorly available complex. Plants also vary in their ability to accumulate selenium from soil. On soils high in available selenium some species of *Astragalus* accumulate much larger amounts of selenium than other plant species grown side by side, but on soils low in available selenium these differences largely disappear.[53] The fungus *Amanita muscaria* may be one of the few plants that accumulate high amounts of selenium from acid soils low in selenium.[251]

Plants and Microorganisms

The selenium content of the common forages and hay is very low. In a large portion of central United States, levels of 0.1 to 0.5 ppm are found in the majority of forage samples, corresponding to a protective but nontoxic level for livestock.[113] Extremely low levels of selenium are found in forages which cause white muscle disease in sheep. Analysis of such plants by a number of workers shows a range of 0.01–0.05 ppm, with an average of approximately 0.02 ppm.[34, 55, 42, 31] Although such levels produce selenium deficiencies in animals consuming forages, it is very significant that the growth of the plants themselves is not impaired and crop yields are considered to be normal in these low selenium regions.[224]

Even lower levels of selenium are found in microorganisms grown on media low in selenium, such as *Torula* yeast grown on spent sulfite liquor. Kelleher and Johnson[113] found 0.004 to 0.007 ppm Se in dried *Torula* yeast, U.S.P. grade. Ewan *et al.*[55] reported a level of 0.009 ppm in *Torula* yeast. Converting these data to a fresh weight basis, and assuming no loss on drying, a concentration of approximately 0.002 ppm would exist in the living cell. Brewers yeast contained 0.1–1.0 ppm Se,[113] reflecting the higher selenium content of the medium used to grow the yeast.

Animal Tissues

Prior to 1960, there was very little information available on selenium levels in tissues of animals other than for those exposed to toxic levels. With the development of sensitive methods for determining selenium in biological materials, data comparing levels of selenium in deficient and normal animals have become available. In the following discussion of these data, selenium levels will be expressed as $\mu gm/gm$ (i.e., ppm) in the fresh tissue.

Early analyses for selenium content of organs from lambs with white muscle disease were reported by Hartley and Grant[88] and Lindberg and Siren.[127] Hartley[89] has summarized numerous analyses performed on tissues of sheep in areas of New Zealand where a variety of conditions responsive to selenium occur in young sheep. From these data it appears that most tissues from lambs having white muscle disease or an "unthriftiness" responsive to selenium contain on the order of 0.01–0.05 ppm selenium.   Kidney cortex is an exception, retaining 0.15 ppm of selenium or more in all cases.   Hidiroglou et al.[91, 93] have reported similar findings in dystrophic lambs and calves, and their data for healthy and dystrophic lambs are shown in Table 9-1.   Heart and skeletal muscle of dystrophic animals contained about 0.02 ppm of selenium.   Most tissues of healthy animals contained 0.1–0.5 ppm, kidney being somewhat higher.

TABLE 9-1   SELENIUM LEVELS IN TISSUES OF
DYSTROPHIC AND HEALTHY
LAMBS [93]

Tissue	ppm Se (fresh basis[a])	
	Healthy	Dystrophic
Kidney cortex	1.10	0.39
Liver	0.33	0.042
Heart	0.19	0.022
Spleen	0.29	0.075
Lung	0.22	0.050
Adrenal	0.17	0.17
Pancreas	0.35	0.08
Quadriceps femoralis	0.11	0.020
Neck muscle	0.10	0.018
Blood	0.33	0.032

[a] Calculated by dividing ppm Se in dried tissues by 4.

Burk et al.[22] have measured tissue selenium levels during the development of liver necrosis in weanling rats fed Torula yeast diets low in selenium and vitamin E for 4 weeks, and also in animals fed the same diet supplemented with 0.25 and 0.5 ppm selenium as selenite.   A level of 0.5 ppm approximates the content of selenium in Purina Chow,[67, 55] a commercial laboratory animal ration widely used in this country.   In liver, the initial level of approximately 0.75 ppm was maintained in animals fed diets supplemented with 0.25 or 0.5 ppm of selenium, but liver selenium declined quite rapidly to a level of 0.07 ppm in animals fed the low selenium diet that caused liver necrosis to develop.   Vitamin E prevented the liver necrosis without preventing the marked decline in liver selenium.   Blood levels were maintained at a fairly constant level of about 0.3 ppm in animals fed the diets supplemented with selenium, but declined to 0.05 ppm in those fed the low-selenium diet.   Kidney, highest in initial selenium content at 0.96 ppm, accumulated selenium during the feeding of 0.25 or 0.5 ppm selenium, in contrast to the fairly constant level maintained in blood and liver.   Kidney levels declined during the low-selenium regimen, but not as much as blood and liver, and still contained 0.27 ppm after 4 weeks.

There is thus a consistent finding that selenium levels in most tissues of deficient animals contain 0.01–0.05 ppm selenium. Normal levels fall in the range of 0.1–0.5 ppm, and up to 1 ppm in kidney cortex. Kidney is exceptional in its ability to accumulate selenium on a normal diet and retain it on a deficient diet. The selenium in kidney is known to have a high biological activity.[206] These characteristics probably explain why the kidney, with an apparent exception in mice,[47] is unaffected in dietary selenium deficiencies that cause liver or muscular lesions.

Although these levels of selenium are usually associated with tissue lesions, low concentrations of selenium that produce liver necrosis in rats not receiving vitamin E may not be detrimental in the presence of vitamin E.

A number of workers have reported data on selenium concentrations in blood and other tissues in humans, as related to geographic factors or to the occurrence of various disorders. Allaway et al.,[2] in 210 samples of whole blood from male donors to blood banks at different locations in the United States, found the mean concentration to be 20.6 μg Se/100ml, with a range of 10–34 μg/100ml. There were regional variations that reflected to some extent the known distribution of selenium in soils and plants. Dickson and Tomlinson[48] had earlier reported extensive data from Canada which show a very similar distribution of values in a population of 254 normal individuals. Burk et al.[21] have found a lower than normal level of selenium in blood of children affected with kwashiorkor.

FORMS OF SELENIUM

The forms of selenium that occur in living systems (see Table 9-2) will depend on the form of selenium supplied to the organism, the amount of selenium supplied, and the species. It is convenient, for the purpose of discussion, to divide the forms of selenium into compounds of low molecular weight occurring in free form, and forms of selenium present in proteins.

Low Molecular Weight Compounds

Se-methylselenocysteine. A few species of *Astragalus* accumulate high levels of selenium, ranging up to several thousand ppm. Se-methylselenocysteine ($CH_3-Se-CH_2-CHNH_2COOH$) is the predominant soluble form of selenium in these accumulator plants. It was first isolated as a crystalline solid from *A. bisulcatus* by ion exchange and partition chromatography.[239]

**TABLE 9-2   ORGANOSELENIUM COMPOUNDS FOUND IN
LIVING ORGANISMS**

Name	Formula
Se-methyl seleno-cysteine	$CH_3-Se-CH_2-CHNH_2-COOH$
Se-methyl seleno-methionine	$(CH_3)_2-\overset{+}{Se}-CH_2-CH_2-CHNH_2-COOH$
Selenocys-tathionine	$HOOC-CHNH_2-CH_2-Se-CH_2-CH_2-CHNH_2-COOH$
Dimethyl selenide	$CH_3-Se-CH_3$
Trimethyl selenonium ion	$(CH_3)_3\overset{+}{Se}$
Selenohomo-cystine	$HOOC-CHNH_2-CH_2-CH_2-Se-Se-CH_2-CH_2-CHNH_2-COOH$
Selenocystine	$HOOC-CHNH_2-CH_2-Se-Se-CH_2-CHNH_2-COOH$
Seleno-methionine	$CH_3-Se-CH_2-CH_2-CHNH_2-COOH$

The compound had an R_f identical with that of synthetic Se-methyseleno-cysteine during paper chromatography in two solvent systems. Its selenium content and infrared spectrum were closely similar but not identical with authentic Se-methylselenocysteine; this was attributed to incomplete separation from the sulfur analogue, S-methylcysteine, which was shown to be present in the same plants. Shrift and Virupaksha[220, 221] demonstrated the biosynthesis of Se-methylselenocysteine from selenite in six species of the genus *Astragalus* as well as *Oonopsis condensata* and *Stanleya pinnata*.

Selenocystathionine. Another compound that was first discovered in the highly seleniferous *Astragalus* species is selenocystathionine, $HOOC-CHNH_2-CH_2-Se-CH_2-CH_2-CHNH_2-COOH$. Horn and Jones[100] obtained a crystalline amino acid containing selenium and sulfur from hot water extracts of the dried plant. Elements analysis yielded the empirical formula $C_{21}H_{42}N_6Se_2SO_{12}$, and this was interpreted to be a 2:1 mixture at $C_7H_{14}N_2O_4Se$ and $C_7H_{14}N_2O_4S$.

The biosynthesis of selenocystathionine from radioactive selenite was demonstrated in the selenium accumulator *S. pinnata* by Virupaksha and Shrift.[244] The compound was obtained by chromatography of a trichloroacetic acid extract on Dowex 50, and comprised about 10% of the radioactive selenium in the extract. After further purification by high voltage electrophoresis, the compound was identified by paper chromatography of the unmodified form and of derivatives obtained by oxidation with hydrogen peroxide or reduction with Raney nickel.

Two groups of investigators have identified selenocystathionine as the cytotoxic compound in seeds of coco de mono (*Lecythis ollaria*), or monkey nut. This tree is widely distributed in Central and South America. Nuts from trees in certain parts of Venezuela are known to produce toxic effects in man, including loss of hair. Aqueous extracts of these nuts were fractionated by ion exchange chromatography to obtain a crystalline material. Infrared and nuclear magnetic resonance spectra, along with elemental analysis and paper chromatography, established the structure.[112]   Tissue culture studies showed that virtually all of the cytotoxic activity present in water extracts of coco de mono could be accounted for by the content of selenocystathionine.[4]

Peterson and Butler[186] have also identified selenocystathionine in a selenium accumulator (*Neptunia amplexicaulis*) of Queensland. The compound was present in seeds collected from seleniferous areas and comprised about 1/3 of the radioactivity in aqueous ethanol extracts of plants grown on ^{75}Se-labeled selenite.

Se-methylselenomethionine.   A radioactive compound of high electrophoretic mobility was first detected by Peterson and Butler[185] in aqueous ethanol extracts of clover and ryegrass roots after growth on radioactive selenite. The compound was identified by paper chromatography as Se-methylselenomethionine $(CH_3)_2\overset{+}{Se}-CH_2-CH_2CHNH_2-COOH$. Virupaksha and Shrift[245] found this to be the predominant soluble organoselenium compound synthesized from selenite by several species of *Astragalus* which lacked the ability to accumulate selenium. The compound was eluted from Dowex 50 with $4N$ HCl and identified by paper chromatography and electrophoresis. The occurrence of Se-methylselenomethionine as the predominant derivative in nonaccumulators apparently distinguishes these species from accumulator species, where Se-methylselenocysteine is the predominant compound.

Dimethyl Selenide.   It has long been known that animals metabolize selenium and tellurium to odorous compounds that can be detected in the breath. Levine[126] lists a number of workers who observed this phenomenon, the earliest report being that of Japha in 1842. Hofmeister[97] obtained evidence that the characteristic garlic odor in dogs injected with sodium tellurite was dimethyl telluride. From the similarity of the garlic odor in the breath of selenium-treated animals to that of dimethyl selenide, Hofmeister reasoned that selenium was also volatilized as the dimethyl derivative.

The biological methylation of selenium, arsenic, and tellurium by fungi has been investigated extensively by Challenger and co-workers, and a review of this very interesting subject is available.[36]

A variety of evidence has been obtained for the formation of dimethyl selenide in animals. McConnell and Portman[144] showed that rats injected with radioactive selenate exhaled a radioactive compound that could be isolated with carrier dimethyl selenide and recrystallized to a constant specific activity as the mecuric chloride derivative $(CH_3)_2SeHgCl_2$. Klug and Froom,[116] using gas chromatography, reported the presence of dimethyl selenide as a respiratory product in rats given selenite, and in plants and seeds of *Astragalus*. Ganther,[66] studying the enzymic synthesis of dimethyl selenide in liver extracts, found that a double labeled product containing ^{14}C and ^{75}Se in a ratio very close to 2:1 was formed after incubation with ^{75}Se-labeled selenite and ^{14}C-labeled S-adenosylmethionine. Although the double labeled study is consistent with CH_3SeCH_3, it would not exclude CH_3SeSCH_3; additional studies with ^{35}S-labeled glutathione, however, for which the system has an absolute and specific requirement, showed that only 1 atom of sulfur was volatilized for every 300 atoms of selenium (Ganther, unpublished data).

Dimethyl Diselenide. Evans et al.[54] obtained four volatile compounds from *Astragalus racemosus* by trapping the volatile compounds on active carbon followed by extraction from the carbon in various solvents. One of these compounds was identified by gas chromatography as dimethyl diselenide, $CH_3SeSeCH_3$. No evidence was obtained for the production of dimethyl selenide. Two unidentified water soluble compounds appeared to have a behavior on ion exchange chromatography identical with two volatile compounds found previously in alfalfa.[6]

Trimethyl Selenonium. One of the urinary selenium compounds excreted by rats was recently identified by Byard[27] as the trimethyl selenonium ion, $(CH_3)_3Se^+$, and confirmed by Palmer et al.[171] The compound was isolated in both cases by adsorption on Dowex 50, followed by elution and precipitation as the Reineckate derivative. Identification was based on comparison with synthetic trimethyl selenonium ion using paper and ion exchange chromatography, nuclear magnetic resonance, infrared, and mass spectrometry, and cocrystallization. Since the compound could be detected in rat tissues[27] and was not formed when normal rat urine was mixed with selenite,[171] it appears to be an actual metabolite and not an artifact of isolation.

Elemental Selenium. It has long been known that microorganisms reduce selenium salts to elemental selenium.[126, 57, 266, 56, 264] The recent study by McCready et al.[153] has conclusively identified amorphous selenium as the

end product of selenite reduction by *Salmonella heidelberg*. The red precipitate from a 48 h culture grown in 0.1 % Na_2SeO_3 was isolated and analyzed
by x-ray fluorescence. Heating the red precipitate for 1 hour at 170°C gave
a crystalline black powder identified as metallic selenium by x-ray diffraction.
Deposition of elemental selenium has been reported to occur in the roots of
plants given toxic levels of selenium, and a high proportion of selenium,
presumably elemental selenium, was extracted from the roots of a selenium
accumulator with bromine water.[185]   It has been suggested that reduction of
dietary selenium to elemental selenium occurs within the rumen of sheep.[38]

Selenohomocystine.   Excised leaves of *Astragalus crotalariae* metabolize
[75]Se-selenomethionine to selenohomocystine, $HOOC-CHNH_2-CH_2-$
$CH_2-Se-Se-CH_2-CH_2-CHNH_2-COOH$.[246]   Column chromatography on Dowex 50 followed by further fractionation by paper chromatography and electrophoresis gave a compound which was shown to yield selenomethionine after reduction and methylation.

Selenomethionine and Selenocystine.   The selenium analogues of methionine
and cystine are largely incorporated into proteins but are found in the free
form to a certain extent.   The identification of selenoamino acids in protein
hydrolyzates is discussed in a later section.

   Free selenoamino acids have been detected in plants.   Onions (*Allium
cepa*) contain a rich variety of sulfur compounds and were therefore used as
starting material in an attempt to isolate selenoamino acids after the injection
of radioactive selenite into the bulb.[227]   After extraction with 70 % ethanol
and isolation of the amino acid fraction by ion exchange chromatography,
two-dimensional paper chromatograms were prepared.   Two of the five
radioactive spots observed moved with selenocystine and selenomethionine.
The authors also concluded that one major radioactive spot was propenyl
β-amino-β-carboxyethyl selenoxide, the selenium analogue of the lachry-
motory precursor.

   Peterson and Butler[185] fractionated the 80 % ethanol extracts of various
plant species given radioactive selenite.   Ratemeter tracings of paper chromatograms showed a substantial proportion of selenocystine, selenomethionine, and their oxidation products in the soluble fraction of red clover and
white clover, with considerably smaller amounts in extracts of rye grass.

Other Compounds.   Evidence for a seleniferous wax containing selenoesters

$$R-\overset{\displaystyle O}{\overset{\|}{C}}-Se-R$$

of the type R—C—Se—R was obtained by McColloch *et al.*[142]   They found
that 5–10 % of the total radioactivity in leaves of *Stanleya bipinnata* grown in

the presence of ^{75}Se-selenite was extracted by a 30 sec rinse in Skelly F, suggesting the presence of seleniferous wax on the external leaf surfaces. The infrared spectrum of the leaf extract contained the absorptions expected for waxes. When the crystalline urea clathrate was prepared by adding urea-saturated methanol to the leaf extract dissolved in benzene, nearly all of the radioactivity was retained in the clathrate, and could be recovered quantitatively in Skelly F after decomposition of the clathrate with hot water. Chromatography on silica gel provided further evidence for the presence of selenium in the hydrocarbon-vegetable wax fraction.

Incompletely characterized selenopeptides have been detected in plants. Virupaksha and Shrift[245] found appreciable amounts of an acidic ninhydrin-positive selenium compound in *Astragalus vasei* (a nonaccumulator) given selenite or selenate, but not in species of *Astragalus* that accumulated selenium. The compound had a mobility slightly lower than that of oxidized glutathione, and yielded glutamic acid after acid hydrolysis. After milder hydrolysis, a spot in the area of serine was detected as well. The material appears to be a glutamyl peptide containing a selenoamino acid which decomposes during hydrolysis. Selenoglutathione was not found in either accumulators or nonaccumulators, even though the plants synthesized glutathione from sulfate. Peterson and Butler[185] detected an unknown compound in ethanol extracts of clover with a chromatographic behavior suggestive of selenoglutathionine.

Proteins

It has long been known that a high proportion of the selenium in the common plants and in animal tissues is closely associated with protein. This selenium remains with the protein during precipitation with trichloroacetic acid or dialysis under gentle conditions but much of it is sufficiently labile that it is removed by more drastic procedures, especially alkaline conditions. On the basis of present knowledge, two generalizations can be made regarding the chemical form of selenium in proteins: First, the covalently-bound selenium is linked to either carbon or sulfur; secondly, the presence of selenium in a protein is always related to the presence of sulfur—the selenium atom is either incorporated in place of the sulfur atom in a sulfur amino acid, or is attached to the sulfur atoms of cysteine residues. Unless careful attention is paid to the chemistry involved, it is possible to draw false conclusions about the form of selenium in a protein. Traces of radioactive selenite adsorbed to proteins may be mistaken for covalently bound selenium, and selenium present in a covalent form may be mistaken for adsorbed selenite through the use of aggressive treatments that decompose the covalent linkage. The failure to recognize these possibilities has led to considerable controversy, especially regarding the synthesis of selenoamino acids from selenite in animal species that do not synthesize sulfur amino acids from inorganic sulfate.

This discussion will consider first the presence of selenium in selenotrisulfide linkages (S—Se—S) of proteins and secondly, the occurrence of selenoamino acids in proteins, including those of ruminant and nonruminant species.

Selenotrisulfide Linkages in Proteins.   In a review article published in 1941, Painter proposed the following reaction involving thiols and selenious acid:

$$4RSH + H_2SeO_3 \longrightarrow RSSeSR + RSSR + 3H_2O$$

Evidence reviewed in a following section has confirmed that this reaction occurs with a great variety of thiols. The possibility of incorporating selenium into proteins through such a reaction was first investigated by Holker and Speakman.[98] They generated cysteine residues in wool by reducing about 65 % of the disulfide bridges with thioglycolic acid. When the washed wool was treated with selenious acid, they observed loss of sulfhydryl groups, increase in dry weight, and recovery of resistance to stretching in the treated wool. The fact that no reduction of selenious acid to elemental selenium occurred during the treatment indicated that the restrengthening resulted from the formation of cross-linkages containing selenium, rather than from an oxidation of cysteine residues to the disulfides. The presence of large amounts of selenium in the treated protein was shown by the liberation of elemental selenium during acid hydrolysis in an amount nearly equal to the increase in dry weight of the treated wool (2.11 %). This increase in dry weight corresponds to an uptake of 26.7 mmoles of Se per 100 gm of wool, but loss of sulfhydryl amounted to only 46 mmoles/100 gm, or 1.72 SH per Se. The authors concluded that each atom of selenium combined with 2 cysteine residues, and that the data for weight gain favored S—Se—S linkages over

$$\overset{\displaystyle O}{\overset{\displaystyle \|}{S—Se—S}}.$$ It is known, however, that the stoichiometry of selenious acid reduction to —S—Se—S— requires not 2, but 4 thiols to be oxidized per selenium, and so the investigators' conclusion is based on data that do not account for the necessary reducing equivalents. Possibly the reducing equivalents were obtained by oxidation of some other groups in the protein in addition to the cysteine residues. On the other hand, the discrepancy may reflect uncertainties in the experimental data, such as incomplete titration of buried SH groups in the protein that might still have been accessible to selenious acid.

In order to evaluate more precisely the role of sulfhydryl groups in the nonenzymic incorporation of selenium into proteins, Ganther and Corcoran[70] chose the enzyme ribonuclease as a model. The native enzyme has four disulfide bonds that can be reduced to yield four pairs of sulfhydryl groups, providing a well-defined system for the study of selenium incorporation into the same polypeptide chain, with or without sulfhydryl groups. The effect on enzyme activity of converting a disulfide linkage to a selenotrisulfide linkage could also be determined.

Reduced RNase treated with selenious acid at 4°C showed a rapid loss of sulfhydryl groups, spectral changes equivalent to selenotrisulfide formation, and the uptake of radioactive selenium. The stoichiometry of the reaction determined spectrophotometrically corresponded to a 4:1 SH/Se combining ratio, and the expected uptake of 2 moles of radioactive selenium per mole of protein (8 SH groups) was observed. To obtain these results, it was necessary to carry out the reaction at a pH between 2 and 3. At higher pH, elemental selenium was liberated during the course of the reaction; this was easily visible at the levels of selenium being used. The elemental selenium remained in the dialysis bag with the protein during dialysis but could be removed by gel filtration chromatography. The protein treated with selenious acid at pH 7 and purified in this manner contained only 0.44 moles of selenium per mole of protein, approximately 20% of the uptake at pH 2. It should be emphasized that the selenium is apparently liberated from an unstable intermediate in the reaction at pH 7 since very little decomposition of the final product occurs under the same conditions. Native ribonuclease did not undergo spectral changes in the presence of selenious acid and could be completely separated from radioactive selenious acid by gel filtration, providing further evidence that selenious acid was reacting only with sulfhydryl groups of the protein.

It was shown by gel filtration and sedimentation velocity studies that the selenium derivative of RNase was a homogeneous, monomeric species, more unfolded than the native enzyme. The activity of the modified protein was only a few percent of that of native enzyme. Is is possible that the loss of activity resulted from the conditions of low pH used to incorporate a maximum amount of selenium, because the conformation of the protein at low pH leads to incorrect pairing of the disulfide sulfurs during reoxidation.[81] Considerations of the probable geometry for selenotrisulfide and disulfide linkages, however, make it likely that the selenotrisulfide modification in itself would alter protein conformation so much that enzyme activity would be lost.

Additional evidence for selenotrisulfide linkages in protein was obtained by Jenkins.[107] The effect of various treatments on the release of ^{75}Se from the serum proteins of chicks given ^{75}Se-selenite suggested that a major portion of the selenium was covalently bound between the sulfurs of half-cystine residues. In serum collected 4 h after the oral administration of selenite, 76–

89% of the selenium was removed by a 2 h dialysis at pH 9, 25°C, in the presence of 0.1 M cysteine, 0.1 M glutathione, 0.25 M 2-mercaptoethanol, or 0.02 M dithiothreitol.   Sulfitolysis (0.05 M Na_2SO_3 plus 8 M urea) or prolonged dialysis against dilute NaOH (pH 11.5) liberated similar quantities. It was also shown that serum proteins reduced with 2-mercaptoethanol incorporated up to 4 ppm of selenium added as selenite when the reaction was carried out at pH 3, incorporation decreasing with increasing pH.   This is in agreement with the work of Ganther and Corcoran[70] on the formation of selenotrisulfide linkages in ribonuclease, although in the latter work the selenium was incorporated stoichiometrically, corresponding to about 11,500 ppm of selenium.   The relatively low incorporation obtained by Jenkins probably resulted in part from the addition of selenite while mercaptoethanol was still present, so that formation of selenodimercaptoethanol competed with protein selenotrisulfide formation.

The form of selenium in serum proteins collected after 4 days was quite different, since the treatments with thiol were ineffective and the selenium could only be released by prolonged dialysis at pH 11.5.   Control experiments made it likely that significant hydrolysis of the protein did not occur under these conditions.   From the known sensitivity of selenotrisulfides to thiols or alkaline conditions, it is clear that much of the selenite incorporated into chick serum proteins during the first few hours involved the nonenzymic formation of selenotrisulfide linkages.   It would appear that a high proportion of the selenium incorporated after longer periods was in a more stable form, such as a selenoamino acid.   Jenkins, however, preferred to interpret such data as indicating an increasing proportion with time of selenotrisulfide linkages that were stabilized or less accessible to the sulfhydryl reagents, but which were accessible to the hydroxyl ion.   There are very few protein disulfides, however, that are inaccessible to 0.1 M mercaptoethanol in 8 M urea, a treatment that removed only 8% of this more resistant form, and so this seems a rather unlikely explanation.

One would suspect that at least part of the selenium removed by the alkaline treatment resulted from the decomposition of selenoamino acids in the protein, through processes such as β-elimination of hydrogen selenide from selenocysteine residues that might be formed by an initial cleavage of the diselenide linkage.   The disulfide bonds of oxidized glutathione are completely cleaved by 0.1 N NaOH in 2h at 30°C[19] and diselenides would be expected to undergo cleavage under milder conditions because of the generally greater reactivity and instability of the selenium analogues of sulfur compounds.[101]   It is therefore surprising that Jenkins could see no evidence for decomposition of selenocystine or selenomethionine held at pH 11.5 for 3 days when these were subjected to paper chromatography.   Selenite was detected by paper chromatography and electrophoresis of the material liberated by the alkaline treatment, the selenite presumably arising from oxidation of selenides or elemental selenium formed initially. A substance tentatively identified as colloidal elemental selenium was probably a material formed from elemental selenium under the alkaline conditions employed, because it was dialyzable in dilute NaOH (pH 11.5), had an R_f of 0.75 in n-butanol:pyridine:water (1:1:1), and moved half the distance to the anode during electrophoresis at pH 7. Since no selenoamino acids were found in the material liberated from the alkaline-treated protein, it was concluded that none existed in the protein. It is unfortunate, however, that no attempt was made to find such compounds in the radioactive protein residue from the alkaline treatment or after enzymic hydrolysis of the original protein. It can be concluded that Jenkin's work provides evidence for the presence of selenotrisulfides in proteins, rather than evidence for the absence of seleno-cystine and selenomethionine.

Indirect evidence for selenium cross-linkages in wool from sheep supplemented with selenium was reported by Demiruren and Slen.[43] Fibers from the wool of selenium-treated sheep (containing 15 ppm Se) contracted less in both length and diameter than those from control wool (3.4 ppm Se) when treated in 50% lithium bromide at 110°C for 3 h. It was suggested that the formation of new cross-linkages, possibly containing selenium, produced a more stable α-keratin in the wool of the animals treated with selenium.

Selenoamino Acids in Proteins. The original investigations on the form of selenium in proteins from seleniferous wheat have been summarized by Rosenfeld and Beath.[193] Selenium was found to be an integral part of the protein and could be released by hydrolysis, although the selenium compounds were less stable to hydrolysis than the sulfur amino acids. Fractionation of the hydrolysates showed that selenium was in the fractions containing cystine and methionine, indicating the probable presence of selenocystine and selenomethionine. On the basis of more recent work, the existence of selenomethionine in plants and microorganisms is generally regarded as established with considerable certainty.

Microorganisms. Tuve and Williams[242,243] grew E. Coli on a sulfur deficient medium containing radioactive selenite and subjected the bacterial protein to enzymic hydrolysis. Radioactivity was associated with the neutral amino acids in chromatography on Dowex 50 and Dowex 2 and selenomethionine was identified by paper chromatography. A second radioactive compound had properties similar, but not identical, to those of selenocystine.

Other workers have consistently found chromatographic evidence for radioactive selenomethionine in protein hydrolyzates of microorganisms grown in the presence of radioactive selenite.[13,90,253] Many of these studies also provide evidence for the formation of the cystine analogue, but its presence is in general less certain than that of selenomethionine.   Huber and Criddle[102] obtained extensive replacement of methionine by seleno-methionine in β-galactosidase isolated from a strain of *E. coli* grown on a medium high in selenate and low in sulfate.   This protein had about 80 of its 150 methionine residues replaced by selenomethionine, but there was apparently no replacement of cystine by selenocystine.

Two papers have appeared which concern the incorporation of inorganic selenium into selenoamino acids by rumen microorganisms.   Paulson *et al.*[182] compared the incorporation of selenate and selenite to that of seleno-methionine and sulfate in rumen fluid.   There was a time dependent in-corporation of selenomethionine into protein which could be inhibited by excess methionine.   Enzymic hydrolysis with Pronase released most of the incorporated selenium and the predominant form was the unchanged sele-nomethionine, as shown by paper and ion exchange chromatography. Selenate appeared to undergo rapid reduction to selenite.   Selenite was rapidly and extensively bound to protein by a nonenzymic process.   A substantial portion of this protein-bound selenium was removed by dialysis in the presence of glutathione.   This indicates incorporation of selenium into protein S—Se—S cross-linkages.   The apparent failure of N-ethyl maleimide to prevent the binding of selenite may have been due to an in-complete blocking of the SH groups of the protein by this reagent at the pH of rumen fluid.   Even though 13 % of the added selenite or selenate could be released from GSH-treated proteins after Pronase digestion, no discrete peaks corresponding to selenoamino acids were observed after chromatography. Possibly the amount of selenoamino acids put on the column were so low that nonspecific adsorption occurred in the absence of carrier.   Hidiroglou *et al.*[92] also found that selenite was incorporated firmly into proteins of rumen bacteria.   Selenomethionine was demonstrated autoradiographically after paper chromatography of the acid-hydrolyzed proteins but the percentage of the total protein-bound selenium present in this form was not reported. Selenocystine was not detected.   The latter compound in free form is quite easily destroyed under the usual conditions of protein hydrolysis[101] but yields of about 40 % are obtained when peptides containing a half-seleno-cystine residue are hydrolyzed.[250]

Plants.   Peterson and Butler[185] used paper chromatography and paper electrophoresis to examine the forms of selenium in aqueous ethanol extracts of various plants and in enzymic digests of proteins from these plants grown in solutions containing radioactive selenite.   Extensive incorporation of selenium into selenoamino acids was observed in the protein portion of ryegrass, wheat, red clover and white clover.   The major compound was selenomethionine, plus a considerable amount of unidentified material that may have been a peptide.   A much smaller incorporation of selenium into proteins was observed in a selenium accumulator plant, *Neptunia amplexicaulis*.   In later work on a small aquatic plant, *Spirodela oligorrhiza*, Butler and Peterson[25] found that 90% of the radioactive selenium in enzymically-hydrolyzed protein was present as selenomethionine or its selenoxide.   The identification of this amino acid included recrystallization from ethanol to constant specific activity with carrier selenomethionine as well as paper chromatography and electrophoresis.   The relatively small amount of selenocystine found in protein was in contrast to the distribution of the selenoamino acids in the aqueous ethanol extracts of the same plant, where selenocystine and oxidation products were always in excess of selenomethionine.   Whether this difference was due to the greater instability of selenocysteine or to enzymic discrimination against its incorporation into protein could not be determined.

Further evidence for selenoamino acid formation from selenite in plants was reported by Jenkins and Hidiroglou.[106]   They found that 59% of the selenite sprayed directly onto brome grass was present as selenomethionine and selenocystine after one week, mostly in polypeptide form.   Surprisingly, the amount of selenocystine was 2-1/2 times greater than the amount of selenomethionine.   This stands in contrast to the results of Peterson and Butler[185] who fractionated their proteins by similar procedures but added the selenite to the nutrient solutions on which the plants were grown.   It is possible that the direct application of selenite to the leaf portion of the plant, the portion used for later analysis, caused a greater nonenzymic formation of selenodicysteine (CyS—Se—SCy).   If this compound survived the extraction procedures, it would have been difficult to differentiate from cystine during paper chromatography in most solvent systems, and might therefore account for the relatively large proportion of selenocystine reported by Jenkins and Hidiroglou.

Animals.   With animals it is convenient to distinguish between the case where selenoamino acids are incorporated into proteins following the administration of the selenoamino acids per se, and the case where selenoamino acids are found in proteins of animals given inorganic forms of selenium.

Incorporation of Preformed Selenoamino acids.   There is considerable evidence that selenium administered in the form of selenomethionine is readily incorporated into proteins.   Hansson and Blau[86] measured the rate of appearance of ^{75}Se in cat pancreatic juice after intravenous injection of ^{75}Se-selenomethionine.   The secreted radioactivity reached a peak at 1 to 2 h, and was mostly in the proteins precipitated by trichloroacetic acid.   After acid hydrolysis of the protein in the presence of carrier methionine, 85% of the ^{75}Se was eluted at the position of selenomethionine after ion exchange chromatography. Control experiments demonstrated 95% recovery of selenomethionine carried through the same procedure. No ^{75}Se-labeled amino acids were observed after the addition of ^{75}Se-selenite to normal pancreatic juice followed by acid hydrolysis and chromatography.

Based on the report of Blau and Bender[15] that the human pancreas could be visualized by external scanning after ^{75}Se-selenomethionine administration, a large number of papers have appeared on the use of this compound in diagnosis. The pancreas concentrates free amino acids from the blood and incorporates them into digestive enzymes to be secreted. Under favorable conditions, a fairly high and selective uptake may be obtained although this may not be the case in the rhesus monkey and in the cat.[268, 5] Uptake of ^{75}Se in the adjacent liver may interfere with visualization of the pancreas. Selective visualization of the parathyroid glands adjacent to the thyroid has also been attempted.

A comparison of the incorporation of ^{35}S-methionine and ^{75}Se-selenomethionine into purified egg white ovalbumin has been made.[164] In an egg laid the day after injection of the tracers, all the ^{35}S and ^{75}Se activity remained bound to protein during gel filtration chromatography, whereas complete separation of the amino acids added to the protein *in vitro* was demonstrated by the same procedure. After acid hydrolysis of the carboxymethylated protein, 85% of the ^{35}S put on the column was recovered as methionine and 15% as S-carboxymethyl cysteine. All of the ^{75}Se was present as selenomethionine, however, as shown by recoveries of 97–98% of the total ^{75}Se applied to the column in that peak. Apparently, the absence of selenocystine did not reflect destruction of selenocystine in the protein because the authors reported no loss of ^{75}Se when ovalbumin was hydrolyzed, although oxidation of the protein with performic acid prior to hydrolysis caused extensive destruction of the ^{75}Se labeled amino acids. Presumably the selenomethionine was incorporated into protein at the positions normally occupied by methionine; the peptides and amino acids liberated from ovalbumin by a series of proteolytic enzymes had a relatively constant ^{35}S/^{75}Se ratio, suggesting that differences in distribution of selenomethionine and methionine in various peptides were not very great. Similarly, the double-labeled proteins were digested *in vivo* in a parallel manner.[165]

Electrophoretic analysis of venom from tarantulas given ^{75}Se-selenomethionine showed a generally parallel distribution of protein and ^{75}Se, indicating a uniform incorporation of the amino acid.[122] On the other hand, there is one report of a differential incorporation of selenomethionine and methionine into rat serum protein.[119]

Incorporation of Selenoamino acids Synthesized From Inorganic Selenium.
The occurrence of selenoamino acids in animals given selenite can be further
divided for discussion in relation to whether the animal is a ruminant or a
monogastric species.

Ruminants.   The synthesis of sulfur amino acids from inorganic sulfate by
rumen microorganisms is a well-established process, supported by both
nutritional and biochemical studies.   For this reason, evidence for the analogous synthesis of selenoamino acids from selenite has not been seriously
disputed for ruminants.

Rosenfeld[194] looked for selenoamino acids in the wool of sheep given
large amounts of ^{75}Se-selenite orally for a 2 month period.   In wool collected
at the end of the experiment, 59% of the ^{75}Se was retained after extraction
with water, detergent, and alcohol: ether.   After hydrolysis of the wool in
6 N HCl, only 38% of protein-^{75}Se was recovered in the filtrate after removal
of humin and treatment with charcoal.   Radioactivity was present in the
cystine fraction obtained by chemical and chromatographic procedures.
Additional chromatography on ion exchange resins showed that ^{75}Se was
eluted in the cystine-selenocystine fraction and a smaller amount in methionine-selenomethionine, but large amounts of ^{75}Se were retained by the column
and the percentage of original protein-^{75}Se radioactivity recovered in the
eluted amino acids was low.

The assumption that selenoamino acids in proteins of ruminants result
from the biosynthetic activity of rumen microorganisms seems reasonable
but direct studies with rumen fluid have given conflicting results.   As noted
earlier, Paulson *et al.*[182] could find no selenoamino acids in rumen microorganisms incubated with approximately $10^{-6} M$ selenious acid, although the
same concentration of selenomethionine led to substantial uptake into protein.   On the other hand, Hidiroglou *et al.*[92] detected selenomethionine by
paper chromatography of acid hydrolyzed rumen bacteria incubated *in vitro*
with selenite (amount not given), whether or not the sheep from which the
fluid was obtained had been previously treated with selenite.   Similar results
were reported for *in vivo* experiments.

It is noteworthy that the evidence for synthesis of selenoamino acids from
selenite in ruminants is qualitatively quite analogous to that in monogastric
animals.   If it is suggested, then, that such a biosynthetic process occurs in
ruminants and not in monogastric animals, this may reflect the incorrect
belief that monogastric animals cannot reduce the selenium analogues of
inorganic sulfur.   The reduction of selenite to dimethyl selenide is in fact a
major metabolic pathway in rats for detoxifying selenite, and the reduction of
selenite to trimethyl selenonium ion provides another example.

Nonruminants. Because a high proportion of selenite becomes firmly bound to proteins, it has long been suspected that selenomethionine and selenocystine are present in proteins isolated from animals given selenite. Since reduction of selenite to the selenide level occurs during the formation of dimethyl selenide in animals given selenite, there is no particular reason to doubt that selenoamino acids could be formed in animals through related processes. Considerable evidence has been obtained for the formation of trace amounts of selenoamino acids in monogastric animals, and evidence to the contrary has also been reported, so the matter is controversial. Fortunately, if selenium occurs in the diet of animals mainly as selenoamino acids, rather than selenite, the question is largely academic.

Two points must be kept in mind in evaluating the experimental evidence regarding the presence of selenoamino acids in animal proteins: (1) protein-bound selenium may or may not be covalently bound selenium; and (2) covalently bound selenium may or may not be the selenium analogue of methionine or cystine. In regard to the first point, the convenience of using ^{75}Se as a tracer has too often led to complete disregard of the actual amount of selenium being handled, leading to the possibility of nonspecific adsorption processes. Nanogram amounts of selenite can and do become adsorbed to organic moieties that normally undergo no type of a stoichiometric reaction with selenite. This was demonstrated by Schwarz and Sweeny[208] who observed that selenite (0.0001 μg) added to various disulfide amino acids (20 μg) traveled with the disulfides during paper chromatography. There is no association of selenite with disulfides when the molar amounts of both compounds are of the same order of magnitude.[68, 70]

McConnell and Wabnitz[146] first demonstrated that trace amounts of radioactivity were present in the amino acids of hydrolyzed dog liver proteins following the injection of radioactive selenite. Paper chromatography of the hydrolyzate revealed that most of the radioactivity was in the region of synthetic cystine and selenocystine, with lesser amounts in the region of methionine-selenomethionine and adjacent to the leucines. Other work by McConnell supporting selenocystine formation was the demonstration that radioactive selenium was present in the p-bromophenylmercapturic acid fraction isolated from dog urine after feeding bromobenzene to dogs previously injected with radioactive selenite.[147] Approximately 11% of the total radioactivity in the urine was extracted by chloroform, and 1.6% of the activity in the chloroform fraction was present in the mercapturic acid fraction after three recrystallizations. McConnell and Kreamer[148] also found radioactive selenium in the S-sulfokeratein protein and cystine isolated from the hair of dogs given radioactive selenite. On the basis of these studies, McConnell believes that trace amounts of selenoamino acids are formed from selenite. The amounts involved are too small to permit direct chemical measurement, so that the identification of selenoamino acids is based on the correspondence of radioactive selenium content and chromatographic properties.

For this reason, Cummins and Martin[40] have suggested that radioactive selenium associated with animal proteins is really inorganic selenite adsorbed in a reversible manner. These workers dosed a rabbit with substantial amounts of ^{75}Se-selenite and carrier selenium over a 5 week period. Most of the radioactive selenite was given 24 h before sacrifice of the animal, however. The liver was homogenized in 0.01 N sodium hydroxide, and the soluble proteins were dialyzed for at least 9 days against dilute sodium hydroxide, pH 11. About 90% of the original ^{75}Se in the homogenate was removed by this treatment; it was not determined what percentage of the radioactive selenium in the homogenate was originally bound to protein and how much was free selenite. The remaining protein was enzymically hydrolyzed, releasing 54% of the protein radioactivity. An aliquot of this hydrolyzate containing about 3000 cpm was subjected to ion exchange chromatography. No radioactivity was detected in fractions that would contain selenocystine or selenomethionine, and the only radioactive peak corresponded to selenite. The authors concluded that selenocystine and selenomethionine were not present in the protein, or were present in concentrations too low to be detected. They made the interesting observation that rabbit urine treated with radioactive selenite yielded 8 radioactive fractions upon ion exchange chromatography, indicating that selenite was associating in some manner with urinary compounds. Some of the peaks could be generated by treating selenite with a mixture of sulfur compounds.

Because of the prolonged alkaline treatment that the liver protein was subjected to, it is difficult to evaluate the experiments of Cummins and Martin regarding the form of selenium in protein. One can presume that selenotrisulfide linkages initially present in the protein would certainly have been destroyed by homogenization in 0.01 N sodium hydroxide. Even disulfide linkages would be cleaved by this treatment. It would be expected that elemental selenium or selenide ion formed during the decomposition would be oxidized to selenite in alkaline solutions exposed to the air. The possibility that selenocystine or selenomethinione decomposed under such conditions does not appear to have been eliminated by the control experiments which were reported. This criticism does not apply to the experiments with urine, which illustrate very well the possibilities for adsorption or for covalent interactions of selenite with sulfur compounds.

Thus, it appears that the major effort in earlier studies on the form of selenium in monogastric animals given selenite was aimed at detecting trace amounts of selenoamino acids. The possible formation of selenotrisulfide linkages in proteins was recognized but not given equivalent attention. The work of Cummins and Martin[40] emphasized the possibilities for adsorption of traces of selenite to proteins and amino acids, but did not deal with the possibility of selenotrisulfide formation. It appears that these workers, and also Jenkins[107] were overzealous in their attempts to prove that all protein-bound selenium was in the form of selenite or selenotrisulfide, rather than selenamino acids. It is likely that selenium is present in all of these forms, and other forms such as $R-S-Se-R$.

Selenium in Mucopolysaccharides. Interest in the possible substitution of selenate for sulfate in the chondroitin sulfate of mucopolysaccharides has prompted several investigations on this matter. Campo et al.[31] compared the *in vitro* incorporation of ^{35}S-sulfate and ^{75}Se-selenate in slices of bovine costal cartilage. The protein polysaccharides were labeled with both ^{35}S and ^{75}Se, but further fractionation revealed that the chondroitin sulfate was not labeled with ^{75}Se, in contrast to the extensive labeling of this fraction with [35]S. The distribution of ^{75}Se instead paralleled the distribution of protein. Treatment of the ^{75}Se-protein-polysaccharide with hyaluronidase to liberate uronic acid released only a small proportion of the [75]Se. Desulfation with HCl-methanol released ^{75}Se but this radioactivity was not found in the precipitate after barium treatment, indicating that the ^{75}Se had not been incorporated into chondroitin-sulfate in ester linkage as selenate. The incorporation of ^{75}Se-selenate into protein seemed to be a time dependent process related to cellular activity, since slices that were boiled or incubated at 4°C incorporated very little [75]Se. Acid hydrolysis of the ^{75}Se-protein-polysaccharide followed by ion exchange and paper chromatography revealed a high proportion of ^{75}Se moving with the solvent front and also in the vicinity of several amino acids, including cystine and methionine, but this was not regarded as conclusive evidence for the presence of selenoamino acids.

Campo et al.[32] also carried out autoradiographic studies in cartilage and long bones, following incubation of ^{75}Se-selenate with cartilage slices or injection into young rats, and compared the distribution of ^{75}Se to that of ^{35}S-sulfate and ^{14}C-amino acids. The greatest concentration of ^{75}Se was seen over chondrocytes. Incubation of the labeled tissue sections with hyaluronidase removed very little ^{75}Se whereas ^{35}S was almost completely removed. Decalcification of sections of bone similarly caused only a very slight loss of [75]Se. These studies indicate that ^{75}Se is associated with the protein of these tissues rather than the polysaccharide or mineral components.

Earlier work by Galambos and Green had shown that ^{35}S, ^{75}Se, and carbazole-positive material had a parallel distribution after barium carbonate precipitation and glass paper chromatography of the nondialyzable components of rabbit urine, obtained after the injection of ^{75}Se-selenate and ^{35}S-sulfate. They concluded that both elements were incorporated into mucopolysaccharides in the nondialyzable fraction of urine.

Selenium in Iron-Sulfur Proteins.   Certain redox proteins contain an acid-labile form of sulfur bound to iron.   Tsibris *et al.*[241] have substituted ^{80}Se for sulfur in putidaredoxin, an iron-sulfur protein from *Pseudomonas putida*, in order to make electron spin resonance studies of the active center of the protein.   The labile sulfur and iron were first removed to form apo-putidaredoxin, which was then treated under anaerobic conditions with ammonium selenide containing ^{80}Se plus a trace of radioactive ^{75}Se, along with ferrous iron and 2-mercaptoethanol.   The reconstituted selenoprotein, after dialysis, contained approximately 2 moles each of iron and selenium per mole of protein, equivalent to the content of iron and sulfur in the native enzyme.   The visible absorption spectrum maxima of the Se-protein were shifted about 20 mμ to longer wavelengths.   The biological activity of the selenoprotein was nearly equal to that of the sulfur protein.   Substitution of ^{80}Se produced a marked change in the electron spin resonance spectrum.   The selenium could be almost completely displaced by ^{32}S, with restoration of the original ultraviolet absorption spectrum.   Similar results were obtained in a subsequent study,[168] in which ^{77}Se and ^{80}Se were substituted for sulfur in both putidaredoxin and in the adrenal iron-sulfur protein, in order to compare the hyperfine structure of the ESR spectra obtained with the two isotopes.   The authors noted that the selenoproteins formed a red precipitate (Se^0) if exposed to air in the absence of 2-mercaptoethanol, even at ice temperature.   The selenium in these proteins is presumably bound to iron in a manner analogous to the sulfur, but the exact structure of the iron-sulfur complexes is not yet established.

METABOLISM OF SELENIUM

There are extensive data available concerning the general patterns of selenium absorption, distribution, and routes of excretion in laboratory animals and in large animal species.   This information was obtained largely by the use of radioactive ^{75}Se, first employed by McConnell.[143]   Most of these studies were carried out after the discovery in 1957 of a nutritional role for selenium, and largely concerned the fate of inorganic selenite or selenate.   This subject was covered in considerable detail in a previous review.[65]   The major emphasis in the present review will therefore be directed to the intermediary metabolism of inorganic selenite to reduced forms, and to studies on the absorption and distribution of organoselenium compounds such as selenomethionine.

Intermediary Metabolism of Selenite

Although the general chemical similarity of sulfur and selenium might suggest that selenium metabolism should parallel the pathways of sulfur metabolism, it is apparent that important differences exist. As a general rule, sulfur follows oxidative pathways while selenium tends to undergo reduction. The contrasting fate of organic forms of sulfur and selenium in the animal is apparent in the final urinary metabolites. Westfall and Smith[256] found that only 15% of the urinary selenium excreted by rabbits fed organic selenium (seleniferous wheat) was present as inorganic selenium, while the remainder was associated with the ethereal and neutral sulfur fractions; the partition of urinary selenium was therefore just the reverse of that for urinary sulfur. The distribution of urinary selenium following inorganic selenite administration showed a similar preponderance of organic forms. In the case of selenate, however, 85% of the urinary selenium was still present in an inorganic form.

These metabolic differences involving selenite and selenate may have their basis in certain chemical properties of these ions. The $+4$ oxidation state of selenium, as found in selenite, may be regarded as a chemical and metabolic dividing point. Thus selenite, in contrast to sulfite, is a good oxidizing agent and has a considerable tendency to go to a more reduced state. Going the other way, oxidation of selenite to selenate is quite difficult and requires a strong oxidizing agent such as hydrogen peroxide. Thus if one begins with selenite, metabolism to reduced organic forms predominates over oxidation to selenate. Selenate tends to remain in the $+6$ oxidation state and does not show such a marked tendency toward reductive metabolism, although reduction of selenate does occur, as in the formation of dimethyl selenide.

Respiratory Excretion of Selenium.   When high levels of selenite or selenate are administered to animals, as much as 50% of the total dose may be exhaled in the form of volatile organoselenium compounds. There is a considerable body of evidence, summarized previously, that the volatile product is dimethyl selenide. The methylation of selenium appears to be a detoxification mechanism; the LD_{50} of dimethyl selenide in the rat is 1.6 g of Se/kg,[145] nearly a thousandfold higher than that for sodium selenite.[58]

The relative importance of volatile selenium formation as an excretory pathway is greatly influenced not only by the amount of selenium administered but also by the form of selenium administered, the nature of the diet fed previously, and the genetic constitution of the animal   Neither sodium selenate nor selenomethionine are converted to volatile selenium as readily as sodium selenite.[95,151,231]   Unidentified factors in crude diets fed to rats prior to an injection of labeled selenite cause a threefold increase in the formation of volatile selenium.[62,67]   A survey of ten inbred strains of mice at the Jackson

Laboratory revealed a tenfold variation in the percentage of selenium exhaled, under standarized conditions of diet and selenite dosage (Ganther, unpublished data).

The usefulness of dimethyl selenide production from inorganic selenite as a model system for studying the biosynthesis of organoselenium compounds has been emphasized.[66] These advantages include the fact that dimethyl selenide is the major excretory product under appropriate conditions, plus the ease of trapping a volatile compound compared to the isolation of labile organoselenium compounds from complex biological systems. The study of this process in cell-free systems[66] revealed some interesting aspects in regard to selenium metabolism which are next described.

When the $9000g$ supernatant of mouse liver homogenates was incubated with 5×10^{-5} M radioactive selenite, as much as 70% of the total ^{75}Se was converted to volatile compounds that could be trapped in $8N$ nitric acid. Activity was proportional to protein concentration and heated extracts were inactive, indicating an enzymic process. When S-adenosyl-L-(methyl ^{14}C) methionine was used in a double-labeled study with ^{75}Se, the volatile compound had a $^{14}C/^{75}Se$ ratio close to 2, as expected for dimethyl selenide. Incubation under nitrogen increased the production of dimethyl selenide about tenfold compared to air, apparently by preventing the oxidation of labile reduced forms of selenium. Reduced nicotinamide-adenine-dinucleotide phosphate, coenzyme A, adenosine-5 triphosphate, and magnesium were also required.

Since the conversion of selenite to dimethyl selenide involves a 6 electron reduction in the oxidation state of selenium and the addition of two methyl groups, the requirement for reduced nicotinamide-adenine-dinucleotide phosphate and S-adenosyl methionine was not surprising. The most interesting finding, however, was that the crude extracts also had a specific requirement for glutathione that could not be eliminated by various monothiols or by dithiothreitol. Because of the specificity of the requirement for glutathione, it was suggested that an enzyme with a high specificity for a substrate containing glutathione and selenium, such as selenodiglutathione (GSSeSG), might be involved in the pathway. Also, the pronounced inhibitory effect of arsenite on dimethyl selenide formation in the presence of a large excess of monothiols implicated a dithiol enzyme. With these two observations in mind, the possible role of glutathione reductase in selenium metabolism has been investigated.[70]

First, selenodiglutathione was prepared to determine if this compound could be converted to dimethyl selenide, and thus meet one criterion for an intermediate in the pathway. It was found to be converted at the same rate and to the same extent as selenite. Next, selenodiglutathione was tested as a substrate for purified glutathione reductase from yeast. In the presence of NADPH and enzyme elemental selenium was rapidly liberated from selenodiglutathione, accompanied by NADPH oxidation, whereas no selenium was liberated in the absence of enzyme. These results suggest that selenodiglutathione or the mixed selenotrisulfide derivative of coenzyme A and glutathione may be the first intermediate in the biosynthesis of dimethyl selenide from selenite and that glutathione reductase or a similar enzyme catalyses the first enzymic step in the pathway. They also provide an explanation for the observation that washed microsomes, in contrast to 9000g supernatant, are not stimulated by NADPH but have an absolute requirement for glutathione. In the case of microsomes, an excess of GSH might nonenzymically convert selenodiglutathione to a labile reduced species identical with that produced enzymically by the action of NADPH and glutathione reductase present in the cytosol. Methylation of the reduced intermediate, probably by a microsomal enzyme, would then produce dimethyl selenide in either system. These studies of dimethyl selenide formation from inorganic selenite suggest that glutathione is involved in the biosynthesis of at least one organoselenium compound. A role for glutathione in the biosynthesis of other organoselenium compounds, such as selenoamino acids, might explain in part the nutritional interrelationships involving selenium and sulfur amino acids.

Urinary Excretion of Trimethyl Selenonium Ion. The trimethyl selenonium ion, $(CH_3)_3 Se^+$, has recently been identified as a major urinary metabolite of selenium in rats. The identification of the compound was first reported by Byard[27] and confirmed by Palmer *et al.*[171] The evidence regarding the chemical structure of the compound has been summarized.

Byard and Baumann[26] noted that the amount of a positively charged metabolite (later identified as trimethyl selenonium ion) excreted relative to a second unidentified metabolite increased with increasing selenite dosage. Palmer *et al.*[171] found that trimethyl selenonium ion comprised about 40% of the urinary selenium (10% of the dose) when rats were injected with either μg quantities or subtoxic doses (0.8 mg Se/kg) of selenite. This percentage could be increased by prefeeding the animals with unlabeled selenium prior to the injection of labeled selenite.

Mixing normal rat urine with selenite did not result in the formation of trimethyl selenonium ion,[171] and Byard[27] found that trimethyl selenonium ion could be detected in bladder urine. These studies indicate that trimethyl selenonium ion is not an artifact, and is a major metabolite of selenite, probably representing a detoxification pathway analogous to the formation of dimethyl selenide.

Just what the relationship of these two methylated metabolites of selenium might be to one another in terms of precursor and product is uncertain. Trimethyl selenonium ion is an important urinary metabolite at doses of selenite insufficient to trigger the respiratory excretion of dimethyl selenide. In regard to the idea that dimethyl selenide is a precursor of trimethyl selenonium ion, Palmer *et al.*[171] studied the effect of arsenite on trimethyl selenonium ion excretion, since arsenite is known to be a very effective inhibitor of dimethyl selenide formation, both *in vivo* and *in vitro*. They found that arsenite caused only a slight decrease in trimethyl selenonium ion excretion.

Enzymatic Reduction of Selenite in Microorganisms.   It has been known for many years that bacteria are capable of reducing selenite to elemental selenium.[126]   Some of the properties of microbial enzyme systems that reduce selenite have been investigated.

Woolfolk and Whiteley[264] studied the stoichiometric relationship of hydrogen uptake to the reduction of selenite or a variety of other inorganic compounds in extracts from *Micrococcus lactilyticus.*   Hydrogen uptake was measured manometrically for comparison with elemental selenium and selenide formation.   With selenite, an initial phase of hydrogen uptake occurred accompanied by the formation of elemental selenium, two moles of hydrogen being consumed per mole of selenite reduced to the elemental state.   A subsequent reduction of selenium to the selenide level then occurred, consuming an additional mole of H_2.   When colloidal selenium was incubated in the system, it was stoichiometrically reduced to selenide in a manner similar to the second phase of selenite reduction.   Selenate was not reduced in this system, but tellurite, tellurate, and a number of other substances were reduced.   There is apparently a low-potential system, possibly involving ferredoxin,[258] which mediates the flow of electrons from hydrogen to a variety of reducible substances in this organism.

The enzymic reduction of selenite to elemental selenium was studied with intact cells and cell-free extracts of yeast by Falcone and Nickerson.[56,162] The optimal pH for reduction of selenite was about 7 in the cell-free system. The activity of cell-free preparations was present in the soluble, nonparticulate fraction, and was heat labile.   Dialysis caused nearly complete loss of activity, which could be restored by the addition of the dialyzable fraction of heated extracts.   The addition of glucose-6-phosphate, nicotinamideadenine-dinucleotide phosphate, oxidized glutathione, and menadione supported extensive reduction of selenite by the dialyzed enzyme source.   Glucose-6-phosphate apparently served as the primary source of reducing equivalents via NADP-linked enzymes.   Menadione was included because extraction with hexane abolished the activity of the boiled cell-free extract that was used as a coenzyme source, and addition of menadione or the hexane-extracted material partially restored activity.   The reconstituted enzyme system was moderately sensitive to arsenite, a concentration of 10^{-3} M producing about 50% inhibition in the presence or absence of 10^{-3} glutathione.

Nickerson and Falcone interpreted the arsenite sensitivity as evidence for the involvement of an essential dithiol enzyme, and suggested that selenite was bound to two vicinal thiol groups of the protein.   To explain the requirement for glutathione, they concluded that glutathione maintained the thiol groups of the protein in the reduced state.   They observed that 10^{-2} M selenite was slowly reduced to elemental selenium by 2×10^{-6} M glutathione, but concluded that this reaction was too sluggish compared to the reaction of selenite with reduced menadione.   They suggested that the electron donor for selenite reduction was a quinone, linked to nicotinamide-adenine-dinucleotide phosphate and glucose-6-phosphate by a specific dehydrogenase.

It should be noted, however, that a strong case for glutathione and glutathione reductase as mediators of selenite reduction in yeast extracts can be made.   It has been observed that selenite reacts nonenzymically with glutathione to form selenodiglutathione, and that this compound undergoes a NADPH-dependent reduction to elemental selenium in the presence of crystalline glutathione reductase isolated from yeast.[68]   This enzyme, moreover, is inhibited by heavy metals and by arsenite and may possess a dithiol prosthetic group.[12, 134]   The only finding not in accord with this interpretation is the apparent requirement for a quinone.   One would like to know whether the thiol requirement of the yeast system is specific for glutathione, as it is in the reduction of selenite in mammalian systems.[66]   Nickerson and Falcone did not test other thiols to see if they were as active as glutathione, which would be the case if the role of glutathione were merely to keep protein thiols in a reduced state.

The reduction of selenite to elemental selenium in *Streptococcus faecalis* and *S. faecium* was investigated by Tilton *et al.*[237, 238] in relation to the use of this criterion in differentiating the species.   The difference in ability to reduce selenite was greatest for whole cells incubated with high concentrations of selenite, and became smaller when low concentrations of selenite were used and when reduction was studied in cell-free extracts.   The selenite-reducing activity of cell-free extracts was very sensitive to oxygen and was inhibited by mercuric chloride and iodoacetate.   Dialysis resulted in complete inactivation, which could be partially overcome by thiols.   Some evidence for the involvement of a flavoprotein was obtained.

Studies with Selenomethionine

It has long been known that the metabolism of organoselenium compounds may differ considerably from that of inorganic selenite.   One of the important forms of organic selenium found in common plants is selenomethionine. The fate of selenomethionine in animals which consume this form of selenium is thus of considerable interest.   ^{75}Se-labeled selenomethionine, first prepared biosynthetically,[13] is readily available and over 50 papers have appeared describing the use of this material, especially applications in radiographic scanning of the pancreas and parathyroid. Only selected papers pertaining to the general tissue distribution and intestinal absorption of selenomethionine will be discussed here. The enzymic incorporation of selenomethionine into proteins is discussed subsequently.

Distribution in Tissues. Selenomethionine is preferentially taken up in organs such as the pancreas and gastrointestinal mucosa that are characterized by a high rate of protein synthesis. Blau and Manske[14] demonstrated that concentrations of selenomethionine in the pancreas of dogs were as much as 8–9 times greater than in the liver. This is in contrast to selenite, for which uptake in the liver is usually far in excess of pancreatic uptake. Other workers have also noted, especially in the first few hours after injection, a consistently greater uptake of selenomethionine in pancreas than in liver.[105, 231, 3] A paper by Jacobsson[105] contains extensive data on the metabolism of selenite and selenocystine, as well as selenomethionine, which illustrate many important aspects of selenium metabolism. Hansson and Jacobsson[87] studied the fate of selenomethionine in mice by means of whole-body autoradiography. They found very high concentrations shortly after injection in the exocrine portion of the pancreas, and high concentrations in liver, kidney, intestinal and gastric mucosa, salivary gland, and bone marrow. Selenomethionine was readily transferred across the placental barrier to the fetus.

These studies have also shown that, as expected, a high proportion of the tissue ^{75}Se-selenomethionine is associated with the protein fraction at all time periods beyond the first 5 minutes. Awwad et al.[7, 8] noted, however, that as much as 23% of the total radioactivity precipitated by trichloroacetic acid could be removed by sulfitolysis, this fraction decreasing with time to only a few percent in most tissues after 24 h. No binding to protein was observed when ^{75}Se-selenomethionine was incubated with plasma in vitro. It was suggested that in the presence of actively metabolizing tissue, a portion of the selenomethionine is converted to a compound such as selenocysteine that might bind to protein through a S—Se bond.

Intestinal Transport of Selenomethionine. Because selenomethionine appears to be an important form of selenium in plants, the absorption of selenomethionine by animals consuming these plants is a pertinent aspect of selenium metabolism.

Spencer and Blau[228] first reported the active transport of L-selenomethionine in everted sacs of hamster intestine. ^{75}Se-selenomethionine and ^{35}S-methionine appeared to be transported identically. This work was confirmed by McConnell and Cho[150, 152] who made more extensive investigations of the transmucosal movement of various selenium compounds in the same system.   Selenomethionine labeled with ^{75}Se, as well as ^{14}C-methyl labeled selenomethionine, were transported against a concentration gradient, but selenocystine and selenite were not actively transported.   The aerobic conditions required for this procedure precluded experiments with selenocysteine to see if this compound might be actively transported.   The rate of transport of selenomethionine varied in different segments of the intestine.   Methionine and selenomethionine appeared to share the same transport system since an excess of one decreased the transport of the other.   Studies with the inhibitors ouabain and 2, 4-dinitrophenol indicated that transport of selenomethionine was dependent on the sodium pump and on energy derived from oxidative pathways.   McConnell and Cho suggest that the inhibitory effect of methionine on the intestinal absorption of selenomethionine might be related to protective effects of dietary protein and methionine in poisoning caused by the ingestion of plants containing selenomethionine.

PROPERTIES OF SELENIUM AND EFFECTS ON BIOLOGICAL SYSTEMS

Oxidation-Reduction Reactions of Selenium Compounds

Antioxidant Properties.   The oxidation-reduction properties of selenium compounds have been extensively investigated in relation to the antioxidant theory of selenium function.   Tappel and his associates have been most active in these studies, and have concluded that selenium compounds have the ability to function as biological antioxidants.[234]

Denison and Condit[44] found that dialkyl diselenides were more effective than their sulfur analogues as antioxidants for lubricating oils.   Woodbridge[262] summarized and extended these studies.   Bieri[11] found that 0.5 ppm of dietary selenite or 0.3% of cystine significantly inhibited *in vitro* lipid peroxidation in certain tissues of vitamin E-deficient chicks, although direct addition of the same compounds to the *in vitro* system had no effect.   Zalkin *et al.*[265] also found that dietary selenite inhibited the formation of lipid peroxidation products in the tissues of vitamin E-deficient chicks *in vivo* and *in vitro*, and demonstrated antioxidant activity of selenoamino acids in model systems composed of an aqueous emulsion of linoleic acid with hemoglobin as a catalyst.   Olcott *et al.*[166] evaluated selenomethionine as an antioxidant for anhydrous fats and oils, and found that it greatly increased the induction period for lard and triolein, although it had no effect with menhaden oil, a highly unsaturated fish oil.   Methionine had little effect.   Selenomethionine was apparently able to decompose peroxides.   Some evidence was obtained that the active antioxidant might be formed from selenomethionine by partial oxidation. Hamilton and Tappel[85] found that the protein fractions from tissues of animals fed high levels of selenium (10–14 ppm as selenite or selenate) had higher activity as lipid antioxidants than proteins from control animals. They discussed possible mechanisms by which selenium could act as a primary antioxidant, reacting with free radicals formed in the initial stages of lipid oxidation.

Caldwell and Tappel[28,30] studied the ability of selenoamino acids or diselenodicarboxylic acids or their sulfur analogues to decompose hydrogen peroxide and several organic peroxides. Methionine reacted rapidly with hydrogen peroxide in strong acid and consumed 0.90 mole of peroxide oxygen per mole of amino acid within 10 minutes, but never exceeded this amount. Selenocystine reacted more slowly, but brought about a more extensive decomposition, up to 3 moles of peroxide oxygen being consumed per mole of selenocystine. Cystine had no effect. Peroxide decomposition in the presence of selenocystine continued for a period of at least 2 h even though most of the selenocystine was transformed to other products within one minute after peroxide addition. It is possible that regeneration of selenocystine, as by dismutation of oxidized derivatives such as the diselenoxide, occurs to a slight extent in such systems. Neither selenocystine nor methionine were very effective in decomposing organic peroxides such as *tert*-butylhydroperoxide, in contrast to the results with hydrogen peroxide. Diselenodicarboxylic acids were shown to decompose hydrogen peroxide at pH 5.6 at rates up to 30 times faster than their sulfur analogues.

The effect of selenium compounds in reducing radiation damage to amino acids and proteins caused by free radical mechanisms has also been investigated. Shimazu and Tappel[213,214] used systems consisting of enzymes (alcohol dehydrogenase and ribonuclease) or amino acids in solutions to which sulfur- or selenoamino acids had been added. These solutions and controls were subjected to γ-radiation and destruction of the compound was measured as the dose of radiation necessary to reduce its original concentration (activity in the case of enzymes) by one half ($D_{1/2}$). In all cases, the selenoamino acids provided greater protection against radiation damage than the sulfur amino acids, usually on the order of a severalfold higher value for the $D_{1/2}$. Selenomethionine itself was also destroyed to a lesser extent than methionine. Protective effects of sodium selenate have also been reported for catalase inactivation by x-rays.[129]

In related studies, Windle et al.[260] investigated free radicals formed from selenium compounds, including alkyl and aryl selenols, selenides, and diselenides, along with their sulfur analogues. These compounds were subjected to ultraviolet irradiation at liquid nitrogen temperatures, and the electron spin resonance spectra were recorded. A spectrum attributed to a selenium radical of g = 2.08 was obtained with di-*n*-octadecyl diselenide and two other compounds.   In contrast to disulfide compounds, ultraviolet irradiation did not cause rupture of the diselenide bond, but it was suggested that a radical was localized elsewhere in the molecule, usually on a methylene carbon.

In other studies on the possible activity of selenium compounds in catalyzing oxidation-reduction reactions of thiol compounds, it has been shown that selenocystine accelerates the oxidation of cysteine, glutathione, and homocysteine by oxygen or peroxides.[29]   A two–threefold increase was obtained in the rate of oxidation of thiols by hydrogen peroxide.   With molecular oxygen, the spontaneous oxidation rate was zero, but oxidation, as measured by loss of thiol, could be induced by the addition of selenocystine.   Cystine had very little effect in any of these experiments.   In contrast to the results with the low molecular weight thiols, selenocystine protected two enzymes which require sulfhydryl groups for activity from the oxidative inactivation induced by hydrogen peroxide.   Since the rate of loss of enzyme activity was measured rather than loss of thiol, the mechanism of the protective action of selenocystine is uncertain.

Selenium and Disulfide Interchange Reactions.   An important difference exists in the relative ease with which thiols cleave selenium-selenium or selenium-sulfur bonds as compared to disulfide bonds.   A number of investigations along various aspects of this topic have been carried out, especially in regard to the catalysis of disulfide interchange by selenium.

Sanger[199] showed that disulfide interchange reactions can proceed in strongly acid media, the overall reaction being:

$$RSSR + R'SSR' \rightleftharpoons 2R'SSR$$

This reaction complicated Sanger's attempts to determine the pairing of the half-cystine residues in insulin, since hydrolysis of the hormone in concentrated HCl produced more disulfide peptides than would be expected from the cystine content of the protein.   Disulfide interchange also occurs under neutral or slightly alkaline conditions, but is minimal in weakly acidic solutions.   In neutral or basic solutions, thiols catalyze disulfide interchange, the reaction being initiated by the attack of a mercaptide ion on the disulfide bond. In strong acid, Benesch and Benesch[10] suggest that sulfenium ions (RS^+) are formed, the probable mechanism being an initial protonation of the disulfide bond followed by heterolytic cleavage to form the thiol and a highly reactive sulfenium ion:

$$RSSR + H^+ \rightleftharpoons (RSSR)^+ \text{H}$$

$$(RSSR)^+ \text{H} \rightleftharpoons RSH + RS^+$$

Thiols inhibit interchange reactions under acidic conditions, in contrast to their catalytic effect on the process at higher pH.

Selenious acid, at a concentration of 10^{-5} M, has been shown to catalyze the disulfide interchange reaction between 10^{-3} M L-cystine and 10^{-4} M N, N'-bis-2, 4-dinitrophenyl-L-cystine in concentrated HCl solutions.[120,121] The catalytic effect decreases to zero in 4.8 N HCl. It was suggested that selenious acid functions catalytically by removing the inhibitory thiol. It has been shown that selenious acid readily reacts with cysteine and other thiols in acid solutions to form selenotrisulfides.[68]

Selenotrisulfides, compounds in which a selenium atom links two sulfur atoms, appear to participate far more readily than disulfides in reactions with thiols. The possibility of such a sulfhydryl-selenotrisulfide exchange reaction similar to that shown in the reaction below was investigated by Ganther and Corcoran:[70]

$$\text{Protein}\Big\langle{\overset{\displaystyle SH}{\underset{\displaystyle SH}{}}} + \text{Se}\Big\langle{\overset{\displaystyle SR}{\underset{\displaystyle SR}{}}} \longrightarrow \text{Protein}\Big\langle{\overset{\displaystyle S}{\underset{\displaystyle S}{}}}\text{Se} + 2RSH$$

It was found that the selenium derivative of 2-mercaptoethanol underwent a reaction with the thiol groups of reduced ribonuclease, resulting in an exchange of selenium into the protein.[70] The reaction occurred readily at 4°C and at a pH as low as 2 or 3, where the rate of exchange between thiol and disulfides is negligible. No exchange occurred with native ribonuclease, which contains disulfide linkages but no thiol groups.

The fact that the presence of selenium between two sulfur atoms confers a much greater sensitivity (compared to disulfides) to attack by thiols may be of importance in connection with disulfide interchange phenomena and the biological role of selenium. Disulfide interchange enjoys a considerable popularity in speculative mechanisms regarding cellular processes such as cell division or membrane permeability changes, and it is naturally attractive to look for a role for selenium in reactions involving sulfur.

Along the same line, Dickson and Tappel[49] have found some interesting effects involving catalysis of disulfide interchange in proteins by selenoamino acids. Selenocystine catalyzed the activation of papain and glyceraldehyde-3-phosphate dehydrogenase and the inactivation of ribonuclease, processes which take place through sulfhydryl-disulfide interchange reactions. The activation of papain with cysteine (10^{-5} M) was studied at pH 7. The presence of 5×10^{-4} M selenocystine increased the rate of activation 15-fold compared to buffer alone. There was no effect of selenocystine in the absence of cysteine. Inactivation of papain through oxidation of the sulfhydryl groups competes with the activation process, so that enzyme activity goes through a maximum and then declines. The effect of selenocystine was to increase this peak height of activity as well as increase the rate of activation. Catalysis of papain activation probably proceeds through the following reactions: (1–3).

$$CySH + CySeSeCy \rightleftharpoons CySSeCy + CySe^- + H^+ \qquad (1)$$

Reaction 1 was studied by Dickson and Tappel[50] in a related paper. They showed that the addition of cysteine or glutathione to nitrogen-saturated solutions of selenocystine produced ultraviolet absorption changes indicative of selenocysteine formation within 2–5 seconds. The equilibrium constant for reaction 1 at pH 7 was about 10^{-2}, so a considerable excess of thiol over selenium would be required to reduce selenocystine to selenocysteine, but there is sufficient glutathione in most tissues to accomplish this. Reactions 2 and 3, when summed, give the overall reaction for the activation of papain by cysteine (PSSCy is papain prior to activation, and PSH is active papain):

$$CySe^- + H^+ + PSSCy \rightleftharpoons PSH + CySSeCy \qquad (2)$$

$$\underline{CySSeCy + CySH \rightleftharpoons CySSCy + CySe^- + H^+} \qquad (3)$$
$$PSSCy + CySH \rightleftharpoons PSH + CySSCy$$

The rate of activation of the sulfhydryl enzyme glyceraldehyde-3-phosphate dehydrogenase by cysteine was also increased by selenocystine. The inactivation of ribonuclease in the presence of cysteine and urea, at pH 7, also involves cleavage of disulfide bonds, and the rate of this reaction also was increased by the presence of 10^{-3} M selenocystine. Selenomethionine, cystine, or methionine were much less effective than selenocystine.

Dickson and Tappel[49, 50] suggest that one of the functions of selenium may be the catalysis of critical sulfhydryl-disulfide exchange reactions involving enzyme proteins. They recognize the necessity of finding a catalytic effect of selenium in biological systems that is commensurate with its extremely high biological activity, and suggest that such a catalytic function involving enzymes, which are themselves catalysts, might provide the necessary degree of biochemical amplification. Taking into consideration the level of total selenium in tissues, and assuming that a portion of this is selenocystine, they conclude that catalytic effects obtained with 10^{-7} M selenocystine may be reasonable in terms of possible biological roles. This degree of activity does not appear to have been obtained in these particular experiments, however, because 10^{-7} M selenocystine produced a detectable but very modest response, and much higher levels were used in the majority of cases.

Walter et al.[250] observed that elution of selenocystine from an ion exchange column with a buffer containing thiodiglycol consistently gave rise to an additional peak, whereas only a single peak of selenocystine was obtained when thiodiglycol was omitted. Cummins and Martin[40] had previously adopted the practice of omitting thiodiglycol from their buffers for similar reasons. Walter et al.[250] showed that 2-mercaptoethanol, present as an impurity in the thiodiglycol reagent, was generating a mixed sulfur-selenium derivative (2-amino-4-selena-5-thia-7-hydroxyheptanoic acid) through a sulfur-selenium interchange reaction with selenocystine. A mixed selenium-sulfur derivative was also formed when selenocystine was treated with cysteine. This work demonstrates how sulfhydryl-diselenide exchange may occur under normal analytical procedures that do not give rise to sulfhydryl-disulfide interchange.

Formation of Selenotrisulfides. When selenious acid or selenite is introduced into a biological system, there is a high probability that it will undergo a reaction with one or another of the sulfhydryl compounds found in biological systems, and this reaction is very important in understanding the effects of selenite. It has been known for over 30 years that aqueous solutions of selenium dioxide react with sulfhydryl compounds to form relatively unstable derivatives. Painter[169] proposed that the reaction between thiols and selenium dioxide took place in the following manner:

$$4RSH + SeO_2 \longrightarrow RSSeSR + RSSR + 2H_2O$$

Compounds of the type RSSeSR are analogues of trisulfides in which one atom of sulfur is replaced by selenium, and therefore may be conveniently referred to as selenotrisulfides. These compounds are quite stable in dilute acid but decompose very quickly in alkaline solution to liberate elemental selenium. Selenium can also be cleaved from these compounds by treatment with thiols. In the literature there are reports that the reaction produces a "tetra" compound in which selenium is bound to 4 atoms of sulfur.[230] These results can be attributed to the difficulty in separating the selenotrisulfide from the disulfide by recrystallization. Subsepuent studies[114,184,240] demonstrated that the reaction produced two products which could be resolved by mild chromatographic procedures. Although these studies all supported the correctness of the reaction as proposed by Painter, the isolation and characterization of such compounds had not been accomplished.

Because of the importance of these compounds to many aspects of the biochemistry of selenium, the reaction of selenious acid with a variety of thiols was investigated.[68] Spectrophotometric analysis confirmed that the combining ratio for thiols and selenious acid was 4:1, and thin layer chromatography of such reaction mixtures invariably demonstrated the presence of two substances, one corresponding to the disulfide, and complete disappearance of the thiol. To separate the reaction products on a larger scale, a new column chromatographic procedure was developed, based on the stronger interaction of selenotrisulfides with metals fixed to a stationary phase. This technique was used to obtain sufficient quantities of the selenotrisulfide derivative of cysteine to permit its characterization by elemental analysis and by amino acid analysis after performic acid oxidation. The ultraviolet absorption spectra of this compound and the derivative of 2-mercaptoethanol were similar, showing a broad band extending to 400 mμ and a peak at 260–265 mμ. The spectra are very similar to the spectrum of a trisulfide;[232] the extinction coefficients are comparable but the spectrum of the selenotrisulfide is displaced to longer wavelengths, as would be expected.

Substitution of Selenium for Sulfur

The effect of substituting selenium for sulfur in biological systems has for some time attracted considerable attention, both for those specializing in research on selenium and for those in other fields. Some of the earliest studies attempted to relate selenium toxicity to a replacement of essential sulfur compounds by their selenium analogues, on the assumption that the latter were inactive. Investigators studying the metabolism of sulfur compounds have frequently extended their studies to include the corresponding selenium compounds. Other studies were initiated in connection with using selenium analogues as antimetabolites and therapeutic agents, and, when labeled with the gamma-emitting isotope, ^{75}Se, as diagnostic tools in medicine. Of considerable interest to protein chemists investigating structure-function relationships are some recent studies with proteins which were extensively substituted with selenium. Most of these topics have been previously discussed at some length.[193,217,223,51,140]

Effects of Extensive Replacements. No one has ever demonstrated that selenium can totally replace sulfur in living organisms. Replacement of certain individual sulfur nutrients with their selenium analogues has been reported, however. Cowie and Cohen[39] observed that a methionine-requiring mutant of *E. coli* could grow on selenomethionine at a reduced but exponential rate when sulfate was supplied to meet the other sulfur needs. Similarly, selenopantethine supported normal growth for the pantethine-requiring organism *Lactobacillus helveticus.*[137]

The most extensive replacement of sulfur with selenium has been obtained using strains of *E. coli* adapted to growth on high levels (0.01 *M*) of inorganic selenate, as first described by Shrift and Kelly.[218] Studies by Huber *et al.*[103] using this approach, indicate that cells utilize selenium only if sulfur is also present as a contaminant and that selenium can replace only 30–40% of the normal sulfur requirement. Growth on the selenate medium was essentially linear rather than exponential and the lag period was longer than with sulfate medium.. The efficiency of growth (grams of cells produced per gram of glucose utilized) on selenate was less than on sulfate.   Total growth on the selenate medium was markedly reduced.

Huber and Criddle[102] then isolated pure β-galactosidase from these organisms in order to compare its properties to the enzyme from cells grown on sulfate.   Amino acid analyses showed that about 20 of the 27 moles of methionine residues found per 135,000 g of protein had been substituted by selenomethionine, but the cysteic acid content of the selenium enzyme was essentially the same as the sulfur enzyme, indicating that there was little or no replacement of cysteine with selenocysteine.   By neutron activation analysis, it was established that 24 g atoms of selenium were present per 135,000 g of protein, in reasonably good agreement with the amino acid data.   Selenium was not detected in the sulfur enzyme.   The content of other amino acids and the extinction coefficient at 280 mμ were comparable for both enzymes.   The catalytic properties of the enzyme were not appreciably changed, both forms having similar values for the Michaelis constant and for maximum velocity.   The optimum pH for both enzymes was the same.   Although the purified enzymes had similar specific activities, the content of enzyme per mg of cells grown on selenate was only 10% of that for cells grown on sulfate.

In properties related to folding of the protein, some differences were noted. The selenium enzyme was less stable to denaturation, induced either by heating or by treatment with urea.   Moreover, when the enzymes were inactivated by treatment with urea and mercaptoethanol, then dialyzed to remove urea, the selenium enzyme regained its activity at a greater rate and to a greater extent than did the denatured sulfur enzyme.   Sedimentation velocity studies gave some indications that the substitution of selenium for sulfur interfered with the spontaneous aggregation of 4S material to the active 16S protein.   With the selenium enzyme, the proportion of inactive 4S protein was higher and the 4S selenium protein had about 29 methionine residues replaced per subunit compared to 20 for the 16S selenium enzyme.   This suggests that there is a limit to the amount of selenomethionine which can be substituted for methionine without loss of enzyme activity.   The authors suggest that substitution of selenocysteine for cysteine in this protein cannot be tolerated because the active site, believed to contain a sulfhydryl group, would be more directly and radically affected.*

The activity of alkaline phosphatase isolated from E. coli grown on a low-sulfate medium containing selenite has been investigated by Ahluwalia and Williams.[1]   This enzyme appeared to be more sensitive to sulfur replacement

---

*On the other hand, selenocysteine is very easily oxidized to selenocystine, and this compound, like cystine, would not be incorporated into proteins enzymically.   The apparent discrimination against *selenocysteine* incorporation might actually reflect discrimination against *selenocystine*, the degree of discrimination depending on the degree to which selenocysteine was oxidized.

by selenium, having a specific activity less than half that of the normal enzyme even though the estimated molar percentage replacement of sulfur by selenium was only 8–10.   A control study showed that incubation of normal enzyme with a high concentration of selenite had no effect on activity.   The pH optimum of the selenium enzyme was the same as that of normal alkaline phosphatase.

The Suitability of Selenium Analogues as Enzyme Substrates.   It now seems quite clear that the older interpretation of selenium toxicity in terms of a general replacement of sulfur compounds by their selenium analogues is not supported by the experimental evidence.   Many enzymes have comparable activity with either the normal substrate or the analogue.

The enzyme from yeast which catalyzes the first step in sulfate activation, ATP-sulfurylase, was shown to be active with selenate as substrate.[259] Adenosine-5'-phosphoselenate (APSe) and pyrophosphate were formed from adenosine triphosphate and selenate in the presence of the enzyme.   APSe was less stable than the sulfate derivative, apparently undergoing hydrolysis to selenate and adenosine monophosphate, so only trace amounts of APSe were recovered.

There have been a number of studies comparing methionine and seleno-methione as substrates for enzymes.   Transmethylation reactions involving methionine require that methionine first be activated in a reaction with ATP to form S-adenosyl methionine.   Mudd and Cantoni[157] have shown that the transferase from yeast which catalyzes this reaction is more active with selenomethionine than with methionine.   They also demonstrated that the Se-adenosylselenomethionine thus formed readily transferred its methyl group to guanidoacetic acid in the presence of a methylpherase isolated from pig liver.   Bremer and Natori[20] found that S-adenosylselenomethionine was at least as good a substrate as its sulfur analogue in choline synthesis catalyzed by rat liver microsomes, and could also transfer its methyl group to sulfhydryl or selenol compounds, including hydrogen selenide and methane-selenol.   A rather appreciable nonenzymic methylation occurred with Se-adenosylselenomethionine in the presence of boiled microsomes.

Both selenomethionine and selenoethionine were evaluated as substrates for rat liver methionine adenosyltransferase.[173]   In contrast to the earlier results with the same enzyme from yeast,[157] the maximum velocity of the enzyme with the selenium analogues as substrates was lower compared to the sulfur substrates, and the velocity at any given concentration of substrate was always lower for the selenium compounds, even though they had a lower K_M.   Pan et al.[172] also studied the incorporation of the alkyl group of seleno-methionine and selenoethionine into various liver constituents, and found that the pattern of incorporation was similar to that obtained with the sulfur analogues.   Incorporation of the ethyl group was less than that of the methyl, but continued for a longer time.

Selenoamino acids appear to be incorporated into proteins by the same enzymic reactions used for other amino acids.   Early studies with crude extracts of *E. coli* showed that both methionine and selenomethionine were incorporated into protein.[163]   The purified methionyl-tRNA synthetase from *Sarcina lutea* was shown to have similar activity with both methionine and selenomethionine, as measured by ATP-pyrophosphate exchange and by hydroxamic acid formation.[82]   Both substrates gave identical values for maximum velocity and for the Michaelis constant.   Since enzyme specificity in the formation of amino acyl-tRNA is usually somewhat higher than for the synthesis of amino acyl-AMP-enzyme, it is quite interesting that *E. coli* methionyl-tRNA synthetase cannot distinguish between methionine and selenomethionine.[96]   This enzyme aminoacylates methionine tRNA with either compound, the K_M being 0.7×10^{-5} M for methionine and 1.1×10^{-5} M for selenomethionine.

Other Selenium Analogues.   Organic chemists and pharmacologists have shown an increasing interest in the synthesis of selenium analogues of biologically important sulfur compounds, and comparisons of the kinetic differences in reactions involving isologous oxo, thio, and seleno compounds. The list of compounds recently synthesized includes selenopantethine, selenocoenzyme A, selenoctic acid, Se-adenosylselenomethionine, selenoglutathionine, selenooxytocin, and various selenopurines and pyrimidines (Table 9-3).   Some pertinent aspects of the comparative chemistry of sulfur and selenium compounds have been reviewed by Mautner,[139] who, along with Günther, has been especially active in this field.

An interesting series of compounds related to acetylcholine, in which sulfur or selenium atoms replace oxygen, were synthesized by Mautner and his

TABLE 9-3   **SOME RECENTLY SYNTHESIZED ORGANOSELENIUM COMPOUNDS OF BIOLOGICAL INTEREST**

Compound	Reference
Selenopurines and selenopyrimidines	136,138
Selenoctic acid	9
Selenopantethine	79
Selenocoenzyme A	80
Se-adenosylselenomethionine	226
Selenoglutathione	59
Selenooxytocin	60,249
Selenolanthionine	267

associates, and extensive studies were then made comparing the pharmacological effects of these compounds in various systems. Because these compounds were essentially isosteric but differed considerably in their electron distribution, they were investigated with the intention of interpreting differences in the biological actions of sulfur and selenium compounds in terms of electron distribution. Some of the results of these studies have been discussed by Mautner.[140]

The initial compounds studied were acetylselenocholine and acetylthiocholine, analogues of acetylcholine in which the ether oxygen is replaced by selenium or sulfur. The biological activity of these esters in the depolarization of excitable membranes is guinea pig ileum and frog rectus abdominis preparations, as well as the monocellular electroplax preparation from eels, decreased progressively as the oxygen atom was replaced by sulfur and selenium.[211] All three esters were hydrolyzed by the enzyme acetylcholine esterase at approximately the same rate. But when the activity of the hydrolysis products was compared, depolarization activity progressively increased as oxygen in choline was replaced by sulfur and selenium, the activity of the thio and seleno analogues of choline being comparable to that of the parent esters. This apparently explains why inhibitors of acetylcholine esterase increase the biological activity of acetylcholine but have no effect with acetylthiocholine and acetylselenocholine. Webb and Mautner[252] have also compared the depolarizing activities of homocholine, homocholinethiol, homocholineselenol, and the acetyl esters or methyl derivatives of these compounds.

It is interesting that, as expected, biological activity of analogues of acetylcholine and its derivatives generally varied in the same manner as the chemical properties, i.e., oxygen followed by sulfur, then selenium, or the same order in reverse. From the standpoint of assessing the effect of replacing sulfur with selenium, the significant point is that generalizations are risky, and substitution of selenium for sulfur may increase, decrease, or leave unchanged the biological activity of a compound in any given case.

Selenium isologues of the sulfur containing hormone oxytocin have been synthesized by Walter and du Vigneaud.[249] They found in a variety of bioassays that the replacement of sulfur by selenium in the 6 position yields highly potent isologues of oxytocin and deamino-oxytocin, nearly as potent as the natural hormone and its comparable deamino analogues.

Selenium and Cellular Processes

Selenium and Enzyme Cofactors. An apparent cofactor role for colloidal selenium in the enzymic oxidation of cysteamine to hypotaurine was reported by Cavallini *et al.*[35]   Selenium was more effective than elemental sulfur or methylene blue in this respect.   It was later shown by Wood and Cavallini[261] that the enzyme did not require any of these substances for activity when the substrate concentration was lowered to 10^{-5} M.   Cystamine, formed by air oxidation of the substrate during the assay, was also found to be an inhibitor.   It is believed that the inhibition results from disulfide interchange reactions involving the inhibitors and the enzyme.   Just how elemental selenium and the other cofactor-like compounds produced their beneficial effect is not known.

An effect of dietary selenium on the oxidation of pyruvate by rat liver was reported by Bull and Oldfield,[23] using whole homogenates or isolated mitochondria from animals fed a *Torula* yeast diet supplemented with selenium (1 ppm as sodium selenite), vitamin E, or a combination.   Dietary selenium produced a statistically significant increase in oxygen consumption when tissue preparations were incubated with pyruvate, although the differences were not large.   The effect was apparently dependent on the simultaneous presence of α-tocopherol, but could not be eliminated by dietary supplementation with Ethoxyquin, suggesting that a general antioxidant effect was not involved.   The oxidation of succinate, in contrast to pyruvate, was not altered by administering selenium in the diet.   The investigators suggested that selenium and probably vitamin E were involved in the oxidation of pyruvate, by some mechanism not involving antioxidant effects or respiratory decline.   It is unfortunate that the oxygen uptakes were not expressed as specific activities based on protein or some other parameter; the uptake was expressed only in terms of milliliters of homogenate, and the relatively small differences noted in oxygen uptake might therefore involve indirect effects of dietary selenium or vitamin E rather than a direct role for selenium in the oxidative processes of the tricarboxylic acid cycle.   As discussed elsewhere, there is insufficient selenium in normal tissues for a stoichiometric relationship with the more abundant cellular enzymes such as those of the pyruvate dehydrogenase complex, so a direct role for selenium as a cofactor for such enzymes is unlikely.

A requirement for selenite, molybdate, and iron for the production of formic dehydrogenase by *E. coli* could be demonstrated when a highly purified growth medium was used.[187]   Oxidation of formate by the organism was minimal in the combined absence of selenite and molybdate, although the rate of glucose oxidation and the rate of growth were normal.   Enzyme activity was fully developed when the medium contained approximately 3×10^{-8} M to 10^{-7} M selenite and 10^{-8} M molybdate (about 5,000 molecules of selenite and 50 molecules of molybdate per cell, according to Pinsent, which is a very noteworthy biological activity).   Formic dehydrogenase is a complex enzyme system[74] and it is not known whether selenium has a direct role in electron transfer or whether it has a more remote function in the biosynthesis of the enzyme. Pinsent regarded the effects of selenite on formic dehydrogenase activity as an example of a nutritional requirement not related to growth. Since the growth medium has been purified to the extent that at least 10^{-8} M selenite was necessary for the response, it can be tentatively concluded that if a selenium requirement for the growth of E. coli exists, it will probably be less than 10^{-8} M.[222]

Effects on Cell Division and Morphology. It was shown long ago by Rapkine[189] that a close temporal relationship existed between cell division and the thiol content of certain cellular components. More recent work by Sakai[198] suggests that sulfhydryl-disulfide interchange reactions between two types of cellular proteins are involved in the formation and breakdown of the microtubules which comprise the filaments of the mitotic apparatus in sea urchins. It may be possible to regulate the tendency of protein subunits of microtubules to aggregate spontaneously through changes in the sulfhydryl: disulfide ratio[141] and thus influence the formation of many cellular structures. It is likely that selenium compounds would have a profound influence on such processes, and a variety of effects have indeed been described.

In certain cases selenium can apparently act to bring about cell division. Nickerson et al.[161] studied a mutant of *Candida albicans* which lacked the ability to divide, so that a filamentous, mycelial type of growth was obtained instead of the normal single-celled budding form. When this filamentous mutant strain was grown on agar containing 10^{-4} M selenite, the yeast grew largely in the single-celled form. On a transfer to medium devoid of selenite, all of the cells reverted to a filamentous form. Selenium at this concentration apparently had little effect on total growth but in some manner caused division to occur in the divisionless strain. An analogous effect was also observed with other types of filamentous fungi. These effects of selenium may be related to the occurrence of certain sulfur-rich proteins in the cell wall, but the mode of action was not established.

An effect of selenium apparently related to cell division in mammalian cells has also been reported.[131] The effect of vitamin E and selenium on the regeneration of liver in rats fed a basal semipurified diet was determined through measurements of liver weight and of mitotic index following subtotal hepatectomy. Sodium selenite given in combination with vitamin E stimulated all processes leading to regeneration. The effect of the two nutrients was much greater when given in combination than when given separately.

Selenium compounds can also inhibit cell division or cause morphological changes in developing organisms, presumably by disrupting processes that involve sulfur. Shrift[216] found that selenomethionine caused the formation of giant cells of the algae *Chlorella vulgaris*, apparently by inhibiting division without affecting growth.   Methionine could overcome the effect of seleno-methionine.   Selenomethionine failed to uncouple growth from cell division in other microorganisms such as *E. coli*.   The giant cells eventually adapted and resumed cell division.   The adaptation persisted for many generations and apparently involved a decreased permeability of the cell to selenomethionine.[219]   Werz[255] obtained evidence that selenate interfered with morphogenesis in the alga *Acetabularia* through effects on sulfur metabolism.

Rulon[196] observed that sodium selenite modified the developmental patterns of newly fertilized eggs and early larval stages of the sand dollar.   Developmental defects included structural abnormalities such as inhibition of skeletal formation and exogastrulation, as well as a high percentage of radially elongated larvae.   Congenital malformations occur in poultry and livestock grown in seleniferous areas in the western United States, and can be induced experimentally with selenium salts.[193]   Sentein[212] has made extensive studies on chromosomal alterations and pole anomalies produced by 0.005 to 0.01 M selenium dioxide during mitotic division in developing salamander eggs.

Deformation of chromatin may be the basis for the decreased rate of cross-over in barley seedlings grown on soil treated with 0.5 to 5 ppm of selenium as sodium selenate.[247]   There was some degree of polyploidy with the highest level of selenium, possibly resulting from destruction of spindle fibers or interference with their formation.   The researchers speculate that incorporation of selenium into protein may cause "relaxation" of the chromosomal protein and thus decrease breakage caused by stress.

Rounds *et al.*[195] have shown that the mitochondria of isolated rabbit endothelial and rat heart cells treated *in vitro* with sodium selenite (100–200 μg per ml) showed marked swelling and fragmentation after 3 h, plus disaggregation of the nucleolus.

Effects on Other Processes.   An interesting report of an antiinflammatory property of selenium has appeared.[192]   In a screening program of potential antitumor compounds it was found that a liver fraction exhibited antiinflammatory properties when tested in rats by the Selye granuloma pouch method.   When the alkaline ash of liver fractions proved to be active, the findings of Patterson *et al.*[180] on retention of factor 3 selenium during alkaline ashing were recalled.   Sodium selenite was tested and proved to be an effective antiinflammatory agent.   A maximum effect was obtained at 9.6 μg of selenium per kg of body weight; this treatment reduced the volume of exudate to 0.7 ml, compared to 6.4 ml in the controls.   Based on the high activity of selenium and its concentration in the liver fractions, it was concluded that selenium accounted for the antiinflammatory activity of liver.

It was postulated by Siren[225] that selenium might be involved in the excitation mechanisms of the visual process.   This suggestion was based in part on the photoelectric properties of selenium and in part on his finding of exceptionally high levels of selenium in the retinal tissue of some species. The selenium content in the retina of the common tern and of the roe deer, species having a high visual acuity, was found to be 630–810 ppm (dry basis) by atomic absorption spectroscopy. The retina of the guinea pig, which has low visual acuity, was found to be only 7 ppm by a fluorometric method. Subsequent analyses of selenium content in a variety of species, including man, have yielded data more in line with the figure for guinea pig retina.[37,235] There is no confirmatory data or supporting evidence for Siren's proposal; it is possible that the unusually high values for tissue selenium which he obtained were related to the difficulty of measuring selenium by atomic absorption, or were due to contamination.

Relative Potency of Selenium Compounds in Preventing Selenium Deficiency, The nutritional value or biological potency of selenium in various organoselenium compounds relative to that of selenite selenium has been investigated.[203,206,210] Some are more active than selenite and a larger number are less active, per g of selenium. In a paper by Schwarz and Fredga, a large number of monoseleno- and diselenodicarboxylic acids and some amides were tested. The compounds found to be most active were monoseleno-11, 11'-di-*n*-undecanoic acid, diseleno-11, 11'-di-*n*-undecanamide, and diseleno-4, 4'-di-*n*-valeric acid, None of these had more than twice the biological activity of selenite, however. In several series of homologous acids the lower three members containing carboxylic acid residues with chain lengths of C_2 to C_4 were inactive. In the series of symmetrical monoselenodicarboxylic acids, $HOOC-(CH_2)_n-Se-(CH_2)_n-COOH$, an interesting alternation in activity was observed from valeric to undecanoic acid; the even-numbered acids were rather inactive and those with uneven numbers were quite active.

INTERRELATIONSHIPS OF SELENIUM WITH OTHER SUBSTANCES

Nutritional Interrelationships with Vitamin E, Sulfur Amino Acids, and Heavy Metals

Efforts to define the biological role of selenium have been greatly complicated by the apparent interchangeability of the two substances in certain nutritional disorders, such as liver necrosis in rats and exudative diathesis in chicks, which could be completely prevented either by vitamin E or by minute amounts of selenium. Of the various mechanisms put forth to explain how two substances that differ so much in their chemical nature can produce similar biological responses, two ideas have received the most attention.

Schwarz[209] has maintained that vitamin E and selenium function independently in alternate pathways of metabolism, so that a deficiency of one would not be deleterious because the alternate route would produce the same product. Only when there was a deficiency of both nutrients would there be a derangement in metabolism. This theory of metabolic redundancy suffers from lack of experimental verification and has not enjoyed wide acceptance.

Others, notably Tappel, have cited the role of vitamin E as a biological antioxidant and have concluded that selenium also functions as an antioxidant. Here there is some experimental support, because some organo-selenium compounds are effective antioxidants. They are often only 2–3 times better than their sulfur analogues, however, and since sulfur is present in the body at a level 1,000–10,000 times higher than selenium, it is difficult to see how selenium could make a significant contribution to the relatively abundant pool of sulfur antioxidants[223] even if this pool were reduced by sulfur amino acid deficiency. The ineffectiveness of selenium in preventing diseases such as encephalomalacia in chicks, that respond to a variety of fat-soluble or water-soluble antioxidants, further weakens this theory of selenium function.

These two proposals have in common the idea that vitamin E or selenium function independently of each other but still bring about the same result, i.e., a metabolic product or an antioxidant effect. Such theories of independent function, however, were based upon the apparent ability of vitamin E to substitute for selenium. The fact that all diets contain selenium, in minute amounts, but nevertheless significant relative to physiologically active levels, has always left open the possibility that supplemental vitamin E was effective by itself only because it interacted with a marginal level of selenium to preserve a critical level of function. Two recent studies have now provided strong evidence that selenium is an irreplacable micronutrient in the chick and the rat, two species in which vitamin E was believed by many to substitute for selenium.

Thompson and Scott[236] investigated the nutritional role of selenium in chicks which were rigorously deprived of selenium, using diets composed of crystalline amino acids which were fed to chicks from selenium-depleted hens. The level of selenium in the basal diet was 0.005 ppm. Chicks fed the diet supplemented with 10 ppm of d-α-tocopheryl acetate and 125 ppm of an antioxidant (Ethoxyquin) grew poorly and died in a few weeks. The addition of 0.02 to 0.05 ppm of selenium as selenite to this diet maintained growth and prevented death. This effect of selenium could not be duplicated by vitamin E at levels as high as 200 ppm. Higher levels of vitamin E prevented mortality, but even 1000 ppm of vitamin E did not duplicate the growth promoting effect of the selenium supplement.

The requirement for selenium was shown to be dependent on the level of vitamin E.   When a generous level of vitamin E was fed, the requirement was 0.02 ppm or less, but increased to about 0.05 ppm when low levels of the vitamin were fed.   Even when attempts were made to eliminate vitamin E from the diet completely, the requirement for selenium did not exceed 0.05 ppm in the presence of the synthetic antioxidant.

It is therefore clear that performance with diets containing a marginal level of selenium (in this case, about 0.02 ppm) can be either adequate or inadequate, depending on the dietary content of vitamin E.   This explains why chicks fed semipurified diets based on *Torula* yeast and soybean protein respond to either vitamin E or selenium.   Thompson and Scott confirmed this point in the same experiments, demonstrating that 10 ppm of d-α-tocopheryl acetate or 0.04 ppm of selenium gave comparable results in a semipurified diet containing 0.02 to 0.035 ppm of selenium.

McCoy and Weswig[154] have provided evidence that selenium is an essential nutrient in the rat.   Female rats maintained on a *Torula* yeast diet (containing 0.02 ppm of selenium) supplemented with 60 ppm of d-α-tocopheryl acetate grew and reproduced normally, but their offspring did not grow well, were nearly hairless, and failed to reproduce   Supplementation of the second generation deficient animals with 0.1 ppm of selenium improved growth and caused improvement in the hair coat within 2 weeks.   These animals were fertile but had small litters that failed to survive more than a few days. Supplementation with higher levels of vitamin E and other vitamins or with methionine did not duplicate the effects of selenium.

It appears from these studies that chicks and rats rigorously deprived of selenium develop nutritional disorders which respond only to selenium and not to vitamin E.   It is interesting, however, that the disorders observed in the presence of substantial quantities of vitamin E are not the classic ones for chicks and rats (exudative diathesis and liver necrosis, respectively).   Thompson and Scott noted that exudative diathesis appeared only in the terminal stages of the deficiency, if at all.   The major effects in both species under these conditions appear to be on growth and on hair or feather development.

Complicating the nutritional relationships between selenium and vitamin E is a third factor which has not received as much attention—a marginal or deficient level of sulfur amino acids is a frequent characteristic of diets used to produce nutritional disorders responsive to vitamin E and selenium.   The level of sulfur amino acids may be especially important when a marginal level of selenium is present in the diet, even though selenium deficiencies can be produced experimentally in the presence of adequate levels of sulfur amino acids, as in the experiments of Thompson and Scott.[236]

Schwarz, in his studies of dietary liver necrosis, identified selenium as the active substance in a dietary factor called "Factor 3," the other two factors being cystine and vitamin E.   It was subsequently shown that part of the effectiveness of cystine resulted from its contamination with selenium.[204] Nevertheless, Schwarz and Foltz[205] have reported that sulfur amino acids free of selenium increase the effectiveness of vitamin E against liver necrosis approximately tenfold.

Nutritional muscular dystrophy in chicks is especially interesting with regard to sulfur amino acids.   Supplements of either cystine or vitamin E completely prevent the muscular dystrophy.   Selenium is also effective when combined with a low level of vitamin E which in itself has little or no effect on the disease.*   The synergistic effect of selenium and vitamin E has been attributed by Scott and co-workers to an improved utilization of the vitamin brought about by selenium.[45]   No explanation has been given for the effectiveness of cystine against muscular dystrophy, however.   The fact that muscular dystrophy develops in chicks within 48 h after the removal of a protective dietary supplement of sulfur amino acids[46] suggests that the rapid loss of a labile sulfur compound, or a compound whose formation is dependent on a sulfur-containing compound, may be a primary factor in this condition.

In the face of these complex interrelationships between vitamin E and antioxidants, selenium, and sulfur amino acids, the introduction of an additional parameter might not at first be welcome.   However, the recent experiments of Diplock et al.[52] brought out the very interesting fact that silver acetate can induce the symptoms of liver necrosis when added to the drinking water of rats consuming a diet that does not by itself produce this lesion, and that vitamin E, antioxidants (DPPD), or selenium give essentially complete protection against the effects of silver on the liver.   These experiments were confirmed by more extensive investigations[24] with both rats and chicks. The induction of liver necrosis in rats was rather specific for silver, since gold chloride produced only slight necrotic changes and none were produced by cupric sulfate.   Arsenic also did not produce liver necrosis when given at a level of 70 ppm as arsanilic acid.   With chicks, the addition of silver had little effect on the liver but produced a high incidence of exudative diathesis and also muscular dystrophy.   The similarity of these patterns to the symptoms of vitamin E and selenium deficiency in the same species is striking. Vitamin E, selenium, sulfur amino acids, and antioxidants, or various combinations of these substances, were effective in preventing the metal-induced lesions in chicks, as in rats.   It should be emphasized that as little as 0.05 ppm of selenium protected against the effects of 130 ppm of silver; even taking into consideration the likelihood that silver is poorly absorbed from the intestinal tract, the effective level of selenium is probably much lower than the level of silver in the tissue.

---

* In a myopathy of skeletal muscle in turkey poults, however, Walter and Jensen found no protection was afforded by cystine, whereas either vitamin E or selenium (1 ppm) prevented the condition.[248]

The experiments of Diplock *et al.* again emphasize the importance of sulfur amino acids to the etiology of selenium- and vitamin E-responsive disorders. Their diets are low in protein and contain only 50–60% of the established requirements for sulfur amino acids.   Parallel experiments with fat-free diets suggest that the lesions induced by silver, originally added in an attempt to accelerate lipid peroxidation and thus create a stress for vitamin E, are not related to gross lipid peroxidation.

Because of the high affinity of metals for organoselenium compounds it is attractive to suggest that silver can induce an apparent deficiency of selenium by complexing some essential organoselenium compound in the liver, leading to liver necrosis.   In this respect, Diplock *et al.*[52] thought it possible that the necrotic lesions produced by silver, on the one hand, and selenium and vitamin E deficiency, on the other, may be biochemically related.   Green and Bunyan,[75] in their recent review, report that electron microscopic studies by Grasso indicate that the nuclear and mitochondrial changes in selenium deficiency and silver poisoning are indeed similar.   Silver poisoning in the chick, although more complex than in the rat, is also felt to be consistent with selenium antagonism by silver.

Sulfate

There is a close chemical similarity of sulfate and selenate which leads to a biological antagonism between these two ions.   Studies with plants[104] and microorganisms[215,254] demonstrated that sulfate effectively reduced the accumulation of selenate in the organisms.   Other studies have shown that selenate metabolism in the rat can also be altered appreciably by the presence of sulfate.   The urinary excretion of selenium following a parenteral dose of sodium selenate was increased nearly threefold in rats given sulfate parentally and in the diet.[63]   Halverson *et al.*[84] also observed that dietary sulfate increased the urinary excretion of selenium by rats fed selenate, and earlier studies demonstrated that injected sulfate decreased blood and liver selenium concentrations following a subacute dose of selenate.[17]   Sulfate has only a slight effect on the urinary excretion of selenium that is administered in the form of selenite.[63]   These specificities also hold true in the case of chronic selenium toxicity; sulfate consistently diminishes the toxicity of selenate but has a smaller effect or none at all on the toxicity of selenite or seleniferous wheat.[83,84,63]

Similarly, the extent to which dietary sulfur may accentuate selenium deficiency would be expected to vary, depending on the forms of sulfur and selenium which are involved.   Studies along this line have indeed given rather diverse results.   Muth *et al.*[159] and Hintz and Hogue[94] reported that the addition of sodium sulfate to the diet reduced the effectiveness of dietary supplements of sodium selenite in preventing white muscle disease. Treatment of alfalfa fields with gypsum increased the sulfate content of the plants and increased the incidence of white muscle disease in sheep consuming these plants.[200] On the other hand, subsequent studies have generally provided little support for a significant effect of sulfur in promoting selenium deficiency.[18,257] Mathias *et al.*[135] found the biological activity of the selenium in alfalfa of low or of high sulfur content was similar, and comparable to that of selenite in preventing liver necrosis in rats, although the selenium in high sulfur alfalfa was slightly less effective against exudative diathesis in chicks. Paulson *et al.*[181] observed that the metabolism of a physiological dose of selenate in lactating ewes was not altered appreciably by supplementing their normal ration with sulfate.

Arsenic

There are a number of interesting and unexplained effects involving arsenic and selenium. It has been known for many years that arsenite can counteract chronic and acute selenium poisoning.[155,170] Earlier attempts to relate the mechanism of the protective action of arsenic to effects on selenium metabolism did not give a consistent picture. In several studies the retention of selenium in tissues was not decreased by protective amounts of arsenic.[190,115,183] In some cases liver selenium was decreased by arsenic,[156,170] but in one case an increase was observed.[33] Nevertheless, there are some striking and well-established effects of arsenic on selenium metabolism.

One example involves the respiratory excretion of dimethyl selenide by rats given sodium selenite, which is almost completely blocked by arsenite.[111,62,167,125] It is paradoxical that arsenic, which prevents selenium toxicity, should inhibit what seems to be a mechanism for detoxifying selenium. This effect of arsenite appears to involve a direct inhibition of the biosynthesis of the volatile compound since arsenite is a remarkably effective inhibitor of the process *in vitro*.[66] A level of 10^{-6} M inhibited the *in vitro* system 50%, and this is especially remarkable because the system contained a 20,000-fold molar excess of monothiols. Moreover, this level of arsenite is at least an order of magnitude less than the level of selenium which might be available in a reduced form, thus the possible trapping of selenium in the form of arsenic-selenium complexes is not an adequate explanation for the arsenite inhibition. It is more likely that arsenic inhibits a catalytic component in the pathway, such as a dithiol enzyme. Inhibition by arsenite or cadmium in the presence of monothiols is a characteristic of enzymes containing a dithiol moiety in accord with the effects of arsenite or cadmium on volatile selenium formation.[66]

The gastrointestinal excretion of injected selenium is greatly increased by arsenic.[62,123]   This effect of arsenic is due to an increased clearance of selenium from the liver into the bile, as shown by direct measurements in rats after cannulation of the bile duct.[124]   One hour after the injection of selenite (0.5 mg Se/kg), 20% of the dose appeared in the bile of arsenite-treated animals compared to only 1% in control animals.   The retention of selenium in the liver of animals given arsenite was correspondingly reduced to only 15% of the dose compared to 41% in the controls.   The stimulatory effect of arsenite (As:Se ratio = 2) on biliary selenium excretion was observed at dosages of selenium ranging from 0.02 to 1.0 mg/kg, although the effect was greatest at high dosages.   It was also shown that arsenite stimulated biliary excretion of selenate (but not of sulfate) and that selenite increased the biliary excretion of arsenite.

The mechanism by which arsenite increases biliary excretion of selenium is unknown.   Possible explanations related to elevation of the blood selenium level by arsenite, or changes in the permeability at the blood/bile barrier, are not consistent with all the data.   The specific ability of arsenic and selenium each to increase the excretion of the other suggests that the formation of a readily excreted seleno-arsenic complex may be involved.[124]

Levander and Baumann suggest that the protective effect of arsenic in selenium poisoning, especially the decrease in liver damage, results from enhanced biliary excretion of selenium.   This view is supported by a recent study[125] which confirmed that arsenic improved the growth of rats chronically poisoned with selenium, decreased the amount of selenium in the liver and kidney, and largely prevented the liver damage caused by selenium.   It was suggested that those studies where an increased fecal excretion of selenium was not observed in animals receiving arsenic[167,183] may reflect a decreasing or marginal effect of arsenic when the amount of selenium administered is decreased.

Heavy Metals

There is considerable evidence that selenium protects animals from the toxic effects of certain heavy metals.   Such antagonisms are interesting in themselves and may prove to be very useful as a probe for investigations of the biological role of selenium.

Cadmium.   The ability of selenium to prevent a cadmium-induced testicular degeneration is a very interesting phenomenon that has been studied by a number of workers.   It is well-known that cadmium chloride administered as a single subtoxic dose selectively damages the rat testis, in contrast to other organs.[175]   Mason *et al.*[132] have concluded that the action of cadmium is to alter the permeability of the capillary endothelium.   It was discovered by Kar *et al.*[108] that selenium protected against the cadmium-induced injury. The remarkable effectiveness of selenium is clearly revealed in the experiments of Mason and Young[133] where the damage produced by cadmium at levels just above the minimum effective dose was prevented by selenium at a level half equimolar to cadmium. Testicular damage from cadmium can also be prevented by zinc or cysteine, but these substances are effective only at considerably higher levels.

Gunn and Gould[76] investigated the mechanism of the protective effect of selenium and found that selenium did not prevent cadmium from reaching the testis, but instead caused a marked and prolonged elevation of testicular cadmium levels. Since cadmium also increased the uptake of selenium by the testis, these authors postulated that cadmium and selenium form a complex that constitutes a nontoxic form of cadmium. Although this appears to be a reasonable explanation, nothing is known about the chemical nature of such a complex. The fact that most of the selenium later leaves the testis while cadmium remains behind suggests a subsequent transfer of cadmium from selenium to some other site in the testis[77] and it follows that cadmium is not irreversibly precipitated in a form such as cadmium selenide.

Although the primary effect of cadmium *in vivo* is believed to be on the circulatory system in the testis rather than a direct effect on the seminiferous tubules, Kar *et al.*[110] have found that very high concentrations (0.02 M) of cadmium chloride cause marked degeneration when incubated *in vitro* with the isolated tubules of the rat testis. These changes were prevented by the addition of an equimolar amount of selenium dioxide or zinc acetate to the incubation medium, or by preincubation of the tubules with selenium prior to the cadmium treatment. Surprisingly, these levels of selenium or zinc were reported to have no effect on the tubules when given alone. Although these effects of selenium and cadmium *in vitro* are interesting, it is doubtful that the high concentration of cadmium required to produce a direct effect on the tubules could be obtained *in vivo* by leakage of cadmium through the damaged capillaries.

The effect of injecting selenium or zinc intratesticularly rather than sub-cutaneously at a remote site was studied by Kar and Kamboj.[109] Both selenium and zinc failed to protect the rat testis from the effects of cadmium chloride when injected locally along with cadmium chloride, whereas both protected against locally injected cadmium chloride when adminstered at the remote site. The authors suggest that failure of selenium and zinc to reach the vulnerable vascular region from an intratesticular location could explain the observations, although this necessarily supposes that cadmium adminis-tered in the same way does reach the vulnerable site.

The same laboratory has investigated the progressive calcification of the seminiferous tubules of the rat testis which occurs between 7 and 90 days after cadmium administration.   A single dose of selenium dioxide injected along with the cadmium prevented the calcification.[233]

In view of the efficacy of selenium for preventing the testicular injury caused by minimal doses of cadmium, Gunn et al.[78] investigated the effect of selenium against the generalized acute toxicity of larger doses of cadmium in mice. They found that the dose of selenium (0.024 mmole/kg) which protected the testis from cadmium (0.012 mmole/kg) also prevented or reduced mortality from larger doses of cadmium (0.044–0.066 mmole/kg).   When higher doses of cadmium were given, a larger dose of selenium (0.072 mmole/kg) was required for protection against lethality.   This dose of selenium had no effect against a dose of mercury, zinc, or cobalt which induced complete mortality, in contrast to its effectiveness against cadmium.   Parizek et al.[179] have also mentioned that rats treated with lethal doses of cadmium survive well when injected simultaneously with selenite or selenomethionine.

The effect of cadmium on the female gonads has been studied in the laboratory of Parizek.   In an early study,[174] no morphological injury to the ovaries could be observed when cadmium was given in a dose known to cause testicular necrosis in males.   A similar dose administered to pregnant rats, however, caused complete destruction of the fetal part of the placenta and death of the fetuses.[176]   Administration of cadmium during the last four days of pregnancy was also lethal for a high percentage of the mothers even though the same dose was well-tolerated by nonpregnant animals.[177]   Parizek et al.[179] have recently investigated the effect of selenium on the toxic effects of cadmium in pregnancy.   Sodium selenite (0.04 mmole/kg) was reported to prevent the placental changes, fetal death, and maternal death produced by a 0.03 mmole/kg dose of cadmium acetate given on the 21st day of pregnancy. Selenium did not decrease the amount of cadmium deposited in the placentae 6 to 10 h after injection, and the cadmium level was in fact markedly elevated.

Mercury.   Selenite has been reported by Parizek and Ostadalova[178] to be a very effective protective agent for rats poisoned with mercury.   A dose of mercuric chloride (0.02 mmole/kg) which killed 39 out of 40 animals caused only one death out of 40 animals given sodium selenite (0.03 mmole/kg) within one hour after the mercury.   Renal necrosis induced by mercuric chloride was also eliminated by a simultaneous dose of sodium selenite.   The mechanism of this protective effect is unknown; it might possibly be associated with some type of mutual antagonism between selenium and mercury although mercury does not protect against a chronic selenosis induced by feeding selenite to rats.[125]   As noted above, Gunn et al.[78] did not observe protection by selenite against lethal doses of mercury (much above LD_{50}).

Thallium.   The prevention of death from thallium poisoning by sodium

selenate was reported by Hollo and Zlatarov.[99]   Sodium selenate administered to animals at a dose of 6–9 mg/kg for several days after a 30 mg/kg dose
of thallium acetate completely prevented death whereas most of the animals
given thallium alone died.   This work was confirmed and extended by
Rusiecki and Brzezinski,[197] who investigated the effect of selenate on thallium
metabolism.   They found that urinary excretion of thallium was not altered
but fecal excretion was increased.   In spite of this effect, thallium storage
in the bones, liver, and kidney was increased 2–3 fold by selenate.   The
finding of an increased level of thallium in tissues of protected animals,
rather than a decrease, is analogous to the affect of selenite on cadmium-
induced testicular necrosis and suggests that selenium compounds combine
with the toxic metals in some way to form a less toxic compound.   Levander
and Argrett[125] found that thallium in turn caused a markedly increased retention of selenium in liver and kidneys of rats given selenite, but did not
alleviate chronic selenium poisoning induced by selenite.

Other Substances

Some evidence exists for an interrelationship between phosphate and selenite.
Bonhorst[16] observed that the inhibitory effect of selenite on yeast respiration
was relieved by phosphate.   Direct evidence for an antagonism between
selenite and phosphate uptake in respiring yeast cells was obtained by Mahl
and Whitehead,[130] who found that phosphate suppressed radioactive selenite uptake, and selenite suppressed radioactive phosphate uptake.   Falcone
and Nickerson[56] also observed competition between phosphate and selenite
for uptake by yeast cells, and uptake of both ions was inhibited by 2,4-
dinitrophenol.   Similarly, an interference of phosphate with selenite transport across the plasma membrane may account for the finding that the
synthesis of dimethyl selenide from selenite in liver slices incubated in
bicarbonate buffer is markedly superior to that obtained in a phosphate
buffer; cell-free systems did not show any indication of an inhibitory effect of
phosphate.[64]

ON THE BIOLOGICAL ROLE OF SELENIUM

An outstanding problem which remains to be elucidated is the biochemical
mechanism by which selenium exerts its very significant beneficial effects on
animals when present in the diet in the parts per billion range.   Beginning
with the identification of selenium as the active substance in a dietary factor
which prevented a fatal liver necrosis in rats[201] and exudative diathesis in
chicks,[180,202] followed by the discovery that selenium prevented white muscle
disease in sheep,[158] extensive evidence has accumulated that selenium may be

an essential trace element for many animal species.   There has been no satisfactory explanation of what selenium does in the decade that has elapsed
since the beneficial effects of selenium were discovered.   Whatever selenium
does, it does it very well, because the amount required to produce its characteristic biological responses seldom exceeds 0.1 ppm in the diet.   This is a
level of activity few of the known micronutrients can match.   The solution of
this interesting problem will undoubtedly clarify the mechanism of some other
outstanding biological processes as well.

Thinking very broadly about possible biological roles for selenium, one
may ask if it has a role in the life processes associated with unicellular organisms, or whether it is involved uniquely with multicellular forms of life.   Such
a classification could direct the search for this role either to aspects of cellular
physiology common to the great majority of cells, such as glucose oxidation or
ATP formation, or to physiological systems characteristic of multicellular
life, such as the circulatory, endocrine, or nervous systems found in higher
animals.

Considerable effort has been made by Schwarz and others to pinpoint a
defect in the citric acid cycle, a metabolic pathway common to almost all
forms of life, as the site affected in selenium deficiency, but this is an unlikely
possibility for several reasons.

In the first place, the ability of selenium to sustain life in animals is well
established, but no comparable evidence for the essentiality of selenium to
microorganisms exists, even though they carry out very similar oxidations.
It is true that *E. coli* cells grown on a rigorously purified medium have low
levels of the enzyme formic dehydrogenase, which can be restored to normal by
the addition of approximately 10^{-8} M selenite to the medium,[187] but this
interesting finding should not obscure a more important fact—glucose
oxidation and the rate of growth were completely normal with or without
added selenite.   If there is a growth requirement for selenium in this organism,
it must be less than 10^{-8} M.

Secondly, *Torula* yeast grows vigorously on a medium very low in selenium,
with the dried cells containing only a few parts per billion of selenium, yet
animals fed diets containing 40% or more of the dried yeast develop a fatal
liver necrosis.   This fact strongly implies that all cellular processes in yeast
occur without the need for selenium, or occur at levels of selenium insufficient to sustain life in higher organisms.   The necrogenic effect, it should be
added, is not due to an innate property of the yeast, since the yeast loses its
necrogenic properties when grown on a different medium containing higher
amounts of selenium.[72]

Finally, serious quantitative difficulties arise when one compares the
concentrations of selenium found *in vivo* to the molarity of the common
enzymes, assuming that a one-for-one stoichiometric relationship is required

for selenium to be considered as a cofactor for these enzymes. From the data described earlier, we may assume that the minimal concentration of selenium consistent with normal function is approximately 0.05 ppm or 1 ppm, for tissues low in selenium (muscle) or high in selenium (kidney), respectively. If all of the selenium in these tissues were present as a single molecular species (which is unlikely) the molarity of this substance would not exceed 6.3×10^{-7} M in muscle and 1.3×10^{-5} M in kidney (Table 9-4). Because of the tendency of selenium to be incorporated into a great variety of compounds, it is likely that the concentration of a given compound is lower by an order of magnitude, giving a probable concentration in the more vulnerable tissues on the order of 10^{-7} to 10^{-8} M. In contrast, when the concentrations of some common enzymes of intermediary metabolism are expressed on the same basis (Table 9-5), values of 10^{-5} to 10^{-6} M are obtained, approximately 100-fold higher than the probable concentration of any selenium cofactor. Even if allowances are made for subcellular compartmentation, it is apparent there is little chance of finding stoichiometric

TABLE 9-4   MOLARITY OF SELENIUM IN TISSUES

Tissue	Total Se (fresh basis) ppm	Total gram atoms Se per kg of tissue
Skeletal muscle	0.05	5.3×10^{-7}
Kidney cortex	1.0	1.3×10^{-5}

TABLE 9-5   APPARENT MOLAR CONCENTRATIONS OF ENZYMES *IN VIVO*[229]

Enzyme	Moles of enzyme per kg (fresh basis)
Glycolytic enzymes (rabbit skeletal muscle)	
Aldolase	4.5×10^{-5}
α-Glycerophosphate dehydrogenase	4×10^{-6}
Phosphoglucomutase	5.5×10^{-6}
Citric acid cycle enzymes (pig heart mitochondria)[a]	
Citrate synthetase	$2.6–6.5 \times 10^{-5}$
Isocitrate dehydrogenase	9.2×10^{-5}
α-Ketoglutarate dehydrogenase	6×10^{-6}
Fumarase	$2.5–8 \times 10^{-6}$
Malate dehydrogenase	$3–5 \times 10^{-5}$
Lipoyl dehydrogenase	1×10^{-5}
Cytochrome a (pig heart mitochondria)[a]	$1.6–2.2 \times 10^{-4}$

[a] The concentrations of these enzymes in moles/kg of tissue, rather than moles/kg of mitochondria, can be estimated by dividing these values by 5.

amounts of selenium in the major cellular proteins.   This line of reasoning is borne out by analyses of purified lipoyl dehydrogenase,[207] or aldolase and myosin,[149] which have failed to reveal selenium in amounts anywhere near that required for it to be a prosthetic group in these proteins.

It may thus be possible to rule out a role for selenium in cellular processes common to most forms of life, and its role as a cofactor for the more abundant enzymes (but not excluding less abundant enzymes).   Attention should therefore be directed to finding a role for selenium in other processes, especially those associated uniquely with higher forms of life.   In the absence of evidence that is essential for plants,[224] we can narrow the search further, concentrating on features unique to animals or which are especially well developed in animal species.   One of the most prominent features of life in animals is the extensive employment of complex membrane systems, and it is natural to suspect a role for selenium in such processes.   It is especially tempting to speculate on roles for selenium involving the plasma membrane of cells, or the "stickiness" of cells in connection with intercellular relationships in multicellular organisms.*

There is one physiological system, the microvascular system, which appears to be prominently involved in many of the pathological conditions induced by dietary deficiencies of vitamin E, selenium, and sulfur amino acids, and in certain specific lesions induced by heavy metals.   Many years ago, Dam and Glavind[41] described a condition in chicks characterized by an exudation of plasma from the capillaries, which they named exudative diathesis.   This was later shown to be preventable by either vitamin E or by very low levels of selenium.   Another disorder of the microcirculatory system occurs in pigs when diets low in vitamin E and selenium are fed which contain oxidatively unstable fat.   This condition, sometimes referred to as "mulberry heart," has been described in detail by Grant,[73] who named it dietetic microangiopathy.   This term refers to vascular lesions in the capillaries and small blood vessels, especially in the myocardium.   Grant produced the condition experimentally in pigs by feeding a diet based on soya meal and showed that either selenium or vitamin E would prevent the disease.

Microcirculatory changes may also be the cause of dietary liver necrosis. In a recent study of the prenecrotic ultrastructural changes in rats fed a basal *Torula* yeast diet, Porta *et al.*[188] have found that a progressive degeneration occurs in the hepatocellular plasma membranes facing the liver sinusoids.

---

*On the other hand, intracellular membrane processes, as in the elaborate compartmentation and shuttle systems operative in mammalian cells, might also be a site of selenium function.   To the extent that these phenomena are more important in animal cells than in cells of microorganisms, an indirect effect of selenium on biological oxidation (for instance, the citric acid cycle) in animals might be imagined even though selenium appears to have no role in the citric acid cycle proper.

This was the only major finding unique to animals in the group fed the basal necrogenic diet.   Changes in mitochondria and endoplasmic reticulum were seen, similar to those reported by earlier investigators, but these changes were nearly as prominent in control animals supplemented with vitamin E or selenium, and therefore could not be regarded as prenecrotic.   Porta *et al.* regard the changes which lead to membrane rupture in the sinusoidal border of hepatocytes as a primary event in the pathogenesis of this form of hepatic necrosis.

The specific effect of cadmium which causes testicular degeneration is believed to involve a mechanism whereby cadmium binds to the capillary endothelium and causes an increase in capillary permeability.   This capillary lesion is prevented by selenium in amounts equimolar with the cadmium. Vitamin E has no effect.[132]   The antagonism of cadmium-induced testicular degeneration might represent only a pharmacological effect of selenium, resulting from the formation of an organoselenium compound with a high affinity for cadmium.   On the other hand, the effects of cadmium on the testis and the effects of silver in producing liver necrosis and exudative diathesis might represent the induction of a selenium-deficient state by heavy metal complexing of an organoselenium compound essential to the integrity of the microvascular system.

Although there is no evidence concerning the possible deterioration of the microcirculatory system as a primary event in the development of white muscle disease, it is not unreasonable to suppose that this selenium-responsive disorder might also involve such a mechanism.   By decreasing the already marginal state of nourishment in the poorly vascularized muscle tissue, a breakdown in microcirculatory function could initiate the pathological changes.

How might one relate vitamin E, selenium, sulfur amino acids, and heavy metals to the function of membranes and tubular structures comprising the microcirculatory system?   In view of the high affinity of heavy metals for sulfur and selenium, the specific biological interactions which occur between selenium and metals, and the prominent effects of metals on the vascular system, it is not difficult to imagine mechanisms in which these substances might be involved.   Presumably the sensitivity of certain membranes to heavy metals is related to the presence of functional components that have an unusually high affinity for heavy metals.   These heavy metal-sensitive compound might be sulfur-rich proteins and organoselenium compounds, functioning together within a membrane matrix, possibly in dynamic processes such as sulfhydryl-disulfide interchange.   Vitamin E, the other factor, could have a secondary but very important role in stabilizing the membrane matrix and its components so that the overall function of a highly ordered system would be preserved.

In conclusion, an attempt has been made to define the biological role of selenium, and focus attention on possible biochemical roles for selenium that seem to be the most likely on the basis of present knowledge. It is suggested that selenium has an essential role in multicellular forms of life, notably animals, involving cellular membranes, especially the membranes of the microvascular system. Although attention has been directed to the most prominent effects of selenium in animals, it was not intended to exclude additional roles in other processes in animals or in other forms of life. Selenium may have a number of distinctly different roles, as different as the role of vitamin A in vision versus its role in the integrity of epithelial tissues.

Acknowledgment   This chapter was largely written while the Author was a member of the Department of Biochemistry at the University of Louisville. The author expresses his appreciation to the staff of the University of Louisville Medical School Library for their splendid cooperation.

REFERENCES

1. Ahluwalia, G. S., and Williams, H. H., *Arch. Biochem. Biophys.*, **117**, 192 (1966).
2. Allaway, W. H., Kubota, J., Losee, F., and Roth, M., *Arch. Environ. Health*, **16**, 342 (1968).
3. Anghileri, L. J., and Marques, R., *Arch. Biochem. Biophys.*, **111**, 580 (1965).
4. Aronow, L., and Kerdel-Vegas, F., *Nature*, **205**, 1185 (1965).
5. Aronsen, K. F., Gynning, I., and Walderskog, B., *Acta Chir. Scand.*, **129**, 624 (1965).
6. Asher, C. J., Evans, C. S., and Johnson, C. M., *Austral. J. Biol. Sci.*, **20**, 737 (1967).
7. Awwad, H. K., Potchen, E. J., Adelstein, S. J., and Dealy. Jr., J. B., *Metabolism*, **15**, 370 (1966).
8. Awwad, H. K., Potchen, E. J., Adelstein, S. J., and Dealy, Jr., J. B., *Metabolism*, **15**, 626 (1966).
9. Bergson, G., *Acta Chem. Scand.*, **11**, 1607 (1957).
10. Benesch, R. E., and Benesch, R., *J. Am. Chem. Soc.*, **80**, 1666 (1958).
11. Bieri, J. G., *Nature*, **184**, 1148 (1959).
12. Black, S., *Ann. Rev. Biochem.*, **32**, 399 (1963).
13. Blau, M., *Biochim. Biophys. Acta*, **49**, 389 (1961).
14. Blau, M., and Manske, R. F., *J. Nuclear Med.*, **2**, 102 (1961).
15. Blau, M., and Bender, M. A., *Radiology*, **78**, 974 (1962).
16. Bonhorst, C. W., *J. Agr. Food Chem.*, **3**, 700 (1955).
17. Bonhorst, C. W., and Palmer, I. S., *J. Agr. Food Chem.*, **5**, 931 (1957).
18. Boyazoglu, P. A., Jordan, R. M., and Meade, R. J., *J. Anim. Sci.*, **26**, 1390 (1967).
19. Boyer, P., in *The Enzymes*, Vol. I, P. Boyer, H. Lardy, and K. Myrbäck, eds., Academic Press, Inc., New York, N.Y., 1959, p. 544.
20. Bremer, J., and Natori, Y., *Biochim. Biophys. Acta*, **44**, 367 (1960).

21. Burk, R. F., Jr., Pearson, W. N., Wood, R. F., and Viteri, F., *Amer. J. Clin. Nutr.* **20**, 723 (1967).
22. Burk R. F., Jr., Whitney, R., Frank, H., and Pearson, W. N., *J. Nutr.*, **95**, 420 (1968).
23. Bull, R. C., and Oldfield, J. E., *J. Nutr.*, **91**, 237 (1967).
24. Bunyan, J., Diplock, A. T., Cawthorne, M. A., and Green, J., *Brit. J. Nutr.*, **22**, 165 (1968).
25. Butler, G. W., and Peterson, P. J., *Austral. J. Biol. Sci.*, **20**, 77 (1967).
26. Byard, J. L., and Baumann, C. A., *Fed. Proc.*, **26**, 476 (1967).
27. Byard, J. L., *Arch. Biochem. Biophys.*, **130**, 556 (1969).
28. Caldwell, K. A., and Tappel, A. L., *Biochemistry*, **3**, 1643 (1964).
29. Caldwell, K. A., and Tappel, A. L., *Arch. Biochem. Biophys.*, **112**, 196 (1965).
30. Caldwell, K. A., and Tappel, A. L., *Arch. Biochem. Biophys.*, **127**, 259 (1968).
31. Campo, R. D., Wengert, P. A., Jr., Tourtellotte, C. D., and Kirsch, M. A., *Biochim. Biophys. Acta*, **124**, 101 (1966).
32. Campo, R. D., Tourtellotte, C. D., and Ledrick, J. W., *Proc. Soc. Exp. Biol. Med.*, **125**, 512 (1967).
33. Carlson, C. W., Guss, P. L., and Olson, O. E., *Poultry Sci.*, **41**, 1987 (1962).
34. Carter, D. L., Brown, M. J., Allaway, W. H., and Cary, E. E., *Agron. J.*, **60**, 532 (1968).
35. Cavallini, D., Scandurra, R., and Piccinini, G., *Ital. J. Biochem.*, **14**, 261 (1965).
36. Challenger, F., *Adv. Enzymol.*, **12**, 429 (1951).
37. Christian, G. D., and Michaelis, M., *Invest. Ophthalmol.*, **5**, 248 (1966).
38. Cousins, F. R., and Cairney, I. M., *Austral. J. Agric. Res.*, **12**, 927 (1961).
39. Cowie, D. B., and Cohen, G. N., *Biochim. Biophys. Acta*, **26**, 252 (1957).
40. Cummins, L. M., and Martin, J. L., *Biochemistry*, **6**, 3162 (1967).
41. Dam, H., and Glavind, J., *Nature*, **142**, 1077 (1938).
42. Davies, E. B., and Watkinson, J. H., *New Zealand J. Agr. Res.*, **9**, 317 (1966).
43. Demiruren, A. S., and Slen, S. B., *Canad. J. Animal Sci.*, **43**, 233 (1963).
44. Denison, G. H., and Condit, P. C., *Indust. Eng. Chem.* **41**, 944 (1949).
45. Desai, I. D., and Scott, M. L., *Arch. Biochem. Biophys.*, **110**, 309 (1965).
46. Desai, I. D., Calvert, C. C., and Scott, M. L., *Arch. Biochem. Biophys.*, **108**, 60 (1964).
47. DeWitt, W. B., and Schwarz, K., *Experientia*, **14**, 28 (1958).
48. Dickson, R. C., and Tomlinson, R. H., *Clin. Chim. Acta*, **16**, 311 (1967).
49. Dickson, R. C., and Tappel, A. L., *Arch. Biochem. Biophys.*, **131**, 100 (1969).
50. Dickson, R. C., and Tappel, A. L., *Arch. Biochem. Biophys.*, **130**, 547 (1969).
51. Dingwall, D., *J. Pharm. Pharmacol.*, **14**, 765 (1962).
52. Diplock, A. T., Green, J., Bunyan, J., McHale, D., and Muthy, I. R., *Brit. J. Nutr.*, **21**, 115 (1967).
53. Ehlig, C. F., Allaway, W. H., Cary, E. E., and Kubota, J., *Agron. J.*, **60**, 43 (1968).
54. Evans, C. S., Asher, C. J., and Johnson, C. M., *Austral. J. Biol. Sci.*, **21**, 13 (1968).
55. Ewan, R. C., Baumann, C. A., and Pope A. L., *Agr. Food Chem.*, **16**, 212, (1968).
56. Falcone, G., and Nickerson, W. J., *J. Bacteriol.*, **85**, 754 (1963).
57. Fels, I. G., and Cheldelin, V. H., *J. Biol. Chem.*, **185**, 803 (1950).
58. Franke, K. W., and Moxon, A. L., *J. Pharmacol. Exptl. Therap.*, **58**, 454 (1936).

59. Frank, W., *Z. Physiol. Chem.*, **339**, 214 (1964).
60. Frank, W., *Z. Physiol. Chem.*, **339**, 222 (1964).
61. Galambos, J. T., and Green, I., *Biochim. Biophys. Acta*, **83**, 204 (1964).
62. Ganther, H. E., and Baumann, C. A., *J. Nutr.*, **77**, 210 (1962).
63. Ganther, H. E., and Baumann, C. A., *J. Nutr.*, **77**, 408 (1962).
64. Ganther, H. E., Ph.D. Thesis, University of Wisconsin, 1963.
65. Ganther, H. E., *World Review of Nutrition and Dietetics*, **5**, 338 (1965).
66. Ganther, H. E., *Biochemistry*, **5**, 1089 (1966).
67. Ganther, H. E., Levander, O. A., and Baumann, C. A., *J. Nutr.*, **88**, 55 (1966).
68. Ganther, H. E., *Biochemistry*, **7**, 2898 (1968).
69. Ganther, H. E., In: *Symposium on Trace Element Metabolism in Animals*, C. F. Mills, ed., Livingston, Edinburgh, 1970, p. 212.
70. Ganther, H. E., and Corcoran, C., *Biochemistry*, **8**, 2557 (1969).
71. Gardner, R. W., and Hogue, D. E., *J. Nutr.*, **93**, 418 (1967).
72. Gitler, C., Sunde, M. L., and Baumann, C. A., *J, Nutr.*, **63**, 399 (1957).
73. Grant, C. A., *Acta Vet. Scand.*, **2** (Suppl. 3), 1961.
74. Gray, C. T., and Gest, H., *Science*, **148**, 186 (1965).
75. Green, J., and Bunyan, J., *Nutr. Abstr. Rev.*, **39**, 321 (1969).
76. Gunn, S. A., and Gould, T. C., in *Selenium in Biomedicine*, O. H. Muth, ed., Avi Publishing Co., Westport, Conn., 1967, pp. 395–413.
77. Gunn, S. A., Gould, T. C., and Anderson, W. A. D., *J. Reprod. Fert.*, **15**, 65 (1968).
78. Gunn, S. A., Gould, T. C., and Anderson, W. A. D., *Proc. Soc. Exp. Biol. Med.*, **128**, 591 (1968).
79. Günther, W. H. H., and Mautner, H. G., *J. Amer. Chem. Soc.*, **82**, 2762 (1960).
80. Günther, W. H. H., and Mautner, H. G., *J. Amer. Chem. Soc.*, **87**, 2708 (1965).
81. Haber, E., and Anfinsen, C. B., *J. Biol. Chem.*, **237**, 1839 (1962).
82. Hahn, G. A., and Brown, J. W., *Biochim. Biophys. Acta*, **146**, 264 (1967).
83. Halverson, A. W., and Monty, K. J., *J. Nutr.*, **70**, 100 (1960).
84. Halverson, A. W., Guss, P. L., and Olson, O. E., *J. Nutr.*, **77**, 459 (1962).
85. Hamilton, J. W., and Tappel, A. L., *J. Nutr.*, **79**, 493 (1963).
86. Hansson, E., and Blau, M., *Biochem. Biophys. Res. Comm.*, **13**, 71 (1963).
87. Hansson, E., and Jacobsson, S. O., *Biochim. Biophys. Acta*, **115**, 285 (1966).
88. Hartley, W. J., and Grant, A. B., *Fed. Proc.*, **20**, 679 (1961).
89. Hartley, W. J., in *Selenium in Biomedicine*, O. H. Muth, ed., Avi Publishing Co., Inc., Westport, Conn., 1967, pp. 79–96.
90. Hedegaard, J., Falcone, G., and Calabro, S., *Compt. Rend. Soc. Biol.*, **187**, 280 (1963).
91. Hidiroglou, M., Jenkins, K., Carson, R. B., and Brossard, G. A., *Canad. J. Physiol. Pharmacol.*, **45**, 568 (1967).
92. Hidiroglou, M., Heaney, D. P., and Jenkins, K. J., *Canad. J. Physiol. Pharmacol.*, **46**, 229 (1968).
93. Hidiroglou, M., Jenkins, K. J., Carson, R. B., and MacKay, R. R., *Canad. J. Animal Sci.*, **48**, 335 (1968).
94. Hintz, H. F., and Hogue, D. E., *J. Nutr.*, **82**, 495 (1964).
95. Hirooka, T., and Galambos, J. T., *Biochim. Biophys. Acta*, **130**, 313 (1966).
96. Hoffman, J. L., McConnell, K. P., and Carpenter, D. R., *Fed. Proc.*, **28**, 860 (1969).
97. Hofmeister, F., *Arch. exp. Path. Pharmakol.*, **33**, 198 (1894).
98. Holker, J. R., and Speakman, J. B., *J. Appl. Chem.*, **8**, 1 (1958).

99. Hollo, Z. M., and S. Zlatarov, *Naturwissenchaften*, **4**, 87 (1960).
100. Horn, M. J., and D. B. Jones, *J. Am. Chem. Soc.*, **62**, 234 (1940).
101. Huber, R. E., and R. S. Criddle, *Arch. Biochem. Biophys.*, **122**, 164 (1967).
102. Huber, R. E., and Criddle, R. S., *Biochim. Biophys. Acta*, **141**, 587 (1967).
103. Huber, R. E., Segel, I. H., and Criddle, R. S., *Biochim. Biophys. Acta*, **141**, 573 (1967).
104. Hurd-Karrer, A. M., *Am. J. Bot.*, **25**, 666 (1938).
105. Jacobsson, S. O., *Acta Vet. Scand.*, **7**, 303 (1966).
106. Jenkins, K. J., and Hidiroglou, M., *Can. J. Biochem.*, **45**, 1027 (1967).
107. Jenkins, K. J., *Canad. J. Biochem.*, **46**, 1417 (1968).
108. Kar, A. B., Das, R. P., and Mukerji, F. N. I., *Proc. Nat. Inst. Sci. India, B.*, **26**, 40 (1960).
109. Kar, A. B., and Kamboj, V. P., *Indian J. Exp. Biol.*, **3**, 45 (1965).
110. Kar, A. B., Dasgupta, P. R., and Jehan, Q., *Acta Biol. Med. Germ.*, **16**, 665 (1966).
111. Kamstra, L. D., and Bonhorst, C. W., *S. Dakota Acad. Sci., Proc.*, **32**, 72 (1953).
112. Kerdel-Vegas, F., Wagner, F., Russell, P. B., Grant, N. H., Alburn, H. E., Clark, D. E., and Miller, J. A., *Nature*, **205**, 1186 (1965).
113. Kelleher, W. J., and Johnson, M. J., *Anal. Chem.*, **33**, 1429 (1961).
114. Klug, H. L., and Peterson, D. F., *S. Dakota Acad. Sci., Proc.*, **28**, 87 (1949).
115. Klug, H. L., Lampson, G. P., and Moxon, A. L., *S. Dakota Acad. Sci., Proc.*, **24**, 57 (1950).
116. Klug, H. L., and Froom, J. D., *S. Dakota Acad. Sci. Proc.*, **64**, 247 (1965).
117. Kubota, J., Allaway, W. H., Carter, D. L., Cary, E. E., and Lazar, V. A., *Agr. Food Chem.*, **15**, 448 (1967).
118. Lakin, H. W., and Davidson, D. F., in *Selenium in Biomedicine*, O. H. Muth, ed., Avi Publishing Co., Inc., Westport, Conn., 1967, pp. 27–56.
119. Laurem, B., Maatela, J., and Juva, K., *Scand. J. Clin. Lab. Invest.*, **21** (101), 13(1968).
120. Lawrence, P. J., and Lardy, H., *Science*, **146**, 427 (1964).
121. Lawrence, P. J., *Biochemistry*, **8**, 1271 (1969).
122. Lebez, D., Maretić, Z., Gubenšek, F., and Kristan, J., *Toxicon*, **5**, 261 (1968).
123. Levander, O. A., and Baumann, C. A., *Toxicol. Appl. Pharmacol.*, **9**, 98 (1966).
124. Levander, O. A., and Baumann, C. A., *Toxicol. Appl. Pharmacol.*, **9**, 106 (1966).
125. Levander, O. A., and Argrett, L. C., *Toxicol Appl. Pharmacol.*, **14**, 308 (1969).
126. Levine, V. E., *J. Bact.*, **10**, 217 (1925).
127. Lindberg, P., and Siren, M., *Life Sci.*, **5**, 326 (1963).
128. Lindberg, P., and Siren, M., *Acta Vet. Scand.*, **6**, 59 (1965).
129. Lohmann, W., and Moss, Jr., A. J., *J. Nucl. Med.*, **7**, 878 (1966).
130. Mahl, M. C., and Whitehead, E. I., *S. Dakota Acad. Sci., Proc.*, **40**, 93 (1961).
131. Maros, T. N., Fodor, G. P., Kovacs, V. V., and Katonai, B., *J. Nutr.*, **90**, 219 (1966).
132. Mason, K. E., Brown, J. A., Young, J. O., and Nesbit, R. R., *Anat. Rec.*, **149**, 135 (1964).
133. Mason, K. E., and Young, J. O., in *Selenium in Biomedicine*, O. H. Muth, ed., Avi Publishing Co., Westport, Conn. 1967, pp. 383–394.
134. Massey, V., and Williams, Jr., C. H., *J. Biol. Chem.*, **240**, 4470 (1965).
135. Mathias, M. M., Allaway, W. H., Hogue, D. E., Marion, M. V., and Gardner, R. W., *J. Nutr.*, **86**, 213 (1965).

136. Mautner, H. G., *J. Am. Chem. Soc.*, **78**, 5292 (1956).
137. Mautner, H. G., and Günther, W. H., *Biochim. Biophys. Acta*, **36**, 561 (1959).
138. Mautner, H. G., Chu, S. H., Jaffe, J. J., and Sartorelli, A. C., *J. Med. Chem.*, **6**, 36 (1963).
139. Mautner, H. G., *Radioact. Pharm.*, **20**, 409 (1965).
140. Mautner, H. G., *Pharmacol. Rev.*, **19**, 107 (1967).
141. Mazia, D., *Sym. Int. Soc. Cell Biol.*, **6**, 39 (1967).
142. McColloch, R. J., Hamilton, J. W., and Brown, S. K., *Biochem. Biophys. Res. Comm.*, **11**, 7 (1963).
143. McConnell, K. P., *J. Biol. Chem.*, **141**, 427 (1941).
144. McConnell, K. P., and Portman, O. W., *J. Biol. Chem.*, **195**, 277 (1952).
145. McConnell, K. P., and Portman, O. W., *Proc. Soc. Exp. Biol. Med.*, **79**, 230 (1952).
146. McConnell, K. P., and Wabnitz, O. H., *J. Biol. Chem.*, **226**, 765 (1957).
147. McConnell, K. P., Kreamer, A. E., and Roth, D. M., *J. Biol. Chem.*, **234**, 2932 (1959).
148. McConnell, K. P., and Kreamer, A. E., *Proc. Soc. Exp. Biol. Med.*, **105**, 170 (1960).
149. McConnell, K. P., and Roth, D. M., *Proc. Soc. Exp. Biol. Med.*, **120**, 88 (1965).
150. McConnell, K. P., and Cho, G. J., *Am. J. Physiol.*, **208**, 1191 (1965).
151. McConnell, K. P., and Roth, D. M., *Proc. Soc. Exp. Biol. Med.*, **123**, 919 (1966).
152. McConnell, K. P., and Cho, G. J., *Am. J. Physiol.*, **213**, 150 (1967).
153. McCready, R. G. L., Campbell, J. N., and Payne, J. I., *Canad. J. Microbiol.*, **12**, 703 (1966).
154. McCoy, K. E. M., and Weswig, P. H., *J. Nutr.*, **98**, 383 (1969).
155. Moxon, A. L., *Science*, **88**, 81 (1938).
156. Moxon, A. L., and Dubois, K. P., *J. Nutr.*, **18**, 447 (1939).
157. Mudd, S. H., and Cantoni, G. L., *Nature*, **180**, 1052 (1957).
158. Muth, O. H., Oldfield, J. E., Remmert, L. F., and Schubert, J. R., *Science*, **128**, 1090 (1958).
159. Muth, O. H., Schubert, J. R., and Oldfield, J. E., *Am. J. Vet. Res.*, **22**, 466 (1961).
160. Muth, O. H., Oldfield, J. E., and Weswig, P. H., eds., *Selenium in Biomedicine*, Avi Publishing Co., Westport, Conn., 1967.
161. Nickerson, W. J., Taber, W. A., and Falcone, G., *Can. J. Microbiol.* **2**, 575 (1956).
162. Nickerson, W. J., and Falcone, G., *J. Bacteriol.*, **85**, 763 (1963).
163. Nisman, B., and Hirsch, M. L., *Ann. Inst. Pasteur.*, **95**, 615 (1958).
164. Ochoa-Solano, A., and Gitler, C., *J. Nutr.*, **94**, 243 (1968).
165. Ochoa-Solano, A., and Gitler, C., *J. Nutr.*, **94**, 249 (1968).
166. Olcott, H. S., Brown, W. D., and Van der Veen, J., *Nature*, **191**, 1201 (1961).
167. Olson, O. E., Schulte, B. M., Whitehead, E. I., and Halverson, A. W., *J. Agr. Food Chem.*, **11**, 531 (1963).
168. Orme-Johnson, W. H., Hansen, R. E., Beinert, H., Tsibris, J. C. M., Bartholomaus, R. C., and Gunsalus, I. C., *Proc. Nat. Acad. Sci.* (U.S.), **60**, 368 (1968).
169. Painter, E. P., *Chem. Rev.*, **28**, 179 (1941).
170. Palmer, I. S., and Bonhorst, C. W., *J. Agr. Food Chem.*, **5**, 928 (1957).
171. Palmer, I. S., Fischer, D. D., Halverson, A. W., and Olson, O. E., *Biochim. Biophys. Acta*, **177**, 336 (1969).

172. Pan, F., Natori, Y., and Tarver, H., *Biochim. Biophys. Acta.* **93**, 521 (1964).
173. Pan, F., and Tarver, H., *Arch. Biochem. Biophys.*, **119**, 429 (1967).
174. Parizek, J., and Zahor, Z., *Nature*, **177**, 1036 (1956).
175. Parizek, J., *J. Endocrinol.*, **15**, 56 (1957).
176. Parizek, J., *J. Reprod. Fert.*, **7**, 263 (1964).
177. Parizek, J., *J. Reprod. Fert.*, **9**, 111 (1965).
178. Parizek, J., and Ostadalova, I., *Experientia*, **23**, 142 (1967).
179. Parizek, J., Ostadalova, I., Benes, I., and Babicky, A., *J. Reprod. Fert.*, **16**, 507 (1968).
180. Patterson, E. L., Milstrey, R., and Stokstad, E. L. R., *Proc. Soc. Exp. Biol. Med.*, **95**, 617 (1957).
181. Paulson, G. D., Baumann, C. A., and Pope, A. L., *J. Anim. Sci.*, **25**, 1054 (1966).
182. Paulson, G. D., Baumann, C. A., and Pope, A. L., *J. Animal Sci.*, **27**, 497 (1968).
183. Petersen, D. F., Klug, H. L., Harschfield, R. D., and Moxon, A. L., *S. Dakota Acad. Sci., Proc.*, **29**, 123 (1950).
184. Petersen, D. F., *S. Dakota Acad. Sci., Proc.*, **30**, 53 (1951).
185. Peterson, P. J., and Butler, G. W., *Austral. J. Biol. Sci.*, **15**, 126 (1962).
186. Peterson, P. J., and Butler, G. W., *Nature*, **213**, 599 (1967).
187. Pinsent, J., *Biochem. J.*, **57**, 10 (1954).
188. Porta, E. A., Iglesia, F. A., and Hartroft, W. S., *Lab. Invest.*, **18**, 283 (1968).
189. Rapkine, L., *Ann. Physiol. Physiochim. Biol.*, **7**, 382 (1931).
190. Rhian, M., and Moxon, A. L., *J. Pharmacol. Exp. Therap.*, **79**, 249 (1943).
191. Roberts, M., *Toxicol. Appl. Pharmacol.*, **5**, 485 (1963A).
192. Roberts, M., *Toxicol. Appl. Pharmacol.*, **5**, 500 (1963B).
193. Rosenfeld, I., and Beath, O. A., *Selenium*, Academic Press, Inc., New York, N.Y., 1964.
194. Rosenfield, I., *Proc. Soc. Exp. Biol. Med.*, **111**, 670 (1962).
195. Rounds, D. E., Raiborn, Jr., C. W., Massey, J. F., and Pomerat, C. M., *Texas Rep. Biol. Med.*, **23**, 402 (1965).
196. Rulon, O., *Physiol. Zool.*, **25**, 333 (1952).
197. Rusiecki, W., and Brzezinski, J., *Acta Pol. Pharmac.*, **23**, 74 (1966).
198. Sakai, H., *J. Biol. Chem.*, **242**, 1458 (1967).
199. Sanger, F., *Nature*, **171**, 1025 (1953).
200. Schubert, J. R., Muth O. H., Oldfield, J. E., and Remmert, L. F., *Fed. Proc.*, **20**, 689 (1961).
201. Schwarz, K., and Foltz, C. M., *J. Amer. Chem. Soc.*, **79**, 3292 (1957).
202. Schwarz, K., Bieri, J. G., Briggs, B. M., and Scott, M. L., *Proc. Soc. Exp. Biol. Med.*, **95**, 621 (1957).
203. Schwarz, K., and Foltz, C. M., *J. Biol. Chem.* **233**, 245 (1958).
204. Schwarz, K., Stesney, J. P., and Foltz, C. M., *Metabolism*, **8**, 88 (1959).
205. Schwarz, K., and Foltz, C. M., *Fed. Proc.*, **19**, 421 (1960).
206. Schwarz, K., *Fed. Proc.*, **20**, 666 (1961).
207. Schwarz, K., *Vitamins and Hormones*, **20**, 463 (1962).
208. Schwarz, K., and Sweeney, E., *Fed. Proc.*, **23**, 421 (1964).
209. Schwarz, K., *Fed. Proc.*, **24**, 58 (1965).
210. Schwarz, K., and Fredga, A., *J. Biol. Chem.*, **244**, 2103 (1969).
211. Scott, K. A., and Mautner, H. G., *Biochem. Pharmacol.*, **13**, 907 (1964).
212. Sentein, P., *Chromosoma*, **23**, 95 (1967).

213. Shimazu, F., and Tappel, A. L., *Radiation Res.*, **23**, 210 (1964).
214. Shimazu, F., and Tappel, A. L., *Science*, **143**, 369 (1964).
215. Shrift, A., *Am. J. Bot.*, **41**, 223 (1954).
216. Shrift, A., *Am. J. Bot.*, **41**, 345 (1954).
217. Shrift, A., *Fed. Proc.*, **20**, 695 (1961).
218. Shrift, A., and Kelly, E., *Nature*, **195**, 732 (1962).
219. Shrift, A., and Sproul, M., *Biochim. Biophys. Acta*, **71**, 332 (1963).
220. Shrift, A., and Virupaksha, T. K., *Biochim. Biophys. Acta*, **71**, 483 (1963).
221. Shrift, A., and Virupaksha, T. K., *Biochim. Biophys. Acta*, **100**, 65 (1965).
222. Shrift, A., in *Selenium in Biomedicine*, O. H. Muth, ed., Avi Publishing Co., Westport, Conn., 1967, pp. 241–271.
223. Shrift. A., in discussion of paper by A. L. Tappel, *Selenium in Biomedicine*, O. H. Muth, ed., Avi Publishing Co., Westport, Conn., 1967, p. 357.
224. Shrift, A., *Ann. Rev. Plant Physiol.*, **20**: 475 (1969).
225. Siren, M. J., *Science Tools*, **11**, 37 (1964).
226. Skupin, J., *Rocz. Chem.*, **36**, 631 (1962).
227. Spare, C. G., and Virtanen, A. I., *Acta Chem. Scand.*, **18**, 280 (1964).
228. Spencer, R. P., and Blau, M., *Science*, **136**, 155 (1962).
229. Srere, P. A., *Science*, **158**, 936 (1967).
230. Stekol, J. A., *J. Am. Chem. Soc.*, **64**, 1742 (1942).
231. Sternberg, J., and Imbach, A., *Intern. J. Appl. Rad. Isotopes.*, **18**, 557 (1967).
232. Szczepkowski, T. W., and Wood, J. L., *Biochim. Biophys. Acta*, **139**, 469 (1967).
233. Takkar, G. L., Chowdhury, S. R., Kar, A. B., and Kamboj, V. P., *Acta Biol. Med. Germ.*, **20**, 97 (1968).
234. Tappel, A. L., and Caldwell, K. A., in *Selenium in Biomedicine*, O. H. Muth, ed., Avi Publishing Co., Westport, Conn., 1967, pp. 345–361.
235. Taussky, H. H., Wahington, A., Zubillaga, E., and Milhorat, A. T., *Nature*, **210**, 949 (1966).
236. Thompson, J. N., and Scott, M. L., *J. Nutr.*, **97**, 335 (1969).
237. Tilton, R. C., Gunner, H. B., and Litsky, W., *Canad. J. Microbiol.*, **13**, 1183 (1967).
238. Tilton, R. C., Gunner, H. B., and Litsky, W., *Canad. J. Microbiol.*, **13**, 1175 (1967).
239. Trelease, S. F., Disomma, A. A., and Jacobs, A. L., *Science*, **132**, 618 (1960).
240. Tsen, C. C., and Tappel, A. L., *J. Biol. Chem.*, **233**, 1230 (1958).
241. Tsibris, J. C. M., Namtvedt, M. J., and Gunsalus, I. C., *Biochem. Biophys. Res. Comm.*, **30**, 323 (1968).
242. Tuve, T. W., and Williams, H. H., *J. Am. Chem. Soc.*, **79**, 5830 (1957).
243. Tuve, T., and Williams, H. H., *J. Biol. Chem.*, **238**, 597 (1961).
244. Virupaksha, T. K., and Shrift, A., *Biochim. Biophys. Acta*, **74**, 791 (1963).
245. Virupaksha, T. K., and Shrift, A., *Biochim. Biophys. Acta*, **107**, 69 (1965).
246. Virupaksha, T. K., Shrift, A., and Tarver, H., *Biochim. Biophys. Acta*, **130**, 45 (1966).
247. Walker, G. W. R., and Ting, K. P., *Canad. J. Genet. Cytol.*, **9**, 314 (1967).
248. Walter, E. D., and Jensen, L. S., *Poultry Sci.*, **43**, 919 (1964).
249. Walter, R., and duVigneaud, V., *J. Amer. Chem. Soc.*, **87**, 4192 (1965).
250. Walter, R., Schlesinger, D. H., and Schwartz, I. L., *Anal. Biochem.*, **27**, 231 (1969).
251. Watkinson, J. H., *Nature*, **202**, 1239 (1964).
252. Webb, G. D., and Mautner, H. G., *Biochem. Pharmacol.*, **15**, 2105 (1966).

253. Weiss, K. F., Ayres, J. C., and Kraft, A. A., *J. Bacteriol.* **90**, 857 (1965).
254. Weissmann, G. S., and Trelease, S. F., *Am. J. Bot.*, **42**, 489 (1955).
255. Werz, G., *Planta*, **57**, 250 (1961).
256. Westfall, B. B., and Smith, M. I., *J. Pharmacol. Exptl. Therap.*, **72**, 245 (1941).
257. Whanger, P. D., Muth, O. H., Oldfield, J. E., and Weswig, P. H., *J. Nutr.*, **97**, 553 (1969).
258. Whiteley, H. R., and Woolfolk, C. A., *Biochem. Biophys. Res. Comm.*, **9**, 517 (1962).
259. Wilson, L. G., and Bandurski, R. S., *J. Biol. Chem.* **233**, 975 (1958).
260. Windle, J. J., Wiersema, A. K., and Tappel, A. L., *J. Chem. Phys.*, **41**, 1996 (1964).
261. Wood, J. L., and Cavallini, D., *Arch. Biochem. Biophys.*, **119**, 368 (1967).
262. Woodbridge, D. T., Thesis, University of London, 1955.
263. Woolfolk, C. A., *J. Bacteriol.*, **84**, 659 (1962).
264. Woolfolk, C. A., and Whiteley, H. R., *J. Bacteriol.*, **84**, 647 (1962).
265. Zalkin, H., Tappel, A. L., and Jordan, J. P., *Arch. Biochem. Biophys.*, **91**, 117 (1960).
266. Zalokar, M., *Arch. Biochem. Biophys.*, **44**, 330 (1953).
267. Zdansky, G., *Arkiv. Kemi*, **29**, 443 (1968).
268. Zuidema, G. D., Kirsch, M., Turcotte, J. G., Gaisford, W. D., Powers, W., and Kowalczyk, R. S., *Ann. Surg.*, **158**, 894 (1963).

ADDENDUM

During the time which has elapsed since this review was written, extraordinary progress has been made in the biochemistry of selenium. It will only be possible in this addendum to summarize some highlights of recent developments.

Selenoproteins

Beyond a doubt, the most significant advance has been the discovery of naturally occurring selenoproteins having catalytic activity. Where there were no such proteins known four years ago, there are now several mammalian or microbial selenoproteins which have been isolated that contain stoichiometric quantities of selenium and have well documented functions in cellular processes.

The work of Rotruck, Hoekstra, and co-workers at the University of Wisconsin led to the recognition of erythrocyte glutathione reductase as a biologically active form of selenium. This represented the first such example. This discovery has greatly clarified the role of selenium in protecting animals against oxidative stress and the nutritional interactions involving selenium, vitamin E, and sulfur amino acids. The enzyme catalyzes the reduction of peroxides by glutathione:

$$2GSH + ROOH \longrightarrow GSSG + ROH + H_2O$$

The enzyme is quite specific for glutathione but either hydrogen peroxide or a variety of lipoperoxides can serve as the peroxide substrate. The importance of glutathione peroxidase in preventing oxidative damage to erythrocytes has been clearly established.[3] A detailed review of this enzyme covers the discovery, properties, and general functions of the enzyme in biological systems.[8]

Reasoning that selenium, on the basis of its nutritional interactions with vitamin E, might logically play a role in preventing oxidative damage, Rotruck *et al.*[23] chose erythrocytes as a convenient system for investigation. It was shown, in contrast to previous reports that selenium was not effective in preventing hemolysis of erythrocytes from vitamin E-deficient animals. that dietary selenium was effective if the cells were incubated with glucose. The glucose-dependent effect of dietary selenium implicated some step in the sequence of enzymic reactions that link glucose oxidation to the destruction of peroxides by reduced glutathione. After it was shown that the GSH concentration was elevated in Se-deficient erythrocytes and was effectively maintained during *in vitro* incubation,[23,24] the defect was clearly not an inability to generate GSH, but the inability to use GSH to protect the cell, presumably by means of glutathone peroxidase. Subsequently, it was shown that erythrocytes from Se-deficient rats had much lower levels of glutathione peroxidase, and that purified glutathione peroxidase from rats given ^{75}Se contained a large portion of the total erythrocyte [75]Se.[25] The highly purified enzyme from sheep erythrocytes was found to contain 0.34% Se, or about 4 g-atoms of selenium per mole[15] This value has been confirmed for the enzyme isolated from bovine erythrocytes.[9] This stoichiometry suggests a 1 : 1 relationship between selenium and each of the four 21,000 molecular weight, apparently equivalent, subunits of the enzyme. The form of selenium in the enzyme is not yet identified.

Glutathione peroxidase provides a conceptual framework to understand how selenium, vitamin E and sulfur amino acids can interact nutritionally. Vitamin E would prevent fatty acid hydroperoxide formation and other types of oxidative stress, while sulfur amino acids (as precursors of glutathione) and selenium are involved in the destruction of peroxides by glutathione peroxidase. It is gratifying that glutathione peroxidase fulfills some expectations in regard to the biological function of Se, as deduced elsewhere in this review. It functions with vitamin E and sulfur amino acids in the preservation of membrane structure. It is unique to animals[8] and no counterpart has thus far been detected in plants or micro-organisms. The concentration of glutathione peroxidase in bovine blood was calculated to be $2.2 \times 10^{-7}M$, compared to $3.3 \times 10^{-6}M$ for catalase.[17] It is thus one of the less abundant proteins and the level is commensurate with tissue levels of selenium.

Of great interest is the recent demonstration by Stadtman and co-workers[29] that a low molecular weight protein containing selenium is part of the glycine reductase system in *Clostridia*. This enzyme system carries out the reductive deamination of glycine, using dithiols or various other electron donors, coupled to the synthesis of ATP:

$$H_2NCH_2COOH + R(SH)_2 + ADP + P_i$$

$$\longrightarrow CH_3COOH + NH_3 + R{\Big\langle}{\overset{S}{\underset{S}{\Big|}}} + ATP$$

The production of protein A in this system was dependent on the presence of selenium in the culture medium used to grow the microorganisms (although growth of the cells was unaffected), and preliminary studies indicate that ^{75}Se-selenite was incorporated stoichiometrically into protein A (1 g-atom Se per 12,000 g of protein). The selenium is tightly bound to the protein but the form of selenium in the protein is not known. After reduction of the labeled protein, alkylation with iodoacetamide destroys its activity and an ^{75}Se-labeled peptide can be isolated from the enzymically hydrolyzed alkylated protein, suggesting the presence of selenium in the form of a selenol group.[27]   It is likely that selenium functions in some oxidation-reduction step in the glycine reductase complex.

Formate dehydrogenase may also be a selenoprotein. Lester and DeMoss[16] have shown that *E. coli* grown on a conventional medium requires selenite and molybdate for the ability to oxidize formate, thus confirming the earlier results of Pinsent with purified media. Activity could not be restored to extracts of Se-deficient cells by the addition of selenite (with or without molybdate) *in vitro*, and the addition of chloramphenicol to cells grown with selenite inhibited the production of formic dehydrogenase, suggesting that the effect of selenium was dependent on protein synthesis. Se deficiency did not affect growth rate, cell yield, or the oxidation of a number of other substrates.[7] Partial purification of formic dehydrogenase from *E. coli* grown in the presence of ^{75}Se-selenite resulted in an enrichment of ^{75}Se in the fraction containing the enzyme activity.[26] The possibility has been raised, however, that in some organisms, the same seleno-protein may be a subunit of both the glycine reductase and the formate dehydro-genase enzyme system.[29]

An additional low molecular weight selenoprotein apparently is present in lamb muscle.[20]   A protein which binds ^{75}Se of approximately 10,000 molecular weight was purified from muscle of lambs given a dose of ^{75}Se-selenite 16 h earlier. The Se-binding protein was not detected in Se-deficient lambs. Until now there has been no method for detecting this protein other than by its binding of ^{75}Se, but it was reported that the purified protein absorbs in the visible region[30] and may be similar to cytochromes.

The presence of trace amounts of selenium in the proteins of rat liver subcellular fractions in a form volatilized (apparently a hydrogen selenide) by acidification under strictly anaerobic conditions has been demonstrated by Diplock and his co-workers.[4] The acid-labile selenium is found in highest amounts when vitamin E is given to the animals, and additional vitamin E plus 2-mercaptoethanol is employed during the isolation of the subcellular organelles. Diplock[5] has discussed the possibility that this selenide may be part of nonheme iron proteins that function in microsomal electron transfer. Levander *et. al*[16a,b] have described interesting effects of selenium catalyzing electron transfer from thiols to cytochrome *c* and possible implications involving mitochondrial electron transport.

Protection by Selenium Against Heavy Metals

The upsurge of interest in heavy metals as environmental toxicants has stimulated further research on the interactions between selenium and heavy metals, particularly mercury.   The remarkable ability of selenium to reduce the toxicity of mercury, first reported for acute mercuric chloride poisoning by Parizek and Ostadalova,[18] has been confirmed in an enrivonmentally significant setting, and there is now much evidence of a true mutual antagonism between mercury and selenium.

With regard to mercury, it has been shown that selenium is present in sufficient levels in fish to modify the toxicity of methylmercury, the major form accumulated in fresh water fish from mercury contaminated waters and by marine food fish from mercury of natural origin in the oceans.   When rats were given 0.5 ppm of selenium (an amount approximately equal to the nutritional requirement) in the form of dietary sodium selenite, the toxicity of methylmercury hydroxide added to the drinking water at levels supplying from 5–20 ppm mercury was greatly reduced,[10] as shown by improved growth and increased survival time.   At the same time it was shown that Japanese quail given 10 or 20 ppm of mercury as methylmercury in the diet survived longer when 17% of tuna (supplying about 0.5 ppm of selenium) was added to the diet.[10,12]   The addition of sodium selenite to the diet of quail given methylmercury duplicates the effect of adding tuna.[6]   The effectiveness of selenium against mercury toxicity was confirmed independently by Stillings et al.[28] who fed rats 25 ppm of mercury as methylmercury with or without 0.6 ppm of selenium as sodium selenite, and by Potter and Matrone,[21] using rats fed various levels of methylmercury or mercuric chloride with or without 5 ppm of selenium as sodium selenite.

Data on tissue levels of total mercury in animals protected by tuna or added selenium suggest that selenium does not reduce the absorption or increase the excretion of mercury, since mercury levels are frequently as high or higher in tissues of protected animals, but selenium may cause an altered distribution of mercury, particularly in the case of mercuric chloride.[11,2,22]   These results fit with the earlier work from Parizek's group[19] which shows that mercury retention in the body is increased by Se and that Se retention is increased by mercury.

Additional evidence suggesting a true antagonism between mercury and selenium comes from the work of Hill.[13,14]   Growth of chicks fed diets containing 10–40 ppm of selenium as selenite was greatly decreased, but the addition of 500 ppm of Hg as mercuric chloride (a level which in itself decreased growth) largely counteracted the growth retardation induced by selenium.

The ability of a nutritional level of selenium (0.5 ppm) to overcome the growth depressing effects of silver nitrate added to the drinking water (75–750 ppm Ag) has also been observed in this laboratory (P. Wagner, M.S. thesis, University of Wisconsin, 1973).   The marked effect of Se on growth of Ag-treated animals is quite comparable to that obtained with methylmercury under the same conditions. The level of glutathione peroxidase in liver of rats given Se was greatly decreased by the chronic administration of silver.   Silver concentrations in the liver were higher in the animals receiving Se compared to those receiving only silver.   No evidence for selenium antagonizing the toxicity of lead acetate (up to 10% in the diet) was noted.   In a preliminary study with $CdCl_2$, there appeared to be no antagonism between Cd and Se comparable to that for Ag and Se, but this question needs to be explored further.[11]

The mechanism by which selenium decreases the toxicity of metals like mercury and silver probably involves some type of complexing between the metal and some form of selenium. This complexing may involve either biologically active forms of selenium, or selenoproteins formed nonspecifically when an excess of selenium is fed. The selenium compounds would have a high affinity for metals and could bind them in a less available form.[11] The effect of silver on glutathione peroxidase is an example of the first type, where the metal reacts with an essential form of Se. An example of the second type may be found in the prevention of Cd-induced testicular injury.[1] In this instance, Se increases total Cd uptake in the testes and testes soluble fraction, but Se gets incorporated into large molecular weight proteins of the testicular soluble fraction and decreases Cd uptake in a low molecular weight protein that seems to be a target of Cd specifically related to Cd-induced testicular injury.[1] The general picture emerging in regard to selenium antagonisms with mercury is also consistent with the idea that when toxic levels of mercury are given, Se alters the biological activity or distribution of the metal in the tissues, rather than affecting its absorption or elimination from the body, or the rate of conversion of methylmercury to the less toxic mercuric ion.[19,2,11,22]

REFERENCES

1. Chen, R., Wagner, P. A., Hoekstra, W. G., and Ganther, H. E., *J. Reprod. Fert.* (in press).
2. Chen, R. W., Ganther, H. E., and Hoekstra, W. G., *Biochem. Biophys. Res. Comm.* **51**, 383 (1973).
3. Cohen, G., and Hochstein, P., *Biochemistry*, **2**, 1420 (1963).
4. Diplock, A. T., Caygill, C. P. J., Jeffrey, E. H., and Thomas, C., *Biochem. J.* **134**, 283 (1973).
5. Diplock, A. T., in *Trace Element Metabolism in Animals*, J. W. Suttie, W. G. Hoekstra, H. E. Ganther, and W. Mertz, eds., University Park Press, Baltimore (in press).
6. El-Begearmi, M. M., Goudie, C., Ganther, H. E., and Sunde, M. L., *Fed Proc.* **32**, 886 (1973).
7. Enoch, H. G., and Lester, R. L., *J. Bacteriol.* **110**, 1032 (1972).
8. Flohe, L., *Klin. Wschr.*, **49**, 669 (1971).
9. Flohe, L., Gunzler, W. A., and Schock, H. H., *FEBS Letters*, **32**, 132 (1973).
10. Ganther, H. E., Goudie, C., Sunde, M. L., Kopecky, M. J., Wagner, P., Oh, S.-H., and Hoekstra, W. G., *Science* **175**; 1122 (1972).
11. Ganther, H. E., Wagner, P. A., Sunde, M. L., and Hoekstra, W. G., in *Trace Substances in Environmental Health*, Vol. VI, D. D. Hemphill, ed., University of Missouri, Columbia, Mo. 1973, p. 247.
12. Ganther, H. E., and Sunde, M. L., Abstracts, Institute Food Technology Symposium, Elements in the Food Chain: Nutrients and Toxicants, Miami, Fla., June 13, 1973 (*J. Food Science*, submitted for publication).
13. Hill, C. H., *Fed. Proc.* **31**, 692 (1972).
14. Hill, C. H., *Environ. Health Perspect.* Experimental Issue **4**, 104 (1973).
15. Hoekstra, W. G., Hafeman, D., Oh, S. H., Sunde, R. A., and Ganther, H. E., *Fed. Proc.* **32**, 885 (1973).
16. Lester, R. L. and DeMoss, J. A., *J. Bacteriol.* **105**, 1006 (1971).

16a. Levander, O. A., Morris, V.C., and Higgs, D. J., *Biochemistry* **12**, 4586 (1973).
16b. Levander, O. A., Morris, V. C., and Higgs, D. J., *Biochemistry* **12**, 4591 (1973).
17. Nicholls, P., *Biochim. Biophys. Acta*, **279**, 306 (1972).
18. Parizek, J., and Ostadalova, I., *Experientia*, **23**, 142 (1967).
19. Parizek, J., Ostadalova, I., Kalouskova, J., Babicky, A., and Benes, J., in *Newer Trace Elements in Nutrition*. W. Mertz and W. E. Cornatzer, eds., Marcel Dekker, New York, 1971, p. 85.
20. Pedersen, N. D., Whanger, P. D., Weswig, P. H., and Muth, O. H., *Bioinorganic Chem.* **2**, 33 (1972).
21. Potter, S. D., and Matrone, G., *Fed Proc.* **32**, 929 (1973).
22. Potter, S. D. and Matrone, G., *Environ. Health Perspect.*, Experimental Issue **4**, 100 (1973).
23. Rotruck, J. T., Hoekstra, W. G., and Pope, A. L., *Nature New Biol.*, **231**, 223 (1971).
24. Rotruck, J. T., Pope, A. L., Ganther, H. E., and Hoekstra, W. G., *J. Nutr.* **102**, 689 (1972).
25. Rotruck, J. T., Pope, A. L., Ganther, H. E., Swanson, A. B., Hafeman, D. G., and Hoekstra, W. G., *Science* **179**, 588 (1973).
26. Shum, A. C. and Murphy, J. C., *J. Bacteriol.* **110**, 447 (1972).
27. Stadtman, T. C., *Fed. Proc.* **32**, 1443 (1973).
28. Stillings, B., Lagally, H., Soares, J., and Miller, D., Ninth International Congress of Nutrition, Mexico City, Summaria, p. 206 (1972).
29. Turner, D. C., and Stadtman, T. C., *Arch. Biochem. Biophys.*, **154**, 366 (1973).
30. Whanger, P. D., Pedersen, N. D., and Weswig, P. H., Abstracts, Second International Symposium on Trace Element Metabolism in Animals, University of Wisconsin June 18–22, 1973, p. 22.

10

Analytical Chemistry of Selenium

W. CHARLES COOPER*

Noranda Research Centre,
Pointe Claire, Quebec, Canada

INTRODUCTION

The analytical chemistry of selenium is quite well defined both with respect to the determination of macro as well as micro amounts of the element. Its common association with tellurium presents no serious problems since it can be readily separated by volatilization or selective precipitation. The traditional gravimetric and volumetric methods are commonly used for macro determinations of selenium. Although numerous instrumental techniques, including neutron activation analysis, have been applied successfully to the estimation of traces of selenium, the use of 3,3'-diaminobenzidine and more recently 2,3-diaminonaphthalene as reagents for the colorimetric or fluorometric determination of as little as 0.02 μg of selenium has gained wide acceptance. In particular, the fluorometric method has

* Present Address: United Nations Development Programme, Casilla 197-D, Santiago, Chile.

been of considerable importance in the determination of small amounts of selenium in biological materials.

In the present chapter, it has not been possible to cover all aspects of the analysis of selenium. However, a reasonably comprehensive treatment is provided including the separation and isolation of selenium, its detection and identification, determination by various methods, as well as the analysis of selenium itself including impurities therein, and the determination of selenium in specific materials. Earlier reviews on the analytical chemistry of selenium are useful in providing more detailed information on analytical procedures and these should be consulted where appropriate.[41,62,127,140]

SEPARATION AND ISOLATION OF SELENIUM

Decomposition, Dissolution and Other Preliminary Treatment of Inorganic Materials

A broad range of inorganic materials have been analyzed for selenium, many of which are associated with the extraction of the element from its ores and its application in metals, alloys, glass, ceramics and semiconducting compounds. The rather limited terrestrial abundance of selenium and its isomorphous association with sulfur, particularly in sulfide ores, have given rise to selective dissolution and separation methods. Although precautions must be taken to prevent the volatilization of selenium if halogen or hydrohalide media are used during sample dissolution, this volatility of selenium, notably as selenium tetrabromide, affords a convenient method for the separation of selenium from complex inorganic materials.

The dissolution of sulfide ores and concentrates is commonly effected by the use of mixed acids such as nitric, sulfuric and hydrochloric acids,[89] nitric, sulfuric and perchloric acids,[117] or nitric and sulfuric acids.[24] Care must be taken in such procedures to ensure that there remains no undissolved sulfur which might entrap selenium. Such procedures are usually limited to samples smaller than 5 g. A strongly oxidizing mixture of potassium chlorate–nitric acid to which mercuric oxide has been added is apparently effective in preventing any loss of selenium due to volatilization since, under these conditions selenium is oxidized to the hexavalent state.[51] Because the selenium content of ores and concentrates, although small, can vary considerably, it is desirable to have dissolution procedures which provide for ease of sample dissolution and the use of larger samples. To this end, the use of 15% fuming sulfuric acid can be recommended.[41]

In a dissolution procedure which has given excellent results for the determination of small amounts of both selenium and tellurium in sulfide ores, the sample is first treated with nitric acid.   Following the initial attack, the solution is warmed gently to maintain the reaction but not exceeding the melting point of sulfur.   When the reaction appears to be complete, the solution is evaporated cautiously to dryness.   Fuming nitric acid is added to dissolve any sulfur, followed by 70% perchloric acid and evaporation of the solution to strong fumes to expel nitric acid.   Dilute (1 : 1) hydrochloric acid is added and the solution is warmed to dissolve soluble salts.   Any insoluble residue is filtered off and washed with 1 : 1 hydrochloric acid to give a total sample volume of about 150 ml.   Selenium and tellurium are then separated from this solution by precipitation using hypophosphorous acid as reducing agent and arsenic as collector.

The separation of selenium, following dissolution of the sample, can be realized by a number of procedures.   In addition to the arsenic collection method noted above, the distillation of selenium as $SeBr_4$ from hydrobromic acid solution and the precipitation of selenium using tin(II) chloride have been successfully employed.

In the determination of selenium in flue dusts, the sample can be dissolved in sulfuric acid in a Knorr arsenic still and the selenium separated by distillation from hydrobromic acid–bromine solution.[41]   During dissolution, sulfur can be formed.   This causes contamination of the distillate.   This effect, if not too severe, can be prevented by the addition of a small quantity of basic tellurium sulfate (selenium-free) prior to dissolution of the sample. A more serious contamination can be prevented through the use of 15% fuming sulfuric acid instead of sulfuric acid, with the addition of the tellurium salt being maintained.

Smelter slags cannot be dissolved in sulfuric acid nor are they appreciably decomposed by hydrochloric or hydrobromic acids.   Dissolution can be effected conveniently by fusion with sodium peroxide.   However, such a procedure gives rise to the formation of colloidal silica which renders difficult a satisfactory selenium determination.   A suitable approach is through nitric acid oxidation followed by sulfuric acid fuming and hydrobromic acid distillation of the selenium.   Detailed procedures for handling ores, concentrates, flue dusts and smelter slags have been reported.[41]

Although sintering procedures are generally time-consuming and are to be avoided, such methods are sometimes advisable in the case of difficultly decomposable minerals.   One such procedure which has been applied successfully to samples of less than 2 g involves the fusion of the sample with a 4 : 1 : 1 mixture of zinc oxide–magnesium oxide–sodium carbonate mixture.[19] A 1 : 4 mixture of sodium carbonate and zinc oxide has been employed in the dissolution of cinnabar prior to selenium determination.[92]

Sulfur, containing traces of selenium, can be decomposed by treatment with a mixture of nitric acid and bromine in which sulfur and selenium are dissolved and oxidized to sulfuric acid and selenious acid, respectively.[132]

In the determination of selenium in electrolytic copper refinery slime, which constitutes the principal commercial source of selenium, dissolution can be most conveniently realized by the use of a 1 : 1 mixed nitric–sulfuric acid[41]. Decomposition of the sample and the destruction of any organic matter can be effected through addition of a few crystals of potassium chlorate. An alternative procedure which is claimed to provide for accelerated decomposition of anode slimes involves the use of a mixture of hydrogen peroxide and hydrochloric acid.[26]

In the determination of selenium in standard rock samples by neutron activation analysis, Brunfelt and Steinnes[25] decomposed the sample after neutron irradiation by fusion with sodium hydroxide following the addition of the selenium carrier. The resulting fusion cake was treated with sulfuric acid and the selenium separated by distillation from a solution of hydrochloric and hydrobromic acids.

Fusion with alkali followed by distillation of selenium has been found to be a convenient procedure for the determination of selenium in decolorized soda–lime–silica glasses.[54]

In general, the dissolution of metals and semiconducting compounds of selenium presents no serious difficulties. It is worth noting that in recent work on the determination of selenium in iron and steels, the dissolution of the sample has been found to result conveniently through the use of 1 : 1 nitric acid or hydrochloric acid–nitric acid for stainless steels followed by evaporation to fumes with perchloric acid.[67]

Decomposition of Organic Materials Containing Selenium

The widespread interest in selenium in biological materials arising originally from the occurrence of significant amounts of selenium in certain plants and their toxic effect on animals, and the extension of this interest to the nutrient role of traces of selenium in numerous animal species, has given rise to a number of procedures for the decomposition of organic materials containing the element. The various dissolution procedures have been reviewed critically.[140]

The two principal methods which have been used for the decomposition of organic materials in the analysis of selenium are the oxygen flask combustion method and wet oxidation using mixed acids. The increased sensitivity of the selenium determination based on the fluorescence of the complex formed with 3,3'-diaminobenzidine or 2,3-diaminonaphthalene has focused attention on the importance of the dissolution procedure. According to Dye et al.,[50] extraneous fluorescence can result following nitric acid or mixed nitric–

perchloric acid treatment. This fluorescence which has been ascribed to lipids or lipoidal material in the sample cannot be removed by organic extraction nor by absorbants such as charcoal or diatomaceous earth. In the case of plant material, the extraneous influence may be due to unoxidized chlorophyll and/or xanthophyll, both of which are known to fluoresce strongly. These possible interferences have necessitated the separation of selenium prior to final estimation. However, Cummins, Martin and Maag[44] have recommended a more direct procedure which has been found to give excellent results in the analysis of both plant and animal materials. In this method the sample is digested in a mixture of concentrated sulfuric acid and 70% perchloric acid to which sodium molybdate has been added. The isolation of selenium prior to determination with 3,3'-diaminobenzidine has been obviated by the use of ethylenediaminetetraacetic acid (EDTA) which masks interfering ions such as iron, chromium and vanadium. Check analyses by neutron activation have indicated that this method is capable of recovering virtually all of the organically bound selenium. In the application of the oxygen flask combustion method, traces of fluorescent compounds may be produced during combustion and selenium may be retained in the ash from samples which are high in inorganic matter.[140]

The use of mixed nitric and perchloric acids is favored by Watkinson,[140] who cites the following advantages for this mixture in the determination of selenium in biological materials: oxidation capability for relatively large samples per unit weight; virtual elimination of charring and attendant loss of selenium; maximum temperature of about 200°C at which volatilization loss of selenium is negligible; and ease of purification. Liquid samples such as blood can be handled more conveniently and accurately by acid digestion.

The present indications are that the drying of plant and animal tissues containing low or normal concentrations of selenium results in negligible volatilization of the element. Drying temperatures which have been used range from freeze drying,[128] up to air drying at 100°C[140] without appreciable loss of selenium. However, prolonged heating at elevated temperatures should be avoided and special precautions must be taken to avoid volatilization loss when higher levels of selenium are present. A summary of the decomposition methods which have been applied to a variety of organic materials is presented in Table 10-1.

Selective Separation Methods

Selenium can be separated selectively from a number of elements including tellurium by volatilization, precipitation, solvent extraction and ion exchange. The selective separation of selenium is particularly important in the determination of small amounts of the element in complex materials.

TABLE 10-1   METHODS FOR DECOMPOSITION OF ORGANIC MATERIALS
CONTAINING SELENIUM

Material	Oxidation Method	Reference
Plant materials and animal tissues and secretions	Oxygen combustion (Parr bomb and Schöniger flask)	50
Plant materials and animal organs	Oxygen flask combustion (addn $Mg(NO_3)_2$ with animal samples)	74
Biological fluids and tissues	Oxygen flask combustion	128
Animal feeds	Oxygen flask combustion	8
Animal tissue	Na_2MoO_4(10 g/150 ml H_2O) H_2SO_4 (150 ml) 70% $HClO_4$ (200 ml)	44
Animal tissue	$HNO_3 \cdot H_2SO_4 \cdot HClO_4$ (10 : 3 : 5)	38
Animal tissue	H_2SO_4–30% H_2O_2 after neutron irradiation	111
Hay, flour, bone, hair, fat	HNO_3–H_2SO_4 (1 : 4) after neutron irradiation	122
Plant material	HNO_3–60% $HClO_4$ (3 : 1)	49
Soil, hair, wool	HNO_3–70% $HClO_4$ (4 : 1)	74
Soil	HNO_3–H_2SO_4 (2 : 1)	53

Volatilization of selenium as selenium tetrabromide from a bromine–hydrobromic acid solution is a separation procedure which has been applied extensively.   The hydrobromic acid distillation of a sulfuric acid solution containing selenium, tellurium and other elements gives a distillate of selenium which is relatively free from tellurium, iron, and copper but which can contain arsenic, antimony, and germanium.

Precipitation procedures can be used for the separation of both micro and macro quantities of selenium.   Larger amounts of selenium can be separated readily from tellurium by Keller's method which is based on the fact that selenium is precipitated quantitatively by sulfur dioxide from concentrated hydrochloric acid solution, while tellurium is not.   Besides sulfur dioxide, sodium hypophosphite in 6–7N hydrochloric acid solution can be used for the selective precipitation of selenium and its separation from tellurium.[9]   In the reverse separation, $Na_2S_2O_5$ acts as a selective reductant for tellurium(IV) in the presence of selenium(IV).   In this case the selenium remains in solution as the selenosulfate ion.[82]

Precipitation methods are very useful for the collection of small amounts of selenium, thereby affording an important preconcentration of the element. Small amounts of selenium can be recovered quantitatively by coprecipitation of selenium and arsenic using as reducing agent, hypophosphorous acid.

Tellurium is also separated quantitatively by this technique. Another preconcentration precipitation method in which both selenium and tellurium are recovered utilizes freshly prepared tin(II) chloride solution. Other reducing agents which have been employed include hydroxylamine hydrochloride and hydrazine hydrochloride.

The separation of selenium (and tellurium) by coprecipitation with metal hydroxides has long been used for the purpose of separating the two elements in their subsequent determination in metals such as copper. In a study of the separation of selenium by this method, Plotnikov[105] demonstrated that sparingly soluble metal selenites are formed such as $Fe_2(SeO_3)_3$ which are entrained in the precipitated metal hydroxide which acts as a carrier.

The separation of selenium by solvent extraction can be effected by extraction into benzene or toluene as elemental selenium or as the complex which is formed between selenium and 3,3'-diaminobenzidine or 2,3-diaminonaphthalene. In some instances selenium can be determined directly in the organic phase by colorimetric methods.

Selenium(IV) and tellurium(IV) as chlorides or bromides can be extracted with diantipyrilpropylmethane.[31]

The separation of selenium(IV) from elements such as iron, aluminum, cobalt, manganese, and zinc can be realized by means of an H^+ cation exchanger prior to the determination of selenium polarographically,[112] or colorimetrically.[81] According to Yoshino,[144] selenite ions enter the effluent quantitatively at pH 0.7–5.0.

Zelyanskaya and Gorshkova[145] separated copper in ammoniacal solution from small amounts of selenium by cation exchange. Following the sorption step, the resin was rinsed with dilute ammonia (1 : 9) and selenium was determined in the combined effluents. Selenium(IV) and tellurium(IV) are retained effectively by strongly basic anion resins. The separation of tellurium and selenium can be effected by elution with $0.5M$ NaOH–$3M$ NH$_3$.[64] Traces of selenium can be separated from telluric acid by the selective sorption of selenium using an anion exchanger such as the acetate form of a strongly basic resin.[137]

DETECTION AND IDENTIFICATION

Selenium is commonly detected by precipitation with sulfur dioxide in hydrochloric acid solution. The cold solution of selenium(IV) or selenium(VI) in concentrated hydrochloric acid, when gassed with sulfur dioxide, gives a red amorphous precipitate of selenium which, on warming, goes over to the grey trigonal crystalline modification. There is no interference on the part of tellurium if the hydrochloric acid concentration exceeds $8.8N$. The hydrogen sulfide gassing of a selenium(IV) solution gives a yellow precipitate of selenium sulfide which, on standing, dissociates in the sulfur and red amorphous selenium.

Other reducing agents which can be used to reduce selenium(IV) in the cold to give red elemental selenium are tin(II) chloride, iron(II) sulfate, hydroxyl-amine hydrochloride, hydrazine sulfate, hypophosphorous acid and ascorbic acid. The reduction of selenium(IV) to elemental selenium by thiocyanate in 6N hydrochloric acid solution is said to permit the detection of as little as 0.05 ppm of selenium.[78] Potassium iodide, when added in excess to a cold hydrochloric acid solution of selenium(IV) or selenium(VI), gives a red elemental selenium together with iodine. On warming, the iodine distills and red selenium goes over to the grey trigonal form.

When gently warmed with concentrated sulfuric acid or when treated with fuming sulfuric acid in the cold, selenium and selenides undergo oxidation to give a green color which has been attributed to the formation of the cations Se_4^{+2} and Se_8^{+2}.[18] The color varies in intensity from a light green to an almost opaque greenish black, depending upon the amount of selenium present. In concentrated sulfuric acid, the green color is destroyed by warming the solution for a few minutes. When the solution is added to water, red ele-mental selenium is precipitated. It should be noted that this test is not applicable to an oxidized selenium compound.

In the detection of selenium in complex inorganic materials, the sample can be treated with aqua regia or with a mixture of hydrochloric acid and potas-sium chlorate. The free chlorine must be expelled at a temperature below boiling to avoid loss of volatile selenium chlorides. Materials which are difficultly soluble in acid media can be solubilized by fusion in a nickel crucible with an alkaline oxidizing flux. Following dilution and filtration to remove insoluble matter, the acid solution is gassed with sulfur dioxide. A precipitate is indicative of the possible presence of selenium, tellurium, or gold. Selenium and tellurium can be separated from the gold by dissolution in concentrated nitric acid. Careful evaporation with hydrochloric acid serves to destroy the nitric acid and subsequent gassing of this hydrochloric acid solution with sulfur dioxide can serve to indicate the presence of selenium via the formation of a red precipitate.

The reaction of selenium with mercury(II) cyanide to form selenium cyanide provides the basis of a spot test for selenium.[52] The selenium cyanide hydrolyzes in the vapor phase to yield hydrocyanic acid, which is detected by the blue color formed with copper acetate and benzidine acetate reagent. This test can be used to detect as little as 8 μg of selenium. How-ever, sulfur and tellurium behave in a similar manner.

By thin layer chromatography on alumina, Gaibakyan and Aturyan[56] were able to detect selenium and tellurium using solutions of methyl, ethyl and propyl alcohols containing hydrochloric acid. For identification, the film was sprayed with tin(II) chloride solution, with selenium being detected by the appearance of a light brown color and tellurium by a black coloration.

A similar procedure has been developed by Larsen[75] for the paper chromatographic identification of selenium in the presence of tellurium, gold, and platinum.

A field test capable of detecting as little as 2 μg of selenium has been developed by Niebuhr and Macmillan.[93] This test is based on the precipitation of elemental selenium with tin(II) chloride following initial repeated extraction of the powdered sample with 30% H_2O_2 and concentrated hydrochloric acid and separation of the hydroxides of iron, nickel, copper, aluminum and bismuth. Tellurium, which forms a black precipitate, can be dissolved by dilute hydrogen peroxide thereby exposing any red elemental selenium.

DETERMINATION OF SELENIUM

Precipitation and Gravimetric Methods of Determination

Selenium is commonly determined by precipitation as the element by reduction in a hydrochloric acid solution of selenium(IV) or selenium(VI) with sulfur dioxide. As mentioned previously, this precipitation, if carried out in hydrochloric acid solution more concentrated than 8.8N, serves to separate selenium from tellurium. Other reducing agents which can be used are hydroxylamine hydrochloride and hydrazine hydrochloride or hydrazine sulfate. The selenium should be filtered on an asbestos padded Gooch crucible, and after washing and drying at 105°C, weighed, ignited, and the weight loss determined. In this way any error from gold and most of the errors due to occlusions can be eliminated.

Although the precipitation of elemental selenium cannot be recommended as a method for the determination of selenium in refined selenium, it has been employed for the microgravimetric determination of small amounts of selenium.[84] The loss upon ignition of the precipitated selenium is determined to 0.01 mg in a 3 ml Gooch crucible. As little as 20 ppm selenium in a 20 g sample can be determined by this method.

Heavy metal selenites or selenates, e.g., barium selenite, have been suggested as a possible means of determining selenium gravimetrically. However, the solubility of many of these compounds, e.g., those of zinc, cadmium, and mercury,[108] is too high for an accurate analysis.

The piazselenols formed by the reaction of selenium(IV) with such compounds as 2,3-diaminonaphthalene,[79] and 4,5-dichloro-o-phenylenediamine[121] can be used for the gravimetric as well as the colorimetric determination of selenium. Adequate masking, such as with EDTA, is necessary to overcome interference by cations, particularly copper(II).   The recovery of selenium is dependent on pH and in the range 0–2 complete recovery should be realized with a 4 to 6-fold excess of the reagent.   The precipitate, which has been washed to remove all traces of acid, is stable on air drying at 100°C.   The reagent 4,5-dichloro-o-phenylenediamine requires 30 to 60 minutes for complete precipitation, whereas in the case of the 2,3-diaminonaphthalene, 150 to 180 minutes are required.   The molar solubilities of products of the two reagents at pH 2 are respectively 4.53×10^{-8} and 4.12×10^{-8}.

Although insoluble selenides can be formed by the reactions of metallic cations with hydrogen selenide, selenium is seldom, if ever, determined by this method.   Nevertheless, it is worth noting that the solubility products and entropies of sulfides, selenides, and tellurides have been correlated.[28]   The relation between the solubility products (K_{sp}) of selenides and the corresponding sulfides is given by the equation:

$$pK_{sp\,Se} = 7.11\ K_{sp\,S}^{0.62} - 17.18$$

Volumetric Methods

Relatively few volumetric procedures for selenium have gained general acceptance, although numerous electrochemical titration methods (see section entitled "Polarography and Other Electrochemical Methods") have been developed in recent years.   Since the volumetric methods which have been employed are redox procedures, it is important that interfering reducing or oxidizing substances be absent from the selenium solution.   Thus, any organic matter should be destroyed, preferably by digestion with mixed sulfuric and nitric acids.   Any residual nitric acid must also be removed. One of the oldest volumetric methods is that of Norris and Fay[95] in which selenium(IV) is reduced by an excess of standard sodium thiosulfate solution. The excess sodium thiosulfate is back titrated with standard iodine solution.

Another approach to the iodometric determination of selenium involves the addition of a moderate excess of potassium iodide solution, with the liberated iodine being titrated with standard sodium thiosulfate solution. This procedure possesses the disadvantage that the red selenium precipitate renders the starch-iodine end-point less sharp, particularly when larger amounts of selenium are being determined.   In addition, iodine may be occluded by the selenium.   This iodometric procedure is better suited to the determination of small amounts of selenium as has been demonstrated by Sill and Peterson[118] and by McNulty et al.[86] who determined as little as 0.002% selenium in copper.

The difficulties inherent in the gravimetric method and in the iodometric-volumetric procedures as applied to the determination of macro amounts of selenium prompted Barabas and Cooper[16] to extend the permanganate method of Schrenk and Browning[113] to the determination of the purity of refined selenium and selenium compounds.   In this method, selenium(IV) is oxidized by an excess of permanganate, the excess being determined by back-titration with ferrous ammonium sulfate.   Disodium phosphate is added to prevent the precipitation of manganese dioxide.   Since tellurium(IV) interferes in this determination, this element must be separated prior to selenium determination or a suitable correction must be applied.

An excellent differential potentiometric method for macro quantities of selenium has been developed which is based on the titration of selenium(IV) with tin(II) chloride following the precipitation of approximately 97% of the selenium with a calculated deficiency of hydrazine sulfate.[15]   No tellurium is precipitated in the presence of as much as 1000 times its amount of selenium in a $3.6N$ sulfuric acid solution when a small deficiency of hydrazine sulfate, relative to the selenium content, is added.   The reduction of the selenium-tellurium ratio to a much smaller value, say 20 : 1, permits the simultaneous potentiometric determination of the two elements.   Of the impurities other than tellurium which may be associated with selenium—copper, gold, and iron can interfere seriously if present in significant amounts.   Iron(III) destroys the end-point and copper(II) and gold(III) cause large positive errors.   Thus, this procedure is well-recommended for the analysis of refined selenium and selenium compounds but should be applied with considerable caution to refinery intermediate products.

Polarography and Other Electrochemical Methods

The polarographic behavior of selenium in its various oxidation states has been studied systematically over a wide range of pH and supporting electrolytes.[77]   For analytical purposes, it appears that the best supporting electrolyte for selenium(IV) reduction is $1M$ NH_4Cl–0.1 to $1M$ NH_3 (pH 8.0 to 9.5) 0.001% in gelatin.[77]   A well-developed single wave is produced denoting reduction to Se^{-2} and the diffusion current is directly proportional to the selenium(IV) concentration.   At pH 8.0, $E_{1/2} = -1.44$ V vs S.C.E. and at pH 9.5, -1.54 V.

A serious limitation in the application of polarography to the determination of selenium is the fact that the apparent height of the selenium(IV) wave can be decreased if metal ions such as copper(II), which form insoluble selenides, are present in the solution.   A similar depressive effect can occur in acid supporting electrolytes, e.g., $3M$ H_2SO_4 in which mercury undergoes dissolution.[37]   In this case, the mercury ions react with hydrogen selenide produced in the primary electrode reaction to form mercury selenide.   Correction for this phenomenon must be made in the determination of selenium in such solutions.

Following a prior separation of selenium by reduction to the element, Nangniot[91] found a sensitivity for selenium(IV) of $2.5 \times 10^{-6}M$ or 0.2 μg selenium per ml in 0.7M HBr. This sensitivity was increased to around 0.01 μg selenium per ml in the same electrolyte using oscillopolarography. It should be noted that measurements were made on the first wave, the second wave being subject to the decrease observed by Christian, Knoblock and Purdy.[37]

An interesting microdetermination of selenium has been developed which is based on the polarographic reduction of diphenylpiazselenol formed by the reaction between selenium(IV) and 3,3'-diaminobenzidine.[76] In 0.1N NH_4ClO_4 at pH 2.5, the heights of the two waves ($E_{1/2} = -0.17$ and -0.63 vs S.C.E.) are proportional to the concentration of the piazselenol up to its solubility limit (3.5–4 mg per ml). The determination of less than 0.5 mg of selenium is facilitated by a prior separation and isolation of the piazselenol by extraction into benzene and evaporation of the solvent.

According to Lingane and Niedrach,[77] selenide ion produces a well-defined single anodic wave in the pH range 0 to 12. The half-wave potential shifts from -0.49 V vs S.C.E. at pH 0 to -0.94 V at pH 12. In 0.1M NaOH, the anodic selenide wave is around 0.2 V more negative than that for the sulfide ion permitting the simultaneous determination of these two species. It appears that the reduction of selenium(VI) in both acid and alkaline medium is too slow to yield polarographic waves.

In the presence of sulfite, the selenosulfate ion in 0.6M NH_4OH–0.6M NH_4Cl gives a square-wave polarogram having one selenium peak at -0.77 V vs S.C.E.[66] It is assumed that the mercury electrode catalyzes the decomposition of selenosulfate to sulfite with the latter being adsorbed on the electrode and reduced at the potential peak of the selenosulfate ion. Thus, the reversible 2 electron reduction is that of selenium to selenide. The pulse-polarographic behavior of selenosulfate has been applied to the determination of traces of selenium in very pure metals such as antimony, indium, gallium and bismuth.

Selenium(IV) (and tellurium(IV)) can be determined by amperometric titration in acid solutions using as reagents hexyl- and cyclopentyldithiocarbamates.[131] The reagents are oxidized at a platinum anode to form slightly soluble compounds with selenium and tellurium. If tellurium is present, selenium must be determined by difference. Thus tellurium is estimated by titration at pH 5.5–4.0 and the sum of selenium and tellurium by titration at pH < 3.7. A similar procedure has been developed using sodium diethyldithiocarbamate.[135] Since the reagent is unstable in acid solutions, the titration is carried out in the presence of lead nitrate, lead(II) forming with the reagent a soluble complex which is oxidized at the platinum electrode.

Several coulometric procedures have been developed for the determination of selenium. Thus, micro amounts of selenium(IV) can be determined by titration with electrolytically generated hypobromite ions.[4] The sensitivity for selenium was 2.2 μg per ml under the following optimum conditions: $1M$ KBr, $0.1M$ NaHCO$_3$, 70°C, and cd 0.5–5 ma per cm^2. Tellurium(IV) behaves similarly so that a prior separation of selenium is mandatory.

Selenium(IV) can be determined coulometrically in the presence of a 50-fold excess of tellurium by electrolytically generated iodide ions.[3] The sensitivity was $1.2 \times 10^{-7}M$ or 0.4 μg selenium per ml. Reproducible results were obtained using a graphite electrode and a platinum counter electrode with an amperometrically indicated end-point.

The reduction of selenium(IV) in $0.2M$ HCl or $0.5M$ citric acid at constant potential until a minimum constant current is reached has been used as the basis of a coulometric method for minor amounts of selenium.[130] In this determination, the quantity of current was measured using a micro hydrogen–oxygen coulometer. It would appear that a prior separation of selenium is necessary since a number of elements, including tellurium(IV), interfere. Electrolytically generated titanium(III) can be employed for the coulometric titration of selenium(IV) with a sensitivity of $1 \times 10^{-5}M$.[5] Tellurium(IV) behaves similarly.

The difficulties inherent in the polarography of selenium and the availability of other more sensitive methods of analysis can be considered as having an important bearing on the limited number of applied polarographic procedures. Although no specific applications of the various amperometric and coulometric methods have been indicated, it would appear that in general these methods lack the necessary specificity and sensitivity for the trace analysis of selenium, particularly in natural products.

Photometric Methods

Spectrophotometry. One of the most sensitive methods for the determination of selenium is furnished by the reaction of selenium(IV) with 3,3'-diamino-benzidine at pH 2.5:[35,138]

The intense yellow compound, diphenylpiazselenol, which is formed, can be extracted quantitatively by toluene at pH above 5. Beer's law is obeyed over the range of 5 to 25 μg of selenium per 10 ml of toluene at the wavelengths 340 and 420 mμ.[35] The use of ethylenediaminotetraacetic acid (EDTA) permits the determination of selenium in the presence of iron, copper, molybdate, titanium, chromium(III), nickel, cobalt, tellurium, arsenic, and up to 5 mg of vanadium(V). Strong oxidizing and reducing agents must be absent. According to Cheng,[35] the limit of sensitivity of the method is 50 ppb with a 1 cm absorption cell.

An important extension of the diaminobenzidine method was devised by Watkinson[138] who utilized the strong fluorescence of the diphenylpiazselenol at 580 mμ to determine as little as 0.02 μg of selenium. Interference from materials, the fluorescence of which is not overcome by EDTA, can be avoided by the extraction of selenium with toluene-3,4-dithiol into a 50% mixture of ethylene chloride and carbon tetrachloride from strong hydrochloric acid solution.

A useful bibliography of 3,3'-diaminobenzidine as an analytical reagent notably for selenium covering the years 1937–1961 has been compiled by Broad and Barnard.[23]

Other reagents such as 2,3-diaminonaphthalene and 4,5-diamino-6-thiopyrimidine which also form piazselenols with selenium(IV) have been investigated. According to Lane,[74] 3,3'-diaminobenzidine is less sensitive than 2,3-diaminonaphthalene, and suffers from the disadvantage of a relatively high and variable blank. Also, the solution containing the diphenylpiazselenol must be neutralized before the complex can be quantitatively extracted into the organic phase.

The reagent, 4,5-diamino-6-thiopyrimidine, can be used to determine selenium(IV) directly in aqueous solution at pH 1.5 to 2.5.[34] The determination can be carried out at 380 mμ or at 410 mμ where the absorbance due to the blank is comparatively low. Measurement must be made within 30 minutes of the addition of the reagent due to the precipitation of elemental selenium. Conformity to Beer's law was observed in the range 0.5 to 2.5 μg per ml. As expected, iron, copper, and redox agents interfere and must be removed or sequestered. It should be noted that reagents forming piazselenols are subject to air oxidation[34] and suitable precautions must be taken in their use.

According to Kirkbright and Ng[72] selenium(IV) can be determined in the presence of tellurium using thioglycolic acid provided the tellurium is oxidized to the hexavalent state. The yellow selenium thioglycollate complex extracted into ethyl acetate followed Beer's law at 260 mμ in the range 6 to 34 ppm. Gold(III), chromium(III), copper, iron(III), germanium, mercury(II), and vanadium(V) can interfere as a result of the consumption of reagent to form nonextractable complexes or through oxidation by the metal ion. Arsenic(III) and (V), bismuth(III), molybdenum(VI), and antimony(III) yield high results due to the formation of extractable complexes.

An older and still very useful photometric method for determining micro amounts of the element is that in which a colloidal suspension of selenium, as the hydrosol, is formed by a reducing agent such as sulfur dioxide, phenyl-hydrazine, tin(II) chloride, or ascorbic acid. A sol stable for 5 days obeying Beer's law in the range 1–10 μg selenium per ml has been reported.[115]   Gum arabic or polyvinyl alcohol can be employed as sol stabilizers. The use of methycellulose permits a nephelometric determination of selenium in the range 0.7–4.5 μg selenium per ml without interference from copper, iron, or tellurium.[41]

Alternatively, the elemental selenium can be extracted into benzene or toluene and determined spectrophotometrically at 276 mμ with a sensitivity of 1 μg selenium per ml.[133]   However, mercury(II), silver, iron(III), bromide, and sulfur interfere.

The reaction of 1-amidino-2-thiourea with micro amounts of selenium(IV) in $6N$ HCl yields a yellow-orange color due to the formation of elemental selenium.[90]   The reaction which leads to the color formation is complete at a selenium : reagent ratio of 1 : 2. Beer's law is obeyed for the concentration range 1–9 μg of selenium per ml. The interferences due to ions such as tellurium(IV), copper(II), and iron(III) which are commonly encountered can be overcome by selective redox procedures. However, a prior separation of selenium from such interfering elements should be undertaken.

Busev[30] has investigated a number of reagents containing sulfhydryl groups such as 2-mercaptobenzimidazole, N-mercaptoacetyl-p-toluidine which react with selenium(IV). The resulting yellow compounds can be extracted by a 1 : 5 mixture of butyl alcohol and chloroform. Beer's law is obeyed at 420 mμ over the concentration range 1–40 μg per ml for the compound with 2-mercaptobenzimidazole and at 340 to 380 mμ over the range 7 to 80 μg per ml for the compounds with the derivatives of thioglycolic acid. The extent to which extractants of the reagents absorb and the interference of tellurium(IV), bismuth and copper limit the usefulness of these reagents.

Atomic Absorption Spectroscopy. Atomic absorption spectroscopy affords a convenient method for the determination of small amounts of selenium. The greatest population of members of the ^3P state of the selenium atom is found at an absorption wavelength of 1960 Å. However, a disadvantage of wavelengths as short as 1960 Å is the absorption of light by the flame gases. Therefore, the use of a pulsed power supply and the shortest practicable absorption path length are important factors in realizing increased sensitivity of selenium determination. Using an acetylene-air fuel-rich flame and a 10 cm slit burner, Rann and Hambly[107] were able to obtain a sensitivity of 1 ppm selenium at 1960 Å.   The same sensitivity (0.9 ppm for 1 % absorption) was found in air-hydrogen (Hetco burner) and air-acetylene (laminar burner) flames although the air-acetylene flame gave more variable readings.[71]   The observed sensitivity was highest in the burner with the lowest adequate temperature and the highest aspiration rate.   The sensitivity for selenium was nil in the nitrous oxide-acetylene flame.

The use of a high intensity hollow-cathode lamp together with selected modulation of the resonance lines, or a resonance monochrometer, or double-beam instrumentation should permit the detection of at least 0.05 ppm of selenium.[140]

Atomic fluorescence measurements of the selenium 2040 Å reasonance line using a microwave electrodeless discharge tube as the spectral source increases the sensitivity for selenium compared with atomic absorption.   Thus, Dagnall, Thompson, and West[46] found the limits of determination and detection for selenium by atomic fluorescence spectroscopy to be 0.25 and 0.15 ppm respectively vs 5 and 1 ppm for atomic absorption spectroscopy. However, it should be noted that these atomic absorption results are by no means optimum values.

Emission Spectroscopy.   Since selenium has no sensitive spectral lines above 2200 Å, the determination of selenium by optical emission spectroscopy is really beyond the range of the ordinary spectrograph and demands a vacuum spectrograph.   The higher excitation potential of selenium relative to the common matrices and its high volatility reduce the spectral sensitivity of the element.

X-Ray Fluorescence.   If a suitable collection or separation method is employed, there is no difficulty in determining small amounts of selenium by x-ray fluorescence spectroscopy.   Thus, selenium can be isolated using arsenic as the collector and hypophosphorous acid as the precipitating agent. The precipitate is collected on a micropore filter and analyzed for selenium by measuring the intensity of the $SeK_{\alpha1}$ peak.   Since tellurium behaves in the same manner as selenium, the procedure permits the simultaneous determination of tellurium.   The detection limits for this method as estimated from the calibration curves are 1 ppm for selenium and 0.1 ppm for tellurium.[70]

A novel approach which has been applied to the x-ray determination of selenium in semiconductor materials, utilizes the formation of a film of elemental selenium produced by the decomposition of the compound formed between selenium(IV) and 4,5-diamino-6-thiopyridine.[33]   A linear relation was observed between concentration and x-ray intensity up to 800 μg of selenium.

As little as 1 μg of selenium can be detected along with elements such as antimony, tin, and arsenic by x-ray fluorescence using 100 μg of copper as a carrier and an internal standard.[143]   The elements being determined together with the carrier are precipitated as sulfides which are collected on a membrane filter, air dried, and analyzed.

Neutron Activation Analysis

A number of techniques have been advanced for the determination of selenium by neutron activation particularly in biological materials.[22,40,45,47,122]   There are only three radionuclides produced by thermal neutron irradiation which are useful for analytical purposes.   These are ^{75}Se(^{74}Se(n, γ)^{75}Se $t_{1/2}$ = 120 days), ^{77m}Se(^{76}Se(n, γ)^{77m}Se $t_{1/2}$ = 17.5 sec), and ^{81}Se(^{80}Se(n, γ)^{81}Se $t_{1/2}$ = 18.6 minutes).

The long half-life of ^{75}Se permits complete radiochemical separation of the element.   However, some workers have elected to permit the decay of shorter-lived nuclides, usually for a period of 5 to 7 days, to facilitate the direct measurement of ^{75}Se gamma radiation in the sample.   Such a procedure has the obvious disadvantage of serious delay in obtaining analytical results. Also, the sensitivity is less than that which is obtained when chemical separation is carried out.

Conventional radiochemical separation can be avoided by a dry distillation procedure following irradiation of the sample by either thermal neutrons or by 14 MeV neutrons each having a flux of 10^{10} neutrons/cm^2/sec.[40] Thermal irradiation for 6 h produced ^{75}Se by the reaction ^{74}Se(n, γ)^{75}Se. Irradiation with 14 MeV neutrons for 30 minutes yielded ^{81m}Se by the reaction ^{82}Se(n, 2n)^{81m}Se.   The calculated detection limit for selenium using both irradiation sites was 50 ppm.   Consequently, a preconcentration of the selenium would be required for samples containing very small amounts of the element.   The procedure was applied successfully to the estimation of selenium in steels (approximately 0.25% Se) using a sample weight of 1.0 to 1.5 grams.[40]

The very short half-life of ^{77m}Se, which affords the opportunity for extremely rapid selenium analyses, demands direct and rapid measurement of the gamma activity.   Unfortunately the presence of ^{19}O ($t_{1/2}$ = 29 sec) can limit the accuracy and sensitivity of the selenium determination.   Dickson and Tomlinson[47] developed a correction procedure for oxygen interference which gave a sensitivity for selenium of 0.01 μg in biological materials such as blood cells and plasma.   The average time for analysis of prepared samples was less than 5 minutes per sample.   Quantities of selenium as low as 0.01 μg could be determined and as little as 0.005 μg of selenium was detectable.

Using a radiochemical separation of ^{81}Se by distillation, Bowen and

Cawse[22] were able to separate selenium from arsenic, bromine, manganese, sodium, and zinc and determine the element rapidly with a sensitivity of 0.005 μg. Dahl and Steinnes[45] used the same nuclide for the determination of selenium in animal tissue, separating the selenium by hydrochloric–hydrobromic acid distillation.

ANALYSIS OF SELENIUM (AND TELLURIUM)

Analysis of Selenium

The accurate estimation of selenium in refined selenium cannot be realized by the conventional gravimetric procedure in which elemental selenium is precipitated by sulfur dioxide.

The permanganate method of Schrenk and Browning[113] has been extended to the determination of selenium, not only in refined selenium, but also in sodium selenite, sodium selenate, and iron selenide.[16] In this procedure, selenium(IV) and tellurium(IV) are oxidized in sulfuric acid medium by an excess of permanganate, the excess being determined by back-titration with ferrous ammonium sulfate. Disodium phosphate is added to prevent the precipitation of manganese dioxide. Since commercial selenium generally analyzes around 99.5% selenium with tellurium as the principal impurity, a correction for the tellurium content of the sample must be made. This can be realized by direct permanganate titration following volatilization of the selenium by sulfuric acid fuming at elevated temperatures.[41]

A procedure which has been used routinely with excellent results involves the potentiometric titration of selenium(IV) with tin(II) chloride solution following the precipitation with hydrazine sulfate.[15] It was established that no tellurium will precipitate in the presence of as much as 1000 times its concentration of selenium for a 3.6N sulfuric acid solution when a small deficiency of hydrazine sulfate, relative to the selenium content, is added. Thus, it was possible to reduce the selenium–tellurium ratio to a favorable value, say 20 : 1, so as to permit the simultaneous potentiometric determination of the two elements. Of the various impurities which are usually encountered in refined selenium, only tellurium and oxygen are apt to be present at concentrations high enough to have any effect on the selenium analysis. Neither of these impurities interferes.

Determination of Impurities in Selenium

The production of high purity selenium and its use in rectifiers, photocells, and xerography has demanded the availability of rapid and accurate procedures for the determination of traces of elements such as mercury, tellurium, lead, iron, nickel, copper, arsenic, and other metals.   The determination of non-metallic elements such as oxygen, sulfur, chlorine, and carbon has also been necessary.   Although chemical procedures have been developed for most of the impurities, these methods are too time-consuming to be suitable for production control.   The most convenient method for metallic elements is afforded by optical emission spectroscopy.

Mellichamp[87] has developed a semiquantitative spectrographic procedure for impurities in selenium in which a 50 mg sample is burnt to completion in a 10 amp dc arc.   Twenty-three elements are determined by this procedure from a spectrogram in the region 2200–4400 Å using the background as internal standard.

A dc arc procedure has been developed by Peterson and Currier[103] for the quantitative determination of aluminum, bismuth, copper, lead, magnesium, silver, and zinc in selenium in which the detection limits are considerably lower than those reported by Mellichamp.   In this procedure a 500 mg sample containing 4 μg of palladium as the internal standard is burnt in a 5 amp dc arc for 35 sec.   The working curves were obtained from standards prepared by adding 0.05 ml of a standard impurity solution to pure selenium which had been previously melted in the electrode crater.   Following evaporation of the solution to dryness, the crater was filled with additional selenium and the entire mass was melted prior to excitation to ensure homogeneity.

A more rapid and convenient procedure can be realized by the use of external comparison standards.   In this method, the spectra of the samples are compared with those of carefully prepared standards which are recorded on the same spectrographic plate.   A 400 mg sample in the form of shot is excited in a 10 amp dc arc for 30 sec following a 5 sec preexposure.   A transmission of 100% with a slit width of 30 μ is employed.   The electrodes are preformed graphite rods with the lower electrode being covered by a boiler cap.   The electrode spacing is 0.3 mm.   To accommodate the impurities determined by this procedure (mercury, tellurium, lead, iron, copper, arsenic, antimony, tin, bismuth, nickel, chromium, aluminum, titanium, zinc, cadmium, silver, sodium, magnesium, silicon) two wavelength ranges are employed, namely, 2275–3030 Å and 2500–3670 Å.   A transmission of 25% is employed for the second wavelength range.   The spectra of both samples and standards on each range are recorded in duplicate using Eastman spectrum analysis No. 1 plates.   The standards may be actual production samples or samples to which known amounts of impurities have been added.   The concentrations of the significant impurities in all standards should be determined by properly established alternative procedures.

The difficulties inherent in the determination of traces of impurities, particularly nonmetallic elements, in selenium can be overcome by the use of spark source mass spectrometry.   The detection limits for impurities, both metallic and nonmetallic, are in the order of 0.1 ppm atomic or less.   In addition to the metallic impurities determined by optical emission spectroscopy, nonmetallics such as carbon, chlorine, sulfur, boron, nitrogen, oxygen, hydrogen, and compounds like Se_2C, SeO_2, and SeO can be determined. Detection limits for the elements oxygen, sulfur, chlorine, and carbon have been reported as 0.03, 0.04, 0.06 and 0.03 ppm atomic, respectively.   It should be noted that elemental selenium is a difficult material to analyze by spark source mass spectrometry unless special precautions are taken.   The high resistance, low melting point, and high vapor pressure of the element can give rise to instrument contamination and a serious memory effect which can persist for some time.   The contamination due to selenium can be overcome by changing those parts of the instrument which have come into direct contact with selenium.   Alternatively, the selenium which remains deposited on instrument parts can be coated with some noninterfering element or destroyed by running the instrument until no selenium shows on the spectrographic plate.   The contamination due to selenium has been overcome successfully by application of the sample cooling system developed by Cherrier and Nalbantoglu.[36]

Although methods have been reported for the determination of as many as 20 elements in selenium by neutron activation analysis,[10,11,12,13] this technique has never received acceptance as a convenient trace analytical method for impurities in selenium.   However, the method is useful when accurate determinations are required for traces of elements such as chlorine and sulfur.   Although nondestructive neutron activation analysis only allows a detection limit for chlorine of 10 ppm,[58] the neutron irradiation of selenium followed by chemical separation using the reaction $^{37}Cl(n, \gamma)^{38}Cl$ permits a limit of detection for chlorine of 0.01 ppm.[59]   In this method the sample flux used was 5.1×10^{11} neutrons cm^{-2} sec^{-1} at an irradiation time of 76 minutes.   The half life of ^{38}Cl of 38 minutes is sufficient to permit the separation of chlorine by distillation and its precipitation as silver chloride prior to the counting of the photo-peak intensity.   In a similar procedure in which 2 g samples were irradiated for 37 minutes at a flux of 8×10^{10} neutrons cm^{-2} sec^{-1}, Ballaux, Dams and Hoste[12] determined amounts of chlorine in the range 0.4–1 ppm in high purity selenium.

The irradiation of selenium by epithermal neutrons followed by chemical separation permitted a limit of detection for sulfur and antimony in selenium of 0.05 and 0.001 ppm, respectively.[60]   The reactions selected, $^{32}S(n, p)^{32}P$ and $^{121}Sb(n, \gamma)^{122}Sb$ were such as to permit a suitable irradiation time and chemical separation of phosphorus as magnesium ammonium phosphate, and antimony as the sulfide.   Sulfur in the range 1.5–4.6 ppm sulfur was found in high purity selenium samples in a similar procedure in which the separation of phosphorus was accomplished by precipitation of ammonium phosphomolybdate.[12]

The neutron activation determination of small amounts of tellurium (15–25 ppb) in high purity selenium can be carried out by irradiating 100 mg samples for 4 days at a flux of 6×10^{12} neutrons $cm^{-2} sec^{-1}$. The daughter isotope of ^{131}Te, ^{131}I, is separated by distillation and extracted into carbon tetrachloride. The iodine is precipitated as silver iodide and its activity measured by β, γ-coincidence counting.[11]

Schuster and Wohlleben[114] have used the reaction $^{16}O(d, n)^{17}F$ for the determination of oxygen in evaporated selenium layers. By varying the deuteron energies it is possible to determine oxygen in layers of varying thickness. Thus with an energy of 3.0 MeV, oxygen could be determined in layers of a thickness of 26 μ. To determine the ^{17}F activity it is necessary to separate the fluorine, and for this purpose the classical Willard-Winter distillation was employed. An oxygen content of between 50 and 100 ppm was determined, and for less heavily oxidized surfaces, the oxygen content ranged between 10 and 90 ppm.

Although chlorine can be determined nephelometrically or turbidometrically as silver chloride following the separation of chlorine from selenium by distillation, such methods are not sufficiently sensitive for small amounts of chlorine. Tada[125] has applied a variation of the thiocyanate-ferric alum method for the colorimetric determination of chlorine as encountered in chlorine-doped selenium used in selenium rectifiers. An interesting feature of this method is the direct anodic dissolution of a sample weight of 15 mg from the selenium surface.

A promising method for the determination of traces of chlorine in selenium is a microcoulometric technique in which the chlorine produced, following combustion of the selenium sample, is titrated with electrolytically-generated silver ions.[14] In the combustion of the sample, the formation of selenium dioxide was avoided by heating the selenium samples in an atmosphere of helium at 400°C. The sensitivity of this method is indicated by the fact that a sample weighing 119.2 mg and containing 17 ppm chlorine gave a recorded current peak having an area of 4.3 in². Thus, it should be possible to determine nanogram quantities of chlorine without difficulty using this method.

To date no rapid, convenient method for the determination of oxygen in selenium has been devised. It is of interest to note that although the determination of oxygen in selenium by infrared spectrophotometry has been claimed,[1,136] in a detailed infrared spectrophotometric study of vitreous selenium doped with selenium dioxide, Burley[29] has shown that oxygen is present in selenium in two distinct forms, the relative amounts of which are dependent on the temperature of the melt from which the specimen was prepared. The infrared spectra of these two forms, as indicated by the presence of bands at 904 and 925 cm^{-1}, undergo changes with time, indicating a transformation from one form to the other.   This rate of transformation is influenced by a number of factors, including temperature and the thermal history of the selenium.

Sulfur, in the range 5 to 1200 ppm in selenium, can be determined by a combustion procedure in which the resulting sulfur dioxide is absorbed in pararosaniline solution.[2]   Since selenium has a decidedly negative effect on the absorbance of the sulfur-pararosaniline compound, it must be absorbed from the stream of combustion gases.   The disadvantages of the pararosaniline method and its limited sensitivity have been overcome in a colorimetric procedure in which sulfate is reduced to sulfide by hydriodic and hypophosphorous acids prior to its determination as methylene blue.[119]   In this method, selenium is removed by distillation from a hydrobromic acid solution. As little as 0.4 ppm of sulfur in selenium has been determined by this technique.

The determination of small amounts of mercury in selenium is of importance in view of the marked effect which this element can have on the asymmetric conduction of selenium.   A convenient procedure which depends upon the extraction of the mercury–dithizone complex into chloroform permits the determination of as little as 0.5 ppm mercury in selenium.[106] Other impurities such as may be encountered in selenium do not interfere with this method.   According to Marshall,[84] the complex formed between mercury and di-β-naphthylthiocarbazone permits the determination of as little as 0.2 ppm mercury in selenium.   However, this method has a disadvantage in the utilization of an unstable reagent blank.

Since tellurium is commonly encountered as an impurity in selenium, its determination in both refined and high purity selenium is of some importance. Therefore, it is appropriate that more detailed information should be given on methods for the determination of this element to supplement the procedure which has already been detailed under the optical emission spectrographic analysis of selenium.

Tellurium, in the order of a few tenths of a percent in selenium can be determined by permanganate titration following the removal of selenium from sulfuric acid solution containing potassium bisulfate by repeated evaporation to dryness and the absence of fumes.[41]

As has been indicated previously in this chapter, the differential potentiometric method of Barabas and Bennett[15] can be applied to the determination of tellurium in refined selenium.   The analyses for tellurium in selenium provided by this method compared very favorably with those given by more lengthy methods such as permanganate titration.

Polarographic procedures have been developed for the determination of tellurium in selenium in the range 0.10–5%,[102] and as low as 0.2 ppm.[7]   In the former procedure following the removal of selenium by repeated evaporation with nitric acid, the residue containing the tellurium is taken up in 20% NaOH and an appropriate aliquot taken for polarographic analysis.   In the latter method, the tellurium is separated by electrodeposition onto a platinum cathode after dissolution of the sample in sodium hydroxide.   The electrodeposit is taken up in nitric acid solution and the evaporation residue is dissolved in $1M$ NaOH–$1M$ NH_4Cl, pH 9 which serves as the supporting electrolyte for polarographic analysis.   Such a procedure requires the careful use of microtechniques and a micro-cell to attain the reported sensitivity.

The spectrophotometric procedure based on the iodotellurite complex has been applied to the estimation of trace amounts of tellurium in selenium.[41] Here again selenium must be removed completely prior to the analysis.   This procedure can be used to determine as little as 0.1 ppm tellurium in high purity selenium.

Tellurium can be determined rapidly and with acceptable accuracy in selenium by spectrographic analysis.   Apart from the spectrographic analysis procedure given for high purity selenium, refined selenium, which generally contains 99.5+% selenium, with tellurium as a principal impurity, can be analyzed by excitation to extinction of a 130 mg sample either as powder or shot in a condensed dc arc.   The tellurium analysis line, 2385.8 Å, is measured at 100% and 10% transmission for concentration ranges 0.006–0.10% and 0.07–1%, respectively, and the selenium internal standard line, 2548.2 Å, at 10% transmission.   The standards can be either actual production samples or samples prepared by the addition of known amounts of tellurium powder to molten selenium, stirring and shotting the melt.   In either case, the standards are analyzed chemically for tellurium, e.g., by permanganate titration or for smaller quantities, spectrophotometrically, using the iodotellurite complex.

A sensitivity of 0.05 ppm tellurium in selenium has been given by Fratkin and Polivanova[55] for a spectrographic ac arc method in which the impurities in a 1 gram sample are concentrated by sublimation of the selenium as SeO_2 and then transferred to the flat surface of a treated carbon electrode following dissolution of the sublimation residue in 1 : 1 HNO_3.   Indium and beryllium are added following dissolution of the original sample but it is not clear if this addition is necessary for the tellurium determination.

ANALYSIS OF SELENIUM COMPOUNDS

The selenium content of compounds such as selenium dioxide, sodium selenite, sodium selenate and iron selenide can be determined by the permanganate titration method,[16] or by the differential potentiometric procedure of

Barabas and Bennett.[15]   Both of these procedures have been discussed in the section entitled "Volumetric Methods."

Selenium dioxide in sulfuric acid solution can be determined by polarographic reduction over the concentration range 0.3–30 mM.[99]   The complex formed between elemental selenium and concentrated sulfuric acid interferes in the determination.

The separate determination of selenium(IV) and selenium(VI) when present together in solution can be realized by a number of approaches, most of which are based on the selective oxidation (permanganometric or bromate titration) or reduction (titration with thiosulfate or precipitation with SO_2 or $SnCl_2$) of selenite ions and a precipitation of total selenium by SO_2 gassing in a strongly hydrochloric acid solution.   It should be noted also that selenium(IV) can be separated from selenium(VI) by coprecipitation with heavy metal hydroxides. A convenient method has been developed for the separate determination of selenites and selenates in solution which is based on the precipitation of selenium(IV) by tin(II) chloride in approximately 5 % (vol) hydrochloric acid solution and the precipitation of selenium(VI) upon boiling the filtrate which has been made 20 % (vol) in hydrochloric acid.[27].   The disadvantages of a gravimetric determination can be overcome by acid dissolution of the precipitate and titration of selenium(IV).

A novel paper chromatographic procedure was developed by Kempe,[68] for the separation, identification, and determination of selenite and selenate. The basis of the method is the formation of a yellow crystalline compound, $3C_5H_5N \cdot SeO_2 \cdot 9MoO_3 \cdot 4H_2O$, from a solution of pyridinium sulfate, selenious acid, and $(NH_4)_6Mo_7O_{24}$.   Selenate, which does not form such an heteropoly compound, is reduced to selenite by treating the paper strip with gaseous HCl.   Selenium-containing spots are treated in boiling water and the extracted selenium is determined spectrophotometrically as the selenium sol following reduction with potassium iodide and stabilization with gum arabic.

In the determination of selenium in organic compounds, sample dissolution is facilitated by the much smaller quantity of organic matter present than is the case with biological materials.   Thus, the oxidation of organoselenium compounds can be effected conveniently and without any significant loss of selenium by a mixture of sulfuric and nitric acids.[48]   The use of perchloric acid is to be avoided to eliminate the oxidation of selenium to selenium(VI). The resulting selenious acid is reduced to elemental selenium by ascorbic acid and the selenium sol stabilized by chlorpromazine hydrochloride.   The absorbance is measured at 420 mμ.   This procedure was used to determine selenium in such compounds as amino-selenazolone, selenothiazine and selenazolidine.

Another dissolution procedure which has been applied successfully involves the use of a sulfuric acid–potassium permanganate mixture.[20]   The selenium(IV) is determined by iodometric titration.   Compounds analyzed by this method included derivatives of selenazole, selenazolone, and selenodiazine.

Heating a sample containing 0.2 to 100 mg of selenium under reflux at 160–180°C with 1 g potassium perchlorate and concentrated sulfuric acid provided a dissolution of organic compounds which was followed by a final determination of selenium by titration with $0.1N$ or $0.01N$ $Na_2S_2O_3$.[73]   The oxygen flask combustion method which has been applied routinely in the determination of selenium in biological materials (see section entitled " Decomposition of Organic Materials Containing Selenium") has been used for the decomposition of organoselenium compounds.[17]   In this procedure the sample containing 0.4 to 4 mg of selenium is combusted in the flask containing oxygen and 5 ml of water.   After what appears to be an unnecessarily involved pretreatment of the resulting solution, the selenium is determined by titration, with thiosulfate, of the iodine liberated from excess iodide.

The polarographic behavior of organic selenium compounds has been studied not so much from the viewpoint of analysis as from a study of the reduction of organic selenides and the nature of the rupture of the diselenide bond.[96]   Thus, in perchlorate solution at pH 1.15 which is 35% (vol) in ethanol, diphenyl diselenide gives a double wave, the more positive of which is diffusion-controlled and is directly proportional to the diselenide concentration.[96]   The following reaction scheme is proposed:

$$C_6H_5Se\text{—}SeC_6H_5 \xrightarrow{\text{fission}} 2C_6H_5Se\cdot$$

$$2C_6H_5Se\cdot + 2Hg \longrightarrow 2C_6H_5SeHg$$

$$C_6H_5SeHg + H^+ + e^- \longrightarrow C_6H_5SeH + Hg$$

In view of the biological interest in the selenium analogues of cystine and cysteine, the polarographic properties of these compounds have been investigated.[97]   In the case of selenocystine, there is no tendency to form separate pre- or post-waves and the reduction is characterized by single, well-defined two-electron waves at all pH values which vary directly as the selenocystine concentration.   There was no indication of double waves of the type noted for cystine.   Selenocysteine undergoes anodic processess at the dropping mercury electrode which resemble closely those for cysteine.   In the case of selenocysteine, a single well-defined diffusion controlled anodic wave is produced at pH > 11 corresponding to the reaction:

$$RSeH + Hg \rightleftharpoons RSeHg + H^+ + e^-$$

Mercury compounds are formed at the electrode surface following fission of the Se—Se bond. For pH < 9 the mercury(II)-selenocystine complex, $(RSe)_2Hg$, is reduced reversibly in a two electron step, whereas for pH > 9, the mercury(I)-selenocystine complex RSeHG is reduced reversibly in a one electron process.

DETERMINATION OF SELENIUM IN SPECIFIC MATERIALS

Iron and Steel

The use of selenium as an additive in ferrous materials, notably as a free-machining agent in stainless steels, has resulted in the development of suitable analytical procedures for selenium.

The determination of selenium in iron and steel can be effected by its extraction into benzene followed by back extraction and iodometric determination.[126]   Sample dissolution is effected by 1 : 1 HNO_3 or $HCl–HNO_3$ in the case of stainless steel, followed by boiling with $HClO_4$ to fumes.   Following the same dissolution procedure, Kammori and Ono[65] using only a 0.1 g sample determined 0.01 to 10 μg of selenium in pure iron using extraction with isobutyl methyl ketone to separate the iron.   The final estimation is carried out colorimetrically with 3,3'-diaminobenzidine in toluene.

A fairly direct method for selenium in steel, which should be applicable to small, but not trace amounts of selenium, involves its precipitation with iodide following sample dissolution in 1 : 5 $H_2SO_4 : H_2O$ containing a small amount of saturated $FeSO_4$ solution.[88]   The precipitated selenium is dissolved in hydrochloric acid–nitric acid medium reduced once again with phenylhydrazine, extracted into $CHCl_3$, and determined colorimetrically.

Selenium, in the order of 0.25% in alloy steels, can be determined readily by means of atomic absorption spectrophotometry following dissolution of a 1 g sample in mixed hydrochloric–nitric acids (5 : 1).[69]   The dissolution flask is connected to a reflux distillation apparatus provided with a flask containing a 1% cupric chloride solution to trap any selenium which may be volatilized.   The sample solution, condenser washings and cupric chloride solution are combined, made up to volume, and aspirated together with standard solutions using the Hetco hydrogen-air burner, 1960 Å wavelength, and a 3-pass optical system.

Selenium in the range 0.01 to 0.20% in cast iron can be determined conveniently by a fluorometric method based on the piazselenol formed with 2,3-diaminonaphthalene.[39]   Inhibition of the formation of the piazselenol by iron(III) is overcome by reduction of iron to the divalent state and interferences due to other elements are counteracted by EDTA.   For a sample containing 10 μg of selenium, the maximum fluorescence is attained in 20 to 30 minutes on a boiling water bath.

Other Metals and Alloys

In addition to its determination in ferrous alloys, the accurate estimation of selenium in both blister and refined copper is of considerable importance. The small amounts of selenium (and tellurium) usually encountered in copper necessitate the separation of determinable quantities of the elements either by direct precipitation as with sodium hypophosphite or by coprecipitation with hydrous ferric oxide. Zirconium sulfate in ammoniacal solution has also been used successfully for the coprecipitation. The method employed for the final estimation depends on the amount of selenium present. Volumetric iodometric methods have been reported to give excellent results for selenium in the range 0.001 to 0.1 % using 5 to 20 g samples.[94]   The thiosulfate titration procedure of Sill and Peterson[118] has been adapted successfully to the determination of microgram quantities of selenium in a 100 g sample of copper.[120]

A sensitive x-ray fluorescence procedure for the determination of traces of selenium in refined copper has been developed in which the selenium is preconcentrated by coprecipitation with arsenic.[80]   The precipitate is collected on a 0.8 μ membrane filter and the SeK_α line is excited from a molybdenum target.

A fairly direct determination of selenium in zirconium–selenium alloys can be realized by extracting into toluene the piazselenol formed between selenium(IV) and 3,3'-diaminobenzidine following nitric acid dissolution of the sample and complexation of the zirconium with fluoride.[32]

In the case of bismuth–arsenic–selenium alloys, the selenium is determined gravimetrically using SO_2 reduction following dissolution of a 0.1 to 0.3 g sample in concentrated sulfuric acid.[109]   The same procedure should be applicable to alloy systems of interest in thermoelectric applications.

An interesting pulse-polarographic procedure has been developed by Kaplan and Sorokovskaya[66] for the determination of as little as 2×10^{-5} % of selenium in very pure antimony, indium, gallium, and bismuth. The selenium, which is collected by sulfur, is converted to selenosulfate which is polarographed in $0.6M$ NH_4OH–$0.6M$ NH_4Cl solution in the range -1.0 to -0.5 V using a square-wave polarograph. In the presence of sulfite, selenosulfate gives one selenium peak at -0.77 V vs S.C.E.

Semiconductor Materials

Although semiconducting selenides have been studied extensively, and selenium has been used as a dopant or alloying constituent, particularly in thermoelectric materials, little attention seems to have been devoted to the determination of selenium in such materials. The determination of selenium as a dopant in gallium phosphide by means of spark source mass spectrometry showed a good correspondence with spectrophotometric 3,3'-diaminobenzidine results in the range 10^{18} to 10^{20} atoms per cm^3.[6]

Small amounts of selenium in silver chloride and uranium oxide (U_3O_8) can be determined colorimetrically with 3,3'-diaminobenzidine following extraction of the piazselenol into toluene.[110] Considering the number of selenium separation procedures which are available, it is difficult to understand the use of aluminum hydroxide as a collector for selenium in this procedure, particularly because of the problems which are presented in the formation and extraction of the piazselenol.

Traces of elemental selenium in cadmium selenide can be separated by volatilization 520°C in a stream of nitrogen and the selenium determined colorimetrically following dissolution, precipitation with ascorbic acid solution, and extraction into benzene.[134] This procedure has a sensitivity for selenium of 1 μg per ml of extract.

Small amounts of selenium in tellurium can be determined spectrophotometrically with 3,3'-diaminobenzidine with the complex being extracted from the reaction medium by $CHCl_3$ at pH > 9.[8a] The original reaction with selenium(IV) must be carried out in acid medium (pH 2–3) with EDTA being used to counteract the effect of oxidizing cations. The detection limit was found to be 0.1 ppm selenium for a 1 g sample.

By preconcentrating the selenium present in tellurium via distillation from a $HCl-H_2SO_4$ solution, Pats and Semochkina[101] were able to determine polarographically 0.2–2 ppm selenium with a loss of 6–7%. A concentration of selenium as low as $5 \times 10^{-9} M$ could be determined with the favored supporting electrolyte being 0.4–1.3M HCl.

Rocks, Minerals, Ores, and Anode Slimes

Since the selenium content of rocks, minerals, ores, and concentrates is small yet subject to considerable variation, close attention must be paid to sample dissolution and to the preconcentration of selenium (see section on "Decomposition, Dissolution and Other Preliminary Treatment of Inorganic Materials"). The final estimation of selenium can be realized by a number of procedures, the more important of which will be discussed.

It should be noted that the earlier procedures for sulfide ores, concentrates, slags, and flue dusts which have been detailed[41] are still applicable in terms of the dissolution and separation steps. However, the development of specific reagents for selenium such as 3,3'-diaminobenzidine and 2,3-diaminonaphthalene, coupled with the use of arsenic as a collector, has resulted in a greatly enhanced accuracy and sensitivity for the determination of traces of selenium in complex materials.

A rapid and convenient method, particularly for concentrates, involves the collection of selenium by arsenic precipitation using hypophosphorous acid as reducing agent.   The precipitate is transferred to a micropore filter and analyzed for selenium by x-ray spectrometry.   Since tellurium behaves in the same manner as selenium, the procedure permits the simultaneous determination of both elements.

Extraction photometric procedures have been applied to a number of mineral and metallurgical materials, including sulfide ores and minerals, slag, and sulfur.   Of the more sensitive reagents for selenium, including o-phenylenediamine, 1,4-diphenylthiosemicarbazide, dithizone, diantipyrinylmethane, and 3,3'-diaminobenzidine, Shcherbov, Ivankova, and Gladysheva[116] found dithizone to be most suitable.   A complex formed with selenium in $6N$ HCl or $11N$ H_2SO_4 was extracted into $CHCl_3$ and the absorbance measured at 415 mμ.   Selenium in the range 10^{-2} to $10^{-5}\%$ in ores could be determined following preconcentration by precipitation of selenium (and tellurium) with arsenic.   The fluorometric determination of selenium in mineral samples using 3,3'-diaminobenzidine was also carried out.

The determination of small amounts of selenium in ore or slag samples can be realized by a benzene extraction of the complex formed with o-phenylenediamine followed by dissolution of the sample in mixed nitric, sulfuric, and perchloric acids.[117]   The absorbance, measured at 335 mμ, provides a sensitivity of 0.5 μg per 5 ml of benzene.

A procedure for selenium in sulfide ores with comparable sensitivity to the foregoing method involves the extraction of the complex formed with 1,4-diphenylthiosemicarbazide into $CHCl_3$ or CCl_4, following dissolution of the sample in mixed nitric, sulfuric, and hydrochloric acids and preconcentration of the selenium (and tellurium) by precipitation with tin(II) chloride.[89]

In an extensive study of the distribution of selenium in sulfide minerals, Tischendorf[129] recommended the determination of selenium spectrophotometrically by means of the complexes formed with 3,3'-diaminobenzide or o-phenylenediamine extracted into toluene.   Although a number of possible interfering ions can be masked, the separation of selenium via bromine–hydrobromic acid distillation and precipitation with tellurium as a collector is proposed.   The sensitivity of the method, 1 μg of selenium per 20 ml of toluene, is superior to that reported for similar procedures.

Selenium in the order of $10^{-3}\%$ in sulfur can be determined by ascorbic acid precipitation of the element and measurement of the absorbance at 276 mμ of the benzene extract.[132]   The sensitivity of this procedure can be enhanced by benzene extraction of the free iodine formed by iodometric reduction by selenium(IV).   The absorbance, measured at 300 mμ, permits the determination of 1×10^{-4} to $5 \times 10^{-5}\%$ selenium in 1 g of sulfur.

Even with preconcentration of the selenium, the procedures described above are incapable of determining the extremely small amounts of selenium found in mineral samples.   The superiority of neutron activation analysis is shown by Brunfelt and Steinnes[25] who found a sensitivity for selenium in rocks of 0.0005 μg.   In this procedure, a 500 mg sample of finely crushed rock is irradiated for 14 days in a thermal neutron flux of 1×10^{12} neutrons/cm^2-sec. Seven days are allowed for the decay of short-lived nuclides.   After this time, the sample, to which 20 mg of selenium carrier has been added, is fused with sodium hydroxide and the selenium is separated from the leached fusion cake by hydrochloric–hydrobromic acid distillation.   The selenium is then precipitated and collected on a membrane filter for the determination of the intensity of the 0.40 MeV γ-ray coincidence peak produced by ^{75}Se.

In the determination of selenium in soil samples, Lane[74] obtained excellent agreement with neutron activation analyses using an extraction-photometric procedure based on measurement of the fluorescence of the selenium-diaminonaphthalene complex in decalin.   A linear relationship between fluorescence and selenium concentration was observed up to 1 μg selenium per ml of decalin.

Selenium in electrolytic copper slimes can be determined conveniently by dissolution of the sample in 1 : 1 mixed nitric–sulfuric acids with the possible addition of a little potassium chlorate.[41]   Any dissolved silver is precipitated by hydrochloric acid and the filtered solution is made up to 80% by volume with hydrochloric acid and gassed with sulfur dioxide.   The precipitated selenium is determined gravimetrically by loss upon ignition.   Alternatively, the selenium can be determined by iodometric titration.[26]

Selenium in the 2–10 ppm range in decolorized soda–lime glass can be determined by fusion of the sample with Na_2CO_3, leaching with sulfuric acid and separation of the selenium by bromine–hydrobromic acid distillation.[54] The selenium is extracted as the selenium–diaminobenzidine complex into toluene following complexation of iron with ethylene diaminetetraacetic acid.

Biological Materials

The biochemistry of selenium in both plants and animals is a field of considerable and growing interest.   The nutrient role which has been noted for traces of selenium in numerous animal species and the toxicity of selenium which has been observed for larger amounts of the same element have given rise to various studies in which the accurate and sensitive determination of selenium has been of critical importance.

In recent years a number of excellent methods have been developed for the estimation of small amounts of selenium in biological material.   An excellent review of analytical methods for selenium in biological materials has been presented by Watkinson.[140]   Because of the volatility of selenium, appropriate consideration must be given to the sample dissolution procedure.

These procedures are discussed in the section of this chapter entitled "Decomposition of Organic Materials Containing Selenium."

Photometric methods based on the piazselenols particularly of 3,3'-diaminobenzidine and 2,3-diaminonaphthalene have gained wide acceptance in the analysis of selenium in both plant and animal materials. The sensitivity of 3,3'-diaminobenzidine has been increased through the use of ethylene diaminetetraacetic acid (EDTA) to complex interfering ions such as copper(II), iron(III), and vanadium(III)[44,138] and by fluorescence rather than absorption measurements.[44] The fluorescence of the selenadiazole permits the determination of as little as 0.02 μg selenium in plant materials.[44] In the analysis of biological material, extraneous fluorescence can be overcome by using a combustion rather than a wet oxidation method of sample dissolution.[50] Such a procedure is superior to the isolation of selenium by coprecipitation with arsenic[42] or extracting the complex formed by selenium with toluene-3,4-dithiol.[44] As pointed out by Watkinson,[44] the separation of selenium by toluene-3,4-dithiol can be circumvented if interferences due to the reaction of iron and copper with 3,3'-diaminobenzidine can be masked by EDTA. However, such a procedure cannot be applied to materials which produce an interfering fluorescence which is not quenched by EDTA, or to soils. The fluorometric diaminobenzidine method has been applied to the determination of selenium in alfalfa, grass, linseed meal, wheat, corn, oats, lettuce, pear, apple, and radish leaves, blood, muscle, liver, brain, and milk;[50] biological fluids and tissues.[128]

Using the oxygen flask combustion method, Allaway and Cary[8] extended the fluorometric 2,3-diaminonaphthalene procedure, as developed by Parker and Harvey[100] and Cukor, Walzcyk, and Lott[43], to the determination of as little as 0.02 μg selenium in biological materials such as dry skim milk, alfalfa, mixed hay, mixed grains, and kidney beans. The results compared favorably with those obtained by neutron activation analyses. Hydroxylamine was used to minimize the oxidation of diaminonaphthalene. A similar procedure was used by Lane[74] for the analysis of plants and animal organs with the modification of a magnesium nitrate addition to the oxygen combustion flask in the case of the animal samples to prevent the formation of a black residue which could absorb selenium. The fluorometric 2,3-diaminonaphthalene procedure of Watkinson[136] has been applied to numerous biological materials including papers and tobaccos.[98]

An interesting variation of the diaminobenzidine method as applied to biological materials such as liver, kidney, spleen, muscle and hair is the polarographic procedure of Christian, Knoblock and Purdy.[38] In this determination the selenium–diaminobenzidine complex is formed directly in the acid digest of the sample, then extracted into a mixture of chloroform and ethylenechloride and back-extracted into 2M perchloric acid. The polarographic behavior of the resulting solution permitted the determination of as little as 0.2 μg selenium in a 1–2 g wet weight sample.

A simple polarographic procedure has been developed by Nangniot[91] for the determination of selenium in plants. Selenium reduced to the element by ascorbic acid is dissolved in hydrobromic acid solution in which medium (0.7M HBr) selenium is estimated polarographically with a sensitivity of 0.2 μg per ml. This sensitivity can be increased to around 0.01 μg selenium per ml using oscillopolarography.

Excellent agreement with alternative data for selenium in South Dakota wheat (16.7 ppm Se) was obtained by Rann and Hambly[107] using atomic absorption analysis. The sample was prepared by the combustion (in 5 stages) of 10 g of roughly ground wheat in a Parr bomb calorimeter under oxygen pressure. The calibration curve was prepared from selenium-free wheat samples to which selenium had been added.

Since, in selenium deficiency studies on animals, the critical selenium concentration appears to be in the range 0.02 to 0.06 ppm,[74] it is not surprising that neutron activation methods of analysis have been developed. Thus, Dickson and Tomlinson,[47] using the nuclide ^{77m}Se ($t_{1/2} = 17.5$ sec), obtained excellent precision in the ppm region and a detection limit of 0.005 μg of selenium in biological materials such as blood cells and plasma using a flux of 10^{13} neutrons/cm^2 sec and a correction procedure for ^{19}O($t_{1/2} = 29$ sec).

The very short half-life of ^{77m}Se and the long half-life of ^{75}Se (120 days), the latter isotope requiring a long activation period, prompted Dahl and Steinnes[45] to develop a neutron activation method for the determination of selenium in animal tissue based on ^{81}Se ($t_{1/2} = 18.5$ minutes). In this procedure the sample is irradiated for 30 minutes at a flux of about 2.7 × 10^{12} neutrons/cm^2 sec. The irradiated sample is dissolved in mixed nitric–sulfuric acids (with selenium carrier added), after which the selenium is separated by hydrochloric–hydrobromic acid distillation and precipitated with NaHSO$_3$. This selenium separation procedure is effected in the remarkably short period of 15 minutes, which is adequate to permit measurement of the β-activity of ^{81m}Se and ^{81}Se. The chemical yield was found to be in the range 60 to 80%. The interference from bromine due to the reaction ^{81}Br(n, p)^{81}Se is negligible if bromine is present in minor amounts (< 10 ppm). Under optimum conditions, this procedure should have a sensitivity of 0.01–0.02 μg selenium at the flux indicated.

The 120 day ^{75}Se which allows time for careful radiochemical separation and counting, has been used by a number of investigators for the analysis of biological tissue.[61,85,122,142] Steinnes[122] has shown that at a flux of 2 × 10^{12} neutrons/cm^2 sec, the limit of detection of selenium can be as low as

0.0005 μg.   In this procedure, as for soil samples, the biological samples were irradiated for 14 days at the above flux, then shelved for 7 days to permit decay of short-lived nuclides.   The separation procedure which followed was the same as that used previously.[45]   The γ-activity of the samples and standard was measured with a 3 inch × 3 inch NaI (Tl) well crystal connected with a 400 channel γ-spectrometer.   The diameter of the well was 15 mm, and the depth 37 mm.   For low activity samples the counting time was generally 30 minutes.   Measurements in the well permitted detection of nearly 100% of the disintegrations in the 0.41 MeV photopeak.   Steinnes applied this procedure successfully to the determination of selenium in several types of biological materials such as hay, flour, bone, hair, and fat. The selenium content of fossil fuel samples from various areas in the United States has been determined by neutron activation analysis based on the ^{75}Se isotope.[104]   Following irradiation for 20 h at a thermal neutron flux of approximately 3×10^{13} neutrons cm^{-2} sec^{-1}, the samples were set aside for 5 days to allow the decay of short-lived nuclides.   The selenium was then separated by hydrochloric–hydrobromic acid distillation.   In the majority of the coal samples, the selenium ranged between 1 and 5 ppm whereas in the oil samples, the selenium content was relatively low (0.06–0.35 ppm).

In the determination of selenium in urine, Glover[57] obtained consistently good results with an accuracy of 0.02 mg of selenium per liter by digesting a 300 ml sample in a sulfuric–nitric acid mixture to which mercuric oxide had been added to prevent any volatilization loss of selenium.   The selenium was separated by hydrobromic acid–bromine distillation and precipitated with sulfur dioxide.   The final estimation was made titrimetrically using the Norris and Fay method[95] although a superior procedure could be realized based on the fluorescence of the piazselenol.

Air and Water

In view of the growing interest in environmental health, the accurate determination of small amounts of selenium in air and water is of some importance.   In an interesting study of selenium in the atmosphere, Hashimoto and Winchester[63] described a neutron activation procedure in which a 1 liter sample of rain or melted snow water was first vacuum evaporated to 20 ml, then reduced to 0.1 ml by gentle heating in a dust-free enclosure and irradiated for up to 14 h in a reactor having a flux of approximately 2×10^{13} neutrons/ cm^2.   In the case of air, samples of 100 m^3 are taken by passing air at about 1 m^3 per h through an aqueous bubbler or through a 1μ pore diameter millipore filter.   The aqueous solutions are treated in the same manner as the water samples, whereas the microfilters are irradiated directly.   Following irradiation, the selenium is separated by distillation from a bromine–hydro-bromic acid solution and precipitated from hydrochloric acid solution by hydrazine using a suitable selenium(IV) addition as carrier. The activity of ^{75}Se (120 day half-life) is determined by measuring the integrated sum of the gamma ray photopeaks. The method is sensitive to a limit of 0.01 μg of selenium. It is noteworthy that the selenium concentration averaged 0.2 μg per liter of water (0.2 ppb) (precipitation samples) or per 200 m^3 of air.

In the estimation of selenium in air Glover[57] employed a simple collection procedure involving the use of two wash bottles containing a 95–5% HBr–Br$_2$ solution. A range of 100–2500 liters of air was taken depending on the selenium concentration found in a trial run (e.g., from 25 to less than 0.1 mg per m^3) at a flow rate of 2 liters per minute.

The colorimetric diaminobenzidine method has been applied to the estimation of selenium in air.[123,124,141] West and Cimerman[141] determined selenium from the color of the piazselenol ring formed on filter paper by the ring oven technique. Copper and iron as possible interfering elements present in airborne particulate matter were sequestered using 1% sodium oxalate solution. The concentration range for selenium was established at 0.1 to 0.5 μg. The lower limit is considerably above that afforded by neutron activation analysis. Unfortunately, West and Cimerman present no analytical data on actual air samples, nor is the sampling method described.

The ring-oven technique has also been applied to the determination of selenium in water.[21] In this method a 10 liter water sample treated with sodium peroxide is evaporated nearly to dryness and the selenium separated by bromine–hydrobromic acid distillation. The selenium ring is developed using a 5% solution of thiourea as reductant and the concentration determined by visual comparison with the intensities of rings formed using suitable drops of a standard solution. Any arsenic codistilled with the selenium is retained at the center of the filter paper by precipitation as magnesium ammonium arsenate. Although Biswas and Dey[21] indicate a determination of selenium at 0.24 ppm, it should be possible to determine selenium below this level. The use of the piazselenol rather than selenium as the basis of the ring-oven method provides for a more sensitive procedure and it would appear that West and Cimerman's procedure could be extended without difficulty to the determination of submicrogram quantities of selenium in water.

As little as 0.005 ppm of selenium in water can be determined by x-ray fluorescence following the separation of selenium from a 1 liter sample by extraction into chloroform of the complex formed with ammonium pyrrolidine dithiocarbamate.[83] The chloroform extract is evaporated to between 0.25 and 0.5 ml, taken up on a 1/2 inch diameter filter paper disc and exposed to radiation from a molybdenum target x-ray tube. A linear counts vs concentration plot was obtained in the range 0.005 to 0.5 ppm of selenium.

REFERENCES

1. Abdullaev, G. B., Mekhtieva, S. I., Aliev, G. M., Abdinov, D. Sh., and Kerimova, T. G., *Phys. Stat. Sol.*, **16**, K31 (1966).
2. Acs, L., and Barabas, S., *Anal. Chem.*, **36**, 1825 (1964).
3. Agasyan, L. B., Agasyan, P. K., and Nikolaeva, E. R., *Vestn. Mosk. Univ.*, *Ser. II*, **21**, 93 (1966).
4. Agasyan, L. B., Nikolaeva, E. R., and Agasyan, P. K., *Vestn. Mosk. Univ.*, *Ser. II*, **21**, 96 (1966).
5. Agasyan, L. B., Nikolaeva, E. R., and Agasyan, P. K., *Zh. Anal. Khim.*, **22**, 904 (1967).
6. Ahearn, A. J., Trumbore, F. A., Frisch, C. J., Luke, C. L., and Malm, D. L., *Anal. Chem.*, **39**, 350 (1967).
7. Alekperov, A. I., and Novryzova, F. S., *Azerb. Khim. Zh.*, **2**, 117 (1966).
8. Allaway, W. H., and Cary, E. E., *Anal. Chem.* **36**, 1359 (1964).
8a. Angermann, W., and Ackermann, G., *Proc. Conf. Appl. Phys.-Chem. Methods Chem. Anal.*, *Budapest*, **3**, 49 (1966).
9. Arstamyan, Zh. M., and Tarayan, V. M., *Arm. Khim. Zh.*, **19**, 590 (1966).
10. Ballaux, C., Dams, R., and Hosti, J., *Anal. Chim. Acta*, **37**, 164 (1967).
11. Ballaux, C., Dams, R., and Hosti, J., *ibid.*, **41**, 147 (1968).
12. Ballaux, C., Dams, R., and Hosti, J., *ibid.*, **43**, 1 (1968).
13. Ballaux, C., Dams, R., and Hosti, J., *ibid.*, **47**, 397 (1969).
14. Barabas, S., Noranda Research Centre, unpublished work.
15. Barabas, S., and Bennett, P. W., *Anal. Chem.*, **35**, 135 (1963).
16. Barabas, S., and Cooper, W. C., *Anal. Chem.*, **28**, 129 (1956).
17. Barcza, L., *Acta Chim. Acad. Sci. Hung.*, **47**, 137 (1966).
18. Barr, J., Gillespie, R. J., Kapoor, R., and Malhotra, K. C., *Can. J. Chem.*, **46**, 149 (1968).
19. Belopol'skaya, T. L., *Materialy po Geol. i Polezn. Iskopaemym Severo-Vostoka SSSR*, *Sb.*, **14**, 165 (1960).
20. Bieling, H., and Wagenknecht, W., *Z. Anal. Chem.*, **201**, 419 (1964).
21. Biswas, S. D., and Dey, A. K., *Analyst*, **90**, 56 (1965).
22. Bowen, H. J. M., and Cawse, P. A., *Analyst*, **88**, 721 (1963).
23. Broad, W. C., and Barnard, A. J., Jr., *Chemist-Analyst,* **50**, 124 (1961).
24. Bruckner, K., *Z. Anal. Chem.*, **94**, 305 (1933).
25. Brunfelt, A. O., and Steinnes, E., *Geochim. Cosmochim. Acta*, **31** (2), 283 (1967).
26. Buketov, E. A., Makhmetov, M. Zh., and Gromakova, Z. I., *Izv. Akad. Nauk Kaz. SSR*, *Ser. Khim. Nauk*, **15**, 41 (1965).
27. Buketov, E. A., Moiseevich, O. Yu., and Ugorets, M. Z., *Zavod. Lab.*, **30**, 787 (1964).
28. Buketov, E. A., Ugorets, M. Z., and Pashinkin, A. S., *Russ. J. Inorg. Chem.*, **9**, 292 (1964).
29. Burley, R. A., *Phys., Stat. Sol.*, **29**, 551 (1968).
30. Busev, A. I., *Talanta*, **11**, 485 (1964).
31. Busev, A. I., Babenko, N. L., and Huang, Min-Ch'iao, *Zh. Anal. Khim.*, **18**, 1094 (1963).
32. Carter, T., and Parker, A., U.K. At. Energy Auth., Res. Group, At. Energy Res. Establ., Rept. No. 5293 (1966).
33. Chan, F. L., *Advances in X-ray Analysis*, Vol. 7, W. M. Mueller, G. Mallett and M. Fay, eds., New York, Plenum Press, 1964, pp. 542–554.

34. Chan, F. L., *Talanta*, **11**, 1019 (1964).
35. Cheng, K. L., *Anal. Chem.*, **28**, 1738 (1956).
36. Cherrier, C., and Nalbantoglu, M., *Anal. Chem.*, **39**, 1640 (1967).
37. Christian, G. D., Knoblock, E. C., and Purdy, W. C., *Anal. Chem.*, **37**, 425 (1965).
38. Christian, G. D., Knoblock, E. C., and Purdy, W. C., *J.A.O.A.C.*, **48**, 877 (1965).
39. Clarke, W. E., *Analyst*, **95**, 65 (1970).
40. Conrad, F. J., and Kenna, B. T., *Anal. Chem.*, **39**, 1001 (1967).
41. Cooper, W. C., *Standard Methods of Chemical Analysis*, 6th ed., Vol. 1, N. H. Furman, ed., Princeton, N.J., van Nostrand, 1962, pp. 925–949.
42. Cousins, F. B., *Australian J. Exptl. Biol. Med. Sci.*, **38**, 11 (1960).
43. Cukor, P., Walzcyk, J., and Lott, P. F., *Anal. Chim. Acta*, **30**, 473 (1964).
44. Cummins, L. M., Martin, J. L., and Maag, D. D., *Anal. Chem.*, **37**, 430 (1965).
45. Dahl, J. B., and Steinnes, E., Kjeller Report 95 (1965), Institutt for Atomenergi Kjeller, Norway.
46. Dagnall, R. M., Thompson, K. C., and West, T. S., *Talanta*, **14**, 557 (1967).
47. Dickson, R. C., and Tomlinson, R. H., *Int. J. Appl. Radiat. Isotopes*, **18**, 153 (1967).
48. Dingwall, D., and Williams, W. D., *J. Pharm. and Pharmacol.*, **13**, 12 (1961).
49. Duff, R., and Chessin, M., *Nature*, **208**, 1001 (1965).
50. Dye, W. B., Bretthauer, E., Seim, H. J., and Blincoe, C., *Anal. Chem.*, **35**, 1687 (1963).
51. Edwards, A. B., and Carlos, G. C., *Proc. Australian Inst. Min. Met.*, 172 (1954).
52. Feigl, F., and Del'Acqua, A., *Chemist-Analyst*, **55**, 15 (1966).
53. Fini, L. O., *Agronomy*, **9**, Pt. 2,1117 (1965).
54. Frackiewicz, J., and Budd, S. M., *Glass Technol.*, **4**, 134 (1963).
55. Fratkin, Z. G., and Polivanova, N. G., *Metody Anal. Khim. Reaktivov Prep.*, *Moscow*, **12**, 21 (1966).
56. Gaibakyan, D. S., and Aturyan, M. M., *Arm. Khim. Zh.*, **21** (12), 1015 (1968).
57. Glover, J. R., *Ann. Occup. Hyg.*, **10**, 3 (1967).
58. Gobrecht, H., Bock-Werthmann, U., Tausend, A., Brätter, P., and Willers, G., *Intern. J. Appl. Radiat. Isotopes*, **16**, 655 (1965).
59. Gobrecht, H., Tausend, A., Brätter, P., and Willers, G., *Solid State Commun.*, **4**, 307 (1966).
60. *Ibid.*, **4**, 311 (1966).
61. Grant, C. A., Thafvelin, B., and Christell, R., *Acta pharmac. tox.*, **18**, 285(1961).
62. Green, T. E., and Turley, M., *Treatise on Analytical Chemistry*, Pt. II, Vol. 7, I. M. Kolthoff and P. J. Elving, eds., New York, Interscience, 1961, pp. 137–205.
63. Hashimoto, Y., and Winchester, J. W., *Environ. Sci. and Tech.*, **1**, 338 (1967).
64. Iguchi, A., *Bull. Chem. Soc. Japan*, **31**, 748 (1958).
65. Kammori, O., and Ono, A., *Bunseki Kagaku*, **15**, 290 (1966).
66. Kaplan, B. Ya., and Sorokovskaya, I. A., *Zavod. Lab.*, **30**, 783 (1964).
67. Katu, T., Nobuyuki, T., and Hiroshi, T., *Bunseki Kagaku*, **18**, 319 (1969).
68. Kempe, G., *Werner Denn. Z. Anal. Chem.*, **217**, 169 (1966).
69. Kerbyson, J. D., and Acs, L., Noranda Research Centre, unpublished work.
70. Kerbyson, J. D., and Givens, B. J., Noranda Research Centre, unpublished work.

71. Kerbyson, J. D., Lemieux, R. R., and Ratzkowski, C., Noranda Research Centre, unpublished work.
72. Kirkbright, G. F., and Ng, W. K., *Anal. Chim. Acta*, **35**, 116 (1966).
73. Kotarski, A., *Chem. Anal.* (*Warsaw*), **10**, 321 (1965).
74. Lane, J. C., *Irish J. Agr. Res.*, **5**, 177 (1966).
75. Larsen, E., *J. Chem. Ed.*, **41**, 435 (1964).
76. Le Peintre, C., *Compt. rend.*, **252**, 1968 (1961).
77. Lingane, J. J., and Niedrach, L. W., *J. Am. Chem. Soc.*, **70**, 4115 (1948).
78. Ljung, H. A., *Ind. Eng. Chem.*, *Anal. Ed.*, **9**, 328 (1937).
79. Lott, P. F., Cukor, P., Moriber, G., and Solga, J., *Anal. Chem.*, **35**, 1159 (1963).
80. Maasson, G., *Z. Erzbergbau Metallhuettenw.*, **18**, 116 (1965).
81. Magin, G. B., Thatcher, L. L., Rettig, S., and Levine, H., *J. Am. Water Works Assoc.*, **52**, 1198 (1960).
82. Makhmetov, M. Zh., and Polukarov, A. N., *Tr. Khim.-Met. Inst., Akad. Nauk Kaz. SSR*, **3**, 52 (1967).
83. Marcie, F. J., *Environ. Sci. and Tech.*, **1**, 164 (1967).
84. Marshall, H., Canadian Copper Refiners Limited, unpublished work.
85. McConnell, K. P., *Proc. Int. Conf. Mod. Trends in Activation Analysis*, College Station, Texas, p. 137 (1961).
86. McNulty, J. S., Center, E. J., and MacIntosh, R. M., *Anal. Chem.*, **23**, 123 (1951).
87. Mellichamp, J. W., *Applied Spectroscopy*, **8**, 114 (1954).
88. Murashova, V. I., and Sushkova, S. G., *Izv. Vyssh. Ucheb. Zavod., Khim. i Khim. Tekhnol.*, **9**, 551 (1966).
89. Murashova, V. I., Sushkova, S. G., and Bakunina, L. T., *Zavod. Lab.*, **33**, 280 (1967).
90. Nadkarni, R. A., and Haldar, B. C., *Indian J. Chem.*, **3**, 539 (1965).
91. Nangniot, P. J., *Electroanal. Chem.*, **12**, 187 (1966).
92. Nepeina, L. A., and Kuznetsova, M. K., *Uch. Zap., Tsentr. Nauchn. Issled. Inst. Olovyan. Prom.*, **1**, 92 (1965).
93. Niebuhr, P. E., and Macmillan, A. H., U.S. Bur. of Mines, Rept. of Invest. No. 6006 (1962).
94. Noakes, F. D. L., *Analyst*, **76**, 542 (1951).
95. Norris, J. F., and Fay, H., *Am. Chem. J.*, **20**, 278 (1898).
96. Nygard, B., *Acta. Chem. Scand.*, **20**, 1710 (1966).
97. Nygard, B., *Ark. Kemi*, **27**, 341 (1967).
98. Olson, O. E., and Frost, D. V., *Environ. Sci. Technol.*, **4**(8), 688 (1970).
99. Pacauskas, E., Janickis, J., and Rinkeviciene, E., *Liet. TSR Mokslu Akad. Darb., Ser. B.*, **3**, 55 (1966).
100. Parker, C. A., and Harvey, L. G., *Analyst*, **87**, 558 (1962).
101. Pats, R. G., and Semochkina, T. V., *Zavod. Lab.*, **33**, 1491 (1967).
102. Pats, R. G., Vasil'eva, L. N., Zaglodina, T. V., and Shuvalova, E. D., *Zavod. Lab.*, **29**, 928 (1963).
103. Peterson, G. E., and Currier, E. W., *Applied Spectroscopy*, **10**, 1 (1956).
104. Pillay, K. K. S., Thomas, C. C., Jr., and Kamenski, J. W., *Nucl. Applns. and Tech.*, **7**, 478 (1969).
105. Plotnikov, V. I., *Russ. J. Inorg. Chem.*, **9**, 245 (1964).
106. Pollock, E. N., *Talanta*, **11**, 1548 (1964).
107. Rann, C. S., and Hambly, A. N., *Anal. Chim. Acta*, **32**, 346 (1965).

108. Redman, M. J., and Harvey, W. W., *J. Less Common Metals*, **12**, 395 (1967).
109. Reshchikova, A. A., and Shorina, E. V., *Izv. Akad. Nauk Moldavsk. SSR, Ser. Biol. i Khim. Nauk*, **11**, 76 (1964).
110. Russell, B. G., Lubbe, W. V., Wilson, A., Jones, E., Taylor, J. D., and Steele, T. W., *Talanta*, **14**, 957 (1967).
111. Samsahl, K., *Anal. Chem.*, **39**, 1480 (1967).
112. Samuelson, O., *IVA*, **17**, 5 (1946).
113. Schrenk, W. T., and Browning, B. L., *J. Am. Chem. Soc.*, **48**, 2550 (1926).
114. Schuster, E., and Wohlleben, K., paper presented to German Physical Society, Berlin, March 21, 1968.
115. Shakhov, A. S., *Zavod. Lab.*, **11**, 893 (1945).
116. Shcherbov, D. P., Ivankova, A. I., and Gladysheva, G. P., *Issled. Razrab. Fotometrich. Metod. Opred. Mikrokolichestv. Elem. Miner. Syr'e*, 10 (1967).
117. Shkrobot, E. P., and Shebarshina, N. I., *Zavod. Lab.*, **35**, 417 (1969).
118. Sill, C. W., and Peterson, H. E., U.S. Bur. Mines, Rept. of Invest. No. 5047 (1954).
119. Sjoborg, B.-L., *Talanta*, **14**, 693 (1967).
120. Skowronski, S., private communication.
121. Starace, C. A., Wiersma, L. D., and Lott, P. F., *Chemist-Analyst*, **55**, 74 (1966).
122. Steinnes, E., *Int. J. Appl. Radiat. Isotopes*, **18**, 731 (1967).
123. Strenge, K., *Arch. Gewerbepathol. Gewerbehyg.*, **16**, 588 (1958).
124. Suzuki, Y., Nishiyama, K., and Matsuka, Y., *Shikoku Igaku Zasshi*, **11**, 77 (1957).
125. Tada, O., *J. Appl. Phys., Japan*, **33**, 823 (1964).
126. Tanaka, K., Takagi, N., and Tsuzimura, H., *Bunseki Kagaku*, **18**, 319 (1969).
127. Tarayan, V. M., Arustamyan, E. M., and Shaposhnikova, G. M., *Sovrem. Metody Khim. Spektral. Anal. Mater.*, 33 (1697).
128. Taussky, H. H., Washington, A., Zubillaga, E., and Milhorat, A. T., *Biochem. J.*, **10**, 470 (1966).
129. Tischendorf, G., *Freiberger Forschungsh.*, **C208**, 162 (1966).
130. Tso, Tsung-Chi, and Yang, Hsin Min., *K'o Hsueh T'ung Poa*, **17**, 75 (1966).
131. Tulyupa, F. M., Barkalov, V. S., and Usatenko, Yu. I., *Zh. Anal. Khim.*, **22**, 399 (1967).
132. Tumanov, A. A., and Shakhverdi, N. M., *Zavod. Lab.*, **33**, 20 (1967).
133. Tumanov, A. A., Shakhverdi, N. M., and Glazunova, Z. I., *Poluch. Anal. Veshchestv. Osoboi Chest., Mater. Vses. Konf., Gorky, USSR*, 255 (1963).
134. Tumanov, A. A., Shakhverdi, N. M., and Glazunova, Z. I., *Zavod. Lab.*, **33**, 8 (1967).
135. Usatenko, Yu. I., Tulyupa, F. M., and Barkalov, V. S., *Zavod. Lab.*, **32**, 787 (1966).
136. Vasko, A., *Phys. Stat. Sol.*, **8**, K41 (1965).
137. Veale, C. R., *J. Inorg. and Nucl. Chem.*, **10**, 333 (1959).
138. Watkinson, J. H., *Anal. Chem.*, **32**, 981 (1960).
139. Watkinson, J. H., *Anal. Chem.*, **38**, 92 (1966).
140. Watkinson, J. H., *Symposium, Selenium in Biomedicine*, Chapt. 6, Muth, O. H., ed., Westport, Conn., AVI Publishing Co., Inc., 1967, p. 97.
141. West, P. W., and Cimerman, C., *Anal. Chem.*, **36**, 2013 (1964).
142. Wester, P. O., Brune, D., and Samsahl, K., *Int. J. Appl. Radiat. Isotopes*, **15**, 59 (1964).

143. Wlotzka, F., *Z. Anal. Chem.*, **215**, 81 (1966).
144. Yoshino, Y., *J. Chem. Soc. Japan, Pure Chem. Sect.*, **71**, 577 (1950).
145. Zelyanskaya, A. I., and Gorshkova, L. S., *Trudy Inst. Metal., Akad. Nauk SSSR, Ural. Filial*, **137**, 141 (1960).

The following additional references, not included in the chapter, will be useful to the interested reader.

C. L. Chakrabarti, "The Atomic Absorption Spectroscopy of Selenium." *Anal. Chim. Acta.*, **42**, 379 (1968).
H. L. Kahn, "Improvement of Detection Limits for Arsenic, Selenium and Other Elements With Argon-Hydrogen Flame." *Atomic Absorption Newsletter*, **7**, 5 (1969).
M. S. Cresser and T. S. West, "Spectrophotometric Determination of Selenium with Cyclohexanone." *Talanta*, **16**, 416 (1969).
P. W. West and T. V. Ramakrishna, "A Catalytic Method for Determining Traces of Selenium." *Anal. Chem.*, **40**, 966 (1968).
R. L. Osburn, A. D. Shendrikar, and P. W. West, "A New Spectrophotometric Method for the Determination of Submicrogram Quantities of Selenium." *Anal. Chem.*, **43**, 594 (1971).
T. Kawashima and M. Tanaka, "Determination of Submicrogram Amounts of Selenium (IV) by Means of the Catalytic Reduction of 1, 4, 6, 11-Tetraazanaphthacene." *Anal. Chem. Acta.*, **40**, 137 (1968).
M. Tanaka and T. Kawashima, "Some Substituted *o*-Phenylenediamines as Reagents for Selenium." *Talanta*, **12**, 211 (1965).
I. Rosenfeld and O. A. Beath, "Selenium," Academic Press Inc., New York, N.Y., 1964.
"Preliminary Air Pollution Survey of Selenium and its Compounds." United States Department of Health, Education and Welfare, NAPCA.

11

The Toxicology of Selenium and its Compounds

W. CHARLES COOPER*

*Noranda Research Centre,
Pointe Claire, Quebec*

J. R. GLOVER

*Welsh National School of Medicine, University of
Wales, Cardiff, Wales*

INTRODUCTION

The toxicology of selenium and its compounds is of considerable interest, particularly in view of the long established selenium poisoning of cattle foraging on seleniferous plants and the more recent extensive research confirming the nutritional essentiality of the element. The widely held belief that selenium is a toxic substance is due, no doubt, to the highly toxic nature of a number of plants, notably of the species *Astragalus* growing in seleniferous soils and able to accumulate as much as several thousand ppm of selenium. Although the toxic effects on humans of vegetation grown in seleniferous areas have been documented, no long-term systemic toxicity has been indicated.[27] Indeed, seleniferous grains have come to be regarded as of higher nutritional value for animals than those containing no selenium and of importance in combatting selenium deficiency diseases.

* Present Address: United Nations Development Programme, Casilla 197-D, Santiago, Chile.

Although selenium itself is relatively nontoxic, certain selenium compounds are definitely toxic towards humans, one of the most seriously toxic being hydrogen selenide.

It is the object of this chapter to review the present knowledge of the toxicity of selenium and its compounds, including selenium-bearing natural products. Although valuable data can be obtained from animal experiments, caution must be used in applying these data to humans. The toxicology of selenium has been reviewed in previous publications,[7,11,25,28,35,50] amongst which the most extensive treatment has been provided by Rosenfeld and Beath.[50] The industrial toxicology of selenium has been reviewed most recently by Glover.[28] The treatment of various effects of selenium intoxication is discussed together with a description of preventive measures.

SELENIUM INTOXICATION IN MAN

As has been indicated by Glover,[27,28] no rural or industrial cases of irreversible chronic disease or death attributable to selenium or its compounds have occurred in humans. Nevertheless, selenium can be toxic to animals and presents a potential hazard to humans in rural farm and industrial situations. The absorption of selenium compounds by the body usually occurs through the lungs by dust or fumes, and through the skin. In the case of intoxication arising in seleniferous areas, the selenium is ingested generally from foods grown in such areas. Although it is difficult to attribute definite symptoms to selenium intoxication, the following factors have been observed in chronic selenium intoxication in man: depression, languor, nervousness, occasional dermatitis, gastrointestinal disturbances, giddiness, and garlic odor of the breath and sweat. Of these, the most common symptom appears to be gastrointestinal disturbances.

The chronic and acute selenium poisoning of livestock in seleniferous regions of the western United States, which has been well documented by Rosenfeld and Beath,[50] has given rise to numerous studies on the health of the rural population in these areas. In such a study, Smith and Westfall[57] attempted to correlate the symptomatology with selenium intake in the diet and selenium excretion. Fifty families in South Dakota and Nebraska in areas known to be highly seleniferous were selected for this survey. The majority of the subjects had lived on seleniferous farms for from 10 to 40 years, and no one had a residence of less than 3 years. The relation of urinary selenium to food selenium and the frequency of observed symptoms and their percentage distribution are shown in Tables 11-1 and 11-2. None of the symptoms listed in Table 11-2 can be regarded as specifically due to selenium. However, the high incidence of gastrointestinal disturbances is of some importance in the light of observations made on lower animals. The incidence

TABLE 11-1   RELATION OF URINARY SELENIUM TO FOOD SELENIUM

Family No.	Urinary Selenium (ppm)	Milk	Eggs	Meat	Vegetables	Cereal Grains
97	0.25, 0.27, 0.32	Trace	0.57	—	Trace	Trace
51	0.20, 0.27	—	1.35	1.60	0.36	1.90
22	0.20, 0.21	0.36	1.40	—	0.41–0.74	Trace
113	0.20, 0.20, 0.24	0.25	1.45	—	0.30–0.82	0
52	0.26	0.25	0.32	—	Trace–0.58	—
83	0.13, 0.38, 0.40	0.34	—	2.19	—	—
27	0.29, 0.56	0.22	—	2.22	—	—
76	0.43, 0.73	0.35	4.08	3.30	0.27–1.05	—
107	0.94	0.39	3.65	—	Trace–0.18	—
47	0.70, 0.80, 0.98	0.57	3.08	—	0.23–2.04	3.30
74	1.03, 1.10	Trace	4.12	—	—	0.45–1.00
78	1.00, 1.14	0.36	5.04	—	1.03–17.80	3.60
16	1.05, 0.36, 1.33	1.14	—	—	2.42	2.50–18.80
19	1.24, 1.98	1.27	—	8.00	1.26	4.20–10.00

TABLE 11-2   FREQUENCY OF OBSERVED SYMPTOMS AND PERCENTAGE DISTRIBUTION[57]

Clinical Group	Frequency of Occurrence	Percent Distribution
No obvious symptoms	24	16.0
Gastrointestinal disturbances	31	20.7
Bad teeth	27	18.0
Icteroid discoloration of the skin	28	18.7
History of recurrent jaundice	5	3.3
Vitiligo	2	1.3
Pigmentation of the skin (chloasma?)	3	2.0
Sallow and pallid color, especially in younger individuals	17	11.3
Dermatitis	5	3.3
Rheumatoid arthritis	3	2.0
Pathological nails	3	2.0
Cardiorenal disease	2	1.3

of discoloration of the skin is higher than normal and may be related to the occurrence of bilirubinemia which has been observed in experimental animals subjected to chronic selenium poisoning.[21]

More recent work on the correlation between bad teeth and selenium intake has demonstrated a positive correlation between malocclusion and caries of the permanent teeth in native children in seleniferous regions in Wyoming. The higher selenium intake was corroborated by the urinary excretion which ranged from 0.2 to 1.12 ppm selenium.[65]   Studies by Hadji-markos and Bonhorst[31] on the effect of seleniferous diets on the incidence of dental caries in school children indicate that selenium may be a factor in increasing the susceptibility to caries. However, no clearly defined relationship between selenium and caries has been established.

From the studies on the public health problems arising from the ingestion of food grown in seleniferous areas, it would appear that humans may be absorbing from 0.01 to 0.1 mg and even as much as 0.2 mg of selenium per kg of body weight per day. A daily amount of 1.0 mg of selenium per kg of body weight may be sufficient to produce chronic selenium intoxication in man.[56] According to Smith and Lillie,[56] the continued ingestion of food containing selenium in an amount as low as 0.2 mg per kg of body weight may be harmful.

A number of cases of selenium intoxication arising in seleniferous regions have been described. In a study of a family living in western South Dakota, Lemley and Merryman[39] noted the following symptoms of selenium intoxication: slight continual dizziness and clouding of the sensorium, and extreme lassitude. There was also a feeling of depression, together with a moderate emotional instability.

A case of acute dermatitis on the lips, eyelids, and bearded regions of a rancher in South Dakota has been reported[38] The water, meat, vegetables, and dairy products from the patient's ranch contained selenium in appreciable quantities and the animals on the ranch suffered from alkali disease. The dermatitis disappeared following treatment and the elimination of seleniferous foods from the diet.

An interesting case of chronic selenium poisoning arising from the consumption of well water containing 9 ppm of selenium has been reported.[4] The 5 children affected, ranging in age from 6 months to 10 years, showed symptoms of lassitude, total or partial loss of hair, and discoloration and dropping off of the fingernails. Regrowth of the nails and hair commenced upon discontinuation of the use of the water. It is of interest to note that this is the first case of selenium intoxication due to the ingestion of drinking water sufficiently high in selenium to produce the well-known symptoms of alkali disease in cattle.

The occurrence of seleniferous soils and vegetation in Colombia, South America, has been known for some time, with Indian children displaying symptoms of chronic selenium intoxication through the loss of hair and nails.[4]

It should be noted that seleniferous grains have been recognized in a number of other countries besides the United States, and that it is most difficult to obtain a selenium-free diet. In view of the nutrient value of traces of selenium in the animal diet, it might well be that a selenium-free diet is, in fact, nutritionally undesirable.

As regards industrial selenium intoxication, Amor and Pringle[3] and Glover[29] believe that a garlic odor of the breath is the earliest and most characteristic sign of selenium absorption.   However, this indication is of much shorter duration and of lower intensity than in the case of tellurium absorption, and the symptom can vary widely from one individual to another More reliable indications of selenium absorption are provided by an examination of the selenium content of the urine and blood.

Although a relatively high urinary selenium has been often associated with gastrointestinal disorders, pathological disturbances of the nails, poor teeth and icteroid skin, it is difficult to draw direct correlations between selenium intake and such disease symptoms or manifestations.   As will be shown in this chapter, the distribution of selenium in body organs differs widely from one individual to another and it is very difficult to establish selenium levels for body organs and fluids which might be considered as normal.   Although during the last 30 years the studies which have been conducted present valuable information on selenium intoxication, the quantitative data may not always be reliable.   The limitations of the analytical methods used in the earlier work have been overcome in more recent years and more reliable data are now becoming available.

Since there is considerable interest in the possible long-term systemic effects of selenium on the human body, it is noteworthy that in a study of selenium intoxication on the part of industrial workers over a period of 10 years, Glover[27] reported 17 natural deaths amongst workers exposed to selenium.   The mortality pattern of these exposed workers corresponded closely to that of the general population of England and Wales.   Of the 6 workers who died due to malignant neoplasms (expected number, 5.1), there was no sarcomata, and the deaths arose from varied organs of origin (2 bronchus, 1 stomach, 1 colon, 1 ovary, and 1 testis).   A comparison of human cancer mortality data in high seleniferous *vs* low seleniferous states of the United States actually shows a higher incidence of deaths in the low seleniferous regions (171.5 male and 142.0 female *vs.* 191 male and 165.5 female).[53]

SELENIUM IN THE ANIMAL BODY

Retention and Distribution of Selenium

Laboratory animals chronically poisoned with small doses of inorganic selenium administered over a long period indicate a wide distribution of selenium in the tissues.[58]   The highest concentrations of selenium are to be found in the liver, kidney, spleen, pancreas, heart, and lungs.   If the feeding of inorganic selenium is discontinued, most of the selenium is excreted within 2 weeks, although small amounts persist for about a month or even longer. Similar results have been found in the administration of inorganic selenium to rats and rabbits.[42] Of the selenium administered, 35–45% was excreted within 48 h and 50–60% within 10–12 days.

The above findings have been corroborated by Gardiner[26] who found that in the case of sheep, the liver and kidney accumulated far more than muscle, serum and wool. The higher concentrations in the liver and kidney are in agreement with other studies in which the chronic administration of selenium was carried out on various animal species. It is noteworthy that in the case of chronically depleted sheep, the concentration of selenium reached peak concentrations in a few hours and then declined over succeeding days. The tissues having a particular tendency to accumulate selenium appeared to be those undergoing pathological changes resulting from selenium administration.

The concentration of selenium in the animal body is dependent on the animal species as well as on the preinjection levels of selenium in the body organs. The form of selenium is also of importance, as has been shown in studies by Cousins and Cairney,[12] who found that the storage of inorganic selenium in sheep is limited, whereas organically bound selenium has a more long-lasting effect. Also, the mode of administration is of importance as demonstrated by Jacobsson[33] who administered doses of ^{75}Se-sodium selenite to sheep subcutaneously and intraruminally. The fraction of the dose retained was larger for small than for large doses. Also, it has been found that in the case of pigs, a smaller proportion of the administered selenium dose is retained if the selenium concentration of the basal diet has fulfilled the physiological requirements of the treated animal. The higher concentration of selenium in the liver and the relatively constant percentage of its retention in this organ, irrespective of the route of selenium administration, suggests an important metabolic difference between the selenium in the liver and other body organs.

Data on the distribution of selenium in human organs and tissues have been reported by Ermakov,[19] (Table 11-3). Although no explanation is advanced for the wide differences in the selenium content of the body organs of different individuals, the significantly higher and almost constant level of selenium in the pancreas of infants compared with that of adults is particularly striking. Also of interest is the consistently higher level of selenium in the body organs of the 64 year old male compared with other adults. Data on a number of body organs for an infant and an adult have been presented by Dickson and Tomlinson.[13] As shown in Table 11-4, these data indicate higher levels of selenium in the kidney and thyroid gland. There is no

TABLE 11-3   SELENIUM CONTENT IN HUMAN ORGANS AND TISSUES[19]

(μg per gram of fresh substance)

Organ and Tissue	Child Seven Months, Extreme Prematurity	Child Nine Months, Intrauterine Asphyxia	Child Nine Months, Intrauterine Asphyxia	Child Six Months, Extreme Prematurity	Girl 15 years, Food Poisoning	Man 29 Years, Skull Fracture (Trauma)	Man 64 Years, Mechanical Asphyxia	Man 75 Years, General Atherosclerosis
Heart	0.083	0.040	0.006	0.204	0.004	0.072	0.114	0.024
Lungs	0.020	0.016	0.006	0.011	0.024	0.138	0.100	0.024
Spleen	0.140	0.071	0.079	0.046	0.018	0.076	0.078	0.014
Liver	0.072	0.026	0.011	0.019	0.079	0.058	0.161	0.072
Kidneys	0.037	0.028	0.031	0.014	0.080	0.271	0.161	0.016
Pancreas	0.260	0.240	0.261	0.260	0.032	0.026	0.100	0.021
Muscles	0.021	0.021	0.024	0.016	0.014	0.028	0.226	0.020
Skin	0.036	0.021	0.025	0.028	0.012	0.028	0.124	0.032
Bone	0.105	0.090	0.064	0.984	0.021	0.031	0.243	—

TABLE 11-4   SELENIUM CONTENT OF TISSUES OTHER THAN BLOOD[13]

Tissue	Selenium Content per gram of Whole Tissue (μg)	
	Infant	Adult
Stomach	0.19	0.17
Liver	0.34	0.39
Pancreas	0.05	0.13
Spleen	0.37	0.27
Kidney	0.92	0.63
Intestine	0.31	0.22
Heart	0.55	0.22
Lung	0.17	0.21
Artery	0.27	0.27
Muscle	0.31	0.40
Fat	0.09	0.12
Trachea	0.14	0.24
Gonad	0.46	0.47
Thyroid gland	0.64	1.24
Brain	0.16	0.27
Adrenal gland	0.21	0.36
Lymph node	0.26	0.10

obvious reason why the selenium levels reported by Ermakov are, in general, so much lower than those given by Dickson and Tomlinson. The latter authors employed a rapid and sensitive neutron activation method of analysis (see Chapter 10) whereas no information is available on the procedure used by Ermakov.

It would be of interest to compare data on the selenium content of body organs of persons resident in seleniferous *vs* nonseleniferous areas. Higher selenium levels in the blood of people living in seleniferous regions have been reported by Allaway et al.[1] who analyzed some 210 blood samples from 19 collection sites in the United States. The blood levels ranged from an average of 25.6 μg per 100 ml for persons living in Rapid City, South Dakota, a seleniferous area, to an average of 15.7 μg per 100 ml for residents of Lima, Ohio, a nonseleniferous region. No selenium level in the blood of humans suffering from selenium toxicity was reported. It is noteworthy that Allaway et al. found the uniformity of the blood selenium concentrations more striking than the geographic differences.

A significant difference in the blood selenium levels in children suffering from kwashiorkor *vs* those who had recovered from the disease has been reported.[6] Thus, a mean selenium value in whole blood of 14 μg per 100 ml was noted for recovered patients *vs* 8 μg per 100 ml for kwashiorkor untreated patients or those undergoing treatment.

Pathology of Selenium Intoxication

As discussed above, in experimental animals the organ most affected by selenium is the liver, which undergoes a reversible fatty degeneration provided the selenium exposure is not prolonged.[54]   Under continued exposure, liver cirrhosis is likely to occur.   In addition, the spleen becomes enlarged and the stomach and intestinal tract show hemorrhages.

Although in humans respiratory ailments and gastrointestinal disturbances are of major significance in the diagnosis of selenium intoxication, the liver can also be affected.   A case of very early cirrhosis of the liver was reported by Lemley and Merryman[39] in a rancher from northwestern South Dakota who had displayed symptoms of marked selenium intoxication, including a high selenium level in the urine.   However, it should be noted that the liver cirrhosis could have been attributed to causes other than selenium intoxication.

Bilirubinemia has been observed to result from selenium intoxication in rabbits.[21]   The loss in the capacity of the liver to dispose of the yellow pigment, bilirubin, in the form of bile in this type of selenium poisoning suggests a possible explanation of the yellow skin coloration sometimes found in humans exposed to selenium over a period of time.   In recent Japanese studies on selenium poisoning, a yellowish color of the facial skin was noticed in addition to discoloration of the nails of the fingers and toes, anemia, and hypertension.[61]

Selenium Detoxification and Elimination

There appear to be a number of factors which may be operative in limiting selenium intoxication or in effecting a detoxification of ingested selenium. Perhaps the most important of these factors is found in the protein content of food.   Numerous studies on experimental animals, particularly rats, have shown that the toxicity of selenium found in natural foods can be affected markedly by dietary factors.   Thus, a ratio of 1 % protein in the diet to about 30 μg or less of selenium per 100 g of diet is of low toxicity.   However, under the same conditions, a ratio of 1 % protein to 100 μg of selenium per 100 g of diet is decidedly toxic.[54]   In these experiments a level of 10 ppm of selenium in the food was employed.

Similar results have been obtained when the selenium has been administered as inorganic selenite.[40]   The most protective protein in this case was found to be casein.   Thus, a diet containing 30 % of protein in the form of casein in the diet of rats counteracted the presence of 35 ppm of selenium as selenite.[30]   These and other experimental results indicate that a high protein diet may be an important factor in protecting humans against chronic intoxication.

Another counteractant to selenium intoxication is arsenic. In experiments on rats, 5 ppm of arsenic as arsenite in drinking water was found to be effective against 11 ppm of selenium regardless of whether the selenium was introduced into the food as selenite or present as organic selenium as in toxic wheat.[14] This protective action of arsenic against selenium was effective if the arsenic treatment was started not later than 20 days after the initial intake of selenium. This effect of arsenic in counteracting selenium toxicity has been shown not only in rats but also in dogs and pigs.[14,46,49,66] As has been shown by Main and Frost,[23,24] the effect of arsenate in combatting selenium toxicity is dependent on the mode of administration of selenium. Thus, when high levels of sodium arsenate and selenate were simultaneously present in the water, the toxicity towards rats was greatly increased. However, when the selenium was in the diet, arsenate in the water counteracted the selenium toxicity.

Studies on laboratory animals reveal that there occur significant decreases in the concentrations of ascorbic acid, vitamin K and glutathione as a result of selenium intoxication. This suggests that a diet which furnishes larger than normal amounts of these substances might be beneficial in counteracting the toxic effects of selenium.[21,37,52,62] However, more definitive studies are required to determine the effect of these substances in counteracting selenium toxicity.

Although Lemley and Merryman[39] claim an increase in the excretion of selenium resulting from the administration of bromobenzene to humans suffering from selenium intoxication in seleniferous regions, Westfall and Smith[67] found that the oral administration of this compound produced no significant increase in the level of urinary selenium. These conflicting results, combined with the fact that bromobenzene itself is a rather toxic substance, indicate that this compound cannot be recommended for the treatment of selenium intoxication.

In studies on the retention and elimination of selenium by growing lambs over a 12 week period, Ewan, Baumann and Pope,[20] found that, in general, the elimination of selenium for most organs showed a continuous decrease with time. The concentration of selenium in the liver decreased markedly during the 12 weeks, but of all the organs sampled, the liver contained the highest concentration of selenium at the end of the period (13 μg per 100 g).

In the case of humans, the elimination of selenium is best followed through analysis of the urine. However, as shown by Dudley,[15] and Glover,[29] it is not possible to establish a definite correlation between the symptoms of selenium intoxication and the concentration of selenium in the urine. In the case of people not known to have been exposed to selenium, Glover,[27] found that in 793 men and women examined, in 90.7% of the samples tested, the selenium content ranged between 10 and 100 µg selenium per liter.   In studies on a much smaller population, Sterner and Lidfelt,[59] found 62% of the samples in this same range with 30% indicating no selenium (less than 10 µg per liter).

In studies on 200–300 selenium workers over a period of 17 years, Glover,[27] found an average selenium urinary level of 84 µg per liter.   The highest figure was 490 µg selenium per liter for a selenium worker who had inhaled selenium dust during a maintenance operation.   Glover's observations are of particular interest in providing an indication of the elimination of selenium through the urine and pointing to a proper maximum allowable concentration of the element in both air and urine.   Glover's data indicate a tendency for urinary excretion to rise towards the end of the work week, indicating a lower level during days off and a higher level during periods of ingestion.   Workers taken off highly seleniferous jobs showed a drop in selenium urinary level after 24 h, with the level usually below 100 µg per liter at the end of the first week.   It is of interest to note that on some seleniferous jobs there were workers who showed no selenium in the urine. From this study Glover suggested that in public, rural, and industrial health situations, the selenium urinary level should be below 100 µg per liter.   This is considerably less than 400 µg per liter maximum allowable concentration of selenium in the urine given by Elkins.[18]

The study of selenium intoxication by Smith, Franke, and Westfall[55] of 127 members of 90 families in seleniferous regions of Wyoming, South Dakota and Nebraska showed that in 89% of the urine samples, the selenium content ranged between 20 and 99 µg per liter.

TOXICITY OF SELENIUM, SELENIUM COMPOUNDS, AND SELENIUM-BEARING NATURAL PRODUCTS

Elemental Selenium

From all of the data available, it appears that elemental selenium is relatively nontoxic.   The only forms of elemental selenium which have given rise to irritation are fine dust and selenium fumes.   The commonest symptoms reported by Glover[27] on the part of selenium-exposed workers were garlic odor of the breath, skin rashes, indigestion, and psychological symptoms.   The symptoms of lassitude and irritability noted on the part of several men exposed to selenium were similar to the symptoms noted on the part of people living in seleniferous regions of the western United States.

These symptoms disappeared when the workers were removed from selenium. The skin rashes were believed to be almost entirely due to the effect of selenium dioxide. It is quite likely that certain of the symptoms observed with selenium fumes are due to selenium dioxide. In the case of workers exposed to fine selenium dust, Amor and Pringle[3] noted that the dust collected in the upper nasal passages and caused catarrh, nose bleeding and a loss of the sense of smell. The same investigators noted some cases of dermatitis on the backs of the hands of workers who had handled selenium.

A case of interest is that reported by Glover[27] in which a worker inhaled a cloud of selenium dust while dismantling a ventilation duct. Although this dust contained some selenium dioxide as well as selenium, and the worker had a high urinary selenium level, he suffered no dyspnea and did not require hospitalization.

There are few accounts of severe selenium intoxication due to the ingestion of selenium fumes. An important case has been reported by Clinton[10] in which workers were exposed to dense selenium fumes which were emitted from selenium rectifier plates which had been charged to a scrap aluminum furnace. The red selenium fumes which had an unpleasant, garlic-like odor caused an intense irritation of the eyes, nose, and throat. The more severely exposed workers noticed a severe burning sensation of the nostrils, immediate sneezing, coughing, nasal congestion, dizziness, and inflammation of the eyes. Severe headaches were experienced 2–4 hours after the initial exposure, which lasted until the next day. Other symptoms noted only in the most severely exposed workers were slight difficulty in breathing, edema of the uvula, and in one case, severe dyspnea. The most common symptoms were irritation of the mucous membranes and frontal headache. Although some workers noted an obnoxious smell on the skin and metallic taste in the mouth, none reported a garlic odor of the breath. No selenium could be detected in the urine of the most severely exposed workers and the white blood count and hemoglobin were normal. The failure to detect selenium in the urine is difficult to reconcile with observations made in other studies. However, no mention is made of the analytical method which was employed. All workers were entirely well in 3 days and no ill effects persisted.

Clinton believed that in this case of selenium fume exposure the intense irritation caused by the fumes limited the exposure sufficiently to prevent the absorption of sufficient amounts of selenium to produce any systemic toxic effects.

Molten selenium was reported to cause a mild first degree burn on the face of an industrial worker.[48] However, there was no toxic reaction whatsoever and no injury apart from the burn caused by the heat of the molten selenium. Normal healing of all affected parts was realized in 3 weeks.

Mild interstitial pneumonitis was found in rats, guinea pigs, and rabbits exposed for periods of up to 16 h to 30 mg of selenium dust per m³.[32]   In studies on laboratory animals, little evidence of toxicity was noted in the case of rats which had been injected intraperitoneally with grey powdered selenium in doses of 200–1000 mg per kg.[32]   Rats exposed to selenium fumes showed a mild to moderate pneumonitis with hemhorrage and edema.[32]

Selenium Dioxide, Selenites and Selenates

Selenium dioxide, which is a white crystalline solid, subliming at 315°C, and selenious acid and its salts can be absorbed by the body through the skin, causing intense local irritation and inflammation.   A number of cases of exposure to selenium dioxide powder which resulted in the development of dermatitis have been reported.[48]   This dermatitis can spread if untreated. Pringle[48] treated this condition successfully by applying calamine lotion several times a day.   It has been noted that some people, especially those with fair complexions, can become allergic to selenium dioxide with the result that their eyes puff up whenever they enter areas where selenium dioxide is being handled and where others can work without untoward effect.[29]

Areas of the skin affected by selenium dioxide can also be treated by a washing of affected areas with a 10% aqueous solution of sodium thiosulfate and by a cream containing 10% sodium thiosulfate in the case of more severe irritation or burns.[29]

The penetration of selenium dioxide under the nails results in an inflammation of the nail beds which is very painful.   A successful treatment of this condition proposed by Glover[29] is to cut the nail and rub in soapy water for 5 minutes, and then to introduce the sodium thiosulfate ointment under the nail.   In this way selenium dioxide is reduced to elemental selenium and the pain is relieved.

A case of a burn of the eye by selenium dioxide has been reported in which the patient flushed the affected area with tap water immediately after exposure.[41]   In spite of this treatment, the patient experienced intense pain, lacrimation, and congestion of the conjunctiva.   Complete recovery from acute symptoms occurred within 10 days, but grafting was necessary.   According to Glover,[29] conjunctivitis rarely occurs if the selenium dioxide is washed out immediately.   Glover has treated eye burns successfully using sodium thiosulfate solution.

Since selenium dioxide and its solutions can cause burns and skin irritations, rubber gloves, and suitable protective clothing should be worn so that the skin will not come into contact with these materials when they are being handled.

Animal experiments have shown that, in general, sodium selenite is more toxic than sodium selenate. A comparison is given in Table 11-5 of the toxicity of sodium selenite and sodium selenate with that of compounds of arsenic, molybdenum, tellerium, and vanadium. Of these compounds, sodium tellurite and sodium selenite showed the highest toxicity toward rats.

TABLE 11-5   MINIMUM DOSES FATAL TO RATS AMONG
            COMPOUNDS OF TELLURIUM, SELENIUM, ARSENIC,
            VANADIUM, AND MOLYBDENUM[22]

Compound		Minimum Fatal Dose[a]
Sodium tellurite	Na_2TeO_3	2.25–2.50 mg Te/kg
Sodium tellurate	Na_2TeO_4	20.0–30.0 mg Te/kg
Sodium selenite	Na_2SeO_3	3.25–3.50 mg Se/kg
Sodium selenate	Na_2SeO_4	5.25–5.75 mg Se/kg
Sodium arsenite	Na_2HAsO_3	4.25–4.75 mg As/kg
Sodium arsenate	Na_2HAsO_4	14.0–18.0 mg As/kg
Sodium vanadate	$NaVO_3$	4.0–5.0 mg V/kg
Ammonium molybdate	$(NH_4)_6Mo_7O_{24}$	Above 160.0 mg Mo/kg

[a] Minimum doses fatal to at least 75% of young rats within 48 h after substances were injected intraperitoneally.

Hydrogen Selenide

Hydrogen selenide is one of the most toxic and irritating selenium compounds. Gaseous hydrogen selenide is formed by the action of acids and in some cases water on inorganic selenides such as aluminum selenide. It can also be formed by the reaction of selenium with organic matter and by the direct combination of the elements.

Berzelius, the discoverer of selenium, first noted the effects of exposure to hydrogen selenide. Since that time a number of cases of subacute and acute hydrogen selenide intoxication have been reported. In an acute case, in which the gas, in concentrated form, entered the nasal passages, a pronounced metallic taste was reported and, after a brief sensation of intoxication, no ill effects were observed for four hours.[45] Then a copious discharge of mucous from the nasal passages commenced which persisted with violent sneezing for 3 or 4 days. No ill effects were noted later and no garlic breath was observed.

The principal symptom of acute hydrogen selenide intoxication observed by Symanski[63] and corroborated elsewhere is acute irritation of the mucous membranes of the respiratory tract. Following a short period of relief, Symanski observed the occurrence of pulmonary edema, severe bronchitis and bronchial pneumonia.   In all such reported cases, the patients recovered completely.

An interesting case of subacute hydrogen selenide intoxication has been reported by Buchan[5] in which a selenious acid solution was used in an etching and imprinting operation on steel strip.   Following imprinting, the strip was treated in an oil bath for rust prevention and it appears that hydrogen selenide was produced in both the oil bath and the etching solution. The observed symptoms were nausea, vomiting, metallic taste in the mouth, dizziness, extreme lassitude and fatigability.   Garlic odor of the breath was also noted.   The selenium level in the urine was found to be within normal limits and no correlation between the symptoms and urinary selenium level could be observed.   It is noteworthy that after cessation of exposure, the selenium content of the urine decreased significantly and no other complaints were registered.

Although hydrogen selenide is a gas with a very offensive smell, it can cause olfactory fatigue.   Thus, at a concentration of 1 μg of hydrogen selenide per liter of air, the odor disappears quickly.   Consequently, the odor of this gas cannot be used as a warning of high and extremely toxic concentrations.   The symptoms induced by hydrogen selenide are such as to limit the exposure which can be tolerated.   Dudley and Miller[17] point out that at 5 μg per liter, hydrogen selenide is intolerable to man, causing eye and nasal irritation.

The symptoms of hydrogen selenide intoxication are, initially, irritation of the mucous membranes followed by slight tightness of the chest.   Once this situation clears, there may be a latent period of 6–8 hours following which pulmonary edema may occur, depending on the quantity of hydrogen selenide which has been inhaled.   If this condition sets in, the patient must be admitted to hospital without delay and oxygen administered, along with diuretics, antibiotics and intravenous administration of hydrocortisone.[28]

The threshold limit value for hydrogen selenide, as given by the American Conference of Governmental Industrial Hygienists,[2] is 0.05 ppm (by vol) of gaseous hydrogen selenide in air for 8 h (0.2 mg of selenium as hydrogen selenide per cubic meter of air or 0.2 μg per liter).   Since hydrogen selenide can be produced readily and is highly toxic, considerable care should be exercised in all sitations where the gas may be encountered.

Selenium Oxychloride

Selenium oxychloride, which is an almost colorless liquid at toom temperature, is a severe vesicant, capable of producing third degree burns which are extremely painful and slow to heal.[16]   Immediate flushing of the affected area with water and treatment with sodium thiosulfate 10% solution or a weak alkali such as sodium bicarbonate or dilute ammonia is the procedure recommended for avoiding burns when this compound comes into contact with the skin.

Organoselenium Compounds and Selenium-Bearing Natural Products

There is an almost complete lack of reliable data on the toxicity of organo-selenium compounds and selenium-bearing natural products towards humans. The data available from animal experiments indicate that there is a marked variation in the toxicity of such materials.   Data on the toxicity towards rats of organoselenium compounds of biological interest are presented in Table 11-6.   None of these compounds appears to have a toxicity greater than that of inorganic selenium in the form of sodium selenite or sodium selenate or of organic selenium found in seleniferous grains.   In a comparison of the toxicity of selenocystine, sodium selenite and seleniferous wheat fed orally to rats, Moxon, Dubois, and Potter[44] found that *l*-selenocystine was the most toxic of the compounds tested and had a toxicity equivalent to that of seleniferous wheat.   Aromatic selenium compounds seem to be less toxic than aliphatic selenium compounds, with the compound *o,o'*-dicarboxyl-diphenyldiselenide giving rise to similar hematological changes as seleniferous wheat and inorganic selenium when administered to rats.[34]

The toxicity of seleniferous plants towards sheep has been noted by a number of investigators.   Thus, 340 mature sheep died within 24 h after consuming the plant *Astragalus bisulcatus*.[51]   A sheep which had been fed 285 g of the same plant died within 30 minutes.   Acute selenium poisoning in sheep was produced by the forced feeding of the plant *Machaeranthera glabriuscula*.[51]   The same plant containing 400–800 ppm selenium was fatal to mature sheep when fed in amounts ranging from 8–76 g per kg of body weight.

Defatted Brazil nut flour containing 51 ppm of selenium was toxic to rats when this material was incorporated in a diet fed over a period of 28 days. After 21 days all animals showed a highly pigmented urine, yellow diarrhea and anorexia.[9]

A 3 ppm level of selenium in a 15.7% protein diet containing seleniferous sesame cake which was fed to rats had an unfavourable effect on growth after 10 days.[8]   This untoward reaction to selenium was counteracted successfully by supplementing the diet with 3.5% of fish flour.

Carbon diselenide, which has been shown to be an important intermediate in the synthesis of a number of organoselenium compounds, is a yellow liquid having a very offensive odor.   Experiments on rats and rabbits have shown this compound to be highly toxic (LD_{50} approximately, 6 mg per kg body weight for rats).[47]

TABLE 11-6 TOXICITY OF ORGANOSELENIUM COMPOUNDS OF BIOLOGICAL INTEREST

Selenium Compound	Mode of Administration	Experimental Animal	Dose	mg Se/kg Body Weight	Reference
d,l-selenocystine[a] (optically inactive)	Intraperitoneal	Rat	Minimum fatal	4	36
Selenomethionine	Intraperitoneal	Rat	Minimum fatal	4.25	36
n-propylseleninic acid	Intraperitoneal	Rat	Minimum fatal	20–25	43
β-seleninopropionic acid	Intraperitoneal	Rat	Minimum fatal	25–30	43
β,β′-diselenodipropionic acid (Na salt)	Intraperitoneal	Rat	Minimum fatal	25–30	43
β-selenodipropionic acid (Na salt)	Intraperitoneal	Rat	Minimum fatal	>40	43

[a] l-selenocystine is significantly more toxic than d-selenocystine.[44]

CONCLUSION

In considering the toxicology of selenium it is important to realize the limitations of the available data, particularly as regards selenium toxicity in man.   The rather intensive studies carried out some 35 years ago on the selenium poisoning of livestock in seleniferous regions in the western United States directed attention to selenium as a possible toxic substance in human nutrition.   The essentiality of selenium as a nutrient for numerous animal species has been indicated more recently and has pointed up the differences which exist between toxic and nutrient selenium levels in the diet.   The occurrence of selenium in a wide variety of foodstuffs[64] demonstrates that it may be difficult to realize a nonseleniferous diet.   The documentation of selenium intoxication in people living in seleniferous regions is insufficient to delineate clearly any long term effects of the ingestion of seleniferous foods. More research should be carried out to determine the harmful and possibly in some cases beneficial effects of higher than normal ingestion of selenium.

The limitations of the analytical methods for selenium which were used in earlier studies on selenium intoxication should be considered in the treatment of these previous data.   The development of sensitive methods for the determination of traces of selenium in more recent years has permitted the accurate determination of small amounts of selenium such as are found in tissues, blood, and urine.   It is to be hoped that work will be continued on the determination of such small amounts of selenium so that a better knowledge of the levels of selenium to be expected in human organs can be available.

Selenium and its compounds can present an industrial hazard as has been indicated and there is a need for the establishment of maximum allowable concentrations for more selenium materials.   The 1969 threshold limit value established by the American Conference of Governmental Industrial Hygienists relates only to selenium compounds as Se (recommended maximum atmospheric concentration for 8 h: 0.2 mg per m^3.[2]   In view of the marked differences in the toxicity of individual selenium compounds and the relative nontoxicity of elemental selenium, the present threshold limit value is inadequate and should be updated in terms of definitive values for various selenium materials.   A revision of the present threshold limit value is under consideration.[60]   It is noteworthy that Glover[27] has suggested that the maximum allowable concentration of urinary selenium should be 0.1 mg or 100 μg per liter.

Exposure to selenium dust and fumes should be avoided through the use of adequate protective clothing, safety goggles and respirators.   Where such dust and fumes are produced, adequate ventilation should be provided and a program of air sampling implemented to ascertain that the selenium concentration is below the threshold limit value.   Particular care should be taken to avoid contact of selenium and selenium compounds with the skin. Where such contact has been made, prompt flushing of the affected areas with water should be carried out, together with the application of calamine lotion and sodium thiosulfate ointment where burns have occurred.

There is a dearth of long term data on selenium exposure and more studies are needed to determine whether or not in various types of exposure there are any long term systemic effects due to selenium intoxication.

REFERENCES

1. Allaway, W. H., Kubota, J., Losee, F., and Roth, M., Arch. *Environ. Health*, **16**, 342 (1968).
2. Amer. Conf. of Governmental Industr. Hygienists, "Threshold Limit Values of Airborne Contaminants Adopted by ACGIH for 1969 and Intended Changes," United States Department of Health, Education and Welfare, Bureau of Occupational Safety and Health, Cincinnati, Ohio.
3. Amor, A. J., and Pringle, P., *Bull. Hyg.*, **20**, 239 (1945).
4. Beath, O. A., *Sci. Newsletter*, **81**, 254 (1962).
5. Buchan, R. F., *Occup. Med.*, **3**, 439 (1947).
6. Burk, R. F., Jr., Pearson, W. N., Wood, R. P., and Vitteri, F., *Am. J. Clin. Nutr.*, **20**, 723 (1967).
7. Cerwenka, E. A. Jr., and Cooper, W. C., *Arch. Environ. Health*, **3**, 189 (1961).
8. Chavez, J. F., *Arch. Latinoamer. Nutr.*, **17**, 69 (1967).
9. Chavez, J. F., *Bol. Soc. Quim. Peru*, **32**, 195 (1966).
10. Clinton, M. Jr., *J. Ind. Hyg. Toxicol.*, **29**, 225 (1947).
11. Cooper, W. Charles, *Symposium: Selenium in Biomedicine*, Muth, O. H., ed., Westport, Conn., AVI Publishing Co. Inc., 1967, p. 185–199.
12. Cousins, F. B., and Cairney, I. M., Aust. *J. Agric. Res.*, **12**, 927 (1961).
13. Dickson, R. C., and Tomlinson, R. H., *Clin. Chem. Acta*, **16**, 311 (1967).
14. Dubois, K. P., Moxon, A. L., and Olson, O. E., *J. Nutrition*, **19**, 477 (1940).
15. Dudley, H. C., *Am. J. Hyg.*, **23**, 181 (1936).
16. Dudley, H. C., *U.S. Pub. Health Rept.*, **53**, 93 (1938).
17. Dudley, H. C., and Miller, J. W., *J. Ind. Hyg. Toxicol.*, **23**, 470 (1941).
18. Elkins, H. B., *The Chemistry of Industrial Toxicology*, New York, Wiley and Sons, Inc., 1950, p. 353.
19. Ermakov, V. V., *Byull. Eksp. Biol. i Med.*, **59**, 61 (1965).
20. Ewan, R. C., Baumann, C. A., and Pope, A. L., *J. Animal Sci.*, **22**, 1119 (1963).
21. Fimiani, R., *Folia. Med.* (Naples), **34**, 260 (1951).
22. Franke, K. W., and Moxon, A. L., *J. Pharmacol. Exp. Ther.*, **61**, 89 (1937).
23. Frost, D. V., *Proc. Cornell Nutr. Conf.*, Cornell Univ., Ithaca, N.Y., pp. 31–40 (1967).
24. Frost, D. V., and Main, B., Abstract, 7th Intern. Cong. on Nutrition, Hamburg, Germany, 1966.
25. Galen, M. L., *Le śelenium—apercu sur son état naturel, sa tocixité, son rôle dans la nutrition*, Faculté des Sciences de Montpellier, Montpellier, France, 1965.

26. Gardiner, M. R., Aust. *Veterinary J.*, **42**, 442 (1966).
27. Glover, J. R., *Ann. Occup. Hyg.*, **10**, 3 (1967).
28. Glover, J. R., *Industr. Med. Surg.*, **39**, 50 (1970).
29. Glover, J. R., *Trans. Assoc. Ind. Med. Officers*, **4**, 94 (1954).
30. Gortner, R. A. Jr., *J. Nutrition*, **19**, 105 (1940).
31. Hadjimarkos, D. M., and Bonhorst, C. W., *J. Pediatrics*, **52**, 274 (1958).
32. Hall, R. H. *et al.*, *Arch. Ind. Hyg. Occup. Med.*, **4**, 458 (1951).
33. Jacobsson, S. O., *Acta vet. scand.*, **7**, 303 (1966).
34. Kando, S., *Japan J. Med. Sci. IV*, **9**, 29 (1935).
35. Klenka, J., *Chem. Listy*, **60**, 1656 (1966).
36. Klug, H. L., Peterson, D. F., and Moxon, A. L., *Proc. S. Dakota Acad. Sci.*, **28**, 117 (1949).
37. Lardy, H. A., and Moxon, A. L., *Proc. S. Dakota Acad. Sci.*, **22**, 39 (1942).
38. Lemley, R. E., *J. Lancet*, **60**, 528 (1940).
39. Lemley, R. E., and Merryman, M. P., *J. Lancet*, **61**, 435 (1941).
40. Lewis, H. B., Schultz, J., and Gortner, R. A. Jr., *J. Pharmacol.*, **68**, 292 (1940).
41. Middleton, J. M., *A.M.A. Arch. Ophthal.*, **38**, 806 (1947).
42. Miura, H., *Kokumin. Eisei*, **27**, 336 (1958).
43. Moxon, A. L., Anderson, H. D., and Painter, E. P., *J. Pharmacol. Exptl. Therap.*, **63**, 357 (1938).
44. Moxon, A. L., Dubois, K. P., and Potter, R. L., *J. Pharmacol. Exptl. Therap.*, **72**, 184 (1941).
45. Painter, E. P., *Chem. Rev.*, **28**, 179 (1941).
46. Palmer, I. S., and Bonhorst, C. W., *J. Agr. Food Chem.*, **5**, 928 (1957).
47. Peters, A. C. and Allton, W. H., Report, Battelle Memorial Institute, Columbus, Ohio, to Selenium-Tellurium Development Association, December 1963.
48. Pringle, P., *Brit. J. Dermat.*, **54**, 54 (1942).
49. Rhian, M., and Moxon, A. L., *J. Pharmacol.*, **78**, 249 (1943).
50. Rosenfeld, I., and Beath, O. A., *Selenium: Geobotany, Biochemistry, Toxicity and Nutrition*, New York, Academic Press Inc., 1964.
51. Rosenfeld, I., and Beath, O. A., *Selenium: Geobotany, Biochemistry, Toxicity and Nutrition*, New York, Academic Press Inc., 1964, p. 142–143.
52. Sessa, T., *Folia Med.*, (Naples), **35**, 572 (1952).
53. Shamberger, R. J., and Frost, D. V., *Can. Med. Assoc. J.*, **100**, 682 (1969).
54. Smith, M. I., *U.S. Pub. Health Rept.*, **54**, 1441 (1939).
55. Smith, M. I., Franke, K. W., and Westfall, B. B., *U.S. Pub. Health Rept.*, **51**, 1496 (1936).
56. Smith, M. I., and Lillie, R. D., *U.S. Pub. Health Serv., Nat. Inst. Health, Bull.*, **174**, 1 (1940).
57. Smith, M. I., and Westfall, B. B., *U.S. Pub. Health Report.*, **52**, 1375 (1937).
58. Smith, M. I., Westfall, B. B., and Stohlman, E. F., *U.S. Pub. Health Rept.*, **52**, 1171 (1937).
59. Sterner, J. H., and Lidfelt, V., *J. Pharmacol. Exptl. Therap.*, **73**, 205 (1941).
60. Stokinger, H. E., United States Department of Health, Education and Welfare, Laboratory of Toxicology and Pathology, Bureau of Occupational Safety and Health, Cincinnati, Ohio, private communication.
61. Suzuki, Y. *et al.*, *Shikoku Acta Medica*, **14**, 846 (1959).
62. Svirbeley, J. L., *Biochem. J.*, **32**, 467 (1938).
63. Symanski, H., *Deut. Med. Wochschr.*, **75**, 1730 (1950).
64. Takano, Y., *Shikoku Acta Medica*, **14**, 35 (1959).

65. Tank, G., and Storvick, C. A., *J. Dental Res.*, **39**, 473 (1960).
66. Wahlstrom, R. C., Kamstra, L. D., and Olson, O. E., *J. Animal Sci.*, **14**, 105 (1955).
67. Westfall, B. B., and Smith, M. I., *J. Pharmacol.*, **72**, 245 (1941).

The following articles, not referred to in the text, may be of use to the interested reader.

Schroeder, H. A., Frost, D. V., and Balassa, J. J., "Essential Trace Metals in Man: Selenium," *J. Chronic Diseases*, **23**, 227 (1970).
Shapiro, J. R., "Selenium and Carcinogenesis—A Review," *Ann. N.Y. Acad. Sci.*, **192**, 215 (1972).

12

Selenium in Agriculture

ALVIN L. MOXON

Department of Animal Science, Ohio Agricultural Research and Development Center, Wooster, Ohio

OSCAR E. OLSON

Department of Experiment Station Biochemistry, South Dakota Agricultural Experiment Station, Brookings, South Dakota

INTRODUCTION

Agricultural interest in selenium first developed in 1931 as a result of its association with the poisoning of farm animals. It was renewed following the discovery of the beneficial effects of this element in animal nutrition during the late 1950s.

Twenty-five years after the discovery of the element by Berzelius and Gahn in 1817, Japha[115] showed that compounds of selenium were toxic to animals. Cameron[37] appears to have been the first to add selenium to soil and note its effects upon vegetation. He reported a chlorosis in plants grown on soil to which he had added selenium and he postulated that the selenium was being absorbed and was interfering with the sulfur metabolism of the plants. Taboury[252] announced the actual qualitative detection of selenium in plant material and soon Robinson[207] reported its quantitative determination in plant material.

Just before the Civil War, Madison,[140] an Army

675

surgeon stationed at Fort Randall in what was then the Territory of Nebraska, described a disease which occurred among the Army horses grazing in a specific area adjacent to the fort. The disease is now known to have been caused by the high selenium content of vegetation grown in soils derived from the seleniferous geological formations of the area.

The territory around Fort Randall, opened for settlement about 1891, is located in the south central part of South Dakota near the Nebraska border. The homesteaders observed symptoms in their horses and other livestock similar to those described by Madison. They called the malady "alkali disease" because they associated it with alkali seeps in the soil and with water of high salt content. Experiments completed in 1913 at the South Dakota Agricultural Experiment Station[131] showed that samples of the suspected water did not cause the syndrome in cattle, but the name "alkali disease" is still associated with the condition.

In 1929, Franke started investigations of the "alkali disease" problem at the South Dakota Agricultural Experiment Station. This led to cooperation with the Bureau of Chemistry and Soils and other bureaus of the United States Department of Agriculture, and resulted in the discovery by Robinson[207] of selenium in some cereal grains from South Dakota which Franke[69,70] had found to be toxic to laboratory animals. Wyoming workers[11,58] soon related selenium to a "blind staggers" type of poisoning in livestock. More detailed accounts of the early history of work on the selenium problem may be found elsewhere.[7,153,212,261]

The rather recent discovery of selenium as an integral part of "Factor 3"[223] created renewed interest in the role of the element in agriculture. It was soon found that a number of nutritional diseases in farm animals responded to therapy with selenium,[62,169,201,248] and most research now being done on the element is concerned with its nutritive properties rather than with its toxicity. In discussing selenium as it relates to agriculture, it is convenient to give separate consideration to its role as a toxicant and as a nutrient.

SELENIUM POISONING IN LIVESTOCK

Selenium is not restricted to any specific area or areas of the earth. It appears to be present, at least in trace amounts, in all soils and thus, in all natural feeds. However, some soils contain an excess of the element in forms available to plants, and the plants absorb it in amounts that make them toxic to animals. Historically, such soils and plants have been referred to as "seleniferous," and the term implies excessive levels of the element. It is used in that sense in the discussion that follows.

Distribution of Seleniferous Soils and Vegetation Over the World

In studying the distribution of selenium poisoning over the world, two types of information has been used, namely: (a) observed cases of toxicity in animals, and (b) analysis of soils and plants.

In the United States, selenium toxicity in livestock has been reported in Nebraska,[194] in South Dakota,[76] and in Wyoming.[13]   Symptoms of selenium toxicity have also been observed in Counties Limerick, Tipperary and Meath in Ireland,[68,186,272,273,279] in the Huleh Valley and other locations in Israel,[203] in North Queensland, Australia,[122,123] in Colombia, South America,[6,16] and in the state of Guanajuato, Mexico.[33]   In South Africa, dangerously high levels of selenium have been reported in the tissues of animals affected with Geeldikkop and enzootic icterus, and a predisposing role for the element has been suggested.[23,24,171]

Seleniferous plants have been found growing in various locations in most of the states of the Great Plains and Rocky Mountain areas of the United States[261,286] as well as in at least three provinces of Canada.[34,259]   High levels of selenium have also been reported for soils and/or plants in Mexico,[33] the U.S.S.R.,[124] Spain,[286] and Venezuela.[114]   Robinson[208] analyzed wheat samples from the major wheat production areas of the world.   He found some selenium in all of them, but the highest value was 1.9 parts per million (ppm) which is not considered to be a toxic level.

Perhaps the best information concerning the location of seleniferous areas will eventually be based on geology.   In a preliminary field survey of " alkali disease " in South Dakota by Franke et al.,[76] the cases observed all occurred in " gumbo " soils derived from Pierre shales.   Beath and his associates[11] studied the geological distribution of selenium as it related to livestock poisoning, and they associated certain poisonous plants with Cretaceous shales even before they found selenium in the plants.

It is now well-established that, in the Great Plains area of the United States, seleniferous soils have derived almost entirely from rocks of Cretaceous age.   These were deposited in a shallow sea which covered most of the area during the Mesozoic era.   These sedimentary rocks are not the primary source of selenium.   Instead, the element must have been injected into the erosion cycle from some other source and deposited with these rocks. It has been shown that selenium occurs in present day volcanic materials,[32] and these may well have been the primary source.   Furthermore, selenite is easily precipitated from solution by combination with iron hydroxide[182] and selenium from volcanic gases or that dissolved from soils or rocks and carried by river waters to the sea could very possibly be oxidized to selenite, precipitated with iron hydroxide, and appear in sea bottom deposits.   Normally only traces of selenium appear in these deposits.   For example, a composite sample from 12 locations in the Gulf of Mexico contained less than 1.0 ppm.[159]   On the other hand, samples from the Gulf of California contained 3 to 5 ppm.[128]   The Colorado River, which carries appreciable quantities of selenium from drainage and seep water, drains into the Gulf of California and may be contributing most of the selenium to its sediments.

The selenium content of sedimentary rocks varies considerably throughout a geological profile.   This suggests that during their formation the selenium was supplied from a primary source at a rate that varied from the rate at which sediments were deposited.   Almost all seleniferous soils in the United States have weathered from sedimentary rocks of the Cretaceous period.   Only a limited number of the such formations contain enough selenium so that they become incorporated into soils which produce toxic vegetation.   This is illustrated by studies in South Dakota where the various geological formations, especially those of the Cretaceous system, have been analyzed for selenium content.   Stratigraphically, the parent materials from which most of the seleniferous soils in the state were derived occur only in the upper portion (Mobridge and upper Virgin Creek members) and at the base (Sharon Springs member) of the Pierre formation, and in the Smoky Hill (Upper) member of the Niobrara formation which lies just below the Pierre.   However, the selenium content of rocks of a particular age vary from location to location.   For example, while the Mobridge member of the Pierre formation has been the parent material for soils over a large area of South Dakota, it has a high selenium content only in the south central portion of the state and this is the only area within the state where the soils derived from it produce toxic vegetation.   The Sharon Springs member outcrops in limited areas only and the soils it forms are not productive.   Consequently, it does not contribute significantly to the selenium problem.   The Smoky Hill member of the Niobrara formation is the parent material for some of the most seleniferous soils in the state, but here again it outcrops in only limited areas and only a very small part of the soils of the state have weathered from it.   The variability in selenium content between and within formations has contributed in making detailed mapping of seleniferous areas complex.

Glacial draft, which is younger than the Cretaceous formations, has been reported to contain rather high levels of selenium in Canada and North Dakota[286] and in South Dakota.[236]   The selenium here probably originated in materials of Cretaceous age which have been reworked by glaciation.

Forms of Selenium in Soils

The bulk of the selenium in the earth's crust is reported to occur in sulfide minerals.[61]   Slimes of sulfide-ore milling operations have been reported to furnish selenium for the production of areas of toxic alluvial soils in Mexico[33,285] and in Utah.[129] In these ores, the element probably occurs as a selenide. Williams and Byers[284] have reported that in addition to the pyritic form, selenium also occurs in elemental form, in organic form, and as the selenite and selenate in soils. Of the various chemical forms of selenium in soils, selenate appears to be the most readily available to plants,[159] and it is probably the form in which most of the selenium is absorbed by grasses and grains.[183] Selenite has a great affinity for iron hydroxide and in soils containing iron oxide it is bound and is relatively unavailable to crop plants.[284]

Surveys have revealed areas of highly seleniferous soils in Hawaii and Puerto Rico which do not produce toxic vegetation.[130] This is apparently because the selenium in these soils occurs largely as a basic iron selenite, which is not readily available to plants.

Selenium in Plants

The selenium content of plants varies over a wide range, depending upon a number of factors. One of the most important of these is the kind of plant. One type has been referred to as indicator plants. These grow on soils containing high levels of selenium and are therefore helpful in locating seleniferous soils in the field.[13] Rosenfeld and Beath[212] have divided the indicator plants into two categories. The first they call primary indicators. These include over 20 species of *Astragalus*, *Machaeranthera*, *Haplopappus*, and *Stanleya* which appear to require selenium for their growth. They normally accumulate selenium at very high levels, often several thousand parts per million. The other category they refer to as secondary indicators. These do not appear to require selenium for their growth, but will accumulate the element when they grow on soils of high available selenium content. They belong to a number of genera, including *Aster*, *Atriplex*, and *Grindelia*. Finally, there is a third group of plants which includes the grains and grasses and which do not normally accumulate selenium in excess of about 50 ppm under field conditions.

The indicator plants are of additional interest in that they appear to absorb from the soil selenium in a form that is not available to other plants, and redepositing it in a form that is available.[9] In view of this, they have occasionally been referred to as "converter" plants. Their actual role in this type of conversion is obscure, but many are deep rooted and they may remove soluble selenium from the subsoil and deposit it on the surface where it is more readily accessible to other plants.

With few exceptions, the selenium content of plants decreases with maturity.[14,31,159,184]   Much of the selenium in crop plants is associated with the protein[70,106] and its distribution in the plant relates in a general way to the distribution of protein, being higher in the seeds than in the straw.[14,160]

The relationship between the selenium in plants and in the soils in which they grow has not been fully explained.   The selenium contents of surface soils do not correlate well with those of the plants growing in them.[31,32,33] On the other hand, the water-soluble selenium in soil used in greenhouse studies[184] or in field soil profiles to a depth of 3 feet[185] has been found, for alkaline soils, to be a fairly good index of the availability of the element to plants.   The water-soluble selenium probably is present as selenate derived through normal weathering processes from the more reduced and insoluble forms present in the parent material.

Hurd-Karrer[107,108,109] found that sulfates would inhibit the uptake of selenate by plants under greenhouse conditions.   On the other hand, Martin[143] found that sulfates did not appreciably affect the uptake of selenite.   Since selenate contributes in an important way to the selenium content of plants under field conditions, the sulfate level of the soil would appear to be significant in determining the selenium content of plants.   However, this has not been found to be the case in the field, and many soils produce vegetation with toxic levels of selenium in spite of the presence of very high levels of naturally occurring sulfates.

Other factors such as thrift of the plant and weather have been mentioned relative to their effect on the selenium content of plants,[159] but the effects in these cases have not been well-defined.

Shortly after selenium was established as the cause of the alkali disease syndrome, it was found that the element was associated with the protein of grains.[72,106]   Early attempts to isolate selenium compounds from protein hydrolysates of grain were abandoned in favor of working with plants of higher selenium content.   Horn and Jones[105] isolated a crystalline material containing both selenium and sulfur from *Astragalus pectinatus*. They tentatively concluded that it was a mixture of cystathionine and seleno-cystathionine.   Selenocystathionine has been identified in plants by other workers, and Se-methylselenocysteine, Se-methylselenomethionine, seleno-cystathionine, and its glutamyl peptide, selenohomocystine, selenocystine, selenomethionine, selenomethionine selenoxide, selenoglutathione, γ-L-glutamyl-Se-methyl-seleno-L-cysteine, selenite, selenate, selenocysteic acid, selenocysteine seleninic acid, dimethyl selenide, and dimethyl diselenide have also been reported.[64,144,195,30,240,243,261,265,266,267,136,175,176]   In addition to organic compounds of selenium, many plants contain a rather high proportion of soluble inorganic selenium.[12,93]   Quite possibly, the chemical form of selenium in plants has some effect on their toxicity.

Selenium in Waters

While there have been isolated reports of waters relatively high in selenium content, there are few documented field cases of selenium poisoning from waters.   Spring and irrigation drainage waters have on occasion been found to contain over 1 ppm of selenium, but as a rule surface, ocean and well waters contain less than 0.05 ppm.[35,128,283]   Stream and dam waters from eight locations on a seleniferous ranch in South Dakota were all found to contain less than 0.005 ppm of selenium throughout the growing season (unpublished data, S. Dak. Agr. Exp. Sta.).   Byers and his co-workers[35] explain the low selenium content of waters as the result of precipitation with iron hydroxide.   However, microbiological action is very possibly involved as well.[1]   Whatever the reason, it appears that waters are of no real consequence in selenium poisoning.

Symptoms of Selenium Poisoning

Three types of selenium poisoning in livestock have been described by Rosenfeld and Beath:[212] (1) acute, (2) chronic, blind staggers, and (3) chronic, alkali disease.

The acute poisoning results when plants of high selenium content, such as the primary indicator plants, are consumed in sufficient quantity.   The symptoms[13] are severe, including abnormal movement and posture, watery diarrhea, elevated temperature and pulse rate, labored breathing, prostration, and often death from respiratory failure.   The most obvious pathologic changes have been reported[209] as necrosis, hemorrhage, edema and congestion of several tissues.   Deaths from the acute type of poisoning are quite rare, since the highly seleniferous plants are not palatable and their consumption is largely limited to conditions where extreme shortage of other forage prevails.   However, a few cases of large losses in individual herds have been reported.[10]

According to Rosenfeld and Beath,[212] the blind staggers type of chronic selenium poisoning results from the consumption of primary indicator plants in limited amounts over a considerable period of time.   In early stages of the poisoning, cattle wander, stumble, appear to have impaired vision, and lose their appetite for food and water.   Later their front legs become weak, and finally paralysis and respiratory failure precede death.   Occasionally, symptoms are delayed and do not appear until after the animals are in the feedlot being fattened for the market.   In sheep, this type of poisoning has a similar pathology, but is not as readily diagnosed.   That this type of poisoning actually results from selenium has been questioned,[138] since alkaloids could be involved.   The matter needs some further study.

Chronic selenium poisoning of the alkali disease type results from the continuous ingestion of feeds that contain over 5 but usually less than 40 ppm of selenium.[153]   The most common gross symptoms of the poisoning in horses, cattle and swine have been reported as follows: rough hair coat, lack of vitality, lameness due to erosion of the joints of the long bones and to inflammation with swelling along the coronary band, loss of appetite and emaciation.   Liver cirrhosis, atrophy of the heart and anemia are reported to occur in advanced cases.   Horses lose the long hair from the main and tail, resulting in a "bobtailed" appearance.   Cattle lose the long hair from their tails, and hogs occasionally lose their body hair.   In growing chicks, depressed rate of gain, roughed feathers and characteristics of nervousness have been observed.   Sheep have not been reported to exhibit symptoms similar to those listed above for cattle, swine and horses, with the possible exception that they may suffer a loss of appetite and reduced rate of gain. Histologically, chronic types of degenerative changes have been reported[211] but are not specific to this type of poisoning.

In addition to the above, chronic selenium poisoning of the alkali disease type has a marked effect on reproduction.   In chickens, egg production is delayed and hatchability is reduced.[77,198]   Decreased hatchability is largely the result of deformed embryos.   The most common deformities are missing or short upper beaks, missing eyes, deformed feet and legs, edema of the head and neck, and a wiry down which gives those chicks that do hatch a greased appearance.   Similar effects have been reported in turkeys.[41]   Selenium has also been reported to have adverse effects on reproduction in swine.[271] In sheep, field observations have indicated that reproduction of sheep on seleniferous range is markedly reduced (unpublished data, S. Dak. Agr. Exp. Sta.), but Glenn et al.[81] did not find inorganic selenium fed over a long period of time to affect reproduction in these animals.   Congenital malformations, presumably resulting from excessive selenium intake by the dam have been reported for a colt[245] and in lambs.[15]   These latter cases are probably unusual.   Observations on cattle on a seleniferous ranch in South Dakota suggest that reduced reproductive performance may be quite common, may be rather severe without the animals showing other typical symptoms of alkali disease, and may be the most significant effect of excessive selenium intake from the standpoint of economics.[55,152]

Visual observations of cattle grazing seleniferous lands or of animals on seleniferous diets indicate that there is considerable individual variation in susceptibility to selenium poisoning.   Diagnosis is sometimes difficult because of this variation, so hair analysis has been used as an aid in diagnosis, especially in cattle.[180]   In these animals, selenium contents of less than 5 ppm suggest that selenium poisoning is no problem, values between 5 and 10 ppm are borderline, and values greater than 10 ppm indicate that excessive selenium has been ingested and that symptoms of "alkali disease" can be expected in at least some of the animals.

Muth and Binns[168] have reviewed the toxicity of inorganic selenium compounds toward various species of animals. Ruminant animals tolerate more selenium as inorganic salts than monogastric animals because of the reactions which may take place while the selenium is in the rumen. Selenium salts may be reduced to elemental selenium in the rumen and as much as 40% of a single dose may be excreted in the feces. This reduction of selenium in the rumen may be influenced by diet constituents and by the nature of the rumen flora.[29]

Maag and his associates[138,139] administered 0.25–0.50 mg per pound of body weight of selenium as sodium selenite to cattle 3 times weekly in gelatin capsules over an extended period of time. They observed few symptoms of alkali disease or blind staggers. They suggest that alkaloids and other toxic substances in the seleniferous plants may cause the syndromes commonly attributed to selenium poisoning. Others[212,138] have suggested that only organic forms of selenium found in plants produce either alkali disease or blind staggers. However, studies with rats,[75] chickens,[71,153] swine,[151,269,270] dogs,[205] and cattle[181] have shown that the most common symptoms of alkali disease can be produced in these animals on prolonged feeding of inorganic salts of selenium. Typical alkali disease symptoms have not been described for sheep and continued feeding of inorganic selenium to these animals at acutely subtoxic levels causes deaths without producing symptoms similar to those found in cattle suffering from alkali disease or blind staggers.[81,82]

Factors Affecting Selenium Toxicity

The toxicity of selenium in animals is influenced by several factors. Among these are the animal species, the composition of the ration, the chemical form of the selenium, the concentration of the element and the duration of intake. In early studies on selenium poisoning, attempts were made to find dietary factors which would reduce or prevent the toxicity.[153] Feeding trials with rats showed that changes in calcium and phosphorus levels and ratios, increases in vitamins A and D, the addition of yeast as a source of B-complex vitamins gave no beneficial responses. The supplemental feeding of orange juice, dry yeast and cod liver oil with a toxic diet gave some increase in growth of rats, probably because of calories present in supplemental materials, but failed to prevent the development of the typical pathological changes. The addition of cystine at levels of 0.6 per cent of the diet gave no protection against the toxicity of selenium,[217] and sulfur at 0.5% of the diet was slightly detrimental to rats and chicks.[153]

Early work[153] indicated that high protein levels in the diet of rats afforded some protection against selenium toxicity, and a number of protein supplements have been studied with respect to this.[155,205,209,210,211,246]   Linseed oil meal has excelled other supplements,[155,163] but Halverson et al.[91] suggest that protein is probably not involved since the protective principle is extractable with hot 50% ethanol and is not precipitated with lead.   In general, a high protein level in the diet appears to offer only limited protection against selenium poisoning.

The effect of adding methyl donors to seleniferous diets has also been studied, since Hofmeister[101] suggested that selenium might be detoxified and eliminated in part as dimethyl selenide.   While in some experiments a degree of protection has been afforded by methionine,[137,179] choline,[179,213] and betaine,[179] this has not been the case in all experiments,[121,237,244,246] and at best the effect of these methyl donors on selenium toxicity has been limited.   Sellers et al.[237] and Levander and Morris[135] have determined the level of Vitamin E required in order for methionine to give protection against liver damage due to selenium in the rat.   The levels of methionine and Vitamin E used are higher than those commonly present in practical livestock rations.

Arsenic has proven to be the most effective factor for the counteraction of selenium toxicity.[154,156,164,166,205]   After arsenic was found to protect against selenium toxicity, a number of other elements were investigated for possible protective action.   Arsenic (sodium arsenite) at a level of 5.0 ppm in the drinking water protected rats against the toxicity of 10.0 ppm of selenium in the ration.[154]   Other elements (chromium, fluorine, vanadium, molybdenum, cadmium, zinc, cobalt, nickel and uranium) at levels of 5.0 ppm in the drinking water failed to give any apparent protection in rats against 11.0 ppm of selenium in the ration and, in fact, cause some increase in mortality.   Tungsten at 5.0 ppm caused some decrease in gross liver pathology and in mortality, and a level of 2.5 ppm reduced mortality slightly, but failed to prevent liver lesions.[157]   Gallium and germanium levels of 10 ppm in the drinking water were ineffective against selenium at 15 ppm in the diet of rats.   Antimony, in the form of antimony trichloride, when added at the rate of 12 ppm to a diet which contained 12 ppm of selenium, gave some protection against the toxicity of selenium, but not as much as arsenic on an equal weight basis.[158]   Thallium and mercury have been found ineffective against selenium toxicity.[134]

Organic as well as inorganic compounds of arsenic have been shown to have some protective effect against selenium.[59,95,269]   Further, arsenic will protect against organic as well as inorganic forms of selenium,[59] and it was found effective in counteracting selenium toxicity regardless of the route of administration (oral or subcutaneous) for either element.[161]   The basis for the protective effect of arsenic against selenium has not been established.

DuBois *et al.*[60] found that intraperitoneal injections of SH-glutathione at a ratio of 10 moles to 1 mole of selenium protected rats against lethal doses of sodium selenite.   The glutathione was administered 2 h before the selenium.   Bersin[17] has stated that selenite reacts with glutathione to form an intermediate compound GS—Se—SG which then decomposes to form oxidized glutathione and elemental selenium which is very much less toxic than selenite.[74]   The reaction of selenite with thiols to form the selenotrisulfide has recently been confirmed by Ganther.[78]

Sulfate has been found to reduce the toxicity of selenate to rats but not of other forms of selenium.[79,89,92]   An increase in urinary selenium has been reported following the administration of bromobenzene,[133,165] but others have failed to confirm this.[281]   It has been suggested[153] that plants absorb more selenium in wet than in dry years, but there is no experimental evidence to verify that weather or climate has any effect on selenium poisoning in animals.   The rate of grazing has also been proposed as having an effect on the severity of selenium poisoning, but again this has not been confirmed and what little work has been done gave inconclusive results.[55]   Hair color or breed of animal has sometimes been suggested as correlating with the susceptibility of an animal to selenium poisoning, but experimental evidence for this is lacking.

Control of Selenium Poisoning in Livestock

The control of selenium poisoning in livestock should obviously be based on prevention.   With the acute and the blind staggers types of poisoning, the avoidance of grazing areas where selenium indicator plants grow in abundance should resolve the problem.   Elimination of these plants through spraying with herbicides or with other weed control programs may be useful in limited situations, but areas involved and the costs of the programs will normally make this practice uneconomical.

In the case of the alkali disease type of poisoning, other control measures are required.   In view of reports by Hurd-Karrer[108,109] that in greenhouse studies the addition of elemental sulfur or of sulfates to soils containing added selenate reduced the selenium uptake by plants, Franke and Painter[73] tried treating soils under field conditions.   They found that adding elemental sulfur or calcium sulfate to the soil in a field producing toxic grain did not reduce the selenium content of the crops grown there and concluded that this type of soil treatment would not prove a practical control measure.

In poultry, organic arsenicals have been found to give some protection against selenium.[40]   However, a simple and practical method of preventing selenium damage to growing or laying poultry is based on the dilution of seleniferous grains with feeds of low selenium content.   Another effective method is to have the individual farmer sell his seleniferous grains on the market, where its dilution during normal commercial grain handling processes renders it safe.   He, in turn, can purchase nonseleniferous grains for feeding.   The problem of the effect of selenium on the hatchability of eggs has been resolved by the advent of the commercial hatchery which can avoid eggs and feeds from seleniferous areas.

Organic arsenicals have also been found effective in at least partially preventing selenium poisoning in swine,[270] but here again the simplest method of prevention involves dilution of seleniferous feeds with those from nonseleniferous areas or the sale and purchase of feed grains as described for poultry.

The most difficult problem of control involves range animals.   Moxon et al.[164] reported that 25 ppm of arsenic as sodium arsenite in the salt of range cattle was effective in giving some protection against selenium poisoning.   This practice was used for a number of years, but following observations by those who used it and additional research findings[56] it was concluded that the amount of arsenic the animals received was too small to be effective, so the practice was discontinued.   Using bromobenzene administration to speed the recovery of cattle showing symptoms of the alkali disease type of selenium poisoning or feeding linseed oil meal to cattle on seleniferous range as a preventive were also found ineffective and present control measures for range animals rely on management rather than on the use of preventives.

The best management of seleniferous range requires locating those soils which are producing toxic vegetation.   It also requires an understanding of the variation in the selenium content of vegetation during the growing season. Seleniferous soils can be located in a general way with a knowledge of the geology of an area.   However, mapping in enough detail for proper management requires either the analysis of soils to a depth of 3 feet for their water-soluble selenium content or the analysis of plant samples.[183]   The latter has been found the most practical, but even it is time-consuming and costly, especially where ranches are large.   Very little seleniferous land has been mapped in sufficient detail for the best management.

The selenium content of range grasses is highest in their early stages of growth,[183] and this has been suggested as the basis for managing seleniferous pastures used for breeding stock.   Dinkel et al.[55] found that when cows on seleniferous ranges were bred prior to or early in the growing season when the cumulative selenium intake was relatively low instead of later when the total selenium intake for the season was high, reproductive performance was much improved.   In addition, since grasses are lowest in their selenium content in the fall and winter, using the most seleniferous land for winter pasture and the least seleniferous for summer pasture can somewhat reduce the damage from selenium.

SELENIUM AS A NUTRIENT

In the 1930s, Trelease and Trelease[264] found that selenium was stimulating and probably essential for the growth of certain "indicator" plants such as *Astragalus racemosus*. Shortly thereafter, Poley *et al.*[199] found that while 5 ppm or more of selenium in the diet depressed the growth rate of chicks, 2 ppm slightly stimulated growth. However, it was when Schwarz and Foltz[223] announced that selenium was an integral part of "Factor 3" and that selenium salts would themselves protect rats against the dietary necrotic liver degeneration prevented by "Factor 3," that interest in the element as a nutrient was really stimulated.

Schwarz[218] had worked with yeast as a source of dietary protein in Germany, and had found that rats fed diets containing yeast as the only source of protein developed a fatal liver necrosis which could be prevented by either vitamin E or supplementary cystine. In the United States, he found that these results could not be duplicated with brewer's yeast but could be with *Torula* yeast. When some brewer's yeast was added to the *Torula* yeast diets, no liver necrosis developed. In 1950, he designated the factor in brewer's yeast that prevented the liver necrosis as "Factor 3" because it was the third natural agent to affect dietary liver necrosis.[220] He later found that other natural products such as casein contained this factor.[219] Then, he and Foltz found that highly potent concentrates of "Factor 3" contained traces of selenium, that there was correlation between selenium content and biopotency, and that the addition of sodium selenite in minute concentrations gave protection against liver necrosis.[223] Schwarz *et al.*[226] further reported that the removal of selenium from commercial cystine, which had long been known to have a protective effect against dietary necrotic liver degeneration,[280] eliminated its "Factor 3" activity.

Selenium-Responsive Diseases

In a relatively few years beyond the discovery of the "Factor 3" activity of selenium, a number of syndromes, especially those previously found to be in some way related to vitamin E deficiency, were found to be prevented or cured by selenium administration. The term, selenium-responsive rather than selenium-deficiency diseases, is commonly used in referring to these, although selenium is now generally considered to be an essential element.

Data are not available which will allow for a comparison of the economic importance of selenium toxicity and selenium responsive diseases to the world's agriculture. At this point, the latter appears to have caused considerably more economic loss because it involves more land area. On the other hand, resolution of the deficiency problem appears simpler than resolution of the toxicity problem.

Recently an economic evaluation[a] has been made of the annual losses to the U.S. poultry industry resulting from "selenium deficiency." This report indicates that over 50% of the poultry (chickens and turkeys) are produced in "selenium deficient" states. It is estimated that poultry producers suffer annual losses of more than $27,000,000 due to selenium deficiency. It is indicated that losses to the swine industry would be equal to the losses in the poultry industry, so the annual losses due to selenium deficiency in swine and poultry in the U.S. are about $55,000,000.

Poultry. The same year that selenium was reported to be associated with "Factor 3," a number of studies on the effect of the element in vitamin E-related nutritional disease of chickens, such as those previously described by Dam[48] and by Scott et al.,[229] were reported. Dam and Søndergaard[52] found that 5 ppm of selenium as selenite in a low casein, vitamin E deficient diet almost completely prevented muscle striation (dystrophy) in chicks. The Lederle group[189,248] found selenite at a level of 0.3 ppm of selenium in a *Torula* yeast diet prevented exudative diathesis. Scott et al.[228] reported similar findings for 0.1 ppm of selenium, again as selenite, and Schwarz et al.[222] found that selenite, selenocystathionine, and elemental selenium stimulated growth and prevented exudative diathesis in chicks on a vitamin E-free *Torula* yeast diet.

Exudative diathesis is characterized by an edema of body tissues, with a marked accumulation of fluid under the skin of the breast and abdomen.[49] Hemorrhages in subcutaneous tissues adjacent to fluid accumulations give a greenish color to the skin. The fluid is often bluish in color and its protein composition is similar to that of plasma.[51] Capillaries in chicks with exudative diathesis are more permeable to trypan blue than those in normal chicks,[50] and their permeability may account for the edema. However, the low blood serum protein, especially the albumin,[20,51,83,204] probably contributes to the edema also. Anemia has also been described as occurring with this syndrome.[202,229]

Nutritional muscular dystrophy in chicks is characterized by white striations in the breast and, less frequently, the leg muscles.[174,231] Histologically, the affected muscles show marked degenerative changes. Walter and Jensen[275] found serum glutamic-oxaloacetic transaminase levels to be re-

[a] Draft Environmental Impact Statement, Rule Making on Selenium in Animal Feeds, Bureau of Veterinary Medicine, Food and Drug Administration, Department of Health, Education and Welfare. C. D. Van Houweling, D.V.M. Director, April 19, 1973.

lated to the incidence of muscular dystrophy in chicks and poults on selenium and vitamin E-deficient diets. Vitamin E alone and the sulfur amino acids or selenium with levels of vitamin E too low to in themselves prevent the dystrophy have all been reported to be effective preventatives.[227]

Exudative diathesis is prevented in chicks by adding as little as 0.1 ppm of selenium as selenite to a vitamin E-deficient diet.[173,189,222] Vitamin E, itself, has been reported to prevent the syndrome,[229] and until recently it appeared that vitamin E would substitute for selenium and so the element could not be classed as essential. Thompson and Scott[256] have, however, shown that on diets prepared with crystalline amino acids and found to contain less than 0.005 ppm of selenium, D-α-tocopherol acetate at levels up to 200 ppm in the diet did not prevent poor growth and mortality in chicks, and at 1000 ppm was inferior to low levels of selenite in terms of growth response. They concluded that a need for selenium in chicks has been demonstrated. Earlier results with chicks, in which 50 IU of vitamin E per pound of diet failed to prevent exudative diathesis completely while 0.1 ppm of selenium did;[174] with poults in which 20 IU of added vitamin E per kilogram of diet failed to completely prevent gizzard muscular dystrophy prevented by 1 ppm of selenite;[275] with chicks severely depleted in vitamin E which were protected against myopathy by selenium;[67] and with quail in which birds given substantial quantities of vitamin E still required selenium[254] support the conclusion that selenium is essential.

Later reports by Scott and Thompson[232,233,234] and Thompson and Scott[257] on the results of studies using highly purified diets containing less than 0.002 ppm of selenium are especially convincing with respect to the essentiality of the element. Although the diet contained 100 ppm of D-α-tocopherol acetate, exudative diathesis and 100% mortality occurred by the time the chicks were 35 days of age, and 0.01 ppm of selenium in the diet produced maximum growth and livability and prevented all symptoms. Chicks on the basal ration showed poor fat and vitamin E absorption and unhydrolyzed fat appeared in the feces. Pancreatic fibrosis was observed and the abnormalities in fat and, consequently, vitamin E absorption were attributed to this lesion. They summarized the most important abnormalities of an uncomplicated selenium deficiency as poor growth and feathering followed by mortality. Exudative diathesis, they suggest, is the result of the secondary vitamin E deficiency resulting from the selenium deficiency. Gries and Scott[88] have studied the pathology of pancreatic degeneration and fibrosis in day-old chicks on a selenium-deficient (.004 ppm) amino acid diet. Deficiency lesions appeared in 6 days even though plasma tocopherol levels were maintained at a high level. Regeneration was detected 4 days after the addition of 0.1 ppm Se to the diet and in 2 weeks recovery from pancreas lesions appeared to be completed.

Vohra *et al.*,[268] in studies on the distribution of ^{75}Se from injected seleno-methionine-^{75}Se in chickens, found the highest specific activity per gram of fresh tissue in the pancreas.

Walter and Jensen[274] found that cystine or methionine would not prevent dystrophy in poults in a *Torula* yeast diet deficient in vitamin E. Selenium at 0.1 ppm was not effective, but at 1.0 ppm it gave complete protection. Vitamin E at 20 IU per kilogram of diet was also effective. Anemia and reduced albumin/globulin ratios were observed on the unsupplemented diet and these were corrected by 0.1 ppm of selenium or with vitamin E. Scott *et al.*[231] produced myopathies in poults on a practical type diet low in selenium and deficient in vitamin E. The addition of vitamin E and methionine to the diet improved growth, but did not prevent gizzard myopathy or give maximum growth unless at least 0.1 ppm of selenium as sodium selenite was also added. In diets low in vitamin E approximately 0.28 ppm Se was required. These authors concluded that selenium appears to be the primary factor concerned in the prevention of the myopathies associated with vitamin E and selenium deficiencies in turkey poults.

Supplee[250] has reported a feather abnormality in poults on a vitamin E and selenium deficient *Torula* yeast diet. The quills of the flight feathers on the affected birds became discolored and atrophied. Either vitamin E or selenium prevented this syndrome.

Feeding hens a *Torula* yeast diet low in its content of selenium and deficient in vitamin E has caused low hatchability of the eggs produced, and early chick mortality.[39] In turkeys, selenium has failed to improve egg hatchability when a *Torula* yeast diet was used and again tocopherol was effective.[45] In Japanese quail, however, different results were obtained. When these birds were reared from one day of age on a diet of very low selenium and vitamin E content, selenium was equal to vitamin E supplementation in maintaining hatchability of the eggs.[117] Young quail which hatch from eggs laid by hens on the selenium-deficient diets had a high incidence of paralysis of the legs and degeneration of the muscle in the gizzard and legs was observed. Recently Cantor and Scott[39] have shown that selenium additions to low Se, low vitamin E diets will improve egg production, hatchability and growth of chicks.

Swine. Until recently, the role of selenium as a nutrient for swine has been given somewhat limited attention. Eggert *et al.*[62] fed pigs on a *Torula* yeast diet deficient in vitamin E and found that while the pigs grew as well on this diet as did those on similar diets supplemented with either vitamin E or selenium, over half of them died suddenly during a 53 day period without previous outward manifestation of ill health. These pigs showed marked necrosis of the liver, and in some cases yellowish-brown discoloration of the fat and hemorrhages of the lymph nodes and gastrointestinal tract.   Either vitamin E or selenium in the diet at a level of 1.0 ppm as the selenite prevented death and other symptoms.

In addition to fat discoloration (yellow fat disease), nutritional muscular dystrophy often occurs with the dietary hepatic necrosis observed in pigs.[177] Selenium appears to protect pigs against the necrosis but is not entirely effective against the dystrophy.[251]   In addition to protecting against the necrosis in pigs on a low selenium, vitamin E-deficient diet, selenium or vitamin E was found to increase the serum amino acids, probably through increasing feed consumption.[276,277]   While there is experimental evidence that vitamin E is effective against the muscular dystrophy, better results have been obtained when both selenium and vitamin E were administered.[187] Yellow fat disease responds to vitamin E therapy, but not to selenium.[86] Mulberry heart (dietetic microangiopathy), which sometimes accompanies dietary hepatic necrosis, also responds to selenium.[46,47,85]   In Australia, a condition called "white pig disease" has been reported to respond to selenium and vitamin E therapy combined, but not to either alone, and sows on a diet with 10 mg of vitamin E added per pound but of low selenium content farrowed prematurely and their pigs were small and weak.[46]

Recently, Michel et al.[150] used pigs from herds with naturally occurring dietary hepatic necrosis to study the nature of the disease.   In both naturally occurring and in experimentally produced cases, lesions of acute dietary hepatic necrosis appeared as red scattered spots on the surface and in the parenchyma of the liver.   On a 6% protein diet with *Torula* yeast as the protein source and containing vitamin E-free lard as a fat source, dietary hepatic necrosis occurred consistently.   Supplementation of the diet with methionine or vitamin E administration at 21.8 IU 3 times weekly did not prevent the snydrome.   Increasing the protein level of the diet to 20% with the *Torula* yeast, the administration of 0.2 mg of selenium as selenite 3 times weekly, or the administration of 150 IU of vitamin E 2 times weekly prevented the dietary hepatic necrosis.   Microscopic lesions of nutritional myopathy were observed.   These were not prevented by selenium administration, but were prevented by vitamin E.   It is interesting that the necrosis occurred spontaneously or was produced experimentally without the inclusion of unsaturated fat in the diet.   Previous reports had suggested that the presence of polyunsaturated fats was essential to the production of the disease.

Ewan et al.[66] have reported that young pigs fed diets low in selenium and vitamin E showed hepatic necrosis, icterus, edema, anemia, pale yellow areas in skeletal and cardiac musculature and prominent histologic lesions in the liver and skeletal muscle tissues.   54% of the pigs died before 14 weeks of age.   Supplementing with selenium, with vitamin E or with both had no significant effect on growth rate, but did reduce mortality to 7% and prevented the other lesions.

Increasing numbers of field cases of vitamin E-selenium deficiency in swine are being reported in the eastern part of the U.S. corn belt[260,142] and in eastern Canada.[116]   Changes in swine production which may be involved in this increase include: confinement of swine with elimination of pasture, diets of corn-soybean meal with rather pure vitamin and mineral additives, faster growing meatier animals, depletion of selenium reserves in the soil, loss of vitamin E during handling and storage of feeds, and better methods of diagnosis.   The present methods for selenium analysis have made possible the accurate determination of selenium in feeds and in animal fluids and tissues to confirm the deficiencies in field cases as they are encountered. Often, the selenium content of practical corn-soybean meal swine diets in use on farms is very little higher than in some *Torula* yeast experimental diets which have been designed to study selenium deficiency in experimental animals.

Patrias and Olson[188] determined the selenium content of corn samples from 11 midwest states with values going as low as 0.01 ppm.   Nineteen samples of corn grown in Ohio[141] averaged only 0.019 ppm selenium.   Corn makes up 75% to 80% of many swine diets with soybean meal as the other main ingredient.   With soybean meals containing 0.070 ppm of less selenium, it is obvious that the final diet will be well under the National Research Council[170] estimated requirement of 0.100 ppm selenium for swine.   Ku et al.[125] determined the selenium content of swine diets and longissimus muscle from pigs fed under controlled conditions in a cooperative study conducted in 13 different states.   The highest concentration of selenium in the diets (0.493 ppm) produced the highest value in the muscle 0.521 (ppm/wet basis) and the lowest diet concentration (0.027 ppm) produced the lowest value in the muscle (0.034 ppm/wet basis).

In the absence of approval for the addition of selenium in livestock diets injectable preparations of selenium and vitamin E have been resorted to for the effective treatment of swine in selenium deficient areas.[260,142]

Ruminants.   Muth et al.[169] found that adding 0.1 ppm of selenium as sodium selenite to dystrophogenic diets fed prepartum to ewes protected the lambs from white muscle disease.   At about the same time, Proctor et al.[201] obtained similar results using 1.0 ppm of selenite in the ration of the ewes, and also by feeding 0.25 pounds of linseed oil meal containing 1.18 ppm of selenium per head per day.   Sharman et al.[238] found that the administration of 0.25 mg of selenium as selenite per day to suckling calves in a region of Scotland where muscular dystrophy was prevalent prevented the disease while the administration of 200 mg of α-tocopherol per week gave incomplete protection.

In New Zealand, where a considerable area has been found to be selenium-deficient, a number of selenium-responsive diseases have been reported. The situation there has been reviewed by Hartley and Grant[94] and more recently by Andrews et al.[8] These authors describe two forms of white muscle disease. One of these is designated as "congenital." Lambs affected with it are either born dead or die suddenly after exertion a few days following birth. Myocardial, liver, and body cavity lesions are observed, but skeletal musculature is rarely affected. The other form is designated as "delayed," and it occurs predominantly in lambs from 3 to 6 weeks of age. Affected animals have a stiff gait and arched back, avoid movement, lose condition and die. Some with severe heart involvement may die suddenly. Skeletal musculature shows symptoms of dystrophy, but cardiac lesions are not always present.

In addition to white muscle disease, other selenium-responsive diseases are described by Andrews et al.[8] An unthriftiness in both sheep and cattle characterized by loss of condition, diarrhea, and often death has been found to respond to selenium. Periodontal disease of ewes, characterized by loosening and shedding of permanent premolars and molars, is associated with the other diseases and responds to selenium therapy. A type of myopathy in hoggets (9–12 month old sheep) which usually develops when the animals have wintered on turnips and swedes, and for which clinical signs including listlessness, stiffness, and often death, appears suddenly during cold and wet weather and seems to respond to selenium therapy although experimental work on this is lacking. These authors also report that in ewes, but not in cattle, a failure to reproduce is often found in areas where white muscle disease occurs, that this responds to selenium treatment, and that reproductive failure is apparently the result of embryonic death. With respect to reproduction, Buchanan-Smith et al.[26] found that on a low-selenium, low-vitamin E diet, selenium administration alone did not give satisfactory reproduction but when accompanied by vitamin E administration it did. The fertility of rams was not affected.

Buchanan-Smith and his co-workers[26,27] have also reported that on low-selenium, low-vitamin E diets growth was poor and sheep died between about 80 and 230 days. Death was sudden and selenium delayed but did not prevent it. Selenium improved growth but only those lambs born of ewes administered both selenium and vitamin E did not exhibit myocardial necrosis. Several plasma enzymes were increased on the basal diet, but with selenium administration the increases were transitory.

There is an additional body of evidence that selenium in the diet improves weight gains[8,65,103,148,149,178,191,206,214] but not under all conditions.[21,239] There is also some experimental evidence that selenium in the diet improves the fleece of sheep.[242] In California, a nonspecific diarrhea of calves in a white muscle disease area responded to selenium therapy.[118]

Increased use of corn silage as a major ingredient replacing hay in beef and dairy cattle and sheep diets in the eastern corn belt may bring about more cases of selenium deficiencies because of the low selenium content of these silages. Many samples analyzed in Ohio are in the range of 0.020 ppm selenium or below, whereas, hay samples from the same areas contain about 0.100 ppm selenium.

Other Animals. In addition to poultry, swine, sheep and cattle, selenium-responsive diseases have been reported in foals and race horses,[8,94] mink,[249] and goats.[288]

The laboratory rat has continued to be very useful for studies on selenium toxicity and selenium metabolism.[111,241,282,215,258]

Distribution of Selenium-Responsive Diseases

In reviewing the problem, Wolf *et al.*[288] list the following countries in which selenium-responsive diseases have been reported: Australia, Canada, New Zealand, South Africa, Turkey, England, Scotland, Scandinavia, Germany, France, Switzerland, Italy, South America, Japan, and the United States. Selenium-responsive diseases have also been observed in Finland.[177]

In New Zealand, where the incidence is among the highest,[288] white muscle disease occurs largely in a restricted area at the center of the North Island and sporadically over much of the South Island, particularly in the southeast area.[8]

In the United States, white muscle disease has been rather widely reported.[288] All but 14 states were reported to have experienced it in lambs or calves. Workers at the U.S. Plant, Soil and Nutrition Laboratory, Ithaca, New York, found evidence that regional patterns of occurrence of the diseases are related to regional differences in the selenium content of feed crops.[4] With their data and that previously published by other workers, they mapped the United States showing in a general way the selenium content of crop plants.[42,127] Their maps indicate that there are extensive areas where crops are generally low in their selenium content in the Pacific Northwest, the Southeastern Seaboard States, an area covering large parts of New England, New York, Pennsylvania, Maryland, West Virginia, Ohio, Michigan, Indiana, Illinois, Wisconsin, and along the Allegheny, Blue Ridge and Smoky Mountains, and the Appalachian Plateau. In addition, an area in southwestern Montana and one in the White Mountains and their foothills in Mexico and Arizona produced crops of low selenium content.

Selenium-responsive diseases in Canada are reported to be prevalent in all Provinces except Manitoba, Saskatchewan and Alberta where they are encountered to a lesser extent.[116,287]

Hoffman *et al.*[100] have determined the selenium content of muscle and kidney tissue from calves and lambs raised in various parts of Canada to assess the prevalence of the element in a selenium-deficient area. They have concluded that selenium levels below 5 ppm in kidney cortex dry matter and 0.5 ppm in muscle dry matter might be used as an indication of selenium deficiency in animals. Using these criteria, they have determined the percentage of selenium-deficient animals in areas of Canada. They report from 100% in areas of Quebec and Ontario to 0% in areas of Saskatchewan and Alberta.

Control Measures

There have been a number of approaches to the control of selenium-responsive diseases suggested. These include (1) soil treatment to raise the selenium content of crops, (2) the administration of selenium to the animal orally, by injection or by diet supplementation and (3) the selection of crops that will efficiently absorb selenium from the soil.

The application of selenium to soils as a means of correcting deficiencies in the plants has been studied by a number of investigators.[5,53,54,63,84,196,278] It is possible to apply selenium to the soil in a number of forms[278] and increase the selenium content of the plants to a level where selenium-responsive diseases would be prevented. This method has, however, several disadvantages.[3] In terms of selenium, it is inefficient. The amount and form of selenium to be added to the soil and the method of addition will vary with a number of factors, especially with the soil type. There is some hazard which can result from the accumulation of excessive levels of selenium in the forage as the result of soil treatment, but this seems minimal. The selenium content of forages or crops rises sharply after selenium addition to the soil and then falls off rather rapidly. All of these matters complicate the control of the level of selenium in the feed or forage being grown. In brief, while this system of control has possibilities, it is not as simple and effective as other systems.

The administration of selenium to the animal in the feed or water, by injection or by oral administration appears to be the most practical method for controlling selenium-responsive diseases, when regulatory agencies will allow such practices. In New Zealand, for example, recommendations for control are based on these methods of selenium administration[8] as described below. For sheep, oral administration of 1–5 mg of selenium as sodium selenite is recommended, depending upon the age of the animal and the nature of its condition. Ewes are dosed at this level about 1 month before mating to prevent infertility and 1 month before lambing to prevent congenital white muscle disease in the lambs. For selenium-responsive unthriftiness in lambs, dosing at marking and at 2 or 3 month intervals thereafter is recommended. For cattle, it is suggested that it is most convenient to use subcutaneous injections of 10 mg of selenite for calves and up to 30 mg for cows at about 3 month intervals. In swine, a manufactured supplement is recommended which, when used according to directions, provides 0.15 ppm of selenium in the diet.

In Finland, following field tests which demonstrated the effectiveness of selenium in preventing or curing muscular dystrophy in calves,[112] a commercial mineral mixture containing 10 ppm of selenium as the selenite and 500 ppm of vitamin E was marketed. A study of the effects of this supplement showed that it greatly reduced muscular dystrophy in calves. It was, therefore, recommended that in areas where this syndrome was common the dry fodder should be supplemented with 0.3 ppm of selenium.[113]

In the United States, an injectable selenium preparation containing vitamin E has been licensed for use in the treatment or prevention of selenium-responsive diseases.[167,288] The use of injectable selenium to prevent the development of a deficiency has been resorted to in the absence of approval for additions of selenium to feeds. The influence of injectable selenium upon blood levels of selenium in cattle and in sheep has been reported by Preston and Moxon[200] and Moxon et al.[162] In steers the peak value of selenium occurred about 12 days after injection and declined gradually. Injections would be required at about 2 month intervals to keep the values significantly above the basal value observed on the corn, corn silage, protein supplement ration.

The American Feed Manufacturers Association has filed a petition with the U.S. Food and Drug Administration for use of selenium as an additive in feeds for chickens, turkeys, and swine. A notice of a proposed Food Additive Regulation for selenium in animal feed was published in the *Federal Register*, Vol. 38, No. 81—Friday, April 27, 1973, pages 10458–10460.

There is a considerable amount of evidence that about 0.1 ppm of selenium in the diet will provide protection against selenium-responsive diseases, but Thompson and Scott[255] suggest that for poultry diets where the vitamin E level very often falls below 10 IU per pound, at least 0.15 ppm and possibly 0.20 ppm of selenium should be present in the diet. Scott et al.[231] have stated that 0.28 ppm was required for turkey poults on diets low in vitamin E. Not being able to add selenium compounds, an adequate level of the element in diets may be provided by purchasing feed ingredients or supplements naturally sufficiently high in selenium. This, of course, requires monitoring since the selenium content of diet ingredients is usually widely variable.[2] Fish meals may be considered as a fairly potent source of selenium. Analyses of a number of samples have been made and the values have fallen between about 1 and 4 ppm of selenium. In using this method of supplementing diets with selenium, the variability in the activity of different forms of selenium[224,225] and in the "availability" of selenium in various feedstuffs[145,146] require some consideration.

Wisconsin workers have found that incorporation of 132 ppm of selenium as the selenite into the salt fed to ewes *ad libitum*, prevented clinical nutritional muscular dystrophy and improved weight gains in lambs.[191] In a 3 year trial, similar results were obtained using levels of 26 and 132 ppm of selenium, but at 264 ppm depressed weight gains were observed in the lambs.[214] The ewes showed no apparent signs of selenium toxicity.

It has been shown that inorganic selenium (sodium selenite) added to feeds in quantities to meet the requirements of swine[126] or poultry[235,132] does not increase the tissue levels of selenium as much as equal amounts of the naturally occurring forms of selenium in corn and soybeans.

In the ruminant, the effectiveness of orally administered inorganic salts of selenium may be reduced by microbial action in the rumen. Cousins and Cairney[43] found that about 40% of an oral dose of selenite was eliminated in the feces of sheep in a very insoluble form. Butler and Peterson[29] reported similar findings, and Peterson and Spedding[197] found that the selenium in feces was only in small part available to plants. Conflicting reports have been made relative to the incorporation of the selenium of inorganic salts into methionine by rumen microorganisms.[96,190]

Ehlig *et al.*[63] and Davies and Watkinson[53] have studied the selenium content of different species of plants growing in soils of low selenium content. It appears that on these soils the selenium concentrations in the various species differ very little. It seems that a shift in crops from one type to another will not usually affect the incidence of selenium-responsive diseases. In marginal cases, however, there may exist the possibility that shifting from a shallow rooted crop like grass to a deep rooted crop like alfalfa will reduce the incidence of selenium-responsive syndromes.

BIOLOGICAL FUNCTION OF SELENIUM

The mode of action of selenium in preventing or curing selenium-responsive diseases has not been resolved. Selenium compounds have antioxidant effects on some *in vitro* biological systems,[20,19,289] and Tappel and Caldwell[253], have summarized evidence which supports the concept that selenium acts as a nonspecific antioxidant and inhibits free radical peroxidation of lipids. Diplock *et al.*[57] and Green *et al.*[87] have suggested that selenium may be concerned in the biosynthesis of ubiquinone. Schwarz[221] found that supplementing the animal diet with selenium enhanced the over-all activity of the α-ketoglutaric acid system and suggested that the element might be involved in the decarboxylation reaction of this system.   The data of Bull and Oldfield[28] are in agreement with this to the extent that they suggest the involvement of selenium in the tricarboxylic acid cycle.   Some work[147,192] suggests the involvement of selenium in long chain fatty acid metabolism.   Scott and Thompson[234] have interpreted their recent results to indicate that, at least for chickens, selenium may be important to pancreatic protein synthesis.   In a deficiency of the element, impaired function of the fibrotic pancreas results in a failure of absorption of fat and thus of vitamin E from the intestine. With respect to this finding, the dystrophogenic properties for sheep of raw kidney beans[80,97,98] not entirely accounted for by their low selenium and vitamin E contents have recently been explained by the prevention of vitamin E absorption by some factor in the uncooked beans.[102]

Rotruck et al.[216] have reported that selenium is an integral part of GSH peroxidase and Hoekstra et al.[99] have reported that red blood cell and liver levels of GSH peroxidase are dependent upon dietary selenium intake. GSH peroxidase prevents hydrogen peroxide or ascorbic acid-induced oxidation of hemoglobin.

Scott and Noguchi[230] have found that the plasma level of GSH peroxidase in chicks fed a selenium-deficient diet will drop to about 10% of the normal value in 5 days.   This drop precedes the onset of exudative diathesis by 1 or 2 days.   Vitamin E has no effect upon the level of GSH peroxidase even though it is effective against exudative diathesis.   There was no relationship between muscular dystrophy development and GSH peroxidase level in chicks.   Selenite is more effective than selenium in selenomethionine in the maintenance of GSH peroxidase activity and for the prevention of fibrotic degeneration of the chick pancreas.[38]

Selenium as an Essential Element for Plants

In view of the role of selenium in animal nutrition, the question of its essentiality to plant growth is pertinent.   In the late 1930s, Trelease and Trelease[264] reported marked stimulation in the growth of indicator plants by selenium and suggested that it might be essential for their growth.   The same year (1938), Perkins and King[193] reported that small additions of selenium to soil stimulated the growth of wheat seedlings.   A year later, Stanford and Olson[247] reported that the growth of certain crop plants was stimulated by 0.05 ppm of the element in culture solution.   Additional studies on the effect of selenium on plant growth have been reviewed by Rosenfeld and Beath.[212] Evidence that selenium is esesential to plants is not convincing.   A few years ago, Broyer et al.[25] reported that selenium added to highly purified culture solutions failed to stimulate plant growth.   The likelihood that selenium deficiencies in soils will be found to have an effect on plant growth under natural conditions appears remote.

Selenium as an Insecticide

Several selenium compounds have been used as insecticides and have proven to be very effective for the control of certain insects.   However, because of the toxicity of most selenium compounds, the use of selenium as an agricultural insecticide will probably be limited to a few nonfood areas. Selenium is especially objectional as an insecticide because of its stability in the soil and its uptake by most plants which may render them toxic to animals and humans.

Certain insects such as red mites[22] are very susceptible to selenium insecticides but, on the other hand, there are insects which live in and reproduce in the seeds of seleniferous plants which contain over 1000 ppm of selenium.[263]

Selenium was found to be an effective insecticide against aphids in 1936.[110] Neiswander and Morris[172] studied the use of selenium as a systemic insecticide for the protection of certain ornamental plants against insects.   Selenium is used by florists for the protection of a number of greenhouse plants,[7] however, care must be taken that the greenhouse soils do not find their way into food-producing gardens or greenhouse beds.   The use and limitations of selenium as an insecticide have been carefully reviewed by Smith[7] and more recently by Rosenfeld and Beath.[212]

REFERENCES

1. Abu-Erreish, G., M. S. Thesis, S. Dak. State Univ. (1967).
2. Allaway, W. H., *Proc. Georgia Nutrition Conf.*, 61 (1969).
3. Allaway, W. H., Cary, E. E., and Ehlig, C. F., in *Selenium in Biomedicine* Muth, O. H., Oldfield, J. E., and Weswig, P. H., eds., AVI Publishing Co., Inc., Westport, Conn., 1967, p. 273.
4. Allaway, W. H., and Hodgson, J. F., *J. Animal Sci.* **23**, 271 (1964).
5. Allaway, W. H., Moore, D. P., Oldfield, J. E., and Muth, O. H., *J. Nutrition* **88**, 411 (1966).
6. Ancizar-Sordo, J., *Soil Sci.* **63**, 437 (1947).
7. Anderson, M. S., Lakin, H. W., Beeson, K. C., Smith, F. F., and Thacker, E., *U.S. Dep. Agr. Handbook* No. 200 (1961).
8. Andrews, E. D., Hartley, W. J., and Grant, A. B., *New Zealand Vet. J.* **16**, 3 (1968).
9. Beath, O. A., *Wyoming Agr. Exp. Sta. Bull.* No. 221 (1937).
10. Beath, O. A., Draize, J. H., and Eppson, H. F., *Wyoming Agr. Exp. Sta. Bull.* No. 189 (1932).
11. Beath, O. A., Draize, J. H., Eppson, H. F., Gilbert, C. S., and McCreary, O.C., *J. Am. Pharm. Assoc.* (Sci. Ed.) **23**, 94 (1934).
12. Beath, O. A., and Eppson, H. F., *Wyoming Agr. Exp. Sta. Bull.* No. 278 (1947).
13. Beath, O. A., Eppson, H. F., and Gilbert, C. S., *Wyoming Agr. Exp. Sta. Bull.* No. 206 (1935).

14. Beath, O. A., Eppson, H. F., and Gilbert, C. S., *J. Am. Pharm. Assoc.* (Sci. Ed.) **26**, 394 (1937).
15. Beath, O. A., Eppson, H. F., Gilbert, C. S., and Bradley, W. B., *Wyoming Agr. Exp. Sta. Bull.* No. 231 (1939).
16. Benavides, S. T., and Mojica, R. F. S., *Inst. Geograf. Colombia Publ.* No. IT-3 (1959).
17. Bersin, T., *Ergebn. Enzymforsch.* **4**, 68 (1935).
18. Bieri, J. G., *Nature* **184**, 1148 (1959).
19. Bieri, J. G., *Am. J. Clin. Nutrition* **9**, 89 (1961).
20. Bieri, J. G., and Pollard, C. J., *J. Nutrition* **69**, 301 (1959).
21. Boyazoglu, P. A., Jordan, R. M., and Meade, R. J., *J. Animal Sci.* **26**, 1390 (1967).
22. Boyce, A. M., *J. Econ. Entomol.* **29**, 125 (1936).
23. Brown, J. M. M., *Ann. N. Y. Acad. Sci.* **104**, 504 (1963).
24. Brown, J. M. M., and de Wet, P. J., *Onderstepoort J. Vet. Research* **29**, 111 (1962).
25. Broyer, T. C., Lee, D. C., and Asher, C. J., *Plant Physiol.* **41**, 1425 (1966).
26. Buchanan-Smith, J. G., Nelson, E. C., Osborn, B. I., Wells, M. E., and Tillman, A. D., *J. Animal Sci.* **29**, 808 (1969).
27. Buchanan-Smith, J. G., Nelson, E. C., and Tillman, A. D., *J. Nutrition* **99**, 387 (1969).
28. Bull, R. C., and Oldfield, J. E., *J. Nutrition* **91**, 237 (1967).
29. Butler, G. W., and Peterson, P. J., *New Zealand J. Agr. Research* **4**, 484 (1961).
30. Butler G. W., and Peterson, P. J., *Aust. Biol. Sci.* **20**, 77 (1967).
31. Byers, H. G., *U.S. Dep. Agr. Tech. Bull.* No. 482 (1935).
32. Byers, H. G., *U.S. Dep. Agr. Tech. Bull.* No. 530 (1936).
33. Byers, H. G., *Ind. Eng. Chem.* **28**, 1200 (1937).
34. Byers, H. G., and Lakin, H. W., *Can. J. Research* **17**, 364 (1939).
35. Byers, H. G., Miller, J. T., Williams, K. T., and Lakin, H. W., *U.S. Dep. Agr. Tech. Bull.* No. 601 (1938).
36. Byers, H. G., Williams, K. T., and Lakin, H. W., *Ind. Eng. Chem.* **28**, 821 (1936).
37. Cameron, C. A., *Proc. Roy. Soc.* (Dublin) **2**, 231 (1880).
38. Cantor, A. H., Langevin, M. L., Noguchi, T., and Scott, M. L. *Federation Proc.* **32**, 885 (1973).
39. Cantor, A. H., and Scott, M. L., *Abstracts of Papers, 62nd Annual Meeting, Poultry Science Assoc. Inc.* p. 16 (1973).
40. Carlson, C. W., Guenthner, E., Kohlmeyer, W., and Olson, O. E., *Poultry Sci.* **33**, 768 (1954).
41. Carlson, C. W., Kohlmeyer, W., and Moxon, A. L., *S. Dakota Farm and Home Research* **3** (No. 1), 20 (1951).
42. Carter, D. L., Brown, M. J., Allaway, W. H., and Cary, E. E., *Agronomy J.* **60**, 532 (1968).
43. Cousins, F. B., and Cairney, I. M., *Australian J. Agr. Research* **12**, 927 (1961).
44. Creech, B. G., Feldman, G. L., Ferguson, T. M., Reid, B. L., and Couch, J. R., *J. Nutrition* **62**, 83 (1967).
45. Creger, C. R., Mitchell, R. H., Atkinson, R. L., Ferguson, T. M., Reid, B. L., and Couch, J. R., *Poultry Sci.* **39**, 59 (1960).
46. Cunha, T. J., *Feedstuffs* **41** (No. 18), 21 (1969).
47. Cunha, T. J., *Feedstuffs* **42**, 22 (1970).

48. Dam, H., *J. Nutrition* **27**, 193 (1944).
49. Dam, H., and Glavind, J., *Nature* **142**, 1077 (1938).
50. Dam, H., and Glavind, J., *Naturwissenschaften* **28**, 207 (1940).
51. Dam, H., Hartman, S., Jacobsen, J. E., and Søndergaard, E., *Acta Physiol. Scand.* **41**, 149 (1957).
52. Dam, H., and Søndergaard, E., *Experimentia* **13**, 494 (1957).
53. Davies, E. B., and Watkinson, J. H., *New Zealand J. Agr. Research* **9**, 317 (1966).
54. Davies, E. B., and Watkinson, J. H., *New Zealand J. Agr. Research* **9**, 641 (1966).
55. Dinkel, C. A., Minyard, J. A., and Ray, D. E., *J. Animal Sci.* **22**, 1043 (1963).
56. Dinkel, C. A., Minyard, J. A., Whitehead, E. I., and Olson, O. E., *S. Dakota Agr. Exp. Sta. Circ.* No. 135 (1957).
57. Diplock, A. T., Bunyan, J., Edwin, E. E., and Green, J., *Brit. J. Nutrition* **16**, 109 (1962).
58. Draize, J. H., and Beath, O. A., *J. Am. Vet. Med. Assoc.* **86**, 753 (1935).
59. DuBois, K. P., Moxon, A. L., and Olson, O. E., *J. Nutrition* **19**, 477 (1940).
60. DuBois, K. P., Rhian, M., and Moxon, A. L., *Proc. S. Dakota Acad. Sci.* **19**, 71 (1939).
61. Earley, J. W., *Am. Miner.* **34**, 337 (1950).
62. Eggert, R. G., Patterson, E., Akers, W. T., and Stokstad, E. L. R., *J. Animal Sci.* **16**, 1037 (1957).
63. Ehlig, C. F., Allaway, W. H., Cary, E. E., and Kubota, J., *Agronomy J.* **60**, 43 (1968).
64 Evans, C. S., Asher, C. J., and Johnson, C. M., *Australian J. Biol. Sci.* **21**, 13 (1968).
65. Ewan, R. C., Baumann, C. A., and Pope, A. L., *J. Animal Sci.* **27**, 751 (1968).
66. Ewan, R. C., Wastell, M. E., Bicknell, E. J., and Speer, V. C., *J. Animal Sci.* **29**, 912 (1969).
67. Ewen, L. M., and Jenkins, K. J., *J. Nutrition* **93**, 470 (1967).
68. Fleming, G. A., and Walsh, T., *Proc. Roy. Irish Acad.* **58**, 151 (1957).
69. Franke, K. W., *J. Nutrition* **8**, 597 (1934).
70. Franke, K. W., *J. Nutrition* **8**, 609 (1934).
71. Franke, K. W., Moxon, A. L., Poley, W. E., and Tully, W. C., *Anat. Record* **65**, 15 (1936).
72. Franke, K. W., and Painter, E. P., *Cereal Chem.* **13**, 67 (1936).
73. Franke, K. W., and Painter, E. P., *Ind. Eng. Chem.* **29**, 591 (1937).
74. Franke, K. W., and Painter, E. P., *Cereal Chem.* **15**, 1 (1938).
75. Franke, K. W., and Potter, V. R., *J. Nutrition* **10**, 213 (1935).
76. Franke, K. W., Rice, T. D., Johnson, A. G., and Schoening, H. W., *U.S. Dep. Agr. Circ.* No. 320 (1934).
77. Franke, K. W., and Tully, W. C., *Poultry Sci.* **14**, 273 (1935).
78. Ganther, H. E., *Biochemistry* **7**, 2898 (1968).
79. Ganther, H. E., and Baumann, C. A., *J. Nutrition* **77**, 408 (1962).
80. Gardner, R. W., and Hogue, D. E., *J. Nutrition* **93**, 418 (1967).
81. Glenn, M. W., Jensen, R., and Griner, L. A., *Am. J. Vet. Research* **25**, 1479 (1964).
82. Glenn, M. W., Jensen, R., and Griner, L. A., *Am. J. Vet. Research* **25**, 1486 (1964).
83. Goldstein, J., and Scott, M. L., *J. Nutrition* **60,** 349 (1956).

84. Grant, A. B., *New Zealand J. Agr. Research* **8**, 681 (1965).
85. Grant, C. A., *Acta Vet. Scand.* **2** (Supplement 3) (1961).
86. Grant, C. A., and Thafvelin, B., *Nord. Veterinärmed.* **10**, 657 (1958).
87. Green, J., Diplock, A. T., Bunyan, J., Edwin, E. E., and McHale, D., *Nature* **190**, 318 (1961).
88. Gries, C. L. and Scott, M. L., *J. Nutrition* **102**, 1287 (1972).
89. Halverson, A. W., Guss, P. L., and Olson, O. E., *J. Nutrition* **77**, 459 (1962).
90. Halverson, A. W., Hendrick, C. M. and Olson, O. E., *J. Nutrition* **56**, 51 (1955).
91. Halverson, A. W., Jeade, L. G., and Hills, E. L., *Pharmacol.* **7**, 675 (1965).
92. Halverson, A. W., and Monty, K. J., *J. Nutrition* **70**, 100 (1960).
93. Hamilton, J. W., and Beath, O. A., *J. Range Management* **16**, 261 (1963).
94. Hartley, W. J., and Grant, A. B., *Federation Proc.* **20**, 679 (1961).
95. Hendrick, C., Klug, H. L., and Olson, O. E., *J. Nutrition* **51**, 131 (1953).
96. Hidiroglou, M., Heaney, D. P., and Jenkins, K. J., *Can. J. Physiol. Pharmacol.* **46**, 229 (1968).
97. Hintz, H. F., and Hogue, D. E., *J. Nutrition* **82**, 495 (1964).
98. Hintz, H. F., Hogue, D. E., and Walker, E. F., Jr., *Proc. Soc. Exp. Biol. Med.* **131**, 447 (1969).
99. Hoekstra, W. G., Hafeman, D., Oh, S. W., Sunde, R. A., and Ganther, H. E., *Federation Proc.* **32** (No. 3) 885 (1973).
100. Hoffman, I., Jenkins, K. J., Meranger, J. C., and Pigden, W. J. *Can. J. Animal Sci.* **53**, 61 (1973).
101. Hofmeister, F., *Arch. Exp. Pathol. Pharmakol. Naunyn-Schmiedeberg's* **33**, 198 (1893–4).
102. Hogue, D. E., Banerjee, G. C., Hintz, H. F., and Walker, E. F., Jr., *Federation Proc.* **29**, 694 (1970).
103. Hopkins, L. L. Jr., Pope, A. L., and Baumann, C. A., *J. Animal Sci.* **23**, 674 (1964).
104. Horn, M. J., and Jones, D. B., *J. Am. Chem. Soc.* **62**, 234 (1940).
105. Horn, M. J., and Jones, D. B., *J. Biol. Chem.* **139**, 649 (1941).
106. Horn, M. J., Nelson, E. M., and Jones, D. B., *Cereal Chem.* **13**, 126 (1936).
107. Hurd-Karrar, A. M., *Science* **78**, 560 (1933).
108. Hurd-Karrar, A. M., *J. Agr. Research* **49**, 343 (1934).
109. Hurd-Karrar, A. M., *J. Agr. Research* **50**, 413 (1935).
110. Hurd-Karrar, A. M., and Poos, F. W., *Science* **84**, 252 (1936).
111. Hurt, H. D., Gary, E. E., and Visek, W. J., *J. Nutrition* **101**, 761 (1971).
112. Hyppola, K., *Feedstuffs* **34** (No. 4), 22 (1962).
113. Hyppola, K., *Feedstuffs* **34** (No. 42), 36 (1962).
114. Jaffe, W. G., Chavez, J. F., and de Mondragon, M. C., *Archivos Latinamericanos de Nutricion* **17**, 59 (1967).
115. Japha, A., *Dissertation, Halle* (1842).
116. Jenkins, K. J. and Hidiroglou, M. *Can. J. Animal Sci.* **52**, 591 (1972).
117. Jensen, L. S., *Proc. Soc. Exp. Biol. Med.* **128**, 970 (1968).
118. Kendall, O. K., *Calif. Veterinarian* **14** (No. 1), 39 (1960).
119. Kifer, R. R., and Payne, W. L., *Feedstuffs* **40** (No. 35), 32 (1968).
120. Kifer, R. R., Payne, W. L., and Ambrose, M. E., *Feedstuffs* **41** (No. 51), 24 (1969).
121. Klug, H. L., Harshfield, R. D., Pengra, R. A., and Moxon, A. L., *J. Nutrition* **48**, 409 (1952).

122. Knott, S. G., McCray, C. W. R., and Hall, W. T. K., *Queensland Department of Agriculture and Stock Division of Animal Industry Bulletin* No. 41 (1958).
123. Knott, S. G., McCray C. W. R., and Hall, W. T. K., *Queensland J. Agr. Sci.* **15**, 43 (1958).
124. Kovalskii, V. V., *Priroda* 4, 11 (1954); (*Nutrition Abstr. Rev.* **25**, 544 (1955)).
125. Ku, P. K., Ely, W. T., Groce, A. W. and Ullery, D. E., *J. Animal Science* **34**, 208 (1972).
126. Ku, P. K., Miller, E. R., Wahlstrom, R. C., Groce, A. W., Hitchcock, J. P. and Ullery, D. E., *J. Animal Sci.* **37**, 501 (1973).
127. Kubota, J., Allaway, W. H., Carter, D. L., Cary, E. E., and Lazar, V. A., *J. Agr. Food Chem.* **15**, 448 (1967).
128. Lakin, H. W., and Byers, H. G., *U.S. Dep. Agr. Tech. Bull.* No. 783 (1941).
129. Lakin, H. W., and Byers, H. G., *U.S. Dep. Agr. Tech. Bull.* No. 950 (1948).
130. Lakin, H. W., Williams, K. T., and Byers, H. G., *Ind. Eng. Chem.* **30**, 599 (1938).
131. Larsen, C., and Bailey, D. E., *S. Dakota Agr. Exp. Sta. Bull.* No. 147, 299 (1913).
132. Latshaw, J. D., *Abstracts of papers*, 62nd Annual Meeting Poultry Sci. Assoc. Inc. p. 60 (1973).
133. Lemley, R. E., *J. Lancet* **60**, 528 (1940).
134. Levander, O. A., and Argett, L. C., *Toxicol. Appl. Pharmacol.* **14**, 308 (1969).
135. Levander, O. A., and Morris, V. C., *J. Nutrition* **100**, 1111 (1970).
136. Lewis, B. G., Johnson, C. H., and Broyer, T. C., *Biochim. Biophys. Acta* **237**, 603 (1971).
137. Lewis, H. B., Schultz, J., and Gortner, R. A., Jr., *J. Pharmacol. Exp. Therap.* **68**, 292 (1940).
138. Maag, D. D., and Glenn, M. W., in *Selenium in Biomedicine* Muth, O. H., Oldfield, J. E., and Weswig, P. H., eds. AVI Publishing Co., Inc., Westport, Conn., 1967, p. 127.
139. Maag, D. D., Orsborn, J. S., and Clopton, J. R., *Am. J. Vet. Research* **21**, 1049 (1960).
140. Madison, T. D., (*U.S.*) *36th Cong. 1st Sess. Senate Exec. Doc.* **52**, 37 (Jan. 1855 to Jan. 1860).
141. Mahan, D. C., *Distillers Feed Research Council Proc.* **28**, 6 (1973).
142. Mahan, D. C., Jones, J. E., Cline, J. H., Cross, R. F., Teague, H. S., and Grifo, A. P., Jr. *J. Animal Science* **36**, 1104 (1973).
143. Martin, A. L., *Am. J. Botany* **23**, 471 (1936).
144. Martin, J. L., and Gerlach, M. L., *Anal. Biochem.* **29**, 257 (1969).
145. Mathias, M. M., Allaway, W. H., Hogue, D. E., Marion, M. V., and Gardner, R. W., *J. Nutrition* **86**, 213 (1965).
146. Mathias, M. M., Hogue, D. E., Loosli, J. K., *J. Nutrition* **93**, 14 (1967).
147. McCoy, K. E. M., and Weswig, P. H., *Proc. N. W. Am. Chem. Soc.* **23**, 34 (1968).
148. McLean, J. W., Thomson, G. G., and Claxton, J. H., *Nature* **184**, 251 (1959).
149. McLean, J. W., Thomson, G. G., and Claxton, J. H., *New Zealand Vet. J.* **7**, 47 (1959).
150. Michel, R. L., Whitehair, C. K., and Keahey, K. K., *J. Am. Vet. Med. Assoc.* **155**, 50 (1969).
151. Miller, W. T., and Schoening, H. W., *J. Agr. Research* **56**, 831 (1938).
152. Minyard, J. A., *S. Dakota Farm and Home Research* **12** (No. 1), 1 (1961).

153. Moxon, A. L., *S. Dakota Agr. Exp. Sta. Bull.* No. 311 (1937).
154. Moxon, A. L., *Science* **88**, 81 (1938).
155. Moxon, A. L., Ph.D. Thesis, Univ. Wisconsin (1941).
156. Moxon, A. L., *Proc. S. Dakota Acad. Sci.* **21**, 34 (1941).
157. Moxon, A. L., and DuBois, K. P., *J. Nutrition* **18**, 447 (1939).
158. Moxon, A. L., Jensen, C. W., and Paynter, C. R., *Proc. S. Dakota Acad. Sci.* **26**, 21 (1946–47).
159. Moxon, A. L., Olson, O. E., and Searight, W. V., *S. Dakota Agr. Exp. Sta. Tech. Bull.* No. 2 (1939).
160. Moxon, A. L., Olson, O. E., Searight, W. V., and Sandals, H. M., *Am. J. Botany* **25**, 794 (1938).
161. Moxon, A. L., Paynter, C. R., and Halverson, A. W., *J. Pharmacol. Exp. Therap.* **84**, 115 (1945).
162. Moxon, A. L., Parker, C. F., and Potter, E. L., *Ohio Agr. Res. and Dev. Center Res. Summary* **67**, 20 (1973).
163. Moxon, A. L., and Rhian, M., *Physiol. Rev.* **23**, 305 (1943).
164. Moxon, A. L., Rhian, M. A., Anderson, H. D., and Olson, O. E., *J. Animal Sci.* **3**, 299 (1944).
165. Moxon, A. L., Schaefer, A. E., Lardy, H. A., DuBois, K. P., and Olson, O. E., *J. Biol. Chem.* **132**, 785 (1940).
166. Moxon, A. L., and Wilson, W. O., *Poultry Sci.* **23**, 149 (1944).
167. Muth, O. H., *J. Am. Vet. Med. Assoc.* **142**, 272 (1963).
168. Muth, O. H., and Binns, W., *Ann. N.Y. Acad. Sci.* **111** (art. 2), 583 (1964).
169. Muth, O. H., Oldfield, J. E., Remmert, L. F., and Schubert, J. R., *Science* **128**, 1090 (1958).
170. National Research Council., *Publication No. 1959*, National Academy of Science, Wash. D.C. (1968).
171. Neethling, L. P., Brown, J. M. M., and DeWet, P. J., *J. S. Afr. Vet. Med. Assoc.* **39**, 93 (1968).
172. Neiswander, C. R., and Morris, V. H., *J. Econ. Ent.* **33**, 517 (1940).
173. Nesheim, M. C., and Scott, M. L., *J. Nutrition* **65**, 601 (1958).
174. Nesheim, M. C., and Scott, M. L., *Federation Proc.* **20**, 674 (1961).
175. Nigam, S. N., and McConnell, W. B., *Biochim. Biophys. Acta* **192**, 185 (1969).
176. Nigam, S. N., and McConnell, W. B., *Federation Proc.* **32** (No. 3) 886 (1973).
177. Oksanen, H. E., in *Selenium in Biomedicine* Muth, O. H., Oldfield, J. E., and Weswig, P. H., eds., AVI Publishing Co., Inc., Westport, Conn.,1967, p. 215.
178. Oldfield, J. E., Muth, O. H., and Schubert, J. R., *Proc. Soc. Exp. Biol. Med.* **103**, 799 (1960).
179. Olson, O. E., Carlson, C. W., and Leitis, E., *S. Dakota Agr. Exp. Sta. Tech. Bull.* No. 20 (1958).
180. Olson, O. E., Dinkel, C. A., and Kamstra, L. D., *S. Dakota Farm and Home Research* **6** (No. 1), 12 (1954).
181. Olson, O. E., and Embry, L. B., *Proc. S. Dak. Acad. Sci.* (in press) (1973).
182. Olson, O. E., and Jensen, C. W., *Proc. S. Dakota Acad. Sci.* **20**, 115 (1940).
183. Olson, O. E., Jornlin, D. F., and Moxon, A. L., *J. Am. Soc. Agron.* **34**, 607 (1942).
184. Olson, O. E., and Moxon, A. L., *Soil Sci.* **47**, 305 (1939).
185. Olson, O. E., Whitehead, E. I., and Moxon, A. L., *Soil Sci.* **54**, 47 (1942).
186. O'Moore, L. B., *Irish Vet. J.* **6**, 392 (1952).
187. Orstadius, K., Nordstrom, G., and Lannek, N., *Cornell Vet.* **53**, 60 (1963).

188. Patrios, G. and Olson, O. E., *Feedstuffs* **41** (No. 43) 29 (1969).
189. Patterson, E. L., Milstrey, R., and Stodstad, E. L. R., *Proc. Soc. Exp. Biol. Med.* **95**, 617 (1957).
190. Paulson, G. D., Baumann, C. A., and Pope, A. L., *J. Animal Sci.* **27**, 497 (1968).
191. Paulson, G. D., Broderick, G. A., Baumann, C. A., and Pope, A. L., *J. Animal Sci.* **27**, 195 (1968).
192. Pendell, H. W., Muth, O. H., and Oldfield, J. E., *Proc. N. W. Am. Chem. Soc.* **22**, 10 (1967).
193. Perkins, A. T., and King, H. H., *J. Am. Soc. Agron.* **30**, 664 (1938).
194. Peters, A. T., *Nebraska Agr. Exp. Sta. 17th Annual Rpt.* p. 13 (1904).
195. Peterson, P. J., and Butler, G. W., *Australian J. Biol. Sci.* **15**, 126 (1962).
196. Peterson, P. J., and Butler, G. W., *Nature* **212**, 961 (1966).
197. Peterson, P. J., and Spedding, D. J., *New Zealand J. Agr. Research* **6**, 13 (1963).
198. Poley, W. E., and Moxon, A. L., *Poultry Sci.* **17**, 72 (1938).
199. Poley, W. E., Wilson, W. O., Moxon, A. L., and Taylor, J. B., *Poultry Sci.* **20**, 171 (1941).
200. Preston, R. L., and Moxon, A. L., *Ohio Agr. Res. and Dev. Center. Res. Summary* **63**, 81 (1972).
201. Proctor, J. F., Hogue, D. E., and Warner, R. G., *J. Animal Sci.* **17**, 1183 (1958).
202. Rahman, M. M., Deyoe, C. W., Davies, R. E., and Couch, J. R., *J. Nutrition* **72**, 71 (1960).
203. Ravikovitch, S., and Margolin, M., *Agr. Research Sta. Rehovot* **7**, 41 (1957).
204. Reid, B. L., Rahman, M. M., Creech, B. G., and Couch, J. R., *Proc. Soc. Exp. Biol. Med.* **97**, 590 (1958).
205. Rhian, M., and Moxon, A. L., *J. Pharmacol. Exp. Therap.* **78**, 249 (1943).
206. Robertson, T. G., and During, C., *New Zealand J. Agr.* **103**, 306 (1961).
207. Robinson, W. O., *J. Assoc. Off. Agr. Chem.* **16**, 423 (1933).
208. Robinson, W. O., *Ind. Eng. Chem.* **28**, 736 (1936).
209. Rosenfeld, I., and Beath, O. A., *Am J. Vet. Research* **7**, 52 (1946).
210. Rosenfeld, I., and Beath, O. A., *J. Agr. Research* **75**, 93 (1947).
211. Rosenfeld, I., and Beath, O. A., *Wyoming Agr. Exp. Sta. Bull.* No. 275 (1957).
212. Rosenfeld, I., and Beath, O. A., *Selenium*, Academic Press, New York, 1964, p. 411.
213. Rosenfeld, I., and Eppson, H. F., *Am. J. Vet. Research* **18**, 693 (1957).
214. Rotruck, J. T., Pope, A. L., Baumann, C. A., Hoekstra, W. G., and Paulson, G. D., *J. Animal Sci.* **29**, 170 (1969).
215. Rotruck, J. T., Pope, A. L., Ganther, H. E., and Hoekstra, W. G., *J. Nutrition* **102**, 689 (1972).
216. Rotruck, J. T., Hoekstra, W. G., Pope, A. L., Ganther, H., Swanson, A., and Hafeman, D., *Federation Proc.* **3** (No. 2) 691 (1972).
217. Schneider, H. A., *Science* **83**, 32 (1936).
218. Schwarz, K., *Z. Physiol. Chem., Hoppe-Seyler's* **281**, 109 (1944).
219. Schwarz, K., *Proc. Soc. Exp. Biol. Med.* **80**, 319 (1952).
220. Schwarz, K., *Federation Proc.* **20**, 666 (1961).
221. Schwarz, K., *Federation Proc.* **24**, 58 (1965).
222. Schwarz, K., Bieri, J. C., Briggs, G. M., and Scott, M. L., *Proc. Soc. Exp. Biol. Med.* **95**, 621 (1957).

223. Schwarz, K., and Foltz, C. M., *J. Am. Chem. Soc.* **79**, 3292 (1957).
224. Schwarz, K., and Foltz, C. M., *J. Biol. Chem.* **233**, 245 (1958).
225. Schwarz, K., and Fredga, A., *J. Biol. Chem.* **244**, 2103 (1969).
226. Schwarz, K., Stesney, J., and Foltz, C. M., *Metabolism* **8**, 88 (1959).
227. Scott, M. L., in *Selenium in Biomedicine* Muth, O. H., Oldfield, J. E., and Weswig, P. H., eds., AVI Publishing Co., Inc., Westport, Conn., 1967, p. 231.
228. Scott, M. L., Bieri, J. G., Briggs, G. M., and Schwarz, K., *Poultry Sci.* **36** 1155 (1957).
229. Scott, M. L., Hill, F. W., Norris, L. C., Dobson, D. C., and Nelson, T. S., *J. Nutrition* **56**, 387 (1955).
230. Scott, M. L., and Noguchi, T., *Federation Proc.* **32** (No. 3) 885 (1973).
231. Scott, M. L., Olson, G., Krook, L., and Brown, W. R., *J. Nutrition* **91**, 573 (1967).
232. Scott, M. L., and Thompson, J. N., *Poultry Sci.* **48**, 1868 (1969).
233. Scott, M. L., and Thompson, J. N., *Proc. Cornell Nutrition Conf.* p. 59 (1969).
234. Scott, M. L., and Thompson, J. N., *N.Y. Food Life Sci. Quart.* **3** (No. 1) 12 (1970).
235. Scott, M. L., and Thompson, J. N., *Poultry Sci.* **50**, 1742 (1971).
236. Searight, W. V., and Moxon, A. L., *S. Dakota Agr. Exp. Sta. Tech. Bull.* No. 5 (1945).
237. Sellers, E. A., You, R. W., and Lucas, C. C., *Proc. Soc. Exp. Biol. Med.* **75**, 118 (1950).
238. Sharman, G. A. M., Blaxter, K. L., and Wilson, R. S., *Vet. Record* **71**, 536 (1959).
239. Shirley, R. L., Koger, M., Chapman, H. L., Jr., Loggins, P. E., Kidder, R. W., and Easley, J. E., *J. Animal Sci.* **25**, 648 (1966).
240. Shrift, A., and Virupaksha, T. K., *Biochim. Biophys, Acta* **100**, 65 (1965).
241. Siami, G., Schulert, A. R., and Neal, R. A., *J. Nutrition* **102**, 857 (1972).
242. Slen, S. B., Demiruren, A. S., and Smith, A. D., *Can. J. Animal Sci.* **41**, 263 (1961).
243. Smith, A. L., M. S. Thesis, S. Dakota State Univ. (1949).
244. Smith, M. I., *Public Health Repts.* (U.S.) **54**, 1441 (1939).
245. Smith, M. I., Franke, K. W., and Westfall, B. B., *Public Health Repts.* (U.S.) **51**, 1496 (1936).
246. Smith, M. I., and Stohlman, E. F., *J. Pharmacol. Exp. Therap.* **70**, 270 (1940).
247. Stanford, G. W., and Olson, O. E., *Proc. S. Dakota Acad. Sci.* **19**, 25 (1939).
248. Stokstad, E. L. R., Patterson, E. L., and Milstrey, R., *Poultry Sci.* **36**, 1160 (1957).
249. Stowe, H. D., and Whitehair, C. K., *J. Nutrition* **81**, 287 (1963).
250. Supplee, W. C., *Poultry Sci.* **45**, 852 (1966).
251. Swahn, O., and Thafvlin, B., *Vitamins and Hormones* **20**, 645 (1962).
252. Taboury, M., *Acad. Sci. Paris. Compt. Rend.* **195**, 171 (1932).
253. Tappel, A. L., and Caldwell, K. A., in *Selenium in Biomedicine* edited by Muth, O. H., Oldfield, J. E., and Weswig, P. H., eds., AVI Publishing Co., Inc., Westport, Conn., 1967, p. 345.
254. Thompson, J. N., and Scott, M. L., *Proc. Cornell Nutrition Conf.* p. 130 (1967).
255. Thompson, J. N., and Scott, M. L., *Proc. Cornell Nutrition Conf.* p. 121 (1968).
256. Thompson, J. N., and Scott, M. L., *J. Nutrition* **97**, 335 (1969).
257. Thompson, J. N., and Scott, M. L., *J. Nutrition* **100**, 797 (1970).

258. Thomson, C. D., and Stewart, R. D. H., *British J. Nutrition* **30**, 139 (1973).
259. Thorvaldson, T., and Johnson, L. R., *Can. J. Research* **18**, 138 (1940).
260. Trapp, A. L., Keahey, K. K., Whitemack, D. L. and Whitehair, C. K., *J. Am. Vet. Med. Assoc.* **157**, 289 (1970).
261. Trelease, S. F., and Beath, O. A., *Selenium*, The Champlin Printers, Burlington, Vt., 1949.
262. Trelease, S. F., DiSomma, A. A., and Jacobs, A. L., *Science* **132**, 618 (1960).
263. Trelease, S. F., and Trelease, H. M., *Science* **85**, 590 (1937).
264. Trelease, S. F., and Trelease, H. M., *Am. J. Botany* **25**, 372 (1938).
265. Virupaksha, T. K., and Shrift, A., *Biochim. Biophys. Acta* **74**, 791 (1963).
266. Virupaksha, T. K., and Shrift, A., *Biochim. Biophys. Acta* **107**, 69 (1965).
267. Virupaksha, T. K., Shrift, A., and Tarver, H., *Biochim. Biophys. Acta* **130**, 45 (1966).
268. Vohra, P., Johnson, C. M., McFarland, L. Z., Siopes, T. D., Wilson, W. O., and Winget, C. M., *Poultry Sci.* **52**, 644 (1973).
269. Wahlstrom, R. C., Kamstra, L. D., and Olson, O. E., *J. Animal Sci.* **14**, 105 (1955).
270. Wahlstrom, R. C., Kamstra, L. D., and Olson, O. E., *S. Dakota Agr. Exp. Sta. Bull.* No. 456 (1956).
271. Wahlstrom, R. C., and Olson, O. E., *J. Animal Sci.* **18**, 141 (1959).
272. Walsh, T., and Fleming, G. A., *Trans. Intern. Soc. Soil Sci. Comm. II and IV* **2**, 178 (1952).
273. Walsh, T., Fleming, G. A., O'Connor, R., and Sweeney, A., *Nature* **168**, 881 (1951).
274. Walter, E. D., and Jensen, L. S., *J. Nutrition* **80**, 327 (1963).
275. Walter, E. D., and Jensen, L. S., *Poultry Sci.* **43**, 919 (1964).
276. Wastell, M. E., and Ewan, R. C., *J. Animal Sci.* **29**, 149 (1969).
277. Wastell, M. E., Ewan, R. C., Bicknell, E. J., Hays, V. W., and Speer, V. C., *J. Animal Sci.* **26**, 1474 (1967).
278. Watkinson, J. H., and Davies, E. B., *New Zealand J. Agr. Research* **10**, 116 (1967).
279. Webb, J. S., and Atkinson, W. J., *Nature* **208**, 1056 (1965).
280. Weichselbaum, T. E., *Quart. J. Exp. Physiol.* **25**, 363 (1935).
281. Westfall, B. B., and Smith, M. I., *J. Pharmacol. Exp. Therap.* **72**, 245 (1941).
282. Whanger, P. D., Pedersen, N. D., Elliot, P. H., Weswig, P. H. and Muth, O. H., *J. Nutrition* **102**, 435 (1972).
283. Williams, K. T., and Byers, H. G., *Ind. Eng. Chem. Anal. Ed.* **7**, 431 (1935).
284. Williams, K. T., and Byers, H. G., *Ind. Eng. Chem.* **28**, 912 (1936).
285. Williams, K. T., Lakin, H. W., and Byers, H. G., *U.S. Dep. Agr. Tech. Bull.* No. 702 (1940).
286. Williams, K. T., Lakin, H. W., and Byers, H. G., *U.S. Dep. Agr. Tech. Bull.* No. 758 (1941).
287. Winter, K. A., Gupta, U. D., Nass, H. G., and Kunelius, H. T., *Can. J. Animal Sci.* **53**, 113 (1973).
288. Wolf, E., Kollonitsch, V., and Kline, C. H., *J. Agr. Food Chem.* **11**, 355 (1963).
289. Zalkin, H., Tappel, A. L., and Jordan, J. P., *Arch. Biochem. Biophys.* **91**, 117 (1960).

13
Selenium in Glass

GEORGE B. HARES

Senior Research Associate, Research and Development Laboratories, Corning Glass Works, Corning, New York

INTRODUCTION

The chemistry of selenium in oxide glasses is very complex and was difficult to fathom by early investigators. Even though the use of selenium to produce colors in glass was mentioned as early as 1865, as late as 1928, A. Silverman writes the following:[71]

Selenium produces a ruby color in zinc–potash glasses in the presence of cadmium sulphide and a reducing agent. An amber shade is obtained in lead glasses in the presence of an oxidizing agent. In lime glasses, a pink color is obtained in the presence of an oxidizing agent without the use of cadmium sulphide, and an orange shade when cadmium sulphide and a reducing agent are added. No mention has been made in the literature of the last three colors in the particular types of glasses so the writer submits this note hoping that chemists and ceramists may ascertain the cause of the color produced by selenium under various conditions.

The red color of the zinc–potash glasses is already attributed to the formation of polyselenides. This would imply the action of a reducing agent. The amber color in lead glasses could be attributed to lead selenide were it not for the presence of an oxidizing agent.   The pink color in lime glasses is certainly formed under oxidizing conditions, for niter is employed, and the glass does not have to be reheated to produce the color.   The orange color in lime glasses forms in the presence of cadmium sulphide and a reducing agent, and changes somewhat on reheating.   It is probably the result of reducing action.

As previously stated, the element selenium produces a number of colors under various conditions.   Can the ceramists and chemists offer any explanation for the behavior of this versatile color-producing element?

The earliest known mention of the use of selenium to impart a color to glass is that of the French chemist, Pelouze,[61] who found that a small proportion of selenium, when added to a glass batch, imparted a pink or rose color to the glass.   Little commercial interest appeared to develop from his observation.   This may be due partly to the pale, delicate coloration of these glasses at a time when new, intense colors were being sought, or, it more probably was due to the fact that the tint was difficult to reproduce, the glass sometimes being nearly colorless, at other times having an off-color brownish tint.

Experimentation and commercial exploitation of selenium as a glass colorant increased markedly during the last decade of the nineteenth century following a number of patents issued to glass chemists in Bohemia, Hungary, and Germany.   Welz[79] was granted a patent for the production of a rose-colored glass using selenium alone, as well as orange-red glasses using a mixture of cadmium sulfide and selenium.   These latter glasses were brilliant in color and subsequently found wide use in signal colors as well as in artware.

Spitzer[73] pointed out the advantage of using selenites or selenates in preparing the rose-colored selenium glasses, and thus avoiding the off-colors caused by iron and copper impurities in the commerically available selenium metal of that period.   As a result of these two patents, comments on the use of selenium as a glass colorant begin to appear in recipe books for the practical glassmaker.[5,7]

A third patent of interest was granted to Hirsch[35] for a means of decolorizing glass with selenium.   The use of selenium as a decolorizer for container glass steadily increased from that time on and today is the largest use of selenium in the glass industry.

Each of these three uses of selenium in glass (selenium pinks, cadmium sulfoselenides, and decolorizers) has its own chemistry and will be discussed separately.

SELENIUM PINK GLASSES

The results of the earliest scientific investigations into the nature of the colorant in selenium pink glasses were published in 1912–1915.   The most important work was that of Fenaroli[21] who postulated that the pink color was due to metallic selenium in the colloidal state. Zsigmondy[86] had already shown that the color of the gold ruby glasses was due to colloidal gold, while it had been shown much earlier that copper ruby glasses were colored by very finely subdivided metallic copper particles.[19] Thus, it is not surprising that Fenaroli assumed that the pink color of selenium glasses was due to colloidal selenium metal particles.

Witt and Frankel[84] regarded the colloidal nature of the colorant as less important than the proportion of selenium that acts as a chromophore in the glass. By a combination of wet chemical analysis to obtain the total selenium content and a colorimetric comparison technique to determine the amount of selenium acting as a chromophore, they found that only 8% of the total selenium content of the glass was responsible for the color. The remainder of the selenium was present in the glass in a colorless form.

Witt and Frankel also reported that the composition of the base glass affected the color. For example, potassium-containing glasses were more highly colored than the sodium analogue glasses. Kraze,[47] who also investigated the color of selenium-containing glasses as a function of the base glass composition, had earlier reported the same observation regarding the effect of a sodium-for-potassium substitution. He also found that decreasing the silica content decreased the color, as did substituting barium, zinc, or lead for calcium in a potash–lime silicate glass.

A definitive investigation into the chemistry of selenium in silicate glasses was reported by Höfler and Dietzel.[37] The primary evidence that led Fenaroli to postulate that the pink color was due to colloidal selenium was the appearance of a red-violet Tyndall cone in the glass when illuminated with an intense light source. Höfler and Dietzel investigated this Tyndall effect and proved that it was fluorescence and not light scattering by colloidal particles. The beam is not polarized, retains its red color when a blue filter is placed over the light source, and disappears when a red filter is used.*

With a series of carefully controlled laboratory experiments, Höfler[36] made one of the most significant contributions to understanding the chemical

---

* This fluorescence phenomenon can be demonstrated very easily by using a glass melted according to the following recipe:

Batch		Oxide Wt%	
Iron-free sand	296	SiO_2	74
Na_2CO_3	82	Na_2O	14
$NaNO_3$	22	CaO	12
$CaCO_3$	87	Se	0.25
Na_2SeO_3	2		

By using a microscope lamp as a source and a series of sharpcut filters,[10] the activating wavelength for this red fluorescence can be found to be about 530 nm.

nature of the color changes in selenium-containing glasses. By melting in different atmospheres, he found great differences in the color of glass melted from the same batch. These colors were reproducible in the same atmosphere, as to the tint and intensity of the color.

Glasses melted in oxygen were colorless in the upper third of the melt and pink in the lower portion; in air, the colorless layer was very shallow, with the pink layer more intense in color. In argon, the glasses were pink with no colorless layer, the intensity of the pink color being somewhat greater than in the melts made in air or oxygen. Three % carbon monoxide in nitrogen produced brownish salmon-colored glasses, 10 % carbon monoxide produced glasses still more brownish, while pure carbon monoxide, as well as hydrogen, produced brown glasses but with a colorless surface layer again.

Arranged in decreasing oxidation conditions the color effects are: (1) colorless I, (2) pink, (3) brown, (4) colorless IV. These are suggested to be due to the following oxidation states of selenium: (1) selenate (Se^{+6}) or selenite (Se^{+4}), (2) elemental selenium (Se^0), (3) polyselenide (Se_n^{-2}), (4) alkali selenide (Se^{-2}).

Since the oxidation state is very dependent on the furnace atmosphere, the temperature of melting, as well as the state of oxidation of selenium in the batch, it is no wonder that pure selenium colors were not popular with the glassmaker. There was a great tendency for the selenium color to "burn out" (become colorless due to over-oxidation) or become brown (because of reduction). Colorless selenium glasses due to over-reduction probably would not occur in practice.

Höfler confirmed observations of earlier investigators—that an increase in alkali content above 24% decreases the color of selenium-containing glasses as does a substitution of sodium for potassium. The reason for this change in color as a function of the base glass is explained by Weyl.[80] According to Weyl, the selenium pink glasses are true solutions of metallic selenium atoms in glass where they form a kind of "frozen in" metal vapor. The high polarizability of the large selenium atoms accounts for the fact that the color of these glasses is greatly influenced by the composition of the base glass. The outer electron orbitals of the selenium atoms are easily deformed by the electrical fields of positively charged alkali atoms. The bluish-pink of potash glasses changes to brown or even yellow colors if the large potassium ion is replaced by smaller ions having a stronger, perturbing effect, such as sodium or lithium. The same effect occurs on increasing the number of alkali ions.

SELENIUM AS A DECOLORIZER FOR GLASS

Selenium was first mentioned as a decolorizer for glass in a German patent issued to Rechter.[64] The British patent issued to Ellis[20] and the U.S.

patent issued to Hirsch[35] are foreign issues of the same patent, but for some unknown reason different names are on the patent.   No mechanism for the decolorization phenomenon was discussed in this patent.

One early investigator explained the decolorization as an additive optical phenomenon, the reddish color of selenium compensating the green color due to iron.[63]   The same view was held by Miskowsky[50] and Frommel.[26] The latter also noted that selenium-decolorized glasses assume a brownish tinge on prolonged exposure to sunlight.

The use of decolorized glass for food containers became more and more general in the period between 1910 and 1925.   One writer of the time states:[2]

The natural tint of the contents shows to advantage in colorless glass; and since, also, people generally like to see what they are buying (especially with food), preference is frequently given to the article put up in colourless glass on account of its more attractive appearance.   For this reason, such glass is gradually replacing pale green.

This is the same period in the history of glass manufacturing when the changeover was nearly complete from the hand manufacture of bottles from glass melted in a pot furnace, to automatic production of bottles by machine from glass melted from a continuous tank.   As a consequence, a number of papers were written describing the practical aspects of selenium as a decolorizer in a continuous tank.   The use of selenium was said to possess three distinct advantages over the use of manganese (the traditional glass decolorizer).   It was cheaper, it was easier to maintain regularity of color from day to day, and it permitted "the use of high temperatures over Sunday, which is essential as an aid in 'plaining up' the glass when the tank is being pulled hard during the week."[3]

Instructions are given for the procedure to be used in converting a continuous tank from manganese decolorizer to selenium decolorizer.[3]   The form in which selenium was introduced was investigated to reduce the volatility loss of selenium during melting.

Elemental selenium volatilizes readily below its boiling point (690°C). Above 220°C, it forms a dense liquid which could segregate at the bottom of the batch pile before it has a chance to react with other batch components. As a consequence, other compounds of selenium were developed for use by the glass industry.   Sodium selenite, theoretically containing 45.7% Se, is hygroscopic and the selenium content of the material falls rapidly after being exposed to air.   Stable, nonhydroscopic selenite compounds of barium and zinc were developed.   However, it was found both in practice[2] and in the laboratory[46] that the selenium loss during melting was independent of the nature of the selenium-containing raw materials.   To minimize the loss of selenium, quick melting was essential so that the outside of the batch pile rapidly glazed over.   Adams also pointed out that the presence of arsenic appeared to be necessary for the best results when using selenium, although he did not understand its function.

An experimental study of selenium decolorization was made by Cousen and Turner,[11] in an attempt to better understand the mechanism. They found that the maximum iron content of the glass which could be decolorized was 0.1 wt%. Above this value, green, yellow, or amber tints occurred. The presence of salt-cake (sodium sulfate) required a larger amount of selenium for decolorization. When selenium-containing glasses were reheated, color changes took place only in those glasses which had an excess of selenium, i.e., in which there was a distinct selenium tint.

The amount of selenium actually needed for decolorizing was found to be minute; 1/2 oz of selenium per 1000 lb of sand. Arsenic trioxide prevented the yellowish-brown color due to selenium, even when an excessive amount of selenium was added to the batch. Arsenic trioxide was the only substance found which made the decolorizing of selenium-containing glasses easy and would give stability to the melting process.[12]

The use of antimony trioxide resulted in a bluish-green glass,† while phosphates had no effect on the color. Arsenic trioxide, therefore, was unique among the oxides of Group V elements for stabilizing the decolorizing process.[13] A solarization effect, in which yellowish selenium glasses became paler in color, while colorless or pink glasses developed a yellow tint, was also noted.[14]

A number of theoretical studies of the mechanism of selenium decolorization was made between 1935 and 1942, a period in which the understanding of the structure of glass expanded rapidly. The random network theory of glass, proposed by Zachariasen[85] and confirmed by x-ray diffraction studies of Warren,[77] helped to explain many of the observations that had puzzled glass technologists for years.

Gooding and Murgatroyd[29] discovered that there was an interaction between selenium and iron rather than a purely physical decolorizing process. By making a quantitative determination of the color of glasses containing iron and/or selenium using a Lovibond tintometer, they discovered that the color of a glass containing both iron and selenium was not the same as that of a combination of two glasses, one containing iron only and the other containing selenium only. For glasses containing iron oxide without selenium present, the higher the temperature of melting, the bluer was the resulting glass. They attributed this increase in the blue color to the following reaction:

$$6Fe_2O_3 \; \underset{\longleftarrow}{\overset{heat}{\longrightarrow}} \; 4Fe_3O_4 + O_2$$

postulating the existence of the compound, ferrous ferrate, in the glass.

† The bluish-green color would be due to the reduction of trivalent iron to divalent iron during cooling.

It is now well-known, that for elements which have more than one oxidation state in glass, the proportion of the element in the lower oxidation state increases with an increase in the melting temperature. The observation of Gooding and Murgatroyd is an example of this principle. However, glasses do not contain compounds in the usual sense. Trivalent iron can replace silicon in the silica network, while divalent iron plays the role of a network modifier. In this manner a chromophore group $[Fe^{+3}-O^--Fe^{+2}]$ is formed. The higher the proportion of Fe^{+2}, the bluer the glass will be. Conversely, a low concentration of Fe^{+2} produces a yellow green glass of higher light transmission.

Gooding and Murgatroyd proposed that the decolorizing effect of selenium was due to the formation of an iron selenide as follows:

$$2Se + 2Fe_3O_4 \longrightarrow 2FeSe + 2Fe_2O_3 + O_2$$
$$\text{red} \quad \text{blue} \qquad \text{yellow-brown} \quad \text{brown}$$

The same view was held by Höfler and Dietzel[38] who also found that the light absorption of a glass is different than that resulting from a combination of glasses containing iron and selenium separately. In a thorough study of the formation of ferrous selenide in glass, Höfler and Dietzel found that the amount of ferrous selenide formed was a function of heat treatment and melting conditions. For example, if a soda–lime–silica glass containing 0.05% Fe_2O_3 and a quantity of selenium in excess of that necessary for decolorizing is melted under strong oxidizing conditions, quickly quenched rods drawn from this melt are yellow-green. Heat treating these pieces at a temperature slightly above 550°C produces a reddish color which becomes more intense as the time of the heat treatment is extended.

Conversely, the color of a glass melted under neutral conditions was found to change only slightly, if at all, upon reheating. Dietzel[16] gives the following explanation of this phenomenon. At melting temperatures the glass contains colorless selenide ions and ferrous ions. On cooling a reaction between these ions takes place:

$$Fe^{+2} + Se^{-2} \rightleftharpoons FeSe$$

This equilibrium, which at the lower temperatures and for a very great excess of iron lies to the right, is very quickly established. At temperatures around 1100°C, the formation of FeSe is nearly instantaneous. Under neutral melting conditions, a large proportion of the selenium in the glass is present as the selenide, allowing appreciable quantity of FeSe to form spontaneously on cooling. In a more strongly oxidizing melt, the selenium is present in the glass as selenite, some elemental selenium, and only a small quantity of selenide. Therefore, the amount of FeSe formed on rapid cooling is correspondingly small. On heat treatment or long annealing, the selenite ions decompose to form selenide ion:

$$2SeO_3^{-2} \longrightarrow 2Se^{-2} + 3O_2$$

This in turn reacts with ferrous ion to form FeSe.

Tillyer[76] took advantage of the fact that large amounts of iron shifted the equilbrium:

$$Fe^{+2} + Se^{-2} \rightleftharpoons FeSe$$

to the right in developing a series of warm rose-smoke tinted glass compositions suitable for use in eye-protective spectacles.

Still another mechanism for the selenium decolorization of iron-containing glasses was proposed by Day and Silverman.[15] Measurements of the near infrared absorption spectra of glasses containing iron and selenium separately, as well as in combination, showed that some ferrous ions are oxidized to ferric when selenium is present in the glass. Since the absorption coefficient of trivalent iron is smaller than that of the divalent iron, the glass would be partially decolorized by simple oxidation. Day and Silverman proposed that the remaining decolorization was the physical compensation of the green iron color with the pink color of elemental selenium.

In spite of the viewpoint of Day and Silverman that the oxidation-reduction relationship:

$$4Fe^{+2} + SeO_3^{-2} \rightleftharpoons 4Fe^{+3} + Se^0 + 3O^{-2}$$

is the only mechanism of decoloration, the actual mechanism is probably a complex one involving not only oxidation of iron by selenium, a physical compensation of the yellow-green iron color by the pink color due to elemental selenium, but also formation of iron selenides as proposed by Höfler and Dietzel.[38] Since the resulting glass has a slight yellowish tint, a small amount of cobalt oxide is added for its blue coloration, resulting in a glass with a neutral tint.

A shortage of the selenium supply for the glass and electronics industries occurred in the early 1950s. With its accompanying price spiral, renewed efforts were made on a practical scale to reduce the amount of selenium used by the glass industry.[48,40,32] Much of this effort was centered on determining the optimum melting conditions and the effect of various batch materials. The level of arsenic was found to be important. Too much arsenic required an excessive amount of selenium, while too little resulted in an unstable decolorizing situation.[32]

A method to minimize the volatilization problems of selenium was patented by Silverman.[72] The invention comprises adding a selenium-containing glass frit to the molten glass in the forehearth of a glass-melting tank just prior to delivery to the glass-forming equipment. Since temperatures in a forehearth are considerably lower than in the melting end of a glass tank, resulting in a lower residence time for the selenium at high temperatures, this invention could result in considerable savings in the amount of selenium used.

A different approach to lowering the amount of selenium needed for de-colorizing is to use cerium oxide instead of arsenic trioxide in the batch.[31] It is claimed that a 30% reduction in the amount of selenium needed to de-colorize a container glass was realized using a cerium oxide concentrate.[53] Care must be taken when using cerium oxide. The glass cannot contain an oxide of arsenic, since a combination of arsenic and cerium produces a glass that is so sensitive to solarization that it will develop a brown color on exposure to light. Also, if an excess of cerium oxide is used, selenium will be oxidized to the colorless selenite state making it of no value as a decolorizer. Indeed, it is too early to tell whether the use of cerium oxide with selenium is a practical means of decolorizing glass. Only after several years of commercial experience at a number of different glass plants will this be known.

SELENIUM RUBY GLASSES

The so-called selenium ruby glasses are the most brilliant red glasses known to the glassmaker. Originally mentioned by Welz,[79] the first appearance of selenium rubies in commercial ware, such as vases and religious objects, occurred around 1895.[67] The coloration is caused by the precipitation of particles composed of a solid solution of cadmium sulfide and cadmium selenide; in other words, cadmium sulfoselenides. The glasses range in hue from the pure yellow of a cadmium sulfide-containing glass, through orange to a deep red as the ratio of selenium to sulfur increases.

The nature of these cadmium sulfoselenide particles was determined by Rooksby[65] using x-ray crystallographic methods, and was confirmed by Bigelow and Silverman[6] using x-ray techniques, both on selenium ruby glass and synthesized solid solutions of cadmium sulfide and cadmium selenide. Rooksby derived the compositions of the solid solutions from the shifts in spacing of the x-ray diffraction lines and compared these calculated compositions with the visual color of the corresponding glass, as follows:

Visual Colour of Glass	Composition of Precipitated Particles	
	CdS	CdSe
Yellow	100	0
Orange	75	25
Red	40	60
Deep Ruby	10	90

Rooksby points out that the values in the last two columns are only approximate. The main point is that the cadmium sulfide percentage decreases, with corresponding increase in the cadmium selenide content, as the color of the glasses changes from yellow through orange and red to deep ruby.

Another important aspect of Rooksby's work was the observation by microscopic examination that there could be a large increase in the size of the particles without a change in the visual color of the glass. Thus, the coloration of selenium rubies is mainly due to the color of the cadmium sulfoselenide particles themselves rather than to their size and nature, as in the case of gold and copper ruby glasses.

The use of cadmium sulfoselenides as pigments for ceramic enamels and paints is not as old as is the use in ruby glasses. The earliest cadmium sulfoselenide pigments were apparently manufactured in Germany just prior to the First World War.[74] At present, the pure cadmium sulfoselenide colors are used mostly for artists' colors and other special color applications because of their high cost. Extended pigments containing barium sulfate are known as cadmium red lithopones and are used in interior latex paints, enamels and lacquers.[28] These pigments are not generally suitable for exterior paints, since they may be converted to carbonates under the influence of moisture and oxygen, resulting in a bleaching of the color.[55]

Cadmium sulfide, cadmium selenide and their solid solutions are semiconductors. As such, their optical absorption is the result of the excitation of valence band electrons to the conduction band. In an ideal semiconductor at absolute zero the valence band would be completely full. Electrons could not be excited to a higher energy state within the valence band. The only absorption is that of light quanta of high enough energy for the electrons to be excited across the forbidden band into the empty conduction band. This results in an absorption spectrum having a region of intense absorption at short wavelengths (high energy), with a more or less steep absorption edge at a characteristic wavelength. For wavelengths beyond this edge, the semiconductor is practically transparent. This absorption edge often, though not necessarily, occurs at $hv = E_g$, where E_g is the width of the forbidden energy gap.

The values of the band gap for cadmium chalcogenides at:[43]

Crystal	E_g(eV)	E_g(nm)
CdS	2.42	512
CdSe	1.74	713
CdTe	1.45	855

Cadmium sulfoselenide crystals maintain the optical absorption characteristics of semiconductors even when precipitated in glasses. A sharp change in transmission from almost total absorption to nearly total transmission is characteristic of the cadmium sulfoselenide colored glasses, resulting in brilliant red, orange, and yellow colors.   As a result of this sharp "cut" in transmission, these glasses were chosen for the red-colored lenses of American Railway signals in 1908.[62]

Around 1900, yellow was introduced as the color of a caution signal indication, supplanting green, which was made the "clear" indication.[9]   It was soon discovered that confusion of yellow and red signals could be avoided only by developing a red glass which would never look orange.   Certain cadmium sulfoselenide glasses were found to completely absorb wavelengths shorter than that of the D line of sodium[8] and were accepted as standards. A cadmium sulfoselenide glass was not used for yellow or amber lenses for railway signals since the intensity of the transmitted light was much greater than for the red.   Balanced light intensities for red, yellow, and green signals were deemed most desirable.   Not until the introduction of motor vehicle traffic signal lights were cadmium sulfoselenide yellow glasses used.

The absorption edge of cadmium chalcogenide semiconductors shifts to longer wavelengths and becomes less steep with increasing temperature.[18] This same effect occurs in the cadmium sulfoselenide yellow and red glasses. This shift in color was first noted by Silverman[70] and examined quantitatively by Gibson.[27]   Other discussions of the effect of temperature on the shift in color include Angstrom and Drummond[4] and Holland and Turner.[39]   The latter include the cadmium sulfoselenide glasses in a comprehensive survey of the effect of temperature on the transmission of colored, optical and miscellaneous commercial glasses.   A knowledge of the effect of temperature on the visual transmission of the cadmium sulfoselenide-containing glasses (as well as other glasses) is of importance whenever a glass is used for transmission of light at elevated temperatures.   Typical uses where this knowledge is necessary include optical pyrometers, signal lenses, electric light bulbs, and other forms of cover for illuminants.

While the color of the cadmium sulfoselenide rubies depends only on the ratio of cadmium, sulfur, and selenium in the mixed crystals, the manufacture of these colored glasses requires a high degree of skill and experience on the part of the glassmaker.   Not only is the compounding of the batch of great importance, but also the melting conditions, and the subsequent heat treatment.   The majority of these glasses require a subsequent heat treatment in order to develop their full color.   This color development is known as "striking" and is defined by the ASTM[1] as the development of color or opacity during cooling or reheating.

In general the higher the subsequent heat treatment the more red is the glass.   The sharpness of the transmission "cut" also increases with increase in the reheat temperature.   The manufacture of multicolored (red and yellow) glass artware vessels makes use of this phenomenon.   A hand-crafted article is attached to the end of a long, hand-held iron rod and placed in a reheating furnace or "glory hole." The part of the vessel opposite to the iron receives more heat than that to which the iron is attached. The color of the finished piece may shade from a deep red to a yellow where it was attached to the iron.

The complexities of the striking of the cadmium sulfoselenide-containing glasses can be explained in the following manner.[81,82] At the melting temperature there is a random distribution of the anions O^{-2}, S^{-2}, and Se^{-2}. Rapid chilling of the glass maintains this random distribution. When the glass is reheated slowly to a temperature where the anions become mobile, a regrouping of O^{-2} and S^{-2} ions leads to the formation of CdS nuclei. Since CdS is only slightly soluble in silicate melts, it is precipitated in the form of subcolloidal particles. If the reheating temperature is in the range where the large Se^{-2} ion has too low a diffusion rate to participate in nucleation and crystal growth, only CdS crystals will grow.

At higher reheating temperatures, CdS tends to dissociate. The cadmium cation can react with both the sulfur and selenium anions to form crystals richer in selenium. The higher the temperature, the greater the proportion of selenium to sulfur in the crystals and the more constant the sulfur-selenium ratio due to a closer approach to equilibrium.

Time cannot be substituted for temperature to increase the selenium content of the particles, as the crystals will grow to the size of the wavelength of light, resulting in a hazy or waxy appearance. In some cases a waxy appearance is desirable as in some yellow tubing used in advertising signs.

Since the color of these glasses is due to a compound whose solubility limit has been exceeded, it is impossible to make pale colors (such as pink) by reducing the concentration of cadmium, sulfur, and selenium. Solution colors in glass, such as those of Co^{+2}, Cu^{+2}, and Ni^{+2} can be diluted to very faint tints.

The melting of cadmium sulfoselenide ruby glasses requires close attention in order to reproducibly obtain a constant proportion of cadmium, sulfur, and selenium in the final glass. Selenium dioxide has a high vapor pressure and sublimes as low as 317°C. Sullivan and Austin[75] found that if no reducing agent was added to the melt, and if the gaseous furnace atmosphere was oxidizing, virtually all of the selenium was lost. On the other hand, the presence of strong reducing agents, such as carbon, produces elemental cadmium which distills from the melt.

A proper balance of mildly reducing conditions must be maintained during the melting process. Most of the selenium is lost during the early stages of the batch melting reactions. Once the selenium is incorporated into the glass matrix, it is retained permanently, provided the temperature of the melt is not raised above 1400°C.[42]

In addition to melting the batch under carefully controlled conditions, the glassmaker often adds one or more characteristic minor ingredients in order to produce a successful ruby glass. These are zinc oxide, cryolite or other fluorine compounds, bone ash, and minute amounts of copper or nickel oxide. Shively and Weyl[69] established the functions of these minor but important constituents. Because of the unique character of this work, their paper will be discussed in some detail.

It has been known for some time that zinc oxide was a desirable constituent of the cadmium sulfoselenide ruby glasses. Weckerle[78] found that with increasing zinc oxide content, the selenium content increased rapidly, giving more brilliant red colors in the glass. However, a high concentration of zinc oxide (18 wt%), while helping to retain a high selenium content in the glass, produced a colorless glass. In this latter case, the equilibrium between the sulfides or silicates of zinc and cadmium sulfide (or selenide) to remain within its solubility limits. Shively and Weyl,[69] by substituting calcium oxide for zinc oxide in a commercial ruby batch, confirmed the selenium retention of zinc-containing batches.

They point out that zinc oxide as a glass constituent resembles silica in many respects. An occasional SiO_4 is probably replaced by a ZnO_4 group, decreasing the viscosity of the glass. The increased mobility of the structural units enhances the striking of the ruby. In addition, the formation of ZnSe stabilizes the selenium and prevents its oxidation. Selenium ions combined with zinc ions are more strongly held in the glass structure than those which replace an O^{-2} of a SiO_4^{-4} group.

Fluorine compounds are known to lower the viscosity of glass by weakening the network structure. Shively and Weyl prepared a series of fluorine-containing ruby glasses and noted that the intensity of the color increased with increasing fluorine content. Analysis showed that fluorine had no effect on the selenium content. They point out that the tendency to form cadmium sulfoselenide increases with decreasing temperature. The presence of fluorine, by lowering the temperature range where molecular rearrangements are possible, permits a larger amount of colorant to be precipitated from the glass.

Bone ash is chiefly a calcium phosphate with small amounts of fluoride. These constituents produce the mineral apatite, which is extremely insoluble in glass even at high temperatures. Thus, the addition of bone ash, particularly in combination with fluorides, produces crystal nuclei of apatite providing sites for the precipitation of cadmium sulfoselenide.

The sulfides and selenides of heavy metals such as copper and nickel form at higher temperatures than do the sulfides and selenides of zinc and cadmium. In addition, nickel and copper sulfides and selenides are extremely insoluble, again producing nuclei upon which cadmium sulfoselenide can precipitate. In certain compositions, these nuclei of copper or nickel sulfides and selenides are sufficient to allow the striking of the cadmium sulfoselenide ruby without requiring an additional heat treatment.

Black selenide-containing glasses were produced by Pavlish and Austin[58] using either copper or cobalt in conjunction with selenium as the colorants and with silicon added to the batch as a reducing agent.

Selenium dioxide has been investigated as an ingredient of the overall glass composition, as opposed to the usual small amounts used for coloring purposes.[54]   The major portion of the selenium dioxide in the batch is vaporized during the melting process.   Using low melting glass compositions, Navias and Gallup were able to retain about 25 wt% SeO_2 in a soda–lime–boric oxide glass, and about 20 wt% SeO_2 in a soda–lead oxide–boric oxide glass.   These glasses were very unstable, disintegrating readily when immersed in water.   The most stable glass they obtained was found to have the following analysis.   The theoretical batch composition is given for comparison:

Class No. 42

	Batch Composition	Analyzed Composition
Na_2O	22.4%	28.3%
CaO	8.3	16.4
B_2O_3	25.2	35.1
SiO_2	3.9	8.7
SeO_2	40.2	10.8
	100.0	99.3

This glass was fairly insoluble, remaining in water for some time without forming a surface film.   However, its resistance to moisture attack was very poor in comparison with commercial soda–lime glasses.

SELENIUM IN CHALCOGENIDE GLASSES

Glass is "an inorganic substance in a condition which is continuous with, and analogous to, the liquid state of that substance, but which, as the result of a reversible change in viscosity during cooling, has attained so high a degree of viscosity as to be for all practical purposes rigid."[51]   A slightly different definition states that glass is "an inorganic product of fusion which has cooled to a rigid condition without crystallizing."[1]   Both of these definitions imply an absence of long-range order in the atomic structure of glasses.

Most commercial glasses are complex mixtures of various metal oxides and may be viewed as derivatives of vitreous silica.   However, glasses containing only one kind of atom can be obtained. The ability to form monatomic glasses is only known for the elements of Group VI of the periodic table, oxygen, sulfur, selenium, and tellurium.[83]

These monatomic glasses seem to keep their glass-forming capability when mixed or chemically combined with each other. The system, S–Se, has been known to form glasses for many years. It was used[49] in preparing immersion media for the study of minerals of high refractive index. These mixtures may be readily melted in a borosilicate glass beaker or test tube and quenched to form glasses. The selenium-rich mixtures will remain as glasses for years, while the sulfur-rich materials crystallize almost immediately.

The ternary system, S–Se–Te, was investigated by Jerger[41] for infrared transmitting properties. Mixtures ranging from about 2 to 20% tellurium, 7.5 to 40% sulfur, and 52.5 to 85% selenium were found to be the most favorable for forming stable glasses. The glasses are prepared by heating the S–Se–Te mixture at 250°C to melt the sulfur and selenium. The temperature is then raised to 500°C to complete the reaction between the molten sulfur–selenium mixture and the solid tellurium.

Other elements which do not form glasses by themselves may be incorporated into a complex vitreous network with the elements of Group VI. The simplest of these glasses contain one element from Group VI and one element from another group. This second element must be from Group III, IV, or V.

Some of these diatomic glasses are well-known. Vitreous silica (SiO_2) is used widely in the optical and electronic fields.[17] Vitreous B_2O_3 and P_2O_5 are largely laboratory curiosities because of their hygroscopicity. Arsenic trisulfide glass was first reported in 1852[23] and its infrared transmitting characteristics were noted as early as 1870.[68] Frerichs[24] is credited with the rediscovery of arsenic trisulfide glass and its infrared transmitting properties. He writes, "Arsenic trisulfide forms a perfectly clear, stable red glass which has remarkable optical properties hitherto unknown." This work represents a significant turning point in the development of infrared optical materials. A subsequent paper[25] mentions the incorporation of selenium into arsenic trisulfide melts.

Fraser[22] investigated the system As_2S_3–As_2Se_3–As_2Te_3 and found that while As_2Te_3 melts crystallized easily, small additions of As_2S_3 and/or As_2Se_3 to the As_2Te_3 melts resulted in stable glasses. While all of these As_2Se_3-modified glasses had excellent transmission in the infrared, the softening points‡ were 200°C or less. Since one need for infrared transmit-

‡ Softening point: For a glass of density near 2.5, the softening point temperature corresponds to a viscosity of $10^{7.6}$ poises.[1] At this viscosity a glass will rapidly deform under its own weight.

ting glasses was in airborne optical systems at temperatures as high as 500°C, these As_2S_3 modified glasses were unsatisfactory.   In order to raise the softening points of these glasses, it was necessary to introduce elements of Group IVA.   An intensive investigation of ternary glasses composed of IVA–VA–VIA elements was undertaken by scientists at Texas Instruments Incorporated. Many useful glass compositions were found, but none had the combination of excellent transmission in the 8–14 μ region and a softening point in the 500°C range.[34]

Extreme precautions must be taken when preparing large quantities of glasses containing the elements S, Se, As, and P to prevent dangerous explosions.   A two-step method is generally used[33]   In the first step, the glass is compounded in large quartz tubes at fairly low temperatures (700–800°C) in 800–1500 g batches.   A small amount of the tube may be left in a cool zone to minimize explosion dangers.   After a melt is obtained, the tube may be slowly cooled, maintaining the glass in a single ingot, which minimizes the oxygen-glass surface reaction when the tube is opened.   Fabrication into the final desired shape is accomplished in the second step.   The required number of ingots are placed in a container inside an oxygen-free furnace and reheated.   The melt is agitated to give homogeneity and either cooled in the container or poured into a preheated mold.   The glass is quickly cooled to below the annealing point and allowed to cool slowly from this point.

A comparison of the infrared transmissions of purified sulfide, selenide, and telluride glasses shows that the useful ranges of infrared transmission of sulfide glasses is 1–10 μ, of selenide glasses 1–20 μ and of telluride glasses 1–30 μ.[33]   Unless great care is taken to eliminate impurities, especially oxides, the infrared transmission of these glasses is substantially diminished. One technique to eliminate the oxide impurities is to redistill the glass in an atmosphere of hydrogen.[66]

The physical properties of the chalcogenide glasses show that they are weak solids when compared to the common oxide glasses:

	Chalcogenide Glasses	Common Oxide Glasses
Knoop Hardness	100–200	400–600
Density (gm/cc)	4.0–5.0	2.5–4.0
Young's Mod. (psi $\times 10^6$)	1.5–4	8–13
Poisson's Ratio	~0.25	0.20–0.25
Soft. Pt. (°C) (Visc $10^{7.6}$ poise)	100–500	450–950
Thermal Expansion (°C $\times 10^7$)	100–300	30–100

The chalcogenide glasses are semiconductors, the conduction process being electronic in nature rather than ionic, as is the case of oxide glasses.[44] Whether the electron transfer process occurs in the conduction or valence bands, or whether it is a hopping mechanism between localized states is unresolved.[57] The electrical conductivities of these glasses range from 10^{-12} to 10^{-3} ohm^{-1} cm^{-1}. The mobility of the charge carriers is very low.[45] Because of this they have found limited use in electronic applications.

The nonohmic voltage-current characteristics of these semiconducting glasses appears to have potential use in electronic switching and memory device applications[56,30] Pearson[60] describes the behavior of one device as follows:

The behavior on initially increasing the voltage is ohmic up to somewhat above 20 V, but above this level the current begins to increase at a faster rate than would be expected for ohmic behavior. At about 23 V an abrupt change in the conduction characteristics of the device takes place, the voltage dropping to somewhat less than 2 V with an increase in current to 35 mA. .... On reducing the voltage to zero, the device switched abruptly back to the high-resistance condition. This whole cycle could be traversed apparently indefinitely.

Two additional states in this device were reported by Pearson.

The first was a so-called memory state of high conductivity apparently coincidental with the ON state of the switching device, but stable at zero bias. This memory state is realized by passing current through the device in the ON state so that a critical current is exceeded. .... The device can be made to switch back from the memory state to the high-resistance OFF state by passing through it a pulse of current greater than a particular critical value for composition and geometry and having a sharp trailing edge. The second additional state which can be realized in these devices consists of a negative resistance condition. This can be obtained by operating the device using a constant current power supply.

The nature of the switching mechanism of these devices is still speculative. It may be due to phase changes dependent on Joule heating or it may depend upon electronic effects. To date neither mechanism has been proven conclusively.

The overall field of amorphous semiconductors, including chalcogenide glasses, has intrigued many investigators, resulting in a voluminous literature on the subject that cannot be covered in this review. Kolomiets[45] and Pearson[59] have written excellent general review articles. Two symposia on amorphous semiconductors are reprinted in the *Journal of Non-Crystalline Solids*, Volumes 2 and 4, 1970. All of these contain numerous literature references.

REFERENCES

1. ASTM Committee C-14, *ASTM Standards on Glass and Glass Products*, 5th ed., The American Society for Testing and Materials, Philadelphia, 1962, p. 84.
2. Adams, F. W., *J. Soc. Glass Technol.* **6**, 205, (1922).
3. Adams, I. E., *Glass Ind.* **3**, 219 (1922).
4. Angstrom, A. K., and Drummond, A. J., *J. Opt. Soc. Amer.* **49**, 1096 (1959).
5. Appert, L., and Henrivaux, J., *Verre at Verrerie*, Gauthier-Villars, Paris, 1894, p. 398.
6. Bigelow, M. H., and Silverman A., *J. Amer. Ceram. Soc.* **16**, 214 (1933).
7. Biser, B. F., *Elements of Glass and Glass Making*, Glass and Pottery Publishing Company, Pittsburgh, 1899, p. 119–120.
8. Churchill W., *The Roundel Problem*, A paper presented at the Ninth Annual Meeting of the Railway Signal Association, Niagara Falls, N.Y., October 10–12, 1905.
9. Churchill, W., *Proceedings of the New York Railroad Club* **24** (4), 1 (1914).
10. Corning, *Color Filter Glasses* Bulletin CFG, Corning Glass Works, 1970, p. 24.
11. Cousen, A., and Turner, W. E. S., *J. Soc. Glass Technol.* **6**, 168 (1922).
12. Cousen, A., and Turner, W. E. S., *J. Soc. Glass Technol.* **7**, 309 (1923).
13. Cousen, A., and Turner, W. E. S., *J. Soc. Glass Technol.* **9**, 119 (1925).
14. Cousen, A., and Turner, W. E. S., *J. Soc. Glass Technol.* **9**, 111 (1925).
15. Day, F., Jr., and Silverman, A., *J Amer. Ceram. Soc.* **25**, 371 (1942).
16. Dietzel, A., *J. Soc. Glass Technol.* **21**, 87 (1937).
17. Dumbaugh, W. H., and Schultz, P. C., Silica (Vitreous) *Kirk-Othmer Encyclopedia of Chemical Technology*, 2nd ed., Vol. 18, Interscience Publishers, New York, 1969, p. 73.
18. Dutton, D., *Phys. Rev.* **112**, 785 (1958).
19. Ebell, P., *Dinglers Polytech. J.* **213**, 53 (1874).
20. Ellis, British Patent 17,931 (September 25, 1895).
21. Fenaroli, P., *Chem. Zt.* **36**, 1149 (1912); *J. Soc. Glass Technol.* **4**, Abs. 11 (1920).
22. Fraser, W. A., (to Servo Corportation of America) U.S. Patent 2,883,291 (April 21, 1959).
23. Fraser, W. A., and Jerger, J. Jr., *J. Opt. Soc. Amer.* **43**, 332 (1953).
24. Frerichs, R. *Phys. Rev.* **78**, 643 (1950).
25. Frerichs, R. *J. Opt. Soc. Amer.* **43**, 1153 (1953).
26. Frommel, W. *Keram. Rundsch.* **25**, 95 (1917).
27. Gibson, K. S. *Phys. Rev.* **7**, 194 (1916).
28. Gloger, W. A., " Pigments (Inorganic)," *Kirk-Othmer Encyclopedia of Chemical Technology*, *2nd ed.*, Vol. 15, Interscience Publishers, New York, 1968, p. 522.
29. Gooding, E. J., and Murgatroyd, *J. Soc. Glass Technol.* **19**, 43 (1935).
30. Henisch, H. K., *Sci. Amer.* **221** (5) 30 (1969).
31. Herring, A. P., Dean R. W., and Drobnick, J. L., *Glass Ind.* **51**, 316 (1970).
32. Hill, D. K., *J. Soc. Glass Technol.* **42**, 3 (1958).
33. Hilton, A. R., *J. Non-Cryst. Solids* **2**, 28 (1970).
34. Hilton, A. R., Jones, C. E., and Brau, M., *Phys. Chem. Glasses* **7**, 105 (1966).
35. Hirsch, M., U.S. Patent 576,312 (February 2, 1897).
36. Höfler, W., *Glastech Ber.* **12**, 117 (1934). Engl. transl. by Scholes, S. R. *Glass Ind.* **15**, 122, (1934).
37. Höfler, W., and Dietzel, A., *Glastech. Ber.* **12**, 297 (1934).

38. Höfler, W., and Dietzel, A., *Glastech. Ber.* **14**, 411 (1936).
39. Holland, A. J., and Turner, W. E. S., *J. Soc. Glass Technol.* **25**, 164 (1941).
40. Howes, H. W., *J. Soc. Glass Technol.* **36**, 181 (1952).
41. Jerger, J., Jr., (to Servo Corporation of America) U.S. Patent 2,883,295 (April 21, 1959).
42. Kirkpatrick, F. A., and Roberts, G. C., *J. Amer. Ceram. Soc.* **2**, 895 (1919).
43. Kittel, C., *Introduction to Solid State Physics*, 3rd ed., John Wiley and Sons, Inc., New York, 1966, p. 302.
44. Kolomiets, B. T., *The Structure of Glass* Vol. 2, Consultants Bureau, New York, 1960, p. 403.
45. Kolomiets, B. T., *Phys. Status Solidi* **7**, 713 (1964).
46. Krak, J. B., *Glass Ind.* **10**, 181 (1929).
47. Kraze, F., *Sprechsaal Z.* **45**, 214 (1912).
48. Manring, W. H., *Ceramic Ind.* **57**, 49 (1951).
49. Merwin, H. E., and Larsen, E. S., *Amer. J. Sci.* **34**, 42 (1912).
50. Miskowsky, J., *Keram. Rundsch.* **18**, 457 (1910).
51. Morey, G. W., *The Properties of Glass* 2nd ed., Reinhold Publishing Company, New York, 1954, p. 28.
52. Moss, T. S., *Optical Properties of Semi-Conductors* Butterworths Scientific Publications, London, 1959, p. 34.
53. Myers, D. D. *Glass Ind.* **51**, 354 (1970).
54. Navias, L. and Gallup, J. *J. Amer. Ceram. Soc.* **41**, 441 (1931).
55. Nylén, P., and Sunderland, E. "*Modern Surface Coatings*," Interscience Publishers, London, 1965, p. 488.
56. Ovshinsky, S. R., U.S. Patent 3,271,591 (September 6, 1966).
57. Owen, A. E., and Robertson, J. M., *J. Non-Cryst. Solids* **2**, 40 (1970).
58. Pavlish, A. E., and Austin, C. R., *J. Amer. Ceram. Soc.* **30**, 1 (1947).
59. Pearson, A. D., in *Modern Aspects of the Vitreous State*, Mackenzie, J. D., ed., Vol. 3, Butterworths, Washington, 1964, p. 29.
60. Pearson, A. D., *J. Non-Cryst. Solids* **2**, 1 (1970).
61. Pelouze, T. S., *C. R. Acad. Sci.* **61**, 615 (1865).
62. Railway Signal Association, "Specifications for Signal Roundels, Lenses, and Glass Slides," *Proceedings of the Railway Signal Association*, **5**, 1908.
63. Rauter, G., *Sprechsaal Z.* **37**, 598 (1904).
64. Rechter, German Patent 88,615 (January 13, 1895).
65. Rooksby, H. P., *J. Soc. Glass Technol.* **16**, 171 (1932).
66. Savage, J. A., and Nielson S., *Seventh International Congress on Glass*, Brussels paper No. 307, Gordon and Breach, New York, 1965.
67. Schofield, M. A., *Glass* **44**, 310 (1967).
68. Schultz-Sellack, C., *Ann. Phys.* (*Leipzig*) **139**, 182 (1870).
69. Shively, R. R. Jr., and Weyl, W. A., *J. Amer. Ceram. Soc.* **30**, 311 (1947).
70. Silverman, A., *Trans. Amer. Ceram. Soc.* **16**, 547 (1914).
71. Silverman, A., *J. Amer. Ceram. Soc.* **11**, 81 (1928).
72. Silverman, W. B., U.S. Patent 2,955,948 (October 11, 1960).
73. Spitzer, A., German Patent 74,565 (1893); U.S. Patent 518,336 (April 17, 1894).
74. Sugie, S., *J. Jap. Ceram. Assoc.* **341**, 152, **342**, 193, **343**, 226 (1921).
75. Sullivan, J. D., and Austin, C. R., *J. Amer. Ceram. Soc.* **25**, 123 (1942).
76. Tillyer, E. D., U.S. Patent 2,524,719, (October 3, 1950).
77. Warren, B. E., *Chem. Rev.* **26**, 237 (1940).
78. Weckerle, H. *Glastech. Ber.* **11**, 273 (1933).

79. Welz, F., German Patent 63,558, (December 6, 1891); U.S. Patent 479,689, (July 26, 1892).   German Patent 73,348, (January 29, 1893).
80. Weyl, W. A., *Coloured Glasses*, Society of Glass Technology Sheffield, 1951, reprinted, Dawson's of Pall Mall, London, 1959, p. 286.
81. Weyl, W. A., *Coloured Glasses*, Society of Glass Technology, Sheffield, 1951, reprinted, Dawson's of Pall Mall, London, 1959, p. 311.
82. Weyl, W. A., and Marboe, E. C., *The Constitution of Glasses, A Dynamic Interpretation*, John Wiley and Sons, New York, 1964, p. 773.
83. Winter, A., *J. Amer. Ceram. Soc.* **40**, 54 (1957).
84. Witt, O. N., and Frankel, E., *Sprechsaal Z.* **47**, 444 (1914).
85. Zachariasen, W. H., *J. Amer. Chem. Soc.* **54**, 3841 (1932).
86. Zsigmondy, R. *Zur Kenntnis der Kolloide*, transl. by J. Alexander as *Colloids and the Ultramicroscope*, John Wiley and Sons, New York, 1909, p. 166.

14

Selenium and Selenium Compounds in Rubber and Plastics

WILLIAM J. MUELLER

Research Chemist, Polymer and Paper Technology,
Battelle Memorial Institute, Columbus, Ohio

INTRODUCTION

Both elemental selenium and several selenium compounds are used by the rubber and plastics industries in a variety of applications. These materials find use primarily because of their particular chemical properties.

Elemental selenium is located just below sulfur in the periodic table, and therefore is a logical alternative material. Since sulfur and sulfur compounds are so widely used by the rubber industry, and to a lesser extent by the plastics industry, we would expect selenium and its compounds to be used in the same general areas.

The apparent use of elemental selenium is to replace sulfur as a vulcanizing agent, where unique properties, such as improved aging, are obtained with this element. The pattern for the use of selenium compounds is somewhat different, as their applications are based either on the selenium and/or the chemical constituents of the molecule other than the selenium.   In this respect, these compounds have found applications as vulcanizing agents, accelerators, antioxidants, UV stabilizers, bonding agents, polymerization additives, modifiers for butyl rubber, carbon black activators, latex preservatives, and flameproofing agents.   These particular applications will be discussed in detail.

CURING-ACCELERATOR SYSTEMS

Selenium and its compounds have found use in two capacities in the curing-accelerator systems of rubber compositions.   Elemental selenium is used as a direct replacement for sulfur as the vulcanizing agent where it is able to impart special properties, such as heat resistance.   However, the large volume in this area is for selenium compounds which find application in the polymer accelerator system to control the rate of vulcanization.   The amount of selenium consumed as an accelerator is believed to be many times as great as that consumed as a vulcanizing agent.   These two functions are so closely related that they will be discussed together.   Also, the effects on both unaged and aged properties, because they cannot logically be separated, will be also discussed together.

Natural Rubber

Probably the most exhaustive review of the use of selenium in rubber was published by Boggs and Follansbee in 1926.[9]   Because of the date of publication, it was, of course, confined to natural rubber.   However, the type of information contained in this review was very basic and most of the ideas presented are probably valid even today.

Early work using selenium as the vulcanizing agent in natural rubber was of no practical importance until it was found that it could be used in combination with organic accelerators to increase the rate of curing.   This basic concept was covered in U.S. Patents 1,249,272[10] and 1,364,055.[11]   Compositions containing suitable accelerators could be vulcanized at temperatures and times comparable to those used in sulfur cures to produce compositions with a good plateau in the curing curve and good aging resistance.   As an example, Figure 14-1 shows the effect on tensile strength of varying amounts of selenium, based on the following composition: pale crepe, 60; zinc oxide, 13; carbon black, 17; litharge, 3; Ozokerite, 2; piperidonium pentamethylene dithiocarbamate, 0.6; selenium, variable.   The data show that the tensile strength and elongation held up very well after 96 h in the Greer oven.

This finding is significant because synthetic rubbers are usually substituted for natural rubber when good aging resistance is required.   To gain this improvement in aging, however, several of the advantages of using natural

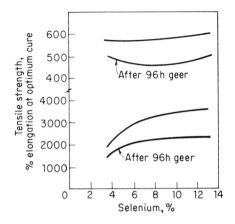

Figure 14-1.

rubber are lost.   By the use of selenium as the vulcanizing agent in natural rubber, it may be possible to obtain both the advantages of natural rubber and the improved aging resistance obtained with synthetic rubber.

Since both selenium and sulfur act as vulcanizing agents for natural rubber, a logical approach would be to determine the effect of adding small amounts of selenium to a composition containing sulfur.   This was done both at normal and higher-than-normal sulfur concentrations.   Figures 14-2 to 14-8[9] show

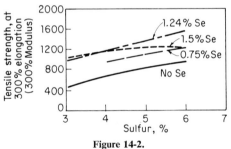

Figure 14-2.

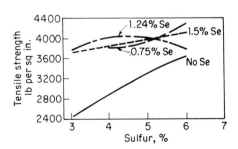

Figure 14-3.

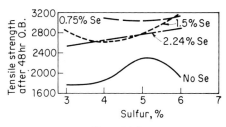

Figure 14-4.

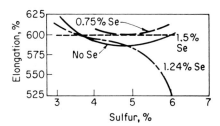

Figure 14-5.

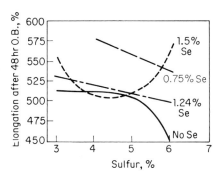

Figure 14-6.

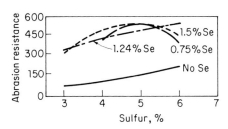

Figure 14-7.

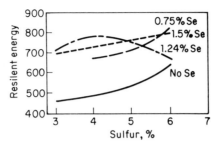

Figure 14-8.

this effect in a tire tread composition containing pale crepe, 30; smoked sheet, 30; zinc oxide, 17; carbon black, 17; Ozokerite, 2; DPG, 0.3; ethylidene aniline, 0.3; sulfur and selenium, variable. Data show that the addition of small amounts of selenium improved the tensile strength before and after aging, abrasion resistance, resilient energy, and generally, the aged elongation. The higher values for tensile stress at 300% elongation shown in Figure 14-2 shows that the added selenium produced a tighter cure.

Compositions with a high carbon black content were also affected in a similar manner. Table 14-1 shows that the inclusion of small amounts of

TABLE 14-1   SELENIUM IN HIGH CARBON BLACK COMPOUNDS[9]

Reference Number	P-43	P-44	P-41	P-42	P-45
Minutes for best cure	80	80	60	60	60
Tensile, lbs per sq in.	3240	4525	4457	4305	4355
Tensile after O.B.	1949	2446	2394	2493	2725
% Elongation	550	606	575	600	556
% Elongation after O.B.	475	494	394	468	481
Resilient energy	553	924	933	881	885
Tensile at 300% elongation (300% Modulus)	1160	1586	1860	1492	2020
Abrasion resistance	18	33	100[a]	75	90
Ingredients	*Parts by Weight*				
Pale crepe	30	30	30	30	30
Smoked sheets	30	30	30	30	30
Zinc oxide	5	5	5	5	5
Carbon black	29	29	29	29	29
Ozokerite	2	2	2	2	2
D.P.G.	0.3	0.45	0.3	0.3	0.45
E.A.	0.3	0.45	0.3	0.3	0.45
Selenium	0	0	0.9	1.12	1.12
Sulfur	2.4	2.4	2.4	1.8	1.8
	99.0	99.3	99.9	98.52	99.82

[a] Arbitrary abrasion resistance of P-41 = 100.

selenium was able to triple the abrasion resistance while significantly improving tensile strength before and after aging. However, the activity observed by the use of small amounts of selenium was not confined to carbon black-filled compositions. When the 17 phr (parts per hundred rubber) of carbon black was replaced with an equal volume of zinc oxide (68 phr), essentially the same results were obtained.[9]

More recent work with smoked sheet has compared the use of selenium compounds in the curing-accelerator system with three basic methods of curing natural rubber. These are: (1) normal sulfur (3 phr)-benzothiazyl disulfide, (2) low sulfur (0.5 phr)-TMTD (tetramethylthiuram disulfide), and (3) TMTD alone. Table 14-2 shows that selenium diethyldithiocarbamate used alone, or in conjunction with tellurium, gave a composition with unaged properties which was quite comparable to that obtained with TMTD as the total curing-accelerator system, but which compared to 0.5 phr sulfur and 2.0 phr TMTD in most aging properties. These compositions, after aging, were significantly better than those prepared from sulfur-benzothiazyl disulfide, again indicating the ability of the selenium compounds to improve aging resistance.

Table 14-3 shows the effect of using selenium diethyldithiocarbamate in combination with low sulfur in a natural rubber insulation composition. Aging resistance was improved, compared to using a combination of benzothiazyl disulfide and Methyl Zimate.

One patent[13] reports on the use of selenides of heterocyclic amines as fast acting vulcanizing agents for natural rubber. The patent cites N,N'-triselenobis(morpholine) as an example.

SBR (Styrene-Butadiene Rubber)

SBR is a copolymer of butadiene and styrene which is used in much greater volume than any other synthetic rubber. This polymer is normally vulcanized with sulfur, and a large number of sulfur-containing accelerators are used to increase the rate of the vulcanization reaction.

Elemental selenium (Vandex, R. T. Vanderbilt Company) has been investigated as a partial or total substitute for sulfur as the vulcanizing agent for SBR-1500. This particular SBR polymer is widely used in many applications and is a major polymer for passenger tire treads. The recipe used was an ASTM standard recipe containing benzothiazyl disulfide as the accelerator. Data in Table 14-4 shows that up to half of the sulfur could be replaced by selenium with no significant effect on unaged stress-strain properties. Complete replacement produced a composition with a badly retarded cure. On the other hand, use of elemental selenium with selenium diethyldithiocarbamate (Ethyl Selenac, R. T. Vanderbilt Company) produced a composition with a high degree of cure, as judged by the 100 and 300% modulus.

TABLE 14-2 EFFECT OF SELENIUM COMPOUNDS ON VULCANIZATION AND AGING OF NATURAL RUBBER (SMOKED SHEET)[4]

Curing-Accelerator System,	Unaged Physical Properties					Physical Properties After Aging 7 Days at 212°F				
	100% Modulus psi	300% Modulus, psi	Tensile Strength, psi	Elong-ation, %	Hardness, Shore A-2	100% Modulus, psi	300% Modulus, psi	Tensile Strength, psi	Elong-ation, %	Hardness, Shore A-2
3.0 Sulfur plus 1.0 benzothiazyl disulfide	400	2020	3820	500	69	980	—	1060	110	80
2.0 TMTD[a]	180	1400	3400	550	62	190	1070	1920	470	63
0.5 Sulfur plus 2.0 TMTD	460	2670	4160	440	71	240	1280	1320	310	67
2.0 Selenium diethyldithio-carbamate	220	1560	3620	530	64	180	890	1100	360	65
1.0 Tellurium plus 2.0 selenium diethyldithio-carbamate	200	1420	3560	530	63	220	1140	1620	400	64

Recipe: *Ingredients*   *Parts by Weight*

Smoked sheet   100
EPC black   50
Zinc oxide   5
Stearic acid   3
PBNA   2
Curing-accelerator system   As shown
Cure: 40 minutes at 284°F

[a] Tetramethylthiuram disulfide

EFFECT OF SELENAC ON AGING OF A NATURAL RUBBER INSULATION COMPOSITION¹

Open Steam Cures at 388°F (198°C)	Accelerator, Altax-Zimate			Accelerator, Ethyl Selenac		
	Stress at 200%, psi	Tensile Strength, psi	Elongation, %	Stress at 200%, psi	Tensile Strength, psi	Elongation, %
Unaged						
5 Sec	595	1700	440	490	1530	415
7–1/2 Sec	775	2100	440	565	1720	440
10 Sec	730	1900	430	635	1800	430
15 Sec	700	1620	410	525	1810	435
20 Sec	645	1610	420	610	1670	430
30 Sec	625	1450	405	580	1580	425
Aged 96 h in Oxygen Bomb at 70°C and 300 lbs Pressure						
5 Sec	850	1380	350	790	1350	375
7–1/2 Sec	820	1500	375	740	1470	390
10 Sec	780	1420	370	730	1430	390
15 Sec	730	1180	360	655	1440	375
20 Sec	675	955	330	730	1200	340
30 Sec	660	710	230	700	1240	370

Recipe

Ingredients	Parts by Weight	Parts by Weight
Smoked Sheets	100	100
Reogen	1	1
Stearic acid	0.5	0.5
Zinc oxide	50	50
Dixie clay	50	50
Kalite	75	75
Rayox	10	10
Sulfur	2	0.75
Altax	0.625	—
Methyl zimate	0.625	—
Ethyl selenac	—	1

TABLE 14-4 EFFECT OF SUBSTITUTING SELENIUM FOR SULFUR IN SBR-1500 COMPOSITIONS[4]

Curing-Accelerator System	Unaged Physical Properties					Physical Properties After Aging 7 Days at 212°F				
	100% Modulus, psi	300% Modulus, psi	Tensile Strength, psi	Elonga-tion, %	Hardness, Shore A-2	100% Modulus, psi	300% Modulus, psi	Tensile Stength, psi	Elong-ation, %	Hardness, Shore A-2
2.0 Sulfur plus 3.0 benzothiazyl disulfide	140	910	3710	660	66	—	—	1080	70	80
1.5 Sulfur plus 0.5 selenium plus 3.0 benzothiazyl disulfide	190	1350	3810	600	67	—	—	1450	90	82
1.0 Sulfur plus 1.0 selenium plus 3.0 benzothiazyl disulfide	150	930	3970	770	65	940	—	1680	140	77
2.0 Selenium plus 3.0 benzothiazyl disulfide	Nil	Nil	130	1080	65	70	250	560	900	64
3.0 Selenium plus 3.0 selenium diethyldithio-carbamate	280	1880	3970	500	68	780	—	2850	220	76

Recipe: *Ingredients*   *Parts by Weight*

SBR-1500	100
EPC black	40
Zinc oxide	5
Stearic acid	1.5
Curing-accelerator system—As shown	

The big advantage for the use of selenium in place of sulfur is seen in the properties after aging 7 days at 212°F. Except for the composition containing 1.5 phr sulfur, replacement of sulfur with selenium increased aged elongation and reduced aged hardness. Both of these are an indication of improved aging.

One method of increasing the aging resistance of SBR is to decrease the sulfur to 0.5 phr or to eliminate it altogether. Table 14-5 shows that the use of 2.0 phr TMTD, alone or with 0.5 phr sulfur, gave a composition which had good aged elongation even after 7 days at 212°F or 2 days at 300°.F The use of selenium diethyldithiocarbamate alone or in combination with elemental tellurium or methyl piaselenol produced compositions with good properties before aging and excellent aged properties. The composition containing methyl piaselenol was the poorest of the group, but even this was far superior after aging to the composition vulcanized with the commonly used 2.0 phr of sulfur.

Nitrile Rubber

This is a copolymer of butadiene and acrylonitrile, and is usually vulcanized with essentially the same systems used for SBR. Data in Table 14-6 show that the composition containing 0.5 phr sulfur plus 2.0 phr TMTD has slightly improved aging properties compared to the composition containing 1.5 phr sulfur plus 1.0 phr benzothiazyl disulfide. This is based primarily on the lower aged hardness, and duplicates the type of result found with SBR. Furthermore, Table 14-6 also shows that the combination of 1.0 phr tellurium plus 2.0 phr selenium diethyldithiocarbamate, or the selenium compound by itself, produced compositions which aged better when the aging temperature was 300°F. This again shows the advantage to be gained in aged properties by using the selenium compounds in the curing-accelerator system.

Butyl Rubber

This is a specialty polymer which is composed primarily of isobutylene with small amounts of isoprene added to provide some unsaturation for vulcanization. The polymer is often vulcanized with sulfur and an accelerator in a manner similar to that used for natural rubber and SBR. However, resin cures are used where maximum heat resistance is required.

The direct substitution of selenium for sulfur in an accelerated composition results in reduced tensile strength and modulus, indicating a lower state of cure with no significant effect on aging. This is shown in Table 14-7. The use of 2.0 phr selenium plus 3.0 phr selenium diethyldithiocarbamate did improve

TABLE 14-5 EFFECT OF SELENIUM COMPOUNDS ON VULCANIZATION AND AGING OF STYRENE-BUTADIENE (SBR) COMPOSITIONS[4]

Curing-Accelerator System, phr	Unaged Physical Properties					Physical Properties After Aging 7 Days at 212°F				
	100% Modulus, psi	300% Modulus, psi	Tensile Strength, psi	Elongation, %	Hardness, Shore A-2	100% Modulus, psi	300% Modulus, psi	Tensile Strength, psi	Elongation, %	Hardness, Shore A-2
2 Sulfur plus 3 benzothiazyl disulfide	160	1230	4190	650	65	—	—	1550	90	81
0.5 Sulfur plus 2 TMTD	280	2200	2800	350	67	520	—	2220	220	73
2 TMTD	80	690	3450	760	62	—	—	—	—	—
2 Selenium diethyldithiocarbamate	120	980	3890	670	63	290	2280	3490	400	70
0.5 Tellurium plus 2 selenium diethyldithiocarbamate	90	750	3890	720	62	—	—	—	—	—
1.0 Tellurium plus 2 selenium diethyldithiocarbamate	50	450	3150	800	53	220	1760	4160	520	67
3 Methyl piaselenol plus 3 selenium diethyldithiocarbamate	210	1340	3640	560	66	930	2520	2610	310	74

Physical Properties After Aging 2 Days at 300°F

Curing-Accelerator phr	100% Modulus, psi	Tensile Strength, psi	Elongation, %	Hardness, Shore A-2
2 Sulfur plus 3 benzothiazyl disulfide	—	—	—	—
0.5 Sulfur plus 2 TMTD	1100	1100	100	75
2 TMTD	480	1140	180	72
2 Selenium diethyl-dithiocarbamate	—	—	—	—
0.5 Tellurium plus 2 selenium diethyldithiocarbamate	510	1640	220	73
1.0 Tellurium plus 2 selenium diethyldithiocarbamate	400	1690	280	70
3 Methyl piaselenol plus 3 selenium diethyldithio-carbamate	—	—	—	—

Recipe:

Ingredients	Parts by Weight
SBR-1500	100
EPC black	40
Zinc oxide	5
Stearic acid	1.5
Curing-Accelerator System	As shown

Cure: 30 minutes at 302°F

TABLE 14-6 EFFECT OF SELENIUM COMPOUNDS ON VULCANIZATION AND AGING OF NITRILE RUBBER COMPOSITIONS[4]

Curing-Accelerator System, phr	Unaged Physical Properties					Physical Properties After Aging 4 Days at 300°F		
	100% Modulus, psi	300% Modulus, psi	Tensile Strength, psi	Elongation, %	Hardness, Shore A-2	Tensile Strength, psi	Elongation, %	Hardness, Shore A-2
1.5 Sulfur plus 1.0 benzothiazyl disulfide	310	1950	2430	420	69	1250	50	87
0.5 Sulfur plus 2.0 TMTD	430	2150	2150	300	69	1270	70	80
1.0 Tellurium plus 2.0 selenium diethyldithiocarbamate	190	930	2640	710	66	1430	110	78
2.0 Selenium diethyldithiocarbamate	210	1140	2530	630	66	1340	110	79

Recipe: *Ingredients* *Parts by Weight*

Hycar 1001   100
SRF black   40
Zinc oxide   5
Stearic acid   1
Curing-accelerator system—As shown
Cure: 40 minutes at 302°F

TABLE 14-7 EFFECT OF SELENIUM COMPOUNDS ON VULCANIZATION AND AGING OF BUTYL RUBBER[4]

Curing-Accelerator System, phr	Unaged Physical Properties				Physical Properties After Aging 14 Days at 250°F			
	100 Modulus, psi	Tensile Strength, psi	Elongation, %	Hardness, Shore A-2	100% Modulus, psi	Tensile Strength, psi	Elongation, %	Hardness, Shore A-2
2 Sulfur plus 0.5 benzothiazyl disulfide plus 1 TMTD	190	2590	630	67	180	600	330	74
2 Selenium plus 0.5 benzothiazyl disulfide plus 1.0 TMTD	70	1800	930	56	80	630	580	63
2 Selenium plus 3 selenium diethyl dithiocarbamate	20	1830	900	56	40	890	620	58
12 Amberol ST-137 plus 3 Hypalon 20	180	2050	430	66	730	1610	170	76
12 Amberol ST-137 plus 3 Hypalon 20 plus 2 selenium	180	2060	430	65	720	1420	160	78
12 Amberol ST-137 plus 2 selenium plus 3 selenium dioxide	120	1630	380	57	260	950	210	63
12 Amberol ST-137 plus 3 selenium dioxide	90	1630	410	57	250	880	210	64

Recipe: *Ingredients*   *Parts by Weight*

Ingredients	Parts by Weight
Butyl 325	100
HAF black	35
Zinc oxide	5
Stearic acid	2
Curing accelerator system—As shown	

Curing Conditions: Resin cures containing Amberol ST-137—30 minutes at 330°F; other compositions 30 minutes at 302°F.

aging compared to the combination of sulfur–benzothiazyl disulfide–TMTD. This is shown by the higher tensile strength and elongation after aging and the small change in hardness during aging.

The resin-cured control composition contained Amberol ST-137 and Hypalon 20 in the curing-accelerator system. The good aging of this composition is shown by the high aged tensile strength. Some small advantage is shown in aged elongation if selenium dioxide is added to the composition. Otherwise, no particular advantage is seen for the use of selenium compounds.

Neoprene

Neoprene or polychloroprene is normally vulcanized with a combination of zinc and magnesium oxides. An organic accelerator such as 2-mercapto-imidazoline is added to accelerate the curing. Table 14-8 shows limited data on the use of selenium compounds as a substitute for the 2-mercapto-imidazoline. No advantage is seen in the use of the selenium compounds, and there is some disadvantage since the compositions containing selenium tended to age more poorly.

Halogenated Butyl

Butyl rubber is modified both by bromination and chlorination to impart special properties which are desirable in particular end items. Table 14-9 shows the effect of using selenium or selenium compounds in several ways in the curing-accelerator system of brominated butyl rubber. The use of 2 phr of selenium diethyldithiocarbamate alone or in combination with tellurium was most effective in improving aged tensile strength, although at a small decrease in the unaged tensile strength.

Limited work on the use of selenium and its compounds in chlorinated butyl rubber indicated that the effects were essentially the same as those observed with brominated butyl rubber.

Hypalon

This specialty polymer is frequently vulcanized with dipentamethylene thiuram tetrasulfide, with typical results shown in Table 14-10. However, 2 phr of selenium diethyldithiocarbamate by itself or with 1 phr of selenium can be used as the vulcanizing system. This had very little effect on the unaged physical properties, but did improve aging resistance. This is best shown by the improved elongation and lower hardness after aging.

TABLE 14-8 EFFECT OF SELENIUM COMPOUNDS ON VULCANIZATION AND AGING OF NEOPRENE W[4]

Curing-Accelerator System, phr	Cure, minutes at 307°F	Unaged Physical Properties					Physical Properties After Aging 7 Days at 250°F			
		100% Modulus, psi	300% Modulus, psi	Tensile Strength, psi	Elong-gation, %	Hardness, Shore A-2	100% Modulus, psi	Tensile Strength, psi	Elong-gation, %	Hardness, Shore A-2
5 Zinc oxide plus 4 magnesium oxide	45	170	1280	3340	570	64	—	1660	50	82
5 Zinc oxide plus 4 magnesium oxide plus 0.5 2-mercaptoimidazoline	30	360	3090	3810	350	67	1390	2890	190	81
5 Zinc oxide plus 4 magnesium oxide plus 0.5 selenium plus 1 selenium diethyldithiocarbamate	30	180	1460	3620	630	67	—	2000	80	86
5 Zinc oxide plus 4 magnesium oxide plus 1 selenium diethyldithiocarbmate	30	190	1450	3860	640	67	1860	2020	110	85

Recipe: Ingredients   Parts by Weight

Neoprene W   100
HAF black   30
PBNA   2
Curing-accelerator system—As shown

TABLE 14-9 EFFECT OF SELENIUM COMPOUNDS ON VULCANIZATION AND AGING OF BROMINATED BUTYL RUBBER[4]

Curing-Accelerator System, phr	Unaged Physical Properties					Physical Properties After Aging 7 Days at 300°F				
	100% Modulus, psi	300% Modulus, psi	Tensile Strength, psi	Elongation, %	Hardness, Shore A-2	100% Modulus, psi	300% Modulus, psi	Tensile Strength, psi	Elongation, %	Hardness, Shore A-2
2 Sulfur plus 0.5 TMTD	370	2250	2430	320	72	Nil	Nil	100	380	77
2 Selenium plus 0.5 TMTD	390	1890	2020	330	73	90	270	280	310	78
2 Sulfur plus 2 Selenium plus 0.5 TMTD	390	1870	2040	340	73	Nil	Nil	100	300	78
2 Selenium diethyldithiocarbamate	250	1450	2280	470	72	130	350	410	350	78
2 Tellurium plus 2 selenium diethyldithiocarbamate	270	1390	1880	450	69	150	570	630	350	78
12 Phenolic resin	350	1320	1920	410	72	290	—	710	220	75
12 Phenolic resin plus 2 Selenium	470	1520	1710	350	80	440	—	750	180	84

Recipe:

Ingredients	Parts by Weight
Hycar 2202	100
HAF black	50
Zinc oxide	5
Curing-accelerator system	As shown

Curing Conditions: Compositions containing phenolic resin cured 30 minutes at 330°F. All others cured 30 minutes at 310°F.

EFFECT OF SELENIUM ON VULCANIZATION AND AGING OF CHLOROSULFONATED POLYETHYLENE (HYPALON) COMPOSITIONS[4]

Curing-Accelerator System, phr	Unaged Physical Properties					Physical Properties After Aging 7 Days at 250°F				
	100% Modulus, psi	300% Modulus, psi	Tensile Strength, psi	Elongation, %	Hardness, Shore A-2	100% Modulus, psi	300% Modulus, psi	Tensile Strength, psi	Elongation, %	Hardness, Shore A-2
2 Dipentamethylene thiuram disulfide	70	440	1450	500	57	280	—	1350	260	61
2 Selenium diethyl-dithiocarbamate	130	400	1450	470	60	—	—	—	—	—
1 Selenium plus 2 selenium diethyldithiocarbamate	90	590	1390	420	54	110	930	1210	350	69
1 Selenium plus 3 selenium diethyldithiocarbamate	130	710	840	310	67	140	840	1230	330	57
3 Selenium plus 3 selenium diethyldithiocarbamate	90	520	840	340	59	180	880	1090	320	58

Curing-Accelerator System, phr	Physical Properties After Aging 4 Days at 300°F			
	100% Modulus, psi	Tensile Strength, psi	Elongation, %	Hardness, Shore A-2
2 Dipentamethylene thiuram disulfide	—	910	50	79
2 Selenium diethyl-dithiocarbamate	—	950	70	68
1 Selenium plus 2 selenium diethyldithiocarbamate	530	970	130	65
1 Selenium plus 3 selenium diethyldithiocarbamate	430	1090	180	64
3 Selenium plus 3 selenium diethyldithiocarbamate	660	990	130	67

Recipe

Ingredients	Parts by Weight
Hypalon 20	100
Magnesia	10
Stearic acid	1
Curing-accelerator system—As shown	

Cure: 30 minutes at 307°F

Viton

This fluorinated polymer is saturated so that it requires materials such as N, N′-dicinnamylidine-1,6-hexanediamine (Diak No. 3, Du Pont) and hexamethylene diamine carbamate (Diak No. 1, Du Pont) for vulcanization. The former is generally preferred because of improved scorch resistance.

Table 14-11 shows the effect of replacing a portion of the Diak with dibenzylselenourea. This substitution was particularly effective with the HMDA carbamate since it resulted in a significant improvement in aged elongation. This would be particularly beneficial under severe aging conditions, where low elongation could limit the service life.

Acrylate Polymer

Saturated acrylate polymers such as Hycar 4021 (B. F. Goodrich) are particularly effective for improved aging in the presence of sulfur-bearing oils. Being saturated, they are not subjected to continuing vulcanization as an unsaturated polymer would be. A recommended vulcanizing agent is Trimene Base (a reaction product of ethyl chloride, formaldehyde, and ammonia; Uniroyal, Inc.) used alone or in combination with a small amount of sulfur.

Table 14-12 shows the effect of using selenium both to replace and supplement the sulfur. The combination of sulfur and selenium was particularly effective in improving aging after 7 days at 350°F. The increased elongation and lower hardness indicates this significant improvement in aging resistance.

Hard Rubber

Because of the large amount of sulfur used in hard rubber compositions, care must be used in choosing accelerators so that the compositions do not become brittle as well as hard. This is particularly important, since many current uses for hard rubber are in applications such as battery boxes, where good impact resistance is required.

Table 14-13[2] shows that the addition of Vandex (elemental selenium) to a hard rubber composition significantly accelerated the curing while maintaining an elongation of at least 5%. This elongation would reduce brittleness and improve shock resistance.

Surface Vulcanization

Some rubber products, such as sheet goods, cannot be vulcanized by the usual methods because of their bulky nature. These items are often vulcanized by contact with sulfur chloride or a similar material. For various

Curing-Accelerator System, phr	Unaged Physical Properties					Physical Properties After Aging 7 Days at 500°F				
	100% Modulus, psi	300% Modulus, psi	Tensile Strength, psi	Elongation, %	Hardness, Shore A-2	100% Modulus, psi	300% Modulus, psi	Tensile Strength, psi	Elongation, %	Hardness, Shore A-2
1.0 Diak No. 1[a]	420	—	2060	280	72	570	—	1290	200	79
0.5 Diak No. 1 plus 0.5 dibenzylselenourea	280	1180	1980	470	72	220	640	1150	700	75
0.25 Diak No. 1 plus 0.75 dibenzylselenourea	220	830	1650	620	72	190	520	1060	980	73
1.0 Diak No. 3[b]	250	1050	2170	530	73	—	—	—	—	—
1.0 Diak No. 3 plus 0.25 dibenzylselenourea	300	1120	2310	600	72	—	—	—	—	—
1.5 Diak No. 3	330	1440	2380	460	72	—	—	—	—	—
2.0 Diak No. 3	460	1660	1880	330	75	—	—	—	—	—

Recipe

Ingredients	Parts by Weight
Viton AHV	100
MT black	25
Magnesia	10

Curing-Accelerator System—As shown

Press-Cure: 30 minutes at 330°F

Postcure: 1 h at 212°F; 1 h at 250°F; 1 hr at 300°F; 1 h at 350°F; 24 h at 400°F.

Physical Properties After Aging 10 Days at 500°F

Curing-Accelerator System, phr	100% Modulus, psi	300% Modulus, psi	Tensile Strength, psi	Elongation, %	Hardness, Shore A-2
1.0 Diak No. 1[a]	—	—	—	—	—
0.5 Diak No. 1 plus 0.5 dibenzylselenourea	—	—	—	—	—
0.25 Diak No. 1 plus 0.75 dibenzylselenourea	—	—	—	—	—
1.0 Diak No. 3[b]	270	830	1400	610	74
1.0 Diak No. 3 plus 0.25 dibenzylselenourea	290	1010	1690	640	75
1.5 Diak No. 3	370	1010	1020	310	78
2.0 Diak No. 3	—	—	3130	Nil	100+

[a] Diak No. 1—Hexamethylenediamine carbamate.
[b] Diak No. 3—N-N′-dicinnamylidene-1,6-hexanediamine.

TABLE 14-12 EFFECT OF ADDED SELENIUM ON VULCANIZATION AND AGING OF AN ACRYLATE POLYMER[4]

Curing-Accelerator System, phr	Postcure	Unaged Physical Properties					Physical Properties After Aging 7 Days at 350°F			
		100% Modulus, psi	300% Modulus, psi	Tensile Strength, psi	Elon- gation, %	Hardness, Shore A-2	100% Modulus, psi	Tensile Strength, psi	Elon- gation, %	Hardness, Shore A-2
3 Trimene base	No	140	860	1490	530	57	—	1250	30	92
	Yes	—	—	1280	70	77	—	990	20	90
3 Trimene base plus 0.5 sulfur	No	560	—	1550	210	66	—	630	80	83
	Yes	860	—	1710	160	70	480	750	150	82
3 Trimene base plus 0.5 selenium	No	150	940	1430	470	72	—	1320	50	91
	Yes	—	—	1120	60	80	—	1510	60	89
3 Trimene base plus 0.5 selenium plus 0.5 sulfur	No	550	—	1770	240	69	370	1050	220	78
	Yes	710	—	1780	170	72	210	810	300	80

Recipe: *Ingredients*   *Parts by Weight*

Hycar 4021   100
SRF black   40
Stearic acid   1
Curing-accelerator system—As shown
Cure: 30 minutes at 310°F
Postcure: 24 h at 300°F

TABLE 14-13 EFFECT OF VANDEX IN HARD RUBBER[1]

Cured Between Tinfoil in Water at 149°C	Tensile Strength, psi	Elongation, %	Shore Hardness, at 100°C	Tensile Strength, psi	Elongation, %	Shore Hardness, at 100°C	Tensile Strength, psi	Elongation, %	Shore Hardness, at 100°C	Tensile Strength, psi	Elongation, %	Shore Hardness at 100°C
3 h	—	Soft	—	3470	60	82	4450	15	82	7600	6	86
4 h	4200	90	77	7200	7	87	6900	6	87	7800	6	92
6 h	7450	7	92	7950	6	94	8050	6	96	8120	6	96
8 h	7900	6	95	8000	6	97	8250	5	97	8200	6	97
10 h	8150	6	96	8160	6	98	8250	5	98	8200	5	98

Recipe:

Ingredients	Parts by Weight	Parts by Weight	Parts by Weight	Parts by Weight
Smoked sheet	100	100	100	100
Cotton seed oil	2	2	2	2
Plastogen	3	3	3	3
Sulfur	45	45	45	45
Vandex	—	1	2	5
	150	151	152	155

reasons, the use of sulfur may be objectionable, and two patents[8,15,16] report that selenium dioxide in solution in an organic solvent may be used to vulcanize such products in place of the sulfur compounds.

ANTIOXIDANTS

Most rubber products contain an antioxidant to retard aging and increase their useful life.   These are normally amines or hindered phenols.   The former are most effective although they often contribute to staining and discoloration.   The latter are used in light-colored rubber goods since they have little effect on the color.

A variety of selenium compounds have been investigated as antioxidants. These materials generally contain an amine or phenolic group, or contained selenium as an alternative to sulfur.   Effects shown in specific polymers are discussed below.

SBR

Table 14-14 shows the effect of using several selenium compounds in SBR as an alternate to PBNA (phenylbetanaphthylamine), a commonly used commercial antioxidant.   Bis(2-hydroxy-1-naphthyl) selenide was about equally effective with the PBNA in minimizing tensile strength and hardness changes during aging.   The dibenzylselenourea was most effective in maintaining properties during aging, although this material significantly decreased the unaged tensile strength.   Other materials investigated were less effective.

Butyl Rubber

Antioxidants are not commonly used with butyl rubber, since this polymer contains very little unsaturation, and is inherently resistant to oxidation. However, data in Table 14-15 show that several of the selenium compounds were more effective than PBNA in maintaining physical properties during extended heat aging at 250°F.   Dibenzylselenourea, phenoxaselenine, and methyl piaselenol were particularly effective in maintaining tensile strength at a high level and minimizing changes in hardness.

Natural Rubber

In natural rubber, the selenium compounds wre not as effective as the PBNA, as shown by the data in Table 14-16.   The most effective of the selenium compounds was methyl piaselenol.

TABLE 14-14 EVALUATION OF SELENIUM COMPOUNDS AS ANTIOXIDANTS IN STYRENE-BUTADIENE RUBBER (SBR)[4]

Antioxidant	Unaged Physical Properties			Physical Properties After Aging 7 Days at 212°F		
	Tensile Strength, psi	Elongation, %	Hardness, Shore A-2	Tensile Strength, psi	Elongation, %	Hardness, Shore A-2
None	2980	760	61	2080	160	77
PBNA	2830	800	60	2010	160	71
Dibenzylselenourea	1780	280	75	1900	140	77
Bis(2-hydroxy-1-naphthyl) selenide	3420	540	69	2150	120	79
Phenoxaselenine	3640	760	63	1070	100	77
Methylpiaselenol	3350	700	63	1580	150	78

Recipe: Ingredients   Parts by Weight

SBR-1500   100
EPC black   40
Zinc oxide   5
Benzothiazyl disulfide   1.75
Sulfur   2
Antioxidant   2.5
Cure: 40 minutes at 302°F

TABLE 14-15   EVALUATION OF SELENIUM COMPOUNDS AS ANTIOXIDANTS IN BUTYL RUBBER[4]

Antioxidant	Unaged Physical Properties			Physical Properties After Aging 7 Days at 250°F		
	Tensile Strength, psi	Elongation, %	Hardness, Shore A-2	Tensile Strength, psi	Elongation, %	Hardness, Shore A-2
None	2130	520	72	1680	400	65
PBNA	2150	560	71	1500	540	66
Dibenzylselenourea	2270	660	70	1960	480	66
Bis(2-hydroxy-1-naphthyl) selenide	2280	680	72	1210	580	63
Phenoxaselenine	2240	540	70	2230	400	75
Methylpiaselenol	2280	550	72	1960	420	68
β-Naphthyl selenocyanate	2150	620	73	1170	560	63

Recipe: *Ingredients*

	Parts by Weight
Butyl 325	100
MPC carbon black	50
Zinc oxide	20
Stearic acid	1
Benzothiazyl disulfide	2
Sulfur	2
Tellurium diethyl-dithiocarbamate	2
Antioxidant	2.5

Cure: 30 minutes at 310°F

TABLE 14-16   EVALUATION OF SELENIUM COMPOUNDS AS ANTIOXIDANTS IN NATURAL RUBBER (SMOKED SHEET)[4]

Antioxidant	Unaged Physical Properties			Physical Properties After Aging 7 Days at 212°F		
	Tensile Strength, psi	Elongation, %	Hardness, Shore A-2	Tensile Strength, psi	Elongation, %	Hardness, Shore A-2
None	2480	360	71	2080	240	61
PBNA	3010	400	71	2840	240	71
Dibenzylselenourea	2830	340	72	1280	140	73
Methylpiaselenol	2940	400	71	1970	240	64
β-Naphthyl selenocyanate	2670	400	72	1570	160	67

Recipe: *Ingredients*   *Parts by Weight*

Smoked Sheet   100
SRF black   50
Zinc oxide   5
Benzothiazyl disulfide   1
Sulfur   1.5
Antioxidant   2.5
Cure: 40 minutes at 302°F

Staining and Discoloration

As mentioned earlier in this section, it is necessary to control staining and discoloration in light-colored rubber products. This staining and discoloration can be caused by the antioxidant as it is used, or by oxidation products which increase in concentration as aging progresses. Table 14-17 shows the effect of using 3 commercial antioxidants and 6 selenium compounds in a pale crepe composition heavily loaded with titanium dioxide. Data show that all compositions containing either control antioxidants or selenium compounds either stained, discolored, or showed a surface effect. The composition containing phenoxaselinine was affected least of the 6 containing selenium compounds.

Miscellaneous Effects

Several patents have been issued which report the use of selenium compounds as antioxidants. For example, British Patent 551,852[26] reports that synthetic or natural rubber has increased oxidation resistance if compounded with a polyvalent metal salt of a dehydroxy dialkyl diaryl sulfide or polysulfide or polymer or tellurium and selenium analogue thereof, at least one of the valencies of the metal being connected to an alkoxy group. British Patent 551,819[25] reports improved aging by the inclusion of a compound of the formula:

Where R is an alkyl group,
   M is a divalent metal,
   X is sulfur, selenium, or tellurium
   n is an integer 1–4.

Two U.S. patents, 2,789,103 and 2,858,325[27,28] show that an organotin compound of the formula $(R_2SnQ)_n$ is capable of stabilizing rubber compositions. Q is selected from the group of elements consisting of sulfur, selenium, and tellurium. Selenium is reported to increase oxidation resistance of silicone fluids[3] and this effect may carry over to silicone rubber.

UV STABILIZERS

Plastics, such as polyethylene, require a UV stabilizer to achieve maximum service life when used outdoors in direct sunlight. Materials used to provide

TABLE 14-17 RESULTS OF NONSTAINING NONDISCOLORING ANTIOXIDANT EVALUATION[a]

Composition	Antioxidant	Condition After 12 Weeks Outdoor Exposure		
		Staining	Discoloration	Appearance
H-1	None	None	None	Surface cracks
H-2	Wingstay S (control)	Slight stain	None	Same as original
H-3	DBPC (control)	Slight stain	None	Same as original
H-4	Ionol (control)	Slight stain	None	Same as original
H-5	Dibenzylselenourea	Slight stain	Gray	Same as original
H-6	Bis(2-hydroxy-1-naphthyl) selenide	Bad staining	Dark tan	Same as original
H-7	Methylpiaselenol	Bad staining	Dark tan	Same as original
H-8	Phenoxaselenine	Slight stain	None	Surface cracks
H-9	β-naphthyl selenocyanate	None	Light tan	Surface cracks
H-10	Selenium dioxide	Slight stain	Light gray	Some pitting

Base composition, parts by weight:

Pale crepe	100	Sulfur	3.25
Zinc oxide	25	Captax	0.5
Titanium dioxide	55	Altax	0.5
Dixie clay	10	TMTD	0.1
Reogen	1		
Stearic acid	1		
Magnesium carbonate	5		
Cure: 50 min at 274°F.			

this stability have the ability to absorb UV energy in the 300–400 mμ range. As a class of materials, derivatives of benzo[2,1,3] selenadiazole have been found to absorb strongly in this region.   Figure 14-9 shows several typical absorption curves from materials in this family, as well as one from a commercial UV stabilizer.

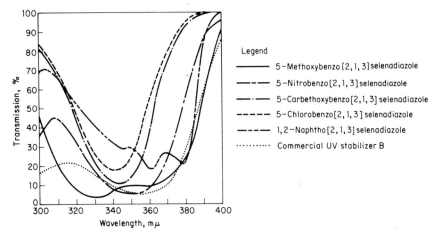

Figure 14-9   Legend

———   5-Methoxybenzo[2,1,3]selenadiazole

– – –   5-Nitrobenzo[2,1,3]selenadiazole

—·—   5-Carbethoxybenzo[2,1,3]selenadiazole

- - - - -   5-Chlorobenzo[2,1,3]selenadiazole

– – – –   1,2-Naphtho[2,1,3]selenadiazole

··········   Commercial UV stabilizer B

Figure 14-9   Ultraviolet absorption spectra of various derivatives of benzo[2,1,3] selenadiazole (solvent: toluene; concentration: 10 mg/liter; cell thickness: 2 cm).

On the basis of the observed absorption, polyethylene was compounded with several benzoselenadiazoles and exposed both to RS sunlamps and to direct sunlight in outdoor exposure.   Deterioration was based both on color and on changes in brittle point.   The latter is known to be a property of polyethylene very sensitive to aging.

Data in Table 14-18 show the effect of this exposure.   The 5-nitrobenzo-[2,1,3] selenadiazole at a level of 0.5 phr was very effective in controlling changes in the brittle point of the polyethylene due to the RS sunlamp exposure.   At the level of 0.1 phr, there was observed a noticeable color change after 21 days of exposure, while the commercial control did not exhibit a change color in comparable exposure time.   This suggests that the use of the benzoselenadiazoles might be restricted to colored compositions, or to applications in which color development is not objectionable.

The use of a selenium compound for improving the heat-oxidative stability of polyolefins has also been suggested.   It is reported[19] that a very effective stabilizer consists of a combination of an organic stabilizer with a compound of a bivalent to a hexavalent metal with an element of main group VI of the periodic table.   Examples of such materials are selenides of cadmium and other metals.

TABLE 14-18   AGING TESTS ON POLYETHYLENE COMPOSITIONS CONTAINING SELENADIAZOLE DERIVATIVES[4]

Stabilizer, parts per 100 polyethylene	Brittle Point After Aging Under RS Sunlamp, C		Color After Outdoor Exposure		
	200 h	300 h	Unexposed	3 days	21 days
0.5 Uvinul 490[a] (control)	< −60	−50	—	—	—
0.1 ,,	> −15	> −15	None	None	None
0.5 5-Nitrobenzo[2,1,3]-selenadiazole	−55	−45	—	—	—
0.1 ,,	> −15	> −15	None	None	Slight yellow
0.3 5-Methoxybenzo[2,1,3]-selenadiazole	> −15	> −15	—	—	—
0.2 ,,	−15	−15	—	—	—
0.1 ,,	> −15	> −15	None	Dark tan	Tan with purple cast

[a] Mixture of tetrasubstituted benzophenones.

BONDING AGENTS

Selenium and selenium compounds are reported to be useful in several ways for improving the bonding of rubber to various substrates.   Improved bonding of rubber to metal or glass is reported if the substrate is first coated with selenium dioxide or selenious acid and the rubber dissolved in an organic solvent.   This procedure is described in identical Canadian[7] and British patents.   A later British patent[17] revised the procedure to report more reproducible bonds when a thermosetting resin was added to the mixture. Another approach is claimed for bonding an olefinic rubber to a substrate which consists of using a bonding phase[14] between the rubber and substrate. This bonding phase incorporates an isocyanate and a selenium dithio-carbamate having the formula:

$$\left(\begin{array}{c} A-C-S \\ \parallel \\ S \end{array} \right)_4 Se$$

Where A is a tertiary amine group

Finally, a very novel method describes the bonding of rubber and steel in which a layer of selenium is deposited on the steel surface.[23]   It is agreed that this process is primarily of theoretical interest.

POLYMERIZATION ADDITIVES

Although the major use for selenium and its compounds in rubber and plastics is in the area of compounding for specific properties, a limited number of uses have been found in polymerization.

In emulsion polymerization, mercaptans are used as chain modifiers at temperatures up through 50°C. At higher temperatures, these materials are often too active to be effective. Under these conditions, selenols with the general formula C_xSeH can be substituted since they are low enough in activity to be effective at temperatures to 150°C. The use of selenols as chain modifiers is described in two patents. One[5] emphasizes use in butadiene-styrene polymerization, while the second[6] is based on the use of butadiene and acrylonitrile.

It is also reported that it is very difficult to effect a *cis–trans* isomerization of polyisoprene. However, this can be readily accomplished in solution at 180–200°C with elemental selenium as the catalyst.[22] This catalyst also causes the *cis–trans* isomerization of polybutadiene at 200°C.[22]

A third reported use for selenium in polymerization relates to apparent control of molecular weight in the preparation of polybutadiene. While conventional processes produced a polymer which was difficult to process, the addition of selenium or tellurium to the polymerizing mixture is reported to lower the plasticity number[29] or increase the plasticity.

MODIFICATION OF BUTYL RUBBER

Butyl rubber compositions containing carbon black are characterized by a lack of resilience. This handicap can be overcome by heat treating the butyl rubber in the presence of carbon black as described in U.S. Patent 2,356,128. The preferred carbon blacks are those which contain a small amount of oxygen on the surface. It has also been found that comparable effects are obtained in the presence of small amounts of sulfur, selenium, and tellurium.[21] When used in equimolar quantities, sulfur is the most effective, while selenium is second in its effectiveness. Figure 14-10[20] compares the effects of sulfur and selenium on stress-strain properties of an SRF-filled butyl rubber composition. The significant differences are seen at extensions greater than 300%. Table 14-19[20] compares the damping properties obtained with the three elements in equimolar quantities. Heat treatment in the presence of sulfur showed the greatest decrease in damping, although the effect with selenium was also very significant.

Another approach to modifying butyl rubber is outlined in a British patent.[18] Normally, it is difficult to adequately disperse fillers in butyl rubber. This can have an adverse effect on electrical properties, one of the main applications for butyl rubber. By first reacting the butyl rubber with one or more nonmetallic aromatic selenenyl halides or nonmetallic aromatic sulfenyl halides, the filler can be more readily dispersed with an improvement in electrical properties.

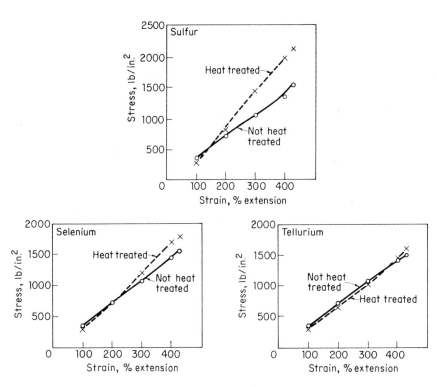

Figure 14-10 Stress-strain properties of butyl-SRF systems heat treated with sulfur, selenium, and tellurium, respectively.

TABLE 14-19 DAMPING PROPERTIES OF COMPOUNDS
            HEAT TREATED WITH SULFUR,
            SELENIUM, AND TELLURIUM[20]

| | $\eta f \times 10^{-6}$, Poises $\times$ CPS | | |
	Sulfur	Selenium	Tellurium
Not heat treated	2.27	2.42	2.38
Heat treated	1.61	2.05	2.25

ACTIVATING CARBON BLACK

Polymeric materials which are essentially saturated can be protected against thermal oxidation by the incorporation of 0.5 to 5 phr of activated carbon black with a maximum particle size of 1000 Å. Activation is achieved by the addition of oxygen, sulfur, or selenium to the surface of the particles.[24]

LATEX PRESERVATIVE

One of the unusual uses of selenium compounds is for stabilizing latex. While ammonia is normally used, it often has to be removed or rendered inactive before use. Small amounts of selenium in the form of selenium compounds of dithiocarbamic acid can significantly reduce the amount of ammonia necessary.[12] For example, 0.15% ammonia and 0.15–0.25% of a selenium compound can replace 0.7% ammonia.

FLAMEPROOFING

Selenium acts as an excellent flameproofing material when applied to the cotton braid of rubber insulated wire, according to Canadian Patent 258,392.[8] This type of application uses the selenium most efficiently, although it also flameproofs when used as a filler in the rubber composition. In the latter use, about 8% selenium must be used.

Acknowledgment   We wish to acknowledge the following organizations who have given permission to reproduce data from their earlier publications: *Rubber Age* for data in Table 14-19 and Figure 14-10; R. T. Vanderbilt Company for data in Tables 14-3 and 14-13; Institution of Rubber Industry for data in Table 14-1 and Figures 14-1 to 14-8. Data in remaining tables were from work at Battelle's Columbus Laboratories sponsored by the Selenium-Tellurium Development Association.

REFERENCES

1. *Vanderbilt News*, **13**, (5) 22 (1947).
2. *Vanderbilt News*, **14**, (1) 9 (1948).
3. Atkins, D. C., Murphy, C. M., and Saunders, C. E., *Ind. Eng. Chem.*, **39**, 1395 (1947).
4. Battelle Memorial Institute reports to Selenium and Tellurium Development Committee on project entitled " Utilization of Selenium and Tellurium."
5. Bennett, Bailey, (to Battelle Development Corporation), U.S. Patent 2,737,507 (March 6, 1956).
6. Bennett, Bailey, (to United States of America, Secretary of the Army), U.S. Patent 2,787,609 (April 2, 1957).

 7. Blackmore, R. H., (to Dunlop Tire and Rubber Goods Company, Ltd.), Canadian Patent 459,471 (Sept. 6, 1949).
 8. Blackmore, R. H., (to Dunlop Tire and Rubber Goods Company, Ltd.), Canadian Patent 465,334 (May 23, 1950).
 9. Boggs, C. R., and Follansbee, E. M., *Trans. IRI*, **2**, 272 (1926).
10. Boggs, C. R., (to Simplex Wire and Cable Company), U.S. Patent 1,249,272 (Dec. 4, 1917).
11. Boggs, C. R., (to Simplex Wire and Cable Company), U.S. Patent 1,364,055 (Dec. 28, 1920).
12. British Rubber Producers Research Association, British Patent 821,872 (Oct. 14, 1959).
13. Christensen, C. W., (to Monsanto Chemical Company), U.S. Patent 2,691,016 (Oct. 5, 1954).
14. Coleman, E. W., (to Lord Manufacturing Company), U.S. Patent 2,837,458 (June 3, 1958).
15. Dunlop Rubber Company, British Patent 604,964 (July 13, 1948).
16. Dunlop Rubber Company, British Patent 604,965 (July 13, 1948).
17. Dunlop Rubber Company, British Patent 738,313 (Oct. 12, 1955).
18. Esso Research and Engineering Company, British Patent 842,557 (July 27, 1960).
19. Fisher, A. and Lenz, H., (to Farbwerke Hoechst Aktiengesellschoft), U.S. Patent 3,029,224 (April 10, 1962).
20. Gessler, A. M. and Ford, F. P., *Rubber Age*, **74**, 397 (1953).
21. Gessler, A. M. and Robinson, S. B., (to Esso Research and Engineering Company), U.S. Patent 2,811,502 (Oct. 29, 1957).
22. Golub, M. A., *J. Poly. Sci.*, **36**, 523 (1959); Rubber Chem. Tech., **32**, 718 (1959).
23. Gurney, W. A., *IRI Proc.*, **7**, 127 (1960).
24. Hawkins, W. L. and Winslow, F. H., (to Bell Telephone Laboratories, Inc.), U.S. Patent 3,042,649 (July 3, 1962).
25. Standard Oil Development Company, British Patent 551,819 (March 11, 1943)
26. Standard Oil Development Company, British Patent 551,852 (March 12, 1943).
27. Weinberg, E. L. and Tomka, L. A., (to Metal and Thermit Corporation), U.S. Patent 2,789,103 (April 10, 1957).
28. Weinberg, E. L. and Tomka, L. A., (to Metal and Thermit Corporation), U.S. Patent 2,858,325 (Oct. 28, 1958).
29. Yunker, M. A., (to E. I. du Pont de Nemours & Company), U.S. Patent 2,333,403 (Nov. 2, 1943).

15

Metallurgical Aspects and Uses of Selenium

S. C. CARAPELLA, Jr.

Research Superintendent, Central Research Laboratories, American Smelting and Refining Company, South Plainfield, New Jersey

R. H. ABORN

Consultant to the Selenium-Tellurium Development Association, Inc.

L. R. CORNWELL

Associate Professor, Department of Mechanical Engineering, Texas A&M University, College Station, Texas

INTRODUCTION

Selenium belongs to Group VI of the Periodic Table which also includes the elements oxygen, sulfur, tellurium, and polonium. With its outer $4s^2 4p^4$ configuration, it is generally electronegative in nature and chemically, relatively reactive. From a metallurgical point-of-view it should be classed as a nonmetal. Some physical and mechanical properties of selenium are listed in Table 15-1. Additional details concerning these properties of selenium are to be found elsewhere in this volume.

Selenium forms a large number of compounds, many of them having interesting semiconductor properties, some of which are discussed in Chapters 5 and 16. Because of its brittle and vitreous nature, elemental selenium and its alloys have not found any application as engineering materials. Its metallurgical uses have thus been limited to small additions to other metals and alloys in order to improve some specific properties.

In this chapter we discuss first the metallurgical behavior of selenium in elemental form and in alloys before turning to its use as indicated above.

TABLE 15-1   PHYSICAL AND MECHANICAL PROPERTIES OF SELENIUM[a,c]

Property		Value
Atomic number		34
Atomic weight		78.96
Density 4.79 gm/cc	(Hexagonal)	4.26 gm/cc (Red amorphous)
Crystal structures	Hexagonal (A8)	$a = 4.346$ Å, $c = 4.954$ Å
	[b]Monoclinic	$a = 12.85$ Å, $b = 8.07$ Å, $c = 9.31$ Å, $\beta = 93°8'$
	[b]Monoclinic	$a = 9.05$ Å, $b = 9.07$ Å, $c = 11.61$ Å, $\beta = 90°46'$
Melting point		217°C (423°F) (Crystalline) red amorphous →black amorphous 40–50°C
Boiling point		685°C. (1265°F) (Crystalline)
Coefficient of linear expansion		37×10^{-6} in/in°C
Electrical resistivity		12μ ohm-cm (0°C) (Hexagonal)
Specific heat		0.084 cal/gm/°C
Thermal conductivity		7–18.3×10^{-4} cal/cm²/cm/sec °C (Hexagonal)
Mohs hardness		2.0 (22 BHN)
Young's modulus		8.4×10^6 psi
Vapor pressure 344°C (651°F) 566°C (1051°F)		1 mm Hg 100 mm Hg

[a] *ASM Metals Handbook* Vol. I.   All values quoted for 20°C unless otherwise noted.
[b] A. Taylor and B. J. Kagle, *Crystallographic Data on Metal Alloy Structures*, Dover, 1963.
[c] See Chapter 1 for additional data on the physical and mechanical properties of Se.

CRYSTAL GROWTH*

Selenium is usually grown as single crystals in the form of thin films on a substrate for semiconductor applications.   However, attempts have been made to grow single crystals large enough for physical and mechanical property studies.   The growth of large single crystals is difficult due to the exceptionally sluggish kinetics of molecular attachment at the liquid–solid interface.[50]   By using pressures of 5 kbar and speeds of 10^{-5} m/sec with 99.999 % pure selenium, Harrison and Tiller[38] were able to obtain single

* See Chapter 3 for a more extensive discussion of this topic.

crystals by spontaneous nucleation from the melt when the tip of the cone was <10°C. These crystals always contained low angle tilt boundaries of 3–4°. They found that the freezing temperature had been increased by 30°C and suggested that the 10^3-fold rise in the rate of crystal growth was due to the effect of this higher temperature on the interface kinetics. It was shown[36] that at atmospheric pressure spherulites are grown, and that either a growth rate of $\leq 10^{-4}$ cm/sec and a pressure of 5 kbar or a growth rate of $\leq 10^{-5}$ cm/sec and a pressure of 4 kbar is required to obtain single crystals.

It is believed that liquid selenium contains long polymer-like chains[24] which result in a high viscosity. Keezer and Moody[44] suggest that it is this high viscosity which causes the slow rate of growth at the solid–liquid interface. The viscosity can be reduced by high pressures as was done by Harrison and Tiller[38] or by adding impurities which break down the long selenium chains. Either method results in a faster growth rate and makes the production of single crystals much easier. Keezer and Moody,[44] adopted the latter technique by growing crystals from melts doped with 500–1000 ppm chlorine. The pulling speeds were 7×10^{-7} to 7×10^{-6} cm/sec with a rotation of 100 rpm. At lower rotational rates or higher pulling speeds, single crystals were not obtained. Attempts to grow crystals oriented along the c-axis always resulted in the formation of polycrystals. The concentration of chlorine in the crystals was found to be 300 ppm. X-ray diffraction by the Laúe technique did not reveal the presence of large defects like low angle grain boundaries. Growth was perpendicular to the c-axis in the [11$\bar{2}$0] direction.

After studying the structure of selenium single crystals, Fitton and Griffiths[26] concluded that growth of large crystals of low defect concentration is unlikely. The chains in the liquid can be terminated by impurities which are introduced by doping the melt.

Harrison[37] used the method of Harrison and Tiller[38] and showed that solidification begins in a polycrystalline manner at the apex of the crucible cone but the crystal best oriented for growth grows fastest to produce a single crystal. He found the c-axis varied randomly with respect to the crystal axis. By using oriented seeds, it was possible to grow crystals parallel to the [0001] and [11$\bar{2}$0] directions but not the [10$\bar{1}$0] direction. These latter crystals did not have a lineage structure.

Shiosaki et $al.$[79] report on a novel method of growing selenium single crystals 2.5 cm long and 1 cm in diameter. Amorphous selenium is put on a cleaved (10$\bar{1}$0) surface of a tellurium single crystal and sealed in a pyrex tube at 10^{-3} torr. The selenium is melted in the high temperature region of a Bridgeman type furnace where the tellurium is still solid and the capsule lowered at $6–12 \times 10^{-6}$ cm/sec to the low temperature region. The composition of the resulting selenium crystals was 2.3% Te near the interface and

0.04 % away from the interface.   Laue x-ray diffraction show the crystals to be of good quality.   The investigators considered that since the initial crystal formed at the surface of the tellurium was an alloy, it had a higher melting point and thus, a lower viscosity than pure selenium.

Single crystals of Se approximately 0.8 cm long and 0.3 mm thick have also been produced by crystallization of Se from a 1.5 N Na_2S solution by slowly cooling a saturated solution from 50°C to 25°C.[47]

MECHANICAL BEHAVIOR

Data on the plastic deformation of selenium is scanty.   Since, however, selenium is very much like tellurium, information on the latter might give some insight into the deformation behavior of selenium.   Stokes et al.[83] have deformed tellurium single crystals and shown that slip occurs on the $\{10\bar{1}0\} < 11\bar{2}0 >$ system.   These planes are also planes of cleavage. In specimens oriented for single slip, the CRSS was 300 psi and gave 20 % elongation.   On the other hand, crystals deformed along the [0001] direction were strong and brittle, fracture occurring at 20,000 psi without plastic deformation.   This behavior was correlated with the chemical bonding between chains and in the chains.

The work of Harrison and Tiller[38] showed that cleavage of selenium single crystals does occur on $(10\bar{1}0)$ planes and sometimes on $(10\bar{1}2)$ planes. Cleavage does not occur on a single prism plane and considerable surface damage was observed.   These authors made hardness measurements on the prism planes, i.e., parallel to the selenium chains and obtained a Knoop hardness of 6 with plastic deformation occurring between the indentations. Hardness measurements perpendicular to the selenium chains gave a Knoop hardness of 29 and produced brittle fracture on the $(10\bar{1}2)$ planes.

Several authors have published data on the elastic constants of selenium. Vedam et al.[87] claim the first measurements of elastic constants on hexagonal and glassy phases.   They grew crystals 2–3 in. long and 0.5 in. diameter by the method of Harrison and Tiller, which always contained low angle tilt boundaries.   They were only able to take measurements along the [0001] direction because of the high attenuation of ultrasonic waves by scattering from the low angle boundaries.   They found that at 25°C, $C_{33} = 8.02 \pm 0.05 \times 10^{11}$ dyn/cm^2 and $C_{44} = 1.83 \pm 0.02 \times 10^{11}$ dyn/cm^2. In the temperature range 15–35°C, $dC_{33}/dT = -2.3 \pm 0.1 \times 10^8$ dyn/cm^2/°C and $dC_{44}/dT = -1.7 \pm 0.1 \times 10^8$ dyn/cm^2/°C.

Mort[64] was able to grow good quality crystals by the method of Keezer[45] and thereby was able to measure the 6 elastic constants.   These are listed in Table 15-2.   The estimated volume compressibility was 1.08×10^{-11} cm^2/dyn.   Mort suggested that the weaker van der Waals forces between

TABLE 15-2   ELASTIC CONSTANTS OF HEXAGONAL SELENIUM AT 300°K

Elastic Stiffness Const	(10^{11} dynes/cm^2)	Calculated Elastic Compliances	(10^{-13} cm^2/dyn)
C_{11}	1.87	S_{11}	143
C_{12}	0.71	S_{12}	-6
C_{13}	2.62	S_{13}	-53
$[C_{14}]$	0.62	$[S_{14}]$	62
C_{33}	7.41	S_{33}	47
C_{44}	1.49	S_{44}	116

chains determine the C_{11} constant while the strong covalent bonds, within the chains, determine the C_{33} constant.

Using the above data in the Voigt-Reuss-Hill approximation, Sirdeshmukh and Sabhadra[80] were able to calculate the elastic constants as follows, K (bulk modulus) $= 1.74 \times 10^{11}$ dyn/cm^2, G (shear modulus) $= 0.92 \times 10^{11}$ dyn/cm^2, E (Young's modulus) $= 2.34 \times 10^{11}$ dyn/cm^2 and v (Poisson's ratio) $= 0.27$. They also found the compressional anisotropy A^* and the shear anisotropy G^* given by Chung and Buessem:[18]

$$A^* = \frac{K_v - K_R}{K_u + K_R} \times 100 = 47\%$$

$$G^* = \frac{G_v - G_R}{G_v + G_R} \times 100 = 16\%$$

Finally they calculated the Debye temperature as 151.9°K which compares well with the value of 151.7°K obtained from specific heat data by Gscheidner.[32]

Compressibility studies by Vereschchagin et al.[88] on selenium showed considerable anisotropic behavior since the a parameter increased while the c parameter decreased in the hexagonal structure. Yet Krishchunas and Mikalkevichyus[51] in a similar investigation found that the a parameter decreased and the c parameter increased.

ALLOYING BEHAVIOR

Selenium Binary Systems

The large ionic radius of selenium suggests a low solid solubility in other elements. A great deal of published data is not available on binary systems with selenium but some have been collected in Table 15-3. The data available indicate a propensity for the formation of compounds. Many of these compounds, particularly of the XSe type, melt congruently at fairly high temperatures.

TABLE 15-3   SELECTED SELENIUM BINARY SYSTEMS[a]

System	Liquid Solubility Wt% Se	Maximum Solid Solubility Wt% Se	Principal Compound and Melting Point °C	Reference
Al–Se	Complete	None	Al_2Se_3 953	17
B–Se		0.4% at 270°C Negligible at 25°C	B_2Se_3 Degenerate eutectic	9 34
Bi–Se	Complete except for miscibility gap between 50–95% Se above 618°C		Bi_2Se_3 706 (melts congruently)	25
Cd–Se	None	None		34
Co–Se	Miscibility gap between 4–37% Se above 1448°C	None	CoSe 1055	25
Cu–Se	Miscibility gap between 5–37% Se above 1107°C No data above 44% Se	0.015% at 800°C	Cu_2Se 1148	34 25
Fe–Se	Miscibility gap between 4–48% Se above 1520°C and 80–99.9 + %Se above 790°C. Sol ~ 3%Se at 1520°C	Sol. of Fe in Se at 220°C 25 ppm	FeSe 1070 (melts congruently)	22 48 56
Ge–Se	Monotectic reaction at 905° between 17–40%Se	11×10^{-6}%	GeSe 675; at 587°C eutectic of GeSe and $GeSe_2$ (melts incongruently)	43 25

Continued

TABLE 15-3—*Continued*

System	Liquid Solubility Wt% Se	Maximum Solid Solubility Wt% Se	Principal Compound and Melting Point °C	Reference
Mn–Se			MnSe 1535	34
Ni–Se	Complete	Probably none	NiSe 980	25
Pb–Se	Miscibility gap 55–97% Se above 681°C	0.0015% at 300°C	PbSe 1076 (melts congruently)	34 25
S–Se	Complete	50% Se in β S at 105°C 25% Se in α S at 75°C		34
Sb–Se	Miscibility gap 12–36% Se above 572°C	None	Sb_2Se_3 617 (melts congruently)	34
Sn–Se	Possible Miscibility gap 12–38% Se	≪0.04% Se at 200°C	SnSe 860	34
Te–Se	Complete	Complete		34
Tl–Se	Miscibility gap 1–15% Se at 354°C and 60–95% Se at 202°C	None	Tl_2 Se 390 TlSe 330	34
Zn–Se	Completely immiscible	None	ZnSe 1515	34
Zr–Se		20.5% Se	$ZrSe_3$	32a

[a] Additional Se binary systems are described by Hansen,[34] Elliott,[25] and Chizhikov and Shchastlivyi.[17]

Selenium Ternary Systems

A considerable amount of work has been done on ternary systems containing selenium, particularly in the Soviet Union.   Most of the work seems to have been done at sections which join two binary compounds.   Systems which have been investigated are given in Table 15-4.

TABLE 15-4   TERNARY SYSTEMS OF SELENIUM

System	Section	Principal Characteristics	Reference
Ag–Pb–Se	Ag_2Se–PbSe	Eutectic type quasi-binary system	68
As–Cu–Se	$AsSe_{1.5}$–Cu		76
Co–Ni–Se	580°C isotherm		35
Fe–Ni–Se	580°C isotherm		74
Mn–S–Se	MnS–MnSe	Complete solid solubility	61
Pb–Hg–Se	PbSe–HgSe	Quasi-binary eutectic system	86
Pb–S–Se	PbS–PbSe		
Pb–Te–Se	PbTe–PbSe	Quasi-binary system with complete solid solubility	31
Ni–Te–Se	580°C isotherm	Continuous range of solid solution between 2-phase regions $NiSe_{1.02}$–$NiSe_{1.3}$ and $NiTe_{1.09}$–$NiTe_2$	75
Sn–Sb–Se	$SnSe$–Sb_2Se_3	Peritectic between β-SnSe and liquid at 572°C to give α-SnSe. Ternary compound $Sn_2Sb_6Se_{11}$ at 563°C	90
Sn–Te–Se	SnTe–SnSe 300–900°C	Quasi-binary eutectic system Eutectic point $SnTe_{.55}Se_{.45}$ and 755°C.	85

OXIDATION AND CORROSION

Very little work appears to have been done on the oxidation of selenium. Dry oxygen or air has no effect on the oxidation of Se at room temperature. One study by Mekhtieva et al.[62] has been published in which the kinetics of reoxidation of previously deoxidised selenium by heating at 215°C are described.   It was found that the rate of oxidation increased sharply for the first 20–40 h after which the rate passed through a maximum and then diminished monotonically.   The height of the maximum and the general appearance of the kinetic curves were influenced by the presence of antimony. The general instability of the physical parameters of selenium was attributed to the diffusion of oxygen through the Se lattice; the number of oxygen atoms diffusing was directly related to the electrical conductivity of the material.

Cramarossa et al.[20] have studied the oxidation of solid and liquid selenium by gaseous oxygen in the temperature range 50–160°C. The rates of selenium oxidation have been determined between 200 and 330°C. The primary reaction product is SeO which undergoes further oxidation in the gas phase to SeO_2. The dependence of the rates on the concentration of oxygen is different below and above the melting point and the temperature dependence is complex. The chemical and the evaporation regimes of the reaction have both been observed.

The corrosion of heat resistant high alloy steels by molten selenium has been studied by Soshnikova and Paikin.[82] They used several steels and found that they were all corroded by molten selenium, the rate of corrosion increasing with temperature. In the range 230–310°C, the minimum corrosion was 0.1 gm/m^2 with a depth of corrosion of 0.1 mm/year. At 600°C, the corrosion depth increased to 50 mm/year.

DIFFUSION

Measurements of the self-diffusion of selenium in the temperature range 142–215°C have been made by Bratter and Gobrecht[14] using ^{75}Se as tracer. They found that in the direction parallel to the c-axis $D^{\parallel} = 0.2 \exp(^{-1.2}/kT)$ cm^2/sec, while in the perpendicular direction $D^{\perp} = 10^2 \exp(-1.4/kT)$ cm^2/sec. The corresponding vacancy migration enthalpies were 0.5 eV and 0.7 eV, respectively. They offered a vacancy diffusion model to explain their observations.

A recent study of the interaction of selenium in contact with an iron wire at 720°C showed that iron selenide (FeSe) formed continuously at the interface by the outward diffusion of iron through the selenide until all of the iron had been transported, leaving a cavity of the same size as that of the original wire.[92] However, the carbon dissolved in a 1% carbon steel wire treated with selenium at 1000°C appeared to retard the outward diffusion of iron so much that some selenium diffused into the steel to form iron selenide inclusions in the surface layer of the wire. When tellurium was used, instead of selenium, its inward diffusion rate through the telluride layer appeared faster than the outward diffusion rate of the iron except in the presence of carbon. The authors attributed these differences to stronger homopolar or metallic bonds in tellurides than in selenides. It should be noted FeSe appears to have a broader homogeneity range than FeTe, thus permitting faster total diffusion of iron atoms through the greater number of iron vacancies in the FeSe lattice.

METALLURGICAL USES

Ferrous Alloys

Casting of Iron and Steel. Trace amounts of selenium lower the surface tension of molten iron and steels. This effect is similar to that observed for oxygen, sulfur, and tellurium. The degree of lowering of the surface tension of iron by selenium is 54,600 dyn/cm/atom % compared to that of sulfur of 15,400 dyn/cm/atom %.[48] A value for the effect of tellerium is lacking, but its influence has been observed to be greater than that of selenium. A

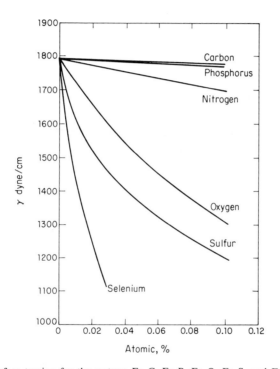

Figure 15-1 Surface tension for the systems Fe–C, Fe–P, Fe–O, Fe–S, and Fe–Se at 1550°C.

graphical comparison[49] of the influence of some of the more common elements on the surface tension of iron is presented in Figure 15–1. Metals, in general, have relatively little influence on the surface tension of iron in contrast to the strong influence of these electronegative elements. The selenium is considered to be adsorbed as a monolayer on the surface of liquid iron or steel. In this situation, it radically alters the nature of the surface and retards the absorption of nitrogen.[65]   Such a reduction of N_2 would be expected to improve the impact properties and reduce strain aging.

The effect of selenium on the structure of cast irons has been extensively investigated.[21]   It was found that the carbide-promoting action of selenium is closely related to the hydrogen content at the time of solidification. In iron of low manganese content (0.02%) selenium has a marked carbide-promoting effect when iron is melted and poured under conditions conducive to normal hydrogen content, as in green sand molds.   On the other hand, when manganese is present, regardless of the hydrogen content, selenium will not suppress the nucleation of graphite.   In contrast to this, manganese does not neutralize the carbide-promoting effect of tellurium.

It has been found that small quantities of selenium added to molten steel will act as a mild deoxidizer and induce grain refinement during solidification. Recent studies reported by the European Selenium-Tellurium Committee and by Almand[3] have also demonstrated the influence of selenium as a grain refiner in alloy steels favoring the formation of equi-axed grains which minimize the directional differences in properties.   The smaller grain size reduces the hardenability, sensitivity to overheating, and tendency to quench cracking.[29,30]   These characteristics suggest improved steels for carburizing and for direct hot rolling without further treatment.[1]

Sulfur in steels usually occurs as manganese sulfide which on hot rolling becomes elongated in shape.   The result of this morphology is an impairment of properties in the transverse direction; e.g., impact strength.   The addition of selenium alters the filamentary shape of the sulfide inclusions to a globular shape with a resultant improvement in transverse properties.[3] This effect is nicely illustrated in Figure 15-2 where two rings made from low alloy steel have been flattened.   The ring made from steel containing an addition of selenium has deformed without fracture whereas the ring with no selenium has fractured after only a small amount of deformation. Selenium also permits more severe cold forging of steel than does sulfur, while conferring improved machinability.[11]

Mannesmann[59] has confirmed this effect and finds that selenium forms selenide or selenide–sulfide inclusions which tend to be larger, more globular, and fewer in number than sulfide inclusions.   More importantly, hot work does not affect this change in inclusion morphology so that longitudinal and transverse mechanical properties become more nearly equal.

Several workers[89,58,33] have reported that pinhole porosity which frequently occurs in high alloy steels is reduced by the addition of 0.005–0.02% selenium.   Foundry practice reveals that elemental selenium, nickel selenide, or iron selenide are satisfactory additives for this purpose.   A recent discus-

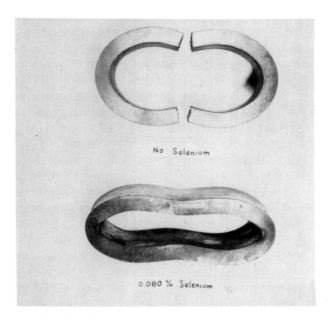

Figure 15-2   Flattening test on transversely cut rings of two 0.36% C, 0.79% Mn, 0.03% S, 0.015% P, 0.90% Cr steels, one without selenium and the other with 0.06% selenium. Original magnification reduced 0.4 times.[3]

sion of the preparation and properties of fused master iron–selenium alloys suggests higher selenium recovery in molten steel from master alloys containing about 40% selenium than from those alloys which contain selenium above 50%.[56]   Recent unpublished work suggests that even lower concentrations may be still more effective.

Free Machining Steels.   Several authors have reported that additions of selenium improve the machinability of both carbon and alloy steels.   Aborn[2] has shown that selenium can permit greater cutting speeds, lower power requirements, increase tool life, improve surface finish, and thus confer greater productivity.   It is generally believed that it is the presence of selenium as selenide–sulfide inclusions in globular form which bring about this improvement.   These inclusions affect the machining process by reducing the chip size, stabilizing the built-up edge, reducing the friction between the rake face of the tool and the chip, and by smearing across the nose of the tool, thereby minimizing or preventing welding.   Poole and Tipnis[71] have made similar observations.   Zaslavsky,[91] however, maintains that the main reason for the improvement in machinability is the formation of a protective coating from decomposed selenides.

TABLE 15-5[a]   FREE-MACHINING LOW CARBON STEELS WITH SELENIUM

Steel	C	Mn	P	S	Pb	Other
J & L type A[c]	0.15 (max)	0.85–1.15	0.04–0.09	0.26–0.35	0.15–0.35	Se added with or without nitrogen
Multicut[b]	0.15 (max)	0.85–1.15	0.04–0.09	0.26–0.35	0.15–0.35	Selenium and nitrogen added

[a] Data from Metal Progress, Oct. 1964.
[b] Republic Steel Corporation
[c] Jones and Laughlin Steel Corp.

Low carbon free-machining steels with selenium additions may also contain other elements such as sulfur, lead, nitrogen, bismuth, and tellurium. Several patents have been issued on these steels[40,70,41,67] and an example of a proprietary free-machining low carbon steel is listed in Table 15-5. This alloy may be considered to be modified AISI 12L14.

In the case of free-machining stainless steels, sulfur is the most popular additive. Selenium, however, is preferred in those applications where corrosion resistance, hot and cold formability, and surface finish are important considerations. Seven selenium-bearing free-machining stainless steels employed commercially in America are listed in Table 15-6.

Selenium has also been added to higher carbon steels such as AISI 1040 and to low alloy steels like 4140. Table 15-7 shows a comparison of their machinibility with the same steels containing other additives.[84]

TABLE 15-6 GRADES OF FREE MACHINING STAINLESS STEELS

Type[a]	C[b]	Mn[b]	Si[b]	Cr	Ni	Se or Other
Austenitic 303 Se	0.15	2.00	1.00	17–19	8–10	0.15 min
Austenitic CF-16F	0.16	1.50	2.00	18–21	9–12	0.20–0.35
Martensitic 416 Se	0.15	1.25	1.00	12–14		0.15 min
Ferritic 430 F Se	0.12	1.25	1.00	14–18		0.15 min
Austenitic 347 F[c]	0.05	1.25	0.50	17.5	9.5	0.13P, 0.30Se, 0.60 Nb
Martensitic 420 F	0.38	0.45	0.35	13.5		0.21Se or 0.18S
Martensitic 440 F	1.00	0.40	0.40	17.0		0.18Se or 0.08S

[a] All wrought except cast CF-16F
[b] Maximum.
[c] Last three nonstandard grades nominal compositions, *ASM Metals Handbook*, Vol. 1, 1961.

TABLE 15-7

Steel	Additive % Retained	Production[a] (Parts per h)	% Increase over Base Steel
1040	None	136	—
1040	0.12Se	193	42
1040	0.09Te	170	25
1040	0.10Bi	157	16
1040	0.19Pb, 0.05Bi	170	25
4140	None	111	—
4140	0.13Se	180	62
4140	0.08Te	160	44
4140	0.19Pb	160	44

[a] Maximum rate of production of standard parts per h.

The beneficial influence of selenium on the machinability of various stainless steels is shown in Figure 15-3 which compares the machinability ratings of selenium-containing steels with various grades of stainless steels based on a rating of 100% for AISI 1112.[8]   From these ratings it will be observed that the addition of selenium to 302 stainless steel (303 Se) improves its machinability by about 25% and to 430 ferritic stainless steel by 50%.   The degree of improvement, however, that can be expected in the machinability of the higher carbon grades of stainless steel by adding selenium is less pronounced due to the presence of carbide particles.

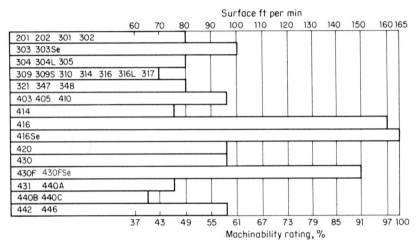

Figure 15-3   Comparison of machinability ratings of the standard stainless steels, based on 100% for AISI B1112, using high speed steel tools.

In France, a 303 austenitic grade has been developed with 0.15% selenium and 0.15% sulfur, thus maintaining the high free cutting quality of sulfur while minimizing its detrimental effect on transverse ductility and corrosion resistance by the selenium addition.[6]

The studies of Goldshtein[29] have shown that the addition of lead in conjunction with selenium further improves the machinability of stainless steel. If the conditions of machining, however, lead to the development of high temperatures, lead sweats will occur on the surface of the part.

Because of the significant influence of selenide compounds on the free machining properties of steels, Kiessling et al.[46] have conducted some basic studies on synthetic selenides and actual inclusions.   They observed that the solid solubility at 1150°C of the metals titanium, vanadium, chromium, iron, cobalt, and nickel in α-MnSe varies in the manner shown in Figure 15-4. Similar trends have been observed for the substitution of these elements for

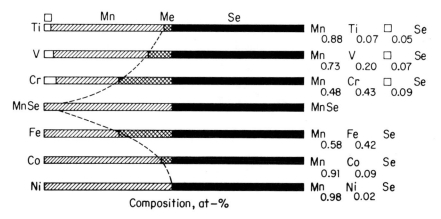

Figure 15-4   Composition of (Mn, Me) Se phases at solubility limit for first long-period transition metals.

manganese in MnS. This is not surprising since both MnSe and MnS possess a NaCl-type structure and form a continuous series of solid solutions.[61] The hardness of the relatively softer MnSe compound is modified by substituting the transition elements for manganese, or sulfur for selenium as shown in Figure 15-5.

The influence of temperature on the hardness of the alloys in the MnS–MnSe system was also studied by Riewald[73] and his results are shown in Figure 15-6. The substitution of the sulfide ion for the selenide ion and vice versa produces some solution hardening. No special significance is attached to the maxima observed other than averaging out of two solid solution effects. Some anisotropy was observed in the hardness of the pure phases and mixtures which was related to the crystallographic orientation. The basic mechanism of plastic deformation, fracture characteristics, and hardness relationships of compounds of manganese with Group VIA elements were also studied by Riewald. This has helped in the understanding of the role of these compounds as inclusions on the hot and cold forming characteristics of the matrix metal and the relationship of their influence on the machinability of steel.

Silicon Steels. The addition of selenium to silicon steels with 2–4% silicon has a significant influence on its secondary recrystallization texture.[41,7] Imai *et al.*[41] showed that with amounts up to 0.1% Se, the (001) [110] cube texture developed in silicon steels is enhanced, thereby improving the magnetic properties for use in transformer cores. The presence of residual selenium is considered undesirable and may be removed by heating the steel to a temperature above 1000°C in a reducing atmosphere such as hydrogen. In

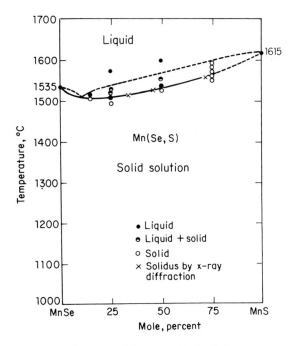

Figure 15-5   The system MnSe–MnS.

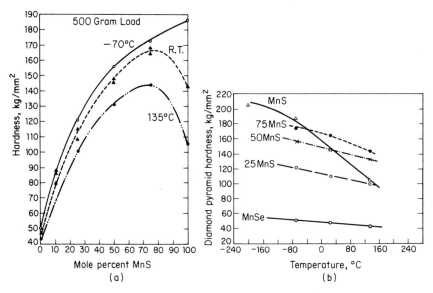

Figure 15-6   (a) Average diamond pyramid hardness of MnSe–MnS solid solutions at various temperatures. (b) Diamond pyramid hardness *vs* temperature for MnSe–MnS solid solutions.

their work on a 3 % Si steel, Benford and Stanley[7] found that on heating to a temperature of 1216°C in an atmosphere containing H_2Se, a (111) [211] texture developed. They suggested that the mechanism for this texture growth is due to surface energy rather than grain boundary energy.

Nonferrous Metals and Alloys

Copper. As with iron, selenium is highly surface-active in liquid copper, forming a monomolecular surface as was noted by Monma.[63] It has been reported by Smith[81] that selenium additions to copper improve its machinability without essentially affecting its strength or conductivity. Lead, bismuth, sulfur, and tellurium may also be used to improve the machinability of copper. However, there are limitations to their use. For example, leaded copper suffers from hot shortness and copper containing bismuth cannot be cold worked due to embrittlement. On the other hand, copper containing either sulfur, tellurium, or selenium is not susceptible to hot shortness or embrittlement. A comparison of the drilling characteristics of copper containing these three elements shows that selenium is the most effective for improving the machinability of copper (Figure 15-7).

The results of additions of selenium, sulfur and tellurium on the properties of copper are compared in Table 15-8. Again, selenium causes a greater improvement in machinability per unit of additive than either sulfur or tellurium. None of the additions has any effect on the strength of cold-drawn copper as might be expected, but reduced its ductility. Both selenium and sulfur slightly increase the yield strength of annealed copper whereas tellurium causes a slight softening. Again, there is a reduction in ductility as the selenium concentration increases. Notice particularly the very small effect on the conductivity.

The microstructure of copper containing selenium consists of primary copper and a eutectic of copper plus Cu_2Se. It is the presence of this copper selenide phase which serves to break up the chips during machining. It is recommended that for extended production runs carbide tools be used.

Historically, tellurium has been the primary additive used in free machining copper. This has been due mainly to the higher cost of selenium during the early stages of commercialization. Finally, it is to be noted that while neither selenium nor tellurium impair the soldering or brazing properties of copper, they do impair the weldability.[19]

Nickel-Iron and Cobalt-Iron Alloys. Selenium improves the machinability of these alloys which are used for electrical purposes, as shown in Table 15-9. It is preferable to sulfur or tellurium, as the latter cause hot brittleness in these alloys.[16] Harrison *et al.*[39] found that the addition of 0.4–0.5 %

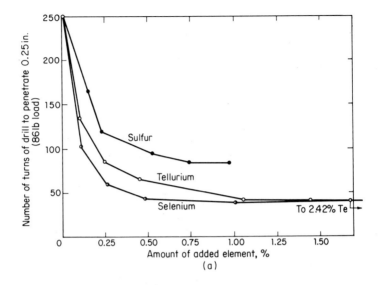

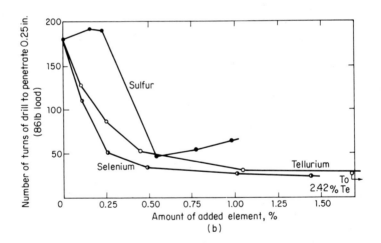

Figure 15-7 (a) The effect of sulfur and its congeners on the machinability of copper (hard-drawn). (b) The effect of sulfur and its congeners on the machinability of copper (annealed 600°C).

TABLE 15-8 PROPERTIES OF COPPER ALLOYS CONTAINING SULFUR, SELENIUM AND TELLURIUM[81 a]

Comp %	Cold Drawn					Annealed 1 h at 600°C				
	YS ksi	UTS ksi	% Elongation in 1.4 in	Cond. IACS	Machinability %[b]	YS ksi	UTS ksi	% Elongation in 1.4 in	Cond. IACS	Machinability %[b]
99.99 Cu	50.7	52.1	21.4	99.7	115	8.7	32.5	60	102.1	22
0.15 S	49.6	52.5	14.3	97.8	24	8.1	32.8	53.6	99.3	20
0.23 S	51.4	52.5	15.0	97.6	33	7.8	33.0	51.4	99.6	21
0.54 S	52.5	54.3	12.1	95.1	41	9.0	33.6	50.7	96.7	81
0.78 S	56.5	56.5	7.9	91.6	45	11.0	34.8	45.7	92.5	70
0.97 S	53.1	56.7	8.6	90.9	45	11.1	34.9	45.7	91.4	70
0.11 Se	51.2	53.2	19.3	98.9	38	8.4	32.8	57.1	99.0	35
0.26 Se	51.6	53.0	15.7	98.0	64	8.9	32.5	55.7	100.1	63
0.48 Se	50.9	53.6	15.7	97.4	89	8.0	32.8	52.9	98.8	112
1.01 Se	49.9	53.4	11.4	94.4	95	8.5	33.3	48.6	95.6	140
1.44 Se	52.0									
0.10 Te	50.7	53.1	15.0	99.5	29	7.9	32.8	53.6	99.8	30
0.25 Te	51.5	53.4	14.3	99.9	45	7.9	33.0	54.3	100.6	45
0.45 Te	50.7	53.1	11.4	98.2	38	8.1	33.0	51.4	99.9	72
1.05 Te	51.2	53.3	10.7	97.1	91	8.3	33.2	41.4	98.0	126
2.42 Te	51.1	54.6	7.1	94.5	91	8.8	33.3		96.1	134

a Alloys were hot rolled from 3 inch to 0.625 inches, annealed and cold-drawn to 0.5 inches in one pass.
b Machinability expressed as % of machinability of leaded brass.

TABLE 15-9   FREE MACHINING Fe–Ni AND Fe–Co ALLOYS[16]

Alloy Use	%Ni	%Co	%Se	Other	%Fe
Low expansion	36	—	0.20	—	Bal
High permeability	49	—	0.15	—	Bal
High magnetostriction	—	49	0.20	2%V	Bal

Se to a 35% Co, 5% Ti iron-base alloy gave a columnar grained structure which more than doubles the coercive force of equi-axed grains in such a permanent magnet alloy.

Lead and Lead Base Alloys.   Information in the literature is scant regarding the influence of selenium on lead and its alloys. It would be anticipated that its influence would be similar to that of tellurium which raises the recrystallization temperature of lead and its alloys, thereby improving the mechanical properties.

It has been shown by Bradhurst *et al.*[15] that selenium reduces the surface tension of liquid lead and the addition of about 0.1% Se together with a similar amount of tellurium is reported to improve the fluidity of lead base solder and avoid coarse dendritic freezing.[66]

Miscellaneous Applications

Use in Plating Solutions.   The addition of selenium in the form of sodium selenate or selenic acid to a chromium plating bath will produce microcracked chromium[77] which when plated over copper and nickel results in the deposition of a corrosion-resistant plating for steel and zinc die castings. A typical bath contains 0.015–0.018 g/liter of selenate ion, 250 g/liter chromic acid, 1.2 g/liter of sulphuric acid and is operated at a temperature of 43–46°C and a current density of 150–200 amp/ft^2. The microcracked chromium plating offers beneficial corrosion protection of the base metal since the deposit contains a network of fine intersecting cracks which distribute the galvanic action uniformly over the surface of the part.

The effect of sodium selenate on both acid and alkaline cyanide plating solutions has been studied by Raub[72] using the radioactive isotope ^{75}Se. He found that in acid chromium, nickel and zinc solutions the deposits contained relatively large amounts of selenium. The quantity of selenium codeposited was dependent on the plating conditions and affected the structure of the deposits formed. In alkaline solutions, selenium is not codeposited with zinc and silver, but is deposited cathodically along with copper and gold, in relatively small quantities. Where selenium is codeposited with another metal from an alkaline bath, the structure of the metal is affected. The structure of the metal deposition from alkaline baths in presence of V_2SeO_3 is, however, unaffected. Similar studies have been conducted by Krusenstjern and Meier[52] who found the conditions under which the best deposits were obtained.

Ostrow[69] has developed the use of a soluble selenium compound as a primary brightener in conjunction with sulfur compounds for silver plating from a cyanide bath.[53]

Lainer et al.[53] have shown that traces of selenium which become mechanically entrapped in the recesses of copper cathodes in the form of slimes during the course of electrolysis can affect the subsequent properties. These slimes are reduced during melting of the cathode and the selenium enters into solid solution upon solidification. Even in amounts as low as 0.001%, selenium can lead to poor welding properties.

Lubricating Agent.   Several studies, by Boes,[10] McConnell,[60] Bowen et al.,[12] and Jamison and Cosgrove,[42] have been made on the lubricating properties of the selenides of tungsten, molybdenum, niobium, and tantalum. They have been found to be superior to graphite and molybdenum disulfide, especially for extended periods of time in vacuum applications beyond 10^{-6} torr. Brainard[13] has shown that the selenides of molybdenum and tungsten are superior to the corresponding sulfides in vacuum up to a temperature of 760°C because of their higher thermal stability. ·

The selenides may be used on surfaces as dry powders, or mixed with other powders, such as copper, silver, nickel, or gold to form machinable hot pressed composites. A composite of silver and niobium diselenide which combines both good electrical conductivity and good lubricity has useful applications in sliding electrical contacts.

Surface Agent.   The surface of steel and cast iron can be coated with selenium by diffusion from a molten salt bath. The surfaces treated in this manner improve the resistance to wear and the resistance to seizure of contact surfaces by a factor of 4.[88a] Surface coatings of selenium on magnesium alloy sheets offer protection against corrosion and provide a good base for paint.[55]

Powder Metallurgy.   The addition of 0.05–2.0% selenium to copper oxide, when reduced and mixed with nonmetallic friction materials, pressed and sintered, was found by Shafer[78] to make a superior friction brake facing with 30% higher pressed green strength, which is important for thin shapes. Selenium reportedly improves the adhesion of the metallic to the nonmetallic particles.

REFERENCES

1. Aborn, R. H., Selenium-Tellurium Symposium, New York, 1965.
2. Aborn, R. H., Amer. Smelting and Refining Co., Machinability Booklet (1969).
3. Almand, E. *Rev. Met.*, **66**, 749 (1969).
4. Asadov, Yal. G. and Bakirov, M. Ya., *Izvest. Akad. Nauk* SSSR, Neorg. Materialy, **5**, 459 (1969).
5. *American Society for Metals Handbook*, Vol. 1, 44 (1961).
6. Bellot, J. and Herzog, E., Symposium on Selenium and Tellurium in Iron and Steel, Swedish Inst. Met. Res., (1969).
7. Benford, J. G. and Stanley, E. B., *Trans. Met. Soc.* AIME, **242**, 1763 (1968).
8. Blott, D. M., *Metal Progress*, **86**, 119 (1964).
9. Boryakova, V. A., Grinberg, Ya. Kh., Zhukov, E. G., Koryazhkin, V. A., and Medvedeva, Z. S., *Izvest. Akad. Nauk.* SSSR, Neorg. Materialy, **5**, 260 (1969).
10. Boes, D. J., ASLE Preprint 65 AM-5C3 (1965).
11. Bougler, F., to be published.
12. Bowen, P. H., Boes, D. J., and McDowell, J. R., *Mach. Design*, **35**, 139 (1963).
13. Brainard, W. A., *NASA Tech. Note* D5141, (1969).
14. Bratter, P. and Gobrecht, H., *Phys. Stat. Sol.*, **37**, 879 (1970).
15. Bradhurst, D. H. and Buchanan, A. S., *J. Phys. Chem.*, **63**, 1486 (1959).
16. Carpenter Technology Corp., "Alloys for Electronic and Magnetic Applications," (1965).
17. Chizhikov, D. M. and Shchastlivyi V. P., *Selenium and Selenides*, translated from the Russian by E. M. Elkin, Collet's Ltd., London, 1968.
18. Chung, D. H. and Buessem, W. R. M., *Anisotropy in Single Crystal Refractory Compounds*, Plenum Press, New York, 1968.
19. Copper Development Association, "Machining of Copper and its Alloys," Pub. 34, London (1961).
20. Cramarossa, F., Molinari, E., and Rio, B., *J. Phys. Chem.*, **72**, 84 (1968).
21. Dawson, J. V., *Trans. AFS*, **77**, 113 (1969); *Cast Metals Res. J.*, 5, 138 (1969); *Iron* and *Steel* **43**, 397 (1970).
22. Dutrizac, J. E., Janjua, M. B. I., and Toguri, J. M., *Can. J. Chem.*, **46**, 1171 (1968); *Can. Met. Quart.*, **7**, 91 (1968).
23. Dzhalilov, S. U., and Khalilov, Kh. M., *Zhur. Fiz. Khim.*, **42**, 1794 (1968).
24. Eisenberg, A., and Tobolsky, A. V., *J. Polymer Sci.*, **46**, 19 (1960).
25. Elliott, R. P., *Constitution of Binary Alloys*, First Supplement, McGraw-Hill, New York, 1965.
26. Fitton, B., and Griffiths, C. H., *J. Appl. Phys.*, **39**, 3663 (1968).
27. Gagnebin, A. P., *Amer. Foundryman*, **12**, 43 (1947).
28. Garber, I., *J. Iron Steel Inst.*, **181**, 291 (1955).
29. Goldstein, Y. E., *Stal.* **2**, 156 (1959); *Mach. and Tooling*, **37**, 34 (1966).
30. Goldstein, Y. E., and Zhizhakina, O. D., *Stal.* **9**, 684 (1961).
31. Grimes, D. E., *Trans. Met. Soc.* AIME, **235**, 1442 (1965).
32. Gschneidner, K. A., *Solid State Phys.*, **16**, 276 (1964).
32a. Hahn, H., and Ness, P., *Z. anorg. u. allgem. Chem.* **302**, 37 (1959).
33. Hall, A. M., and Sims. C. E. *Metals Eng. Quart.* **10** (1), 53 (1970).
34. Hansen, M., *Constitution of Binary Alloys*, McGraw-Hill, New York, 1958.
35. Haraldsen, H., Mollerud, R., and Rost, E., *Acta Chem. Scand.*, **21**, 1727 (1967).
36. Harrison, D. E., *J. Appl. Phys.*, **36** 3150 (1965).
37. Harrison, J. D., *Ibid.*, **39**, 3672 (1968).

38. Harrison, D. E., and Tiller, W. A., *Ibid.*, **36**, 1680 (1965).
39. Harrison, J., and Wright, W., *Cobalt*, **35**, 63 (1967).
40. Holowaty, M. O., U.S. Patent 3,152,890 (1964).
41. Imai, M., *et al.*, U.S. Patent 3,157,538 (1964).
42. Jamison, W. E., and Cosgrove, S. L., Thesis, University of Cincinnati, Cincinnati, Ohio (1969).
43. Karbonov, S. G., Zlomanov, V. P., and Novoselova, A. Y., *Vestn. Moskva. Univ.*, **3**, 93 (1968).
44. Keezer, R. C. and Moody, J. W., *Appl. Phys. Letters*, **8**, 233 (1966).
45. Keezer, R. C., and Wood, C., *Proc. Intl. Conf. on Cryst. Growth*, Boston, Pergamon Press, 1967, p. 119.
46. Kiessling, R., Hassler, B., and Westman, C., *J. Iron and Steel Inst.*, **205**, 531 (1967).
47. Kolb, E. D., *The Physics of Selenium and Tellurium*, Cooper, W. Charles ed., Pergamon Press, 1969, p. 155.
48. Kozakevitch, P. and Urbain, G., *Mem. Sci. Rev. Met.*, **58**, 517, 931 (1961).
49. Kozakevitch, P., ASM Symposium "Bismuth, Selenium and Tellurium in Iron and Steel," 1969.
50. Kozyrev, P. T., *Sov. Phys. Tech. Papers*, **28**, 470 (1958).
51. Krishchunas, V. Yu, and Mikalkevichyus, M. P., *Fizika Tverdogo Tela*, **8**, 3429 (1966).
52. Krusenstjern, A. V., and Meier, M., *Metall.*, **22**, 688 (1968).
53. Lainer, D. I., Cherkashina, N. V., and Brik, L. M., *Tsvet. Metally*, **5**, 86 (1968).
54. Lange, V. N., Lange, T. I., and Tikov, V. A., *Sov. Phys. Cryst.*, **12**, 456 (1967).
55. LeBrocq, L. F., and Cole, H. G., *Protection and Electrodeposition of Metals*, His Majesty's Stationery Office, London, 1951, p. 259.
56. Lundquist, S., Symposium on Selenium and Tellurium in Iron and Steel, Swedish Inst. Met. Res., 1969.
57. Mamedov, K. P., and Nurieva, Z. D., *Sov. Phys. Cryst.*, **12**, 605 (1968).
58. Mangone, R. J., Hall, A. M., Bryan, W. T., and Sims, C. E., *J. Metals*, **10**, 810 (1958).
59. Mannesmann, A. G., Belgian Patent 615,573 (1962).
60. McConnell, B. D., *Prod. Engr.*, **32**, 70 (1961).
61. Mehta, J. M., Riewald, P. G., and VanVlack, L. M., *J. Amer. Cer. Soc.*, **50**, 164 (1967); *Ibid.*, **50**, 1522 (1969).
62. Mekhtieva, S. I., Abdinov, D. Sh., and Alliev, G. M., *Izvest. Akad. Nauk* SSSR, Neorg. Mat., **5**, 1522 (1969).
63. Monma, K. *et al.*, *Nippon Kinzoku, Gakkati*, **24**, 544 (1960).
64. Mort, J., *J. Appl. Phys.*, **38**, 3414 (1967).
65. Mowers, R. G., and Pehlke, R. D., *Met. Trans.*, **1**, 51 (1970).
66. Mustzel, P. S., Brit. Patent, 1,073,861 (1967).
67. Nachtman, E. S., U.S. Patent, 3,250,647 (1966).
68. Novoselova, A. V., Shleifman, Zh. G., Zlomanov, V. P., and Sloma, R. K., *Izvest. Akad. Nauk* SSSR, Neorg. Materialy, **3**, 1143 (1967).
69. Ostrow, B. D., U.S. Patent 2,777,810 (1956).
70. Palmer, F. R., U.S. Patent 2,009,713 (1935); 2,009,714 (1935).
71. Poole, S. W., and Tipnis, V. A., ASM Symposium "Bismuth, Selenium and Tellurium in Iron and Steel," 1969.
72. Raub, E., *Metalloberflache*, **22**, 1 (1968).
73. Riewald, P. G., thesis, University of Michigan, Ann Arbor, Michigan (1968).

74. Rost, E., and Haugsten, K., *Acta. Chem. Scand.*, **23**, 388 (1969).
75. Rost, E., and Vestersjo, E., *Ibid.*, **22**, 2118 (1968).
76. Savan, Ya., Kozhina, I. I., Orlova, G. M., and Binder, Kh., *Izvest. Akad. Nauk SSSR, Neorg. Materialy*, **5**, 492 (1969).
77. Safranek, W. H., and Hardy, R. W., *Plating*, **4**, 1027 (1960).
78. Shafer, W. M., U.S. Patent 3, 383, 198 (1968).
79. Shiosaki, T., Kawabata, A., and Tanaka, T., *J. Appl. Phys.*, **39**, 3502 (1968).
80. Sirdeshmukh, D. B., and Subhadra, K. E., *Ibid.*, **40**, 5404 (1969).
81. Smith, C. S., *Trans. AIME*, **128**, 325 (1938).
82. Soshnikova, L. A., and Paikin, V. A., *Tsvet. Metally.* **2**, 75 (1969).
83. Stokes, R. J., Johnston, T. L., and Li, C. H., *Acta Met.*, **9**, 415 (1961).
84. Tata, H. J., ASM Symposium, Bismuth, Selenium and Tellurium in Iron and Steel, 1969.
85. Totani, A., Akazaki, H., and Nakajima, S., *Trans. Met. Soc.* AIME, **242**, 709 (1968).
86. Vanyarkho, V. G., Zlomanov, V. P., and Novaselova, A. V., *Vestn. Moskva Univ.*, **6**, 108 (1968).
87. Vedam, K., Miller, D. L., and Roy, R., *J. Appl. Phys.*, **37**, 3432 (1966).
88. Vereshehagin, L. F., Kabalkinz, S., and Shulenin, B. M., *Doklady Akad Nauk SSSR*, **165**, 297 (1965).
88a. Vinogradov, Iu. M., *Mettalloved. Term. Obrab. Metal.*, **10**, 36 (1965).
89. Wilcox, R. J., *Alloy Casting Bull.*, No. 15 (1951).
90. Wobst, M., *J. Less Common Metals*, **14**, 77 (1968).
91. Zaslavskii, A. Ya., Gol'dshtein, Ya. E. and Shenk, R. I., *Metalloved. Term. Obrab. Metal.*, **9**, 52 (1967).
92. Zaslavskii, A. Ya., Gorokh, A. V., Goldstein, Ya. E., and Shenk, R. I., *Fiz. Khim. Mekh. Mat.*, **5**, 243 (1969).

The following articles, not referred to in the chapter, will be useful to the interested reader:

Bellot, J., "Influence des inclusions de composés du soufre, du selenium et du tellure sur la déformabilité a chaud et les propriétés mécaniques des aciers", Thesis Docteur-Ingénieur, University of Nancy, France, 1972.

Belov, A. D., Alferov, V. P., Vilim Yu, V., Goryachev, A. D., Kirillova, V. P., Sedov, V. V., and Kosobokov, E. A., "Cast, easily machinable carbon steels alloyed with sulfur, selenium, and tellurium," *Liteinoe Proizvod*, **5**, 14 (1970); *Chem. Abstr.*, **73**, 47751 (1970).

Boulger, F. W., Becker, J. R., Olofson, C. T., and Henning, H. J., "How selenium and tellurium affect formability and machinability of low-carbon steels for cold forging," *Metals Eng. Quart.*, **12** (1) 17 (1972).

Dutton, W. A., Janjua M. B. I., Van den Steen, A. J., and Watkinson, A. P., "Vapour Phase oxidation of Selenium", *Can. Met. Quart.*, **10**, 97 (1971).

Herzog, E., Bellot, J., Gillot, P., and Rolin, M., "Austenitic stainless steel having improved machinability and good corrosion resistance," Fr. 1, 602, 444, 31 Dec 1970, *Chem. Abstr.*, **75**, 39472 (1971).

Janjua, M. B. I., Toguri, J. M., and Cooper, W. Charles, "Nature of Colored Films produced on Copper by Selenious Acid," *Can. Met. Quart.*, **7**, 133 (1968).

Janjua, M. B. I., Yannopoulos, J. C., and Cooper, W. Charles, "The Corrosive Action of Selenium towards Various Materials in the temperature Range 300 to 700°C," *Corrosion by Liquid Metals*, Plenum Press, New York, 1970, p. 339.

Larina, L. S., and Chernov, V. M., "Alloying magnico alloys with selenium additions," *Liteinoe Proizvod*, **9**, 34 (1969); *Chem. Abstr.*, **72**, 34942 (1970).

Malagari, F. A., Jr., "Silicon steels containing selenium," U.S. 3, 556, 873, 19 Jan. 1971; *Chem. Abstr.*, **74**, 69552 (1971).

16

Selenium in Electrophotography

GERALD LUCOVSKY / MARK D. TABAK

Xerox Research Laboratories, Rochester, N.Y.

INTRODUCTION: THE XEROGRAPHIC PROCESS AND THE IDEAL PHOTORECEPTOR

This chapter considers the microscopic processes of free carrier photogeneration and transport which occur in evaporated films of vitreous Se and which determine the performance of that material as a photoreceptor in the xerographic electrophotographic process. Numerous review articles[1,2] and papers[3,4] have been written describing the xerographic process and the photoelectronic properties of vitreous Se. However, a review of this subject matter is in order, owing to recent work[5,6] which shows that the rate controlling process in the photodischarge is the field controlled generation of free carriers[5,6,7] rather than a field dependent carrier range.[3] Before discussing the role of the photoreceptor in xerography and the properties of an idealized photoreceptor, we will first briefly describe the essential elements of the xerographic process.

Xerography is the most highly developed form of electrophotography in present commercial use.   Figure 16-1 illustrates the sequential steps involved in generating a copy.   These process steps are essentially those that take place in commercial machines which employ reuseable photoreceptors.

Step 1:   Sensitization of the Photoreceptor.   The photoreceptor consists of a vacuum-deposited film of Se, approximately 50 μ thick, on an Al substrate.   Sensitization is accomplished by electrostatic charging from a corona discharge.   Fields required for sensitization are typically 10^5 V/cm.

Step 2:   Exposure and Latent Image Formation.   The sensitized photoreceptor is exposed to a light and dark image pattern.   In light-struck areas, the surface potential of the photoconductor is reduced due to a photoconductive discharge.   Since current can only flow normal to the film surface, this step produces an electrostatic potential distribution which replicates the light-dark pattern of the image.

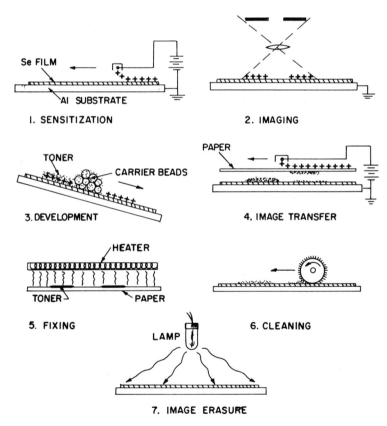

Figure 16-1   Process steps in reuseable xerography.

Step 3: Development of the Image. This is accomplished using a developer which consists of a mixture of black (or colored) toner particles, typically 10μ in diameter, and carrier beads, typically 100 μ in diameter. Toner particles are charged triboelectrically in this process and are preferentially attracted by either the surface fringe field at light-dark boundaries, or in systems employing a development electrode, by the absolute potential in the dark areas. They adhere to the photoreceptor, forming a visible image corresponding to the latent electrostatic image.

Step 4: Image Transfer. This is most simply accomplished using electrostatic transfer, wherein the paper is appropriately charged and the toner particles are attracted to the paper.

Step 5: Print Fixing. The powder image is made permanent by fusing or melting the toner particles into the surface of the paper. This can be accomplished by heat, heat and pressure, or by solvent vapors.

Step 6: Cleaning. Following the transfer process, there is some toner still left on the photoreceptor. This is removed by either mechanical, cloth web or brush, or mechanical and electrostatic means.

Step 7: Image Erasure. This step removes any potential differences due to latent image formation by flooding the photoreceptor with sufficiently intense light to drive the surface potential to some uniformly low value (typically ~ 100 V corresponding to fields $\sim 10^4$ V/cm). At this point, the photoreceptor is ready for another print cycle.

As shown above, the primary function of the photoreceptor is that of a transducer which converts an optical image into an electrostatic image. To perform this task, the photoreceptor must be capable of being charged to sufficiently high fields as required in the sensitization step. The discharge or decay of this potential in the dark must be small in order to proceed through the process steps, preserving differences in surface potential between light-struck and dark areas.

At this point we can identify the electronic properties of an idealized photoreceptor. A simplified specification of idealized behavior provides a convenient basis for comparison with actual photoreceptors.[7] Referring to Figure 16-2, the model behavior is characterized in the following way:

(1) The surface charge density required to reach a given voltage is proportional to the voltage, with the geometric capacitance per unit area as the constant of proportionality.

(2) Each photon absorbed results in the transport of one electronic charge unit completely through the photoreceptor.

(3) There are no mobile carriers that are not photogenerated. Thus, there is no dark decay and no surface conductivity to neutralize the electrostatic image.

(4) The interface to the back conducting electrode is a perfect blocking

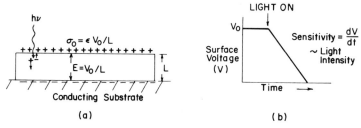

Figure 16-2   (a) Schematic of the electrical properties of an ideal photoreceptor.   Surface charge density is σ_0 and the initial voltage across the photoreceptor is V_0.   Photons of energy $h\nu$ each create one hole-electron pair, the hole being transported under the action of the applied electric field E to the conducting substrate and the electron reaching the top surface.   (b) Light discharge characteristic of the ideal photoreceptor.   There is zero dark decay until the light is turned on. The sensitivity or the rate of discharge is independent of the surface potential and directly proportional to the light intensity.   The total voltage discharge is directly proportional to the exposure; i.e. the light intensity multiplied by the time of exposure.

contact and does not allow charge injection into the photoreceptor, or the accumulation of an interfacial potential drop.

(5) The top surface is blocking and does not permit charge placed on the top surface to be injected into the bulk.   Also the charge cannot move laterally on the surface.

(6) There is no change in the photoreceptor characteristics with cycling.

(7) The properties of the photoreceptor are the same everywhere on the photoreceptor surface.

In this ideal photoreceptor, the potential drop is in direct proportion to the exposure.   The sensitivity defined as the slope of the photoreceptor voltage-exposure curve is constant and independent of the field, as shown in Figure 16-2.

First, we compare the characteristics of amorphous Se as a photoreceptor with those of the idealized photoreceptor discussed above.   Secondly, we summarize the physical properties of amorphous Se that are relevant to its operation as a photoreceptor, focusing on a photoelectronic characterization of the microscopic electrical and optical properties.   Thirdly, we develop a physical model for amorphous Se as a photoreceptor, defining the intrinsic limitations in this material and identifying their effects on its application in technology.   Finally, we treat some current trends in photoreceptor technology in which Se or Se alloys are the active elements.

AMORPHOUS SELENIUM AS A XEROGRAPHIC PHOTORECEPTOR

A comparison of the performance of amorphous Se as a xerographic photoreceptor with that of the ideal photoreceptor indicates that it deviates significantly from the ideal on three counts.   The most important of these deviations from ideal performance is related to the second characteristic (each absorbed photon results in the transport of one electronic charge completely through the photoreceptor) described in the list in the previous section. The deviations characteristic of Se are as follows:

(1) There is a significant spectral region, wherein there is strong optical absorption, but no measurable photodischarge.[4] The region extends from ~ 1.7 eV to ~ 2.3 eV.

(2) For all photon energies (or wavelengths) where there is photoconductivity, the photogeneration of free carriers is electric field dependent.[5]

(3) Every carrier, and here we include both holes and electrons, that is photogenerated is not transported completely through the photoreceptor. Failure on the third count also results in changes in the photoreceptor characteristics with cycling.

Figure 16-3 illustrates the photon energy dependence of the xerographic photosensitivity and the optical absorption. To emphasize the failure in photosensitivity at photon energies where there is strong absorption, we have included a plot of the fractional light absorption in a 40 μ thick Se film. Whereas there is very strong absorption throughout the red, orange, and yellow (1.7 eV to 2.2 eV), the photoconductivity or xerographic sensitivity exhibits a threshold well into the green (~ 2.4 eV). In terms of a practical limitation in copying machines using a Se photoreceptor, one cannot produce acceptable copy from colored originals having black characters on a red or orange background.

Figure 16-4 indicates the voltage sensitivity of the xerographic discharge (xerographic sensitivity is defined in terms of the rate of change of surface potential dV/dt at $t = 0$) for blue light. Under the conditions shown in Figure 16-4, $dV/dt|_{t=0}$ is a direct measure of the photogeneration rate of free carriers. For fields less than 10^5 V/cm the dependence of the discharge rate is approximately the same for all wavelengths of incident light. The inherent voltage sensitivity of the photoconductivity (through the carrier generation step) manifests itself in a system performance in which the photodischarge rate is not solely dependent on the incident light flux, but in addition depends on the electric field. The ideal photoreceptor would give a horizontal line at a level of dV/dt, corresponding to unity gain. The unity gain point is calculated by finding the rate of discharge corresponding to creation and complete transport of one electronic charge per absorbed photon.

With regard to transport limitations (items 3 on the list of deviations), there is trapping of both holes and electrons in the photodischarge process. In practice, electron trapping has a more deleterious effect on cycled xerographic performance since the range of the hole in Se is usually significantly longer than that of the electron. There are two cases of photodischarge to

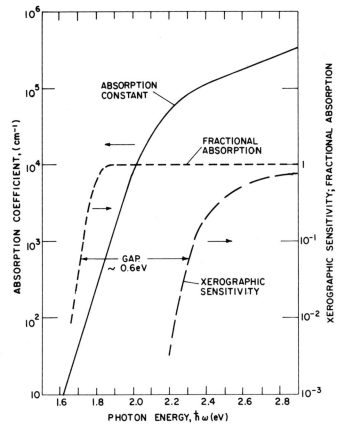

Figure 16-3 The photon energy dependence of (1) the optical absorption constant,[4] (2) the high field (10^5 V/cm) xerographic sensitivity in a 40 μ thick sample,[5] (3) the fractional light absorption in a 40 μ thick sample.

consider; i.e., those corresponding to sensitizing the photoreceptor with (1) positive or (2) negative surface charge.   In almost all machine applications, the photoreceptor is charged positively and hole transport is the dominant step in the photodischarge process.   Light of energy high enough to produce free carriers is always absorbed close to the illuminated surface.

On the other hand, in the negative charge mode, the situation is quite different.   The desirability of photodischarge for both positive and negative charging is prompted by the advantage of producing black on white copy from either a black on white or white on black original.   The simplest way to effect this is to reverse the sense of the charging step and then use all the other xerographic subsystems, i.e., development, transfer and erase, in the same mode.[1]

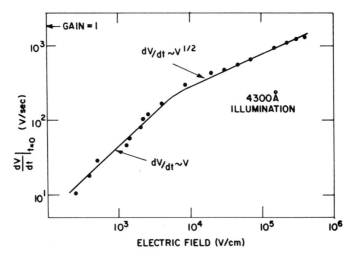

Figure 16-4   The voltage dependence of the xerographic sensitivity for a 44 μ thick amorphous selenium sample, charged positively under 4300 Å illumination.   The xerographic sensitivity is defined in terms of the rate of change of surface potential when the light is first turned on.

However, it is not so much the trapping, but rather the trap release time that is the important factor in cycled xerographic performance.   Holes are released from deep traps in times on the order of seconds or times comparable to machine cycle times.   As such, hole trapping does not contribute to a time-dependent build up of residual potential.   Residual potential is the voltage that remains even after the light erase step.   Electron release times from deep traps are of the order of minutes and therefore the residual potential build-up due to deep trapping is cumulative.   After a very low number of cycles in a negative charge mode, further photodischarge is not possible.   Since trapping effects are spatially preferential; i.e., they occur in the light struck regions, residual build-up is nonuniform spatially and produces ghosting effects when originals are changed.

ELECTRICAL AND OPTICAL PROPERTIES OF AMORPHOUS Se

The electronic properties of an ideal photoreceptor have been described and so has a comparison between this idealized behavior and the actual xerographic properties of amorphous Se.   It is the purpose of this section to introduce the physics of the optical and electrical properties of amorphous Se that have a direct effect on the xerographic behavior as described in the previous section.

Optical Properties Due to Electronic Transitions

The optical properties of a semiconductor in the vicinity of the band edge are normally quite sensitive to the band structure. Careful and reproducible optical absorption measurements[4,8-13] illustrated in Figure 16-5, show both an absorption edge and a high value of the optical absorption constant which persists over a considerable energy range. While these properties are characteristic of interband transitions, the width in energy of the absorption edge is far in excess of that expected for strong interband transitions; e.g., in a crystalline solid. Spectral response measurements[4,8,14-17] of quantum efficiency (quantum efficiency is directly proportional to xerographic sensitivity as long

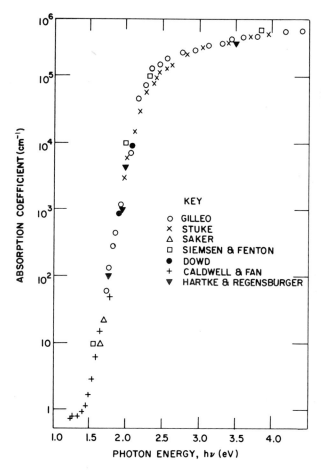

Figure 16-5 Collection of optical absorption measurements on amorphous selenium showing the reproducibility and the exponential edge.

as no transport limitation exists) is consistent with a band model since the number of free carriers generated per absorbed photon is close to unity[4,5,14] at short wavelengths and high fields ($\sim 10^5$ V/cm). However, as shown in Figure 16-6 and Figure 16-3, there is a gap of approximately 0.6 eV between the optical absorption edge and the photoconductive edge. Several models

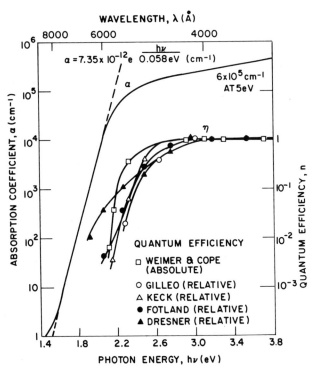

Figure 16-6 Optical absorption and quantum efficiency for amorphous selenium at room temperature. After Hartke and Regensburger.[4]

for this effect have been proposed. Laynon[18] has postulated that this behavior is due to transitions from filled states lying below the Fermi level to the conduction band. These transitions would produce optical absorption and electron photoconductivity but no hole photoconductivity, since the hole would be trapped in a bound state. However, Hartke and Regensburger[4] found equal quantum efficiencies for holes and electrons over the whole spectral range. They postulated that the photoconductive component of the absorption is due to interband transitions, while the nonphotoconductive component is attributed to a separate mechanism similar to intrinsic exciton absorption. This idea has since been corroborated by further optical measurements by Siemsen and Fenton[13] and field-dependent quantum efficiency measurements by Tabak and Warter[5] and Pai and Ing.[6] Possible evidence for the existence of an excitonic absorption has been found by Drews[19] from a peak in the reflectance spectrum of amorphous selenium.

This gap between the optical absorption edge and photoconductive edge has important xerographic implications. This particular limitation has stimulated interest in extending the photoconductive spectral response by addition of other elements to Se; for example, Te and As. These trends will be discussed in the section on Se alloy photoreceptors.

Electrical and Photoelectrical Properties of Amorphous Selenium

In this section we will review the pertinent electrical and photoelectrical properties of amorphous selenium, including dark conductivity, carrier injection, photogeneration, and transport. Fundamental solid state concepts such as carrier mobility, range and trapping lifetime can be used to describe the aforementioned properties. These electrical and photoelectrical microscopic properties can in turn be related to the xerographic performance of the amorphous selenium photoreceptor, including such properties as dark decay, residual potential and light discharge. The remainder of this section will be used to describe these fundamental electrical properties.

Dark Conductivity and Injection. Measurements of dark conductivity in amorphous selenium have been reported by Lanyon[18] and by Hartke.[20] In both cases, a low field ohmic characteristic was obtained, with Lanyon reporting resistivities ranging from 10^{11} to 10^{14} Ω-cm and Hartke reporting values of 10^{14} to 10^{16} Ω-cm. Xerographic dark decay measurements are more consistent with Hartke's generally higher values of resistivity; the higher values should be closer to the intrinsic resistivity since impurities would tend to decrease the resistivity in the conventional semiconductor sense. More recent measurements of steady state dark currents by Rossiter and Warfield[21] also are in agreement with the higher range of dark resistivities. The dark conductivity is thermally activated; Hartke reported values ranging from 0.96 to 1.10 eV. As the electric field is increased, the I-V curve starts to deviate from a linear relationship and the current becomes space charge limited. Both Hartke and Lanyon calculated a distribution of trapping states near the band edges from a study of the injected space charge limited currents. They both found a high density of localized levels (10^{19}–10^{21}/cm^3) near the band edges; such a high density might be expected in an amorphous material lacking long range order.

Photogeneration. By careful control of the experimental variables, Tabak and Warter[5] were able to separate transport and surface recombination from the photogeneration step.   They found that the rate of photogeneration is a function of the applied electric field and temperature   This is quite different from the normal band-to-band, electric-field-independent transition one expects in a covalently bonded semiconductor.   In amorphous selenium, an incoming photon may not produce a band-to-band transition but rather produces a localized excited state that may not contribute to the conductivity.

A detailed discussion of this effect occurs later, where a model describing the nonphotoconductive absorption is presented.   This model explains not only the gap between the photoconductive edge and the absorption edge, but the electric field and temperature dependence of the photogeneration step as well.

Electronic Transport. Both holes and electrons are mobile in amorphous selenium and their transport can be characterized by a drift mobility and a carrier lifetime.   Mobility is defined as the average carrier velocity per unit field, and carrier lifetime by the average time a free carrier remains in the band.   If shallow trapping is present, that is, where the carriers in the band are in equilibrium with some shallow localized level, the simple definitions have to be applied carefully.   If an experiment can be performed very quickly (before the free carriers come into equilibrium with the shallow traps) or if it is inherently sensitive to only those carriers which are free, then the traps will have no measurable effect, the mobility will correspond to the "microscopic" mobility, and the carrier lifetime will be the actual free time

If, however, the drift mobility is measured by finding a transit time and shallow trapping is present, then the measured drift mobility will be reduced from the microscopic mobility by the ratio of the free time to the time the carrier is immobilized in shallow traps.   The measured carrier lifetime will also be increased by this same factor.

Direct drift mobility measurements can be easily performed by photo-injecting a narrow sheet of free charge (narrow with respect to sample thickness) at one end of the sample and measuring the time it takes for the sheet of charge to reach the opposite electrode.   Because of the long dielectric relaxation time, holes and electrons can be separated under the influence of a moderate electric field and their contribution to the total current measured individually.   If the light used for creating the free charge is absorbed near the surface, then only the carrier of the same polarity as that of the illuminated surface will contribute to the signal.   As the sheet of charge drifts across the sample, it will induce a charge on the electrodes which is proportional to the product of the number of charges in the sheet and the distance they have moved.   When the charge sheet hits the opposite electrode, there will be an abrupt change in the slope of the voltage-time trace which defines the transit time.   Knowing the sample thickness and the dc applied voltage enables one to calculate the average velocity, and hence the drift mobility from a simple distance over time formula.

If carriers are lost during transit because of deep trapping, the transit time will become less well-defined and also cause the pulse height (total change of voltage across the sample) to become electric field dependent.   The theoretical dependence of the pulse height or field due to trapping is termed a Hecht curve.   If one assumes that the field dependence of the experimental pulse height curve is due entirely to bulk trapping, then by matching the data to a Hecht curve one can calculate a free carrier range.   The range is defined as the mean distance a carrier drifts before trapping per unit field and is given by the product of the drift mobility and the carrier lifetime.   Unfortunately, the carrier generation rate is also electric field dependent so that the range measured from a Hecht curve-type analysis can be seriously in error.   The extra electric field dependence caused by the photogeneration term will cause this Hecht-type technique to underestimate the range.   Accordingly, Hecht curve analysis of pulse height vs voltage data by Fotland[16] gave a hole range of 10^{-8}cm^2/V and 2×10^{-9} cm^2/V for the electron range.   Similar measurements by Hartke[20] gave a hole range of 3×10^{-8} cm^2/V and 10^{-7} cm^2/V for electrons.

By measuring the drift mobility and carrier lifetime separately, Tabak and Warter[5] were able to show much higher hole ranges ($1.0-6 \times 10^{-6}$ cm^2/V) and normal electron ranges of 10^{-7} cm^2/V.   These higher values of the hole range are consistent with the low residual voltage found in the xerographic mode of operation.

Both the holes and electrons in amorphous Se have well-defined, thermally-activated drift mobilities.   Values of hole drift mobility of 0.10–0.16 cm^2/V-sec at room temperature with a thermal activation energy of 0.14 eV and electron drift mobilities of $4-8 \times 10^{-3}$ cm^2/V-sec at room temperature with activation energies of 0.25–0.33 eV have been reported.[5,20,22,23]   These thermally activated mobilities are characteristic of a shallow trap controlled drift mobility.   By using Spear and Lanyon's[24] estimate of the microscopic mobility of 10 cm^2/V-sec for electrons and 60 cm^2/V-sec for holes and the measured drift mobilities, Hartke[20] estimated shallow trap densities of 10^{19}cm^{-3} for the electron traps of depth 0.28 eV and about 10^{21} cm^{-3} for the hole traps of depth 0.14 eV.

MODELS FOR THE PHOTOELECTRONIC PROPERTIES OF AMORPHOUS Se

In this section we develop two models to describe photoelectronic processes in Se that have a strong bearing on the xerographic performance of Se photoreceptors.   The first is a model that describes the mechanism for the nonphotoconductive energy gap and includes a description of field-controlled carrier generation.   The second is a model for bulk recombination and space charge neutralization.

Model for Nonphotoconductive Absorption

To explain the occurrence of a nonphotoconductive energy gap, Hartke and Regensburger[4] proposed a model based on competition between two sets of electronic energy states.   In their model, the limiting value of the quantum efficiency, η, at high fields is given by:

$$\eta = \frac{\varepsilon_2^P}{\varepsilon_2^P + \varepsilon_2^N} \tag{1}$$

ε_2^P and ε_2^N are the imaginary parts of the components of the complex dielectric constant that characterize respectively the photoconductive (P), or band to band transitions, and the nonphotoconductive (N), or excitonic transitions (see Figure 16-3).   At lower fields, they attributed a field dependence they observed to a range limitation, so that the quantum yield, QY, is

$$QY = G(E, \mu\tau)_\eta \tag{2}$$

where G is the Hecht range limitation function, E is the applied electric field and $\mu\tau$ is the carrier range.

We now modify Equation 2 to include the effects of field-controlled photo-generation.   Tabak and Warter[5] showed that the functional form of the expression defining field-controlled photogeneration was independent of photon energy for fields less than 10^5 V/cm.   We therefore can define a function $F_p(E, T)$ that characterizes the photogeneration step and operates only on photon absorbed through P, so that the quantum yield is given by

$$QY = G(E, \mu\tau)F_p(E, T)_\eta \approx F_p(E, T)_\eta \tag{3}$$

The approximation reflects the fact that a range limitation is not generally applicable for the field range most conveniently studied; i.e., 5×10^3 V/cm $< E < 5 \times 10^5$ V/cm.

We now interpret Equation 3 in terms of specific mechanisms. η gives the fraction of absorptive events that can contribute to the photoconductivity. The assumption here is that any photon absorbed through N can never contribute to the photoconductivity; e.g., this transition can be viewed as an exciton in which the pair becomes self-trapped[4] and cannot be ionized by the electric field and/or temperature.   On the other hand, the fact that $F_p(E, T)$ is independent of wavelength, implies that a pair generated in process P degrades in energy so that the field-controlled generation step is always initiated from the same initial state.   This last statement may seem at odds with the results of Pai and Ing:[6] however, we shall demonstrate that the wavelength-dependent activation energy they reported is an artifact and not an inherent part of the field-controlled generation process.

In the model, the term that embodies the wavelength dependence is η. In the most general case, this term is also dependent on field and temperature. For Se, at temperatures below T_g (the glass transition), it has been shown,[4] that both the absorption and quantum efficiency edges are approximately exponential, so that:

$$\varepsilon_2^{P,N} = \varepsilon_2^{P,N}(0)\exp \beta^{P,N}(\hbar\omega - \hbar\omega_0^{P,N}) \tag{4}$$

where $\varepsilon_2^{P,N}(0)$ and $\beta^{P,N}$ are both constants, and

$$\hbar\omega_0^{P,N} = \hbar\omega_0^{P,N}(300°K, 10^4 \text{ V/cm}) + \left.\frac{\partial(\hbar\omega_0)}{\partial T}\right]^{P,N} (T - 300°K) - \frac{(\beta^{P,N})^2 e^2 E^2}{12(m_0^*)^{P,N}} \tag{5}$$

$\partial(\hbar\omega_0)/\partial T]^{P,N}$ is the temperature coefficient and the third term corresponds to the Franz-Keldysh shift for an exponential edge; $(m_0^*)^{P,N}$ is the effective mass and e the electronic charge.

If we restrict our attention to photon energies characterizing the onset of photoconductivity, then a further approximation is possible:

$$QY \approx F(E, T)\frac{\varepsilon_2^P}{A} \tag{6}$$

where A is a constant. This last approximation follows from the results of Hartke and Regensburger,[4] and applies at low temperature as well.

We now show how this model can explain (1) the apparent wavelength dependent activation energy[6,15] and (2) the increased quantum efficiency at higher fields and lower photon energies. Figure 16-7 indicates the form of the absorption edge and quantum efficiency curves at two different temperatures.[4] Therefore, if one examines the QY at a given wavelength as a function of temperature, the QY will always decrease as the temperature is decreased. The decrease in QY will be smallest for those photon energies where β^P is small at both temperatures ($\hbar\omega_1$), and will be greatest at those photon energies which occur on the steepest part of the edge ($\hbar\omega_3$); for intermediate photon energies ($\hbar\omega_2$), the decrease will be proportionate. These decreases in η, at a given $\hbar\omega$, can also be described in terms of a wavelength-dependent activation energy. A comparison of the activation energies as measured by Mort and Lakatos[25] and those obtained from the translation of the QY curves with temperature as measured by Knights[26] indicate quantitative agreement. The reason that this effect has been reported for Se, but not for other well-known photoconductors, for example, CdS,

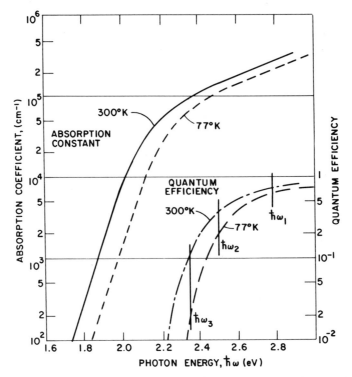

Figure 16-7 The photon energy dependence of the absorption constant and quantum efficiency at 300°K and 77°K.

is related to the fact that in Se the photoconductivity is not proportional to the total absorption, but rather is related to competition between two processes.

The increased field dependence of quantum efficiency (at higher fields and lower photon energies) as reported by Tabak and Warter,[5] can be interpreted in terms of a Franz-Keldysh shift. Since $(\varepsilon_2^P + \varepsilon_2^N)$ is essentially constant in the region of the photoconductivity edge, the field effect on η is approximated by a shift of the photoconductivity edge ε_2^P. Support for this interpretation is based on an analysis of the increased QY which shows that (1) it is proportional to E^2 and that (2) the increased QY is largest at those photon energies at which β^P is largest. The effective mass obtained from the analysis is $1.8\ m_0^*$ as compared to an effective mass of $4.5\ m_0^*$, as obtained by Drews[27] who analyzed the absorption edge shift in terms of the Franz-Keldysh model.

Photoemission studies[25,28] support the assertion that ε_2^N is associated with excitonic transitions and that ε_2^P is associated with one electron transitions.

Exciton states are always observable in absorption and exist only in an occupied state; as such, the photoemission of a single carrier, a hole or electron, into these pair states is not possible.   On the other hand, photoemission into unoccupied one electron states is always possible.

Finally, the analysis of Scher and Mort[29] for their Faraday rotation work is consistent with the model discussed above.   The main contribution to their reported rotation is associated with the photoconductive transitions, ε_2^P and gives an effective energy gap of 2.1 eV.   A more slowly varying component to the dispersion is contributed by the nonphotoconductive transitions ε_2^N.

Molecular Origin of Nonphotoconductive States: Theory and Discussion

It is well established that vitreous Se is a mixture of polymeric chains and Se_8 ring molecules,[30,31] with approximately half of the atoms making up each of the structural components.   Chen[32] studied the molecular orbitals and energy levels of the Se_8 and S_8 molecules and has extended his calculations to a description of the energy states of the associated molecular crystals, α-monoclinic $Se(Se_8)$ and orthorhombic $S(S_8)$.   For each crystal, he found a localized exciton lying below the continuum of one electron transitions.   The transfer integral for this exciton vanishes independently of the intermolecular geometry, implying that the same special type of excitonic state would be present in a disordered array of Se_8 molecules; e.g., in amorphous Se.   ·Both α-monoclinic Se[33] and orthorhombic S[34] have nonphotoconductive energy gaps.

The theoretical calculations and the photoconductivity of the molecular crystals, as well as the structural model for vitreous Se support a model wherein the nonphotoconductive transition is associated with a localized exciton of the Se_8 molecule.   The recent work of Schottmiller, Tabak, Lucovsky, and Ward[35] has demonstrated that the electron transport states in Se are also associated with the Se_8 molecule.   The occurrence of equal temperature coefficients for both ε_2^N and ε_2^P (or equivalently a temperature independent nonphotoconductive energy gap), also supports the assertion that both N and P are associated with the same structural member.

Studies of the photoconductivity in Se-based glasses[35,36] show that as the Se_8 population is depleted, those processes that have been attributed to that species disappear.   It should be noted that in vitreous As_2Se_3, a branched structure with no monomer component,[37] there is a range of energies where there is strong absorption, but only very weak photoconductivity.   However, it is not possible to define a photoconductive edge for As_2Se_3 as we have done for Se.   A better description of the photoconductivity in this system, and other compounds[38] and alloy glasses,[39] is embodied in a model that considers one electron states extending well into the forbidden energy gap.[40]

The elemental nature of amorphous Se, and the fact that it is a mixture of polymeric and molecular components differentiates its energy band configuration from those of the totally branched compound and alloy chalcogenide glasses.

Finally, the model invoked here is not in disagreement with transport studies in which the hole and electron drift mobilities are described in terms of shallow trap controlled models.[22,25] The densities of trapping states required to account for the magnitudes of the carrier mobilities are smaller, by orders of magnitude, than the effective density of localized excitonic states. Therefore, even though it is energetically possible for these trapping states to take part in absorption, their density precludes any significant contribution to absorption.

Trapping Kinetics

As discussed earlier, although the range of the electron in amorphous Se is smaller than that of the hole, it is not so small as to exclude the possibility of xerographic discharge in the negative surface charge mode. Rather, it is the long trap release time of the electrons compared to the cycle time that is responsible for a cumulative build-up of polarization or residual potential. A recent experiment by Scharfe and Tabak[41] has shed considerable light on trapping effects in Se. Here we discuss some additional aspects of that type of experiment that relates to xerography in the negative charge mode.

Figure 16-8 illustrates the experiment in question. It is performed on

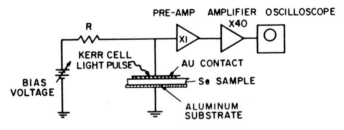

Figure 16-8 Schematic of experimental apparatus used to investigate trapping effects. After Scharfe and Tabak.[41]

a sandwich cell using a semitransparent gold electrode. The results in a xerographic mode of operation would be similar. The Au electrode is biased negatively and highly absorbed light is incident on the electrode. This induces a photocurrent of electrons flowing from the Au to the substrate. After a short exposure to a relatively high flux and at a moderate field, the sample is polarized by electrons which are uniformly trapped throughout the bulk. The existence of a uniform density of bulk negative space charge is determined by probing the sample field with a small signal pulse of photo-generated holes. It was postulated that a large hole photocurrent would result in a cancellation of the negative space charge; the mechanism in question would involve a direct recombination of free holes with the trapped electrons. After subjecting the fatigued sample to a large hole photocurrent, the field profile, as determined by a small signal probe, is found to be uniform. However, in a short period of time, the sample returns to its original fatigued condition, with bulk negative space charge. This implies that in the neutralization step, that holes did not recombine with the trapped electrons, but rather are trapped nearby, thus cancelling the net negative space charge. Due to the asymmetry in release times, the sample returned to a negative space charge condition when holes were released from deep traps and were collected at the electrodes.

Since the hole lifetime against deep trapping is of the order of 10^{-5} sec, this implies that the recombination lifetime of a trapped electron and free hole is longer, $\sim 10^{-3}$ sec. The voltage on the sandwich cell and the geometric capacity gives a measure of the maximum density of trapped electronic charge. This will be of the order of $10^{13}/cm^3$. If we assume a thermal velocity of 10^6 cm/sec for the holes, then the cross section of the electron trap for hole capture is of the order of 10^{-16} cm^2. This is on the order of the recombination cross section for a neutral trap and implies that electron traps in the bulk were positively charged before capture of the electron.

RECENT DEVELOPMENTS IN XEROGRAPHIC PHOTORECEPTORS EMPLOYING Se

There are two recent developments in photoreceptors for electrophotography that utilize Se. The first are alloy photoreceptors. These are vitreous materials in which Se is alloyed with one or more components. The principle motivation for the alloy work was twofold. First, to increase the stability of Se against crystallization, and second to extend its photoresponse into the red; i.e., produce a more panchromatic photoreceptor. The second type of photoconductor is one in which Se is used to sensitize an organic photoconductor; i.e., to move the spectral response out of the ultraviolet into the visible.

Alloy Photoreceptors

Alloying of Se with Te, As, and Bi achieves the goals stated above, namely the spectral response is extended further into the red and the stability against crystallization is increased.[16,42,43] However, alloying invariably produces two other side effects which are not desirable.   The first of these is a degrada-tion of the hole and electron drift mobilities.   Depending on the effects of trapping, this may also be accompanied by a decrease in range.[35,36]   In general, the electron range is reduced whereas increases in the hole lifetime are sufficient to compensate for decreases in the mobility.   The absence of elec-tron transport, as, for example, in As–Se alloys with an As concentration of greater than 10%, introduces fatigue problems associated with a one carrier system.   The problems associated with a one carrier system have been discussed by Warter.[7]   The second effect is associated with an increase of bulk-generated dark decay.   This in turn is a manifestation of the decreased energy gap that promotes the increased red response.

Sensitized Organic Photoconductors

Regensburger[44] has discussed the activation of PVK (polyvinylcarbazole) with a thin layer of amorphous Se on the top surface.   The layer of Se is thick enough to produce complete absorption of short wavelength light (~ 4000–5000 Å).   The free carriers generated in the Se surface layer are then injected into the PVK which acts as a transport medium.   In this sense, the Se sensitizes the PVK film.

Note Added In Proof: This review article was written in July, 1970.   Our understanding of photogeneration in selenium has been expanded since that time.   More recent developments can be found in the following articles:

H. Seki, *Phys. Rev.* **B2**, 4877 (1970).
H. Seki and Schechtman, B. H. *J. Appl. Phys.* **43**, 523 (1972).
H. Seki, *J. Appl. Phys.* **43**, 1144 (1972).
J. C. Knights and Davis, E. A., to be published in *J. Phys. Chem. Solids*, 1974.

REFERENCES

1. Dessauer, J. H. and Clark, H. E., *Xerography and Related Processes*, The Focal Press, London, 1965.
2. Schaffert, R. M., *Electrophotography*, The Focal Press, London, 1965.
3. Li, H. T. and Regensburger, P. J., *J. Appl. Phys.* **34**, 1730 (1963).
4. Hartke, J. L. and Regensburger, P. J., *Phys. Rev.* **139**, A970 (1965).
5. Tabak, M. D., and Warter, P. J., Jr., *Phys. Rev.* **173**, 899 (1968).
6. Pai, D. M. and Ing., S. W., *Phys. Rev.* **173**, 729 (1968).
7. Warter, P. J., Jr., *Appl. Optics* **Supp. 3**, 65 (1969).
8. Gilleo, M. A., *J. Chem. Phys.* **19**, 1291 (1951).
9. Stuke, J., *Z. Physik* **134**, 194 (1953).
10. Saker, E. W., *Proc. Phys. Soc.* (*London*) **B65**, 785 (1952).

11. Dowd, J. J., *Proc. Phys. Soc.* (*London*) **B69**, 70 (1951).
12. Caldwell, R. S. and Fan, H. Y., *Phys. Rev.* **114**, 664 (1959).
13. Siemsen, K. J. and Fenton, E. W., *Phys. Rev.* **161**, 632 (1967).
14. Weimer, P. K. and Cope, A. D., *RCA Rev.* **12**, 314 (1951).
15. Keck, P. H., *J. Opt. Soc. Am.* **42**, 221 (1952).
16. Fotland, R. A., *J. Appl. Phys.* **31**, 1558 (1960).
17. Dresner, J., *J. Chem. Phys.* **35**, 1628 (1961).
18. Lanyon, H. P. D., *Phys. Rev.* **130**, 134 (1963).
19. Drews, R. E., unpublished.
20. Hartke, J. L., *Phys. Rev.* **125**, 1177 (1962).
21. Rossiter, E. L. and Warfield, G., *J. Appl. Phys.* **42**, 2527 (1971).
22. Spear, W. E., *Proc. Phys. Soc.* (*London*), **B76**, 826 (1960); **B70**, 669 (1957).
23. Blakney, R. M. and Grunwald, H. P., *Phys. Rev.* **159**, 664 (1967).
24. Spear, W. E. and Lanyon, H. P. D., *Proc. of Int. Conf. on Semicond. Physics, Prague 1960*, Czech. Acad. of Sciences, Prague, 1961, p. 987.
25. Mort, J. and Lakatos, A. I., *J. Non-Crystalline Solids* **4**, 117 (1970).
26. Knights, J., unpublished.
27. Drews, R. E., *Appl. Phys. Letters* **9**, 347 (1966).
28. Lakatos, A. I. and Mort, J., *Bull. Am. Phys. Soc.* **14**, 737 (1969); Mort, J., Lucovsky, G., and Lakatos, A. I., *Bull. Am. Phys. Soc.* **14**, 783 (1969).
29. Scher, H. and Mort, J., *Bull. Am. Phys. Soc.* **13**, 1414 (1968).
30. Lucovsky, G., Mooradian, A., Taylor, W., Wright, G. B. and Keezer, R. C., *Solid State Comm.* **5**, 113 (1967).
31. Lucovsky, G., *The Physics of Selenium and Tellurium*, Cooper W. Charles ed., Pergamon Press, Oxford, 1969, p. 255.
32. Chen, I., *Phys. Rev.*, **B2**, 1052 (1970).
33. Prosser, V., *Proc. of Int. Conf. on Semicond. Physics, Prague, 1960*, Czech. Acad. of Sciences, Prague, 1961, p. 943.
34. Spear, W. E. and Adams, A. R., *J. Phys. Chem. Solids* **27**, 281 (1966).
35. Schottmiller, J., Tabak, M. D., Lucovsky, G., and Ward, A., *J. Non-Crystalline Solids* **4**, 80 (1970).
36. Tabak, M. D., *Proc. of 3rd Int. Photoconductivity Conf., Stanford, 1969*. Pergamon Press, Oxford, 1970, p. 8.
37. Myers, M. B. and Felty, E. J., *Mat. Res. Bull.* **2**, 535 (1967).
38. Weiser, K. and Brodsky, M. H., *Phys. Rev.*, **B1**, 791 (1970).
39. Fagen, E. A. and Fritzsche, H., *Semiconductor Effects in Amorphous Solids*, North Holland Pub., Amsterdam, 1970, p. 180.
40. Cohen, M. H., Fritzsche, H. and Ovshinsky, S. R., *Phys. Rev. Letters* **22**, 1065 (1969).
41. Scharfe, M. E. and Tabak, M. D., *J. Appl. Physics* **40**, 3230 (1969).
42. Neyhart, J., *Phot. Sci. Eng.* **10**, 126 (1966).
43. Schottmiller, J. C., Bowman, D. L., and Wood, C., *J. Appl. Phys.* **39**, 1663 (1968).
44. Regensburger, P. J., *Photochem. Photobiol.* **8**, 429 (1968).

Index

Absorption, 15, 89, 657, 658, 803, 804
Absorption coefficient, 197, 208, 293, 715
Absorption constant, 285, 795
Absorption discontinuity, 112
Absorption edge, 191–204, 249, 288, 718, 795
  behavior, 203
  electronic absorption, 203–204
  indirect, 200
Absorption, nonphotoconductive, 798, 800–803
Absorption spectrophotometry, 640, 717
Absorption wavelength, 629
Absorptive events, 800
Abundance, 2–25
Accelerator composition, 737
Accelerators, 729, 746
Acceptors, 207, 238, 239, 241, 279
Accumulation, 659
Accumulators, *see* Plants
Acetabularia, 587
ac-Field, 222

Acetylamino group, 499
Acetylcholine, 584
Acetylene, 323, 461, 476, 478, 499
Acetylenic systems, 463
Acetylselenocholine, 525, 584
Acetylthiocholine, 584
Acid amides, 448, 448
Acid chlorides, 425, 479
Acidic solutions, 60
Acids, 56, 60, 68, 423
  esters, 425
  preparation, 423–425
  reduction, 425
Acoustoelectric current saturation, 235–236
Acrylate polymers, 746, 748
Acrylonitrile, 737, 758
Activation, 133
Activation energy, 134, 136, 137, 218, 221, 227, 799, 801
Actual free time, 798
Addition compounds, 367
Addition reactions, 462
Additive optical phenomenon, 712

Additives, 775
Adducts, 317, 321–322, 323, 324, 502
Adenosine triphosphate, 582, 583
Adenosyl transferase, 582
Adsorbents, 77
Adsorption, 69–70, 110
Agglomeration, 43
Aging of Rubber, 733, 742, 747
  effect of added selenium on, 748
  effect of selenium compounds on, 738–739, 740 741, 743, 744, 745
  resistance, 733, 742
Aglycone, 487, 491
Agriculture
  economic evaluation, 688
  economic loss, 687
  selenium in, 675–699
Aguilerite, 37
Air 11–12, 41, 51, 65, 647–648, 769
Air-acetylene flame, 629–630
Airborne dust, 11

810   INDEX

AISI 1040 steel, 775
AISI 1112 steel, 776
AISI 12L14 steel, 775
AISI 4140 steel, 775
Albumin/globulin ratio, 690
Alcohols, 409, 622
Aldehydes, 439
Alfalfa, 697
Algae, 586–587
Aliphatic group, reactions
    modifying, 463
Aliphatic halides, 422
Aliphatic selenium
    trihalides, 427
Aliphatic selenoglucosides,
    481–488
    preparation, 481, 484–488
Alkali, 669
    content, 711
Alkali disease, 676, 677, 680,
    681, 682, 683, 685
Alkali, fusion with, 618
Alkali halides, 117, 153, 159
Alkali metal acetylides, 476
Alkali metal chloride
    solution, 43
Alkali metals, 69
Alkali metal selenites, 311
Alkali soluble selenites, 42
Alkaline ash, 587
Alkaline metals, 239
Alkaline phosphatase,
    581–582
Alkaline phosphate, 56
Alkalines, 51, 53, 57, 60
Alkaline solutions, 782–783
Alkalinity, 555, 558, 559
Alkalized cyanides, 37
Alkaloids, 681, 683
Alkaneselenolates, 516–517
Alkoxides, 478
Alkoxy group, 499, 754
Alkyl aryl selenides, 456–475,
    506
    from diaryl diselenides,
        462–463
    modification of organic
        groups, 463–474
Alkylation, 397–398, 443,
    456–458, 506, 516–518
    of 1-carboxy-2-aminoalk-
        eneselenolates,
        517–518
Alkyl carboxyalkyl selenides,
    448
Alkyl group, 754

Alkyl halides, 476
Alkyl iodides, 419
Alkyl selenides, 393
Alklylselenium halides,
    312–313
4-Alkylselenobenzoic acids,
    469
4-Alkylselenopheny1
    2-chloroethyl ketones,
    472
4-Alkylselenophenyl methyl
    ketones, 472
Allergy, 666
Allotropes, 13, 88
    amorphous, 109, 118
    crystalline, 119–125
    miscellaneous, 124–125
Allotropic modification, 113
Alloying behavior, 766–769
    binary systems, 766–769
    ternary systems, 766–769
Alloy photoreceptors,
    805–806
Alloys, 641, 763
    As–Se, 806
Altrose, 480
Alumina, 622
Aluminum, 38, 117, 131, 789
    hydroxide, 60
Amanita muscaria, 548
Amberlite, 68
Amberol ST-137, 742
American Conference of
    Governmental Industrial
    Hygienists, 668, 671
American Feed
    Manufacturers
    Association, 696
American Metal Climax, Inc.
    44
American Smelting and
    Refining Co., 44
Amides, 478
Amines, 476, 478
Aminoacids, 515, 518, 689
    crystalline, 589
4-Aminophenyl tetracetyl-
    α-selenoglucoside, 492
Amino-selenazolone, 638
Aminoselenides, 478
Ammonia, 50, 51, 476, 746,
    760
Ammonium nitrate, 69
Amorphous films, 117–119
Amorphous modification,
    118

Amorphous selenium, 87, 91,
    94, 97, 99, 109, 119, 133,
    167–171, 293–294
    electrical properties,
        795–799
    elemental nature, 804
    models for photoelectronic
        properties, 797–799
    molecular structure,
        169–171
    optical properties, 795–799
    photoelectrical properties,
        797–799
Amperometric titration, 626
Amplifiers, 138
Analogues, 582–584
Analysis, 632–640, see also
    Determination
    compositional, 32
    of compounds, 637–640
    impurities, determination
        of, 632–637
    phase, 32
Analytical chemistry,
    615–648
    separation and isolation,
        616–621
Anemia, 662, 682, 688, 690
Angle tilt boundaries, low,
    764
Anhydrous sodium diselenide
    304
Anilines, 475
Animals, 26, 515, 547,;
    548–550, 561, 567, 597,
    598, 602, 655, 656
    administration of oral
        selenium, 695
    breed, 685
    poisoning, 675, 681–686
Anionic species, 67
Anisotropic bonding, 184
Anisotropic materials, 175,
    181
Anisotropy, 184, 209, 212,
    224–225, 777
Annealing, 134, 219, 229,
    238, 249, 251, 253, 257
Annealing point, 723
Anode slimes, 3, 31, 47,
    644–647
Anodic wave, 626
Anomeric products, 485
α-Anomers, 497
β-Anomers, 497
Anticorrosives, 452

Antiinflammatory property, 587
Antimetabolites, 580
Antimony, 238, 634, 684
Antimony trioxide, 713
Antioxidants, 589, 697, 729, 750–754
  evaluation of selenium compounds as, 751, 752, 753
  miscellaneous effects, 754
  natural rubber, 750
  nonstaining nondiscoloring evaluation, 755
  properties, 574–576
  staining and discoloration, 754
  styrene-butadiene rubber, 750
Aphids, 699
  electronic, 140
Aqueous bubbles, 647
Arabinose, 480
Argon, 711
Aromatic aliphatic selenides, 456–57
  addition of Ar—Se—X, 459–462
  alkylation of aromatic selenols, 456–475
  alkylation of aryl selenium halides, 459
  bromine addition to olefins 464
  dehydrobromination, 464
  from diaryl diselenides, 462–463
  formation of hydrazones, 464–465
  introduction of substituents into the aromatic ring, 465–466
  isomerization of olefinic and acetylenic systems, 463
  modification of organic groups, 463–474
  modifications in aromatic group, 466–474
  physical properties, 474–475
  preparation, 456–474
  reactions, 474–475
  reactions modifying aliphatic groups, 463
  synthesis, 456–474
Aromatic group,
  modifications, 466–474
Aromatic halides, 485
Aromatic selenium trihalides, 427
Aromatic selenoglucosides, 481–488
  preparation, 481, 484–488
Array, orderly, 138
Arsenic (arsenite), 53, 54, 128, 135, 136, 552, 593–594, 617, 663, 684–686, 713, 797, 803, 805, 806
Arsenic trioxide, 713, 716
Arsenic trisulfide, 722
Arylation, 397–398
Aryloxy group, 499
Aryl selenides, 393
Aryl selenium acetates, 460
Aryl selenium halides, 312–313, 453
Aryl selenocyanates, 458
Aryl trifluoromethyl selenides, 475–476
Ascorbic acid, 663
  precipitation, 643
Assaying, 61
As_2S_3-As_2Se_3-As_2Te_3 system 722
Asymetric conduction, 25
Aster, 679
Astragalus, 547, 548, 551–553, 654, 679
  bisulcatus, 9, 10, 550, 669
  crotalariae, 554
  pectinatus, 680
  racemoses, 553, 687
  vasei, 555
A_2-symmetry optical mode, 155
Atmosphere, 711
Atomic abundance, 5
Atomic fluorescence, 630
Atomic number, 13, 763
Atomic populations, 171
Atomic properties, 13–15
Atomic weight, 13, 763
Atoms, 100, 102, 105
  distance detween, 178
  donor, 337–399
  interaction between, 178
  parametric, 178
  position, 507
  reciprocal, 179–181
  selenium, 449
  singly bonded, 332
unsubstituted carbon, 460
  vibration, 149
Atriplex, 679
  nuttallii, 10
Australia, 552
Autoradiography, 560
Axes, 139, 149, 152, 171, 189
Azeotropes, 70–71

Background, colored, 792
Bacteria, 563
Baia Mare, 43
Bands, 170
  absorption, 164
  dominant, 168
  edges, 797
  energies, 187
  flat, 183
  gap, 200 717
  models, 183, 189, 210, 796
  relation, 185
  structure, 177–190, 795
  structure calculations, 181
  valence and conduction, 182
  weak absorption, 167–168, 170
  weak emission, 161
  width, 188
Band-to-band transitions, 248, 798, 800
Bardeen-Shockley equation, 247
Barite, 3
Barite–carbonate, 3
Barium, 712
  chloride, 56
  selenate, 60
Barley 9, 587
Barrier model, 241, 244–245, 265, 269
Barriers, 243, 255
Battery boxes, 746
b^*–Axis, 124
Beer's law, 628, 629
Behavior, high-spin and low-spin, 381
Bentonite, 44, 52
Benzene, 68, 505
  extraction, 643
Benzeneselenic acid, 310
Benzeneselenols, 459–461, 476, 484

Benzoselenadiazoles, 756
Benzothiazyl disulfide, 733
Benzylselenoaminocarboxylic acids, 447
3-Benzylseleno-2-amino-propionic acid, 521
Benzylseleno group, 522
Berzelionite, 2
Bile, 594
Bilirubinemia, 656, 662
Binary selenide systems, 89
Binary systems, 766–769
Biochemistry, 546–602, 609–613
Biological activity, 546
Biological function, 697–699
  selenium as essential element for plants, 698
  selenium as insecticide, 699
Biologically important compounds, 515–529
  bis(aminocarboxyalkyl) diselenides and selenides, 520–521
  miscellaneous compounds, 525–529
  organylselenoamino-carboxylic acids, 516–519
  Se-aryl aminomonoseleno-carboxylates, 519–520
  selenium containing peptides, 521–525
Biological role, 597–602
Biological systems
  effects of selenium on, 574–588
  sulfhydryl compounds in, 579
Biosynthesis, 547
Bis(aminocarboxyalkyl) diselenides, 520–521
Bis(aminocarboxyalkyl) selenides, 520–521
Bis(6-deoxyglucose) 6-(di)selenides, 493
1,3-Bis(ethylseleno) propane, 370
Bis(2-hydroxy-1-naphthyl) selenide, 750
1,2-Bis(isopropylseleno) ethane, 370
1,2-Bis(methylseleno) ethane, 370
1,3-Bis(methylseleno) propane, 370

Bis(2-methylvinyl) selenide, 446
1,3-Bis(phenylseleno) propane, 370
Bismuth, 779, 805
Bismuth–arsenic–selenium alloys, 641
Black amorphous selenium, 88, 112–113
Black on white copy, 793
Blocking control, 790–791
Blood, 549, 550, 658, 661, 696
"Bobtailed" appearance, 682
Body
  disribution, 658–661
  retention, 658–661
  selenium in animals, 658–664
Bohr radius, 264
Boiling point, 16, 409, 763
Bolidens Gruvaktiebolag, 45
Bolidens process, 60
Bolidens soda roast process, 57
Bonds and bonding agents, 729, 757
  angle, 15, 299–300, 302, 307, 316
  π-bonding, 373, 375
  covalent forces, 163, 167
  distances, 111, 299–300, 302, 316, 333
  energies, 12, 178
  fission, 398
  forces, 171
  homopolar, 770
  isomerism, 351
  length, 14
  links, 316, 318
  metallic, 770
  reproducible, 757
  strength, 175
  van der Waals, 167
Bone ash, 720
Bornite, 3
Borosilicate, 75
Bottles, 712
Bound state, 796
Brazing, 779
Breath, 552, 657–658, 664, 668
Brillouin zone, 179–181, 190, 195, 196, 201, 210
Brine, 54
Briquettes, 44, 119
Brittleness, 746, 762

Bromides, 460, 507–508
Bromination, 742
Bromine–hydrobromic acid distillation, 648
Bromines, 140
Bromobenzene, 663, 685, 686
Bubblers and bubbling, 47, 56
Built-up edge stabilizing, 773
Bulk, 791
Bulk modulus, 21
Burners, 630
Burns, 665, 668
Burstein effect, 281
Butadiene, 461, 737
Butadiene–styrene poly-merization, 758
But-3-enyl butyl selenide, 370
But-3-enyl phenyl selenide, 370
Butyl pent-4-enyl selenide, 370
Butyl rubber, 737–742, 752
  brominated, 744
  modification of, 758–759

Cadmium, 75, 275, 593, 594–596, 601
  chalcogenides, 717, 718
  chlorides, 595
  red lithopones, 717
  selenides, 271, 272, 273, 279, 284–285, 716
  sulfides, 709. 716
  sulfoselenides, 25, 709, 716–720
  sulfoselenides, as pig-ments, 717
Calamine lotion, 672
Calcium, 38
  oxychloride, 37
  phosphate, 721
  selenate, 10
  sulfate, 685
Calculated detection limit, 631
Calculated ε_2-spectrum, 196
California, Gulf of, 678
Calorimetric determination, 615
Calves, 696
Cambridge, Mass., 11
Canada, 5, 27, 550, 678, 692, 694–695
Canadian Copper Refineries, Ltd., 62

Cancer, 658
Candida albicans, 586
Carbanions, 478
Carbide particles, 776
Carbide-promoting action, 772
Carbohydrates
    properties, 480
    selenium derivatives, 480–480–497
Carbon, 52, 129, 304
    activated, 70
    atoms, 1, 6, 495
    dioxide, 56
    diselenide, 669
    disulfide, 96, 120, 133, 141
    monoxide, 52, 711
    tetrachloride, 498
Carbonate leach powders, 59
Carbonates, 38
Carbon black, 732, 758
    activators, 729, 760
Carbon-carbon bonds, 444, 445, 446, 459–462
Carbon-carbon double bonds, 518
Carbon-carbon multiple bonds, 449
Carbonic acid derivatives, 409
Carbon-selenium double bonds, 509–515
    miscellaneous seleno-carboxylic acid derivatives, 515
    selenoaldehydes, 509–511
    selenocarboxylic acid amides, 512–515
    selenoketones, 511–512
Carbonyls, 392
    hexacarbonyls, 394
    selenocyanates, 353
    tetracarbonyls, 393, 394
    tricarbonyls, 393
Carboxylic acid chlorides, 467
Carboxylic acids, selenium-containing, 478
4-Carboxyphenyl alkyl selenides, 467
2-Carboxyphenylmethyl benzyl selenides, 449
Cardiac musculature, 691
Carrier beads, 790
Carrier distribution, 241

Casein, 662, 668
Cast irons, 772, 783
Catalysis, 26, 59, 478, 577
Cathode slimes, 783
Cation-exchanger, 67, 68
Cation exchange resins, 68
Cattle, *see* Livestock
Caustic soda, 56
c-Axis, 764, 770
Cellobiose, 480
Cells
    division and morphology, 586
    physiology, 598–602
Cerium oxide, 716
Cerro process, 39 40
Chain breaker, 98
Chain diffusion process, 137
Chain length, 98, 101, 102, 103, 104, 105, 107
    number-average, 106, 108
Chain modifiers, 758
Chains, 94, 95, 111, 112, 139
    ends, 128, 134, 136
    helical, 171
    interaction, 182
    intermolecular bonding, 150
    long, 115
    long polymer-like, 764
    parallel and perpendicular orientation, 227
    polymeric, 171, 803
    rotation, 216
    saturated, 137
    spiral, 215
    vibrations, 170–171
Chalcogenation, 353,398
Chalcogenides, 721–724
Chalcopyrite, 31
Chalcopyrite–pentlandite pyrrhotite, 3
Characterization, 547
Charcoal, 619
Charge transfer, 367
Charging, 790, 791, 793
Charring, 619
Chelation, 391
Chemical refining, 61–69
Chemical shift, 509
Chemicals, activating, 133
Chest, tightness, 668
Chicks, 588, 589, 590, 591 682
    vitamin E-related nuti-tional disease, 688–690
Chip size, 773

Chlorella vulgaris, 586
Chloride purification, 66–7
Chlorination, 52 742
    partial, 106
Chlorine, 49, 67, 103, 109, 129, 136, 139, 276, 634,; 635, 764,
p-Chlorobenzeneselenic acid, 310
2-Chlorodiethylaminoethane hydrochloride, 467
Chloroethylamine, 473
Chloroform, 498, 648
Chlorophyll, 619
Chlorosis, 765
Chlorosulfonated polyethy-lene (Hypalon) composi-tions, 745
Choline, 515
Chondritic materials, 5
Chromatography, 559
    column, 554
    ion exchange, 554
    paper, 554, 558
Chromic poisoning, 657
Chromium plating bath, 782
Chromophore, 710, 714
Cinnabar, 617
Circulating or transport method, 120
Cirrhosis, 662, 682
Cis-1,4-adducts, 461
Cis configurations, 348, 478
Cis-and *trans*-carboxychloroe-thene, 460
Cis and *trans* compounds, 444, 461, 462
Cis and *trans* isomerization, 758
C-labeled S-adenosylmethio-nine, 553
Clausthalite, 37
Cleaning, 790
Clemmensen reduction, 449
Clover, 552,554
[14]C-methyl labeled seleno-methionine, 574
Coagulation, 58
Coal, 11
Cobalt, 596
Cobalt-iron, 779–782
Coco de mono, 552
Coefficient of linear expan-sion, 763
Coefficient of thermal expan-sion, 18

Coenzyme A, 515
Coils, helical, 149
Cold-drawn copper, 779
Cold finger, 138
Cold working, 779
Collection, 53, 59
Collodion, 118
Colloidal selenium, 571
Collodial state, 710
Colombia, 9, 657
Colorado, 11
Colorado River, 678
Colors
  amber tint, 713
  burning out, 711
  change, 756
  pink, 709–711
  ruby, 708
  solution, 719
  yellow and red, 718
Commercial-grade selenium,
    89
Compensation, 241
Complex membrane systems,
    600, 602
Components, photo- and
    nonphotoconductive, 796
Compounds, 728–760
  analysis of, 637–640
  containing X—Se—Se—
    Se—X linkages,
    305–306
  damping properties, 759
  toxicity, 655
Compressibility, 20, 766
Cooling, 114, 723
  slow, 141
Coordination, 115
Coordination, compounds
    337–399
Coordination, covalent, 149
Coordination number, 99
Concentrates, 642–643
Concentrations, 3, 599,
    658–659, 660, 661, 664,
    681, 683
Condensation, 63, 110,
    498–499
Conducting electrode, 790
Conduction band, 101
Conduction mechanism,
    239–248
Conductivity, 176, 222–225,
    230, 270, 779
  corresponding values, 223
  dark, 218, 227

electrical, 217–228
  frequency dependence, 222,
    243
  ratio of, 226, 227
  tensor, 232
Cone angle, 139
Congenital malformation,
    587, 682
Constant carrier concentra-
    tion, 229
Constant of proportionality,
    790
Consumption, 25
Contact, 272, 317
Contaminants, absorption of,
    110
Contamination, 70, 71
  by metals, 75
"Converter" plants, 679
Copolymer, 737
Copper, 3, 31–70 *passim*, 125
    239, 621, 641, 709, 710,
    720, 721, 779, 782
  alloys containing sulfur,
    selenium, and tellur-
    rium, 781
  amine sulphate, 38
  annealed, 779, 780
  cold-drawn, 779
  deoxidized, 55
  electrolytic, 38
  hard-drawn, 780
  microstructure, 779
  molybdenum, 3
  oxide, 720, 783
  selenide, 42, 51, 58, 60
  sulphate, 48
  sulphide, 26
  telluride, 58
Copper refinery slimes, 618
  see also Anode slimes
Coprecipitation, 641
Corn, 10, 692, 694
Corn–soybean meal, 692
Coronary band, 682
Corrosion, 64, 66, 72, 75,
    769–770
Cosmic abundance, 5
Cottrell precipitator, 47, 54
Coughing, 665
Coulometric diaminobenze-
    dine method, 648
Coulometric procedures,
    627
Counter electrode, 280
Coupled electromagnetic-

lattice waves, 157
Coupled-wave form, 158
Coupling, spin-orbit, 184–188
Covalent bonding, 564
Cracking, 772
Cr(III) complexes, 361
Cretaceous systems, 677–678
Critical constants, 19
Critical surface tension, 24
Crops
  selection, 695
  selenium content, 695
Cross-links, 136
  forming, 128
Cross-linkages, 559, 560
Crude selenium, 63
Cryolite, 720
Crystal-field splitting, 163
Crystalline urea clathrate,
    555
Crystallinity, 111
Crystallites, 133
Crystallization, 75, 113, 119,
    123, 125–142, 805
  factors affecting behavior,
    127–133
  inhibition, 129
  kinetics, 133–137
  rate of, 127–128, 129
  temperature dependent,
    127–128
Crystallochemistry, 3
Crystallographic orientation,
    77
Crystals, 120–121
  chain-folded, 135
  elemental, 152–153
  geometry, 134
  growth, 131, 135, 139–140,
    763–765
  growth of single, 137–142
  ionic radii, 14
  large, 141, 764
  large single, 163
  molecular, 149, 150, 803
  morphology, 124, 131, 133
  photoelectrical properties,
    265
  prismatic, 138
  properties of single, 176
  seed grown, 139
  shapes, 138
  sheaf-like, 131
  single, 160, 242, 763–766
  structure, 121, 177–179
  α-Cubic selenium, 124

β-Cubic selenium, 124–125
Cupellation, 39
Cuprammine, complex 349
Cupric oxide, 51
Cupric salts, 51
Curing, 729, 746
  resin, 737, 742
  retarded, 733
Curing-accelerator systems,
  729–750 passim
Currents, 627
  density, 69
  forward density, 279
  limited, 218
  orientation, 226–227
  ratio, 225
  saturation, 244, 269
Current voltage characteris-
  tic, 272–273, 281, 282
  forward, 274, 276, 277
  temperature dependence,
  272
Curves, 127
Cyanidation, 37–38, 54
Cyanides, 37, 53
  acid and alkalines, 782
Cycle, geochemical, 8
Cyclic selenides, 364, 366
Cyclic systems, 321–332
  miscellaneous, 329–332
Cycling, 791, 792, 794
Cyclopentadienyls, 394
Cystamine, 515
Cystathionine, 680
Cysteamine, 515
Cysteine, 564, 577, 578, 639
  residues, 555, 556
Cystine, 575, 591, 639, 683
Cytotoxic compounds, 552
Czochralski technique, 139,
  140

Damping, 758
Dark conductivity, 291, 797
Dark decay, 790
  bulk-generated, 806
  measurement, 797
dc-Ambient light, 266, 267
D-carbohydrates, synthesis,
  480
dc Arc, 633, 637
dc-Field, 221, 222
dc-Voltage, 243
Deacetylation, 488
Deamino derivatives, 524
Death, see Mortality

Debye temperature, 18
Decantation, 63
Decarboxylation, 448
Decolorizers, 709, 711–716
Decomposition, 8
  of inorganic materials,
  616–618
  methods, 620
  of organic materials,
  618–619
Deficiency, 654, 688, 693,
  694, 696
  prevention, 588
  vitamin E-selenium, 692
Degeneration, 180
Dehydration, 474, 504
Dehydrobromination, 464
Dehydroxy dialkyl diaryl
  polysulfide, 754
Dehydroxy dialkl diaryl
  sulfide, 754
Δ-Axis, 181, 182, 183
Demethylation, 375–376
Denaturation, 581
Dendrite form, 130
Density, 22, 763
Deoxidation, 769
Deoxidizer, mild, 772
Department of Health, 11
Depletion layers, 242, 243,
  244, 249,250 283
Depolarization, 584
Deposition, 117
Deposits, 6–7
  upper horizons, 3
Dermatitis, 26, 657
Detection, 621–623
Determination, 623–632
  neutron activation analysis,
  631–632
  photometric methods,
  627–631
  polarography and other
  electrochemical
  methods, 625–627
  precipitation and
  gravimetric methods,
  623–624
  volumetric methods,
  624–625
Determination in specific
  materials, 640–648
  air and water, 647–648
  biological materials,
  644–647
  iron and steel, 641

  miscellaneous metals and
  alloys, 641
  rocks, minerals, ores, and
  anode slimes, 642–644
  semiconductors, 641–642
Detoxification, 568, 662–664
  products, 547
Development electrode, 790
Deviations of Se as a
  photoreceptor, 792
Dewatering, 48
Diacetyl selenosemicarbazone
  oxime, 377
Diak no. 1, 746
Diak no. 3, 746
Dialkyl diselenides, 574
Dialkyl
  diselenodicarbamates,
  378–382
Dialkyldiselenophosphates,
  354–363
  electronic spectra-
  spectrochemical effect,
  354–357
  infrared spectra, 357—358
  magnetic susceptibility
  measurements, 358
  nephelauxetic effect, 354,
  355
  optical electronegativities,
  356
Dialkyldithiocarbamates, 378
Dialkyl dithiophosphates, 355
Dialkyl selenides, 364, 506
Dialysis, 558, 571, 572
Diamagnetism, 101
Diamino compounds, 455
3,3'-Diaminobenzidine, 615,
  618, 619, 626, 627, 628,
  642, 645
2,3-Diaminonaphthalene, 615,
  618, 628, 645
Diaryl selenides, 364, 507
Diaryl selenium halides, 325
Diatomaceous earth, 619
Diazonium salts, 454
Diazotization, 469
Dibenzoselenophene, 328
Dibenzylselenourea, 750
N, N'-dicinnamylidine-1,6-
  hexanediamine, 746
Dielectric constant, 22, 193
Diet, 585, 586, 589–590, 597,
  662–663, 693
  adequate level of selenium,
  696

basal necrogenic, 601
dystrophogenic, 692
low vitamin E, 693, 696
nonseleniferous, 57, 671
protein, 669, 691
purified, 689
selenium-deficient, 8
vitamin E deficient,
    689–690
Dietetic microangiopathy, 691
Diethyl ether, 497–498
Diethyldiselenocarbamate
    complex, 379–380
Diethyldiselenophosphate,
    354
Diethyldithiocarbamate, 380,
    733, 737, 742
Differences, interband energy,
    188
Differential potentiometric
    method, 636
Diffraction, 115
Diffusion, 770
Digestion, 46, 47, 110, 112
Digestive enzymes, 562
Digital computer, 99
Diglucosyl selenide, 489
Diglycosyl diselenides, 488,
    489
Dihalides, 497–502
Diisocyanato derivative, 455
Diketones, 473
Dilatometric study, 134
Dilute solutions, 58
Dimers, 317, 370–371, 393
    planar, 328
Dimethyl diselenide, 553
Dimethyl selenide, 504,
    552–553, 569, 684
Dimethyl selenium difluoride,
    498
Dimethyl sulfoxide, 369
Dimethylthioselenocarbamate
    381
2,4-Dinitrophenol, 574
Diorganyl diselenides,
    427–436
    symmetric, 428–435
    see also Diselenides
Diorganyl selenides, 437–478,
    476–478, 497–498
    oxidation, 504
    symmetric dialkyl, 437–441
    synthesis, 476, 478
Diorganyl selenide sulfides,
    436–437

Diorganyl selenium
    dialkoxides, 501–502
Diorganyl selenium
    dicarboxylates, 501–502
Diorganyl selenium dihalides,
    497–502
    addition of selenium
        tetrachloride and
        oxychloride to carbon-
        carbon multiple bonds,
        499
    condensation of selenium
        tetrachloride and
        oxychloride with
        hydrocarbons, 498–499
    dialkoxides, 501–502
    dicarboxylates, 501–502
    from diorganyl selenoxides,
        500
    dissociation, 501
    by halogen exchange, 500
    imines, 501–502
    insolubility, 501
    physical properties,
        500–501
    reaction of diorganyl
        selenides with halogen,
        497–498
    reactions, 500–501
    reduction, 501
    thermal stability, 501
    transformation into triaryl
        salts, 501
Diorganyl selenium imines,
    501–502
Diorganyl selenones, 504–505
    oxidation, 505
    by oxidation of diorganyl
        selenides or
        selenoxides, 504
    properties and reactions,
        505
Diorganyl selenoxides, 500,
    502–504
    dehydration, 504
    properties, 504
    synthesis, 502–503
Diorganyl triselenides, 437
Dioxane solution, 770
Dioxides, 770
Diphenyl diselenide, 370, 437
Diphenylpiazselenol, 626, 628
Diphenyl triselenides, 437
Dipole moments, 452
Dipeptides, 525
Diradical chains, 93

Diradicals, 106
Direct edge, 211
Discoloration, 754
Discovery, 1–2
Diseases
    control measures, 695–697
    deficiency, 26
    distribution, 694–695
    nutritional, 676
    selenium-responsive,
        687–695
1,4-Diselenacyclohexane, 366
1,2-Diselenane-3,
    6-dicarboxylic acid, 330
1,4-Diselenane, 321–323
Diselenide ion, 303
Diselenides, 370–371, 421,
    423, 441, 442, 476, 575
    reactions, 435, 436
1,2-Diselenolates, 389–390
1,1-Diselenolates, 389
Diselenobisformamidinium
    ion, 304
Diselenocarbamates, 378, 379
Diselenocyanates, 424
Diselenodicarboxylic acids,
    575, 588
Diselenoglutathione, 522
Diselenoxytocin, 523
Diselenophosphates, 358–363
    crystallographic studies,
        361–362
    electronic spectra, 360–361
    infrared spectra, 361
    phosphinato complexes,
        359–360
Diselenophosphoric diamides,
    358
Disk, 113, 114
Dislocations, 220
Dispersion curve, 158
Dissemination, 3
Dissociation, 41, 501
    constants, 93, 95–96 367
Dissolution of inorganic
    materials, 616–618, 640
Distances, interatomic, 121,
    122
Distillation, 52, 63, 70–77,
    127, 617
    with another compound,
        76–77
    simple, 70, 75–76
Distribution, 547–550,
    658–661
    of symptoms, 656

Disulfides, 556, 557
and selenium interchange reactions, 576–579
Dithiocarbamates, 379, 380, 381
Dithiocarbamic acid, 760
1,4-Dithiacyclohexane, 366
"Dithizone," 374, 643
Divinyl selenide, 445
Dizziness, 665
Dogs, 564, 573, 663
Donors, 238, 241, 243, 245
hard and soft, 343
Doping, 61, 103, 105, 106, 129, 134, 135, 136, 137, 139, 238
Doré furnace, 39, 48, 57
Double waves, 639
Dowex 2, 559
Dowex 50, 553, 554, 559
Dowfroth, 37
DPG, 732
Drift mobility, 292–293
defined, 798
degradation of, 806
electrons, 804
holes, 799, 804
measurment, 798, 799
shallow trap controlled, 799
Drift velocity, 236
Drilling, characteristics, 779
Drossing, 54
Dry distillation procedures, 631,
Drying, 110, 619
D-selenopantethine, 529
Dusts, 33, 664–665, 671
Dynamic charge, 152–156
Dyspnea, 665
Dystrophic lambs, 549

East Germany, 42, 76
Edema, 688
Effective charge, 247
Eggs, 587, 682
Eigenfrequency, 156–157
Eigenvectors, 151, 155
Elastic constants, 114, 765, 766
Elastic deformation, 225
Electrical conductivity, 770
Electrical properties, 174–294, 759
Electrical resistivity, 23, 763
Electric dipole, 154
transitions, 186

Electric field, 263
applied, 800
dependence 212, 792, 799, 800
independence transition, 798
orientation, 214
Electric moment, 151, 154
second order, 162
Electroabsorption, 208–209
Electrochemical methods, 625–627
Electrode distance, 219
Electrodeposition, 117, 637
Electrodes, 798
Electrodialysis, 109
Electrolysis, 69
Electrolytes, 625, 626
Electrolytic copper slimes, 644
Electrolytic leaching, 55
Electromagnetic portions, 158
Electron beam intensity, 132
Electron diffraction, 88
Electron diffraction pattern, 118, 141
Electron diffraction studies, 118
Dowex 50, 553, 554, 559
Electronegativity, 380, 762
Electron irradiation, 131–132
Electron paramagnetic resonance, 100–101
Electron-phonon interaction, 163
Electon spin resonance spectrum, 567
Electron traps, 252–255, 258, 260, 262, 264, 265, 267, 293
depth, 254–255, 268–269
Electronic charge, transport, 790
Electronic configuration, 13
Electronic spectra, 360–361, 380
measurments, 344
Electronic switching, 724
Electronic transfer process, 724
Electronic transport, 798–799
Electons, 92, 792
distribution and biological action, 584
free, 101
mobility, 267

photoconductivity, 796
range, 806
release times, 794
states, 803
transitions, 802
transmission or reflection, 114
transport, 803
traps, 799
Electrophoresis, 68, 559
Electrophoretic mobility, 552
Electrophotography, 128, 788–806
amorphous selenium as a xerographic photoreceptor, 791–794
photoreceptor, ideal, 788–794
xerography, 788–794
Electrorefining, 69
Electroreflectance, 194, 201, 210–213, 214
peaks, 212
signal, 247
temperature dependence, 213
Electrostatic image, 790
Electrostatic transfer, 790
Elemental selenium, 8, 9, 299–301, 410–411, 429, 438, 476, 553–554, 622, 623, 629, 630, 664–666
as vulcanizing agent, 728–729
Elimination, 65, 662–664
Elongation, 737, 742, 765
Elution, 68
Embrittlement, 779
Embryonic death, 693
Empty conduction band, 717
Emulsion polymerization, 758
Enamels, 717
Endothermic heat, 112, 113
Energy, 249
gap, 184
higher, 289
levels, 803
region, 192
states, 800, 803
values, 344
Engineering, 762
Enthalpies of transition, 112
Entropy, 18
Enzootic icterus, 677
Enzymes, 578, 582–583, 585, 598–600

cofactors, 584–586, 600
molar concentrations, 599
substrates, selenium
analogues, 582–583
systems, 571
Epoxides, 443
Equilibrium, 798
constants, 62, 64, 90–91,
104
curve, 72
Escherichia coli, 559, 560,
580, 581, 583, 585, 586,
587, 598
esr spectrum, 101, 102
esr studies, 128
Esterification, 448
Esters, 479–480, 515, 584
N-protected, 520
Estimation, 107
Etching, 668
Etch pits, 246
Ethoxyquin, 585
Ethyl chloride, 746
Ethylenediaminetetraacetic
acid, 619, 628, 640, 642, 645
Ethylidene aniline, 732
Ethyl iodide, 485
Ethyl Selenac, 733
Ethylselenoglucoside, 487
Eucairite, 2, 37
European Selenium-
Tellurium Commitee,
772
Eutectics, 779
Evaporation, 90, 120, 138
Ewes, 693
Excessive intake, 547
Excitation, 250
Excitonic transitions, 800,
802–803
Excitons, 194–197, 211, 212
Excretion, 597, 663
biliary, 594
gastrointestinal, 593–594
Exhalation, 568, 571, 593
Exothermic oxidation, 44
Exothermic reaction, 66
Experimental ε_2 curves, 194
Experimental distribution
function, 115
Experimental error, 91
Explosives, 723
Exponential edge, 795
Exposure, 789, 791
time, 111
Expulsion, 47

Extensive replacement,
580–582
Extraction photometric
procedures, 643
Exudative diathesis, 588, 590,
597, 600, 688–689, 698
Eyes
burns, 666
inflammation, 665

Faces, prism and
rhombohedral, 138
Factor group, 150
Factor 3 selenium, 587, 676,
687, 688
Faraday rotation work, 803
Far-infrared, 156
Feathering, 689
Fe–Co alloys, 782
Feedstuffs, 682, 685, 697
Fe–Ni alloys, 782
Fermi level, 265, 796
Ferric chloride, 49, 51
Ferric hydroxide, 60
Ferric salt, 51
Ferroselenium, 26
Ferrous alloys, 771–779
casting of iron and steel,
771–773
free machining, 782
free machining steels,
773–777
silicon, 777–779
Ferrous ferrate, 713
Ferrous selenide, 77,
714–715
Field, 801
effect, 802
high, 796
Field strength, 242
effective, 221
high, 218
Field splitting parameter,
365, 381
Fillers, 758
Films, 129, 140, 788, 792,
806
amorphous, 128
crystallization, 132, 133
evaporated, 788
single crystal, 138, 141
structure, 131
thickness, 141
Filtration, 63, 110
First-order electric moment,
153, 154

Fish meals, 696
Five coordinated complexes,
368, 376
Five membered ring system,
475
Flameproofing agents, 729,
760
Flattening, 793
Flory's equation, 97
Flotation, 32, 37
Flue dusts, 617
Fluidized beds, 46
Fluids, 658
Fluorocarbon (Viton)
compositions, 747
Fluorescence, 618, 619, 645,
710
Fluorine, 720
Fluorometric determination,
615, 640
Fluorometric
2,3-diaminonaphthalene
procedure, 645
Flushing, 668
Fluxing, 39
Food containers, 712
Food and Drug
Administration, 696
Foodstuffs, 656, 671
Forages, 548
selenium content, 695
Forbidden band, 717
Forbidden energy gap, 803
Force constants, 388
Formaldehyde, 476, 746
Formic dehydrogenase, 585
Forms, 550–567
low molecular weight
compounds, 550,–555
Formvar, 117
Fossil fuels, 8, 10–11, 647
Fourier analysis, 99
430 Ferritic stainless steel,
776
Fowl, *see* Poultry
Fractionation, 554
Fracture, 772
characteristics, 777
section, 125
Fragmentation process, 93
Franz-Keldysh shift, 208,
801, 802
Fredga's method, 518–519
Free carriers, 792, 793, 798
generation, 799, 806
lifetime, 799

measured lifetime, 798
mobility, 797
one carrier system, 806
photogeneration of, 792
range, 799, 800
Free charge, narrow sheet, 798–799
Free cutting quality, 776
Free-machining agent, 640
Free radical ending. 102
Free radical mechanisms, 575
Freezing point, 97, 105, 138, 139
Frequencies, 159
ring and chain vibrational, 167
vibrational, 167
zone-center optical phonon, 157
Frequency dependence, 221, 243
Friction, 773
Friction brakes, 783
Frothers, 53
Fumes, 664–665, 671
Fundamental absorption, 190–197
Fungi, 552, 586
Furanose forms, 480
Furnace atmosphere, 711
Furyl alkyl selenide, 458

Galactose, 480
β-Galactosidase, 581
Galena, 31
Galisteo deposits, 37
Gallium, 684
Gallium phosphides, 641
Galvanomagnetic properties, 232–234
Gamma-emitting isotope, 580
Γ point, 151
Gamma radiation, 132
Gamma ray photopeaks, 648
Gaps, 187, 196, 796
direct and indirect, 203, 209
energy, 800
smallest, 188
Gases
roaster, 33, 58
roaster dust, 32
Gas phase, 770
Gassing, 59
Gastrointestinal disorders, 658

Gauche configuration, 372
Geeldikkop, 677
Gelatinous hydroxides, 70
Gel filtration, 557
Generation step, field controlled, 800
Geobotany, 10
Geographic areas, 676, 694
Geographic differences, 661
Geographic factors, 550
Geometric capacitants, 790
Germanium, 37, 38, 158, 175 181, 684
Germany, 69
GET model, 106
Glasses, 117, 131, 708–724
alloy, 803
amber tint, 713
amount of selenium as a decolorizer, 711–716
base, 710
black selenide, 721
cadmium sulfoselenide-containing, 719
chalcogenide, 721–724, 804
commercial, 721–722
decolorizer, 25, 709
diatomic, 722
electrical conductivity, 724
explosions, 723
iron-containing, 713–715
light absorption, 714
melting temperatures, 714
monatomic, 722
nonohmic voltage-current characteristics, 724
oxidation state in, 714
oxide, 708, 723–724
potassium-containing, 710
pure selenium colors, 711
random network theory, 713
ruby, 708
Se-based, 803
selenium pink, 709–711
selenium ruby, 716–721
semiconductors, 724
silicate, 710
soda-lime silica, 714
sodium analogue, 710
softening point, 723
solution colors, 719
stable, 722
ternary, 723
transition temperature, 20
unstable, 721

viscosity, 720, 721
yellow and red, 718
zinc oxide and selenium content, 720
zinc-potash, 708
Glassmaking, 718
heat treatment, 718–719
Glassy selenium, 88, 301
Globular shape, 772, 773
Glucofuranosides, 495
Glucose, 480, 485
derivatives, 485
Glucose–6–phosphate, 571
Glucosides, 482–483
free, 488
Glucosyl derivatives, 489, 491
2-(β-Glucoseleno)-(α-glucosylamino)benzene, 492
Glutathione, 522, 555, 569, 570, 571, 578, 663
reductase, 569, 572
Glyceraldehyde-3-phosphate dehydrogenase, 578
Glycosyl groups, 488, 489, 491, 493
Gold, 37, 49, 75, 710, 782
telluride, 42
Gonads, female, 596
Grades, 27, 61, technical, 62
Gradient technique, 139
Grain, 680
boundary edge, 779
refinement, 772
size, 772
toxic, 685–686
Graphite, 783
Grass, 10
range, 686
Gravimetric methods, 615
Green strength, 783
Greer oven, 729
Grey selenium, 89, 117
Grignard reagent, 452, 458, 459, 475
Grindelia, 679
Gripsholm, Sweden, 1, 2
Group III–V, 245, 270, 723
Group IVA, 723
Group VA, 723
Group VI, 722, 723, 756, 762
Group VIA, 723, 777
Growth, 586, 590, 687, 689
epitaxial, 131, 140

Grüneisen parameter, 19
GSH peroxides, 698
"g" value, 101

Hair, 657, 682, 685
Half-life, 631, 646
Halides, 59, 65, 506
Hall effect, 218, 232–233, 242
Hall mobility, 243
Haloalkyl selenium chlorides, 419
Haloaminocarboxylic acids, 516–517
Halogenated butyl rubber, 742
Halogen exchange, 500
Halogens, 128, 129, 140, 238, 239, 497–498
Happlopappus, 679
Hardness, 737, 742, 750, 778
Hard rubber, 746
  effect of Vandex on, 749
Hartree-Fock wave functions, 182
Hatchability, 682, 690
Hawaii, 679
Hay, 548, 694
H-bands, 211–212
Heart atrophy, 682
Heat, 790
  capacity, 17
  of combustion, 17, 112
  of dissociation, 17
  of evaporation, 17
  of formation, 66
  of fusion, 17
  resistance, 729, 737
  of transition, 112
  treatment, 718–719, 758
Heat-oxidation stability, 756
Hecht-type technique, 799, 800
  range limitation function, 800
Hemorrhage, 688
Heterocyclic amines, 733
Heterocyclic ring systems, 473
Heterocyclic selenium compounds, 409
Heterojunction, 273, 274, 278–279
Heteropolyacids, 340–341
Hexafluorides, 320
Hexagonal selenium, 89, 511
Hexahaloselenates, 320

Hexamethylene diamine carbamate, 746
Hexose derivatives, 480
Higher order phonon-involving optical processes, 159–160
History, 1–2, 675–676
Holes, 792
  concentration, 240, 242–243, 245, 248
  density, 228–229, 230, 242
  draft mobility, 799
  free, 101
  lack of photoconductivity, 196
  mobility, 227, 269
  mobility, temperature dependence, 269
  range, 799
  transport, 793, 798
  trapping, 794
Hollow-cathode lamp, 630
Hopping, 237, 263
  probability, 264–264
Hornbeck-Haynes bridge, 267, 268
Hot brittleness, 779
Hot pressed composites, 783
Hot shortness, 779
Howard-Sondheimer formula, 247
Hybridization, 182
Hycar 4021, 746
Hydantoins, 518–519
Hydrazine hydrate, 109, 110
Hydrazones, 464–465
Hydride purification, 63–66
Hydrides, 65, 396, 397
  reactivities, 397
Hydrobromic acid, 617, 618
Hydrocarbons, 498–499
Hydrochloric acid, 49, 59, 67, 68, 109, 617, 618, 620–621, 622, 623
Hydrochloric-hydrobromic acid distillation, 647
Hydrolysis, 432, 485, 488, 497, 502–503, 518–519, 555
  enzymic, 559
Hydrogen, 60, 711, 772, 777
  bonding, 310
  chloride, 479, 502
  peroxide, 62, 65, 66, 502–503, 568
  selenide, 56, 63–66,

341–342, 582, 655
  selenide intoxication, 667–668
  selenide toxicity, 667–668
  sulfide, 8
  uptake, 571
Hydrogenation, 65, 455
Hydrolyzation, 67
Hydroselenide group, 396–397
Hydroselenoglucosides, 485
Hydrosol, 629
Hydrostatic pressure, 226–227
Hydroxy compounds, 474
Hydroxy group, 499
Hydroxylamine, 645
3-Hydroxyphenyl methyl selenide, 467–469
4-Hydroxyphenyl methyl selenide, 467–469
Hypalon 20, 742
Hypertension, 662
Hypophosphorous acid, 617

Icteroid skin, 658
Identification, 1, 621–623
Igneous rocks, 2
Illumination, 221, 223, 225, 229, 257–259
Image development, 790, 793
Image erasure, 790, 793
Image transfer, 790, 793
Imidazolidine ring, 514
Impact properties, 772
Impact strength, 772
Impurities, 53, 56, 57, 61, 70, 110, 128–129, 139, 237–239, 723
  detection limits, 634
  determination of, 632–637
  traces, 633–634
Incident light, 162
Incident light flux, 792
Indicator plants, 10, 679, 681, 685, 687, 698
Indigestion, 664
Industrial hazard, 671
Industrial selenium, 62, 70
Industrial toxicology, 655
Infants, 659
Infrared absorption, 164–165, 167–168, 169
Infrared active vibrations, 155
Infrared data, 346, 372

Infrared, photoremittance,
    207–208
Infrared spectrophotometry,
    357–358, 414, 635
Infrared transmission, 162,
    722–723
Ingestion, 657
Injected space charge limited
    currents, 797
Injection, 797
Inorganic materials, 616–618
Inorganic salts, 683
Inorganic selenium, 563–566
Inorganic symbiosis, 343
Insecticides, 699
Insects, 699
Insolubility, 501
Insulin, 576
Intake, 686
    duration of, 683
Interactions, 319
Interatomic bonding, 149
Interband transitions, 795,
    796
Interface, 764–765
    kinetics, 763
Interference diagrams, 98
International Nickel
    Company, 33, 55
Interrelationships with other
    substances, 588–597
    miscellaneous, 597
    nutritional, 588–592
Intestines, 573–574, 662
Intoxication, 655–658
    pathology, 662
    symptoms, 655, 666, 667,
        668
Intrinsic excitonic absorption
    796–797
Iodides, 505, 507
Iodine, 97, 98, 103, 129, 323
    324
    monochloride, 324
Iodometric determination,
    624, 640
Iodometric titration, 639
Ion currents, 92, 93
Ion-exchange resins, 63, 67
Ion-exchange techniques, 67
Ionic charge, 152
Ionic oxychloride systems, 317
Ionic phenomena, 159
Ionization, 800
    cross sections, 94
    potentials, 13

Ireland, 9
Iron, 5, 38, 44, 70, 392, 548,
    567, 640, 709, 771–773
    divalent and trivalent, 714
        715
    hydroxide, 677, 679, 681
    oxide, 679
    maximum content, 713
    selenide, 26, 770
Iron-selenium alloys, 773
Iron-sulfur potential, 567
Irradiation, 631, 634–635,
    647
    time, 634
Irreducible wedge, 180, 190
Irritability, 664
Irritants, 62
Isobutene, 460
Isobutylene, 737
Isochrome annealing, 220
Isolation, 1, 547
Isologues, 584
Isomerization, 422, 463
Isomers, 461
Isoprene, 737
Isopropyl acetate, 133
Isoselenocyanates, 345–346
    349
    N bonded, 350
Isoselenuronium salt, 488
Isotopes, 631
Isotopic materials, 181

Japan, 8, 12, 27
Jobs, seleniferous, 664

Kennecott Copper
    Corporation, 48
Kerosene, 37
α-Ketoglutaric acid, 697
Ketones, 469
Kidney, 549–550, 594, 658
Kinetic curves, 769
Knudsen cell, 88, 90
KU-1-2, see Cation
    exchanger
Kwashiorkor, 550, 661

Lability, 555
Lactobacillus helveticus, 580
Lactones, 443–444
Lamellae, 125, 132, 137
Larvae, 587
Laser, 161
Lassitude, 664
Latex preservatives, 729, 760

Lattice, 138, 177–178, 770
    amplification of modes,
        236
    constant, 15, 125
    contribution, 157
    defects, 218–221 passim,
        237–239, 240, 257
    parameters, 178–179
    reciprocal, 179–180
    ring molecules in, 287
    strong and optically active
        modes, 245
    waves, see Phonons
Layers
    details, 242
    thickness, 131
LCAO method, 182
Leaching, 37–38, 48–52, 54,
    56
    chemistry of, 49–50
    process, 50–52
Lead, 31, 33, 39, 54, 776
    sulfate, 53, 54, 57
Least squares analysis, 121,
    122
Leaves, 554–555
Leningrad Mining Institute
    process, 44, 57
Lesions, liver or muscle, 550
Life processes, 515, 598
Lifetime differential, 261
    long, 97
Ligands
    ambidentate, 342, 369
    bidenate, 350, 370–376,
        378, 387–388
    binegative, 386, 387–391
    inorganic, 338–364
    mixed, 358–363
    monodenate, 364–370, 378
    neutral, 364–378
    nonselenium bearing, 360
    organic, 364–391
    with phosphorous-selenium
        bonds, 354
    polydentate, 376–378
    terdentate, 377, 385–387
    tridentate, 390–391
    uninegative, 378–387
Light
    absorption, 792
    discharge, 797
    flux, 131
    intensity, 258, 282
    interaction with crystal
        lattices, 158

interaction with phonons, 148–171
irradiation, 131–132
polarization, 191–214 *passim*
transmitted, 117
waves, *see* Photons
Light-dark boundaries, 790
Lignosulfonic acid, 48, 58
Lime, 55
Limestone, 9
Linear polymers, 511
Lines, Stokes and anti-Stokes 160
Linkage isomerism, 350
Lipids, 697
Liquid air, 96
Liquid phase, 96–109
Liquid selenium, 65, 66, 90, 167, 764
Litharge, 729
Lithium, 410, 452
Liver, 549, 550, 564–565, 569, 573, 586, 591, 594, 659, 662, 663
  cirrhosis, 662
  fatty degeneration, 662
  necrosis, 549, 550, 588, 590, 592, 597, 600, 687, 690–691
Livestock, 8, 548, 655
  selenium poisoning, 676–686
Long chain fatty acid metabolism, 698
Long-chain polymeric molecules, 97
Long-period transition, 777
Long-term systemic effects, 658
Long waves, 149
Loss, 32
Lovibond tintometer, 713
Low angle boundaries, 765
Low defect concentration, 764
Lower limit, 92
Low field ohmic characteristic, 797
Low molecular weight compounds, 550–555
  dimethyl diselenide, 553
  dimethyl selenide, 552–553
  elemental selenium, 553–554
  miscellaneous compounds, 554–555

selenocystathionine, 551–552
selenocystine, 554
selenohomocystine, 554
selenomethionine, 554
Se-methylselenocysteine, 550–551
Se-methylselenomethionine, 552
trimethyl selenonium, 553
Low sulfur (0.5 phr) TMTD, 733
Lubricants, 26, 452, 783
Luminescence, 204–207

Machaeranthera, 679
Machaeranthera glabriuscula, 669
Machinability, 773, 776, 779
Macro determinations, 615, 625
Macroporous bed of microporous masses, 46
Magma, 2
Magnesium
  oxide, 131, 742
  tellurate, 60
Magnetic nonequivalence, 509
Magnetic properties, 777
Magnetic susceptibility, 23, 88, 100–101
  measurments, 358
Magnetoconductivity, 233–234, 269
  temperature dependence, 233
Magnetoresistance, 232, 233, 237
Manganese, 772, 777
  decolorizer, 712
  oxide, 2
  sulfide, 772
Mannich condensation, 459, 472, 478
Mannose, 480
Mansfield, A. G., 42–43
Mass ratio, 213, 247
Mass spectrometry, 88–90, 634
Matte, 55
Maxima, 195, 204, 294
Mechanical behavior, 765–766
Mechanical properties, 763
Medicine, 580
Melting, 40

Melting point, 16, 139, 409, 763
Melts, 138, 139, 140
  supercooled, 125
Melt temperature, 114, 115, 138
Memory device, 724
Memory effects, 110
Menadione, 571
Mercaptans, 758
2-Mercapto imidazoline, 742
Mercuric chloride, 553, 596
Mercury, 33, 53, 68, 75, 596, 636, 684
Mercury selenide, 53
Mercury (II) cyanide, 622
Metabolism, 547, 567–574, 589
Metallic selenium, 97
Metal-ligand bonding, 350
Metallurgical aspects, 762–783
  alloying behavior, 766–769
  crystal growth, 763–765
  diffusion, 770
  ferrous alloys, 771–779
  mechanical behavior, 765–766
  miscellaneous applications, 782–783
  nonferrous metals and alloys, 779–782
  oxidation and corrosion, 769
  uses, 771–783
Métallurgie Hoboken N. V., 39
Metallurgy, 25–26
Metals, 344, 347, 378, 384, 398–399, 618, 641;
  bivalent to hexavalent, 756;
  deposition, 783
  divalent, 754
  heavy, 341–342, 588, 591, 594–597
  nontransition compounds, 395–398
Metal–selenium stretching, 239
Meteorites, 5, 8
Methaneselenol, 582
Methionine, 564, 574, 575, 581, 583, 587, 684
Methoxides, 460
Methoxy compounds, 467
Methyl aryl selenides, 506

Methylated metabolites, 570–571
Methylation, 568
  biological, 552
Methyl donors, 684
Methylene iodide, 120
5-Methyl-2-methylselenophenyl methyl ketone, 471
Methyl phenyl selenides, 416
Methyl piaselenol, 737, 750
Methyl salicylate, 138
Methylseleno derivatives, 514
4-Methyl selenodithiobenzoic acid, 467
2-Methyl selenophenylmethyl ketone, 471
4-(4-Methylselenophenyl)-phenyl bromoethyl ketone, 471
Methyl sulfate, 467
Methyl Zimate, 733
Mexico, 37
Mica, 131, 132
Mice, 550
Micelles, 117
Microbiological action, 681
Microcirculatory changes, 600–601
Micrococcus actilyticus, 571
Microcoulometric technique, 635
Microcracking, 782
Micro determinations, 623, 629
Microhardness, 20
Microorganisms, 548, 559–560, 571–572, 592, 598
"Microscopic" mobility, 798
Microscopic processes of free carrier photogeneration, 788
Microvascular system, 600, 602
Millipore filter, 647
Mineralization, hydrothermal stage, 3
Minerals, 37, 642–644
  names, 4–5
  number, 2
  sulfide, 2
Minority carrier diffusion length, 267
Miscellaneous properties, 22

Mixed acids, 616, 618, 619, 638
Mixtures, 3, 76
Mn Se system, 778
Mobility, 227, 234, 239, 244
  of charge carriers, 724
  drift, 237
  high, 176
  ratio, 247, 269
  temperature dependance, 235
  trap-controlled, 237
  values, 239
Mobridge member, 678
Model behavior, 790
Modes, vibrational, 171
Modification, 125, 133
Modulation technique, 208–215
Modulus, 737
Mohs hardness, 763
Molar solubilities, 624
Molar volume, 23
Mole fraction, 94
Molecular attachment, 763
Molecular orbital (MO) diagrams, 388
Molecular orbitals, 803
Molecular rotation, 485
Molecular species, 89, 94–95
Molecular transport, 136, 137
Molecular weight, 103, 114
Molecules, 111, 112, 166
  cyclic, 93
  dissociation, 502
  eight-membered ring, 163
  organic, 503
  ring, 803
Molten polymers, 97
Molten selenium, 70
Molybdenum
  disulfide, 783
  trioxide, 240–241
Monkey nut, 552
Monoclinic selenium, 141–142, 287–294
  crystal structure, 287–288
  electical properties, 291–293
  optical properties, 288–291
  photoconductivity, 293–294
α-Monoclinic selenium, 88, 89, 90, 94, 115, 116, 119, 120–122, 133, 141, 163–167, 287, 289, 300, 300, 803

infrared absorption, 164–165
  Raman scattering, 165–166
  structural parameters, 121
  structure, 163
  vibrational assignments, 166–167
β-Monoclinic selenium, 88, 90, 115, 119, 120–122, 133, 141, 287, 300–301
β structure, 122
β'-Monoclinic selenium, 124
Monocrystals, 141–142
  conductivity, 238
  dimensions, 141
  melt-grown, 236
Monogastric animals, 563, 565, 683
Monomers, 370, 455, 803
Monoselenodicarboxylic acids 588
Monoselenooxytocins, 524
Monothiols, 593
Monotropism, 123
Montana, 11
Monte Carlo procedure, 115
Mortality, 596, 658
Mucopolysaccharides, 566–567
Mucous membranes, irritation, 665, 667, 668
Mulberry heart, 691
Muscle, 692
Muscular dystrophy, 591, 688, 691, 692, 696, 697
Myopathy, 689, 690, 691

NADP, 569, 571, 570–573, Nails, 657
  discoloration, 662
  inflammation, 666
Naming, 2
Naphthalene, 138
Nasal symptoms, 665
Native enzymes, 557
Natural products, selenium-bearing toxicity, 669
Natural rubber, 729–733, 750, 753
  affect of Selenac on aging, 735
  affect of selenium compounds on vulcanization and aging, 734
  insulation composition, 735

Natural state, 480
Naumanite, 37
N bonding, 350, 351
Nebraska, 677
Needles, 138
Neighbor, nearest, 14–15
Neoprene, 742
Neoprene W⁴, 743
Nephelauxetic effect, 354–357
Nephelometric
    determination, 635
Neptunia amplexicaulis, 552,
    561
Neutron activation analysis,
    11, 618, 631–632, 644,
    645, 646, 647, 661
New Mexico, 11, 37
New Zealand, 26, 693, 695
Nickel, 3, 51, 55, 782
    cathode, 117
    selenide, 272
Nickel-iron, 779–782
Nikko Electric Copper
    Works, 50
Niobara formation, 678
Nitrate, fusion with, 69
Nitration, 424
Nitric acid, 54, 56, 63, 617
Nitrile rubber, 737, 740
Nitriles, 513
Nitroaryl 2-amino-
    5-methylphenyl selenides,
    454
Nitroaryl 4-aminophenyl, 454
Nitro aryl selenides, 453
5-Nitrobenzo [2,1,3]
    selenadiazole, 756
Nitrocellulose, 141
Nitrogen
    absorption of, 772
    boiling, 110, 111
    liquid, 113
Nitro substituted
    derivatives, 415
Nonbonded approach
    distance, 306
Noncyclic compounds with
    X—Se—Se—X linkages,
    303–305
Noncyclic X—Se—X
    linkages, 301–303
Nonferrous metals and
    alloys, 779–782
    cobalt-iron, 779–782
    copper, 779
    lead, 782

lead base, 782
    nickel-iron, 779–782
Nonmetallic aromatic
    selenenyl halides, 759
Nonmetallic aromatic
    sulfenyl halides, 759
Nonphotoconductive energy
    gaps, 803
Nonphotoconductive states,
    803–804
    molecular origin, 803–804
Nonruminants, 564–566
Noodles, 44
Norddeutsche Affinerie, 39,
    45
Normal sulfur(3phr)-
    benzophiazyl disulfide,
    733
North Sea, 12
N,Se-diglycosides, 493
n-type, 274, 280
Nuclear irradiation, 131–132
Nuclear magnetic resonance
    spectrum, 525
Nuclear magnetic resonance
    studies, 449
Nucleation and nuclei, 117,
    124
    behavior, 135
    rate, 128
    rate curve, 129
    rate of formation, 127–128
    sites, 128
    spontaneous, 139
Nutrient, selenium as,
    687–697
Nutrition, 567, 654
    human, 671

O-alkyldiselenocarbonates,
    382
Occurence, 2–25, 547
Odor, 2, 10
Ohmic conductivity, 273
Ohmic resistance, 273
Oil, 11, 574
    sulfur–bearing, 746
Olefinic systems, 463
Olefins, 464, 474, 499
Olen (Belgium), 39
Olfactory fatigue, 668
ON-OFF state, 724
Onions, 554
Oonopsis condensata, 551
Open-ended chain, 99
Optical absorption, 792

edge, 796, 797
    measurements, 795
Optical activity, 215–217
Optical constants, 191–197
Optical electronegativities,
    354–357
Optical image, 790
Optical lattice modes, 247
Optical measurements,
    190–197
Optical modulations, 138
Optical properties, 174–294
    passim
    history, 190–191
Optical quenching, 249, 251,
    252, 256–257, 263
    energy dependence,
    253–254, 255
    rate, 257–258
    temperature dependence,
    254
Optical transitions
    energy, 193
    polymerization, 193
Orbital diagram, 339–340
Ore microscopy, 3
Ores, 642, 644
    base-metal sulfide, 39
    gold–selenium, 38
    recovery from, 37–38
    sulfide, 3, 616–617, 643,
    678–679
Organic accelerators, 742
Organic arsenicals, 686
Organic compounds, 680
    electronic properties, 175
Organic dihalides, 476
Organic halides, 485
Organic polymers, 125
Organometallic compounds,
    351–354, 391–398
Organoselenium compounds,
    546–602 *passim*, 638, 669,
    670
    amount of toxicity, 669,
    670
    high affinity of metals for,
    592
    lack of toxic data, 669, 671
    recently synthesized, 583
Organs, 658–659, 660
    elimination of selenium,
    663
Organyl isoselenocyanates,
    421–423
    amines, 422

isomerization, 422
ultraviolet spectra, 422
Organyl selenium halides,
    415–416, 417, 433
    nuclear magnetic resonance
        investigation, 416
    synthesis, 416
Organyl selenium
    isocyanates, 417
Organyl selenium trihalides,
    424
Organylselenoamino-
    carboxylic acids, 516–519
    addition of
        phenylmethaneselenol,
        518
    alkylation of
        alkaneselenolates with
        haloaminocarboxylic
        acids, 516–517
    alkylation of 1-carboxy-
        2-aminoalkaneseleno-
        lates, 517–518
Organyl selenocyanates, 417,
    418–421, 423
    amines, 421
    esters, 421
    hydrolysis, 430–431
    reaction, 420–421
    synthesis, 418–420
Organyl selenosulfates, 417,
    429
Organyl thiocyanates, 417
Oriented seeds, 764
Ortho, *meta*, and *para*
    monohalophenyl
    derivatives, 508, 509
Orthorhombic form, 511
Oscillators
    parameters, 156–158, 159
    strength, 188, 190
Oscillopolography, 626
Ouabain, 574
Ovalbumin, 562
Overheating, 772
1,4-Oxaselenane, 324
Oxidation, 41, 43, 53, 331,
    331, 485, 502, 568,
    574–580, 619, 711, 715,
    769–770
    resistance, 750, 754
    states, 625
Oxidation-reduction, 715
Oxides, 710n, 770
    metal, 721
    purification, 62–63

structures, 307- 308
Oxyacids, 309–312
Oxyanions, 309–312
Oxygen, 12, 32, 51, 63, 102,
    110, 128, 238, 316, 584,
    635, 711, 758, 760, 769
    atmospheric, 40, 49
    ring atoms, 495
Oxygen flask combination
    method, 618, 639, 645
Oxygen-glass surface reaction,
    723
Oxytocin, 522, 523, 584
Ozokerite, 729, 732
Ozone, 503

Pacific Ocean, 12
Paints, 717
Pair states, 803
Pale crepe, 729, 732, 754
Panchromatic photoreceptor,
    805
Pancreas, 562, 573, 659, 690
    degeneration, 689
    fibrosis, 689
Pancreatic protein synthesis,
    698
Pantetheine, 515
Papain, 577–578
Paper, 645, 790
Paper chromatography, 638
Parallel chains, 100
Parameters, thermal and
    position, 123
Parametric oscillators, 138
Para-position, 499
Particles, 133, 716
PBNA, *see* Phenyl-β-
    naphthylamine
2-Character, 181
Peaks, 192, 194–196, 205, 211
Pelletization, 44, 46, 61
Pent-4-enyl phenyl selenide,
    370
Pentose derivatives, 480
Peptides, 562
    3-benzylseleno-2-aminopro-
        prionic acid (Sebenzyl-
        selenocysteine), 521
    disulfide, 576
    selenium containing, 522
    sulfur containing, 522
Perchlorates, 506
Periodicity, 12
Periodic position, 13
Periodontal disease, 693

Permanganate method, 625
Permanganate titration, 636,
    637
Peroxides, 574, 575, 576
Peroxycarboxylic acids, 503
Perturbation, 115
pH, 67, 68, 557, 558, 559, 624
    diagrams, 49
    range, 49, 57–58
Phase transformation, 134
Phenol, 68
Phenoxaselenine, 750, 754
Phenyl-β-naphthylamine,
    750
Phenylmethaneselenol, 518
Phenylselenoacetic acids,
    383–385
Phenylselenoglucosides,
    482–483, 497
Phenylseleno group, 460
Phonon-assisted electronic
    optical processes,
    162–163
Phonon drag, temperature
    dependence, 231
Phonons, 158
    energies, 209
    scattering, 245, 247
    structure, 198, 200–201
Phonons, interaction with
    light, 148–171
    infrared active, 158
    long wavelength, 151
    zone-center, 151
Phosphates, 597, 713
Phosphinato complexes,
    359–360
Phosphine selenides, 363–364
Phosphinic acids, 359
Phosphorus, 305–306
Phosphorylation, 529
Phosphoselenopantetheine,
    529
Phosphoselenopantethine,
    529
Photocells, 128
Photoconductive gain, 269
    temperature dependence,
    270
Photoconductive spectral,
    response, 797
Photoconductivity, 25, 174,
    248–270, 792, 800, 801
    decay, 266
    energy dependence, 291
    impurity, 250–251

quenching, 251–258
spectral distribution, 249, 250
Photoconductor edges, 796, 797, 802, 803
Photoconductors, sensitized organic, 806
Photocurrent, 251, 259, 281
amplitude, of pulse, 267
decay of, 259–260, 262, 266
temperature dependence of decay, 261
weak dependence on light intensity, 263
Photo discharge, 792, 794
Photoelectric work function, 23
Photoemission studies, 802–803
Photoemittance, thermal, 207–208
Photogeneration, 798
field controlled, 800
Photographic toners, 26
Photography, 25
Photoinjection, 798
Photometric methods, 627–631, 645
atomic absorption spectroscopy, 629–630
emission spectroscopy, 630
spectrophotometry, 627–629
x-ray fluorescence, 630–631
Photon absorption, 790
Photon energies, 290, 792, 800, 801, 802
dependence, 793
Photon-phonon interaction, 152, 157
Photons, 158, 207–208, 258
emission, 204–207
far-infared, 159
Photoreceptors
amorphous selenium as, 791–794
cleaning, 790
development of the image, 790
employing Se, 805–806
exposure, 789, 791
ideal, 788–794
image erasure, 790
latent image, 789
print fixing, 790

selenium as, 791
sensitization, 789
Photo response, 805
Photovoltage, 282
spectral distribution, 253
Photovoltaic cells, 280–286
spectral sensitivity, 283
structure, 281
Physical properties, 22–24, 72, 128, 763
Physical refining, 69–77
Physics of selenium, 98
Piazselenols, 632, 640, 645, 648
Picrates, 505
Pierre formation, 677–678
Piezoelectric effect, 324–235
Piezoreflectance, 194, 212, 213–215
Pigments, 25, 717
Pigs, 659, 663, 686, 690–692
Pinhole porosity, 772
Pink glass, 709–711
Piperidonium pentamethylene dithiocarbamate, 729
Plaining up, 712
Planes, 138, 765
Plants, 8, 9–10, 26, 547, 548, 560–561, 592, 646
analysis, 686
greenhouse, 699
growth, 698
qualitative and quantative detection of selenium, 675
seleniferous, 654
selenium in, 679–680
selenium as essential element for, 698
selenium toxicity, 669
Plastic constants, 21
Plastic deformation, 219–221, 224, 225, 229, 240, 249, 251, 253, 257, 765, 777
Plastics, 455, 754–757
selenium in, 728–760
Platelets, 120, 133, 138,
Plate-type still, 73
Plating, 782–783
Plating solutions, 782–783
Platinum screen electrodes, 69
Pneumonia, 668
Pneumonitis, 666
p-n junction, 278
Point group, 150

Point h, 184–185
Poisoning, see Toxicity
Poisson's ration, 21
Polariton dispersion curves, 157, 158
Polarization dependence, 192
Polarizations, 191, 195, 288, 289, 384
macroscopic, 155
parallel, 211
rotation, 176, 216
Polarized incident and scattered radiation, 160–161
Polarized waves, 217
Polagraphic behavior, 639
Polarography, 625–627, 636–637, 645
Polaron coupling constant, 247
Polybutadiene, 758
Polychloroprene, 742
Polycrystalline selenium, 239, 240
Polyethylene, 754, 756
aging tests with compositions containing selenadiazole derivatives, 757
Polyisoprene, 758
Polymerization, 102, 167, 511
additives, 729, 758
number-average degree, 104–105, 107, 108
Polymers, 128, 737
low mobile weight, 98
Polyploidy, 587
Polyselenide, 57
Polyurethan coating system, 455
Polyvinylcarbazole (PVK), 806
Populations of ring and chain components, 116
Potassium, 710, 711
amyl xanthate, 37
dichromate, 503
permanganate, 503
Potential drop, 791
Potentiometric method, 625
Poultry, 682, 685, 688-890
Powder image, 790
Powder metallurgy, 783
Precipitation, 58, 63, 109–110 112, 620
Predominant species, 91
Pressing, 144

Pressure, 73, 95, 104, 790
  atmospheric, 75
  orientation, 203
  partial, 93
  reduced, 75
Pressure bomb, 103, 104
Pressure coefficient, 215, 228
Pressure dependence, 184,
  201–203, 225–228
Pressure leaching, 39
Pretreatment, 127
Price, 28
Print fixing, 790
Proceedings of the sympo-
  sium on selenium, Oregon
  State University (1966),
  547
Processes I and II, 260–262,
  264
Production, 26–27
Propenyl β-amino-β-carbo-
  xyethyl selenide, 554
Properties, 12–25, 87–88,
  574–588
Protective coating, 773
Proteins, 554, 555–567, 575,
  583, 592, 662
  disulfides, 558
  hydrolysates, 680
  iron-sulfur, 567
  levels, 684
  mucopolysaccharides,
    566–567
  selenoamino acids in, 559
  selenomethione in, 564
  serum, 557–558
  sulfur-rich, 601
Protonation constants, 502
Pseudohalogen analogue, 306
Pseudohalogen groups, 371
Pseudomonas purida, 567
^{3}P state, 629
P-type, 274, 278, 279
Puckering, 121, 122, 166, 287
Puerto Rico, 679
Pulling speed, 764
Pulmonary edema, 668
Pulse height, 799
Pulse-polagraphic procedure
  641
Pure monomer selenium, 105
Purification, 61, 557, 619
Purification by ion-exchange
  reaction, 67–68
Purification by solvent extrac-
  tion, 67–68

Purina Chow, 549
Purines, 515
Purity, 27, 60, 102, 110, 134
  high, 62–63, 65, 66, 68, 72,
    75, 76, 77
  ultrahigh, 61
Putidaredoxin, 567
Pyramid hardness, 778
Pyranose form, 480
Pyranosides, 495
Pyrex, 129
2-(2-Pyridyl)benzo[b]seleno-
  phene, 373
Pyridyl selenoglucosides, 484
Pyrimidines, selenium deriva-
  tives of, 515
Pyrite, 3, 31, 52
Pyrrhotite–chalcopyrite, 3
Pyruvate, 585

Quadratic current-voltage
  dependence, 218
Quantum efficiency, 795, 796
  802
  field dependence, 802
  field dependent measure-
    ments, 797
  increased, 801
  limiting value, 800
Quantum yield, 800
Quartz, 75, 76
Quartz–barite, 3
Quartz–chalcopyrite, 3
Quartz–chalcopyrite–molyb-
  denite, 3
Quartz labyrinth, 110
Quartz molybdenite, 3
Quartz–pyrite–chalcopyrite–
  polymetallic formations,
  3
Quartz sieve-plate column, 73
Quartz tubes, 139, 723
Quenching, 96, 113, 114, 115
Quinoline-8-selenol, 382–383
Quinone, 572

Rabbits, 565, 662
Racah parameter, 381
Radial distribution curves,
  99–100, 111, 114, 115, 118
Radiation damage, 575
Radioactive fractions, 565
Radioactive isotopes, 13
Radioactivity, 552, 554–555,
  557, 561, 563, 564–565,
  567, 573

Radiochemical separation,
  631, 646–647
Radionuclides, 631
Rain, 11
Raman scattering, 148, 152,
  159–162, 165–166, 168
Raman spectra, 116, 414
Random network theory, 713
Random orientation, 139
Ration, 683
Rats, 549, 586, 587, 590, 591,
  596, 662–663, 669, 684,
  685
  doses fatal to, 667
R, configuration, 499, 519,
  520
Recombination centers,
  265
Recombination mechanism,
  258–270
Recombination, transport
  and surface, 798
Recovery, 31–60
  from aqueous solutions,
    56–60
  from the -2 valence state,
    57
  from miscellaneous metal-
    lurgical products,
    54–56
  from miscellaneous solu-
    tions, 60
  from the $+4$ valence state,
    57–59
  from the $+6$ valence state,
    59–60
Recrystallization, 62
Rectification, 70, 71–75
Rectifiers, 25, 33, 56, 88, 128,
  176–177, 270–280
  current-voltage character-
    istic, 273, 275, 277
  polycrystalline, 271–274
  single crystal, 244–277
  theory, 277–280
Red amorphous selenium, 88,
  109–112, 133
Red mites, 699
Red monoclinic selenium, 53,
  91
Redox procedures, 624
Redox proteins, 567
Red precipitate, 567
Red region, 805
Reduction, 51, 60, 69, 109–
  110, 111, 425, 426, 433,

441, 442, 501, 568,
    574–580
Refineries, 34–35
Refining, 60–77
Reflectance spectrum, 797
Reflectivity, 191, 192, 196,
    210–213, 289, 290
Reformatsky reaction, 469
Refractory index
    extraordinary, 23
    ordinary, 23
Refractory materials, 72
Relative potency, 588
Relaxation time, 247
Renal necrosis, 596
Reproducibility, 195
Reproduction, 682
Residence times, 43
Residual potential, 794,
    797
Residual selenium, 777
Resilience, 758
Resins, 455
    thermosetting, 757
Resistance, 281
Resistivity, intrinsic, 797
Respiratory excretion, see
    Exhalation
Reststrahlen, 149, 152–153,
    154–156, 158, 176
Retention, 658–661
Retorting, 47, 70, 75–76
Reverse ratio, 247
Ribonuclease, 557, 577, 578
Rigid-ion value, 153
Ring-chain equilibrium, 104,
    125, 138
Ring-chain theory, 104
Ring-oven technique, 648
Ring molecules, 287
Rings, 94, 114, 116, 118, 119,
    121
    closed, 122
    disordered, 295
    eight-membered, 166
    five-membered, 324, 325
        329
    making and breaking of, 98
    open, 122
    planar, 328
    small, 108
    vibrations, 170–171
Roasting, 38, 40–46
    chemistry of, 41–42
    temperatures, 47
Roasting processes

low temperature, 42–43
    three groups, 42
Rock formations, 8
Rocks, 618, 642–644
    Cretaceous, 677
    sedimentary, 678
Rods, 115, 119
Romania, 43
Rongalite, 441, 442,
Rotary kilns, 47
Rotatory power, 216
Ruby glass, 716–721
    melting of, 719–720
Rubber, 26
    acrylate, 746
    butyl, 729, 737–742, 752
    halogenated butyl, 742
    hard, 746, 749
    hypalon, 742
    natural, 729–733, 734–735,
        750, 753
    neoprene, 742
    nitrile, 737, 740
    selenium in, 728–760
    styrene-butadiene, 733–736,
        751
    synthetic, 730, 733
    viton, 746
Rumen microorganisms, 560,
    563–566
Ruminants, 563–566, 683,
    692–694, 697
Ryegrass, 554, 555

S-adenosyl methionine, 569
(Salicylideneselenosemicarba-
    zidatosalicylideneseleno-
    semicarbazidic)cobalt
    (III), 385
Salmonella heidelberg, 554
Salt-cake, 713
Salts, 340–431, 687
    polyvalent metal, 754
Sand dollar, 587
Sandstone, 37
Sarcina lutea, 583
Saturated vapor pressure, 89
Saturation, 129
SBR-1500 composition, 736
    see also Styrene-butadiene
    rubber
Scattering, 15
Scattering processes, 161
Scorch resistance, 746
Screws, right-hand or left-
    hand, 178

Scrubbing, 44, 47, 58
^{75}Se, 580, 631, 646, 647
^{77m}Se, 631, 646
^{81}Se, 631, 646
^{81m}Se, 646
Se bonding, 351
Se-adenosylselenomethionine,
    582, 583
Se-aryl aminomonoseleno-
    carboxylates, 519–520
Se-benzylselenocysteine, 521
Secondary selenium, 56
Sedimentation velocity
    studies, 557
Se-esters of selenocarboxylic
    acids, 479–480
^{75}Se-labeled selenite, 552
Selection rules, 149–152, 185–
    186, 190
Selective bonding, 62
Selective precipitation, 61
Selective separation methods,
    619–621
1-Selenacyclohexane, 366
1-Selena-4-oxacyclohexane,
    366, 367
1-Selena-4-thiacyclohexane,
    366
Selenates, 9, 10, 51, 59, 60,
    399, 560, 568, 582, 592,
    638, 663, 667, 679, 680,
    685, 709
    heavy metal, 623
    inorganic, 567, 580, 581
    radioactive, 552
Selenazole, 639
Selenazolidine, 638
Selenenic acids, 417–418
    amides, 418
    anhydrides, 417
    esters, 417
    reduction, 433–434
Selenenyl halides, 415–416
Selenic acid, 310, 782
Selenide ion, 626
Selenides, 33, 40, 49, 51, 105,
    438–478, 506,–507 564,
    575, 772, 783
    conversion of diorganyl
        diselenides, 451
    double, 41
    heavy metal, 720–721
    oxidation, 502
    synthesis, 776
Selenide-sulfide inclusions,
    772, 773

Selenide sulfides, 421
Selenfereous areas, 677
Seleniferous diets, 684
Seleniferous range, 686
Seleniferous soils distribution, 677–678
Seleniferous vegetation distribution, 677–678
Seleniferous wheat, 669
Seleninic acids, 423–425, 426 443
  adducts, 423
  amphoteric character, 423
  aromatic derivatives, 423
  colorless, 423
  cyclic ester, 443
  infrared spectra, 423
  odorless, 423
  reduction, 433–434
  solubility, 423
  titration, 423
Seleninyl difluoride, 319
Seleninyl dihalides, 314–320
Selenious acid, 38, 58, 63, 65, 68, 109, 110, 310, 556, 557, 668
  reaction with thiols, 579
Selenites, 9, 10, 33, 51, 57, 60, 399, 547, 549, 560, 563, 565, 586, 638, 662, 667, 692, 698, 709
  adsorption, 565
  enzymatic reduction in microorganisms, 571–572
  heavy metal, 623
  inorganic, 567
  intermediary metabolism, 568–572
  radioactive, 551, 552, 554, 555, 564
  reduction, 554, 564, 572
  roasting to form, 42–43
  transport, 597
Selenium-carbon bonds, 445, 449, 495
Selenium-cell, 174
Selenium
  chlorides, 61
  cyanide, 622
  dichlorides, 66, 439, 498–499
  diethyldithiocarbamate, 26
  dioxide, 33, 38, 41, 42, 57, 58, 61, 62–63, 109, 240–241, 438, 507, 595, 638,

665, 666–667, 721
  dioxide, roasting to expel, 43–45
  dioxide, recovery, 45
  disulfide, 26
  (IV), 621–637 passim
  hexahalide, 320
  hydride, 61
  monochloride, 66
  oxychloride, 314–320, 498, 500, 668, 669, (VI), 621, 638
  sulfide, 2
  tetrabromide, 620
  tetrachloride, 66, 67, 498–500
Selenium-selenium bonds, 576
Selenium-sulfur bonds, 576
Selenoaldehydes, 509–511
Selenoaminoacids, 408, 518–519, 554–602 passim
  formation, 561, 564
  incorporation, 563–566
  inorganic selenium in, 560
  preformed, 561–562
  in proteins, 559, 583
  synthesis from inorganic selenium, 563–566
  synthesis from selenite, 563
Selenoamino carboxylic acids, 515
Selenoantipyrine, 364
Selenocarbohydrates, 408–409, 487, 488–491
Selenocarboxylic acids, 479–480
  amides, 512–515
  esters, 515
Selenocoenzyme A, 529, 583
Selenoctic acid, 583
Selenocyanates, 342–345
  carbonyl, 353
  infrared and electronic spectra, 343–344
  mixed complexes, 347–351
  N bonded, 245–346, 353
  organometallic, 351–354
  Se bonded, 346–347, 351, 353
Selenocyanation, 420
Seleno-cyanide complex, 53
Selenocystathionine, 551–552, 680
  biosynthesis, 551
  from radioactive selenite, 551

Selenocysteine, 581
Selenocystine, 554, 561, 564, 565, 576, 577, 578, 579, 669
l-Selenocystine, 669
Selenodiazine, 639
Selenodicysteine, 561
Seleno esters, 479–480
Selenoethionone, 582
Selenoformaldehyde, 511
Selenoglucosides, 481–488
α-Selenoglucosides, 485, 497
β-Selenoglucosides, 484, 497
Selenoglutathione, 555, 569–570, 583
Selenoglycosides, 487
Selenohomocystine, 554
Selenoindigo, 329
Selenoketones, 382, 511–512
Selenolates, 442
Selenols, 409–414, 444, 476, 479. 481, 575, 758
  alkylation, 456–458
  aromatic, 411
  from carbon diselenide, 411–412
  from carbon oxide selenide, 411–412
  from elemental selenium, 410–411
  from hydrogen selenide, 411–412
  by hydrolysis, 432
  nuclear magnetic resonance 414
  oxidization, 431
  physical properties, 414
  reactions, 414
  as reagents, 414
  by reduction of diselenides, 413–414
  by reduction of seleninic acids, 413
  by reduction of seleno-cyanates, 413
  from sodium hydrogen selenide, 411–412
  spectra, 414
  synthesis, 410–441
Selenomethionine, 554, 559–567 passim, 572–574, 582, 583, 587, 698
  decomposition, 565
  radioactive, 559–560
  studies with, 572–574
  transport, 562, 573–574

Selenomethionine-^{75}Se, 690
Selenonic acids, 426
  decomposition, 426
  mother liquor, 426
  oxidization, 426
  reduction, 426
Selenonium bromides, 506
Selenonium iodoides, 509
Selenooxytocin, 583
Selenopantetheine derivatives, 526
Selenopantethine, 526, 529, 580, 583
Selenopeptides, 408, 555
Selenophthene, 328
Selenoprotein, 567
Selenopurines, 583
Selenopyrimidine derivatives, 526, 583
Selenosemicarbazides, 373
Selenosulfate, 62, 117
Selenosulfate ion, 626, 641
Selenosulfite, 60
Selenothiazine, 638
Selenotrisulfide linkages, 556–559, 565
  sensitivity, 558
Selenotrisulfides, 577, 579–580
  formation, 579–580
Selenourea, 304, 364, 368–369
4-Selenoxoimidazolidines, 513–514
Selenazolone, 639
Self-diffusion, 770
Self-diffusion coefficient, 24
Self-trapping, 800
Selye granuloma pouch method, 587
Se-methyl organylselenoacetic acids, 480
Se-methylselenomethionine, 552
Semiconductors, 60, 159, 174–175, 231, 630, 641–642, 717, 718, 724, 763, 795
  amorphous, 724
  covalently bonded, 798
  cubic, 188
  energy spectrum of amorphous, 290
  transport mechanism, 232
Seminiferous tubules, 595
  calcification, 595–596

Semiquantitative spectrographic procedure, 633
Sensitivity, 630
Separation, 8, 616–621
  decomposition and discoloration of inorganic materials, 616–618
  decomposition of organic materials, 618–619
  selective separation methods, 619–621
Serum glutamic-oxaloacetic transmission level, 689
Serum proteins, 557–558
^{75}Se-selenomethionine, 573
^{75}Se-sodium, 659
Shaft furnaces, 46
Sharon Springs, 678
Shear modulus, 21
Sheep, 548–549, 597, 659, 663, 669, 681, 682, 693–694, 695
Sheet goods, 746
Shell model, 153, 155, 162
SH-glutathione, 685
Shock resistance, 746
Shot, 61, 113
Shoulders, 170
Signal, 798
Silica, 73, 710, 720
  vitreous, 721, 722
Silicon, 138, 175, 714
Silicone rubber, 141
Silver, 32, 37, 49, 50, 51, 52, 75, 691
  chloride, 43, 642
  salts, 473
  selenide, 42
  selenite, 40, 41, 43
  telluride, 42
Single bond values, 319
Sink, 113
Sintering, 617, 783
Six-coordinate complexes, 345, 377
6-4-2 group, 206
Six-membered ring system, 475
Skin, 672
  discoloration, 656
  rashes, 664, 665
Slab, 113, 114
S-labeled glutathione, 553
Slag, 39, 40, 48, 55
Slimes, copper refinery, 32–33, 38–52,

composition, 34–35
  elementary analysis, 36
  leaching, 48–52
  miscellaneous treatment, 52
  phase analysis, 36
  sulfatizing, 46–48
Slip, single, 765
Sludge, 31, 52
  treatment of sulfuric acid plant, 52–54
Small-ring molecules, 96, 97
Smearing, 773
Smelter slags, 617
Smelting, 32,33, 39–40
  soda, 40
Smoked sheet, 732, 733, 753
Sneezing, 665
Snow, 11
Soda, 38
Soda ash, 53
Soda glass, 129, 130
Soda-lime glass, 644
Soda, ratio to slimes, 46
Soda roasting, 42, 44, 45–46
  temperature, 41
Sodium, 54, 710–711
  acetylides, 410
  arsenate, 663
  bisulfate, 48
  carbonate, 41, 44, 57
  carbonate monohydrate, 46
  chloride, 59, 131, 152, 153
  diselenide, 428–429
  D line, 718
  hydroxide, 43, 49, 51, 55, 57, 565
  selenate, 26, 38, 42, 44, 575, 587, 592, 596–597 667, 782
  selenite, 38, 42, 44, 52, 568, 587, 596, 667, 669, 685, 692, 695, 697, 712
  silicate, 59
  sulfate, 48
  sulfite, 57, 62,
  thiosulfate, 66, 672
Soils, 8–9, 546–548, 644
  alluvial, 679
  analysis, 686
  forms of selenium in, 678–679
  "gumbo," 677
  and plants, 680

seleniferous, 654
Soldering, 779
Solid phase, 99, 109–119, 783
Solids, crystalline, 500
Solid state, 60–611, 797
Solubility, 37, 49, 62, 96, 120
limit, 777
solid, 776
Soluble inorganic selenium, 681
Soluble selenium, 783
Solution chemistry, 8
Solution hardening, 777
Solutions
saturated, 120, 138
supersaturated, 139
Solution techniques, 141
Solvation, 316
Solvent extraction, 59
Solvents, 120, 132–133
types, 132
vapors, 790
Source, 696
commercial, 3
South Dakota, 665, 657, 661, 662, 664, 676, 677, 678, 681
Soviet Union, 2, 5, 27, 45, 46, 47, 52, 769
Soybean meal, 692
Space charge, theory of, 218
Space group, 149–150
Sparging, 39
Specific heat, 763
Spectra, 90, 152, 187, 191, 193, 449, 462, 474,
first order infrared, 166, 170, 171
luminescence, 205–206
optical, 190
partition, 205
Raman, 166, 170, 171
reflectivity, 192, 214
second order, 161
second order infrared, 166–167
stress modulated reflectance, 214–215
Spectral region, 792
Spectral response measurement, 795
Spectral sensitivity, 248–250
curve, 282–283, 284
Spectra restrahlen, 156–159
far-infrared reflectivity, 156

polariton, 156–159
Spectrographic analysis, 637
Spectrophotometric procedure, 637
Spectroscopy, 88
infrared, 167
Raman, 167
Spectrum, 89
absorption, 164–165, 169
infrared, 169
low temperature, 165
reflectance, 797
survey, 160
transmission, 164
Spherulites, 125, 126, 129, 134, 136, 137
Spinning, 184–188
Spiradela oligorrhiza, 561
Spleen, 662
Splitting, 185
spin-orbit, 188
S–Se system, 722
S–Se–Te ternary system, 722
Stable isotopes, 13
Stability, 111, 112, 754, 756
Stability constants, 374, 384
Stacking patterns, 122
Staining, 754
Standard reduction potential, 24
Stanleya, 679
Stanleya bipinnata, 554–555
Stanleya pinnata, 10, 551
Static dielectric constant, 156–157
Static displacement, 115
Stationary crystal-stationary counter technique, 123
Steels, 640, 757, 771, 782, 783
free machining, 773–777
free-machining low carbon, 774–775
grades of free machining stainless, 775
heat resistant high-alloy, 770
silicon, 777–779
stainless, 775–776
Stereospecificity, 461
Steric factors, 360
Steroids, selenium containing, 529
Sticks, 61
Still, 73
Stirring shaft, 138
Stomach, 662

Strain aging, 772
Strength, 779
Stress, 215
Stress-strain properties, 733, 759
Streptococcus faecalis, 572
Streptococcus faecium, 572
Structures, 89–142, 298–333
selenium-bridged, 392
Styrene-butadiene rubber, 733–736, 737, 738–739, 751
evaluation of selenium compounds as antioxidants in, 751
Sublimation, 53, 62, 63, 91, 93, 94, 117, 118, 637
Substitution reactions, 351
Substrate, 117, 118, 129–131, 757, 763
temperature, 128
Succinate, 585
Sulfate, 592–593, 680, 685
^{35}Sulfate, 566
Sulfatizing, 45, 47–48, 53–54, 58
chemistry of, 47
Sulfenium ion, 576
Sulfhydryl-disulfide interchange reactions, 577, 578, 601
Sulfhydryl groups, 556–557
Sulfides, 52, 53, 783
heavy metal, 720–721
hydrothermal, 3
ion, 777
Sulfites, 53
Sulfitolysis, 558
Sulfur, 2, 3, 12, 31, 32, 53, 55, 59, 65, 77, 104, 105, 106, 116, 128, 138, 175, 384, 392, 393, 480, 547, 565, 577, 584, 586, 592, 616, 634, 636, 685, 746, 754, 758, 760, 772, 775, 776, 779, 780, 781
amino acid, 555, 588, 589, 590–592, 600
dioxide, 8, 38, 51, 57, 58, 59, 61, 62, 63, 109, 620, 621
effects of substituting selenium, 736, 737
elemental, 59
enzyme, 581
liquid, 107

metabolism, 568
orthorhombic, 166, 171, 803
orthorhombic, stability and
    growth, 163
substitution of selenium
    for, in biological
    systems, 580–584
volcanic, 8
as vulcanizing agent, 728–
    729, 730
Sulfur-benzothiazyl disul-
    fide—TMTD, 733, 742
Sulfuric acid, 32, 38, 46–48,
    49, 51, 53, 57, 58, 59, 63,
    622, 782
concentrated, 54
Sulfur–selenium interchange
    reactions, 579
Sulfur–selenium systems, 70,
    71
Sulphur rubeum, 1
Sunlight, 754, 756
Superconduction, 226
Supercooling, 134
Superohmism, 218
Supplements, 696
Support rod, 138
Surface, 791
Surface-activity, 779
Surface agents, 783
Surface charge, 790
    density, 790
    positive or negative, 793
Surface conductivity, 790
Surface energy, 779
Surface layer, 806
Surface mobility, 133
Surface tension, 24, 771
Swine, see Pigs
Symmetric dialkyl selenides,
    437–441
by alkylation of elemental
    selenium, 438
from inorganic selenium
    compounds, 438–439
miscellaneous preparation,
    440
modifications of organic
    group, 440
physical properties, 440
reactions, 441
by reduction of dialkyl
    selenium dihalides,
    439–440
Symmetric diaryl selenides,
    450–452

from aryldiazonium salts,
    450
conversion of diorganyl
    diselenides, 145
dipole moments, 542
from diselenium dichloride,
    450
from elemental selenium,
    450
from hydrogen selenide,
    450
modification of organic
    groups, 451
physical properties, 452
reactions, 452
from selenium dioxide, 450
from selenium oxychloride,
    450
from selenium tetrachlor-
    ide, 450
sources, 450–452
Symmetric diselenides, 428–
    435
by hydrolysis of seleno-
    cyanates, 430–431
from inorganic selenium
from organic selenium
    compounds, 430–434
and organyl selenium
    halides, 433–434
by oxidization of selenols,
    431
physical properties, 434–
    435
by reduction of selenium
    and selenic acids,
    433–434
spectra, 434–435
synthesis, 428
Symmetry, 124, 180, 321, 354
Symmetry assignments, 160,
    165–166
Symmetry rules, 149–152
Symptoms, 655, 656, 657,
    658, 664–670 passim,
    681–683
Synthesis, 409, 456–475
Synthetic rubber, 730, 733

Tailing, 37
Tarantulas, 562
Technical selenium, 63, 67,
    69, 75
Teeth, 656
Telluric acid, 621
Telluride layer, 770

Tellurides, 40, 51
Tellurium, 2, 3, 12, 40, 45–46
    47, 50, 51, 54, 55, 59, 60,
    67, 68, 69, 70, 75, 77,
    101, 117, 128, 140, 150,
    175, 176, 204, 216, 238,
    247, 275, 280, 362–363,
    552, 617, 620–621, 636–
    637, 642, 733, 737, 742,
    754, 758, 765, 772, 779,
    781, 797, 805
behavior, 42
(IV), 621–637 passim
ratio, 625
water-soluble, 42
Tellurous acid, 58
Temperature, 71, 72, 91, 93,
    96, 101, 102, 105, 108,
    110, 111, 118, 121, 127–
    128, 134, 135, 191, 234,
    293, 801, 803
anisotropic, 120
and color, 718
controlled, 46, 138
freezing, 764
high, 225, 712
higher reheating, 719
isotropic, 122
low, 219, 230
melting, 711
transition, 112–113
Temperature coefficient, 120
Temperature dependence
    137, 176, 197–201, 205,
    217–224, 770
Temperature shift, 212
Tensile strength, 729, 737,
    742, 750
aged and unaged, 742
Ternary systems, 769
Terrestrial abundance, 616
Tert-butyl selenide, 460
Testis, 595–596, 601
Tetraammine copper salt, 311
Tetraacetylglucoside, 482–483
Tetrachalcogenometallates,
    339
Tetrafluoroborates, 506
Tetrahalides, 312–313
Tetrahydroselenophene, 366
Tetramethylthiuram disulfide,
    see TMTD
2-(2,3,4,6-Tetra-0-acetyl-β-D-
    glucopyranosyl)-2-
    selenopseudourea hydro-
    bromide, 485

Tetraorganyl selenium compounds, 509
Tetraoxometallates, selenium analogue of, 338–340
Tetrapeptides, 534
Tetraselenium tetranitride, 330
Tetravalent selenium, 57–59
Texas, 11
Texture growth, 779
Thallium, 105, 109, 128, 137, 139, 140, 239, 240, 596–597, 684
 storage, 597
Therapy with selenium, 676, 687–694
Thermal activation energy, 254–255
Thermal conductivity, 19, 763
Thermal decomposition, 64, 509
Thermal history, 127, 135
Thermal neutron cross sections, 15
Thermal properties, 16–19
Thermal shock, 114
Thermodynamic constants, 388
Thermoelectric emf, 24
Thermoelectric power, 228–232, 242
1-Thia-4-oxacyclohexane, 366
Thienyl alkyl selenide, 458
Thiocyanate, 622
Thiols, 556, 558, 572, 576–577
Thiourea, 59
III–V compounds, 175, 182
303 austenitic grade steel
302 stainless steel, 776
Threshold limit value, 668, 671
Thyroid gland, 659
Tight-binding calculation, 182
Time reversal, 180
Tin(II) chloride, 617, 622, 623
Tissues, 574–575, 548–550, 659, 660, 661, 681
 molarity of selenium, 599
 selenium content, 60–61
 selenium levels in, 549, 550, 573
 skeletal muscle, 691
Titanium dioxide, 754
TMTD, 733, 742

Tobacco, 645
α-Tocopherol, 585, 692
D-α-tocopherol acetate, 590, 689
Toluene extraction, 643
Toner particles, 790
Torsion effusion, 88, 93,
Torula yeast, 548, 549, 585, 590, 598, 600, 687, 690, 691
Tosylate group, 522
Total content, 547–550
Toxicity, 8, 546, 552, 582, 593, 597
 acute, 681
 alkali disease, 676, 677, 680, 681, 682
 blind staggers, 681, 683
 of compounds, 666–669
 control of, 685
 of elemental selenium, 664–666
 factors affecting, 683–685
 incidence in livestock, 677
 lack of data, 671
 prevention, 685
 of selenium-bearing natural products, 669–670
 see also Toxicology
Toxicology, 654–672
 detoxification and elimination, 662–664
 industrial, 655
 intoxication, 655–658
 lack of data, 761
 pathology, 662
 toxicity of compounds and natural products, 664–671
Trace element, 515, 546, 598
Traces, determination of, 633–634
Trans bonding, 318
Trans configuration, 371, 376
Transducer, 790
Transesterification, 425
Trans-ethanediseleninic anhydrides, 329
Transformer cores, 777
Trans isomers, 365
Transistors, 175
Transition elements, 777
Transition temperature, 136
Transition type, 113
Transitions, 212–213, 344, 347, 356, 796

band-band, 195, 196–199
d-d types, 375
 indirect, 198, 209, 211
Transit time, 799
Transmucosal movement, 573–574
Trans N bonding, 348
Trans octahedral structures, 348–349
Trans-1,4-adducts, 461
Transport, 8
 limitation, 796
 mechanism, 239
Trans position, 348–349, 369, 372
Transverse properties, 792
Trapping
 bulk, 799
 effects of, 806
 electron, 799
 kinetics, 804–805
 lifetime, 797
 shallow, 798, 799, 804
 static, 797, 804
Trap release time, 794
Travelling solvent technique 139
Triaryl selenonium salts, 501
Tricarboxylic acid, 698
Trichlorethylene, 133
Trifluoromethyl selenium chloride, 475
Trigonal band structure, 177–190
Trigonal selenium, 88, 89, 90, 113, 116, 117, 118, 119, 123, 132, 133, 137–141, 149–163, 177–287
 electrical properties, 217–248
 electronic properties, 175
 morphology and growth of crystals, 125
 optical properties, 190–217
 selection rules, 149–152
 structural parameters, 123, 124
 symmetry rules, 149–152
Trimene Base, 746
Trimethyl selenonium, 553 563
 iodides, 312, 314
 ion, 553, 570–571
Triorganyl selenonium compounds, 505–509
 preparation, 505–508

salts, 507, 508–509
Trioxide tetramer, 307
Tripeptides, 525
Triphenyl selenium chloride, 312
Triselenides, 305–306
N,N'-triseleno-bis(morpholine), 733
Tris metal ion compounds, 354, 355–356
TSC-curves, 223, 225, 253, 254–255
Tungsten, 684, 783
Turbidometric determination, 635
Tyndall effect, 710

Ubiquinone, 697
Ultrasonic waves, 765
Ultraviolet absorption, 756
Ultraviolet absorption bands, 339
Ultraviolet stabilizers, 729, 754–756
Uniaxial deformation, 202
Uniaxial stress, 228
Unit cell, 121, 123, 124
Unit-cell electric moment, 154
United Kingdom, 658
United States, 5, 9, 11, 27, 550, 694, 696
Unsaturated selenides, 372
Unsaturation, 750
Unsymmetric aliphatic diorganyl selenides, 441–452
  modification of organic groups, 445–448
  organyl selenium halides as starting materials, 445
  physical properties, 449
  reactions, 449
  from selenols, 441–445
Unsymmetric diaryl selenides, 452–455
  from aryl alkyl selenides, 452–454
  from aryl selenium halides, 452–454
  from aryl selenium thiocyanates, 452–454
  from aryl selenocyanates, 452–454
  modification of functional groups, 455

physical properties, 455
reactions, 455
Unsymmetric diaryl sulfones, 455
Unsymmetric diselenides, 435
Upper limit, 92
Uranium, 37, 38
  oxide, 642
Urbach-rule edge, 162
Urea, 558, 581
Urinary excretion, 570–571
Urine, 553, 565, 568, 592, 597, 647, 656, 658, 663–664, 665
  level of selenium, 663–664
UV, see Ultraviolet

Vacancy migration enthalpies, 770
Vacuum, 52, 114, 138
Vacuum deposition, 117, 118
Vacuum distillation, 53, 70, 89
Vacuum evaporation, 128
Vacuum refining, 74
Valence states, 13, 56–60
Value, 763
van der Waals forces, 163, 318, 765–766
Vandex, 733, 746, 749
Vapor deposited form, 94
Vapor deposition, 118
Vapor-liquid equilibrium, 71
Vapor phase, 88–96, 110
Vapors, 62, 132–133
  chemical, 132–133
  composition, 92
  density, 88
  pressure, 16, 763
  saturated, 91
  thermodynamics, 93
Vegetation, see Plants
Venom, 562
Vesicant, 668
Vibrational assignments, 166–167
Viscoelastic properties, 20–21
Viscometer, 103, 104
Viscosity, 20, 97, 102–106, 136, 139, 720–721, 764
Visible absorption bands, 339
Vision, 587–588
Vitamin A, 602, 683
Vitamin B-complex, 683
Vitamin D, 683
Vitamin E, 549, 550, 574,

585, 586, 588–592, 600, 601, 684–695 passim, 698
  deficiency, 687–688, 689
Vitamin E selenium, 696
Vitamin K, 663
Viton, 746
Vitreousness, 762
Vitreous selenium, 88, 89, 91, 94, 96, 97, 99, 101, 109, 112, 113–117, 118, 167, 803
  properties, 788
VLS technique, 139
Volatility, 62, 76, 568, 569
Volatilization, 42, 52, 55, 619, 620, 712, 715
Voltage, 262, 790, 791
  characteristic, 277
  dependence, 794
  forward, 276
  reverse, 277
Voltage-time trace, 798
Volume change, 134
Volumetric methods, 615, 624–625
Vulcanization, 746
  effect of added selenium on, 748
  effect of selenium compounds on, 738–739, 740, 741, 744, 745, 747
  surface, 746, 750
Vulcanizing agent, 728, 729, 730

Walden's inversion, 484
Waste solutions, 59
Water, 11–12, 54, 647–648
  de-ionized, 62
  drinking, 11
  ground, 11
  ice, 113
  sea, 12
Water-soluble selenium, 40, 45, 680
Wavelength, 116, 717
  dependency, 801
  short, 796, 806
  temperature dependent, 285
Wave number, 216
Wear, resistance to, 783
Weathering, 8
Weight, 110, 693
Weight fraction, 101
Welding, 783
West, 655, 671

Wet oxidation, 33, 618
Wheat, 9, 559, 646, 698
White metal, 55
White muscle disease, 548, 597, 692, 694
White pig disease, 691
Whole-body autoradiography, 573
Wire, 770
Wool, 559, 563
World War I, 25
Wyoming, 676, 677

Xanthophyll, 619
X-band frequencies, 128
Xerographic sensitivity, 792, 794

high field, 793
Xerography, 25, 33, 56, 88, 788–794, 799
  amorphous selenium in, 791, 794
  cycled performance, 794
  discharge, 804
  photoreceptors employing Se, 805–806
  reuseable, 789
  steps, 789–790
X-ray diffraction, 97, 98, 111, 113, 114, 115, 119, 121, 554, 716
X-ray fluorescence, 648
X-ray reflection, 99
X-ray scattering, 99, 301

Xylose, 480, 495,

Yeast, 572, 597, 687
Yellow fat disease, 691
Yields, 409, 432, 443, 449, 450, 458, 469, 476, 478, 487, 500, 505
Young's modulus, 21, 763

Zambia, 45
Zinc, 33, 37, 52, 59, 596, 712, 782
  acetate, 595
  oxide, 720, 729, 732, 742
Zirconium–selenium alloys, 641
Zone center, 153
Zone refining, 70